Biomineralization and Biomaterials

Related titles

Novel biomaterials for bone regeneration
(ISBN 978 0 85709 804 7)

Biomedical foams for tissue engineering applications
(ISBN 978 0 85709 696 8)

Bone substitute biomaterials
(ISBN 9780857094971)

Tissue engineering using ceramics and polymers
(ISBN 9780857097125)

Woodhead Publishing Series in Biomaterials:
Number 104

Biomineralization and Biomaterials

Fundamentals and Applications

Edited by

Conrado Aparicio and
Maria-Pau Ginebra

ELSEVIER

AMSTERDAM • BOSTON • CAMBRIDGE • HEIDELBERG
LONDON • NEW YORK • OXFORD • PARIS • SAN DIEGO
SAN FRANCISCO • SINGAPORE • SYDNEY • TOKYO
Woodhead Publishing is an imprint of Elsevier

WP
WOODHEAD
PUBLISHING

Woodhead Publishing is an imprint of Elsevier
80 High Street, Sawston, Cambridge, CB22 3HJ, UK
225 Wyman Street, Waltham, MA 02451, USA
Langford Lane, Kidlington, OX5 1GB, UK

ISBN: 978-1-78242-338-6 (print)
ISBN: 978-1-78242-356-0 (online)

British Library Cataloguing-in-Publication Data
A catalogue record for this book is available from the British Library

Library of Congress Control Number: 2015942354

For information on all Woodhead Publishing publications visit our website at http://store.elsevier.com/

Contents

Contributors

J. An Radboud University Nijmegen Medical Center, Biomaterials, Nijmegen, The Netherlands

A. Bigi University of Bologna, Bologna, Italy

E. Boanini University of Bologna, Bologna, Italy

F. Buchanan Queen's University, Belfast, UK

C.A. Carrera University of Minnesota, Minneapolis, MN, USA

H. Cölfen University of Konstanz, Konstanz, Germany

C. Combes Institut National Polytechnique de Toulouse, Toulouse, France

E. Cunningham Queen's University, Belfast, UK

Timothy E.L. Douglas Ghent University, Ghent, Belgium

N. Dunne Queen's University, Belfast, UK

A. El-Fiqi Institute of Tissue Regeneration Engineering (ITREN), Dankook University, Cheonan, South Korea

M. Gazzano ISOF-CNR, Bologna, Italy

R.A. Giacaman University of Talca, Talca, Chile

F.J. Gil Mur Technical University of Catalonia, Barcelona, Spain

L.B. Gower University of Florida, Gainesville, FL, USA

J. Jansen Radboud University Nijmegen Medical Center, Biomaterials, Nijmegen, The Netherlands

H.-W. Kim Institute of Tissue Regeneration Engineering (ITREN), Dankook University, Cheonan, South Korea

T. Kokubo Chubu University, Kasugai, Japan

S. Leeuwenburgh Radboud University Nijmegen Medical Center, Biomaterials, Nijmegen, The Netherlands

I.B. Leonor University of Minho, Guimarães, and ICVS/3B's—PT Government Associate Laboratory, Braga/Guimarães, Portugal

M. Navarro International Center for Numerical Methods in Engineering (CIMNE), Barcelona, Spain

R. Osorio University of Granada, Granada, Spain

V.A. Perez University of Talca, Talca, Chile, and University of Melbourne, Melbourne, VIC, Australia

A. Rao University of Konstanz, Konstanz, Germany

R.L. Reis University of Minho, Guimarães, and ICVS/3B's—PT Government Associate Laboratory, Braga/Guimarães, Portugal

C. Rey Institut National Polytechnique de Toulouse, Toulouse, France

A.I. Rodrigues University of Minho, Guimarães, and ICVS/3B's—PT Government Associate Laboratory, Braga/Guimarães, Portugal

P.A. Romero University of California, Los Angeles, CA, USA

S.-J. Seo Institute of Tissue Regeneration Engineering (ITREN), Dankook University, Cheonan, South Korea

T. Serra University College London (UCL), London, UK

J. Seto University of California, San Francisco, CA, USA, and École Normale Supérieure, Paris, France

M. Toledano University of Granada, Granada, Spain

G. Walker University of Limerick, Limerick, Ireland

J. Wolke Radboud University Nijmegen Medical Center, Biomaterials, Nijmegen, The Netherlands

S. Yamaguchi Chubu University, Kasugai, Japan

Preface

Bone, tooth enamel, nacre, coral, mollusk shells, stems of horsetail grass, diatom cell walls, etc., are examples of biominerals formed through biomineralization processes that result in hierarchical structures with unique and, in many cases, outstanding properties. The processes that occur to obtain the exact composition, size, shape, and spatial distribution of the minerals and organic components in these biological materials are controlled by living organisms, from prokaryotes to humans. Control of these processes is exerted via cell-orchestrated production and release of organic materials that interact with inorganic components, inside or outside of the organism that produced them. A biomineralization process is therefore essentially defined by the direct or indirect control of the process by the living organisms and the mediation of the process by organic molecular templates, such as proteins and other biomacromolecules. In many instances, the crystals and structures formed through the biomineralization process incorporate organic macromolecules. This is accomplished as biomineralization processes follow bottom-up strategies that organize defects and matter in such a way that the formed hybrid biomaterials display specific and unique surface, optical, chemical, and mechanical properties that exceed those of their individual counterparts.

The description and understanding of biomineralization and related processes, as well as the assessment of the structure–property relationships in biominerals, have been topics of intense research for many years. In the last 20 years, scientists and engineers have made significant advances in the field not only to the aforementioned understanding of mechanisms of formation, structure, and properties of the biominerals but also to the development of new characterization techniques. As a result of this, we know that key aspects of the process of biomineralization are the interactions at the organic–inorganic interface, the hydration states of the minerals at different stages of mineral formation, the properties of the aqueous solution in which the process is taking place, and the fact that biomineralization occurs at ambient or near-ambient conditions through the formation of precursor transient phases. However, much is still to be investigated and discerned about these sophisticated and elegant mechanisms of mineral formation, which keeps the field open as a hot topic for research with intense discussion and passionate dedication to validate new hypotheses. For instance, the role of organic molecules in polymorph selection, that of lipids and sugars in biomineralization, or of the organic matrix in the structure of some biominerals at different scales are areas that are still not fully understood.

The main motivation for most researchers to understand biomineralization is that it is regarded as a paradigm for the development of new routes for the synthesis of new functional materials with structural precision at different dimensional levels.

The examples that one can find in Nature can lead to the development of biomimetic processes and/or structures that incorporate at least some of the exceptional properties of the biological materials which inspired them. Many different synthetic materials can benefit from this approach, namely optical materials, materials for storage of energy, magnetic and electric sensors, micropatterned surfaces, and biomaterials for implants and scaffolds in biomedical applications. The latter is undoubtedly a field that has drawn much inspiration from the natural processes of biomineralization to either obtain or modify synthetic materials to mimic human body mineralized tissues from a physicochemical, mechanical, and biological perspectives.

Biometals, synthetic and natural biopolymers, bioceramics, and biocomposites have been mineralized most often with calcium phosphate minerals, i.e. the inorganic component of hard tissues in the human body, following synthetic routes that mimic in full or in part natural biomineralization processes and that, in the best-case scenario, result in structures closely resembling those found in Nature. The goal is to produce materials that will be implanted to substitute and/or regenerate bone, dentin or teeth enamel with exceptional ability to perform these functions. Cells and tissues that interact with these biomaterials are responsive to the specific physical, chemical, and mechanical properties of the biomimetic mineralized constructs. The minerals obtained can be formed in any of a series of configurations depending on the selected biomaterial, the route of implantation, and/or the clinical strategy to restore tissue and organ functions. Examples of the different approaches are: the development of calcium phosphate coatings on flat surfaces, micro- and nanofibers, or scaffolds; the intrafibrillar mineralization of scaffolds; mold infiltration at different scales; and bulk, solid–liquid mixture, or solution preparation of calcium phosphate-based materials.

In this book, leading experts in their fields have provided detailed chapters on a range of subjects directly related to the processes of biomineralization, and the design, development, and use of biomineral-based or biomineralized biomaterials. The subjects covered are divided into two parts. Part 1 explores the fundamentals of biomineralization, with an emphasis on describing and discussing new theories about the process of mineralization of hard tissues. Part 2 focuses on the wide range of biomaterials—ceramics, hydrogels, metals, collagen, etc.—that have been mineralized using inspiration from our fundamental knowledge of biomineralization.

The book starts with two chapters that provide the most basic knowledge on the processes of biomineralization. Basic introductions to crystals and biominerals as well as methods for their characterization are provided. The main goal of these two chapters is complementary describing the increasingly better understood nonclassical pathways for nucleation and crystallization of biominerals and how organic templates participate in this process and control crystal polymorphism in Nature. A selection of different natural biomineralized structures and tissues are introduced to exemplify the aforementioned topics.

The rest of the first part of the book focuses on describing the biominerals and the biomineralization processes that form natural human tissues. These are the main mineralized structures that are both mimicked and targeted to be restored by the biomaterials covered in the second part of the book. The chapter introducing the physical chemistry of biological apatites opens this section, including a discussion on bone mineral precursors and transient mineral phases. The properties of bone mineral apatites are correlated with the mineral-ion reservoir function of bone and the interactions between apatite nanocrystals and the organic constituents of the extracellular matrix. Then, the next two chapters provide an excellent overview of the biological and biochemical mechanisms of the traditional and newly described processes of mineralization of bone, tooth enamel, and dentin. Embryogenic and histological considerations for the formation of these tissues are also introduced, as well as the pathological processes associated with their mineralization.

The last chapter in the first part of the book is unique in that it presents a critical analysis on the understanding of the physicochemical mechanisms involved in the intrafibrillar mineralization of collagen with hydroxyapatite using a biomimetic system that induces a nonclassical transient precursor crystallization process. The in-depth, critical, and somewhat controversial discussion is valuable, among other considerations, to better analyze how the processes described in the two previous chapters can be studied, interpreted, and utilized by mimicking them to explore applications such as those presented in the second part of the book.

The second part opens with a chapter on the relevance of ion substitution in biological and synthetic apatites. As a matter of fact, one of the most remarkable distinctive features of biological apatites is the presence of foreign ions, which play an essential role in the functional behavior of mineralized tissues. For this reason, ionic substitution is being explored as a strategy to develop synthetic apatites with enhanced performance, in an approach that conceives the synthetic material as a bioinorganic controlled release system. The chapter reviews the main characteristics of biological apatites derived from the presence of foreign ions, followed by a description of the structural and chemical modifications induced by ionic substitution in synthetic apatites.

This following block of four chapters introduce biomimetic approaches to mineralize different generic types of materials—polymers/hydrogels, ceramics, metals, and surfaces. Each of them introduces a thorough description of the different biomimetic mineralization approaches pursued to advance the functions of the materials to be used in biomedical applications. The rationale for the route of mineralization and/or the consensus mechanism by which the biomineralization is achieved are also discussed. The authors complement the chapters with a more detailed presentation of their research contribution to the subject of interest.

The final four chapters of the book cover specific biomineralized biomaterials and biomineral delivery strategies to treat hard body tissues. This block of content introduces the most advanced approaches to apply inspiration in biomineralization to develop new generations of biomaterials to substitute, repair, engineer, and regenerate enamel, dentin, and bone. The first and second of these chapters provide comprehensive information on strategies to facilitate dentin remineralization, both for products on the market and for those currently under research and development; and how to use and modify glasses and polymers to promote biomineralization to substitute and regenerate bone, respectively. The evolution of biomimetic approaches for the development of calcium phosphate-based bone substitute scaffolds is addressed in the next chapter, with particular emphasis on the development of complex inorganic forms, inspired by naturally occurring architectures. Finally, the last chapter provides an insight into the different strategies used for the development of mineralized fibrous scaffolds, which aim to mimic the native extracellular matrix of mineralized tissues, and have great potential for tissue engineering applications. Again, as in the preceding chapters, the authors complement and illustrate the information with a review of the most relevant research contributions to the subject of interest.

It is clear from the contents presented in this book that our knowledge of the processes of biomineralization and the structure–property relations of the biominerals formed constitute a broad and versatile field that provides continuous inspiration to scientists and engineers to design, develop, or improve the performance of biomaterials for tissue repair and regeneration.

The editors
Conrado Aparicio and
Maria-Pau Ginebra

Woodhead Publishing Series in Biomaterials

1 **Sterilisation of tissues using ionising radiations**
Edited by J. F. Kennedy, G. O. Phillips and P. A. Williams
2 **Surfaces and interfaces for biomaterials**
Edited by P. Vadgama
3 **Molecular interfacial phenomena of polymers and biopolymers**
Edited by C. Chen
4 **Biomaterials, artificial organs and tissue engineering**
Edited by L. Hench and J. Jones
5 **Medical modelling**
R. Bibb
6 **Artificial cells, cell engineering and therapy**
Edited by S. Prakash
7 **Biomedical polymers**
Edited by M. Jenkins
8 **Tissue engineering using ceramics and polymers**
Edited by A. R. Boccaccini and J. Gough
9 **Bioceramics and their clinical applications**
Edited by T. Kokubo
10 **Dental biomaterials**
Edited by R. V. Curtis and T. F. Watson
11 **Joint replacement technology**
Edited by P. A. Revell
12 **Natural-based polymers for biomedical applications**
Edited by R. L. Reiss et al
13 **Degradation rate of bioresorbable materials**
Edited by F. J. Buchanan
14 **Orthopaedic bone cements**
Edited by S. Deb
15 **Shape memory alloys for biomedical applications**
Edited by T. Yoneyama and S.Miyazaki
16 **Cellular response to biomaterials**
Edited by L. Di Silvio
17 **Biomaterials for treating skin loss**
Edited by D. P. Orgill and C. Blanco
18 **Biomaterials and tissue engineering in urology**
Edited by J.Denstedt and A. Atala
19 **Materials science for dentistry**
B. W. Darvell

44 **Durability and reliability of medical polymers**
Edited by M. Jenkins and A. Stamboulis
45 **Biosensors for medical applications**
Edited by S. Higson
46 **Sterilisation of biomaterials and medical devices**
Edited by S. Lerouge and A. Simmons
47 **The hip resurfacing handbook: A practical guide to the use and management of modern hip resurfacings**
Edited by K. De Smet, P. Campbell and C. Van Der Straeten
48 **Developments in tissue engineered and regenerative medicine products**
J. Basu and J. W. Ludlow
49 **Nanomedicine: Technologies and applications**
Edited by T. J. Webster
50 **Biocompatibility and performance of medical devices**
Edited by J-P. Boutrand
51 **Medical robotics: Minimally invasive surgery**
Edited by P. Gomes
52 **Implantable sensor systems for medical applications**
Edited by A. Inmann and D. Hodgins
53 **Non-metallic biomaterials for tooth repair and replacement**
Edited by P. Vallittu
54 **Joining and assembly of medical materials and devices**
Edited by Y. (Norman) Zhou and M. D. Breyen
55 **Diamond-based materials for biomedical applications**
Edited by R.Narayan
56 **Nanomaterials in tissue engineering: Fabrication and applications**
Edited by A. K. Gaharwar, S. Sant, M. J. Hancock and S. A. Hacking
57 **Biomimetic biomaterials: Structure and applications**
Edited by A. J. Ruys
58 **Standardisation in cell and tissue engineering: Methods and protocols**
Edited by V. Salih
59 **Inhaler devices: Fundamentals, design and drug delivery**
Edited by P. Prokopovich
60 **Bio-tribocorrosion in biomaterials and medical implants**
Edited by Y. Yan
61 **Microfluidic devices for biomedical applications**
Edited by X-J. James Li and Y. Zhou
62 **Decontamination in hospitals and healthcare**
Edited by J. T. Walker
63 **Biomedical imaging: Applications and advances**
Edited by P. Morris
64 **Characterization of biomaterials**
Edited by M. Jaffe, W. Hammond, P. Tolias and T. Arinzeh
65 **Biomaterials and medical tribology**
Edited by J. Paolo Davim
66 **Biomaterials for cancer therapeutics: Diagnosis, prevention and therapy**
Edited by K. Park
67 **New functional biomaterials for medicine and healthcare**
E.P. Ivanova, K.Bazaka and R. J. Crawford

68 **Porous silicon for biomedical applications**
Edited by H. A. Santos
69 **A practical approach to spinal trauma**
Edited by H. N. Bajaj and S. Katoch
70 **Rapid prototyping of biomaterials: Principles and applications**
Edited by R.Narayan
71 **Cardiac regeneration and repair Volume 1: Pathology and therapies**
Edited by R-K. Li and R. D. Weisel
72 **Cardiac regeneration and repair Volume 2: Biomaterials and tissue engineering**
Edited by R-K. Li and R. D. Weisel
73 **Semiconducting silicon nanowires for biomedical applications**
Edited by J.L. Coffer
74 **Silk biomaterials for tissue engineering and regenerative medicine**
Edited by S. Kundu
75 **Biomaterials for bone regeneration: Novel techniques and applications**
Edited by P.Dubruel and S. Van Vlierberghe
76 **Biomedical foams for tissue engineering applications**
Edited by P. Netti
77 **Precious metals for biomedical applications**
Edited by N. Baltzer and T. Copponnex
78 **Bone substitute biomaterials**
Edited by K. Mallick
79 **Regulatory affairs for biomaterials and medical devices**
Edited by S. F. Amato and R. Ezzell
80 **Joint replacement technology Second edition**
Edited by P. A. Revell
81 **Computational modelling of biomechanics and biotribology in the musculoskeletal system: Biomaterials and tissues**
Edited by Z. Jin
82 **Biophotonics for medical applications**
Edited by I. Meglinski
83 **Modelling degradation of bioresorbable polymeric medical devices**
Edited by J. Pan
84 **Perspectives in total hip arthroplasty: Advances in biomaterials and their tribological interactions**
S. Affatato
85 **Tissue engineering using ceramics and polymers Second edition**
Edited by A. R. Boccaccini and P. X. Ma
86 **Biomaterials and medical-device associated infections**
Edited by L. Barnes and I. R. Cooper
87 **Surgical techniques in total knee arthroplasty (TKA) and alternative procedures**
Edited by S. Affatato
88 **Lanthanide oxide nanoparticles for molecular imaging and therapeutics**
G. H. Lee
89 **Surface modification of magnesium and its alloys for biomedical applications Volume 1: Biological interactions, mechanical properties and testing**
Edited by T .S. N. Sankara Narayanan, I. S. Park and M. H. Lee

90 **Surface modification of magnesium and its alloys for biomedical applications Volume 2: Modification and coating techniques**
Edited by T .S. N. Sankara Narayanan, I. S. Park and M. H. Lee
91 **Medical modelling: the application of advanced design and rapid prototyping techniques in medicine Second Edition**
Edited by R. Bibb, D. Eggbeer and A. Paterson
92 **Switchable and responsive surfaces and materials for biomedical applications**
Edited by Z. Zhang
93 **Biomedical textiles for orthopaedic and surgical applications: fundamentals, applications and tissue engineering**
Edited by T. Blair
94 **Surface coating and modification of metallic biomaterials**
Edited by C. Wen
95 **Hydroxyapatite (HAP) for biomedical applications**
Edited by M. Mucalo
96 **Implantable neuroprostheses for restoring function**
Edited by K. Kilgore
97 **Shape memory polymers for biomedical applications**
Edited by L. Yahia
98 **Regenerative engineering of musculoskeletal tissues and interfaces**
Edited by S.P. Nukavarapu, J.W. Freeman and C.T. Laurencin
99 **Advanced cardiac imaging**
Edited by K. Nieman, O. Gaemperli, P. Lancellotti and S. Plein
100 **Functional Marine Biomaterials: Properties and Applications**
Edited by Se-Kwon Kim
101 **Shoulder and elbow trauma and its complications: Volume 1: The Shoulder**
Edited by R. M. Greiwe
102 **Nanotechnology-Enhanced Orthopedic Materials: Fabrications, Applications and Future Trends**
Edited by L. Yang
103 **Medical devices: Regulations, standards and practices**
Edited by S. Ramakrishna, L. Tian, C. Wang, S. L. and T. Wee Eong
104 **Biomineralization and biomaterials: fundamentals and applications**
Edited by C. Aparicio and M. Ginebra
105 **Shoulder and elbow trauma and its complications: Volume 2: The Elbow**
Edited by R. M. Greiwe
106 **Characterisation and Design of Tissue Scaffolds**
Edited by P. Tomlins

Part One

Fundamentals

Shaping it up: Design and engineering of biominerals and crystalline materials from the bottom up

1

J. Seto[1,2], *P.A. Romero*[3]
[1]University of California, San Francisco, CA, USA, [2]École Normale Supérieure, Paris, France, [3]University of California, Los Angeles, CA, USA

1.1 Introduction

Humpty Dumpty sat on a wall,
Humpty Dumpty had a great fall,
All the King's horses and all the King's men,
[Finally re-crystallized] Humpty together again.

Unknown

We can search Nature's abyss and find that we are, in fact, surrounded by crystalline-based materials. From the $CaCO_3$ spines of sea urchins that populate the tide pools at the shoreline to the silica granules found in the body plans of sea sponges occupying the vast seas, these biogenic crystals have been sources of mineral that create geological landmarks such as cliffs and seafloors for eons of time. More impressively, each organism that produces these crystalline materials has its own "toolkit" of organic molecules that are tailored to create the structures and functions that allow these organisms to propagate onward into the future. The biological control and regulation involved in the formation and growth of these materials are also flexible enough to enable for additional revision as well as continued evolution, when making new biological mineral components.

Some examples of the spatial, temporal control and precision involved in the regulation of crystal formation and growth can be witnessed through parallels with the basic premise of an embryo in biology. A mineral crystal forms from a nucleation event where a cation and anion pair bonds and creates a nuclei for crystal growth; an embryo forms from the union of female and male gamete cells and cellular developmental processes enable it to grow into an organism (Figure 1.1). In a similar fashion, the pathway a crystal nuclei will take will depend on its chemical and physical environments to determine its trajectory to becoming a fully fledged crystal with specific properties (i.e., polymorph, orientation, lattice spacings, and size) such as the path of crystal growth it will take and whether it will become one crystal polymorph versus another polymorph. Many of these environments can be controlled by biological macromolecules—proteins, lipids, and carbohydrates, as well as nucleic acids (Burke et al., 2000; Fantner et al., 2005; Vos et al., 2007; Knapp et al., 2012; Rao et al., 2014; Chow et al., 2014; Figure 1.2). By taking the simple scenario whereby the presence

Biomineralization and Biomaterials. http://dx.doi.org/10.1016/B978-1-78242-338-6.00002-8

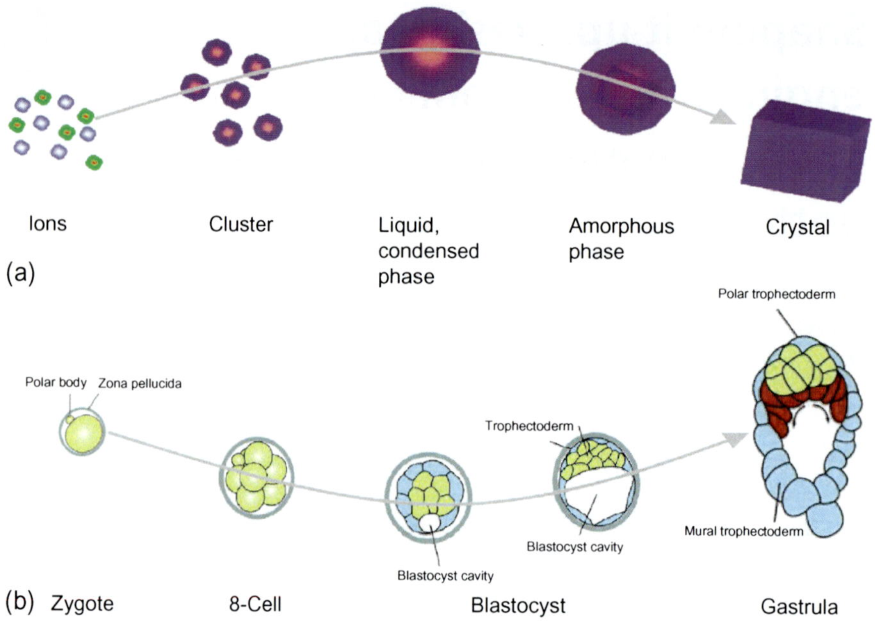

Figure 1.1 Processes of crystallization and embryogenesis compared. (a) Ion pair bonding occurs to initiate nucleation and, subsequently, crystal growth. (b) Sperm and egg cells join to form a zygote that will undergo various stages of growth and development.
Adapted from (a) Cölfen and Antonietti (2008) and (b) Beddington and Robertson (1999).

of biomolecules like nucleic acids, such as deoxyribonucleic acid (DNA), the carboxyl acid group (–COOH) can alter the bulk pH of solutions such that the local supersaturation of ion concentrations are slightly altered to induce a specific polymorph of a mineral (Xu et al., 2008; Figure 1.2). The presence of these molecules has been studied to a great degree and sheds light on how these molecules affect liquid–liquid as well as liquid–solid interfaces in crystal formation as well as the final crystalline product too (Li et al., 2012; Gebauer et al., 2008; Bewernitz et al., 2012; Gower, 2008; Burke et al., 2000; Lu et al., 2005).

It may seem surprising to the reader that such "minor" alterations in the chemical environment, in this case a slight decrease of pH, can inflict such a dramatic outcome in crystallization. The pH, an indicator of the number of H^+ ions in solution, more precisely related to the log of the concentration of H^+ ions, inhibits the kinetics of the anion and thus affects the local supersaturations of the ion pairs involved in crystallization (Figure 1.3). By varying the pH in a crystallizing solution, whether it be lower (more H^+ ions) or higher (more OH^- groups), the cation and anion pair will be obstructed by the additional H^+ or OH^- ions in solution. As a result, the association and dissociation constants, respectively, will be affected above or below the pK of the cation and anion bonding pair (Figure 1.3). Thus, the degree and controls required to keep crystallization in biology in check necessitate regulatory schemes that can handle large amounts of ions as well as various potentials created by the ions.

In biology, regulation comes in the form of compartmentalization such that sensitive elements are localized and surrounded by protective membranes in order to create a

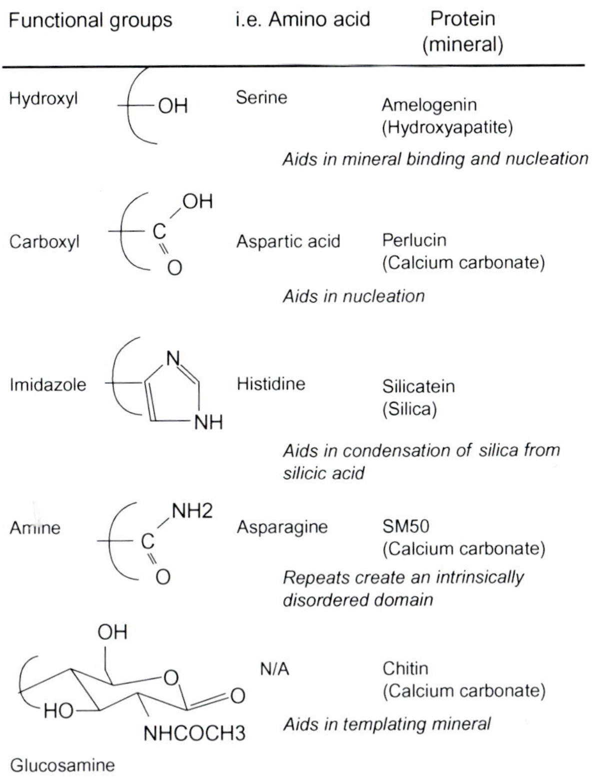

Figure 1.2 Typical functional groups found in biomineralization proteins and the amino acids that form them, known to alter solution properties during crystallization.

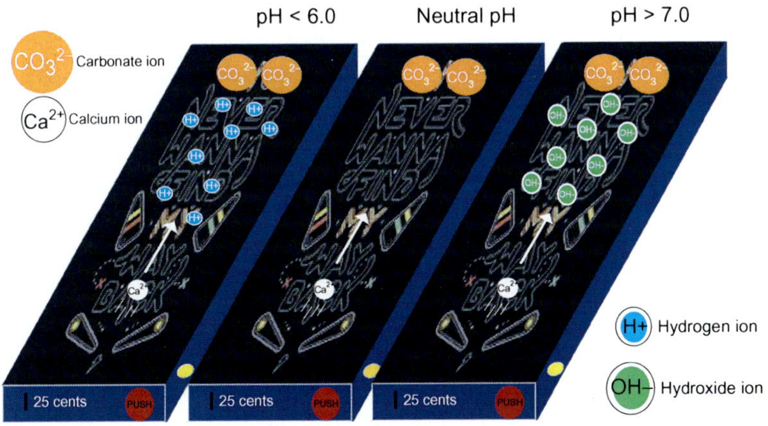

Figure 1.3 Pinball schematic demonstrating the gamut of ions at low pH, neutral pH, and high pH conditions, showing problematic routes to mineralization in the low and high pH conditions.

buffer space separated from the surrounding cellular environment. Some examples of this compartmentalization in Nature can be witnessed in the formation of vesicles that are able to deliver extracellular payloads to specific intracellular organelles, as well as the redox processes occurring in the double membrane of the mitochondria, and finally, the membrane-mediated mineralization processes that occur in a growing sea urchin larvae spicule (Wilt et al., 2003), the growing tunic spicules in ascidians (Aizenberg et al., 2005), and the growing bone (Currey, 2005; Figures 1.4 and 1.5). Specifically, in the case of the growth of a sea urchin larvae spicule, the Ca^{2+} and CO_3^{2-} ions are transported into the primary mesenchymal cells during gastrulation along with matrix

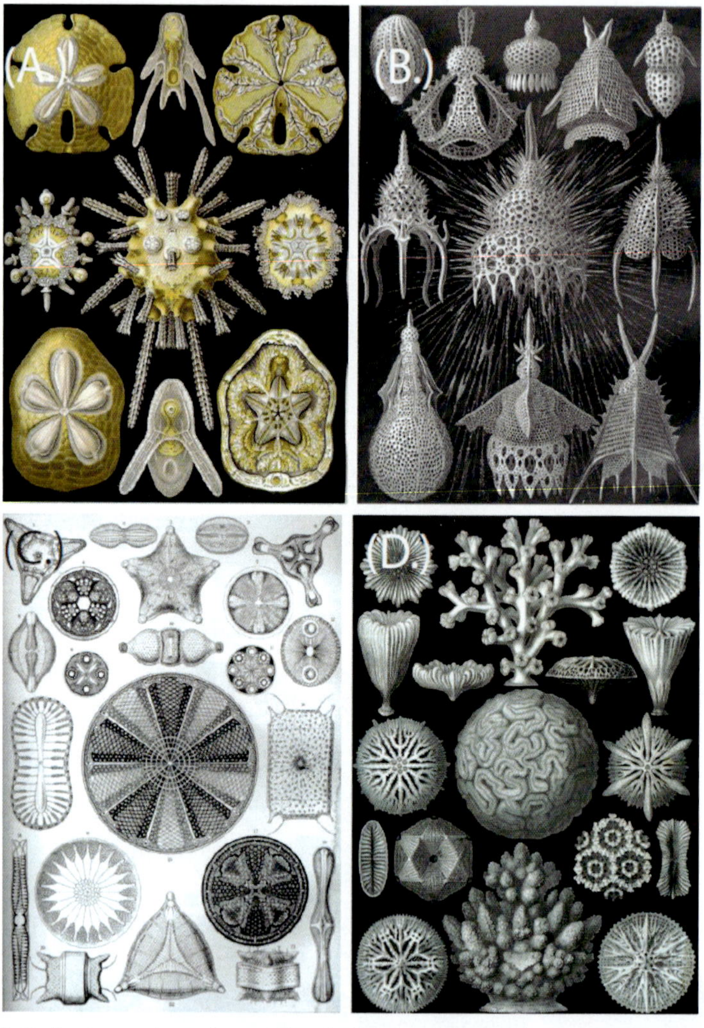

Figure 1.4 Early observations of crystalline body plans in (a) echinoderms, (b) cyrtoidea, (c) diatom, and (d) hexacoralla in lithographic form.
Adapted from Haeckel (1904).

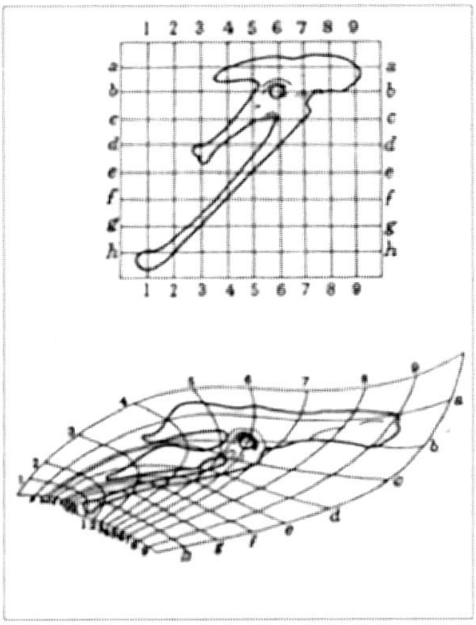

(a)

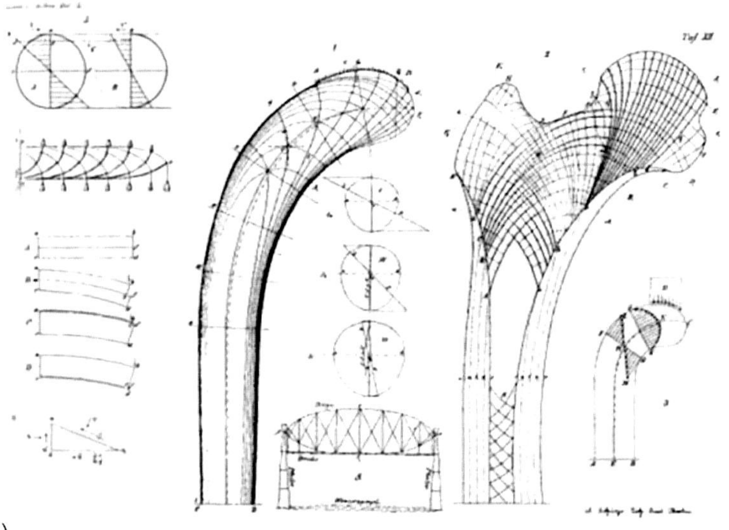

(b)

Figure 1.5 Relation of macroscale forces and counter forces from architectural design schemes to the architecture found in the growth and development of bones. (a) Pelvises of *Archaeopteryx* and *Apatornis*. (b) Femoral head.
Adapted from Thompson (1917) and Wolff (1892).

proteins into a syncytial space to enable nucleation of a precursor mineral that eventually grows into a single crystalline spicule made of $CaCO_3$; subsequently, the same mechanism is adapted by the organism to form the large spines we often see at the beach (Wilt et al., 2003; Seto et al., 2004; Killian et al., 2009; Gong et al., 2012). In the formation and growth of the tunic spicules in ascidians, specific cells release localized silicatein proteins into an extracellular space that is surrounded by a proteinaceous membrane in order to convert silic acid into silica (Aizenberg et al., 2004, 2005). Beyond calcium carbonate and silica, a similar mechanism is utilized for hydroxyapatite in the initial mineralization and development of bone. In the formation of bone, osteoblasts, the cells that secrete minerals, and osteoclasts, the cells that resorb minerals, reach a secretion–resorption equilibrium such that osteoblasts secrete more minerals than are resorbed by osteoclasts (Cowin, 2001; Currey, 2002). The osteoblasts initially absorb minerals in the form of an amorphous precursor and, after intracellular processing, the minerals are secreted into a space called an osteoid that serves as the site of bone mineralization, mineral organization, and growth (Boskey, 1997). Whether it is an echinoderm, mollusk, or vertebrate, the mineralization mechanisms involved in all these diverse organisms share a common theme of membrane-mediated mineralization even though the components involved in each case, the proteins and macromolecules that interact with the mineral ions, are themselves different and share no homology (Boskey et al., 1998, 2002). This suggests that mineralization is a divergent trait that has split off from each genus early in evolution and each genus has "redeveloped" mineralization in its own fashion. Recent work looking at these genes in sea urchin development has found that the biomineralization genes themselves are co-adapted from the set of genes that are responsible for the immune system (Barsi et al., 2014; Davidson, 2014; Tu et al., 2014). Using minerals to trap and sequester dangerous infections or viruses from the environment may have served as the most unlikely way for organisms to begin to evolve crystalline features that now serve in critical functions from mechanical loading to focusing in vision as well as tuning in sound and balance perception in many organisms.

Nature has fashioned crystalline materials with a diverse range of colors and morphologies; each with specific functionalities and properties (Figures 1.4 and 1.5). These features are derived from the properties of the constituent elements: the nanocrystalline stem all the way down to the basic subunit—the single nanocrystalline unit. In this chapter, we will demonstrate several ways of controlling the features of nanocrystals and how these effects can affect assembly schemes of bulk crystalline materials.

1.2 Crystalline properties

> *Since every solid substance contains parts that are crystalline, and since in many of them the whole is an aggregation of crystals, it will be readily understood that a knowledge of crystal structure often affords an explanation of the properties of the substance.*
>
> *William H. Bragg*

1.2.1 What is a crystal?

According to the Third Law of Thermodynamics, crystalline materials are at their most stable physical states. The atoms in a single crystal are all closely packed together in such a way that the total potential energy is at a minimum. The packing order or structure is determined by the features of the atoms involved: their charge, the bond order, as well as their sizes (Torquato, 2002; Torquato and Stillinger, 2002). As an example, when we put quarters and dimes into a jar, a specific packing order will be assumed by these two types of coins, due to their sizes (Seto et al., 2014b; Figure 1.6).

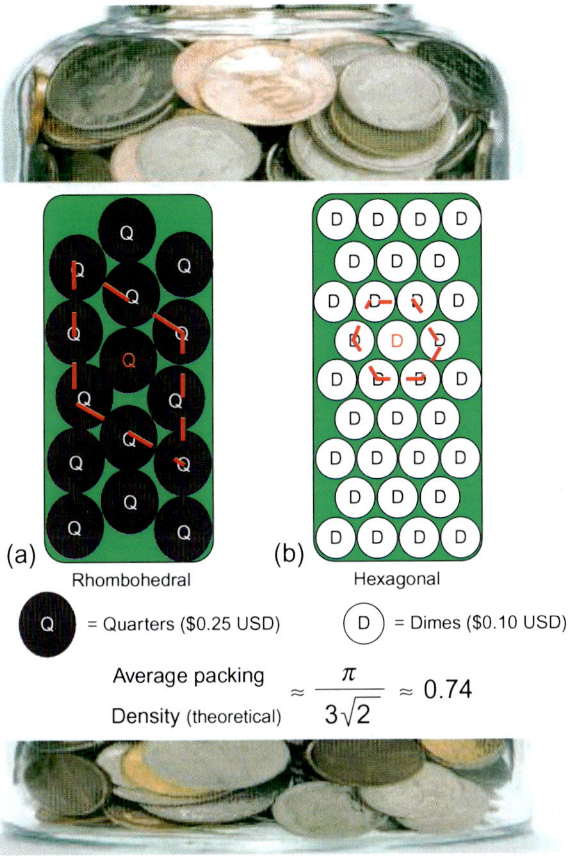

Figure 1.6 Packing density of dimes and quarters in the same volume as a demonstration of how crystalline biological materials use similar strategies to organize constituent crystals. (a) Quarters display more interfaces and void spaces due to the constraint of their size. (b) Better packing of the dimes due to their ability to reduce the availability of spaces. (Note the packing around a specific center denoted by a central coin in each case.) Adapted from Seto et al. (2014b).

The approximate packing density can be estimated as: $\pi/3\sqrt{2}$. In a similar way, when the components of the crystal are known, their packing order as well as the type of crystalline structure can be determined.

What exactly does it mean to be crystalline? In this section, the basic crystalline unit will be defined and the properties related to these structural motifs explored.

1.2.1.1 *Bravais lattice*

Crystalline materials display a specific packing order or organization of a constituent unit cell. This unit cell is the basic configuration of a crystal based on its principal composition. As described in Figure 1.6, the constituent chemical composition, in this case the atomic composition, determines the unit cell organization and the packing/organization within

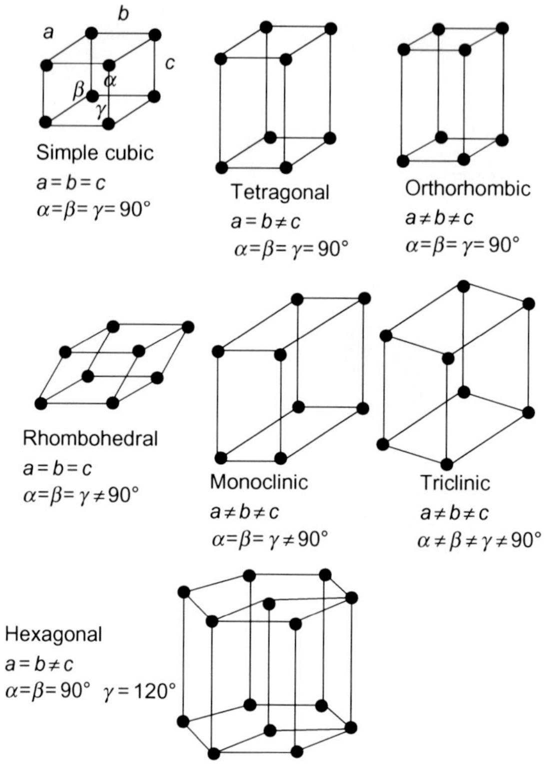

Figure 1.7 The seven common crystallographic Bravais lattices found in unit cells of crystals. Adapted from Chang (2000).

the unit cell. Due to the variety of bonding orders available, the most common types of unit cell configurations include: Simple cubic, Tetragonal, Rhombohedral, Orthorhombic, Monoclinic, Triclinic, and Hexagonal. These types are often represented as Bravais lattices, representing the packing order of the atoms in the unit cell (Morris, 2005; Chang, 2000). Specifically, as shown in Figure 1.7, these seven Bravais lattices show the structural relationship and organization between all the known atoms in the crystal and reveal the basic packing order that determines some of the properties of the bulk crystal.

From this unit cell, as the name indicates, growth of the crystal can be extended in every direction and still maintain the integral unit cell configuration. As is often the case, the fundamental unit cell appears as a nucleation seed and initiates crystal growth as well as other nucleation sites with like unit cell configurations, growing and nucleating until it is physically no longer able to grow, typically meeting an adjoining unit cell to form a macroscale crystal.

1.2.1.2 Grain boundaries

In the formation of macroscale crystals, nucleation of crystals can occur rapidly and at many various locations simultaneously. This leads to multiple single crystals that may seed the crystal and grow to create a larger crystal. As you may imagine, this leads to a problem of registry. Naturally, you might ask, "how do all these crystal nucleates recognize that there are other similar centers of crystal growth and whether there's some mechanism to coordinate cooperative growth?" Simply put, very seldomly do crystal nucleates align together and undergo formation and growth processes in synchronization—leading to a disarray of crystal domains throughout the bulk material. These adjoining crystals form an interface between the neighboring crystalline configurations. These interfaces, albeit small and having negligible volume in the crystals, have a considerable effect on the bulk crystal properties (Greer, 2013; Wang et al., 2013a). These interfaces are better known in materials science as grain boundaries (Morris, 2005).

The interest in grain boundaries arises from the fact that these are sites where adjoining crystals are located and, as a result, it is more likely that mechanical stress concentrations build up. Grain boundaries are also sites where chemically active processes typically start in a crystal. These include reduction–oxidation reactions with a chemical environment whereby redox species can form and precipitate at the grain boundaries, causing grain boundaries to be the regions of interest in strengthening as well as anticorrosion activities of crystalline materials (Wang et al., 2013a).

1.2.1.3 Dislocations

As the name implies, these sites are locations in the crystal where there are breaks from the crystalline structure and organization caused by a shift of atoms from a plane or into a plane of a crystal. These defects are typically a result of stress concentrations in the crystal or the inclusion of voids and impurities that may intercalate into a crystal structure during formation or extreme mechanical loading (Morris, 2005). These dislocations disrupt the organization and bonding order to such an extent that the deformations are irreversible and will eventually be the cause of mechanical failure in the bulk material. Types of dislocations include edge and screw dislocations. As observed in Figure 1.8, these dislocations are not static, but undergo movement and allow deformation at much lower stress than in a perfect single crystal.

Specifically, in crystalline materials, these dislocations cannot move through different planes due to the different charge densities between the current plane and the next location

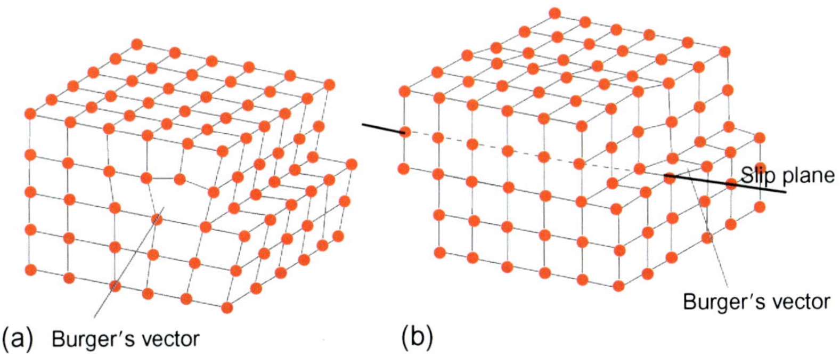

Figure 1.8 Typical crystalline dislocations include: (a) edge and (b) screw dislocations. Adapted from Morris (2005).

of the same charge. As a result, these dislocations will aggregate in the same plane and over a period of time, and the deformation caused by dislocations will be concentrated in a particular plane (Morris, 2005). This contributes to the sensitivity of crystalline materials to shear forces as opposed to failure in compression/tension (Koester et al., 2008; Lakes, 1987, 1993; Liu et al., 1999; Ritchie, 2011). Some mechanisms that have been used to blunt the effects of dislocations include designing grain boundaries or impurities to impede the movement of these dislocations. Such strategies have their pitfalls, as having too many grain boundaries or impurities affect the material properties of the bulk crystal.

1.2.1.4 Atomic impurities

The presence of "contaminants" in the form of other molecules or atoms can affect the properties of the resulting crystal significantly. Normally, due to the environmental conditions, like the exchange or addition of an ion, the unit cell configuration can be slightly skewed such that dislocations in a plane can occur. The mechanism via which this happens is diffusion; a substitutional or interstitial atom can overcome the bond energy due to nearby atoms and achieves movement via diffusion to different parts of the crystal lattice (Morris, 2005). These planar defects themselves can be the cause of materials effects such as optical distortions or even color gradients in a crystal. Additionally, with the adsorption of molecules or atoms at the surfaces of crystals, subsequent crystal growth on these surfaces will have effects on the grain boundaries. These grain boundaries themselves are of great interest since most reactions will occur at these sites because, by definition, these are interfaces that are subject to more exposure and subsequently more reactions with the environment. And with more grain boundaries, the more that crystallite sizes and number can be regulated, as these interfaces directly interact with the crystallite surfaces (Chang, 2000). Due to the dimensions of these impurities, these "flaws" can be structured such that the features of these impurities can be incorporated upon final assembly, but their weaknesses and "flaws" are mitigated by design schemes that arrange these weaknesses away from sensitive points in the crystal assembly. As a result, through strategies of design and organization of the constituent elements, crystalline materials can be engineered from the "bottom up" to have specific functions as regards their optical properties, strength, and toughness.

1.3 Characterization and construction methods of crystals

The important thing in science is not so much to obtain new facts as to discover new ways of thinking about them.

W. L. Bragg

Most of the biominerals that we will discuss here will be in the range of 1 nm to several tens of millimeters and will have been derived from a biological component. To be able to accurately investigate these biological crystalline materials, several techniques will be discussed in order to give the reader a better understanding of the diverse crystalline properties. These methods range from simple inverted microscopy for optical measurements to ion beam-based techniques for spectroscopic investigations. Because of their biological origins, conventional techniques are adapted to enable for these samples to be examined while hydrated and subjected to low doses of electromagnetic radiation (due to "charging" effects). These are some of the tools that are currently utilized to evaluate the multitude of materials properties in biological crystalline materials.

1.3.1 Optical methods

1.3.1.1 Polarized light microscopy

The simplest, most inexpensive way to investigate crystalline materials is through bright-field light microscopy. Without much sample preparation, micron-sized features can be observed in the bulk crystal. By adding specific types of filters, such that polarized light can be utilized, phase information on the crystalline constituents can be obtained. This phase information provides insight on the direction and alignment of the constituent crystals within the bulk material. This provides details on the orientation of crystalline features in the framework of the material as well as the relative organization between crystals (Figure 1.9). By further processing the sample, embedding, and sectioning, as well as using higher objectives, micron-sized features such as grain boundaries and structural defects can be observed.

1.3.1.2 Confocal light scanning microscopy

In a similar manner as optical light microscopy, a light from a laser is emitted at a predefined wavelength to illuminate the structures in the sample and a detector collects the excitation light by rastering an area of the sample at a specific focal plane. This collection of rastered focal plane measurements can be reconstructed with software to obtain a 3D projection (Figure 1.9). Obviously, the quality of these images and projections is limited by the light penetration of the emitted light into the sample, which is determined by the density and the thickness of the sample.

1.3.2 High-voltage, electron-based methods

The diffraction limits of optical white light precludes features smaller than 1 μm from being distinguished and determined. Due to the discovery that electrons can be emitted through special metal filaments and accelerated in vacuum tubes, and the development of electromagnetic lenses to focus electrons, smaller and smaller features at nanometer length scales can be spatially resolved. The way that electrons interact with matter is dependent on the matter. In this chapter, when the matter is biological in nature, consisting of lots of C–H bonding, there is a problem of charge insulation—when electrons are not charge dissipated and the sample begins to burn from the energy of the beam. To make imaging possible with electrons, the sample can be coated with a thin layer of metal, that is, Ag, Au, Pt, C, etc., such that electrons can be dissipated away from the sample and can be imaged without

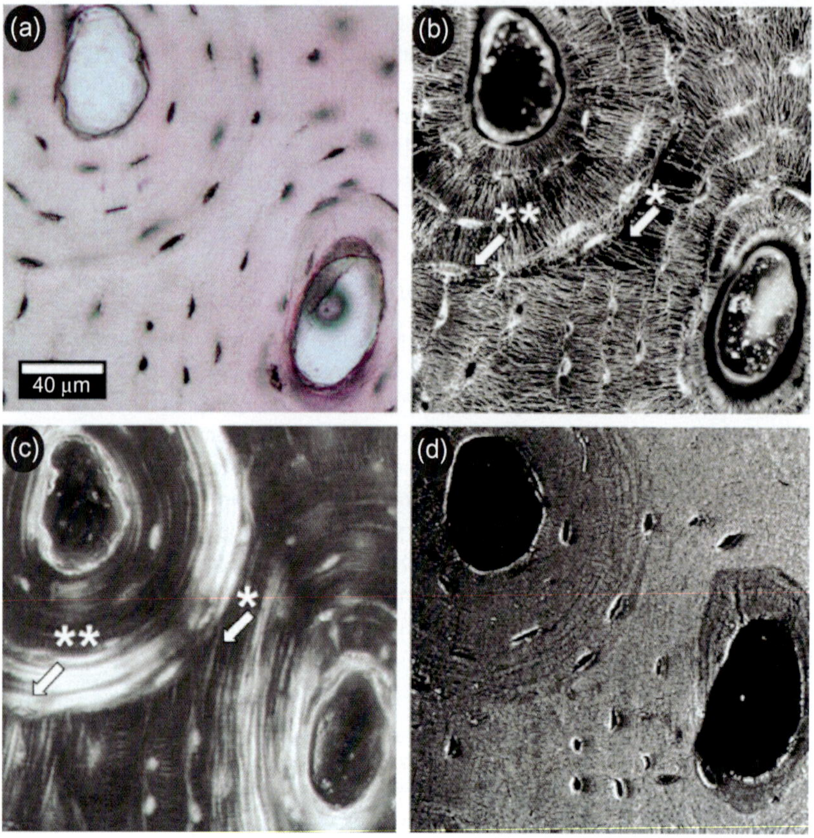

Figure 1.9 Optical imaging of osteons in equine bone. (a) Bright field of rhodamine-B stained sample. (b) Confocal imaging (* and ** represent lacunae–canaliculae networks from neighboring osteons). (c) Polarized light microscopy. (d) Electron micrograph of the same sample.
From Kerschnitzki et al. (2011).

inducing any artifacts (Figure 1.10). This is a surface-sensitive technique and is termed scanning electron microscopy (SEM) from the rastering nature of the electron beam used to produce a 2D image. Another method includes thinning the sample down such that electrons can be transmitted through the sample and provide a "shadow" image of the sample. This method is appropriately termed transmission electron microscopy (TEM). In this way, the electrons can be accelerated to higher voltages and higher spatial resolution images can be imaged (Figure 1.10). The spatial resolutions that have been achieved toward the molecular scale, able to image single atomic bonds. The inherent technique detects electrons and by measuring the energies that the electrons have after interacting with the sample, the specific elements of the sample can be identified. In this way, these electronic techniques can be utilized for elemental spectroscopy and analyses; in particular, they are able to measure the distribution of the different elements in a sample (Figure 1.10).

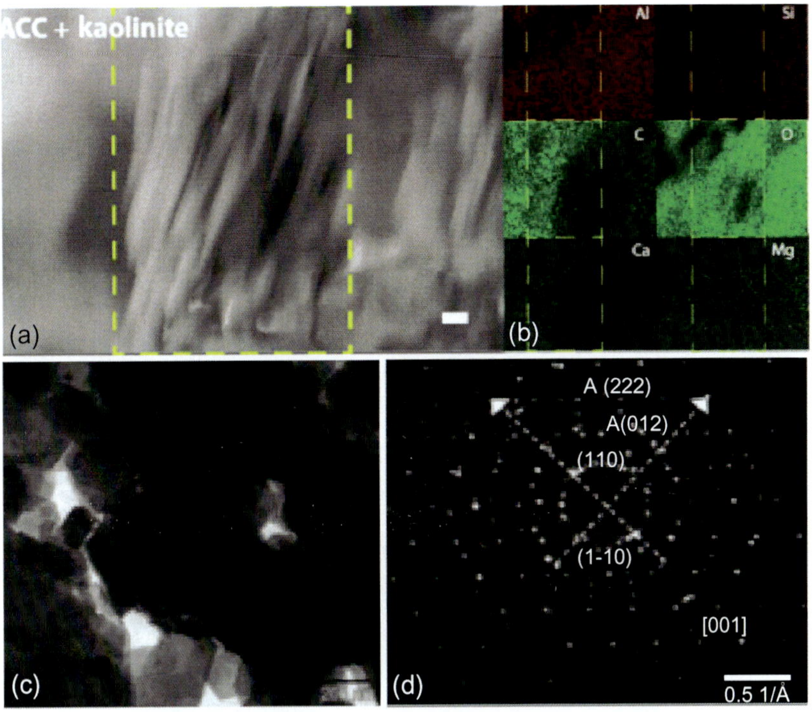

Figure 1.10 Electron micrographs of an amorphous calcium carbonate and kaolinite composite. (a) Scanning electron micrograph (SEM). (b) Corresponding electron dispersive spectroscopy. (c) Transmission electron micrograph (TEM). (d) Electron diffraction pattern to differentiate crystalline domains.
From Seto et al. (2013).

1.3.2.1 Cryo-electron microscopy

Due to the high content of the organic component in some of these biominerals and the subsequent sensitivity of these mineral–organic interactions, utilizing electron-based techniques for these materials require more sample preparation. Preparation in this section will concern itself with cryo-perseveration: using liquid ethane to rapidly plunge freeze the sample such that the liquid and hydration-sensitive organic component is frozen and "fixed" by the surrounding vitreous ice. Depending on the type of electron microscopy to be performed, the sample preparation can diverge at this point.

In the case of cryo-TEM, the sample is frozen with a thin layer of vitreous ice. It is kept frozen with constant exposure to liquid ethane and cold stages before transfer to the TEM with a cryo-holder for imaging. Since the sample is itself surrounded by a frozen network of vitreous ice not only in 2D, the sample stage can be tilted to also image the sample at specific angles relative to the incident electron beam. Using appropriate software, several images of the sample can be collected and an "average" reconstructed image in 3D can be created. Typical spatial resolutions of several tens of angstroms can be achieved with the use of computer-aided image refinements.

For bulk samples such as tissue samples that also have a disproportionate organic content (i.e., proteins, lipids, cells, etc.) with respect to minerals, a technique such as cryo-SEM is required. A major difference in the preparation of material for this technique compared to cryo-TEM is the additional sectioning or "freeze fracturing" or "freeze etching" of the sample while it is frozen with liquid ethane to reveal the sample surface of interest. Since cryo-SEM is a surface-sensitive technique, only the surface matters. Typically, after the sample is frozen, the sample is transferred to a cryo-microtome for "freeze fracturing" large sections of the samples to reveal the bulk interior structure of the sample. In the case of "freeze etching" the sample is placed in a vacuum dryer and under specific pressure, and part of the frozen ice is sublimated—revealing more of the buried frozen sample in the ice. This therefore requires an SEM equipped with a cryo-stage to keep the sample frozen in order to investigate these samples with similar spatial resolution as can be achieved in a typical SEM (Figure 1.11).

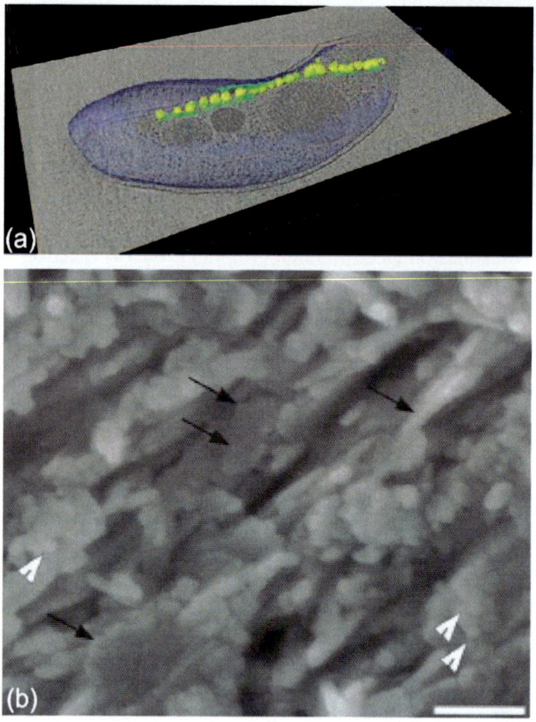

Figure 1.11 Cryo-electron microscopy (cryo-EM) as a tool to image sensitive, hydrated samples. (a) Cryo-TEM is able to visualize and perform tomography showing the process of magnetosome invaginations in bacterial cells. (b) Cryo-SEM shows the presence of amorphous mineral phases in zebrafish fins (black arrows show plate-like mineral hydroxyapatite, white arrowheads show decorations of round amorphous mineral phases). Scale bar = 100 nm.
(a) From Komelli et al. (2006) and (b) from Mahamid et al. (2008).

1.3.2.2 Atom probe tomography

Instead of using accelerated electrons, ions can be used as a source to deposit as well as etch away samples. In the case of ion beam mass spectroscopy, a beam of gallium arsenide (GaAS) are often used to investigate surface and bulk material properties. A series of spectroscopic analyses can also be performed depending on the types of detectors that are coupled with the ion milling processes. An example is time-of-flight mass spectroscopy, whereby the incident GaAs beam induces the sputtering of charged ions from the sample. These charged ions have a typical velocity that is dependent on their mass and the initial energies of the GaAs beam (Figure 1.12). By having a detector that can determine the amount of energies these ion species have at a specific distance away from the GaAs milling process, elemental analyses can be performed to determine the chemical composition of the sample. Furthermore, since the GaAs ion beam can also be used continuously through the bulk thickness of the material, an elemental analysis from the surface to the bulk of the sample can be carried out as a function of depth of the ion milling process. Using appropriate software, the sputtered ions can be reconstructed with their respective spatial information and a tomographic image with combined spectroscopic information can be created. This technique is also dubbed secondary ion mass spectroscopy or ion probe mass spectroscopy. Other techniques that also use a similar method for detection and spectroscopy include Auger and X-ray photoelectron spectroscopies.

1.3.3 Atomic force microscopy

By mechanically tapping a sample surface with a sharp metal tip with defined dimensions, with known elastic modulus and deformation properties, one can determine the hardness properties of the sample surface when observing the changes of the tip during the tapping process. This is the main mechanism in atomic force microscopy (AFM). With AFM, one can observe events on relatively flat surfaces and find defects or planar imperfections at the sample surface (Figure 1.13). The spatial resolution of AFM can be at the single $1 \times 1\ \mu m^2$ area. Additionally, the tip can be chemically functionalized with small molecules such that interactions with a specific chemical

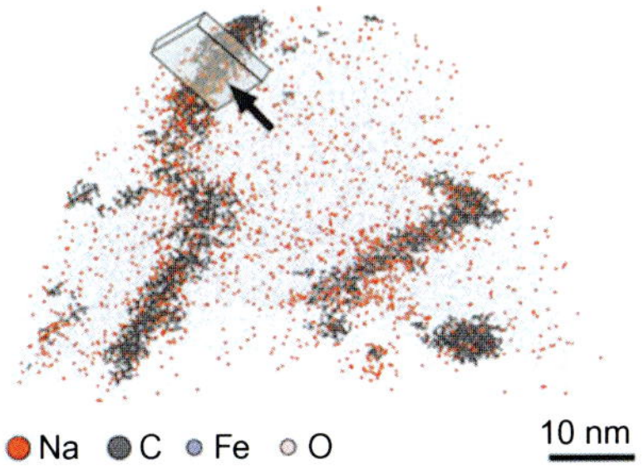

Figure 1.12 Nanoprobe measurement of a radula tooth revealing the spatial distribution of the intrafibrillar organic component and interactions with Na^+ ions in the bulk crystalline material. From Gordon and Joester (2011).

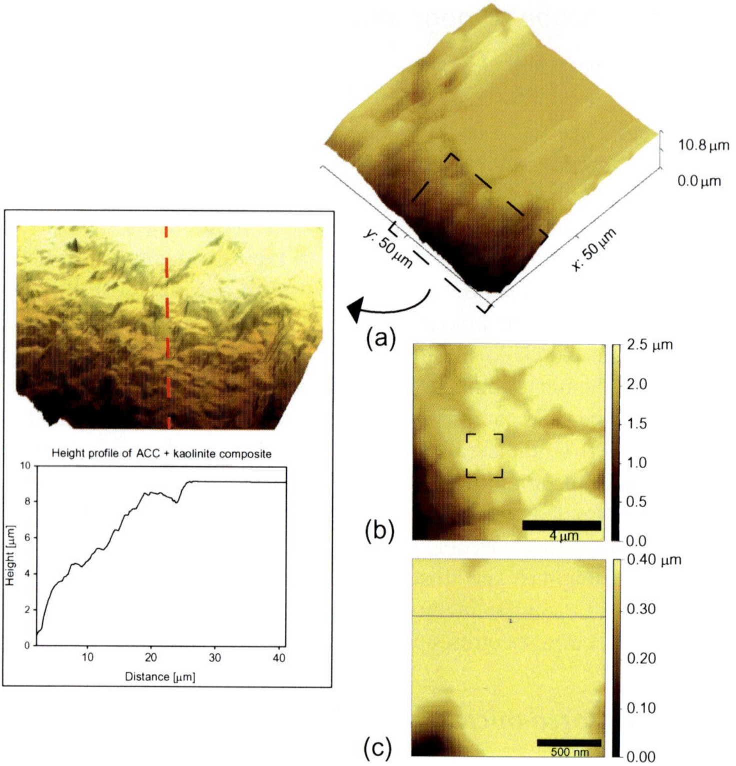

Figure 1.13 Atomic force microscopy (AFM) of an ACC–kaolinite hybrid composite. (a) Three-dimensional height profile of the ACC–kaolinite composite on a slide with subsequent higher magnifications. (b) $10 \times 10\,\mu m^2$ area scan of (a). (c) $1.5 \times 1.5\,\mu m^2$ area scan of the region of interest highlighted in (b).
From Seto et al. (2013).

group on the surface can also be mapped. These include amine-carboxyl side group interactions or examining differences of charge distribution in a material. Additionally, by measuring the pulling force and the displacement on the cantilever of the tip force displacement curves can be obtained to determine the energies that are involved in a specific interaction with the chemically modified tip.

1.3.4 X-ray and high-energy imaging methods

1.3.4.1 X-ray diffraction

The Bragg equation represents the relationship where an incident monochromatic X-ray beam of wavelength λ interacts with a crystal sample and the resulting emitted X-ray beam can be detected for use in delineating the structural properties of the sample. This relationship is summed up in Figure 1.14, demonstrating the behavior of X-rays when interacting with a crystal and the scattering points within the crystalline sample.

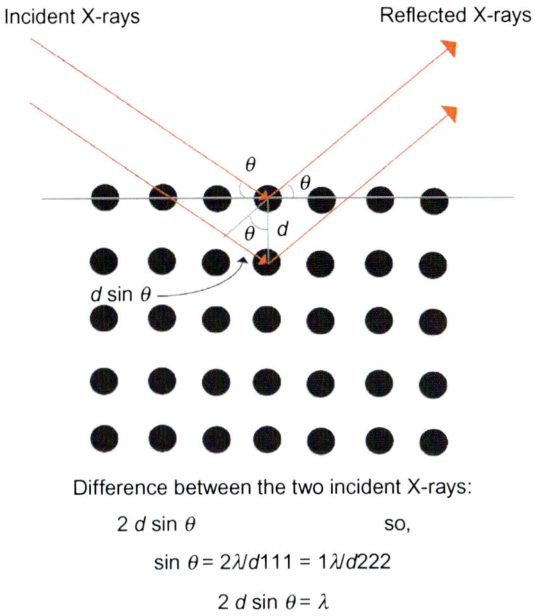

Figure 1.14 Schematic showing the interaction of X-rays with a crystal and the derivation of Bragg's law.

Thus, with information like the wavelength of the incident X-ray beam, the *d*-spacing, and a diffraction pattern, one can measure the cell dimensions of a crystal. Specifically, the dimensions of the interacting scattering centers in the crystalline sample can be pieced together and the unit cell configuration can be determined as a result. Often, it is the case that the X-ray diffraction (XRD) pattern is characteristic of specific doped elements and polymorphs of crystals (i.e., calcite, aragonite, and vaterite). Thus, this technique is invaluable in determining the presence of elements and crystal arrangements in the crystalline sample.

1.3.4.2 Small-angle X-ray scattering and wide-angle X-ray scattering (SAXS and WAXS)

Similarly, by moving the sample-to-detector distance one can focus on specific regions of 2θ. In the case of small sample-to-detector distances, one focuses on the scattering intensities at larger 2θ values, and sometimes this provides overlap with the XRD measurements as it gives a broad 2θ distribution (Figure 1.15). In contrast, at larger sample-to-detector distances the focus is on the smaller 2θ region and this can provide information on the types of phases and their orientations involved in the sample.

1.3.5 Solid-state nuclear magnetic resonance

Yet another method to determine the phase and orientation of the constituents in a crystalline sample is to apply electromagnetic pulses such that specific behaviors of

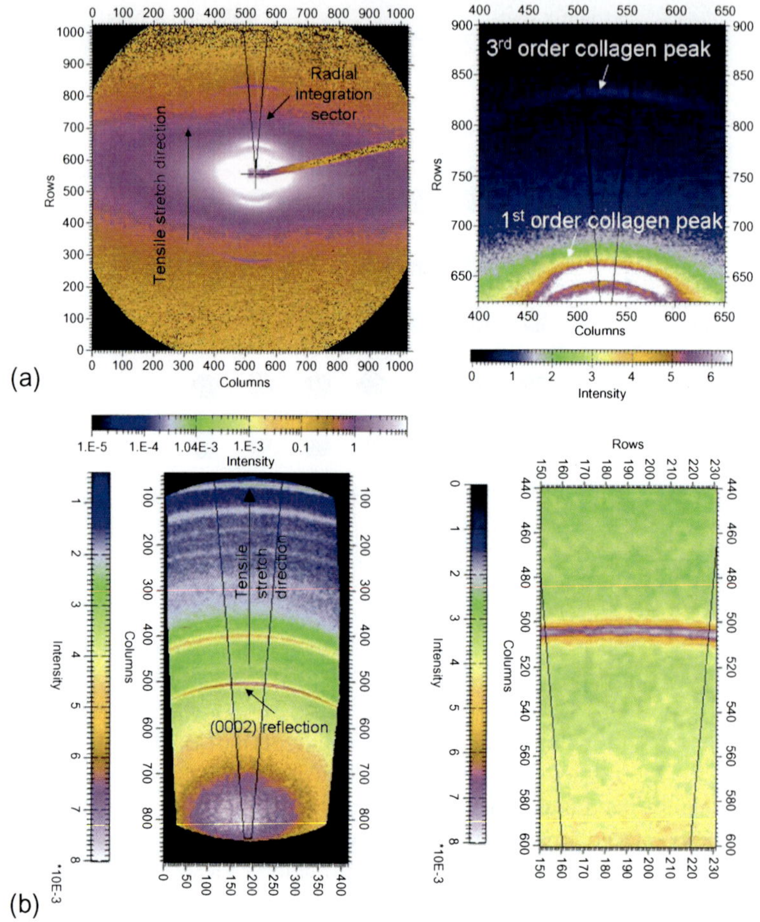

Figure 1.15 SAXS and WAXS measurements of fibrolamellar bone under tensile loading. (a) SAXS measurements showing the *d*-spacing of the mineralized collagen fibers. (b) By tilting the sample, the (0002) reflection from the WAXS measurement of the mineral can be observed and measured.
From Gupta et al. (2006a).

the sample can be recorded and analyzed to infer the types of elements and phases present in the sample. The responses are atomic-level nuclei spin responses to transfers of energies from the applied electromagnetic signals; a coupling of anisotropic spins between bonded atoms can result, such that cross-polarization measurements between adjacent atoms can be detected or a decoupling such that certain spin interactions can be removed to increase the resolution of nuclear magnetic resonance (NMR) spectra and isolate specific spin interactions (Figure 1.16). The advantage of solid-state NMR (ssNMR) is the ability to probe specific atoms at any point in time and, through developments in scan frequencies, be able to resolve neighboring molecular residues with high spatial resolution (Wang et al., 2013). In terms of materials properties, ssNMR is

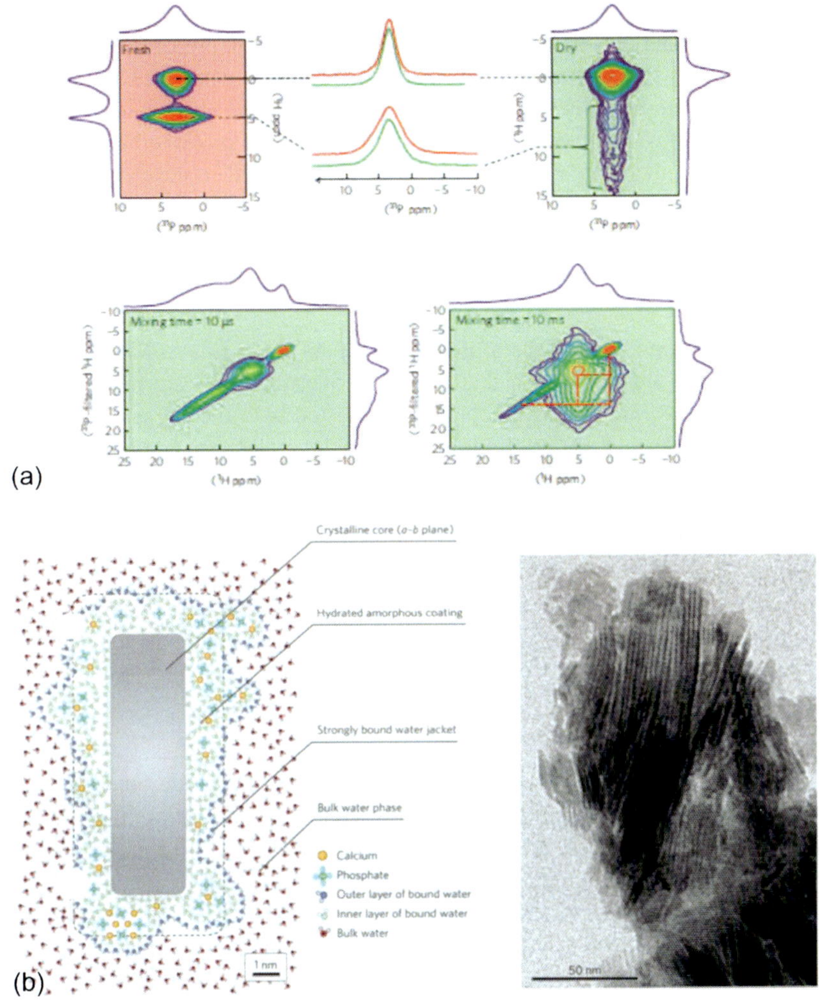

Figure 1.16 NMR measurements of the structural water at the surface. (a) Between fresh and dry bone, indicating that the mineral has many interfaces with an amorphous layer. (b) A model that utilizes the data to unify the NMR measurements and compares it to an electron micrograph. From Wang et al. (2013b).

still the only method able to quantify the amount of an amorphous content in a sample, making it an invaluable tool to study phase transitions during mineralization.

1.3.6 In situ *measurements*

With the presence of proteins and other organic macromolecules, the use of high-vacuum, high energy-based techniques for characterization of these biological hybrid materials may not be the best situation for these samples. Too often these investigations into these biological crystalline materials are clustered with artifacts and correspondingly, the results are skewed and inaccurate as a result of insufficient sample preparation or instrument-induced features that throw the investigations off at a tangent. Instead, the development of microfluidic cells that keep these materials close to their original environments, usually physiologically, hydrated conditions, may prove to be better than the current array of methods. Some of these methods are still in development and may not be readily available, but are likely to be available coupled with AFM and TEM facilities soon.

1.3.6.1 In situ *AFM*

Using a similar approach to the aforementioned AFM, *in situ* AFM encloses the tip and sample in a fluid cell such that the AFM measurements are made in solution. This can occur as long as all the optics are adjusted with the refractive index of the solution (typically ddH_2O) in mind and the sensitive electronics are separated from the wet sample. With the cell, the tip can measure samples while submerged in solution and provide similar types of measurements on the elastic modulus and hardness (Figure 1.17). Additionally, by combining the wet cell with temperature control and the raster speed of the AFM, dynamic processes in solution can be observed in real time (Long et al., 2013; Cho and De Yoreo, 2013). This can include investigations into a protein's denaturation process as a function of temperature for temperature-regulated processes like mineralization.

1.3.6.2 In situ *TEM*

As for TEM, a microfluidic cell made of silicon nitride windows that encapsulates a solution volume is used for *in situ* TEM investigations. The windows themselves have single-figure nanometer thickness, thin enough that the electron beam is able to pass through without any interaction with the cell. The advantage of this method is the ability to probe at nanometer length scales in hydrated conditions without inducing any artifacts from sample preparation or during measurements (Li et al., 2012; Nielsen et al., 2014; Smeets et al., 2015). Processes that require high spatial resolution in solution can be investigated via *in situ* methods, including processes of crystal nucleation, growth, and interactions with small molecules (Figure 1.17).

1.3.7 Formation of crystals

We have exhaustingly gone through the various methods that allow us to quantify some material properties of crystalline materials, but have not yet considered methods to synthesize crystals. Crystal engineering has been on the minds of many research groups for many years—each with specific focuses and using various methods to make crystals. With the increasing interest in how biological macromolecules can interact with crystals to make samples with defined properties, and the use of X-ray crystallography to investigate how crystalline defects can cause disease, more people have begun to investigate ways to examine these interactions. Below are a few areas in which research is active and continuing.

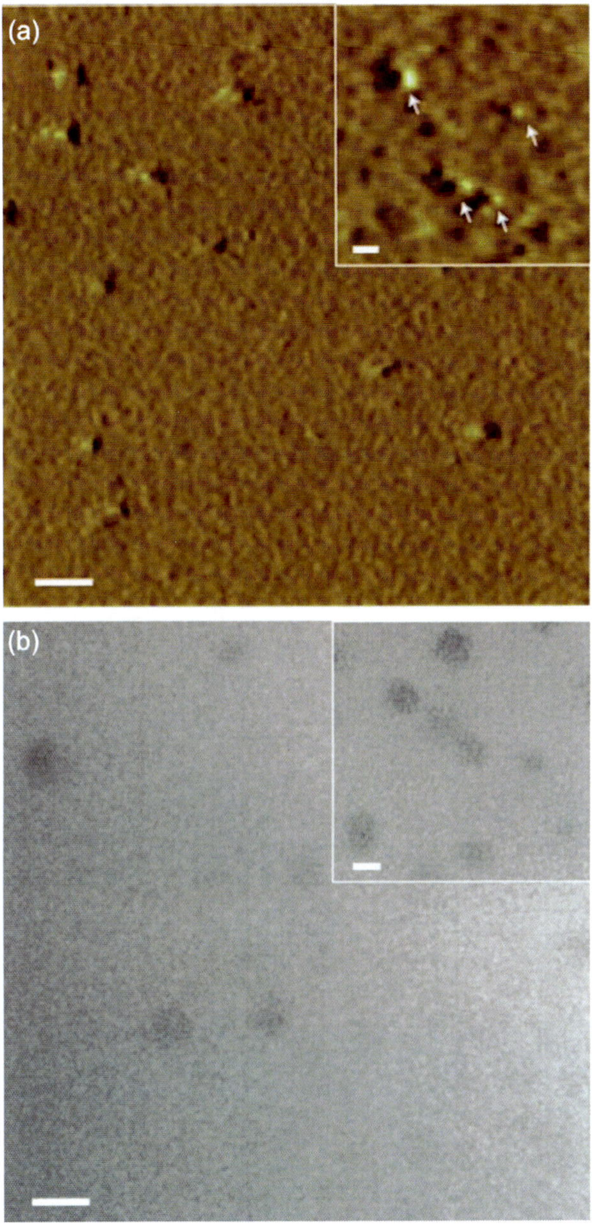

Figure 1.17 Utilizing *in situ* fluid cell AFM and TEM to probe the interactions of a polymer during crystallization. (a) *In situ* AFM demonstrates the physical structure of the polymer–mineral assembly in solution. (b) The same samples are investigated using *in situ* TEM to show the specific interactions of the polymer PSS with mineral.
From Smeets et al. (2015).

1.3.7.1 Vapor diffusion

This method has been utilized for decades to make large crystalline samples for investigations. In the case of calcium carbonate, people know that the use of $CaCl_2$ salts can be combined with a $-CO_3$ group to induce mineralization. When this $-CO_3$ is attached to an ammonium molecule, the vapor pressure at room temperature is high and the compound exists as a gas. In a closed dessicator, a drop of $CaCl_2$ can be placed on an objective slide and placed inside in the presence of powdered NH_4CO_3. Over the course of several hours, the NH_4CO_3 will diffuse into the $CaCl_2$ drop and with time Ca^- will interact with a $-CO_3$ group to form a $CaCO_3$ seed crystal (Chang, 2000; Figure 1.18). This method is the classical method used to investigate crystal nucleation and growth.

1.3.7.2 In solution

By mixing stoichiometric amounts of solutions of the anions and cations of salts, crystals can also be produced (Faatz et al., 2004). However, due to the concentrations and the effect of water solvation, the resulting crystals are much smaller than those produced by the vapor deposition method. However, this method is more controllable as pH, ion activity, and solution volumes can be varied with ease using the solution method (Figure 1.18). Especially in low-volume and complicated flow situations, microfluidic chips can be used to perform the experiments. This enables yet another way to control flow and processes in solution.

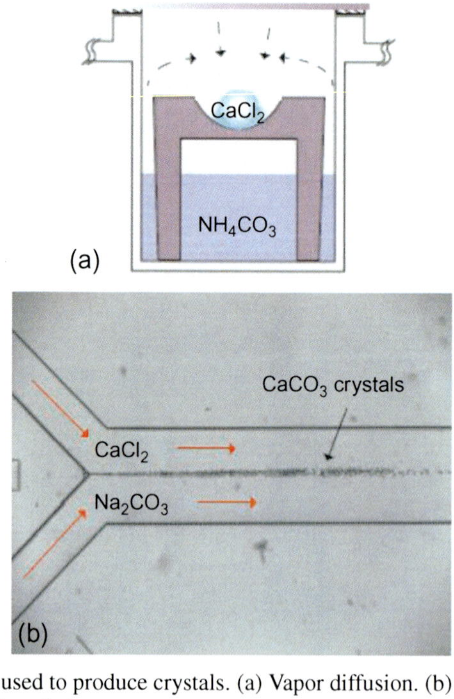

Figure 1.18 Methods used to produce crystals. (a) Vapor diffusion. (b) Mixing with microfluidic channels in solution.

1.3.7.3 Organic additives to drive mineralization

In addition to being able to constrain flow and change flow rates in microfluidic channels to control the diffusion-limited mineralization products, several groups have been using synthetic polymers to regulate local supersaturated ion concentrations in bulk solutions. Among these polymers utilized are polyaspartic acid (pAsp), polyacrylic acid (pAA), and polysulfononesulfate (pSS) (Figures 1.17 and 1.19). A clear observation just from

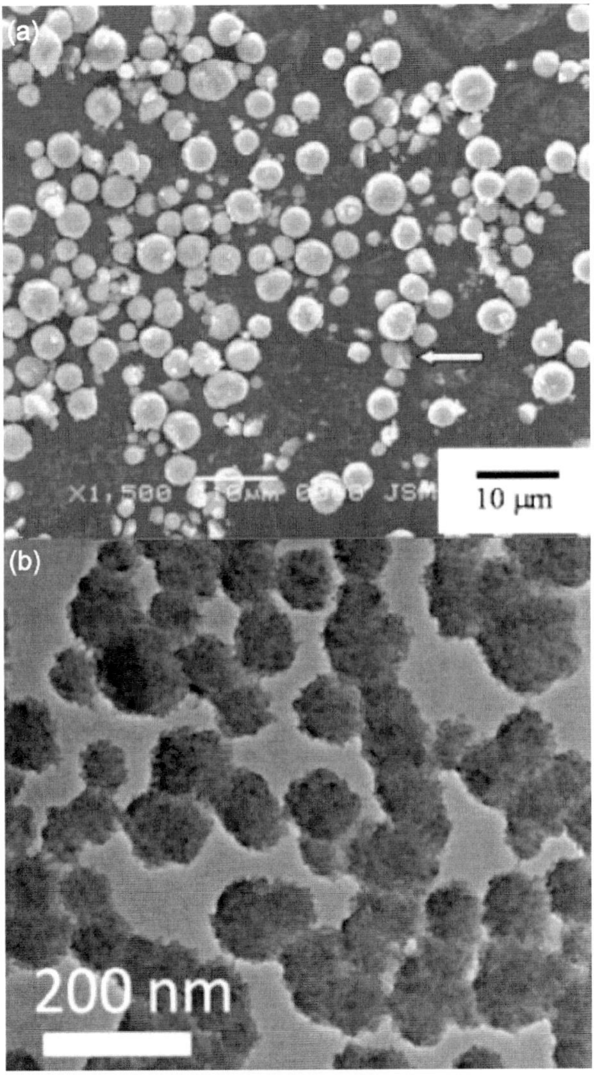

Figure 1.19 The effects of various polymer additives during mineralization. (a) The presence of pAA in calcium carbonate mineralization. (b) The presence of pAsp in calcium phosphate mineralization. (a) From Huang et al. (2008) and (b) from Cantaert et al. (2013).

the names indicates that many of these polymers themselves have acidic characteristics. As briefly described in Figure 1.3, charged functional groups can interfere with anion and cation binding interactions. In the case of pAsp and pAA, the carboxylic acid groups have a predisposition to bind cations. These polymers in fact then mediate the cation and anion interaction processes (Figure 1.19). Specifically, in the presence of pAsp polymers in a mineralizing calcium carbonate or calcium phosphate solution, the pAsp is able to interact weakly with the Ca^{2+} and produces a polymer-induced liquid precursor phase (PILP) that stops the Ca^{2+} from participating in mineral formation (Figure 1.19). Although the use of pAsp may not be directly an equivalent of a biomolecule found in biology and used by organisms, there are parallels that can be drawn from the activity of pAsp in mineralization. Indeed, there are biomolecules that function in a very similar way to stabilize mineral precursor phases. The use of pAsp as well as other synthetic polymers that mimic the function of various mineralization proteins by binding ions to prevent mineralization from progressing and producing amorphous mineral phases is a powerful model system to study the mineralization processes *in vitro*.

1.4 Biological crystalline materials

Crystals are like people: it is the defects in them that make them interesting.
Sir Frederick Charles Frank

The marvelous shapes and forms of crystalline materials are not borne out of only single nanocrystals, but rather by complex assemblies of nanocrystalline subunits. These assemblies are themselves a result of processes that are biological or cellular in nature and require control at various length scales. By observing the diversity of shapes and colors that can be constructed biologically, these materials can serve as an inspiration for future materials with similar functional properties. In this section, the diverse mineral–organic interactions that result in the diversity of crystalline materials in the biological world will be discussed and examples will be highlighted in order to provide the reader with a basic understanding of the mechanisms involved in the formation and assembly of these materials.

The various mineral systems that are biologically controlled are shown below, demonstrating the various mechanisms of mineral–organic interactions including organic molecules as templates, in encapsulation, at interfaces, as well as use in inhibition of crystal nucleation and/or growth. Some examples of these mechanisms as observed in their native biological model systems are given below.

1.4.1 Magnetite

When we pull out our trusty compass, we rely on the Earth's magnetic field to rightfully show us magnetic "North" so we can find our way. The magnetic properties of the compass needle allow reorientation with the orientation of magnetic "North" to allow the needle head to consistently point in the right orientation. Many organisms like migratory birds and fish that depend on having spawning grounds continents away have evolved a magnetic sensory system that enables them to orient their migratory patterns according to magnetic "North." However, even simpler organisms like bacteria have come up with "apparatuses" to be able to orient to this geological magnetic field. These bacteria have developed a network of magnetosomes whereby the magnetic field provides depth information. Being too deep in a lake may be less advantageous to survival due to the lack of light and nutrients; equally disadvantageous is being in water that is too shallow, making these bacteria vulnerable to competition and dramatic changes in environmental conditions.

The formation of the magnetosome is dependent on a family of Mam proteins that transport Fe_3O_4 and act as templates in the growth of these small magnets (Figure 1.20). The orientation of these small magnets is important since there

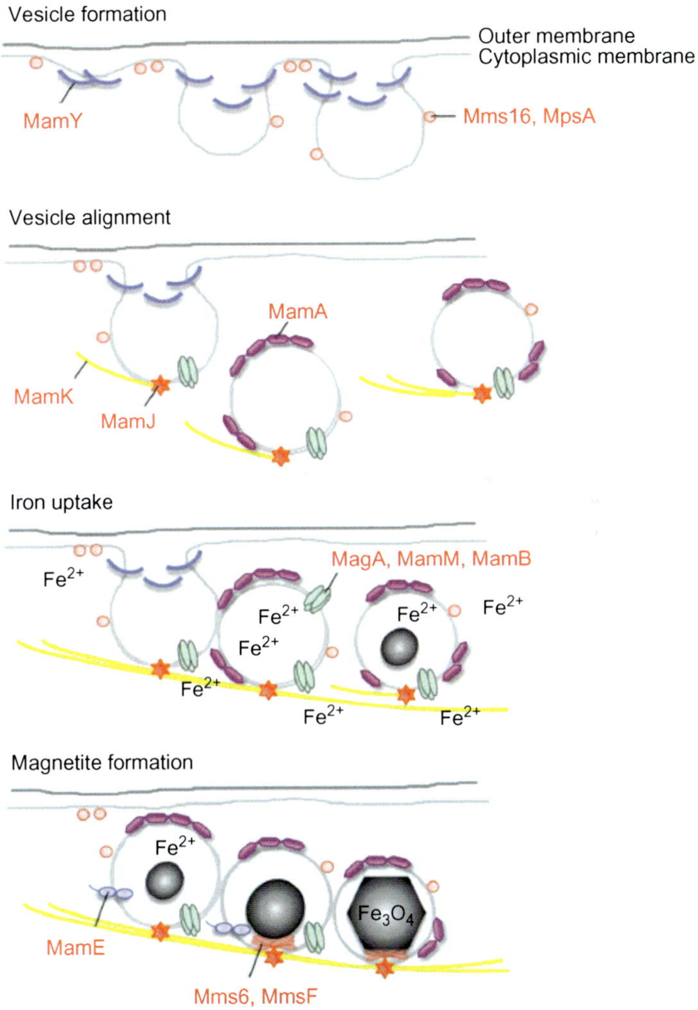

Figure 1.20 Processes involved in the production of magnetite, from vesicle formation to iron uptake and finally magnetite formation regulated by Mam, Mms, and Mps proteins.
Arakaki et al. (2015).

needs to be a net average magnetic moment in the entire magnetosome to be able to detect weak geological magnetic fields. When the individual magnetites are randomly oriented, the net average magnetic dipole is zero. Thus, the MamK protein has a role in positioning the epitaxial group of the magnetite along the same orientation where the magnetic dipole is maximized (Komelli et al., 2006). These proteins have varied functions and have transmembrane segments that form stable oligomers (Gruenberg et al., 2001; Baumgartner et al., 2013a,b). The most abundant MamC and MamF are small, approximately 12 kDa, and have transmembrane domains that can form stable oligomers. Other proteins like MamD and MamG have hydrophobic domains and contain Leu-Gly repeat domains consistent with the characteristics of framework proteins that form scaffolds and act as templates for mineral growth. How all these proteins act in concert and build a magnetosome that can provide the host bacteria with a sense of the Earth's magnetic field is very impressive.

1.4.2 Silica

Silica is found in the form of sand, beach glass, as well as mountains, making up the 10% of silica found in the Earth's crust. A small amount of this silica comes from biologically derived sources. These include plants as well as marine organisms, which use silica to construct their intricate body structures or provide the conditions for symbiotic relationships.

1.4.2.1 Plant-formed silica

Among the first uses of silica in biology came from flora that littered the Earth. At that time 23 million years ago, during the Neogene period, silica was abundant in the soil from the diagenesis of many sources of biogenic silica. Typically, plants are able to absorb elemental Si in the form of silicic acid and via several steps of condensation in the processing of silicic acid such as the formation of amorphous silica, silica mineral is formed and incorporated into the physiology of the plant (Figure 1.21). Through the formation of several types of aquaporin-like membrane proteins that can transport silicic acid (Ma and Yamaji, 2006; Currie and Perry, 2007), specific influx and efflux transporters have evolved to handle the amounts of elemental Si in plants. In many cases, the silica can be a component of the cell wall or an external growth and a depository for silica like a phytolith can be formed as a component of the plant. This is a very common mechanism among grasses and a specific example is horsetail grass, which has nodules of millimeter-sized silica that are precipitated as a result of conversion from silicic acid from the soil.

1.4.2.2 Silicatein sourced silica

Another method utilized in biology to integrate silica into the physiology of organisms is the utilization of proteins called silicateins. These silicateins are a family of enzymatic proteins which can bind soluble silicates, assemble, and precipitate silica to form the intricate mineralized structures found in sponges and corals. Unlike plants,

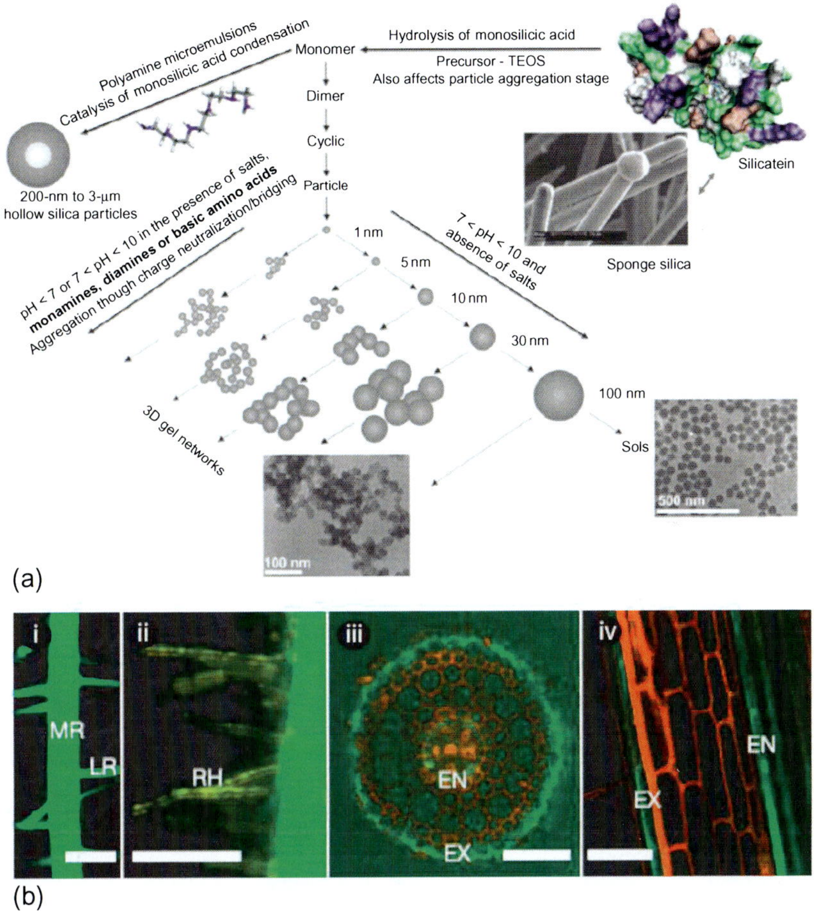

Figure 1.21 The formation of silica via silicic acid-derived compounds in plants. (a) General formation mechanism of silica from orthosilicic acid. (b) Localization of a silicon-related transporter in plants, Lsi1, shown in green in various tissues of rice (i and ii) root and (iii and iv) stem counterstained with propidium iodide in red. (i) MR = Main root, LR = Lateral root, (ii) RH = Root hair, and (iii and iv) EN = Endodermis, EX = Exodermis.
From Belton et al. (2012) and Ma et al. (2006).

which have a cell wall, these organisms can separate sensitive cells away from the high concentrations of elemental Si in order to be able to undergo a long condensation process. In animal cells, all these processes occur in enclosed spaces like vacuoles and require active processes to recruit mineral and package the mineral. These processes take advantage of silicateins, which are a family of proteins that mediate the formation of biosilica from monomeric silicon compounds.

The silicatein family of proteins is highly regulated and undergoes posttranslational modifications such as phosphorylation and glycosylation, indicating that these domains in these proteins are responsible for processes that maintain interactions with other macromolecules or maintain features that enable processes to occur in certain environmental conditions (Weaver and Morse, 2003; Mueller et al., 2007a). There exist three isoforms of silicateins, α/β/γ, each with similar putative sequences to the enzyme cathepsin (Weaver and Morse, 2003). Much of the sequence is composed of hydroxylated amino acids like serine, threonine, and tyrosine, and conspicuously maintains a domain of poly-serine and hydroxyl groups toward the active binding sites to Si mineral. Specifically, the Ser–His hydrogen bond in silicatein is the key active site for nucleophilic attack on the elemental Si. Through an Sn2-type reaction, the side-chain oxygen covalently binds with the Si atom with release of a hydroxyl group. By hydrolysis of the Si–silicatein intermediate, the Ser–His hydrogen bond is reformed and the Si group is released (Figure 1.22). This interaction is reversible and can continue in perpetuity. Other groups have proposed alternative routes for the interaction of silicateins, silicatein-like proteins, and precursor forms of silica in order to form the mineral silica, suggesting that these proteins undergo various routes of evolution to accommodate the various species of precursors to drive the critical formation of silica in organisms (Kroger et al., 2002; Poulsen et al., 2003; Mueller et al., 2007b).

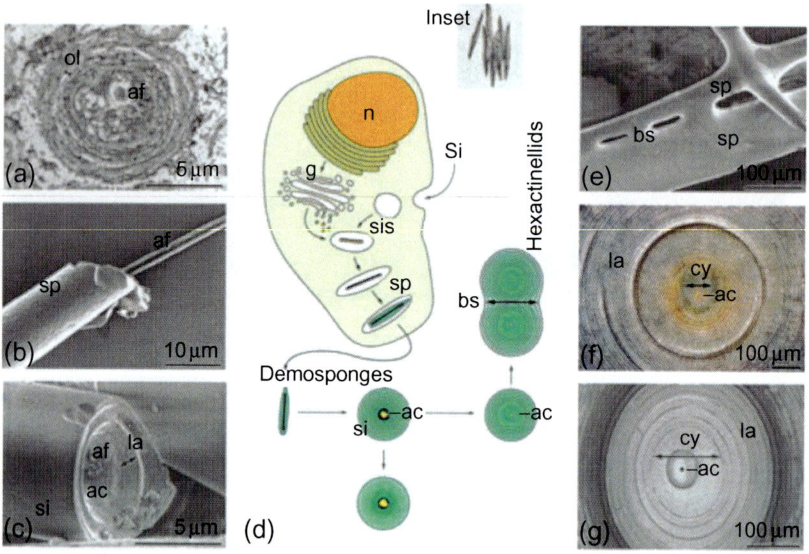

Figure 1.22 The formation of a silica sponge from schlerocytes secreting silicateins. (a–c) The concentric layers of silica around an axial filament (af) of silicatein proteinacous component. (d) Mechanism of cellular secretion of the sponge spicules. (e and f) Post formation development of the sponge spicules can include sintering of two filaments as well as massive production of concentric layers to produce large sponge spicules. (a) ol = organic layer, af = axial filament; (b) sp = tylostyle spicule, af = axial filament; (c) si = siliceous, ac = axial canal, af = axial filament, la = lamellae; (d) Si = silicate, n = nucleus, g = Golgi, sis = silicasomes, sp = spicule, ac = axial canal, bs = biosinter; (e) bs = biosinter, sp = spicule; and (f and g) la = lamellae, ac = axial canal, cy = cylinder.
From Wang et al. (2012).

1.4.3 Calcium carbonate

As one of the most geologically abundant compounds, calcium carbonate can be found everywhere from limestone at the beach and mountains to the surfaces of paper coated with particles of $CaCO_3$. We can also find $CaCO_3$ in our own physiology. Our own physiology contains otoliths, which are composed of $CaCO_3$, are found in the inner ear canals to allow spatial orientation and an upright posture. $CaCO_3$ was predominantly used by organisms from the oceans; as elemental Ca became more geologically abundant and the abundance of CO_2 from the atmosphere was increasing, $CaCO_3$ became plentiful in the seas.

1.4.3.1 Mollusks

Among the organisms which used $CaCO_3$ were mollusks, in the formation of their shells. Mollusk shells are made up of mainly two types of $CaCO_3$—the outer sides contain calcite and the inner sides are made of aragonite polymorphs that form nacre (Barthelat et al., 2006, 2007; Barthelat and Zhu, 2011). There are specialized cells that line the inside and outside of the growing shell. These cells create a treadmill-like conveyer belt that transfers mineral from the outside environment into an interior mineralization space that is surrounded by a membrane (Figure 1.23). This mineralization interior is also subdivided into smaller domains to compartmentalize individual mineral nucleation sites. Eventually, the organic membranes are overgrown by the mineral tablets, which grow until a continuous shell is formed. The cells themselves are able to direct the nucleation and crystal growth processes of the tablets by constructing an extracellular matrix that is formed by various numbers of proteins (Evans, 2008, 2012). One of these proteins, N16, is a framework protein that has regions of long repeats of proline and arginine forming a domain that is intrinsically disordered, as well as domains that are highly charged (Collino and Evans, 2008; Ponce and Evans, 2011). These features presumably enable this framework protein to accomplish its function by self-assembling into large assemblies, as well as bind specific polymorphs of mineral (Seto et al., 2014a). Indeed, as observed by several groups, the N16 protein is required to facilitate organization and growth of an aragonite and calcite polymorph through a strategy of compartmentalization (Finnemore et al., 2012). Through continued nucleation and growth, the shell grows by accretion with new mineral utilizing the same cell-regulated mechanism.

1.4.3.2 Sea urchin larvae

Another example of a mineralized tissue based on $CaCO_3$ is the echinoderm spines. Specifically, in sea urchin larvae over 200 proteins are found to be able to intercalate into micron-sized spicules that are found to provide the larvae their structure and shape. The importance of these proteins can be realized when one by one these proteins can be "deleted" using gene silencing techniques like morpholino antisense oligonucleotides. In one instance, the protein SM50 found in this mineralized spicule is found to stabilize intermediate mineral phases (Gong et al., 2012).

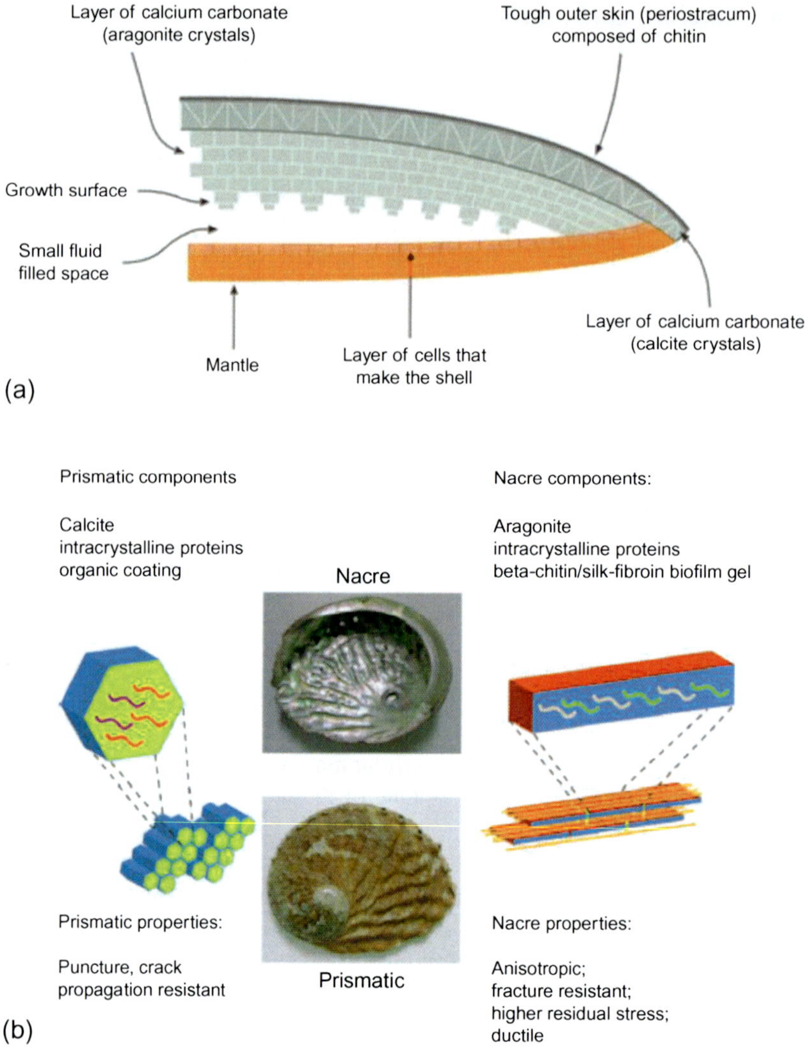

Figure 1.23 Formation of the mollusk shell involves two sets of cells and proteins.
(a) A schematic of the cells and the treadmilling of minerals from the extracellular to the intercellular side. (b) The resulting tissues from the concerted actions of cells and proteins inside and outside of the forming shell.
From Evans (2008).

These hybrid mineral–organic intermediates enable the cell to transport mineral intracellularly without "clogging" other pathways in the cell (Figure 1.24). This protein is the most abundant in the mineralized spicule (Seto et al., 2004). The proposed function of this protein is to bind a mineral precursor, transport these minute amounts of mineral packages through the primary mesenchymal cells, and assemble these mineral packages into coherent mineralized structures used for support.

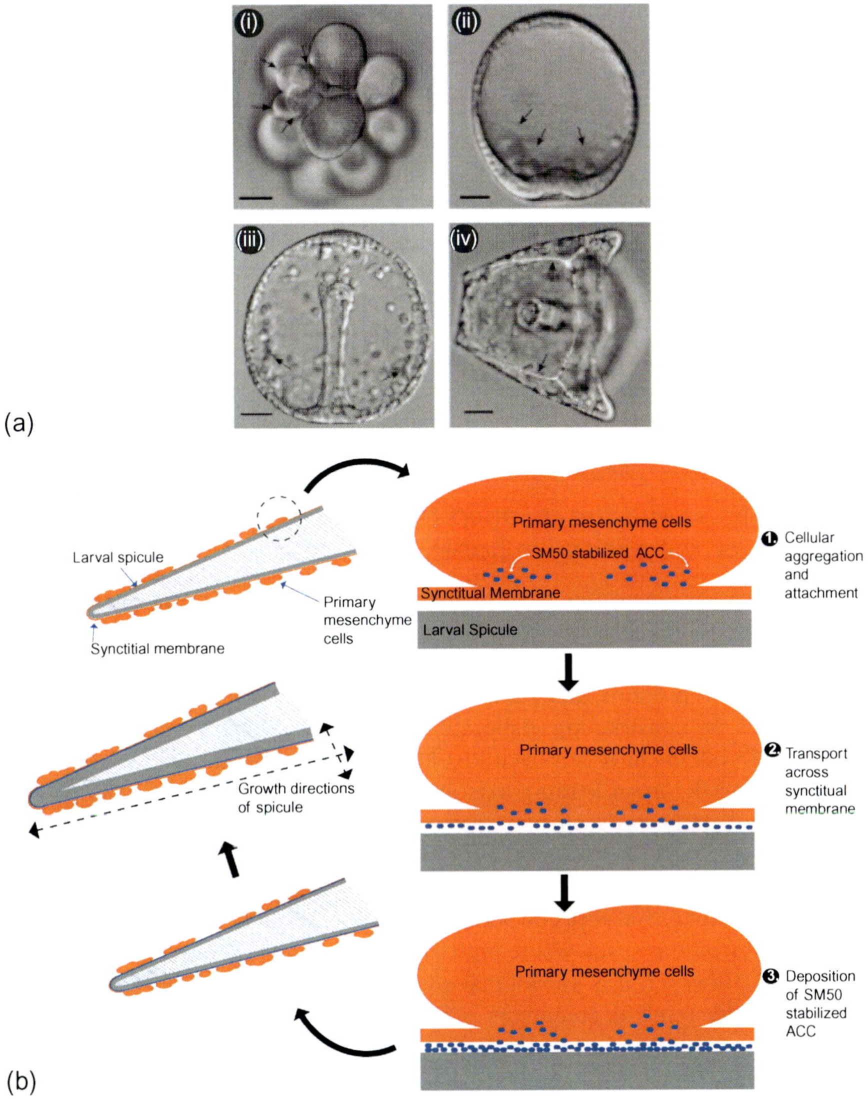

Figure 1.24 The formation of spicules in sea urchin larvae. (a) Light microscopy of the sea urchin larva undergoing: (i) cell division, (ii) blastulation, (iii) gastrulation, and (iv) spicule formation during the pluteus stage. (b) Proposed mechanism of spiculogenesis from primary mesenchymal cells and spicule matrix proteins like SM50.
From Killian and Wilt (2008).

The exact mechanism by which this protein interacts with the precursor phases of $CaCO_3$ is not entirely known, but it does contain hallmarks of proteins that interact with charged mineral ion species. SM50 has intrinsically disordered protein (IDP) regions toward the N-terminal end, as well as a C-type lectin domain, that is assumed to play a role in the self-assembly process, and repeats of asparagine and proline amino acids (Zhang et al., 2000). Additionally, when homologous SM50 proteins are depleted in various other species of sea urchins, no spicules are formed (Peled-Kamar et al., 2002).

With the evolution of organisms out of the oceans and onto terrestrial land, $CaCO_3$ as a mineral for support and higher loads proved to be insufficient. The adaptation to calcium phosphates became more sustainable and better suited for the terrestrial requirements of organisms.

1.4.4 Calcium phosphate

Calcium phosphate in our bodies can be found in abundance in our bones and teeth, two tissues which undergo tremendous wear and tear during our lifetimes. It is due to the structure and organization of calcium phosphate, in the form of hydroxyapatite, that we can enjoy chewing on macadamia nuts and a game of tennis into our 90s. We will discuss some of the schemes below.

1.4.4.1 Bone and the collagen that act as a template for the mineral in bone

Our bones made up of this type of mineral and, without it, our posture would be more horizontal than upright. The mineralized component represents about 50% of the weight/volume of bone tissue, with the remainder being collagen, noncollagenous proteins, and water (Termine, 1988; Cowin, 2001; Currey, 2002). The lessons on constructing elaborate mineralized tissues from the other mineral model systems have not been lost here in bone. The mineral crystals are all aligned along the main loading axis and are able to resist the mechanical loads exerted on a daily basis on bone as well as this tissue continues to grow and develop, while it is also able to be strong and tough (Currey, 1977, 1984, 2005; Weiner and Wagner, 1998).

There is still no consensus on whether the collagen and the mineral are secreted into an extracellular space where both the collagen and the mineral co-align with each other. However, it is known that collagen fibers' gap regions are spaces that hydroxyapatite mineral precursors occupy and where they eventually nucleate, and hydroxyapatite crystal plates grow (Weiner and Wagner, 1998; Gupta et al., 2006a, 2006b; Fratzl, 2008; Figure 1.25). These gap regions are formed from the discontinuous collagen fibrils that assemble into the collagen fiber. These gap regions are periodic and hence, enable crystalline domains that repeat through the length of the collagen. Due to the dimensions of collagen fibers and the orientation of the gap regions, the hydroxyapatite crystal plates that are formed are also oriented along the long axis of the collagen fiber main axis (Fratzl and Weinkamer, 2007). These mineralized collagen fibers themselves assemble into many tissues, including bone and many connective tissues as well (Fratzl and Weinkamer, 2007). With the interaction with other proteins, the mineral content in the collagen can be varied as well as the orientation of the collagen (Fantner et al., 2005; Hassenkam et al., 2004).

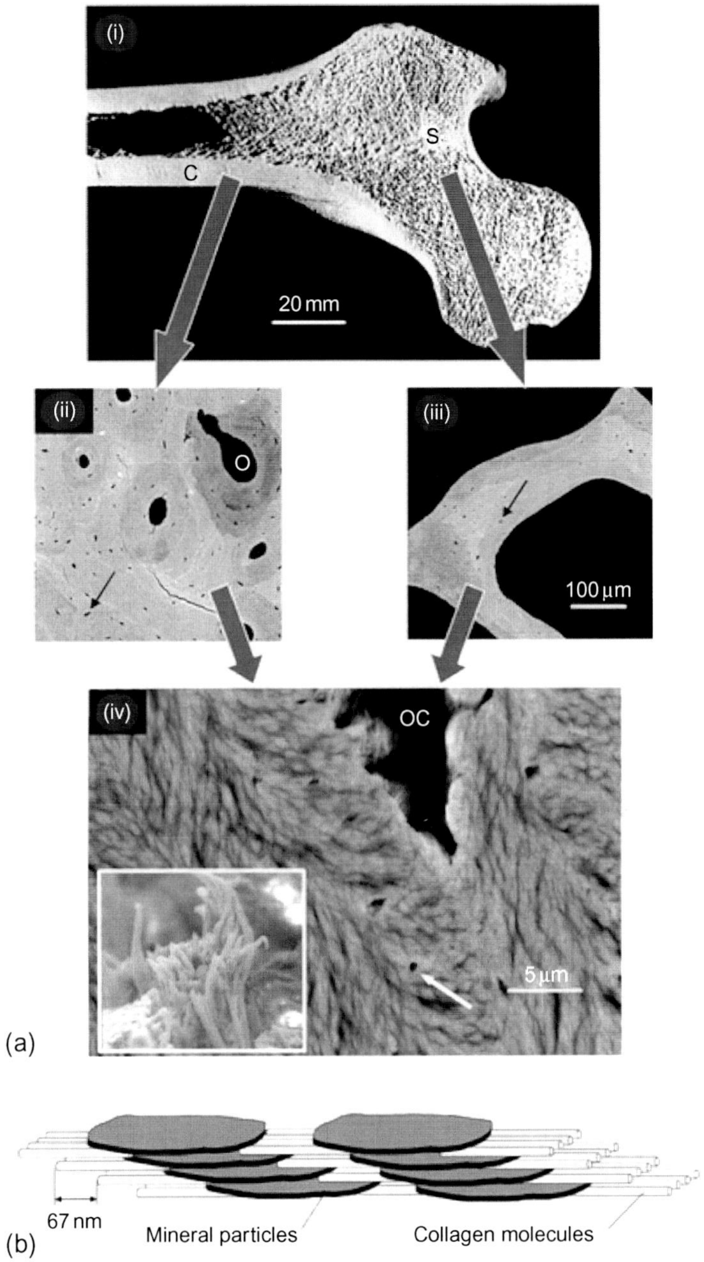

Figure 1.25 Schematic of the hierarchical architecture of bone. (a) From trabecular and lamellar tissues (ii and iii) to mineralized collagen (iv) in a femoral bone (i). (b) The mineral platelets are templated by the collagen gap regions, which are approximately 67 nm apart. (a) (ii) O = Osteon, (iv) OC = Osteocyte lacunae, white arrow indicate area of inset. From Fratzl et al. (2004).

1.4.4.2 Teeth and the amelogenin acting as a template

Adult teeth are made up of two types of mineralized tissues, dentin and enamel. Dentin is made of mineralized tubules that are able to transport fluid and provide a path for nerves to transmit signals—the same signals that are the root cause of tooth pain. Unlike dentin, enamel is actually made of over 90% mineral and it is highly oriented. The process in which the tooth forms is still a debate, but there is agreement that before the tooth emerges, there are cells called odontoblasts which are columnar in nature and initiate the formation of dentin by creating a scaffold of extracellular proteins. These proteins include dentin matrix protein 1 (DMP1), which is found in the juvenile tooth (He et al., 2003; Figure 1.26). Another predominant extracellular

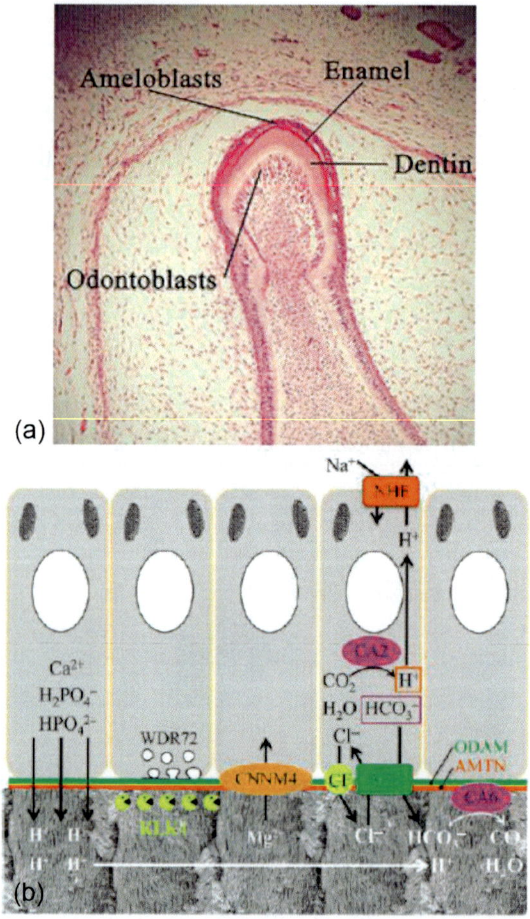

Figure 1.26 Enamel formation. (a) Enamel is formed from the ameloblasts after the formation of the dentin tissue by odontoblasts. (b) Schematic of the ameloblast interacting with the enamel matrix proteins and enzymes to create the conditions for hydroxyapatite mineralization. From Nanci (2013) and Simmer et al. (2010).

protein is amelogenin, which is secreted into the extracellular space by ameloblasts. Ameloblast activity only begins when the dentin is laid down and complete. It is believed that the amelogenin is the matrix that acts as a template in the formation of the enamel and is highly regulated (Le Norcy et al., 2011). Interestingly, over time the amount of amelogenin is reduced while hydroxyapatite increases until amelogenin is found only at the boundaries between neighboring hydroxyapatite crystals (Martinez-Avila et al., 2012; Figure 1.26). Through site-directed mutagenesis as well as knockdown experiments, amelogenin has been demonstrated to be highly phosphorylated and required to stabilize hydroxyapatite as well as transport it to the correct site of mineralization (Le Norcy et al., 2011; Martinez-Avila et al., 2012). It behaves as a mineral chaperone for the proper growth and development of enamel.

1.4.5 Mesocrystals and the materials they make up

Many of the aforementioned cases involve uses of macromolecules, specifically proteins, that bind minerals and act as templates—this may occur via transport of proteins to a site where they act as templates or via simultaneous assembly piece by piece with proteins inside a vesicle connected with a larger construct; the intermediate precursor that is often formed from these interactions is termed "mesocrystalline." By definition, a mesocrystal is an assembly that comprises a 3D array of iso-oriented single-crystal particles of size 1–1000 nm (mesoscale dimensions) (Seto et al., 2012; Figure 1.27). We find that many of these biological mineralized tissues maintain mesocrystalline subunits even when they undergo completely diverse mechanisms of mineralization (i.e. one may be template controlled while another is controlled by an encapsulation and stabilizing effect). Whether this is via a template or interfacial effects, all biogenic crystals come from the "bottom up"; that is, all mineralized tissues are built by the respective nano- and microcrystal subunits (Cölfen and Antonietti, 2005; Cölfen, 2008). Creating mesocrystalline domains may have been a strategy to address a common problem among all organisms that make biological crystals—the ability to create long range order from short range. When examining the microcrystalline platelets of hydroxyapatite in bone or the calcium carbonate crystals in a sea urchin spine, all the individual crystallites are oriented along a main axis from the level of the single crystal to the macro tissue length scale. Biology has a way to register the orientation of single crystals in order to build massive assemblies which themselves have a particular orientation as well.

In all the aforementioned mineralization model systems, genome sequencing and high throughput screening have simplified the investigation of specific molecule–mineral interactions. By having known sequences and sequence motifs that interact with inorganic components, the mechanisms that these domains as well as whole proteins use to control mineralization processes can be delineated and adopted to improve functionalities like better mineral binding or inhibition abilities. Through the understanding of these natural mechanisms to make consistent, strong materials, protein engineering methods can be utilized from molecular biology to design and synthesize materials with more efficacy as well as functionality.

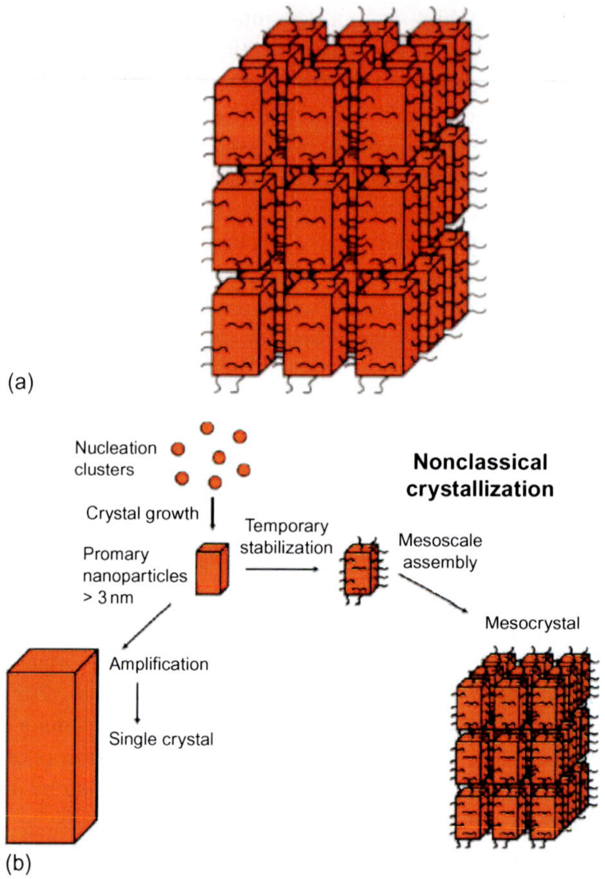

Figure 1.27 Mesocrystals. (a) Mesocrystalline domains in a crystal. (b) The assembly process involved in the formation of bulk crystalline materials.
Niederberger and Cölfen (2006).

1.5 Directing the evolution of mineralization-related proteins

If every one were cast in the same mould, there would be no such thing as beauty.
Charles Darwin in The Descent of Man (1871)

Natural biological systems use proteins as a programmable scaffold for specifying complex, nanostructured materials. As shown in Section 1.3, we can see the considerable influence that organic components have in regulating mineralization processes—from stages of nucleation to crystal growth as well as inhibition, these specific proteins are found to be the control that keeps crystals from being too large or too small. By incorporating protein engineering methods into the "toolkit" to modify these interactions, specific crystalline features can also be screened for or enhanced. The ability to design new proteins with customized functionalities would present great potential for crystalline materials engineering. However, the relationship between a protein's sequence and its biochemical properties is extraordinarily complex, and therefore most efforts to rationally design functional proteins have encountered serious difficulties.

While designing proteins is extremely difficult using a forward engineering approach, Nature has had no difficulty generating an extraordinary array of diverse biological and chemical functionalities. This diversity is the product of natural evolution, whereby small functional changes are made to generate novel functions. Proteins are highly evolvable molecules that can adapt their function through small sequence variations. Protein engineers have learned to exploit this "evolvability" by using iterative rounds of mutation and artificial selection to generate new proteins. Directed evolution is one of the most effective and reliable approaches to engineer new and useful proteins.

A directed evolution experiment involves mutating an initial sequence, assaying the sequence variants for a target biochemical function, selecting the beneficial variants, and repeating this process over multiple generations (Figure 1.28). This sequence optimization can be envisioned as a walk on a high-dimensional "fitness landscape," in which regions of higher elevation represent desirable proteins (Romero and Arnold, 2009; Figure 1.28). Mutations can move you uphill, downhill, or maintain the same elevation, and selection only accepts those mutational steps that go uphill. The combination of mutation and selection over multiple steps enables the gradual optimization of protein function. This simple design algorithm has been used to engineer numerous classes of proteins with varying folds and functions. Some applications include fluorescent proteins for bioimaging (Shaner et al., 2004), therapeutic antibodies (Ho et al., 2005), and industrial enzymes (You and Arnold, 1996).

While the concept of directed evolution is simple, there are several important considerations that affect the ease and speed of the protein optimization process. The first involves choosing a suitable starting sequence. This parent sequence should display low but measurable levels of target activity. If the parent has no measurable activity, then variants that improve upon this activity might also not be detectable. Next, a high-throughput functional assay is chosen. This could be a selection that is coupled to cellular growth and survival, or a screen where the target function is assayed directly (Arnold and Georgiou, 2003). This functional assay must have sufficient precision to identify variants with small improvements in target activity, and sufficient throughput to analyze hundreds to thousands of sequence variants. Finally, a sequence diversification method must be chosen. This could be random, through error-prone polymerase chain reaction (PCR) (Rasila et al., 2009), or if structural or mechanistic information is available, mutations can be made in a more directed fashion (Treynor et al., 2007). Since screening is often the most difficult experimental step, a library should be created to generate the highest probability of finding improved proteins given the screening capacity.

As a proof of concept, directed evolution was recently applied to engineer silicatein enzymes with new mineralization properties (Bawazer et al., 2012). Silicatein libraries were generated by randomly shuffling the α and β silicatein genes from the

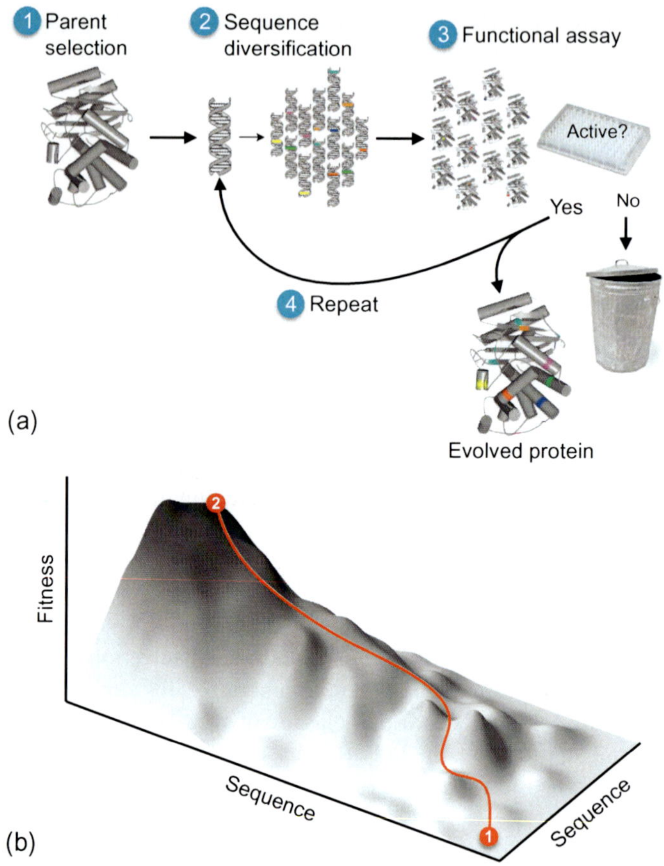

Figure 1.28 Evolution-based optimization of protein function. (a) A typical directed evolution experiment consists of four general steps. The first involves choosing an initial gene for the evolution experiment. Next, this gene's sequence is diversified to generate a library of gene variants. The protein variants encoded by the gene variants are then expressed *in vivo* or *in vitro*, and the expressed proteins are assayed for their function. Sequences that do not display the desired function are discarded. Sequences that display the desired function are used as input sequences for the next round of sequence diversification. This process is repeated over multiple generations until the target specifications are achieved. (b) Directed evolution can be imagined as an uphill walk on a high-dimensional fitness landscape. In this landscape, different sequences are represented in the *XY*-plane and closely related sequences are near to each other. A sequence's "fitness" is represented in the *Z*-direction and is a measure of how well that sequence performs a target function. The initial sequence starts at a low elevation, but the accumulation of uphill mutational steps can uncover highly optimized sequences.

marine sponge *Tethya aurantia*, and also introducing random nucleotide substitutions (Figure 1.29). These libraries were screened using an *in vitro* screening approach that involved expressing silicatein variants in synthetic vesicles, allowing the expressed enzymes to mineralize a polystyrene bead, and sorting the mineralized beads based on the size using a fluorescence-activated cell-sorting (FACS) system (Figure 1.29).

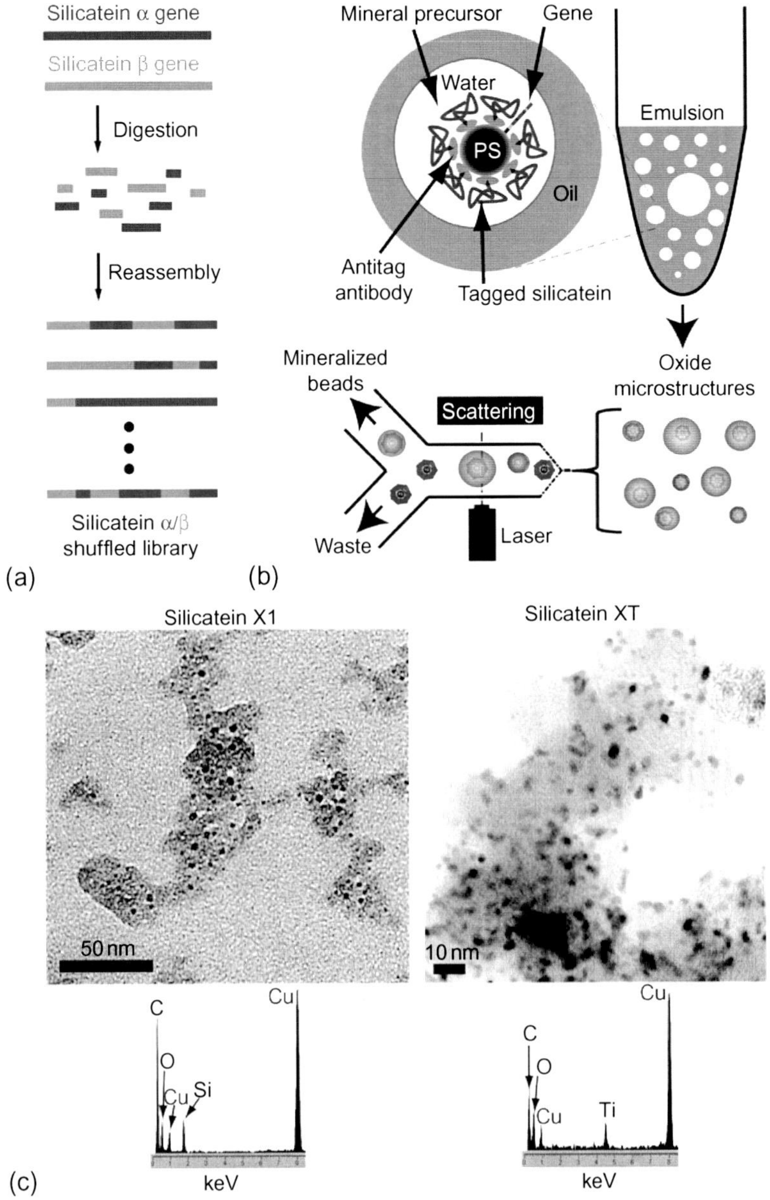

Figure 1.29 Silicatein directed evolution. (a) A chimeric silicatein library was generated by shuffling the sequences of the α and β silicatein genes from *T. aurantia*. (b) Polystyrene microbeads with a silicatein gene attached are encapsulated with a cell-free expression mix. The expressed proteins attach to the bead and serve as a template for mineralization. FACS is used to sort the mineralized beads. (c) Mineralization of evolved silicatein variants. Silicatein X1 was reacted with TEOS and XT reacted with TiBALDH. A TEM image and the associated EDS spectrum of the mineralization products are shown.
From Bawazer et al. (2012).

Silicatein variants were selected based on their ability to mineralize either silica or titania. The evolved mutants were able to produce crystalline inorganic products, while the wild-type enzymes produce amorphous products.

Bacteriophages have also proven to be a versatile platform for template materials. A long-time workhorse of molecular biology, phages have been used as a model system to study self-assembly and have a well-established directed evolution strategy known as phage display. Phage coat proteins assemble into large, supramolecular complexes. These coat proteins can be engineered to bind a wide variety of molecules with high affinity and specificity. A phage display experiment involves inserting a randomized sequence library into a coat protein, binding the coat protein library to a target molecule that is affixed to a solid support, washing away the nonbinding variants, and then eluting the binding variants (Smith and Petrenko, 1997; Figure 1.30). Multiple rounds of positive and negative selection can be used to evolve phages with high affinity and specificity for the target molecule.

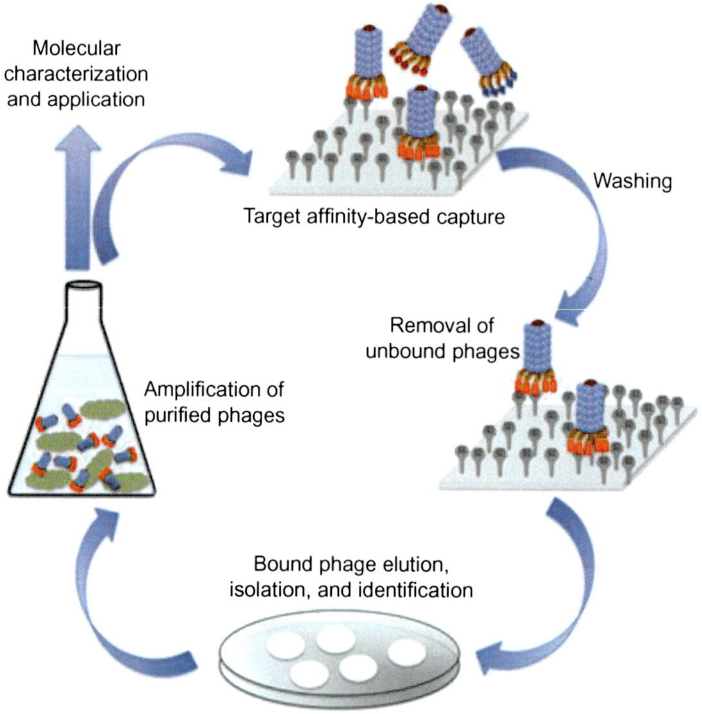

Figure 1.30 Phage display for the selection of improved binding. A target molecule is immobilized on a solid support. A library of phages displaying a randomized binding module on their surface is applied to this target, and the unbound phages are washed off. The tightly bound phages are then eluted and isolated. The selected phages can then be enriched further, or mutagenized again for additional rounds of directed evolution.
From Singh et al. (2013).

Phage display has been used to evolve coat proteins that bind inorganic substances, such as semiconductor, magnetic, and metallic materials (Whaley et al., 2000; Mao et al., 2004; Sano and Shiba, 2003). These phages can be assembled into larger complexes that can serve as templates for inorganic material growth (Flynn et al., 2003). A recent study used phage display to evolve peptides that bound hydroxyapatite (Chung et al., 2011). These phages were then used as templates in the nucleation and growth of crystalline bone-like materials. Phages can also be used for the bottom-up synthesis of nanomaterials because they are composed of multiple coat proteins with a precise spatial localization. Other groups have evolved phages that bind to single-walled carbon nanotubes at one end, and are able to nucleate amorphous iron phosphate at the other end. These phage-templated nanomaterials can serve as cathodes for high-power lithium–ion batteries (Lee et al., 2009).

1.6 Outlook and conclusions

I know that in the study of material things, number, order and position are the threefold clue to exact knowledge; that these three, in the mathematician's hands, furnish the 'first outlines for a sketch of the Universe'.

D'Archy Thompson in On Growth and Form (1917)

By having Nature and its constructions as points of inspiration in designing and engineering novel materials that are stronger, tougher, and more durable, the future can be littered with crystalline materials that can heal and regenerate when damage and wear begins to appear from overuse, as well as materials that maintain the optical properties of crystalline materials, but have high tensile and compressive strengths like steel, such that these materials can be incorporated into works of contemporary architecture.

Nature has figured out strategies to organize defects and weaknesses such that proteins and other macromolecules can endow surface, optical, as well as chemical properties into crystalline assemblies without mechanically weakening the materials assembly. Only recently have we begun to understand and exploit some of these strategies to make better bone and teeth implants. We began to approach silicon-based microelectronics with a "bottom-up" approach little more than 50 years ago. Nature has constructed natural materials with this approach since the beginning of time. Only recently has man begun creating the basic building blocks to make it possible to endow synthetic man-made materials with similar structures and properties (Figure 1.31). The hierarchical structures found in many of these materials provide them with the functional properties we hope to endow in some man-made materials. Perhaps in the not too distant future, civil engineers will also be experts in protein design and engineering for constructing custom building materials. Also, perhaps, we will soon walk by a construction site of the future, where the materials self-assemble into their precise positions.

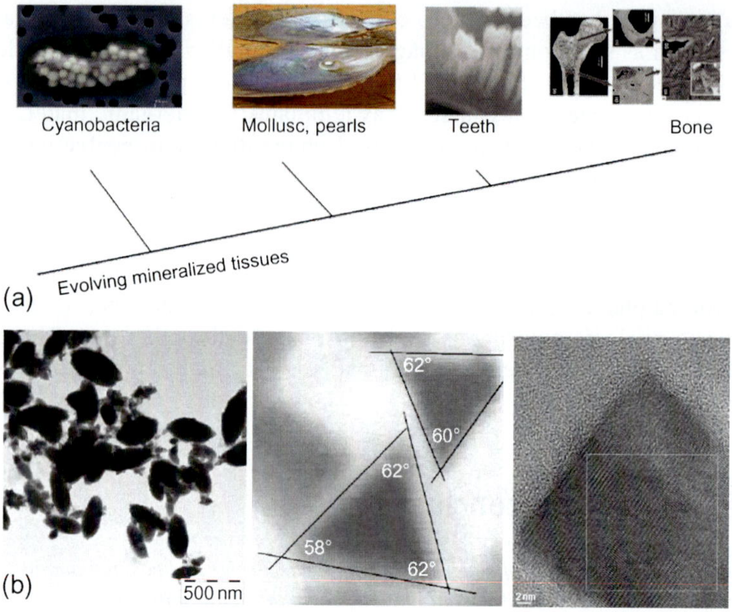

Figure 1.31 Evolution of biomineralized tissues. (a) The evolutionary course in constructing higher strength, high-toughness biomineralized engineered materials. (b) Mineralized constituents designed and engineered as "nanobricks" with various morphologies and functions tuned for future applications.

Acknowledgments

This work would not have been possible without numerous collaborations with the following people: Profs. Fred H. Wilt, Peter Fratzl, Helmut Cölfen, Yong Chen, Stefan Habelitz, Malcolm Snead, Adam Abate, and Micheal Jewett. J. S. would like to acknowledge his family and Dr. Ozlem Sel for their unwavering support and continued presence in his professional life. P. R. acknowledges support from UCLA start-up funds.

References

Aizenberg, J., Sundar, V.C., Yablon, A.D., Weaver, J.C., Chen, G., 2004. Biological glass fibers: correlation between optical and structural properties. Proc. Natl. Acad. Sci. U. S. A. 101 (10), 3358–3363.

Aizenberg, J., Weaver, J.C., Thanawala, M.S., Sundar, V.C., Morse, D.E., Fratzl, P., 2005. Skeleton of *Euplectella* sp.: structural hierarchy from the nanoscale to the macroscale. Science 309 (5732), 275–278.

Arakaki, A., Shimizu, K., Oda, M., Sakamoto, T., Nishimura, T., Kato, T., 2015. Biomineralization-inspired synthesis of functional organic/inorganic hybrid materials: organic molecular control of self-organization of hybrids. Org. Biomol. Chem. 13 (4), 957–1266.

Arnold, F., Georgiou, G., 2003. Directed Enzyme Evolution: Screening and Selection Methods. Humana Press, Totowa, NJ. 383 pp.

Barsi, J., Tu, Q., Davidson, E., 2014. General approach for *in vivo* recovery of cell type-specific effector gene sets. Genome Res. 24 (5), 860–868.

Barthelat, F., Zhu, D., 2011. A novel biomimetic material duplicating the structure and mechanics of natural nacre. J. Mater. Res. 26 (10), 1203–1215.

Barthelat, F., Li, C.-M., Comi, C., Espinosa, H.D., 2006. Mechanical properties of nacre constituents and their impact on mechanical performance. J. Mater. Res. 21 (8), 1977–1986.

Barthelat, F., Tang, H., Zavattieri, P.D., Li, C.M., Espinosa, H.D., 2007. On the mechanics of mother-of-pearl: a key feature in the material hierarchical structure. J. Mech. Phys. Solids 55 (2), 306–337.

Baumgartner, J., Dey, A., Bomans, P., Coadou, C., Fratzl, P., Sommerdijk, N., Faivre, D., 2013a. Nucleation and growth of magnetite from solution. Nat. Mater. 12, 310–314.

Baumgartner, J., Morin, G., Menguy, N., Perez Gonzalez, T., Widdrat, M., Cosmidis, J., Faivre, D., 2013b. Magnetotactic bacteria form magnetite from a phosphate-rich ferric hydroxide via nanometric ferric (oxyhydr) oxide intermediates. Proc. Natl. Acad. Sci. U. S. A. 110 (37), 14883–14888.

Bawazer, L., Izumi, M., Kolodin, D., Neilson, J., Schwenzer, B., Morse, D., 2012. Evolutionary selection of enzymatically synthesized semiconductors from biomimetic mineralization vesicles. Proc. Natl. Acad. Sci. U. S. A. 109 (26), 1705–1714.

Beddington, R., Robertson, E., 1999. Axis development and early asymmetry in mammals. Cell 96 (2), 195–209.

Belton, D., Deschaume, O., Perry, C., 2012. An overview of the fundamentals of the chemistry of silica with relevance to biosilicification and technological advances. FEBS J. 279 (10), 1710–1720.

Bewernitz, M., Gebauer, D., Long, J., Cölfen, H., Gower, L., 2012. A metastable liquid precursor phase of calcium carbonate and its interactions with polyaspartate. Faraday Discuss. 159, 291–312.

Boskey, A., 1997. Amorphous calcium phosphate: the contention of bone. J. Dent. Res. 76, 1433–1436.

Boskey, A.L., Gadaleta, S., Gundberg, C., Doty, S.B., Ducy, P., Karsenty, G., 1998. Fourier transform infrared microspectroscopic analysis of bones of osteocalcin-deficient mice provides insight into the function of osteocalcin. Bone 23 (3), 187–196.

Boskey, A.L., Spevak, L., Paschalis, E., Doty, S.B., McKee, M.D., 2002. Osteopontin deficiency increases mineral content and mineral crystallinity in mouse bone. Calcif. Tissue Int. 71 (2), 145–154.

Burke, E., Guo, Y., Colon, L., Rahima, M., Veis, A., Nancollas, G., 2000. Influence of polyaspartic acid and phosphophoryn on octacalcium phosphaste growth kinetics. Colloids Surf. B: Biointerfaces 17, 49–57.

Cantaert, B., Beniash, E., Meldrum, F.C., 2013. The role of poly(aspartic acid) in the precipitation of calcium phosphate in confinement. J. Mater. Chem. B 1 (48), 1–20.

Chang, R., 2000. Physical Chemistry for the Chemical and Biological Sciences. University Science Books, Sausalito, CA.

Cho, K., De Yoreo, J., 2013. Probing protein assembly, biomineralization, and biomolecular interactions by atomic force microscopy. In: Jena, B., Taatjes, D. (Eds.), NanoCellBiology—Multimodal Imaging in Biology and Medicine. CRC Press, Boca Raton, FL.

Chow, W., Rajan, R., Muller, K., Reid, D., Skepper, J., Wong, W., Brooks, R., Green, M., Bihan, D., Farndale, R., Slatter, D., Shanahan, C., Duer, M., 2014. NMR spectroscopy of native and *in vitro* tissues implicates polyADP ribose in biomineralization. Science 344 (6185), 742–746.

Chung, W., Kwon, K., Song, J., Lee, S., 2011. Evolutionary screening of collagen-like peptides that nucleate hydroxyapatite crystals. Langmuir 27, 7620–7628.

Cölfen, H.A.M., 2008. Mesocrystals and Nonclassical Crystallization. Wiley, Chichester.
Cölfen, H., Antonietti, M., 2005. Mesocrystals: inorganic superstructures made by highly parallel crystallization and controlled alignment. Angew. Chem. Int. Ed. 44 (35), 5576–5591.
Collino, S., Evans, J., 2008. Molecular specifications of a mineral modulation sequence derived from the aragonite-promoting n16. Biomacromolecules 9 (7), 1909–1918.
Cowin, S.C. (Ed.), 2001. Bone Mechanics Handbook. CRC Press, Baca Raton, FL.
Currey, J.D., 1977. Mechanical-properties of mother of pearl in tension. Proc. R. Soc. B Bio. Sci. 196 (1125), 443.
Currey, J., 1984. The Mechanical Adaptations of Bones. Princeton University Press, Princeton, NJ.
Currey, J.D., 2002. Bones: Structure and Mechanics. Princeton University Press, Princeton, NJ.
Currey, J.D., 2005. Materials science—hierarchies in biomineral structures. Science 309 (5732), 253–254.
Currie, H., Perry, C., 2007. Silica in plants: biological, biochemical and chemical studies. Ann. Bot. 100 (7), 1383–1389.
Davidson, E., 2014. The uncommon roles of common gene regulatory factors in the genomes of differentiating cells. EMBO J. 33 (11), 1193–1194.
Evans, J.S., 2008. "Tuning in" to mollusk shell nacre- and prismatic-associated protein terminal sequences. Implications for biomineralization and the construction of high performance inorganic–organic composites. Chem. Rev. 108 (11), 4455–4462.
Evans, J.S., 2012. Aragonite-associated biomineralization proteins are disordered and contain interactive motifs. Bioinformatics 28 (24), 3182–3185.
Faatz, M., Groehn, F., Wegner, G., 2004. Amorphous calcium carbonate: synthesis and potential intermediate in biomineralization. Adv. Mater. 16 (12), 996–1000.
Fantner, G.E., Hassenkam, T., Kindt, J.H., Weaver, J.C., Birkedal, H., Pechenik, L., Cutroni, J.A., Cidade, G.A.G., Stucky, G.D., Morse, D.E., Hansma, P.K., 2005. Sacrificial bonds and hidden length dissipate energy as mineralized fibrils separate during bone fracture. Nat. Mater. 4 (8), 612–616.
Finnemore, A., Cunha, P., Shean, T., Vignolini, S., Guldin, S., Oyen, M., Steiner, U., 2012. Biomimetic layer-by-layer assembly of artificial nacre. Nat. Commun. 3, 1–6.
Flynn, C., Lee, S., Peelle, B., Belcher, A., 2003. Viruses as vehicles for growth, organization and assembly of materials. Acta Mater. 51, 5867–5880.
Fratzl, P. (Ed.), 2008. Collagen: Structure and Mechanics. Springer, New York.
Fratzl, P., Weinkamer, R., 2007. Nature's hierarchical materials. Prog. Mater. Sci. 52 (8), 1263–1334.
Fratzl, P., Gupta, H.S., Paschalis, E.P., Roschger, P., 2004. Structure and mechanical quality of the collagen-mineral nano-composite in bone. J. Mater. Chem. 14 (14), 2115–2123.
Gebauer, D., Volkel, A., Cölfen, H., 2008. Stable prenucleation calcium carbonate clusters. Science 322 (5909), 1819–1822.
Gong, Y.U.T., Killian, C.E., Olson, I.C., Appathurai, N.P., Amasino, A.L., Martin, M.C., Holt, L.J., Wilt, F.H., Gilbert, P.U.P.A., 2012. Phase transitions in biogenic amorphous calcium carbonate. Proc. Natl. Acad. Sci. U. S. A. 109 (16), 6088–6093.
Gordon, L., Joester, D., 2011. Nanoscale chemical tomography of buried organic–inorganic interfaces in the chiton tooth. Nature 469 (7329), 194–197.
Gower, L., 2008. Biomimetic model systems for investigating the amorphous precursor pathway and its role in biomineralization. Chem. Rev. 108 (11), 4551–4627.
Greer, J., 2013. Nanotwinned metals: it's all about imperfections. Nat. Mater. 12, 689–690.
Gruenberg, K., Wawer, C., Tebo, B., Schueler, D., 2001. A large gene cluster encoding several magnetosome proteins is conserved in different species of magnetotatic bacteria. Appl. Environ. Microbiol. 67 (10), 4573–4582.

Gupta, H.S., Seto, J., Wagermaier, W., Zaslansky, P., Boesecke, P., Fratzl, P., 2006a. Cooperative deformation of mineral and collagen in bone at the nanoscale. Proc. Natl. Acad. Sci. U. S. A. 103 (47), 17741–17746.

Gupta, H.S., Wagermaier, W., Zickler, G.A., Hartmann, J., Funari, S.S., Roschger, P., Wagner, H.D., Fratzl, P., 2006b. Fibrillar level fracture in bone beyond the yield point. Int. J. Fracture 139 (3–4), 425–436.

Haeckel, E., 1904. Art forms in Nature. Verlag des Bibliographischen Instituts, Leipzig, Germany.

Hassenkam, T., Fantner, G.E., Cutroni, J.A., Weaver, J.C., Morse, D.E., Hansma, P.K., 2004. High-resolution AFM imaging of intact and fractured trabecular bone. Bone 35 (1), 4–10.

He, G., Dahl, T., Veis, A., George, A., 2003. Nucleation of apatite crystals *in vitro* by self-assembled dentin matrix protein 1. Nat. Mater. 8, 552–558.

Ho, M., Kreitman, R., Onda, M., Pastan, I., 2005. *In vitro* antibody evolution targeting germ-line hot spots to increase activity of an anti-CD22 immunotoxin. J. Biol. Chem. 280 (1), 607–617.

Huang, S.C., Naka, K., Chujo, Y., 2008. Effect of molecular weights of poly(acrylic acid) on crystallization of calcium carbonate by the delayed addition method. Polym. J. 40 (2), 154–162.

Kerschnitzki, M., Wagermeier, W., Roschger, P., Seto, J., Shahar, R., Duda, G., Mundlos, S., Fratzl, P., 2011. The organization of the osteocyte network mirrors the extracellular matrix orientation in bone. J. Struct. Biol. 173 (2), 303–311.

Killian, C., Wilt, F., 2008. Molecular aspects of biomineralization of the echinoderm endoskeleton. Chem. Rev. 108 (11), 4463–4474.

Killian, C.E., Metzler, R.A., Gong, Y.U.T., Olson, I.C., Aizenberg, J., Politi, Y., Wilt, F.H., Scholl, A., Young, A., Doran, A., Kunz, M., Tamura, N., Coppersmith, S.N., Gilbert, P.U.P.A., 2009. Mechanism of calcite co-orientation in the sea urchin tooth. J. Am. Chem. Soc. 131 (51), 18404–18409.

Knapp, R., Wu, C., Mobilia, K., Joester, D., 2012. Recombinant sea urchin vascular endothelial growth factor directs single-crystal growth and branching *in vitro*. J. Am. Chem. Soc. 134 (43), 17908–17911.

Koester, K.J., Ager III, J.W., Ritchie, R.O., 2008. The true toughness of human cortical bone measured with realistically short cracks. Nat. Mater. 7 (8), 672–677.

Komelli, A., Li, Z., Newman, D., Jensen, G., 2006. Magnetosomes are cell membrane invaginations organized by the actin-like protein MamK. Science 311, 242–245.

Kroger, N., Lorenz, S., Brunner, E., Sumper, M., 2002. Self-assembly of highly phosphorylated silaffins and their function in biosilica morphogenesis. Science 298 (5593), 584–586.

Lakes, R.S., 1987. Foam structures with a negative Poisson's ratio. Science 235, 1038–1040.

Lakes, R., 1993. Materials with structural hierarchy. Nature 361, 511–515.

Le Norcy, E., Kwak, S., Allaire, M., Fratzl, P., Yamakoshi, Y., Simmer, J., Margolis, H., 2011. Effect of phosphorylation on the interaction of calcium with leucine-rich amelogenin peptide. Eur. J. Oral Sci. 119 (S1), 97–102.

Lee, Y., Yi, H., Kim, W., Kang, K., Yun, D., Strano, M., Ceder, G., Belcher, A., 2009. Fabricating genetically engineered high-power lithium–ion batteries using multiple virus genes. Science 324 (5930), 1051–1055.

Li, D., Nielsen, M., Lee, J., Frandsen, C., Banfield, J., De Yoreo, J., 2012. Direction-specific interactions control crystal growth by oriented attachment. Science 336, 1014–1018.

Liu, D.M., Weiner, S., Wagner, H.D., 1999. Anisotropic mechanical properties of lamellar bone using miniature cantilever bending specimens. J. Biomech. 32 (7), 647–654.

Long, X., Ma, Y., Cho, K., Li, D., De Yoreo, J., Limin, Q., 2013. Oriented calcite micropillars and prisms formed through aggregation and recrystallization of poly(acrylic acid) stabilized nanoparticles. Cryst. Growth Des. 13, 3856–3863.

Lu, C., Qi, L., Cong, H., Wang, X., Yang, J., Yang, L., Zhang, D., Ma, J., Cao, W., 2005. Synthesis of calcite single crystals with porous surface by templating of polymer latex particles. Chem. Mater. 17 (20), 5218–5224.

Ma, J., Yamaji, N., 2006. Silicon uptake and accumulation in higher plants. Trends Plant Sci. 11 (8), 1–6.

Ma, J., Tamai, K., Yamaji, N., Mitani, N., Konishi, S., Katsuhara, M., Ishiguro, M., Murata, Y., Yano, M., 2006. A silicon transporter in rice. Nature 440, 688–691.

Mahamid, J., Sharir, A., Addadi, L., Weiner, S., 2008. Amorphous calcium phosphate is a major component of the forming fin bones of zebrafish: indications for an amorphous precursor phase. Proc. Natl. Acad. Sci. U. S. A. 105 (35), 12748–12753.

Mao, C., Solis, D., Reiss, B., Kottmann, S., Sweeney, R., Hayhurst, A., Georgiou, G., Iverson, B., Belcher, A., 2004. Virus-based toolkit for the directed synthesis of magnetic and semiconducting nanowires. Science 303 (5655), 213–217.

Martinez-Avila, O., Wu, S., Kim, S., Cheng, Y., Khan, F., Samudrala, R., Sali, A., Horst, J., Habelitz, S., 2012. Self-assembly of filamentous amelogenin requires calcium and phosphate: from dimers via nanoribbons to fibrils. Biomacromolecules 13 (11), 3494–3502.

Morris, J.W., 2005. The Structure and Properties of Materials. McGraw Hill, New York, NY.

Mueller, W., Boreiko, A., Wang, X., Belikov, S., Wiens, M., Grebenjuk, V., Schlossmacher, U., Schroeder, H., 2007a. Silicateins, the major biosilica forming enzymes present in demosponges: protein analysis and phylogenetic relationship. Gene 395 (1–2), 62–71.

Mueller, W., Schlossmacher, U., Wang, X., Boreiko, A., Brandt, D., Wolf, S., Tremel, W., Schroeder, H., 2007b. Poly(silicate)-metabolizing silicatein in siliceous spicules and silicasomes of demosponges comprises dual enzymatic activities (silica polymerase and silica esterase). FEBS J. 275 (2), 362–370.

Nanci, A., 2013. Ten Cate's Oral Histology: Development, Structure, and Function. Mosby, St. Louis, MO.

Niederberger, M., Cölfen, H., 2006. Oriented attachment and mesocrystals: non-classical crystallization mechanisms based on nanoparticle assembly. Phys. Chem. Chem. Phys. 8 (28), 3261–3360.

Nielsen, M., Aloni, S., De Yoreo, J., 2014. *In situ* TEM imaging of $CaCO_3$ nucleation reveals coexistence of direct and indirect pathways. Science 345 (6201), 1158–1162.

Peled-Kamar, M., Hamilton, P., Wilt, F., 2002. Spicule matrix protein LSM34 is essential for biomineralization of the sea urchin spicule. Exp. Cell Res. 272 (1), 56–61.

Ponce, C., Evans, J., 2011. Polymorph crystal selection by n16, an intrinsically disordered nacre framework protein. Cryst. Growth Des. 11 (10), 4690–4696.

Poulsen, N., Sumper, M., Kroger, N., 2003. Biosilica formation in diatoms: characterization of native silaffin-2 and its role in silica morphogenesis. Proc. Natl. Acad. Sci. U. S. A. 100 (21), 12075–12080.

Rao, A., Berg, J., Kellermeier, M., Gebauer, D., 2014. Sweet on biomineralization: effects of carbohydrates on the early stages of calcium carbonate crystallization. Eur. J. Mineral. 26 (4), 537–552.

Rasila, T., Pajunen, M., Savilahti, H., 2009. Critical evaluation of random mutagenesis by error-prone polymerase chain reaction protocols, *Escherichia coli* mutator strain, and hydroxylamine treatment. Anal. Biochem. 388 (1), 71–80.

Ritchie, R.O., 2011. The conflicts between strength and toughness. Nat. Mater. 10, 817–822.

Romero, P., Arnold, F., 2009. Exploring protein fitness landscapes by directed evolution. Nat. Rev. Mol. Cell Biol. 10 (12), 866–876.

Sano, K., Shiba, K., 2003. A hexapeptide motif that electrostatically binds to the surface of titanium. J. Am. Chem. Soc. 125 (47), 14234–14235.

Seto, J., Zhang, Y., Hamilton, P., Wilt, F., 2004. The localization of occluded matrix proteins in calcareous spicules of sea urchin larvae. J. Struct. Biol. 148 (1), 123–130.

Seto, J., Ma, Y., Davis, S.A., Meldrum, F., Gourrier, A., Kim, Y.Y., Schilde, U., Sztucki, M., Burghammer, M., Maltsev, S., Jaeger, C., Cölfen, H., 2012. Structure–property relationships of a biological mesocrystal in the adult sea urchin spine. Proc. Natl. Acad. Sci. U. S. A. 109 (10), 3699–3704.

Seto, J., Azais, T., Cölfen, H., 2013. Formation of aragonitic layered structures from kaolinite and amorphous calcium carbonate precursors. Langmuir 29 (24), 7521–7528.

Seto, J., Picker, A., Chen, Y., Rao, A., Evans, J., Cölfen, H., 2014a. Nacre protein sequence compartmentalizes mineral polymorphs in solution. Cryst. Growth Des. 14 (4), 1501–1505.

Seto, J., Rao, A., Cölfen, H., 2014b. Hierarchical nanostructured biological materials. In: Zhang, Q., Wei, F. (Eds.), Advanced Hierarchical Nanostructured Materials. Wiley, Weinheim.

Shaner, N., Campbell, R., Steinbach, P., Giepmans, B., Palmer, A., Tsien, R., 2004. Improved monomeric red, orange and yellow fluorescent proteins derived from *Discosoma* sp. red fluorescent protein. Nat. Biotechnol. 22 (12), 1567–1572.

Simmer, J., Papagerakis, P., Smith, C., Fisher, D., Rountrey, A., Zheng, L., Hu, J., 2010. Regulation of dental enamel shape and hardness. J. Dent. Res. 89 (10), 1024–1038.

Singh, A., Poshtiban, S., Evoy, S., 2013. Recent advances in bacteriophage based biosensors for food-borne pathogen. Sensors 13 (2), 1763–1786.

Smeets, P., Cho, K., Kempen, G., Sommerdijk, N., De Yoreo, J., 2015. Calcium carbonate nucleation driven by ion binding in a biomimetic matrix revealed by *in situ* electron microscopy. Nat. Mater. 14, 394–399.

Smith, G., Petrenko, V., 1997. Phage display. Chem. Rev. 97 (2), 391–410.

Termine, J.D., 1988. Non-collagen proteins in bone. CIBA Found. Symp. 136, 178–202.

Thompson, D., 1917. On Growth and Form. Cambridge University Press, Cambridge.

Torquato, S., 2002. Random Heterogeneous Materials: Microstructure and Macroscopic Properties. Springer-Verlag, New York.

Torquato, S., Stillinger, F.H., 2002. Controlling the short-range order and packing densities of many-particle systems. J. Phys. Chem. B 106.

Treynor, T., Vizcarra, C., Nedelcu, D., Mayo, S., 2007. Computationally designed libraries of fluorescent proteins evaluated by preservation and diversity of function. Proc. Natl. Acad. Sci. U. S. A. 104 (1), 48–53.

Tu, Q., Cameron, R., Davidson, E., 2014. Quantitative developmental transcriptomes of the sea urchin *Strongylocentrotus purpuratus*. Dev. Biol. 385 (2), 160–167.

Vos, M., Bomans, P., de Haas, F., Frederik, P., Jansen, J., Nolte, R., Sommerdijk, N., 2007. Insights in the organization of DNA-surfactant monolayers using cryo-electron tomography. J. Am. Chem. Soc. 129 (39), 11894–11895.

Wang, X., Schlossmacher, U., Wiens, M., Batel, R., Schroeder, H., Mueller, W., 2012. Silicateins, silicatein interactors and cellular interplay in sponge skeletogenesis: formation of glass fiber-like spicules. FEBS J. 279 (10), 1721–1736.

Wang, Y., Sansoz, F., LaGrange, T., Ott, R., Marian, J., Barbee, T., Hamza, A., 2013a. Defective twin boundaries in nanotwinned metals. Nat. Mater. 12, 697–702.

Wang, Y., Von Euw, S., Fernandes, F., Cassaignon, S., Selmane, M., Laurent, G., Pehau-Arnaudet, G., Coelho, C., Bonhomme-Coury, L., Giraud-Guille, M., Babonneau, F., Azais, T., Nassif, N., 2013b. Water-mediated structuring of bone apatite. Nat. Mater. 12, 1144–1153.

Weaver, J., Morse, D., 2003. Molecular biology of demosponge axial filaments and their roles in biosilification. Microsc. Res. Tech. 62 (4), 356–367.

Weiner, S., Wagner, H.D., 1998. The material bone: structure mechanical function relations. Annu. Rev. Mater. Sci. 28, 271–298.

Whaley, S., English, D., Hu, E., Barbara, P., Belcher, A., 2000. Selection of peptides with semiconductor binding specificity for directed nanocrystal assembly. Nature 405 (6787), 665–668.

Wilt, F.H., Killian, C.E., Livingston, B.T., 2003. Development of calcareous skeletal elements in invertebrates. Differentiation 71 (4–5), 237–250.

Wolff, J., 1892. Das Gesetz der Transformation der Knochen. Verlag von, Berlin.

Xu, A., Dong, D., Antonietti, M., Cölfen, H., 2008. Polymorph switching of calcium carbonate crystals by polymer-controlled crystallization. Adv. Funct. Mater. 18, 1307–1313.

You, L., Arnold, F., 1996. Directed evolution of subtilisin E in *Bacillus subtilis* to enhance total activity in aqueous dimethylformamide. Protein Eng. 9 (1), 77–83.

Zhang, B., Xu, G., Evans, J., 2000. Model peptide studies of sequence repeats derived from the intracrystalline biomineralization protein, SM50 II. Pro, Asn-rich tandem repeats. Biopolymers 54 (6), 464–475.

Morphology control and molecular templates in biomineralization

2

A. Rao, H. Cölfen
University of Konstanz, Konstanz, Germany

2.1 Introduction

Biominerals are biogenic mineral phases that include single crystals, composites, and amorphous materials (Lowenstam and Weiner, 1989). About 60 biominerals have been identified based on their inorganic constituents (Lowenstam, 1981). These include minerals of carbonates, phosphates, halides, sulfates, silica, iron oxides, manganese oxides, sulfides, citrates, and oxalates scattered across the branches of life, i.e., eukaryotes, bacteria, and archaea (Lowenstam and Weiner, 1989; Schüler and Frankel, 1999; Inskeep et al., 2004). Early descriptions of biomineral architectures utilized geometric principles to explain the beautiful patterns and underlying growth mechanisms in Nature (Figure 2.1).

Recent analyses of biominerals indicate the role of proteins, glycans as well as lipids in formative processes at the micro- to nano-length scales. Considering the diversity of the inorganic and organic components in biominerals (Lowenstam and Weiner, 1989; Evans, 2008; Mann et al., 2010), the study of interactions at the nanoscale and their subsequent effects in the manifestation of unique physical properties is of fundamental and industrial importance (Bensaude-Vincent et al., 2002). Therefore, the present focus on biominerals is on account of their fascinating physical properties as well as the need for new green materials based on a blueprint optimized by evolutionary processes. The study of certain biominerals has elucidated the complex relations between structure–function and the emergence of superior physical properties. The nacreous layer present in the shells of certain mollusks and gastropods exemplifies the emergence of remarkable mechanical properties on account of a hierarchical structure reinforced by biomolecules (Sun and Bhushan, 2012). This structural complexity results in a fracture toughness of nacre that is 3000 times that of bulk aragonite (Jackson et al., 1988). Another fascinating example of structural hierarchy is the siliceous skeletal system of the glass sponge, *Euplectella* sp. (Aizenberg et al., 2004). The cylindrical skeleton is a lattice of fused spicules which in turn are composed of amorphous, hydrated silica nanoparticles, and organic interlayers (Weaver et al., 2007). The composite nature of these concentric lamellae leads to a unique nonuniform refractive index profile, thus endowing these spicules with wave-guiding properties. This biological hybrid material is thus an evolutionary optimized counterpart of industrial optic fibers (Sundar et al., 2003). Given that certain biominerals can display fascinating properties of mechanics and optics, the study of biominerals and their formation mechanisms has attracted attention from various disciplines. Such

Biomineralization and Biomaterials. http://dx.doi.org/10.1016/B978-1-78242-338-6.00003-X

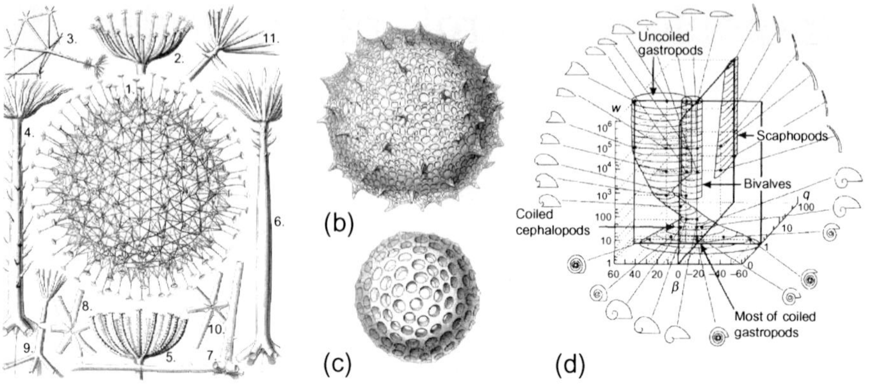

Figure 2.1 Remarkable structures exhibited by skeletons of radiolarians (a) symmetrical structures from the Aulosphaerida family, dissimilar perforations on the shell surfaces of (b) Conosphaera sp. and (c) Ethmosphaera sp. (d) Simulation of different mollusk shell morphologies inspired by early geometric models.
Reproduced from (a–c) the classical works of Ernst Häckel and D'Arcy Wentworth Thompson and (d) Tursch (1997), with permission from Oxford University Press.

biomaterials have motivated interesting discoveries at the interface of biology and physics, such as the role of prenucleation clusters (PNCs) in mineralization (Gebauer et al., 2008, 2009b), the polyamorphous nature of mineralization precursors (Gebauer et al., 2010; Gower and Odom, 2000; Cartwright et al., 2012), stabilization of clusters and transient phases (Tester et al., 2011; Kellermeier et al., 2010, 2012b; Aizenberg et al., 1996; Bentov et al., 2010), occurrence of liquid-like phases (Gower and Odom, 2000; Wohlrab et al., 2005; Bewernitz et al., 2012; Gower and Tirrell, 1998) as well as the emergence of unique material properties in oriented superstructures (Cölfen and Antonietti, 2005; Luz and Mano, 2010; Kunkel et al., 2012). In addition, processes such as the spatiotemporal dynamics between organic and inorganic molecules are under genetic and proteomic regulation. Studying these pathways is vital toward understanding biomineralization. Moreover, considering that certain biominerals possess a hybrid structure with structural hierarchy at different length scales, the nature of physical forces dominant at each length scale cannot be ignored (Mann, 1997; Mann and Ozin, 1996). For example, at the macroscale, gravity, mechanical loading, and external environmental conditions such as temperature, salinity, pH, and humidity play a role in mineral growth and development. Structure at the micron scale is influenced by local gradients in physical parameters, including pH and ionic strength, cellular activities as well as the presence of extracellular matrices. The nanostructure is, however, more affected by molecular structure and composition, curvature, self-organization as well as interfacial properties such as charge and hydrophobicity. During biomineralization, these forces spatiotemporally act together to create fascinating materials that can supersede most synthetic materials in terms of morphological beauty, architectural sophistication, and formative elegance (Figure 2.2).

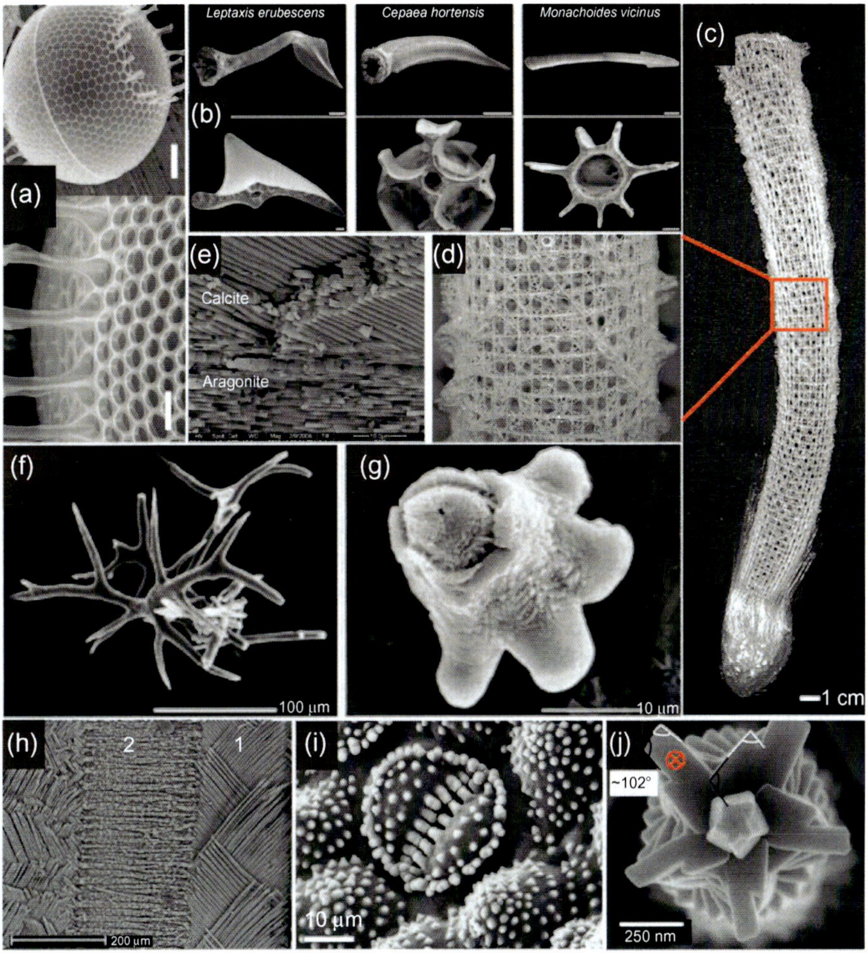

Figure 2.2 Compositional and structural diversity at different length scales illustrated by: (a) the patterned siliceous cell wall of a diatom *Stephanopyxis turris*, (b) calcareous love darts in certain species of land snails, (c and d) siliceous cylindrical skeletal lattice of the sponge *Euplectella aspergillum* with arrays of vertical, horizontal, and diagonal strut, (e) a backscattered electron image of the fractured shell of the bivalve *Mytilus edulis* calcite prisms and aragonite nacre (Copyright 2008 American Chemical Society), (f and g) amorphous skeletal elements from *Pyura pachydermatina*, (h) cross-lamellar layers in the shell of gastropod *Strombus decorus persicus*, (i) cell surface stomata decorated with silicon oxide structures in the common horsetail (*Equisetum arvensis*), and (j) tip of a spine from the coccolith *Rhabdosphaera clavigera* (Copyright 2014 American Chemical Society). Adapted from (a) Sumper and Brunner (2006) with permission from Wiley; (b) Koene and Schulenburg (2005), (c and d) Weaver et al. (2007) with permission from Elsevier; (e) Cusack and Freer (2008) with permission from American Chemical Society; (f and g) Addadi et al. (2003) with permission from Wiley; (h) Pokroy and Zolotoyabko (2003) with permission from Royal Society of Chemistry; (i) Koch et al. (2008) with permission from Royal Society of Chemistry; and (j) van de Locht et al. (2014) with permission from American Chemical Society.

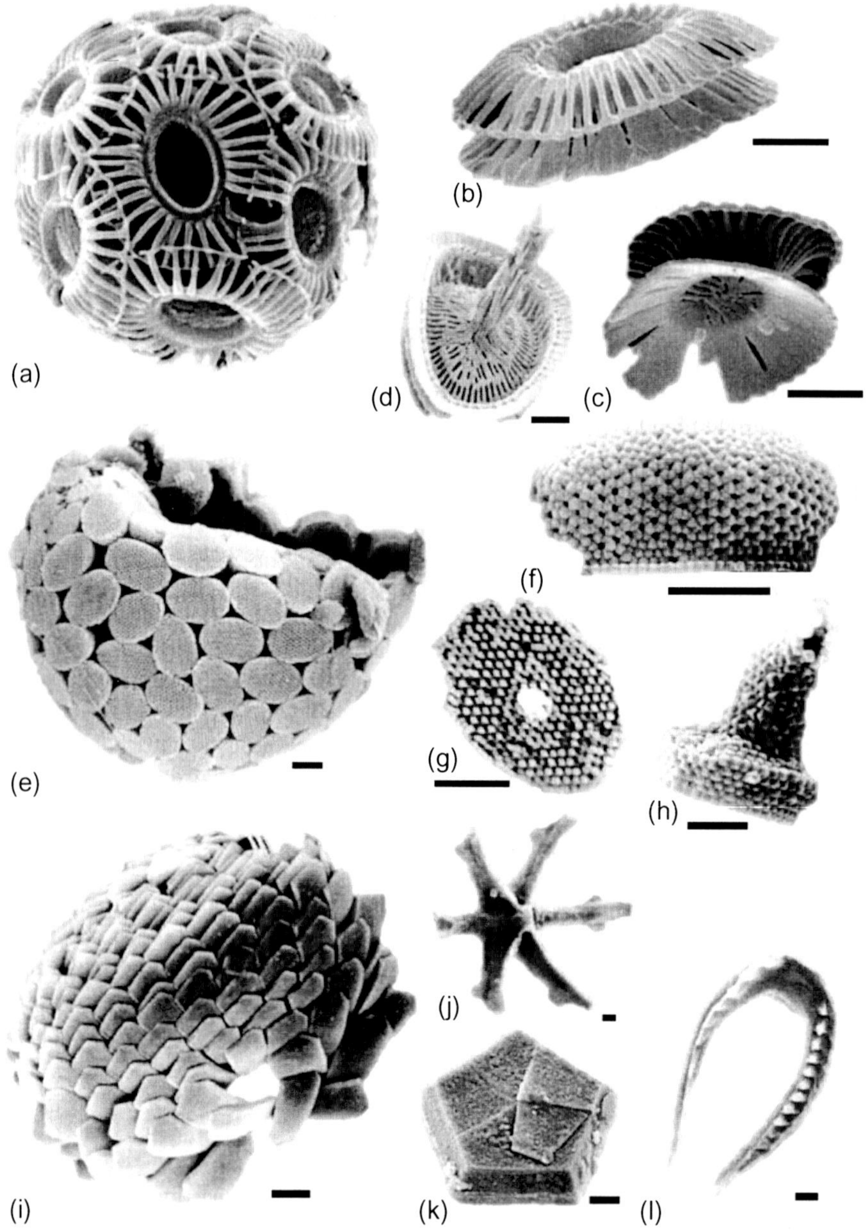

Figure 2.4 Structurally diverse calcareous minerals produced by haptophytes. Heterococcoliths formed of radial arrays of complex crystals: (a–c) complete coccosphere as well as isolated coccoliths from *Emiliania huxleyi* and (d) *Syracosphaera pulchra*. Holococcoliths made of rhombohedral crystal units: (e) *Calyptrolithophora papillifera* coccosphere, (f) *Calyptrosphaera oblonga*, (g) *Syracolithus catilliferus*, and (h) *Anfractus youngii*. Nannoliths from: (i) *Florisphaera profunda*, (j) *Discoaster surculus*, (k) *Braarudosphaera bigelowii*, and (l) *Ceratolithus cristatus*. Scale bars represent 1 μm. Reproduced from Young et al. (1999) with permission from Elsevier.

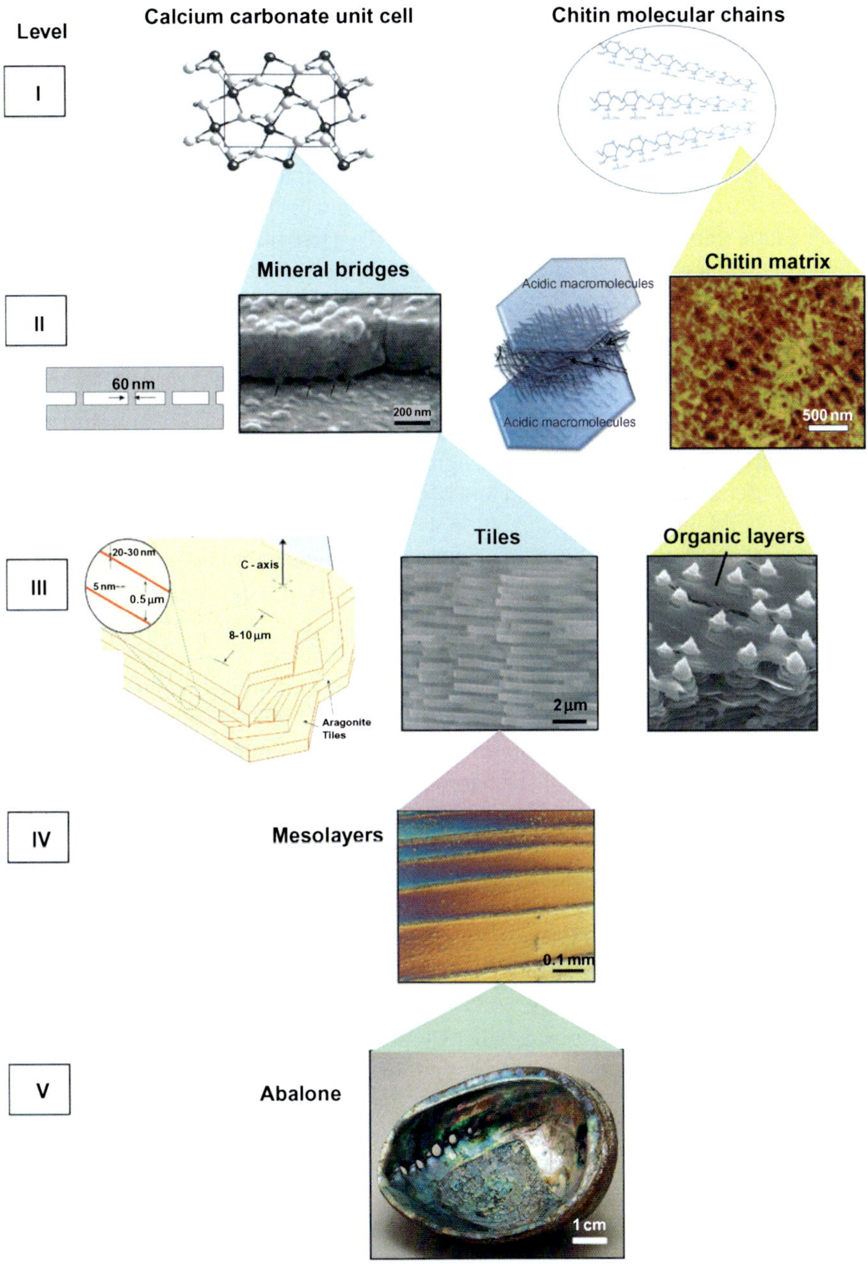

Figure 2.5 Hierarchical organization of abalone nacre represented by: (I) nanoscale association of aragonite crystallites with organic layers of protein and chitin fibers, (II) organic–inorganic interfaces composed of mineral bridges and an inter-tile organic matrix, (III) arrays of interlocked aragonite tiles decorated with nano-asperites, (IV) mesolayers composed of tile assemblages separated by thick organic layers, and (V) macrostructure nacreous layer optimized for strength and toughness.
Reproduced from Meyers et al. (2011) with permission from Elsevier.

of geological aragonite (Jackson et al., 1988; Sarikaya et al., 1990). This property emerges from a hierarchical structure based on aragonite tablets organized in a matrix of biomolecules similar to a brick-and-mortar arrangement of a lamellar ceramic (Wise, 1970; Rousseau et al., 2005) with unique design aspects (Wise, 1970; Currey, 1977; Wang et al., 2001). Firstly, the viscoelastic organic matrix surrounding the aragonite tablets helps in energy dissipation, thereby imparting material toughness and better crack deflection (Smith et al., 1999; Sumitomo et al., 2008). The layered hybrid organization imparts mechanical anisotropy to the nacre material wherein propagation of perpendicular cracks is more difficult than in a direction parallel to the mineral tiles (Meyers et al., 2011). Secondly, under a mechanical load, an aragonite tile can rotate through a small angle to interlock with the lower tablet (Katti et al., 2005). This provides additional deformation. Thirdly, the surfaces of these tiles are covered by nanoprotrusions termed asperities. Analysis of the contacts between adjacent tiles indicates interposition and interlocking of asperities (Wang et al., 2001; Evans et al., 2001). This imparts shear resistance that is augmented by mineral bridges connecting adjacent tiles (Schäffer et al., 1997; Lin and Meyers, 2005). Furthermore, the fundamental inorganic building blocks of nacre are polygonal nanograins that make up the aragonite tiles (Barthelat et al., 2006). The deformation-induced rotation and deformation of nanograins in the surrounding hydrated organic medium can lead to energy dissipation, thereby enhancing the ductile properties of this composite biomineral (Li et al., 2006; Verma et al., 2007). Thus nacre material presents remarkable material design at different length scales and has inspired the fabrication of similar synthetic composites (Tang et al., 2003; Das et al., 2013; Siglreitmeier et al., 2015; Fantner et al., 2006).

Bone tissue is a composite biomineral with unique material properties as well as physiological relevance (US Department of Health and Human Services, 2004; Weiner and Wagner, 1998; Skinner, 1973; Pritchard and Bourne, 1956). Similar to nacre material, bone structure has a hierarchical organization that encompasses several length scales (Weiner and Traub, 1992). At the nanoscale, plate-like hydroxyapatite nanoparticles associate with a matrix of Type I collagen and noncollagenous proteins (Robinson and Watson, 1952; Weiner and Traub, 1986). Type I collagen typically forms fibrils with a periodicity of 67 nm and with gaps of 35 nm between neighboring molecules (Maynes, 2012; Parry, 1988). Furthermore, these fibrils are built up from a triple helix structure of three α-helices (Parry, 1988). In bone, the collagen fibrils are mineralized with plate-like hydroxyapatite nanoparticles about 25–50 nm in size and 2–3 nm thick with the crystallographic *c*-axes aligned well with the fibril long axis (Jackson, 1957; Schmidt, 1936). The mineralized fibrils are in turn assembled as lamellae, 1–10 μm thick. These fibrils and lamellae exhibit diverse tissue-dependent organization such as parallel stacks in tendons of certain flightless birds, woven structures in human fetal femur, plywood-like structure in certain fish scales, and radial fibrillar arrays in dentin (Pritchard and Bourne, 1956; Francillon-Vieillot et al., 1990; Weiss and Ferris, 1954; Kramer, 1951). At the centimeter scale, bone tissue can be comprised of either osteons wherein fibrils are arranged in a compact, concentric manner, or microporous spongy or cancellous bone (Francillon-Vieillot et al., 1990; Leeuwenhoek, 1693; Gibson, 1985). The high stiffness and toughness of bone tissue can be attributed to its intricate structure at different length scales (Weiner and Traub, 1992;

Peterlik et al., 2006). The nanoscale regime of the hydroxyapatite platelets permits elastic deformation corresponding to about two times the failure strain of the bulk apatite (Gupta et al., 2006). At the micron and submicron scale, mechanical anisotropy emerges on account of crystallographic co-orientation of the mineral nanoparticles and unique integration with the collagen matrix. This leads to periodic, weak interfaces at such length scales (Seto et al., 2008; Hasegawa et al., 1994). Moreover, similar to nacre's properties, biomolecules in bone tissue prevent severe irreversible deformation by helping with energy dissipation (Gupta et al., 2005). At the micron scale, fibrils in the lamellar architecture can slide along each other under load. The consequent breakage of sacrificial bonds between glue-like proteins and mineral particles contributes to shear stiffness and reversible deformability (Fantner et al., 2005). Fracture resistance is also imparted due to mechanisms such as crack bridging, crack deflection, and microcrack formation (Peterlik et al., 2006). Such structural complexities and associated properties arise due to spatiotemporally intricate interactions between mineral precursors and organic molecules (Hunter, 1996). These interactions and the underlying physical forces in the formation of fascinating morphologies and structure in biogenic minerals are gradually being understood (Mann, 1997). In the next section, we therefore briefly introduce recent concepts associated with crystal nucleation and growth as well as their possible role in shaping biominerals.

2.3 Pathways of crystal formation and growth

In living organisms, genetic information and regulatory pathways combine to produce the complex processes responsible for transport and deposition of mineral precursors as well as the nucleation and growth of mineralization products. Such mechanisms often incorporate dynamic inorganic and organic molecules that are integrated in a spatiotemporally precise manner. Moreover, since the physical properties of biominerals can depend on hierarchical structuring, the regulation of such materials must also occur at different length scales. These levels of complexity can be in the context of (i) physical concepts describing crystal formation and growth, (ii) principles underlying macromolecular self-assembly, as well as (iii) interactions of biomolecules with mineral precursors and crystal faces. Current developments in the subject of particle formation and growth and transformation are briefly discussed below.

2.3.1 Classical and nonclassical nucleation

According to the classical nucleation theory (CNT), a crystal nucleus can form or disintegrate based on the interplay between energy associated with formation of a new surface and the bulk energy gained from creation of a crystal lattice (Kashchiev, 2000; Becker and Döring, 1935). At a critical size, the surface energy is counterbalanced by the bulk energy. If the nucleus grows beyond this size, the lattice energy gain overcomes the energy loss due to surface formation and a negative change in the free enthalpy of the system is attained, driving growth of the nucleus by ion attachment. However, contemporary investigations probing the early stages of mineral growth indicate that crystallization does not necessarily occur via attachment of ions or molecules to primary particles (Gebauer et al., 2008, 2014; Gebauer and Cölfen, 2011). This fundamentally changes the conventional perception of biomineral growth as well as the nature of mineral precursors. The *in vivo* significance of this crystallization pathway is gradually being understood (Dey et al., 2010; Nudelman et al., 2010; Vidavsky et al., 2014; Politi et al., 2004; Olszta et al., 2007). Hence, we briefly summarize these concepts because of their role in elucidating complex morphologies and hierarchical structures of minerals in the biological world.

The role of stable solute species called PNCs present in under- and supersaturated solutions has been reported for several minerals such as calcium phosphates, calcium carbonates, iron (oxy)(hydr)oxides, silica as well as organic molecules such as amino acids (Gebauer et al., 2008; Kellermeier et al., 2012a; Carcouët et al., 2014; Murphy et al., 1976; Spiro et al., 1966; Dousma and de Bruyn, 1978). PNCs are small, thermodynamically stable solute precursors with subsequent products that form via phase separation (Gebauer et al., 2014). Although PNCs exhibit highly dynamic configurations, they may also exhibit certain short-range structuring that may correspond to crystalline products (Gebauer et al., 2010, 2014; Demichelis et al., 2011). In the context of biomineralization, cluster species are important because PNCs are precursors to liquid and amorphous phases and in Nature amorphous phases are typical precursors for the final crystalline minerals (Politi et al., 2004; Olszta et al., 2007; Nassif et al., 2005). The direct identification of such precursors during biomineralization is challenging due to the transient nature of mineral precursors and the dynamic nature of mineralization processes. Nevertheless, the *in vitro* existence of PNCs has been validated by potentiometric titration, transmission electron microscopy, analytical ultracentrifugation, mass spectrometry, and molecular simulations (Gebauer et al., 2008; Wang et al., 2001; Dey et al., 2010; Kellermeier et al., 2012a; Demichelis et al., 2011). Furthermore, potentiometric titration assays reveal that certain additives including biomolecules significantly affect the early stages of mineralization, including the stability of PNCs, the time required for nucleation of a solid phase as well as the nature of initially precipitated phases (Raiteri et al., 2012; Verch et al., 2011; Picker et al., 2012; Rao et al., 2014; Perovic et al., 2014a). Titration assays also reflect the pH-dependent stability of PNCs (Gebauer et al., 2014). Since the pH of biochemical environments is fine-tuned and biomineralization processes can proceed in confined environments, the solution pH is a primary factor for stability of PNCs and other transient phases during transport of mineralization precursors (Gebauer et al., 2008; Demichelis et al., 2011). Recent studies employing biological additives in such titrations demonstrate quantitatively diverse effects dependent on multiple factors ranging from configuration, charge, position of functional groups, and stereochemistry for simple additives to secondary structure and self-assembly properties for macromolecular additives (Picker et al., 2012; Rao et al., 2014; Perovic et al., 2014a; Seto et al., 2014). For example, D-glucose does not show any significant effects on nascent stages of calcium carbonate mineralization wherein amorphous calcium carbonate (ACC) is the usual postnucleation product. Its enantiomer L-glucose, however, exhibits dissimilar properties such as enhancing the stability of PNCs as well as inducing the formation of vaterite, a crystalline form of calcium carbonate after nucleation (Rao et al., 2014).

Thus a simple switch in stereochemistry can have a substantial influence on mineralization. By assaying other carbohydrates, molecular properties such as glycosyl residue configuration, charge, degree of branching, position of functional groups, and the nature of the glycosyl linkages are found to influence mineralization (Rao et al., 2014; Nielsen et al., 2012). In similar experiments, the stereochemistry and charge of amino acids are found to affect early stages of mineralization (Picker et al., 2012). Thus for large macromolecules modulating mineralization, the effects seem dependent on multiple properties. Such interactions can be significant during the transport and deposition of mineral precursors wherein strict control over the nucleation events in terms of temporal stability as well as type of polymorph nucleated is necessary.

2.3.2 Kinetic control of crystal nucleation

Biological processes often exhibit control over the polymorph of nucleated particles by controlling reaction kinetics, i.e., by changing the barriers of activation energy for mineral nucleation, growth, and phase transformation (Meldrum and Cölfen, 2008). Usually crystallization proceeds by sequential steps that involve structural and compositional changes in mineral precursors and intermediates along the well-known Ostwald rule of stages (Gebauer and Cölfen, 2011; Mann, 2001). In a biological environment, the presence of certain molecules or complex interfaces can potentially influence crystal nucleation and growth by modulating the corresponding free energies of activation (Meldrum and Cölfen, 2008; Xiao et al., 2009; DiMasi et al., 2003; Cölfen and Qi, 2001; Xu et al., 2008). Such effects cannot be described directly from the thermodynamic standpoint provided by CNT. For example, ACC is not stable and can rapidly transform to any of its crystalline forms calcite, vaterite, and aragonite. However, it can be stabilized by ions such as Mg^{2+} as well as certain acidic proteins (Aizenberg et al., 1996; Bentov et al., 2010; Raz et al., 2003; Boskey and Posner, 1974; Politi et al., 2009; Loste et al., 2003). Stabilized ACC has also been reported from several organisms including sea urchins, crustacean exoskeletons, and plant cystoliths (Politi et al., 2004; Becker et al., 2005; Addadi et al., 2003). Additives can also modulate the formation of particular crystalline phases from amorphous precursors (Gotliv et al., 2003). Such controlled transformations are demonstrated by the formation of aragonite in the presence of Mg^{2+} ions (Ma and Feng, 2015). Moreover, proteins isolated from the calcitic or aragonitic layers of shells can induce the precipitation of calcite or aragonite, respectively, from supersaturated solutions (Gotliv et al., 2003; Ma and Feng, 2015; Falini et al., 1996). Kinetic control over crystallization can also be exerted by the interaction of mineral precursors with structured organic surfaces. The use of self-assembled monolayers and biomineral surfaces for crystal growth shows that functional groups and their stereochemistry can lower the activation energy for nucleation of particular faces (Aizenberg, 2010; Aizenberg et al., 1999). In the presence of polymer additives, such surfaces can also mediate selection of a particular polymorph (Balz et al., 2005; Aizenberg, 2004). Thus, considering the prevalence of amorphous phases and interfaces associated with biominerals, the kinetic control of crystal nucleation is an important part of Nature's construction toolbox. Such processes can effectively modulate the shape, stability, structure, and composition of the mineral precursors as well as the final products.

2.3.3 Amorphous precursors

The initial report of amorphous precursors in biomineral formation was by Towe and Lowenstam (1967), who identified iron hydroxide (ferrihydrite) as a precursor phase to magnetite in chiton teeth. Subsequently, the presence of amorphous phases as precursors of biogenic minerals has been noted in several organisms (Weiner et al., 2009). The advantages associated with this strategy include circumventing osmotic stress by high ionic strengths, avoiding transport of large volumes for the formation of crystals with a low solubility product, and adaptability to complex shapes after molding the amorphous phase by a template prior to crystallization (Meldrum and Cölfen, 2008). With respect to such amorphous phases, more than one amorphous state of a substance can exist and this phenomenon is referred to as polyamorphism (i.e., amorphous polymorphism) (Cartwright et al., 2012). Polyamorphism is not limited to solid forms of materials but is also exhibited by liquids such as phosphorus and triphenyl phosphite (Katayama et al., 2000; Kurita and Tanaka, 2004). In the context of biomineralization, studies on polyamorphism are presently focused on calcium carbonate (Figure 2.6),

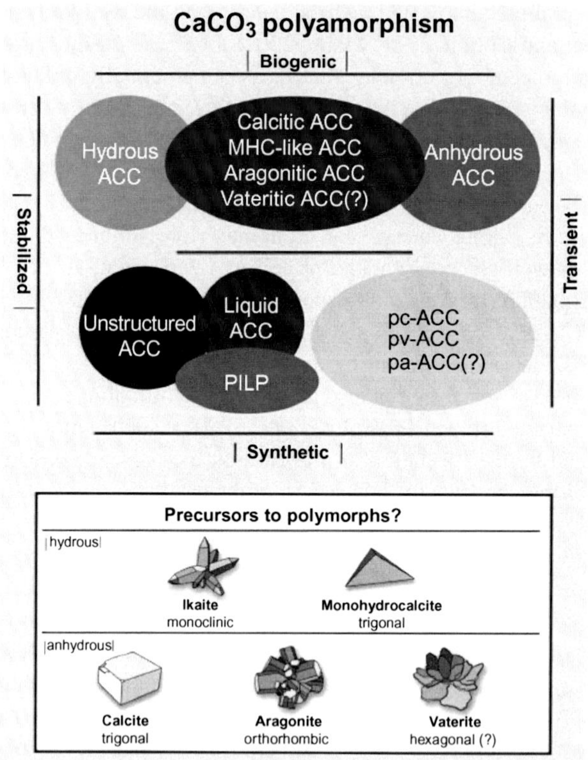

Figure 2.6 Extensive polymorphism and polyamorphism of calcium carbonate. Amorphous forms include hydrous and anhydrous ACC as well as liquid-like mineral precursors, of which certain synthetic and biogenic phases exhibit proto-structures corresponding to specific crystalline forms. Reproduced from Cartwright et al. (2012) with permission from Wiley.

which is the most common constituent of invertebrate biominerals (Gebauer et al., 2010). Amorphous precursors are observed in biominerals such as ascidian spicules, crustaceans, plant cystoliths, sea urchin spine, and sponge spicules, wherein the final products of biomineralization may differ (Politi et al., 2004; Becker et al., 2005; Addadi et al., 2003; Lambert and Lambert, 1997). An interesting observation is the presence of distinct short-range order similar to aragonite in ACC from the aragonitic snail *Biomphalaria glabrata* (Hasse et al., 2000). Similar evidence from structural investigation on amorphous precursors of aragonitic larval mollusks and calcitic sea urchin skeletal elements shows that ACC produced under biological control can exhibit distinct short-range order (Raz et al., 2003; Addadi et al., 2003; Politi et al., 2006). This raises important questions on the existence of diverse short-range orders in ACC of different origins, the nature of phase transformations to particular crystalline forms as well as the impact of additives on such transformations.

Structural analysis of synthetic ACC shows that such short-range structural features exist in samples that are prepared by quenching slightly supersaturated, equilibrated solutions of calcium carbonate even in the absence of additives (Gebauer et al., 2010). Depending on solution pH, the short-range order of amorphous ACC samples exhibits similarities to calcite and vaterite, respectively. Due to the pH-dependent variation in ACC solubility products identified in a previous study, the protocrystalline amorphous phases are termed proto-calcite ACC (pc-ACC/ACC I) and proto-vaterite ACC (pv-ACC/ACC II), corresponding to the solution pH values of 8.75 and 9.8 before quenching (Gebauer et al., 2008, 2010). This nomenclature only indicates structural similarities and not that a particular polyamorphous phase will only transform to a specific crystalline phase. Such transformations are significantly influenced by solution conditions such as the temperature and the presence of additives that can have kinetic effects on crystallization (Cartwright et al., 2012; Kellermeier et al., 2010; Xu et al., 2008; Rodriguez-Blanco et al., 2011). Present studies show the prevalence of polyamorphism among biogenic and synthetic mineral precursors. However, the influence of such proto-structures on mineralization processes as well as whether these proto-structures are distinct enough to be individually recognized by biomolecules and its consequences on mineralization requires further elucidation.

Generally, stabilized or transient ACC correspond to its hydrous or anhydrous form, respectively (Cartwright et al., 2012; Addadi et al., 2003). Under conditions of stabilization, ACC predominantly exhibits a $CaCO_3 \cdot H_2O$ stoichiometry (Gower, 2008). During crystallization, dehydration of ACC generates an anhydrous phase that gradually follows an energetically downhill pathway toward the crystalline phases (Radha et al., 2010). Recent studies also suggest that amorphous forms can present a variation in the amount of hydration (Gower and Odom, 2000; Wolf et al., 2011a,b). For example, liquid forms of calcium carbonate can be synthesized in the presence of acid polymer or certain proteins (Gower and Odom, 2000; Wolf et al., 2011b). These are referred to as polymer-induced liquid precursors (PILPs) and liquid ACC, respectively, and exhibit behavior typical of liquids such as droplet formation and capillary effects (Gower and Tirrell, 1998; Olszta et al., 2007). Classically, PILPs are generated by mixing small quantities of short-chained polyanionic polymers with precursor ions (Gower and Odom, 2000). During mineralization processes, such phases form a metastable emulsion with droplets that play a role in sequestering precursor ions as well as inhibiting nucleation. In addition, complex crystal morphologies can be attained *in vitro* by depositing such phases on films or templates (Gower and Odom, 2000; Gower and Tirrell, 1998). A major advantage of these liquid phases is shape adaptability. Although the role of liquid precursor phases in biomineralization requires evidence, the complex morphologies of biogenic crystals, and their formation in almost inaccessible sites might reflect their transient *in vivo* presence. The mineralization of Type I collagen fibers with calcite disks using PILP supports this notion as well as the infiltration of collagen fibrils with liquid calcium phosphate precursors (Nudelman et al., 2010; Jee et al., 2011; Olszta et al., 2003). Interestingly, these precursors are shown to enter gap zones of collagen fibrils and consequently form a structure analogous to mineralized collagen in bone tissue (Nudelman et al., 2010).

2.3.4 Nonclassical crystallization

The classical perspective on crystallization typically describes crystal growth occurring through the association of ions, atoms, or molecules with certain "active sites" on the crystal surface. Thus nucleation of a new layer on the crystal surface is initiated by the formation of a small island on the growth surface. Further attachment of ions, atoms, or molecules to the steps or kinks leads to formation of a new layer, leading to crystal growth. However, crystal growth can also occur via the aggregation or fusion of nanoparticles. In a slow, thermodynamically controlled regime, precursor nanoparticles can fuse in crystallographic register, forming a large single crystal via a "nonclassical crystallization" process (Cölfen and Antonietti, 2008; Niederberger and Cölfen, 2006). This necessitates that the primary nanoparticles become aligned, connect via a common crystallographic axis, and fuse (Niederberger and Cölfen, 2006). Such oriented attachment phenomena have been described for TiO_2, ZnO, and certain iron oxide particles, and many others in one, two, and three dimensions (Penn and Banfield, 1999; Polleux et al., 2004; Burleson and Penn, 2005; Zhang et al., 2009, 2010). In the case of calcium carbonate, crystal growth in the presence of ACC-stabilizing additives such as phosphate ions, certain proteins, or gel-forming molecules can lead to aggregation of amorphous particles to form assemblies with single-crystal-like properties (Gal et al., 2014). In contrast, kinetically controlled fast processes and crystallographic fusion generally lead to polycrystalline products (Meldrum and Cölfen, 2008). In such scenarios, docking processes that include the rotation of primary nanoparticles for crystallographic alignment and molecular restructuring at particle interfaces are limited. A special case of nonclassical crystallization is when the binding affinity of organic ligands associated with the nanoparticles is high enough so that smaller crystallites cannot grow or fuse and the nanoparticles are aligned by interfacial interactions (Cölfen and Antonietti, 2005; Niederberger and Cölfen, 2006). This leads to the formation of 1D linear aggregates, 2D arrays, or 3D superstructures when the primary nanoparticles, anisotropic in shape, are crystallographically coaligned (Figure 2.7). Such structures are termed as mesocrystals, i.e., mesoscopically structured crystals (Cölfen and Antonietti, 2005). These crystalline assemblies are of relevance to the structure and property of biominerals as discussed in further sections.

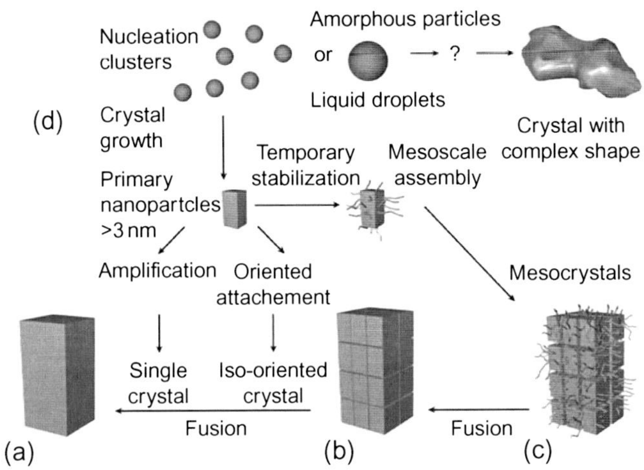

Figure 2.7 Representation of classical and nonclassical crystallization pathways. Pathway (a) shows the classical scheme wherein nucleation clusters grow until they reach a critical size and further grow to a primary nanoparticle, which then matures into a single crystal. Alternatively, the primary nanoparticles can fuse forming iso-oriented crystals by oriented attachment via pathway (b). When stabilized by additives, the primary nanoparticles can assemble to form a mesocrystal as depicted by pathway (c). Pathway (d) involves intermediate amorphous particles that can coalesce and transform to complex morphologies. Reproduced from Cölfen and Antonietti (2008) with permission from Wiley.

2.3.5 Controlling crystal morphology

The equilibrium shapes of crystalline materials are usually attributed to atomic arrangements of the characteristic unit cell that are periodically repeated in the lattice. Thus, single crystals should ideally display well-defined faces and edges. Certain minerals deposited by organisms, however, show intricate morphologies with curved surfaces devoid of typical crystal faces (Lowenstam and Weiner, 1989). For example, the shape of biominerals associated with sea urchins and coccolithophorids deviate considerably from the usual rhombohedral form of calcite, their constituent inorganic phase (Lowenstam and Weiner, 1989; Young et al., 1999; Wilt, 1999). This can be explained by modulation of crystal shape by thermodynamic or kinetic control. Under thermodynamic considerations, the surface energies of individual crystal faces are dictated by external growth conditions. As a generalization of Wulff's rule, the growth rate of faces with high surface energy is higher than that of faces with relatively lower surface energies (Wulff, 1901). Thus, a crystal face with high surface energies will grow relatively faster and vanish whereas the other faces will dominate the final crystal morphology. The equilibrium morphology hence corresponds to the minimum surface free energy, which in turn depends on the growth environment. For example, the calcite (012) face is a high-energy, highly polar surface (Luo and Sonnenberg, 2008). In aqueous conditions, this crystal face is transformed from a smooth to a rough surface and is quickly reconstructed to the neutral (104) face. A similar experiment in the presence of ethanol shows no surface reconstruction. A second strategy for applying Wulff's rule in shaping crystals is utilizing additives that selectively adsorb to certain crystal faces (Wang et al., 2010; Yu et al., 2003). On account of a lower surface energy, crystal growth can be quenched along these faces. However, Wulff's rule is limited to short-range interface energies and excludes long-range interactions (Wang et al., 2006). The prevalence of kinetic control over mineral morphology in biomineralization processes cannot be undermined. As discussed in the early sections, kinetically controlled processes can effectively regulate the nature of the mineral precursors and final products. At high supersaturation, i.e., conditions of fast precipitation, several crystal nuclei are generated. In the absence of an appropriate stabilizing additive, these initially nucleated particles can fuse in crystallographic register to larger particles, leading to conditions that favor nonclassical crystallization (Meldrum and Cölfen, 2008). In a biological context, conditions of pH, temperature, and redox potential have a marked effect on the shape of magnetite crystals in magnetotactic bacteria (Meldrum et al., 1993; Faivre et al., 2008). Gel-like matrices associated with biominerals also offer control over crystal growth by decreasing the diffusion rate of ions and mineral precursors and suppression of convection (Henisch, 2005). Control over gel density and in turn diffusion can lead to complex crystal morphologies such as dendrites and branches (Oaki and Imai, 2003). Moreover, in the presence of gel–mineral interactions, ring-like precipitates named Liesegang rings are observed (Henisch, 2005; Suda et al., 1996). Extracellular environments are typically rich in gel-forming biomolecules such as long-chain proteoglycans and thus can potentially modulate mineralization processes in a diffusion-controlled manner (Nielsen et al., 2012; Arias and Fernández, 2008).

2.3.6 Polymorph control

The ability of a crystalline system to exist as more than one crystal structure is recognized as polymorphism. For example, calcium carbonate exists as three anhydrous (calcite, aragonite, and vaterite) and two hydrated (monoclinic hexahydrate ikaite and calcium carbonate hexahydrate) crystalline forms. Experimentally, synthesizing each of these forms with high purity and reproducibility is tedious because this requires precise modulation of thermodynamic and kinetic effects during crystallization. Nature, however, makes these polymorphs with consistent composition as well as morphology using biomolecules and ions as additives under ambient conditions. For instance, nacre-associated proteins AP7 and AP24 when selectively adsorbed on a surface induce nucleation of aragonite (Kim et al., 2006a). Also, the Pif protein has been thought to induce aragonite formation (Suzuki et al., 2009). Another example of polymorph selection is the ability of ferritin, a shell-like protein complex, to selectively stabilize ferrihydrite ($5Fe_2O_3 \cdot 9H_2O$) (Meldrum et al., 1992; Wong et al., 1998; Uchida et al., 2010). In the absence of this protein, the generated phase is lepidocrocite (γ-FeOOH). Such interactions raise interesting questions pertaining to organic–inorganic interactions that lead to phase selection. Oswald's step rule allows for empirical predictions of the sequence of

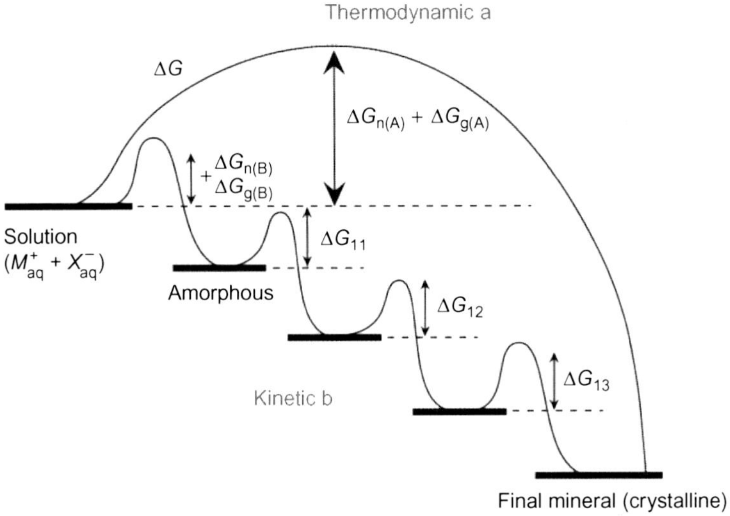

Figure 2.8 Schematic of crystallization pathways under thermodynamic and kinetic control. The free energy of activation associated with nucleation (n), growth (g), and phase transformation(s) determines whether (A) a one-step pathway to the final mineral form takes place or (B) a series of sequential transformations is followed.
Reproduced from Cölfen and Mann (2003) with permission from Wiley.

phases formed during a crystallization event. Although all possible polymorphs may not form, the kinetic transformation sequence strictly follows increasing thermodynamic stability (Figure 2.8). However, by controlling reaction conditions, intermediate polymorphs can be stabilized against further transformation and recrystallization. This can be achieved by optimizing temperature, concentration of precursors, and mixing parameters (Nebel and Epple, 2008; Kotachi et al., 2006). Alternatively, by using additives, particular crystalline forms can be stabilized. For instance, silk nanoparticles can be used to generate vaterite microspheres stable for at least a week (Liu et al., 2015) and in the case of a double-hydrophilic block copolymer (DHBCs) with a strongly chelating PolyEDTA block for even more than a year (Cölfen and Antonietti, 1998). An interesting example is the additive-mediated kinetic stabilization of biphasic calcium carbonate structures composed of vaterite cores surrounded by equatorial loops composed of calcitic nanoparticles. The formation mechanism is attributed to initial temporary stabilization of vaterite by poly(sodium 4-styrenesulfonate) and subsequent folic acid-induced calcite nucleation (Wang et al., 2013). In Nature, however, such effects are difficult to directly capture on account of multiple biomolecules and spatiotemporally complex interactions that are involved in biomineralization processes. Nevertheless, the application of biochemistry and recombinant biomolecules in mineralization assays suggests that organic molecules and interfaces effectively modulate polymorph selection (Kotachi et al., 2006; Ponce and Evans, 2011).

2.4 Molecular templates and morphology control in Nature

2.4.1 Nacre: protein-modulated crystal growth on chitin layers

As discussed above, the unique mechanical properties of nacre emerge on account of the intricate organization of aragonite tablets held together with an organic matrix. Organic sheets situated between the aragonite layers are composed of insoluble, hydrophobic proteins with amino acid compositions similar to those of silk fibers as well as certain soluble, acidic macromolecules. In certain cases, these proteins are associated to fibrils of β-chitin, a linear polysaccharide composed of β-1,4-linked *N*-acetylglucosamine units. Several proteins and polysaccharides associated with nacre are inherently insoluble and thus not easy to investigate. However, recent investigations shed light on the complex interplay between some of these biomolecules and mineralization processes.

In the presence of mineral precursors, soluble molecules extracted from the nacreous layer of certain shells can induce formation of aragonite (Falini et al., 1996). Soluble proteins from abalone nacre can modify the (1014) atomic lattice of calcite crystals and mediate the formation of an aragonitic (001) face (Thompson et al., 2000). This indicates that biomineral-associated macromolecules have functions in polymorph selection. Considering the aragonite-associated proteins present in nacre, all of these are intrinsically disordered and about 95% of these have amyloid-like or cross-β-strand aggregation-prone motifs (Evans, 2012). Given the prevalence of intrinsically disordered and aggregated proteins associated with nacre, the structural dynamics of such proteins may depend on varying mineralization conditions that present precursor ions/clusters, transient mineral precursors, or crystalline faces. In other words, a single protein may influence different stages of mineralization on account of a dynamic, solution-dependent structure. Indeed, recent studies show that the individual proteins have distinct effects on different stages of mineralization (Figure 2.9). For example, nacre-associated proteins AP7 and AP24 effectively modulate the mineralization of calcium carbonate despite having partially unfolded conformations (Michenfelder et al., 2003; Perovic et al., 2014b). The N-terminal of both proteins harbors acidic amino acid residues that may interact with Ca^{2+} ions (Wustman et al., 2004). Also, AP7 can form aggregates that provide a suitable environment for aragonite formation and stabilization in the absence of Mg^{2+} ions (Amos and Evans, 2009). Earlier shown to be induced by protein mixtures, polymorph selection can be mediated by the nacre-associated AP7 protein alone. Moreover, AP7 has a remarkable effect on crystal morphology by inducing the formation of nanometer-sized porosities or voids on calcite surfaces (Chang et al., 2014; Kim et al., 2006b). Thus, proteins associated with nacre can have remarkable effects on the different stages of mineralization. Common sequence features of such proteins are: (i) a high structural disorder propensity, (ii) limited acidic amino acid residues (10–25%), and (iii) copolymer, block-like arrangement of proton donor/acceptor and hydrogen-bonding amino acid residues (Kim et al., 2004). These properties are also exhibited by synthetic DHBCs, which have fascinating effects on colloidal stabilization and crystal growth modification (Cölfen, 2001). Thus, for modulation of

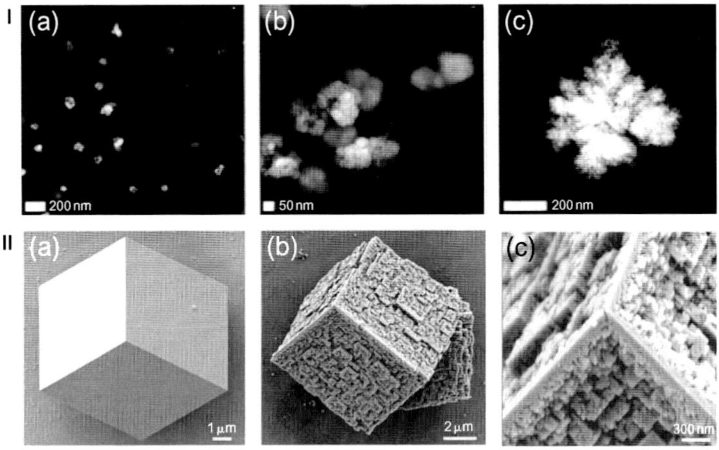

Figure 2.9 Multiple effects of AP7 on calcium carbonate mineralization. (I) Time series of scanning transmission electron micrographs show that: (a and b) initial mineral nanodeposits are localized around ring-like structures and (c) later on a branching morphology is evident. Reproduced from Perovic et al. (2014b) with permission from American Chemical Society. (II) Calcite crystals grown using gas diffusion assays: (a) in the absence of AP7 and (b and c) in the presence of AP7. AP7 introduces surface nanopatterning and subsurface nanoporosities into the crystal.
Reproduced from Chang et al. (2014) with permission from American Chemical Society.

mineralization processes by using additives, an abundance of acidic functional groups, i.e. charge, may not be the primary factor necessary (Cölfen, 2001). Properties such as self-assembly, pH/temperature/ion strength/surface-interaction-dependent structural dynamism and fine-tuned hydrophobicity/hydrophobicity appear to be equally important in biomineralization. This idea is further supported by mineralization studies in the presence of a unique isoform of n16 (specifically n16.3), a nacre-associated protein. Analysis of its primary sequence indicates a tendency toward structural disorder, the presence of a few acidic residues and a low complexity patch abundant in glycine and asparagine. In the presence of Ca^{2+} ions, n16.3 self-assembles into nanometer-sized complexes (Ponce and Evans, 2011). Moreover, with this protein, analysis of the mineralization products atypically indicated aragonite formation. Thus, the self-assembly properties of this protein might be related to binding with mineral precursors as well as crystal nucleation. A similar study indicates that polymorph selection is mediated by formation of unique interfaces in the presence of n16 and β-chitin, which are dependent on the intrinsic disorder of the protein, aggregation propensity, and the presence of certain anionic amino acids (Keene et al., 2010; Amos et al., 2011). Potentiometric titration studies show that although n16.3 does not affect the stability of PNCs, it can affect phases formed after nucleation in a concentration-dependent manner (Perovic et al., 2014a). As in the case of AP7, this protein can also change mineral morphology to exhibit terraced surfaces (Kim et al., 2004). In summary, these studies show that nacre-derived supramolecular protein assemblies modulate prenucleation and postnucleation processes as well as crystal organizing and shaping mechanisms.

Regarding the microstructure of nacre, early crystal growth models assume heteroepitaxial nucleation on organic surfaces to explain the formation of aragonite platelets (Weiner and Traub, 1980). As discussed in earlier sections, adjacent aragonite platelets are connected by mineral bridges that travel through pores in the organic sheets. An alternative hypothesis is that these pores are vital in determining the position of the subsequently forming mineral platelets. This idea is supported by the distribution of pores over the organic sheets observed after proteinase K and collagenase treatment (Schäffer et al., 1997). Thus, mineral bridges were proposed to be determinants of sites of epitaxial growth of aragonite platelets during the biosynthesis of nacre (Checa and Rodríguez-Navarro, 2005). Based on this model, multiple nucleation events may not be required (Checa et al., 2011). This is further validated by recent polarization-dependent imaging contrast studies, which propose one mineral bridge per tablet that serves as a seed for near-epitaxial growth of the new tablet (Olson et al., 2013). However, the presence of continuous layers of amorphous calcium carbonate surrounding the aragonite platelets appears contradictory to epitaxial nacre formation (Nassif et al., 2005). The typical (001) face of aragonite is highly polar and its charge interaction can extend through the surrounding ACC up to 3–5 nm. This face can then interact with an oppositely charged organic matrix. This scenario presents a facile explanation for crystallographic co-orientation. Moreover, amorphous phases of calcium carbonate typically exhibit high-solubility products and could be dissolved during sample preparation procedures for transmission electron microscopy (TEM). In order to refine nacre growth models, further information is required on the formation and transport of mineral precursors, nucleation, and phase transformation mechanisms, as well as the role and dynamics of nacre-associated biomolecules. Recent studies indicate the role of oriented attachment during ACC to aragonite transformations in nacre formation (Huang and Li, 2012; Zhang and Xu, 2013).

2.4.2 Bone: Role of collagen and other proteins

The organic matrix of bone tissue comprises Type I collagen fibrils associated with noncollagenous proteins. Type I collagen has a low-complexity sequence rich in glycine (Gly), proline (Pro) as well as the rare amino acids, 4-hydroxyproline (Hyp) and 5-hydroxyproline. Typically, this collagen type contains several repeats of (Gly-Pro-Hyp) that have important consequences on the protein structure. The periodic presence of proline induces certain structural constraints on the secondary structure by introducing intermittent kinks and twisting into a helical structure. Moreover, three noncoaxial helices pack together forming a superhelical assembly termed tropocollagen. This macromolecular assembly is stabilized by hydrogen bonding between glycine residues of adjacent polypeptide chains and covalent bonds between certain lysine side chains mediated by aldol crosslinks. The tropocollagen superhelix has a distance per turn of 0.3 nm, length of about 280 nm, and width of 1.5 nm. Several superhelices can subsequently assemble to form collagen fibrils with periodic structural features relevant to mineralization mechanisms. Chains of tropocollagen are arranged in a staggered manner such that the resultant array exhibits spaces called hole zones, which are about 40 nm long and 5 nm wide. Several adjacent hole zones decorate the collagen fibril to present regularly spaced parallel grooves. In bone material, plate-like crystals of hydroxyapatite associated with matrix proteins are the inorganic constituents in bone tissue. The platelets are about 45–20 nm in size with a thickness of 3 nm and their flat surfaces are perpendicular to the $[1\bar{1}0]$ direction. The associated organics probably contribute to the thermodynamic stability of these small particles. These crystals have a unique crystallographic orientation wherein the *a* and *c* axes (corresponding to the [110] and [001] axes) are aligned toward the groove direction and collagen axis, respectively. Thus, mineral nucleation and growth are associated with the periodicity presented by the collagen assembly. The unique periodicity in the collagen fibrillar structure is translated to periodicity in the arrangement of the inorganic building units via mechanisms of crystal nucleation and growth.

Although the structure of bone has been well investigated, fundamental aspects of its formation are still not clear (Glimcher et al., 1981). Studies on the nature of the mineral precursors show that calcium phosphate formation in simulated body fluid can proceed by aggregation of PNCs to form an amorphous phase that is subsequently nucleated to crystalline apatite (Dey et al., 2010). Other studies indicate that inhibitors of mineralization may play a role in the infiltration of amorphous liquid-like precursors and nucleation of hydroxyapatite within the collagen fibrils (Nudelman et al., 2010; Olszta et al., 2007). Thus, in addition to self-assembly, charge distribution properties of collagen fibrils also affect mineralization. Simulated models of collagen indicate positively charged junctions near the gap and overlap zones, which are possibly involved in charge-based attraction of anionic mineral precursor–additive complexes (Figure 2.10). Several aspects of nonclassical nucleation including mineralization via PNCs and transient amorphous phases are evident here (Cölfen, 2010). It has also been suggested that interactions between amorphous precursors and collagen are sufficient to obtain crystallographic co-orientation of the hydroxyapatite platelets (Olszta et al., 2007). This would imply that certain biomolecules interact with amorphous phases and modulate secondary nucleation in terms of composition, shape, and orientation of crystalline products. More evidence comes from mineralization in zebrafish fin rays, during which the transport and deposition of amorphous calcium phosphate nanoparticles as well as their subsequent crystallization to apatite are observed (Mahamid et al., 2010). Although such experiments address the inorganic phase transformations during bone formations, the *in vivo* scenario is more complex. For instance, current studies show that bone crystallites are associated with polysaccharides, the functions of which are not clear (Wise et al., 2007). Moreover, the role of noncollagenous proteins is emerging (Marini et al., 2014). Although studies indicate that noncollagenous proteins do not play a primary role in the growth of crystals (Olszta et al., 2007), certain proteins do affect bone mechanics by forming interfaces between mineralized collagen fibrils and enhancing interfacial strength and fracture energy (Hang et al., 2014). Given the recent advances in analytical and biochemical capabilities (Gebauer et al., 2009b; Nudelman et al., 2010; Mahamid et al., 2008; Kellermeier et al., 2013b; Schinke et al., 1996), their joint application as regards investigating the roles of biomolecules such as glycan residues and noncollagenous proteins in stabilization, transport, nucleation, and growth of specific mineral phases appears promising.

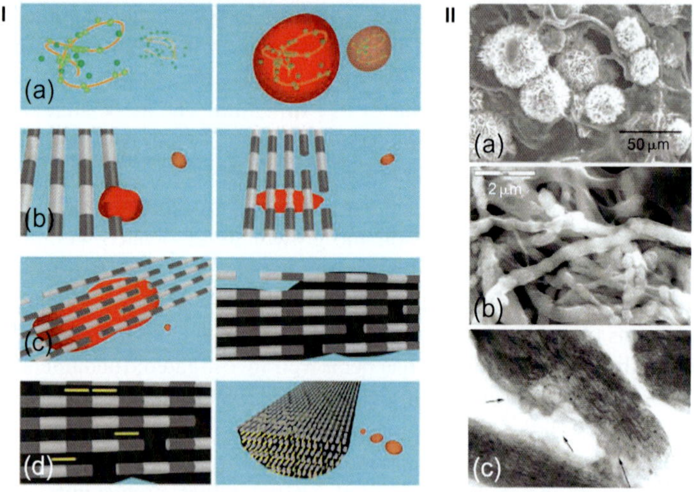

Figure 2.10 (I) Schematic overview of collagen mineralization. (a) Calcium phosphate clusters (red) complex with acidic polymers (yellow) to form droplets of a liquid-like mineral precursor phase. (b) Such droplets associate with distinct sites on the collagen fibril and enter the collagen structure. (c) Inside the collagen network, the liquid-like phase diffuses and solidifies to an amorphous phase. (d) Directed by collagen, oriented apatite crystals are formed from the amorphous phase. (II) Electron micrographs of collagen mineralized (a) in the absence of poly(Asp) show surface growth of hydroxyapatite, whereas (b) in the presence of the polymers, homogeneous mineralization is observed. (c) Amorphous droplets indicative of a PILP phase are associated with the collagen fibrils.
Reproduced from (I) Cölfen (2010) with permission from Nature publishing group and (II) Olszta et al. (2007) with permission from Elsevier.

2.4.3 Sea urchin spicules: Building calcite mesocrystals

Structurally, the adult sea urchin spine is a composite built of calcite containing about 5% Mg (magnesium calcite) reinforced with an organic matrix that constitutes approximately 0.1% of the total mass (Wilt, 2002). This material exhibits interesting properties that are typical of crystalline as well as amorphous materials (Wilt, 2002). X-ray diffraction patterns and intense birefringence of the sea urchin spine show that the constituent crystalline particles are co-oriented. However, the fractured surface of the spine is conchoidal and typical of amorphous materials. This feature is attributed to the mesocrystalline structure of the spine (Seto et al., 2012). Typically, the coherence length of geological calcite is about 800 nm with a domain spread of about 0.003°. In contrast, the calcitic sea urchin spine exhibits a coherence length of about 130 nm and domain spread of 0.07° (Aizenberg et al., 1997). At the nanoscale, the spine consists of nanoparticles that are crystallographically coaligned (Seto et al., 2012; TsrpunsrY and Busncr, 1993; Oaki and Imai, 2006; Berman et al., 1990). The 3D array of nanoparticles is held together by a mixture of ACC and organic molecules (Figure 2.11). Thus, a fracture through the spine cannot only proceed along the calcite {104} cleavage planes but also has to travel through the ACC, resulting in a glass-like fracture surface.

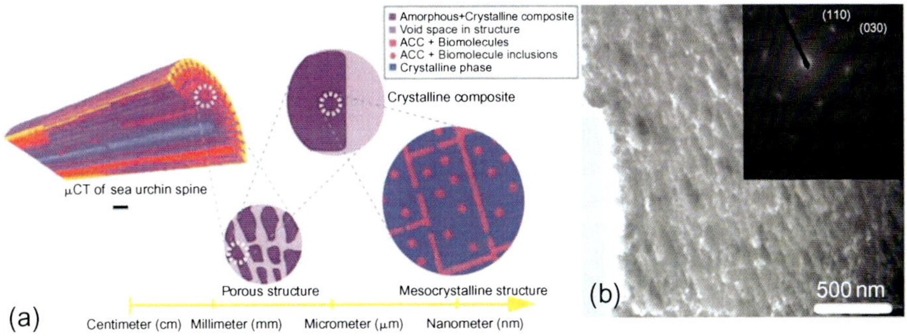

Figure 2.11 (a) Representation of the hierarchical structure in the sea urchin spine. At the centimeter to millimeter scale, the spine exhibits a complex, porous structure. At the nanoscale, the mesocrystalline architecture is composed of calcite nanocrystals (blue) associated with ACC and organic molecules (pink). (b) TEM and selected area electron diffraction (SAED) of a thin section of the spine indicating iso-orientated calcitic nanoparticles.
Reproduced from Seto et al. (2012) with permission from National Academy of Sciences.

Further evidence for this superstructure comes from the homogeneous incorporation of dyes such as Eosin Y in the spine as well as the use of nanoscale interspaces in the spine to act as templates for organic crystals in structures with hierarchical morphologies (Oaki and Imai, 2006; Munekawa et al., 2015; Oaki et al., 2006).

The formation mechanisms of the sea urchin spine are as yet unclear. However, the role of amorphous precursors, proteins, and self-assembly processes in the formation of this intricate structure is evident. Spine formation proceeds by the deposition of an amorphous phase of calcium carbonate (Politi et al., 2004; Beniash et al., 1997). Intracellular vesicles appear to be involved in the stabilization and transport of this typically transient phase (Vidavsky et al., 2014). This suggests a confinement mechanism of phase stabilization (Kellermeier et al., 2010). In addition, three closely located phases of calcium carbonate have been identified in the spicule, including a short-lived, hydrated ACC phase, an intermediate transient ACC form, and biogenic calcite (Politi et al., 2008). Although the exact mechanism of ACC stabilization and transformation is unknown, the role of certain proteins and Mg^{2+} ions is evident (Loste et al., 2003; Gong et al., 2012; Rao et al., 2013). Biochemical studies show that certain proteins such as SM50 and LSM34 are vital for sea urchin spiculogenesis (Peled-Kamar et al., 2002; Wilt et al., 2008). The role of SM50 in the stabilization of ACC is also identified (Raz et al., 2003; Gong et al., 2012; Rao et al., 2013). Furthermore, protein self-assembly processes seem to be involved in spine formation. After demineralization of sea urchin spicules, fibrous lamellae roughly arranged in a concentric manner are observed (Wilt, 2002; Benson et al., 1983; Okazaki, 1960). Mineral dissolution also exposes matrix proteins occluded within the mineral structure (Benson et al., 1986). Although the synchronization of all these processes during spine formation is as yet unclear, the interactions between organic molecules, self-assembly processes, and mineral precursors are intricately regulated and coordinated during spine formation.

2.4.4 Red coral

Among biologically produced mesocrystals, the structure of the axial Mg–calcite skeleton of the red coral, *Corallium rubrum*, is quite intriguing. Although the mechanism of formation is presently under investigation, the fascinating structural complexity of this biomineral is noteworthy. This biomineral exhibits a modular mesocrystalline architecture (Vielzeuf et al., 2008, 2010). The axial skeleton displays six modules of hierarchy. At the macroscale, the inorganic skeleton consists of radially organized units about 200–300 μm wide called "herringbone units." These units can be identified in a section perpendicular to the skeletal axis. When observed with polarized light, a radial pattern of alternating light and dark strips emerges. Electron backscatter diffraction (EBSD) analysis shows that the *c*-axes of the Mg–calcite units in adjacent light and dark bands are oriented ~44° from each other. Each herringbone unit is composed of highly co-oriented, irregular spindles about 100–300 μm long and 10–50 μm wide The crystallographic orientational spread of spindles within a parent strip is about 15°. The spindles are composed of smaller lozenges about 5 μm wide and 10 μm long. The lozenges in turn have a substructure of micrometer-scale crystalline fibers. At lower length scales, the fibers are composed of submicrometer units (200–500 nm) with a stacking arrangement. Lattice fringe analyses show that adjacent domains are nearly crystallographically co-oriented. This remarkable multilevel structure raises interesting questions regarding the underlying mechanisms of formation. Structural analysis of mesocrystalline sclerites from this organism suggests that mineral growth occurs by alignment and fusion of nanoparticles (Sethmann et al., 2007). Current investigations seek to identify biological additives that control the associated formation processes.

2.4.5 Tooth enamel

In reptiles and mammals, a product of biomineralization processes regulated by ameloblast cells is a hierarchically organized crystalline structure known as enamel (Satchell et al., 2002; Sander, 2000). The hydroxyapatite content of human mature enamel is about 96% (w/w) (Gwinnett, 1991). The biomineral is reinforced by an organic matrix, which also plays an important role in crystal growth and structure. A spatially and structurally related tissue called enameloid is produced by dental epithelial and odontoblast cells in sharks, rays, bony fish, and early larval stages of certain amphibians such as newts and salamanders (Davit-Béal et al., 2007). The morphological features of these mineralized tissues vary in a species-dependent manner. Similar to bone tissue, enamel exhibits a hierarchical structure that consists of arrays of hydroxyapatite crystallites aligned along the *c*-axes. Both materials consist of apatite nanofibers about 30–100 nm in diameter; however, at lower length scales, enamel and enameloid structures differ in structure, being composed of dense arrays of nanorods and domains of occasionally curved parallel fibers, respectively (Habelitz, 2015; Diekwisch et al., 2002). Mechanically, enamel is an anisotropic material with values of hardness and Young's moduli depending on the micro/nanostructural features, orientation of crystallites as well as chemical composition (Cuy et al., 2002; Habelitz et al., 2001; White et al., 2001). Consequently, the cracks through enamel material preferentially follow interfaces along inter-rod spaces parallel to the long axes (Rasmussen et al., 1976). Such unique mechanical properties are also contributed to by "sacrificial bonds" between organic molecules and mineral particles that impart higher energy absorption properties and inelastic recovery of the deformed material (He and Swain, 2008). Due to these structural features, dental enamel is the hardest tissue in the human body. Given the physiological relevance of this material, current research focus is toward enamel-based biomimetic materials for clinical applications (Palmer et al., 2008). Although recent developments have been made in this field, enamel formation is not completely understood because of the spatiotemporally dynamic processes involved, which are presently challenging to probe *in vivo*.

The current understanding of enamel formation based on the biochemical and animal models indicates that cellular control over biomineralization processes is mediated by physical association of crystallites with the cell membrane and associated matrix proteins. Studies on amelogenin, a primary constituent of enamel matrix proteins, show that this protein can oligomerize to form nanospheres that can mediate the formation of needle-shaped apatite crystals (Bromley et al., 2011; Du et al., 2005; Fincham et al., 1995). Furthermore, this protein can also self-assemble to form nanoribbons that act as a template in the growth of apatite particles (Du et al., 2005). An interesting study investigating the enamel structure in amelogenin double knockout mice shows a fascinating relationship between phenotypic manifestation and genetic evolution. The role of this protein in the emergence of features typical of mammalian enamel in comparison to reptilian enamel was revealed. The presence of amelogenin is crucial for development of Tomes' process, rod, and inter-rod spaces. The absence of these features is typical of reptilian enamel. The roles of other proteins such as ameloblastin, enamelin, biglycan protein, and tuftelin are unclear (Habelitz, 2015; Diekwisch et al., 2002). An emerging feature in the mechanisms of formation of several biominerals including enamel is the proteolytic processing of the organic constituents. MMP-2, a proteinase associated with the enamel matrix, mediates the proteolysis of full-length amelogenin, which is essential for normal enamel development (Simmer and Hu, 2002; Caterina et al., 2002). Kallikrein-4, a serine protease associated with enamel tissue, can cleave amelogenins and nonamelogenic proteins (Simmer et al., 1998). Cleavage products formed by such proteolytic activity are considered important in protein interactions and biomineral maturation (Simmer and Hu, 2002). Although the components involved in enamel formation have been identified, their concerted role in mineral formation requires attention. Recently an amorphous precursor to enamel apatite has also been identified (Beniash et al., 2009). Considering the interactions of biomolecules with amorphous precursors could help further elucidate the mineralization processes. Secondly, the recent focus on structural disorder in certain enamel-associated proteins and its role in protein function also appears promising (Delak et al., 2009; Ndao et al., 2011). Thus, the overall scheme of enamel development appears to depend on the transport, interactions, and deposition of inorganic precursors mediated by biomolecules under regulation of localized activities of proteases and other biomolecules (Figure 2.12).

There are several other remarkable examples of morphology control and template-driven crystal growth in biomineralization processes, including collagen

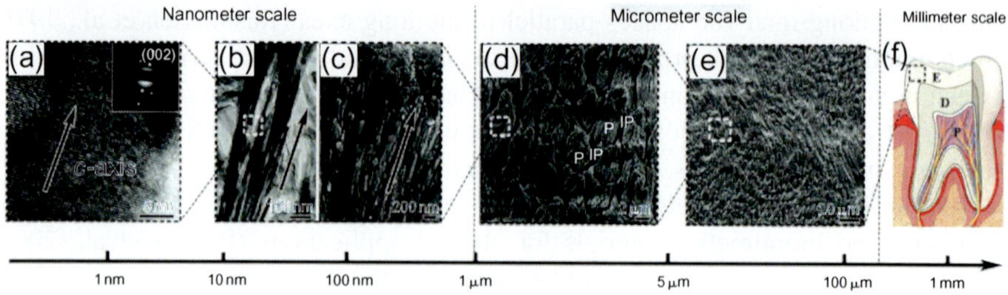

Figure 2.12 Hierarchical organization of enamel depicted (a) with the lattice structure and orientation of the hydroxyapatite crystal aligned along the *c*-axes at the nanoscale and (b and c) a typical bundle of parallel enamel crystals. At the micron scale, the enamel surface shows (d) enamel rods and (e) prism decussation (P, prism; IP, interprism). (f) Schematic image of basic tooth anatomy with enamel (E), dentin (D), and pulp (P).
Reproduced from Ruan and Moradian-Oldak (2015) with permission from Royal Society of Chemistry.

membrane-assisted egg shell formation (Fernandez et al., 1997; Panheleux et al., 1999), intricate composite cells walls of diatoms that are composites of silica and biomolecules (Hildebrand, 2008), and spicules of ACC and calcite in ascidians (Lambert and Lambert, 1997; Aizenberg et al., 2003). Biominerals thus exhibit an immense structural diversity on the nano- to microscale and thus represent Nature's extraordinary architectural and engineering capabilities.

2.5 Summary: Scope of biomineralization—Applications and challenges

The diversity of biominerals certainly provides great opportunities to explore the intricate crystal growth mechanisms and structure–property relationships in Nature. Generally absent in synthetic materials, the properties unique to biologically driven material design are:

(i) Growth of biominerals follows a bottom-up strategy. This offers better control over the synthesis, transport, and composition of precursor transient phases. Moreover, the growth as well as integration of organic and inorganic components is more efficient in comparison to top-to-bottom routes of synthesis.

(ii) Structural hierarchy at different length scales is exhibited by several eukaryotic biominerals. The self-assembly of biomolecules and organization of inorganic elements from the nano- to the macroscale lead to efficient rendering of structural organization into the final material properties.

(iii) By intelligently combining organic and inorganic components, the resultant properties are also complemented with respect to their individual strengths. With respect to mechanical properties, the inherent brittleness of minerals and low stiffness associated with organic molecules are overcome in a composite material. This is evident from an Ashby plot describing the toughness and elastic modulus of diverse materials (Wegst and Ashby, 2004).

(iv) The organic–inorganic interface is a crucial aspect that governs the interplay between organic and inorganic players as well as their influence over each other in terms of compositional as well as conformational stability (Vidavsky et al., 2014; Wilt, 1999; Duchstein et al., 2013; Addadi and Geva, 2003).

(v) The role of hydration in biomineral formation as well as the emergence of properties is significant. At smaller length scales, water molecules determine the structure and stability of transient mineral precursors and also determine the conformation of biomolecules (Cartwright et al., 2012; Demichelis et al., 2011; Sethmann et al., 2007; Raiteri and Gale, 2010; Fenter and Sturchio, 2004; Bryant, 1996; Rupley and Careri, 1991). Moreover, water-associated properties such as ionic strength, solubility, pH, and viscosity are significant determinants in mineralization.

(vi) Like several biological tissues, biominerals can also exhibit regulated growth. Such materials are thus intimately associated with cells and vasculature that can sense the extent of damage and respond by initiating appropriate repair processes.

(vii) Unlike their synthetic counterparts, biological materials are mostly produced under ambient conditions of pH and temperature in aqueous solutions.

It is probably impossible to incorporate all the above-listed properties in a synthetic material. However, by investigating dynamic interactions and structure–property relations between inorganic and organic molecules during biomineralization, concepts can be applied to functional materials. One can be inspired by the beautiful morphologies in Nature and develop new strategies to artificially reproduce these structures and properties. For example, the morphology of calcitic micro-lenses present on brittlestar arms can be imitated by an elegant method utilizing mineral precursors and a commercial surfactant (Figure 2.13; Aizenberg et al., 2001; Lee et al., 2012). Although the formation mechanisms may differ in biogenic and synthetic routes, the macrostructural and optical properties are quite similar.

Alternatively, after investigating *in vivo* the formation processes of biominerals, this knowledge can be applied to create hybrid materials. An interesting example is the exploitation of self-assembly properties of α-silicatein, a protein involved in the formation of siliceous spicules in sponges to produce flexible mesocrystals composed of calcite nanoparticles (Natalio et al., 2013; Gebauer, 2013). In addition, a layer-by-layer growth process has been applied to fabricate a nacre-inspired material with similar mechanical and optical properties (Finnemore et al., 2012) (Figure 2.14).

Figure 2.13 An array of amorphous calcium carbonate microlenses formed by the agglomeration of nanoparticles. Scale bars are (a) 5 and (b) 1 μm.
Reproduced from Habelitz (2015) and Diekwisch et al. (2002) with permission from Nature publishing group.

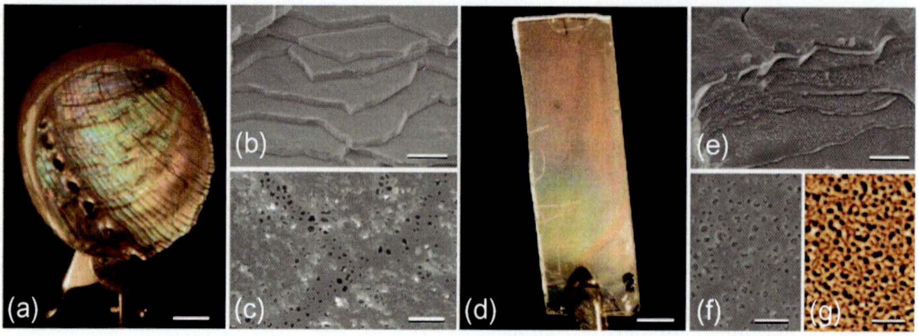

Figure 2.14 (a) Biogenic nacre and (d) its synthetic counterpart compared at the (a and d) macro, (b and e) micron, and (c, f and g) nano scales. Note the bright iridescence, aligned tablets, and porosity in both materials on account of their hierarchical organization. Reproduced from Finnemore et al. (2012) with permission from Nature publishing group.

A third strategy is to use an abundant biological template for *in vitro* mineralization growth under controlled conditions. For example, naturally abundant wood material mineralized with either calcium carbonate or magnetite exhibits fire-retardant or actuator properties, respectively (Merk et al., 2014, 2015). Finally, a more fundamental approach is to understand the multiple interactions of additives and surfaces with mineralization species and products (Figure 2.15; Gebauer et al., 2008, 2009a; Verch et al., 2011). This information can be applied to control different regimes of material

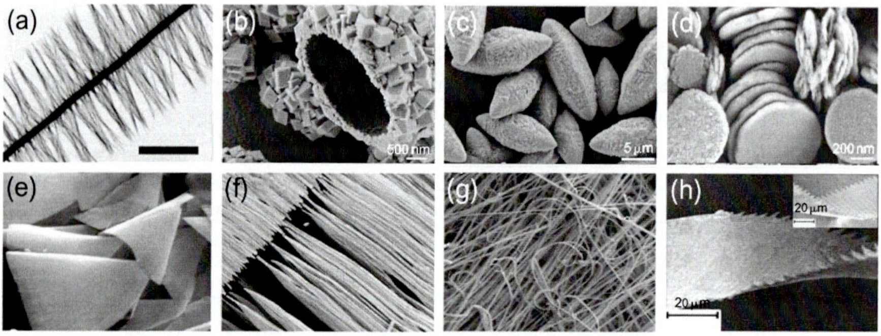

Figure 2.15 Morphologies of synthetic crystals grown in the presence of additives: (a) $BaWO_4$ nanostructures (PEG-*b*-PMAA), (b) spherical/hollow $CaCO_3$ particles (PEG-*b*-PMAA), (c) oval $BaCO_3$ crystals (PEG-*b*-PHEE), (d) $PbCO_3$ platelets (PEG-*b*-PMAA), (e) $BaCrO_4$ funnel-like superstructures (sodium polyacrylate), (f) $BaSO_4$ nanobundles (sodium polyacrylate), (g) $BaSO_4$ nanofibers (sodium polyacrylate). (h) Counterclockwise spiral morphology of $CaCO_3$ superstructures in the presence of copoly(DSer(P)^{75}DAsp^{25}). Clockwise twisting of structures in the presence of copoly(LSer(P)^{75}LAsp^{25}).
Reproduced from (a) Shi et al. (2003), (b–d) Yu et al. (2003), (e–g) Yu et al. (2003b) with permissions from American Chemical Society; and (h) Sugawara et al. (2003) with permission from Wiley.

formation, which is desirable for applications such as inhibiting scale formation as well as for producing micropatterned crystals, implant materials, and other functional materials (Aizenberg, 2004; Kellermeier et al., 2013a; Kikuchi et al., 2001; Ma and Cölfen, 2014; Pouget et al., 2007; Grunenfelder et al., 2014).

Over the past years, significant progress has been made in understanding biomineralization processes; however, there are many questions still unanswered. With better resolution of analytical techniques, lesser addressed features of crystallization such as PNCs, polyamorphism, and oriented attachment are gradually gaining importance (Cartwright et al., 2012; Gebauer and Cölfen, 2011; Politi et al., 2004; Addadi et al., 2003). The importance of water molecules associated with mineralization species is also emerging (Raiteri and Gale, 2010). Moreover, the existence of liquid-like phases of calcium carbonate under physiological conditions reflects a void in the present knowledge of biomineralization (Bewernitz et al., 2012). The mechanisms underlying the role of organic molecules in polymorph selection are unresolved and may hint at macromolecular recognition of short-range order in amorphous phases. Furthermore, the impact of molecular stereochemistry and conformation on mineralization processes is gradually being recognized (Rao et al., 2014; Addadi and Weiner, 1985; Orme et al., 2001). Molecular interactions between transient precursors and biomolecules still remain unknown. The role of lipids and sugars in biomineralization requires more attention (Arias and Fernández, 2008; Wise et al., 2007). Answers to such exciting questions will help to better elucidate the processes of biomineralization and provide new ideas for synthetic materials. Thus, studying the structure, formation, and properties of biominerals not only allows us to understand Nature's constructions in terms of biophysical principles operating at different length scales, but it also provides insights into the design principles, complex interfaces, and fascinating hierarchies yet to be achieved in synthetic materials.

References

Addadi, L., Geva, M., 2003. Molecular recognition at the interface between crystals and biology: generation, manifestation and detection of chirality at crystal surfaces. CrystEngComm 5 (26), 140–146.

Addadi, L., Weiner, S., 1985. Interactions between acidic proteins and crystals: stereochemical requirements in biomineralization. Proc. Natl. Acad. Sci. U.S.A. 82 (12), 4110–4114.

Addadi, L., Raz, S., Weiner, S., 2003. Taking advantage of disorder: amorphous calcium carbonate and its roles in biomineralization. Adv. Mater. 15 (12), 959–970.

Aizenberg, J., 2004. Crystallization in patterns: a bio-inspired approach. Adv. Mater. 16 (15), 1295–1302.

Aizenberg, J., 2010. New nanofabrication strategies: inspired by biomineralization. MRS Bull. 35 (04), 323–330.

Aizenberg, J., Hendler, G., 2004. Designing efficient microlens arrays: lessons from Nature. J. Mater. Chem. 14 (14), 2066–2072.

Aizenberg, J., Addadi, L., Weiner, S., Lambert, G., 1996. Stabilization of amorphous calcium carbonate by specialized macromolecules in biological and synthetic precipitates. Adv. Mater. 8 (3), 222–226.

Aizenberg, J., Hanson, J., Koetzle, T.F., Weiner, S., Addadi, L., 1997. Control of macromolecule distribution within synthetic and biogenic single calcite crystals. J. Am. Chem. Soc. 119 (5), 881–886.

Aizenberg, J., Black, A.J., Whitesides, G.M., 1999. Control of crystal nucleation by patterned self-assembled monolayers. Nature 398 (6727), 495–498.

Aizenberg, J., Tkachenko, A., Weiner, S., Addadi, L., Hendler, G., 2001. Calcitic microlenses as part of the photoreceptor system in brittlestars. Nature 412 (6849), 819–822.

Aizenberg, J., Weiner, S., Addadi, L., 2003. Coexistence of amorphous and crystalline calcium carbonate in skeletal tissues. Connect. Tissue Res. 44 (1), 20–25.

Aizenberg, J., Sundar, V.C., Yablon, A.D., Weaver, J.C., Chen, G., 2004. Biological glass fibers: correlation between optical and structural properties. Proc. Natl. Acad. Sci. U.S.A. 101 (10), 3358–3363.

Amos, F.F., Evans, J.S., 2009. AP7, a partially disordered pseudo C-RING protein, is capable of forming stabilized aragonite in vitro. Biochemistry 48 (6), 1332–1339.

Amos, F.F., Ponce, C.B., Evans, J.S., 2011. Formation of framework nacre polypeptide supramolecular assemblies that nucleate polymorphs. Biomacromolecules 12 (5), 1883–1890.

Arakaki, A., Webb, J., Matsunaga, T., 2003. A novel protein tightly bound to bacterial magnetic particles in Magnetospirillum magneticum strain AMB-1. J. Biol. Chem. 278 (10), 8745–8750.

Arias, J.L., Fernández, M.S., 2008. Polysaccharides and proteoglycans in calcium carbonate-based biomineralization. Chem. Rev. 108 (11), 4475–4482.

Balz, M., Barriau, E., Istratov, V., Frey, H., Tremel, W., 2005. Controlled crystallization of $CaCO_3$ on hyperbranched polyglycerol adsorbed to self-assembled monolayers. Langmuir 21 (9), 3987–3991.

Barker, S., Higgins, J., Elderfield, H., 2003. The future of the carbon cycle: review, calcification response, ballast and feedback on atmospheric CO_2. Philos. Trans. R. Soc. London, Ser. A 361 (1810), 1977–1999.

Barthelat, F., Li, C.-M., Comi, C., Espinosa, H.D., 2006. Mechanical properties of nacre constituents and their impact on mechanical performance. J. Mater. Res. 21 (08), 1977–1986.

Bazylinski, D.A., Frankel, R.B., 2004. Magnetosome formation in prokaryotes. Nat. Rev. Microbiol. 2 (3), 217–230.

Bazylinski, D.A., Garratt-Reed, A.J., Frankel, R.B., 1994. Electron microscopic studies of magnetosomes in magnetotactic bacteria. Microsc. Res. Tech. 27 (5), 389–401.

Becker, R., Döring, W., 1935. Kinetische behandlung der keimbildung in übersättigten dämpfen. Ann. Phys. 416 (8), 719–752.

Becker, A., Ziegler, A., Epple, M., 2005. The mineral phase in the cuticles of two species of Crustacea consists of magnesium calcite, amorphous calcium carbonate, and amorphous calcium phosphate. Dalton Trans. 10, 1814–1820.

Beniash, E., Aizenberg, J., Addadi, L., Weiner, S., 1997. Amorphous calcium carbonate transforms into calcite during sea urchin larval spicule growth. Proc. R. Soc. Lond. Ser. B Biol. Sci. 264 (1380), 461–465.

Beniash, E., Metzler, R.A., Lam, R.S.K., Gilbert, P.U.P.A., 2009. Transient amorphous calcium phosphate in forming enamel. J. Struct. Biol. 166 (2), 133–143.

Bensaude-Vincent, B., Arribart, H., Bouligand, Y., Sanchez, C., 2002. Chemists and the school of nature. New J. Chem. 26 (1), 1–5.

Benson, S., Jones, E.M.E., Crise-Benson, N., Wilt, F., 1983. Morphology of the organic matrix of the spicule of the sea urchin larva. Exp. Cell Res. 148 (1), 249–253.

Benson, S.C., Benson, N.C., Wilt, F., 1986. The organic matrix of the skeletal spicule of sea urchin embryos. J. Cell Biol. 102 (5), 1878–1886.

Bentov, S., Weil, S., Glazer, L., Sagi, A., Berman, A., 2010. Stabilization of amorphous calcium carbonate by phosphate rich organic matrix proteins and by single phosphoamino acids. J. Struct. Biol. 171 (2), 207–215.
Berman, A., Addadi, L., Kvick, Å., Leiserowitz, L., Nelson, M., Weiner, S., 1990. Intercalation of sea urchin proteins in calcite: study of a crystalline composite material. Science 250 (4981), 664–667.
Bewernitz, M.A., Gebauer, D., Long, J., Cölfen, H., Gower, L.B., 2012. A metastable liquid precursor phase of calcium carbonate and its interactions with polyaspartate. Faraday Discuss. 159 (1), 291–312.
Blakemore, R., 1975. Magnetotactic bacteria. Science 190 (4212), 377–379.
Boskey, A., Posner, A., 1974. Magnesium stabilization of amorphous calcium phosphate: a kinetic study. Mater. Res. Bull. 9 (7), 907–916.
Bromley, K.M., Kiss, A.S., Lokappa, S.B., Lakshminarayanan, R., Fan, D., Ndao, M., Evans, J.S., Moradian-Oldak, J., 2011. Dissecting amelogenin protein nanospheres: characterization of metastable oligomers. J. Biol. Chem. 286 (40), 34643–34653.
Bryant, R.G., 1996. The dynamics of water–protein interactions. Annu. Rev. Biophys. Biomol. Struct. 25 (1), 29–53.
Burleson, D.J., Penn, R.L., 2005. Two-step growth of goethite from ferrihydrite. Langmuir 22 (1), 402–409.
Carcouët, C.C.M.C., van de Put, M.W.P., Mezari, B., Magusin, P.C.M.M., Laven, J., Bomans, P.H.H., Friedrich, H., Esteves, A.C.C., Sommerdijk, N.A.J.M., van Benthem, R.A.T.M., de With, G., 2014. Nucleation and growth of monodisperse silica nanoparticles. Nano Lett. 14 (3), 1433–1438.
Cartwright, J.H., Checa, A.G., Gale, J.D., Gebauer, D., Sainz-Díaz, C.I., 2012. Calcium carbonate polyamorphism and its role in biomineralization: how many amorphous calcium carbonates are there? Angew. Chem. Int. Ed. 51 (48), 11960–11970.
Caterina, J.J., Skobe, Z., Shi, J., Ding, Y., Simmer, J.P., Birkedal-Hansen, H., Bartlett, J.D., 2002. Enamelysin (matrix metalloproteinase 20)-deficient mice display an amelogenesis imperfecta phenotype. J. Biol. Chem. 277 (51), 49598–49604.
Chang, E.P., Russ, J.A., Verch, A., Kröger, R., Estroff, L.A., Evans, J.S., 2014. The intrinsically disordered C-RING biomineralization protein, AP7, creates protein phases that introduce nanopatterning and nanoporosities into mineral crystals. Biochemistry 53 (27), 4317–4319.
Checa, A.G., Rodríguez-Navarro, A.B., 2005. Self-organisation of nacre in the shells of Pterioida (Bivalvia: Mollusca). Biomaterials 26 (9), 1071–1079.
Checa, A.G., Cartwright, J.H.E., Willinger, M.-G., 2011. Mineral bridges in nacre. J. Struct. Biol. 176 (3), 330–339.
Cölfen, H., Antonietti, M., 2008. Mesocrystals and Nonclassical Crystallization. Wiley, New York.
Cölfen, H., 2001. Double-hydrophilic block copolymers: synthesis and application as novel surfactants and crystal growth modifiers. Macromol. Rapid Commun. 22 (4), 219–252.
Cölfen, H., 2010. Biomineralization: a crystal-clear view. Nat. Mater. 9 (12), 960–961.
Cölfen, H., Antonietti, M., 1998. Crystal design of calcium carbonate microparticles using double-hydrophilic block copolymers. Langmuir 14 (3), 582–589.
Cölfen, H., Antonietti, M., 2005. Mesocrystals: inorganic superstructures made by highly parallel crystallization and controlled alignment. Angew. Chem. Int. Ed. 44 (35), 5576–5591.
Cölfen, H., Mann, S., 2003. Higher-order organization by mesoscale self-assembly and transformation of hybrid nanostructures. Angew. Chem. Int. Ed. 42, 2350–2365.

Cölfen, H., Qi, L., 2001. A systematic examination of the morphogenesis of calcium carbonate in the presence of a double-hydrophilic block copolymer. Chem. Eur. J. 7 (1), 106–116.

Currey, J., 1977. Mechanical properties of mother of pearl in tension. Proc. R. Soc. Lond. Ser. B Biol. Sci. 196 (1125), 443–463.

Cusack, M., Freer, A., 2008. Biomineralization: elemental and organic influence in carbonate systems. Chem. Rev. 108, 4433–4454.

Cuy, J.L., Mann, A.B., Livi, K.J., Teaford, M.F., Weihs, T.P., 2002. Nanoindentation mapping of the mechanical properties of human molar tooth enamel. Arch. Oral Biol. 47 (4), 281–291.

Das, P., Schipmann, S., Malho, J.-M., Zhu, B., Klemradt, U., Walther, A., 2013. Facile access to large-scale, self-assembled, nacre-inspired, high-performance materials with tunable nanoscale periodicities. ACS Appl. Mater. Interfaces 5 (9), 3738–3747.

Davit-Béal, T., Allizard, F., Sire, J., 2007. Enameloid/enamel transition through successive tooth replacements in Pleurodeles waltl (Lissamphibia, Caudata). Cell Tissue Res. 328 (1), 167–183.

Delak, K., Harcup, C., Lakshminarayanan, R., Sun, Z., Fan, Y., Moradian-Oldak, J., Evans, J.S., 2009. The tooth enamel protein, porcine amelogenin, is an intrinsically disordered protein with an extended molecular configuration in the monomeric form. Biochemistry 48 (10), 2272–2281.

DeMaster, D.J., Leynaert, A., Queguiner, B., 1995. The silica balance in the world ocean: a reestimate. Science 268 (5209), 375–379.

Demichelis, R., Raiteri, P., Gale, J.D., Quigley, D., Gebauer, D., 2011. Stable prenucleation mineral clusters are liquid-like ionic polymers. Nat. Commun. 2, 590.

Dey, A., Bomans, P.H.H., Müller, F.A., Will, J., Frederik, P.M., de With, G., Sommerdijk, N.A.J.M., 2010. The role of prenucleation clusters in surface-induced calcium phosphate crystallization. Nat. Mater. 9 (12), 1010–1014.

Didymus, J.M., Young, J.R., Mann, S., 1994. Construction and morphogenesis of the chiral ultrastructure of coccoliths from the marine alga Emiliania huxleyi. Proc. R. Soc. Lond. B 258, 237–245.

Diekwisch, T.G., Berman, B.J., Anderton, X., Gurinsky, B., Ortega, A.J., Satchell, P.G., Williams, M., Arumugham, C., Luan, X., McIntosh, J.E., 2002. Membranes, minerals, and proteins of developing vertebrate enamel. Microsc. Res. Tech. 59 (5), 373–395.

DiMasi, E., Olszta, M.J., Patel, V.M., Gower, L.B., 2003. When is template directed mineralization really template directed? CrystEngComm 5 (61), 346–350.

Dousma, J., de Bruyn, P.L., 1978. Hydrolysis—precipitation studies of iron solutions. II. Aging studies and the model for precipitation from Fe(III) nitrate solutions. J. Colloid Interface Sci. 64 (1), 154–170.

Du, C., Falini, G., Fermani, S., Abbott, C., Moradian-Oldak, J., 2005. Supramolecular assembly of amelogenin nanospheres into birefringent microribbons. Science 307 (5714), 1450–1454.

Duchstein, P., Kniep, R.D., Zahn, D., 2013. On the function of saccharides during the nucleation of calcium carbonate–protein biocomposites. Cryst. Growth Des. 13 (11), 4885–4889.

Evans, J.S., 2008. "Tuning in" to Mollusk shell nacre- and prismatic-associated protein terminal sequences. Implications for biomineralization and the construction of high performance inorganic–organic composites. Chem. Rev. 108 (11), 4455–4462.

Evans, J.S., 2012. Aragonite-associated biomineralization proteins are disordered and contain interactive motifs. Bioinformatics 28 (24), 3182–3185.

Evans, A.G., Suo, Z., Wang, R.Z., Aksay, I.A., He, M.Y., Hutchinson, J.W., 2001. Model for the robust mechanical behavior of nacre. J. Mater. Res. 16 (09), 2475–2484.

Faivre, D., Schüler, D., 2008. Magnetotactic bacteria and magnetosomes. Chem. Rev. 108 (11), 4875–4898.

Faivre, D., Menguy, N., Pósfai, M., Schüler, D., 2008. Environmental parameters affect the physical properties of fast-growing magnetosomes. Am. Mineral. 93 (2–3), 463–469.

Falini, G., Albeck, S., Weiner, S., Addadi, L., 1996. Control of aragonite or calcite polymorphism by mollusk shell macromolecules. Science 271 (5245), 67–69.

Fantner, G.E., Hassenkam, T., Kindt, J.H., Weaver, J.C., Birkedal, H., Pechenik, L., Cutroni, J.A., Cidade, G.A.G., Stucky, G.D., Morse, D.E., Hansma, P.K., 2005. Sacrificial bonds and hidden length dissipate energy as mineralized fibrils separate during bone fracture. Nat. Mater. 4 (8), 612–616.

Fantner, G.E., Oroudjev, E., Schitter, G., Golde, L.S., Thurner, P., Finch, M.M., Turner, P., Gutsmann, T., Morse, D.E., Hansma, H., 2006. Sacrificial bonds and hidden length: unraveling molecular mesostructures in tough materials. Biophys. J. 90 (4), 1411–1418.

Fenter, P., Sturchio, N.C., 2004. Mineral–water interfacial structures revealed by synchrotron X-ray scattering. Prog. Surf. Sci. 77 (5–8), 171–258.

Fernandez, M.S., Araya, M., Arias, J.L., 1997. Eggshells are shaped by a precise spatiotemporal arrangement of sequentially deposited macromolecules. Matrix Biol. 16 (1), 13–20.

Fincham, A.G., Moradian-Oldak, J., Diekwisch, T.G.H., Lyaruu, D.M., Wright, J.T., Bringas Jr., P., Slavkin, H.C., 1995. Evidence for amelogenin "nanospheres" as functional components of secretory-stage enamel matrix. J. Struct. Biol. 115 (1), 50–59.

Finnemore, A., Cunha, P., Shean, T., Vignolini, S., Guldin, S., Oyen, M., Steiner, U., 2012. Biomimetic layer-by-layer assembly of artificial nacre. Nat. Commun. 3, 966.

Francillon-Vieillot, H., De Buffrénil, V., Castanet, J., Géraudie, J., Meunier, F., Sire, J., Zylberberg, L., De Ricqlés, A., 1990. Skeletal biomineralization: patterns, processes and evolutionary trends. In: Carter, J.G. (Ed.), Microstructure and Mineralization of Vertebrate Skeletal Tissues. Van Nostrand, New York, pp. 471–548.

Gal, A., Kahil, K., Vidavsky, N., DeVol, R.T., Gilbert, P.U.P.A., Fratzl, P., Weiner, S., Addadi, L., 2014. Particle accretion mechanism underlies biological crystal growth from an amorphous precursor phase. Adv. Funct. Mater. 24 (34), 5420–5426.

Gebauer, D., 2013. Bio-inspired materials science at its best—flexible mesocrystals of calcite. Angew. Chem. Int. Ed. 52 (32), 8208–8209.

Gebauer, D., Cölfen, H., 2011. Prenucleation clusters and non-classical nucleation. Nano Today 6 (6), 564–584.

Gebauer, D., Völkel, A., Cölfen, H., 2008. Stable prenucleation calcium carbonate clusters. Science 322 (5909), 1819–1822.

Gebauer, D., Verch, A., Borner, H.G., Cölfen, H., 2009a. Influence of selected artificial peptides on calcium carbonate precipitation—a quantitative study. Cryst. Growth Des. 9 (5), 2398–2403.

Gebauer, D., Cölfen, H., Verch, A., Antonietti, M., 2009b. The multiple roles of additives in $CaCO_3$ crystallization: a quantitative case study. Adv. Mater. 21 (4), 435–439.

Gebauer, D., Gunawidjaja, P.N., Ko, J., Bacsik, Z., Aziz, B., Liu, L., Hu, Y., Bergström, L., Tai, C.W., Sham, T.K., 2010. Proto-calcite and proto-vaterite in amorphous calcium carbonates. Angew. Chem. 122 (47), 9073–9075.

Gebauer, D., Kellermeier, M., Gale, J.D., Bergstrom, L., Cölfen, H., 2014. Pre-nucleation clusters as solute precursors in crystallisation. Chem. Soc. Rev. 43 (7), 2348–2371.

Gibson, L.J., 1985. The mechanical behaviour of cancellous bone. J. Biomech. 18 (5), 317–328.

Glimcher, M.J., Bonar, L.C., Grynpas, M.D., Landis, W.J., Roufosse, A.H., 1981. Recent studies of bone mineral: is the amorphous calcium phosphate theory valid? J. Cryst. Growth 53 (1), 100–119.

Gong, Y.U., Killian, C.E., Olson, I.C., Appathurai, N.P., Amasino, A.L., Martin, M.C., Holt, L.J., Wilt, F.H., Gilbert, P., 2012. Phase transitions in biogenic amorphous calcium carbonate. Proc. Natl. Acad. Sci. U.S.A. 109 (16), 6088–6093.

Gorzelak, P., Salamon, M.A., Lach, R., Loba, M., Ferré, B., 2014. Microlens arrays in the complex visual system of Cretaceous echinoderms. Nat. Commun. 5.

Gotliv, B.-A., Addadi, L., Weiner, S., 2003. Mollusk shell acidic proteins: in search of individual functions. ChemBioChem 4 (6), 522–529.

Gower, L.B., 2008. Biomimetic model systems for investigating the amorphous precursor pathway and its role in biomineralization. Chem. Rev. 108 (11), 4551–4627.

Gower, L.B., Odom, D.J., 2000. Deposition of calcium carbonate films by a polymer-induced liquid-precursor (PILP) process. J. Cryst. Growth 210 (4), 719–734.

Gower, L., Tirrell, D., 1998. Calcium carbonate films and helices grown in solutions of poly(aspartate). J. Cryst. Growth 191 (1), 153–160.

Grunenfelder, L.K., Herrera, S., Kisailus, D., 2014. Crustacean-derived biomimetic components and nanostructured composites. Small 10 (16), 3207–3232.

Gupta, H.S., Wagermaier, W., Zickler, G.A., Raz-Ben Aroush, D., Funari, S.S., Roschger, P., Wagner, H.D., Fratzl, P., 2005. Nanoscale deformation mechanisms in bone. Nano Lett. 5 (10), 2108–2111.

Gupta, H.S., Seto, J., Wagermaier, W., Zaslansky, P., Boesecke, P., Fratzl, P., 2006. Cooperative deformation of mineral and collagen in bone at the nanoscale. Proc. Natl. Acad. Sci. U.S.A. 103 (47), 17741–17746.

Gwinnett, A., 1991. Structure and composition of enamel. Oper. Dent. 5, 10–17.

Habelitz, S., 2015. Materials engineering by ameloblasts. J. Dent. Res. 94 (6), 759–767.

Habelitz, S., Marshall, S.J., Marshall Jr., G.W., Balooch, M., 2001. Mechanical properties of human dental enamel on the nanometre scale. Arch. Oral Biol. 46 (2), 173–183.

Hang, F., Gupta, H.S., Barber, A.H., 2014. Nanointerfacial strength between non-collagenous protein and collagen fibrils in antler bone. J. R. Soc. Interface 11 (92), 20130993.

Hasegawa, K., Turner, C., Burr, D., 1994. Contribution of collagen and mineral to the elastic anisotropy of bone. Calcif. Tissue Int. 55 (5), 381–386.

Hasse, B., Ehrenberg, H., Marxen, J.C., Becker, W., Epple, M., 2000. Calcium carbonate modifications in the mineralized shell of the freshwater snail *Biomphalaria glabrata*. Chem. Eur. J. 6 (20), 3679–3685.

He, L.H., Swain, M.V., 2008. Understanding the mechanical behaviour of human enamel from its structural and compositional characteristics. J. Mech. Behav. Biomed. Mater. 1 (1), 18–29.

Hendler, G., Byrne, M., 1987. Fine structure of the dorsal arm plate of Ophiocoma wendti: evidence for a photoreceptor system (Echinodermata, Ophiuroidea). Zoomorphology 107 (5), 261–272.

Henisch, H.K., 2005. Crystals in Gels and Liesegang Rings. Cambridge University Press, Cambridge.

Hildebrand, M., 2008. Diatoms, biomineralization processes, and genomics. Chem. Rev. 108 (11), 4855–4874.

Huang, Z., Li, X., 2012. Order–disorder transition of aragonite nanoparticles in nacre. Phys. Rev. Lett. 109 (2), 025501.

Hunter, G.K., 1996. Interfacial aspects of biomineralization. Curr. Opin. Solid State Mater. Sci. 1 (3), 430–435.

Inskeep, W.P., Macur, R.E., Harrison, G., Bostick, B.C., Fendorf, S., 2004. Biomineralization of As(V)-hydrous ferric oxyhydroxide in microbial mats of an acid-sulfate-chloride geothermal spring, Yellowstone National Park. Geochim. Cosmochim. Acta 68 (15), 3141–3155.

Jackson, S.F., 1957. The fine structure of developing bone in the embryonic fowl. Proc. R. Soc. Lond. Ser. B Biol. Sci. 173, 270–280.

Jackson, A.P., Vincent, J.F.V., Turner, R.M., 1988. The mechanical design of nacre. Proc. R. Soc. Lond. 234 (1277), 415–440.

Jee, S.S., Kasinath, R.K., DiMasi, E., Kim, Y.-Y., Gower, L., 2011. Oriented hydroxyapatite in turkey tendon mineralized via the polymer-induced liquid-precursor (PILP) process. CrystEngComm 13 (6), 2077–2083.

Kashchiev, D., 2000. Nucleation. Butterworth-Heinemann, Oxford.

Katayama, Y., Mizutani, T., Utsumi, W., Shimomura, O., Yamakata, M., Funakoshi, K.-I., 2000. A first-order liquid-liquid phase transition in phosphorus. Nature 403 (6766), 170–173.

Katti, K.S., Katti, D.R., Pradhan, S.M., Bhosle, A., 2005. Platelet interlocks are the key to toughness and strength in nacre. J. Mater. Res. 20 (05), 1097–1100.

Keene, E.C., Evans, J.S., Estroff, L.A., 2010. Matrix interactions in biomineralization: aragonite nucleation by an intrinsically disordered nacre polypeptide, n16N, associated with a β-chitin substrate. Cryst. Growth Des. 10 (3), 1383–1389.

Kellermeier, M., Melero-Garcia, E., Glaab, F., Klein, R., Drechsler, M., Rachel, R., García-Ruiz, J.M., Kunz, W., 2010. Stabilization of amorphous calcium carbonate in inorganic silica-rich environments. J. Am. Chem. Soc. 132 (50), 17859–17866.

Kellermeier, M., Rosenberg, R., Moise, A., Anders, U., Przybylski, M., Cölfen, H., 2012a. Amino acids form prenucleation clusters: ESI-MS as a fast detection method in comparison to analytical ultracentrifugation. Faraday Discuss. 159, 23–45.

Kellermeier, M., Gebauer, D., Melero-García, E., Drechsler, M., Talmon, Y., Kienle, L., Cölfen, H., García-Ruiz, J.M., Kunz, W., 2012b. Colloidal stabilization of calcium carbonate prenucleation clusters with silica. Adv. Funct. Mater. 22 (20), 4301–4311.

Kellermeier, M., Gebauer, D., Cölfen, H., 2013a. Non-classical nucleation of calcium sulfate: a novel view on the origin of scale formation. In: 17th International Conference on Crystal Growth and Epitaxy, Warsaw, Poland.

Kellermeier, M., Cölfen, H., Gebauer, D., 2013b. Investigating the early stages of mineral precipitation by potentiometric titration and analytical ultracentrifugation. Methods Enzymol. 532, 45–69, Research Methods in Biomineralization Science.

Kikuchi, M., Itoh, S., Ichinose, S., Shinomiya, K., Tanaka, J., 2001. Self-organization mechanism in a bone-like hydroxyapatite/collagen nanocomposite synthesized in vitro and its biological reaction in vivo. Biomaterials 22 (13), 1705–1711.

Kim, I.W., DiMasi, E., Evans, J.S., 2004. Identification of mineral modulation sequences within the nacre-associated oyster shell protein, n16. Cryst. Growth Des. 4 (6), 1113–1118.

Kim, I.W., Darragh, M.R., Orme, C., Evans, J.S., 2006a. Molecular "tuning" of crystal growth by nacre-associated polypeptides. Cryst. Growth Des. 6 (1), 5–10.

Kim, I.W., Collino, S., Morse, D.E., Evans, J.S., 2006b. A crystal modulating protein from molluscan nacre that limits the growth of calcite in vitro. Cryst. Growth Des. 6 (5), 1078–1082.

Koch, K., Bhushan, B., Barthlott, W., 2008. Diversity of structure, morphology and wetting of plant surfaces. Soft Matter 4, 1943–1963.

Koene, J.M., Schulenburg, H., 2005. Shooting darts: co-evolution and counter-adaptation in hermaphroditic snails. BMC Evol. Biol. 5, 25.

Kotachi, A., Miura, T., Imai, H., 2006. Polymorph control of calcium carbonate films in a poly(acrylic acid)-chitosan system. Cryst. Growth Des. 6 (7), 1636–1641.

Kramer, I., 1951. The distribution of collagen fibrils in the dentine matrix. Br. Dent. J. 91 (1), 1–7.

Nielsen, J.W., Sand, K.K., Pedersen, C.S., Lakshtanov, L., Winther, J.R., Willemoës, M., Stipp, S.L.S., 2012. Polysaccharide effects on calcite growth: the influence of composition and branching. Cryst. Growth Des. 12 (10), 4906–4910.

Nudelman, F., Pieterse, K., George, A., Bomans, P.H.H., Friedrich, H., Brylka, L.J., Hilbers, P.A.J., de With, G., Sommerdijk, N.A.J.M., 2010. The role of collagen in bone apatite formation in the presence of hydroxyapatite nucleation inhibitors. Nat. Mater. 9 (12), 1004–1009.

Oaki, Y., Imai, H., 2003. Experimental demonstration for the morphological evolution of crystals grown in gel media. Cryst. Growth Des. 3 (5), 711–716.

Oaki, Y., Imai, H., 2006. Nanoengineering in echinoderms: the emergence of morphology from nanobricks. Small 2 (1), 66–70.

Oaki, Y., Kotachi, A., Miura, T., Imai, H., 2006. Bridged nanocrystals in biominerals and their biomimetics: classical yet modern crystal growth on the nanoscale. Adv. Funct. Mater. 16 (12), 1633–1639.

Okazaki, K., 1960. Skeleton formation of sea urchin larvae. Embryologia 5 (3), 283–320.

Olson, I.C., Blonsky, A.Z., Tamura, N., Kunz, M., Pokroy, B., Romao, C.P., White, M.A., Gilbert, P.U.P.A., 2013. Crystal nucleation and near-epitaxial growth in nacre. J. Struct. Biol. 184 (3), 454–463.

Olszta, M.J., Douglas, E.P., Gower, L.B., 2003. Scanning electron microscopic analysis of the mineralization of type I collagen via a polymer-induced liquid-precursor (PILP) process. Calcif. Tissue Int. 72 (5), 583–591.

Olszta, M.J., Cheng, X., Jee, S.S., Kumar, R., Kim, Y.-Y., Kaufman, M.J., Douglas, E.P., Gower, L.B., 2007. Bone structure and formation: a new perspective. Mater. Sci. Eng. R. Rep. 58 (3–5), 77–116.

Orme, C.A., Noy, A., Wierzbicki, A., McBride, M.T., Grantham, M., Teng, H.H., Dove, P.M., DeYoreo, J.J., 2001. Formation of chiral morphologies through selective binding of amino acids to calcite surface steps. Nature 411 (6839), 775–779.

Palmer, L.C., Newcomb, C.J., Kaltz, S.R., Spoerke, E.D., Stupp, S.I., 2008. Biomimetic systems for hydroxyapatite mineralization inspired by bone and enamel. Chem. Rev. 108 (11), 4754–4783.

Panheleux, M., Bain, M., Fernandez, M.S., Morales, I., Gautron, J., Arias, J.L., Solomon, S.E., Hincke, M., Nys, Y., 1999. Organic matrix composition and ultrastructure of eggshell: a comparative study. Br. Poult. Sci. 40 (2), 240–252.

Parry, D.A., 1988. The molecular fibrillar structure of collagen and its relationship to the mechanical properties of connective tissue. Biophys. Chem. 29 (1), 195–209.

Peled-Kamar, M., Hamilton, P., Wilt, F.H., 2002. Spicule matrix protein LSM34 is essential for biomineralization of the sea urchin spicule. Exp. Cell Res. 272 (1), 56–61.

Penn, R.L., Banfield, J.F., 1999. Morphology development and crystal growth in nanocrystalline aggregates under hydrothermal conditions: insights from titania. Geochim. Cosmochim. Acta 63 (10), 1549–1557.

Perovic, I., Chang, E.P., Lui, M., Rao, A., Cölfen, H., Evans, J.S., 2014a. A nacre protein, n16.3, self-assembles to form protein oligomers that dimensionally limit and organize mineral deposits. Biochemistry 53 (16), 2739–2748.

Perovic, I., Verch, A., Chang, E.P., Rao, A., Cölfen, H., Kröger, R., Evans, J.S., 2014b. An oligomeric C-RING nacre protein influences prenucleation events and organizes mineral nanoparticles. Biochemistry 53 (46), 7259–7268.

Peterlik, H., Roschger, P., Klaushofer, K., Fratzl, P., 2006. From brittle to ductile fracture of bone. Nat. Mater. 5 (1), 52–55.

Picker, A., Kellermeier, M., Seto, J., Gebauer, D., Cölfen, H., 2012. The multiple effects of amino acids on the early stages of calcium carbonate crystallization. Z. Kristallogr., Cryst. Mater. 227 (11), 744–757.

Pokroy, B., Zolotoyabko, E.J., 2003. Microstructure of natural plywood-like ceramics: a study by high-resolution electron microscopy and energy-variable X-ray diffraction. Mater. Chem. 13, 682–688.

Politi, Y., Arad, T., Klein, E., Weiner, S., Addadi, L., 2004. Sea urchin spine calcite forms via a transient amorphous calcium carbonate phase. Science 306 (5699), 1161–1164.

Politi, Y., Levi-Kalisman, Y., Raz, S., Wilt, F., Addadi, L., Weiner, S., Sagi, I., 2006. Structural characterization of the transient amorphous calcium carbonate precursor phase in sea urchin embryos. Adv. Funct. Mater. 16 (10), 1289–1298.

Politi, Y., Metzler, R.A., Abrecht, M., Gilbert, B., Wilt, F.H., Sagi, I., Addadi, L., Weiner, S., Gilbert, P.U.P.A., 2008. Transformation mechanism of amorphous calcium carbonate into calcite in the sea urchin larval spicule. Proc. Natl. Acad. Sci. U.S.A. 105 (45), 17362–17366.

Politi, Y., Batchelor, D.R., Zaslansky, P., Chmelka, B.F., Weaver, J.C., Sagi, I., Weiner, S., Addadi, L., 2009. Role of magnesium ion in the stabilization of biogenic amorphous calcium carbonate: a structure–function investigation. Chem. Mater. 22 (1), 161–166.

Polleux, J., Pinna, N., Antonietti, M., Niederberger, M., 2004. Ligand-directed assembly of preformed titania nanocrystals into highly anisotropic nanostructures. Adv. Mater. 16 (5), 436–439.

Ponce, C.B., Evans, J.S., 2011. Polymorph crystal selection by n16, an intrinsically disordered nacre framework protein. Cryst. Growth Des. 11 (10), 4690–4696.

Pouget, E., Dujardin, E., Cavalier, A., Moreac, A., Valery, C., Marchi-Artzner, V., Weiss, T., Renault, A., Paternostre, M., Artzner, F., 2007. Hierarchical architectures by synergy between dynamical template self-assembly and biomineralization. Nat. Mater. 6 (6), 434–439.

Pritchard, J., Bourne, G., 1956. The Biochemistry and Physiology of Bone. Academic Press, New York, p. 179.

Radha, A.V., Forbes, T.Z., Killian, C.E., Gilbert, P.U.P.A., Navrotsky, A., 2010. Transformation and crystallization energetics of synthetic and biogenic amorphous calcium carbonate. Proc. Natl. Acad. Sci. U.S.A. 107 (38), 16438–16443.

Raiteri, P., Gale, J.D., 2010. Water is the key to nonclassical nucleation of amorphous calcium carbonate. J. Am. Chem. Soc. 132 (49), 17623–17634.

Raiteri, P., Demichelis, R., Gale, J.D., Kellermeier, M., Gebauer, D., Quigley, D., Wright, L.B., Walsh, T.R., 2012. Exploring the influence of organic species on pre- and post-nucleation calcium carbonate. Faraday Discuss. 159, 61–85.

Rao, A., Seto, J., Berg, J.K., Kreft, S.G., Scheffner, M., Cölfen, H., 2013. Roles of larval sea urchin spicule SM50 domains in organic matrix self-assembly and calcium carbonate mineralization. J. Struct. Biol. 183 (2), 205–215.

Rao, A., Berg, J.K., Kellermeier, M., Gebauer, D., 2014. Sweet on biomineralization: effects of carbohydrates on the early stages of calcium carbonate crystallization. Eur. J. Mineral. 26 (4), 537–552.

Rasmussen, S.T., Patchin, R.E., Scott, D.B., Heuer, A.H., 1976. Fracture properties of human enamel and dentin. J. Dent. Res. 55 (1), 154–164.

Raz, S., Hamilton, P.C., Wilt, F.H., Weiner, S., Addadi, L., 2003. The transient phase of amorphous calcium carbonate in sea urchin larval spicules: the involvement of proteins and magnesium ions in its formation and stabilization. Adv. Funct. Mater. 13 (6), 480–486.

Robinson, R.A., Watson, M.L., 1952. Collagen-crystal relationships in bone as seen in the electron microscope. Anat. Rec. 114 (3), 383–409.

Rodriguez-Blanco, J.D., Shaw, S., Benning, L.G., 2011. The kinetics and mechanisms of amorphous calcium carbonate (ACC) crystallization to calcite, viavaterite. Nanoscale 3 (1), 265–271.

Rousseau, M., Lopez, E., Stempflé, P., Brendlé, M., Franke, L., Guette, A., Naslain, R., Bourrat, X., 2005. Multiscale structure of sheet nacre. Biomaterials 26 (31), 6254–6262.

Ruan, Q., Moradian-Oldak, J., 2015. Amelogenin and enamel biomimetics. J. Mat. Chem. B 3 (16), 3112–3129.

Rupley, J.A., Careri, G., 1991. Protein hydration and function. Adv. Protein Chem. 41, 37–172.

Sander, P., 2000. Prismless enamel in amniotes: terminology, function, and evolution. In: Development, Function and Evolution of Teeth. Cambridge University Press, Cambridge, pp. 92–106.

Sarikaya, M., Gunnison, K., Yasrebi, M., Aksay, I., 1989. Mechanical property-microstructural relationships in abalone shell. In: MRS Proceedings. Cambridge University Press, Cambridge, pp. 109.

Satchell, P.G., Anderton, X., Ryu, O.H., Luan, X., Ortega, A.J., Opamen, R., Berman, B.J., Witherspoon, D.E., Gutmann, J.L., Yamane, A., 2002. Conservation and variation in enamel protein distribution during vertebrate tooth development. J. Exp. Zool. 294 (2), 91–106.

Schäffer, T.E., Ionescu-Zanetti, C., Proksch, R., Fritz, M., Walters, D.A., Almqvist, N., Zaremba, C.M., Belcher, A.M., Smith, B.L., Stucky, G.D., 1997. Does abalone nacre form by heteroepitaxial nucleation or by growth through mineral bridges? Chem. Mater. 9 (8), 1731–1740.

Schinke, T., Amendt, C., Trindl, A., Pöschke, O., Müller-Esterl, W., Jahnen-Dechent, W., 1996. The serum protein α2-HS glycoprotein/fetuin inhibits apatite formation in vitro and in mineralizing calvaria cells: a possible role in mineralization and calcium homeostasis. J. Biol. Chem. 271 (34), 20789–20796.

Schmidt, W.I., 1936. Über die orientierung der kristallite im zahnschmelz. Naturwissenschaften 24 (23), 361.

Schüler, D., 2002. The biomineralization of magnetosomes in Magnetospirillum gryphiswaldense. Int. Microbiol. 5, 209–214.

Schüler, D., Frankel, R.B., 1999. Bacterial magnetosomes: microbiology, biomineralization and biotechnological applications. Appl. Microbiol. Biotechnol. 52 (4), 464–473.

Sethmann, I., Helbig, U., Worheide, G., 2007. Octocoral sclerite ultrastructures and experimental approach to underlying biomineralisation principles. CrystEngComm 9 (12), 1262–1268.

Seto, J., Gupta, H.S., Zaslansky, P., Wagner, H.D., Fratzl, P., 2008. Tough lessons from bone: extreme mechanical anisotropy at the mesoscale. Adv. Funct. Mater. 18 (13), 1905–1911.

Seto, J., Ma, Y., Davis, S.A., Meldrum, F., Gourrier, A., Kim, Y.-Y., Schilde, U., Sztucki, M., Burghammer, M., Maltsev, S., Jäger, C., Cölfen, H., 2012. Structure–property relationships of a biological mesocrystal in the adult sea urchin spine. Proc. Natl. Acad. Sci. U.S.A. 109 (10), 3699–3704.

Seto, J., Picker, A., Chen, Y., Rao, A., Evans, J.S., Cölfen, H., 2014. Nacre protein sequence compartmentalizes mineral polymorphs in solution. Cryst. Growth Des. 14 (4), 1501–1505.

Shi, H.T., et al., 2003. Polymer-directed synthesis of penniform $BaWO_4$ nanostructures in reverse micelles. J. Am. Chem. Soc. 125, 3450–3451.

Siglreitmeier, M., Wu, B., Kollmann, T., Neubauer, M., Nagy, G., Schwahn, D., Pipich, V., Faivre, D., Zahn, D., Fery, A., 2015. Multifunctional layered magnetic composites. Beilstein J. Nanotechnol. 6 (1), 134–148.

Simmer, J.P., Hu, J.C.-C., 2002. Expression, structure, and function of enamel proteinases. Connect. Tissue Res. 43 (2–3), 441–449.

Simmer, J., Fukae, M., Tanabe, T., Yamakoshi, Y., Uchida, T., Xue, J., Margolis, H., Shimizu, M., DeHart, B., Hu, C.-C., 1998. Purification, characterization, and cloning of enamel matrix serine proteinase 1. J. Dent. Res. 77 (2), 377–386.

Skinner, C., 1973. The biochemistry and physiology of bone. Yale J. Biol. Med. 46 (3), 251.

Smith, B.L., Schaffer, T.E., Viani, M., Thompson, J.B., Frederick, N.A., Kindt, J., Belcher, A., Stucky, G.D., Morse, D.E., Hansma, P.K., 1999. Molecular mechanistic origin of the toughness of natural adhesives, fibres and composites. Nature 399 (6738), 761–763.
Spiro, T.G., Allerton, S.E., Renner, J., Terzis, A., Bils, R., Saltman, P., 1966. The hydrolytic polymerization of iron(III). J. Am. Chem. Soc. 88 (12), 2721–2726.
Suda, J.-I., Nakayama, T., Nakahara, A., Matsushita, M., 1996. Morphological diversity in the crystal growth of potassium dichromate in gelatin gel. J. Phys. Soc. Jpn. 65 (3), 771–777.
Sugawara, T., et al., 2003. Chiral biomineralization: mirror-imaged helical growth of calcite with chiral phosphoserine copolypeptides. Macromol. Rapid Commun. 24, 847–851.
Sumitomo, T., Kakisawa, H., Owaki, Y., Kagawa, Y., 2008. In situ transmission electron microscopy observation of reversible deformation in nacre organic matrix. J. Mater. Res. 23 (05), 1466–1471.
Sumper, M., Brunner, E., 2006. Learning from diatoms: nature's tools for the production of nanostructured silica. Adv. Funct. Mater. 16, 17–26.
Sun, J., Bhushan, B., 2012. Hierarchical structure and mechanical properties of nacre: a review. RSC Adv. 2 (20), 7617–7632.
Sundar, V.C., Yablon, A.D., Grazul, J.L., Ilan, M., Aizenberg, J., 2003. Fibre-optical features of a glass sponge. Nature 424 (6951), 899–900.
Suzuki, M., Saruwatari, K., Kogure, T., Yamamoto, Y., Nishimura, T., Kato, T., Nagasawa, H., 2009. An acidic matrix protein, Pif, is a key macromolecule for nacre formation. Science 325 (5946), 1388–1390.
Tang, Z., Kotov, N.A., Magonov, S., Ozturk, B., 2003. Nanostructured artificial nacre. Nat. Mater. 2 (6), 413–418.
Tester, C.C., Brock, R.E., Wu, C.-H., Krejci, M.R., Weigand, S., Joester, D., 2011. In vitro synthesis and stabilization of amorphous calcium carbonate (ACC) nanoparticles within liposomes. CrystEngComm 13 (12), 3975–3978.
Thompson, J.B., Paloczi, G.T., Kindt, J.H., Michenfelder, M., Smith, B.L., Stucky, G., Morse, D.E., Hansma, P.K., 2000. Direct observation of the transition from calcite to aragonite growth as induced by abalone shell proteins. Biophys. J. 79 (6), 3307–3312.
Towe, K.M., Lowenstam, H.A., 1967. Ultrastructure and development of iron mineralization in the radular teeth of *Cryptochiton stelleri* (Mollusca). J. Ultrastruct. Res. 17 (1), 1–13.
TsrpunsrY, S.J., Busncr, P.R., 1993. Structure of magnesian calcite from sea urchins. Am. Mineral. 78, 775–781.
Tursch, Bernard, 1997. Spiral growth: the 'museum of all shells' revisited. J. Molluscan Stud. 63, 547–554.
Uchida, M., Kang, S., Reichhardt, C., Harlen, K., Douglas, T., 2010. The ferritin superfamily: supramolecular templates for materials synthesis. Biochim. Biophys. Acta, Gen. Subj. 1800 (8), 834–845.
US Department of Health and Human Services, 2004. Bone Health and Osteoporosis: A Report of the Surgeon General. US Department of Health and Human Services, Office of the Surgeon General, Rockville, MD.
van de Locht, R., Slater, T.J.A., Verch, A., Young, J.R., Haigh, S.J., Kröger, R., 2014. Ultrastructure and crystallography of nanoscale calcite building blocks in Rhabdosphaera clavigera coccolith spines. Cryst. Growth Des. 14, 1710–1718.
Verch, A., Gebauer, D., Antonietti, M., Cölfen, H., 2011. How to control the scaling of $CaCO_3$: a "fingerprinting technique" to classify additives. Phys. Chem. Chem. Phys. 13 (37), 16811–16820.
Verma, D., Katti, K., Katti, D., 2007. Nature of water in nacre: a 2D Fourier transform infrared spectroscopic study. Spectrochim. Acta A Mol. Biomol. Spectrosc. 67 (3–4), 784–788.

Vidavsky, N., Addadi, S., Mahamid, J., Shimoni, E., Ben-Ezra, D., Shpigel, M., Weiner, S., Addadi, L., 2014. Initial stages of calcium uptake and mineral deposition in sea urchin embryos. Proc. Natl. Acad. Sci. U.S.A. 111 (1), 39–44.

Vielzeuf, D., Garrabou, J., Baronnet, A., Grauby, O., Marschal, C., 2008. Nano to macroscale biomineral architecture of red coral (*Corallium rubrum*). Am. Mineral. 93 (11–12), 1799–1815.

Vielzeuf, D., Floquet, N., Chatain, D., Bonneté, F., Ferry, D., Garrabou, J., Stolper, E.M., 2010. Multilevel modular mesocrystalline organization in red coral. Am. Mineral. 95 (2–3), 242–248.

Wang, R., Gupta, H.S., 2011. Deformation and fracture mechanisms of bone and nacre. Annu. Rev. Mater. Res. 41, 41–73.

Wang, R., Suo, Z., Evans, A., Yao, N., Aksay, I., 2001. Deformation mechanisms in nacre. J. Mater. Res. 16 (09), 2485–2493.

Wang, T., Antonietti, M., Cölfen, H., 2006. Calcite mesocrystals: "Morphing" crystals by a polyelectrolyte. Chem. Eur. J. 12 (22), 5722–5730.

Wang, T., Mitchell, J., Borner, H., Cölfen, H., Antonietti, M., 2010. $BaCO_3$ mesocrystals: new morphologies using peptide-polymer conjugates as crystallization modifiers. Phys. Chem. Chem. Phys. 12 (38), 11984–11992.

Wang, S.-S., Picker, A., Cölfen, H., Xu, A.-W., 2013. Heterostructured calcium carbonate microspheres with calcite equatorial loops and vaterite spherical cores. Angew. Chem. Int. Ed. 52 (24), 6317–6321.

Weaver, J.C., Aizenberg, J., Fantner, G.E., Kisailus, D., Woesz, A., Allen, P., Fields, K., Porter, M.J., Zok, F.W., Hansma, P.K., Fratzl, P., Morse, D.E., 2007. Hierarchical assembly of the siliceous skeletal lattice of the hexactinellid sponge Euplectella aspergillum. J. Struct. Biol. 158 (1), 93–106.

Wegst, U., Ashby, M., 2004. The mechanical efficiency of natural materials. Philos. Mag. 84 (21), 2167–2186.

Weiner, S., Traub, W., 1980. X-ray diffraction study of the insoluble organic matrix of mollusk shells. FEBS Lett. 111 (2), 311–316.

Weiner, S., Traub, W., 1986. Organization of hydroxyapatite crystals within collagen fibrils. FEBS Lett. 206 (2), 262–266.

Weiner, S., Traub, W., 1992. Bone structure: from angstroms to microns. FASEB J. 6 (3), 879–885.

Weiner, S., Wagner, H.D., 1998. The material bone: structure–mechanical function relations. Annu. Rev. Mater. Sci. 28 (1), 271–298.

Weiner, S., Mahamid, J., Politi, Y., Ma, Y., Addadi, L., 2009. Overview of the amorphous precursor phase strategy in biomineralization. Front. Mater. Sci. Chin. 3 (2), 104–108.

Weiss, P., Ferris, W., 1954. Electronmicrograms of larval amphibian epidermis. Exp. Cell Res. 6 (2), 546–549.

White, S., Luo, W., Paine, M., Fong, H., Sarikaya, M., Snead, M., 2001. Biological organization of hydroxyapatite crystallites into a fibrous continuum toughens and controls anisotropy in human enamel. J. Dent. Res. 80 (1), 321–326.

Wilt, F.H., 1999. Matrix and mineral in the sea urchin larval skeleton. J. Struct. Biol. 126 (3), 216–226.

Wilt, F.H., 2002. Biomineralization of the spicules of sea urchin embryos. Zool. Sci. 19 (3), 253–261.

Wilt, F., Croker, L., Killian, C.E., McDonald, K., 2008. Role of LSM34/SpSM50 proteins in endoskeletal spicule formation in sea urchin embryos. Invertebr. Biol. 127 (4), 452–459.

Wise, S.W., 1970. Microarchitecture and Mode of Formation of Nacre (Mother of Pearl) in Pelecypods, Gastropods, and Cephalopods. Birkhäuser, Basel.

Wise, E.R., Maltsev, S., Davies, M.E., Duer, M.J., Jaeger, C., Loveridge, N., Murray, R.C., Reid, D.G., 2007. The organic–mineral interface in bone is predominantly polysaccharide. Chem. Mater. 19 (21), 5055–5057.

Wohlrab, S., Cölfen, H., Antonietti, M., 2005. Crystalline, porous microspheres made from amino acids by using polymer-induced liquid precursor phases. Angew. Chem. Int. Ed. 44 (26), 4087–4092.

Wolf, S.E., Muller, L., Barrea, R., Kampf, C.J., Leiterer, J., Panne, U., Hoffmann, T., Emmerling, F., Tremel, W., 2011a. Carbonate-coordinated metal complexes precede the formation of liquid amorphous mineral emulsions of divalent metal carbonates. Nanoscale 3 (3), 1158–1165.

Wolf, S.E., Leiterer, J., Pipich, V., Barrea, R., Emmerling, F., Tremel, W., 2011b. Strong stabilization of amorphous calcium carbonate emulsion by ovalbumin: gaining insight into the mechanism of 'polymer-induced liquid precursor' processes. J. Am. Chem. Soc. 133 (32), 12642–12649.

Wong, K.K., Douglas, T., Gider, S., Awschalom, D.D., Mann, S., 1998. Biomimetic synthesis and characterization of magnetic proteins (magnetoferritin). Chem. Mater. 10 (1), 279–285.

Wulff, G., 1901. On the question of speed of growth and dissolution of crystal surfaces. Z. Kristallogr. 34, 449–530.

Wustman, B.A., Morse, D.E., Evans, J.S., 2004. Structural characterization of the N-terminal mineral modification domains from the molluscan crystal-modulating biomineralization proteins, AP7 and AP24. Biopolymers 74 (5), 363–376.

Xiao, J., Wang, Z., Tang, Y., Yang, S., 2009. Biomimetic mineralization of $CaCO_3$ on a phospholipid monolayer: from an amorphous calcium carbonate precursor to calcite via vaterite. Langmuir 26 (7), 4977–4983.

Xu, A.-W., Dong, W.-F., Antonietti, M., Cölfen, H., 2008. Polymorph switching of calcium carbonate crystals by polymer-controlled crystallization. Adv. Funct. Mater. 18 (8), 1307–1313.

Young, J.R., Bown, P.R., 1991. An ontogenetic sequence of coccoliths from the Late Jurassic Kimmeridge Clay of England. Palaeontology 34 (4), 843–850.

Young, J.R., Didymus, J.M., Brown, P.R., Prins, B., Mann, S., 1992. Crystal assembly and phylogenetic evolution in heterococcoliths. Nature 356 (6369), 516–518.

Young, J.R., Davis, S.A., Bown, P.R., Mann, S., 1999. Coccolith ultrastructure and biomineralisation. J. Struct. Biol. 126 (3), 195–215.

Yu, S.-H., Cölfen, H., Antonietti, M., 2003a. Polymer-controlled morphosynthesis and mineralization of metal carbonate superstructures. J. Phys. Chem. B 107 (30), 7396–7405.

Yu, S.H., Antonietti, M., Cölfen, H., Hartmann, J., 2003b. Growth and self-assembly of $BaCrO_4$ and $BaSO_4$ nanofibers toward hierarchical and repetitive superstructures by polymer-controlled mineralization reactions. Nano Lett. 3, 379–382.

Zhang, G., Xu, J., 2013. From colloidal nanoparticles to a single crystal: new insights into the formation of nacre's aragonite tablets. J. Struct. Biol. 182 (1), 36–43.

Zhang, Q., Liu, S.-J., Yu, S.-H., 2009. Recent advances in oriented attachment growth and synthesis of functional materials: concept, evidence, mechanism, and future. J. Mater. Chem. 19 (2), 191–207.

Zhang, J., Huang, F., Lin, Z., 2010. Progress of nanocrystalline growth kinetics based on oriented attachment. Nanoscale 2 (1), 18–34.

Physical chemistry of biological apatites

3

C. Rey, C. Combes
Institut National Polytechnique de Toulouse, Toulouse, France

3.1 Introduction

Bone is often defined from a material point of view as a nanocomposite constituted of calcium phosphate apatite nanocrystals embedded in a collagen matrix. Its architecture and structure have been the subject of extensive studies using reliable and complementary techniques (Wopenka and Pasteris, 2005; Rey et al., 2009; Currey, 2002; Olszta et al., 2007; Cardoso et al., 2013; Boskey, 2003; Ruppel et al., 2008; Weiner and Traub, 1992), and they seem relatively well documented except, however, at the molecular level, regarding the constitution of bone mineral nanocrystals and the interactions between the different bone components. Some of the most significant progress regarding bone mineral crystal structure, composition, and properties originated from research on synthetic apatites believed, rightly or wrongly, to represent bone mineral. As the original biological material and the conditions of its formation and evolution are still largely unknown, considerations on bone mineral are often highly speculative. The aim of this review is to summarize the current views on bone mineral in order to show the input of apatite nanocrystal studies on the understanding of bone mineral and, in particular, to analyze and discuss different conceptions.

In the first part, the elemental and chemical compositions of bone mineral will be reviewed, followed by a description of the bone apatite structure including the effect of aging and more basic considerations about the hydroxide-ion content, calcium vacancies, and the heterogeneity of the mineral-ion distribution at the crystal level. In the second part, the actual conceptions of the fine structure of bone mineral will be presented, essentially focused on the existence of a hydrated layer at the surface of the nanocrystals, and the hypothesis of the existence of apatite precursors will be discussed. The evolution of bone mineral during aging will be examined and compared to that of synthetic analogs in order to distinguish physicochemical and biological effects. The third part of this chapter deals with the properties of bone mineral apatites and synthetic analogs. Ion exchange and adsorption behavior are detailed and their consequences on the mineral-ion reservoir function of bone and the interactions between bone crystals and matrix constituents are discussed.

3.2 Composition of bone

Bone contains numerous elements that are essential for life in addition to toxic elements. As it is very often difficult to precisely localize the different elements,

Biomineralization and Biomaterials. http://dx.doi.org/10.1016/B978-1-78242-338-6.00004-1

especially trace elements, in the organic and/or the mineral parts of the tissue, a first evaluation of bone composition can be given at the global elemental level, whereas more specific descriptions can be derived based on different hypotheses and considerations. There is no single bone composition as several variations may occur depending on diet, age, bone type, and bone location, and in the same bone, variations can occur within the tissue depending on the stage of remodeling or the zone of analysis (lamellae and cement zone). Herein, we report on a global average composition of bone.

3.2.1 Global elemental composition of bone

The mean values of the main elements present in dry human bones are presented in Table 3.1, and these values are based on the extended review published by Iyengar and Tandon (1999).

Table 3.1 Global elemental composition of bone

Major elements	Weight (%)	Observations; *N*: number of samples
O	31.9	Rather large variations according to authors, depends on sample preparation, especially on drying conditions; *N*=92
C	16.7	(Yoshinaga et al., 1989), *N*=30
N	4.9	Represents the protein content; *N*=55
Ca	25.4	Represents the mineralization ratio, lower Ca content at early ages; *N*=298
P	11.6	Ca/P atomic ratio depends on age (1.58 neonatal to 1.73 in adults (Quelch et al., 1983); *N*=268
Minor elements	**Weight (%)**	**Observations; *N*: number of samples**
Cl	0.13	*N*=7
K	0.0081	(Yoshinaga et al., 1989); *N*=42
Mg	0.269	*N*=298
Na	0.526	*N*=94
S	0.08	*N*=1
Main trace elements	**Weight (ppm)**	**Observations; *N*: number of samples**
Al	28.9	Large variations, considered to be an accumulative element; *N*=90
B	22.3	*N*=32
F	1540	Large variations; *N*=36
Fe	76	Large variations related to bone cleaning; *N*=212
Pb	4.4	Large variations; *N*=117
Sr	69.7	*N*=294
Zn	205	*N*=324

Other trace elements have been detected (less than 10 ppm): Ag, As, Au, Ba, Bi, Br, Cd, Co, Cr, Cs, Cu, Dy, Er, Eu, Ga, Gd, Ge, Hf, Hg, Ho, In, La, Li, Lu, Mn, Mo, Nb, Nd, Ni, Po, Pr, Pt, Ra, Rb, Re, Sb, Sc, Si, Sm, Sn, Ta, Tb, Te, Th, Tl, Ti, Tm, U, V, W, Y, Yb, Zr.
Adapted from Iyengar and Tandon (1999).

The elements are categorized as major, minor, and trace, depending on their concentration. In addition to oxygen, nitrogen, and carbon, the major elements are calcium and phosphorus. The elemental Ca and P contents of mature dry human bone appear to be relatively stable. Among the most important minor elements are Mg, S, Na, Cl, and K. Animal studies have shown that the Mg content varies with age, being higher in younger animals (Burnell et al., 1980). Na and Cl are also found in blood serum and, by extension, in the "bone fluid." The average atomic ratio of these elements (Na/Cl) is around 6 in bone (instead of around 1 in body fluids), suggesting that Na is taken up in bone. Potassium concentrations vary to a large extent and this element, like Cl, is considered to belong to the extracellular fluid of bone (Agna et al., 1958), although high amounts of K have been suggested to be caused by the pollution of bone extracellular fluid (BEF) from the intracellular fluid (Talmage and Talmage, 2006).

Many essential trace elements, such as As, Fe, Cu, Ni, Zn, Se, V, Sr, F, and Mo, are of importance and can be present in bone at variable concentrations. Toxic elements also belong to the category of trace elements, such as Al, Pb, and radioactive elements; furthermore, some essential elements can also become toxic when their concentration is too high. This is the case with F, which can lead to fluorosis at elevated concentrations, and many other elements can cause detrimental effects, such as As or Sr, for example.

For several trace elements, the bone content is directly related to the dietary intake, although diseases, malabsorption, and competitive absorption anomalies may also interfere. In a few cases, Sr content of bone for example, a linear relationship between the blood and bone concentrations of this element has been established, although different slopes were reported depending on the bone type (Cazalbou, 2000). The elimination of some elements can be rather slow (e.g., Al, F, and Pb) and their accumulation in bone is considered to be additive (Cazalbou et al., 1999).

Elemental content is one aspect of bone composition, and a second feature is related to the location of these elements in the bone composite and their chemical molecular structure.

3.2.2 Location of the elements and their molecular associations

Nitrogen belongs exclusively to bone proteins, principally the collagen matrix, and nitrogen can therefore be used as an indicator of the collagen content of bone.

The oxygen within bone has several origins; it is found in the mineral part (carbonate and phosphate ions), water (which is even present in dried bone), and within the organic constituents. The presence of hydroxide ions, OH^-, in bone mineral has been the subject of several divergent communications (Cho et al., 2003; Pasteris et al., 2004; Taylor et al., 2003; Rey et al., 1995a). The detection of OH^- ions using vibrational spectroscopies can lead to the identification of very small OH^- contents, down to between 0% and 5% of the theoretical content of a stoichiometric hydroxyapatite (HA), $Ca_{10}(PO_4)_6(OH)_2$ (Table 3.2).

However, higher contents have been reported when using solid-state nuclear magnetic resonance (SSNMR) evaluations (20% of the HA content) (Cho et al., 2003) and inelastic neutron scattering (INS) experiments (50% of the HA content) (Taylor et al., 2003), although other data obtained using this same technique mention no

Table 3.2 OH^- content in bone mineral of several vertebrate species, as determined from Raman spectral line intensities

Bone	OH^- content (ion percent[a]) (±1)
Hog	2
Embryonic pig	0
Cow	4
Calf	3
Turkey	2

[a] Relative to stoichiometric hydroxyapatite.

OH^- detection (Loong et al., 2000). These divergences can most probably be related to the reactivity of bone mineral and its propensity to be altered by an internal hydrolysis process, thereby increasing the amount of OH^- ions (Rey et al., 2011).

The largest part of bone is carbon, which is of course attributed to the organic matrix; however, a small part corresponds to carbonate ions that are present at various concentrations in the mineral. The amount of carbonate varies with age and increases in older individuals (Pellegrino and Biltz, 1972; Legros et al., 1987; Rey et al., 1991b).

Phosphorus is found mainly as orthophosphate ions (PO_4^{3-}) in the mineral, but is also present in a relatively important amount as hydrogen phosphate ions (HPO_4^{2-}) (Wu et al., 1994), which are often difficult to evaluate with accuracy and are thus ignored. Phosphorus may also exist in a minor amount as pyrophosphate ions and in association with various organic molecules (bone sialoprotein (BSP), phosphorylated fractions of collagen, phospholipids, and cells constituents).

Calcium is mostly located in bone mineral and is considered to represent the mineralization ratio of the tissue. In mammals, local variations can be observed in a tissue depending on the remodeling stage and the mineralization ratio of the collagen matrix. Considering vertebrates, larger global variations are always related to the mineral content. Some bones in fish, for example, are poorly mineralized, which is also the case in embryonic bones of all vertebrates at early stages (Cazalbou et al., 2004).

The determination of the structural environment of minor and trace elements is more difficult and often based on the chemical considerations or correlation studies. The possibility of an element entering the apatite structure is sometimes considered as a justification for its location in the mineral, but if this is a prerequisite, it is not a sufficient condition. The possibility of extracting elements by selective dissolution of the organic matrix or mineral has been explored to assess their location; however, several artifacts may occur. The identification of the element environments using different techniques is also possible. For example, the localization of Si at the mineralization front, in the organic matrix, and not in the mineral has been evidenced using EDX spectroscopy (Landis et al., 1986); the localization of sodium in the apatite structure has also been assessed using SSNMR (Laurencin et al., 2010). Thus, if some minor and trace elements belong primarily to the apatite mineral phase (e.g., Mg, Na, F, Sr, and Pb), others are mainly associated with the organic matrix (e.g., Si, Mn, and Cu); a third category could represent the elements distributed in both the mineral phase and the organic moiety (e.g., Zn). Most often, however, trace elements cannot be precisely localized.

3.2.3 Global composition of bone apatites

Although the identification of bone mineral as an apatite phase has been well established since the first XRD pattern recorded by de Jong (1926), the identification of bone mineral as stoichiometric hydroxyapatite, $Ca_{10}(PO_4)_6(OH)_2$, often reported in publications and textbooks, can no longer be considered as adequate. Several reports have insisted on this incorrect statement (LeGeros, 1991; Wopenka and Pasteris, 2005) for two main reasons: that is, bone mineral only contains a few hydroxide ions that are often barely detectable and, more importantly, bone mineral always contains a rather large amount of carbonate ions. Thus, it can better be defined as a carbonate apatite. In fact, the composition of the apatite phase of bone has been the subject of many debates and, in terms of most of its properties, bone mineral considerably departs from stoichiometric hydroxyapatite, although this phase remains a reference in many studies, reports, and reviews related to bone.

First attempts to derive a bone apatite composition were based on the study of synthetic carbonate apatites. The composition of biological apatites is rather complex, owing to the multiple cationic and anionic substitution possibilities and to the capacity of the apatite structure to host large amounts of ionic vacancies (Wopenka and Pasteris, 2005; Elliott, 1994; Rey et al., 2011). The specific question of ion substitutions in apatites is treated in Chapter 8 of this book. If minor and trace element substituents are neglected, the most relevant substitution concerning bone apatites is the replacement of trivalent PO_4^{3-} anions of the hydroxyapatite structure with bivalent species such as CO_3^{2-} (referred to as type B carbonate) and HPO_4^{2-} ions, which have been detected in large and variable amounts in bone mineral using several spectroscopic and chemical techniques (Pellegrino and Biltz, 1972; Legros et al., 1987; Rey et al., 1990; Wu et al., 1994). The charge compensation mechanism for synthetic apatites containing these bivalent ions involves the formation of two ionic vacancies: one on a cationic site and one on a monovalent anionic site (Montel et al., 1981; Labarthe et al., 1973), which lead to the general formula:

$$Ca_{10-x}(PO_4)_{6-x}(HPO_4 \quad \text{or} \quad CO_3)_x(OH)_{2-x}$$

The departure from stoichiometry can conveniently be determined by the calculation of the atomic ratio: $Ca/(P+C^*)=(10-x)/6$ (where C* is the carbon in carbonate ions). Each divalent ion replacing PO_4^{3-} is thus associated with a missing OH^-, which explains why bone apatite is depleted in hydroxide ions. The limit for this type of apatite non-stoichiometry corresponds to empty OH^- sites ($x=2$). Another explanation for the reduction in OH^- content can be related to the existence of a small amount of carbonate ions in the monovalent anionic sites (type A carbonate) (Rey et al., 1989). The bone mineral composition can thus be correctly approximated by the following chemical composition for a variety of vertebrate species, in which x corresponds to a value close to 1.7 (Legros et al., 1987):

$$Ca_{8.3}(PO_4)_{4.3}(HPO_4 \quad or \quad CO_3)_{1.7}(OH \quad or \quad 1/2CO_3)_{0.3}$$

From a structural point of view, this charge compensation mechanism has been confirmed by Rietveld analyses of X-ray diffraction patterns of substituted synthetic apatites (Wilson et al., 1999, 2005), indicating that cationic vacancies are localized on the Ca(II) sites of the apatitic structure, which are in close proximity to monovalent anionic sites (Figure 3.1). This suggests, as proposed by Labarthe et al. (1973), the existence of vacancy clusters (which are often observed in compounds with a large number of vacancies) in accordance with Pauling's rules, considering that electric charge compensation should occur in a reduced volume of the structure (Figure 3.1). Notably, the very few OH^- ions found in bone using several techniques (SSNMR, Fourier transform infrared (FTIR), and Raman spectroscopies) (Cho et al., 2003)

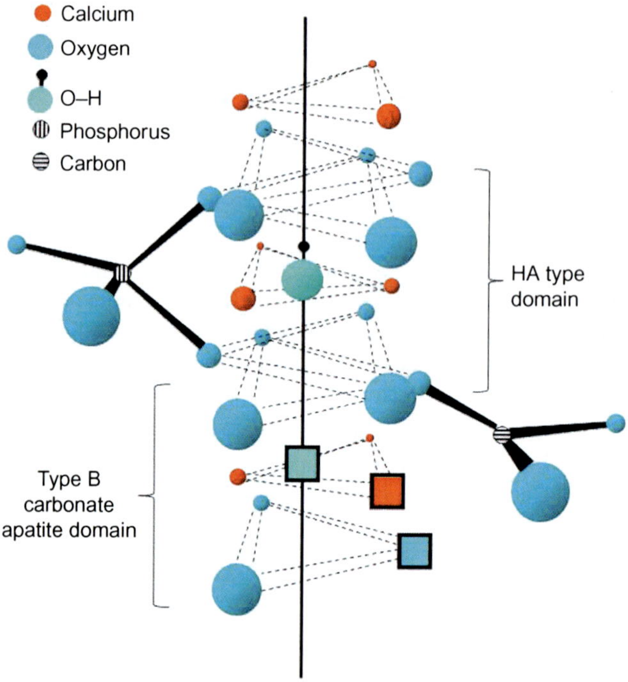

Figure 3.1 Charge compensation mechanism in type B carbonate- and HPO_4^{2-}-containing apatites, explaining the OH^- deficit. The replacement of PO_4^{3-} with a bivalent species (here a carbonate ion) is associated with a cationic vacancy in Ca^{2+} sites (type II) and a vacancy in the monovalent OH^- site, forming vacancy clusters. The vacancies are believed to be occupied, in part, by water molecules. Such vacancy clusters and HA-type environments can coexist in the same crystal.
Reproduced with permission from Rey et al. (2009).

always exhibit the spectroscopic characteristics of hydroxyapatite (exactly the same line positions). Considering that the spectra of these ions are rather sensitive to their environment, it can be concluded that the hydroxylated domains of bone apatite show similar OH^- environments to stoichiometric hydroxyapatite. Thus, a heterogeneous distribution of ions inside bone apatite crystallites can be predicted, depending on the local composition, including Na^+, F^-, and other elements. This intracrystalline heterogeneity, as well as the intercrystalline heterogeneity, has some effect on the dissolution properties of bone apatites, which show abnormal behavior and lead to the concept of metastable equilibrium solubility that is characterized by a variable solubility product depending on the amount dissolved (Baig et al., 1999). Bone crystal heterogeneity could also explain the preferential acidic dissolution of carbonated domains, as reported in earlier papers (Arends and Jongebloed, 1981; Biltz and Pellegrino, 1981). The global composition of bone apatite does not, however, take into account the existence of hydrated domains with non-apatitic environments of mineral ions, which appears as a major source of heterogeneity in bone mineral.

3.3 Constitution of biological nanocrystals

3.3.1 Bone mineral formation

The formation of bone mineral has been extensively studied in *in vivo*, in embryos, in bone defects, in slow mineralizing systems (turkey tendon or fish bones), or in *in vitro*, in osteoblast cell cultures (Glimcher, 2006). In all cases, bone tissue formation begins with the extracellular deposition of a structured collagen matrix that exhibits a specific arrangement of the fibrils, known as quarter stagger array, resulting in the creation of hole zones between aligned fibrils and interfibrillar space for parallel fibrils (Hodge, 1989). In remodeled bone, a specific structure appear, that is, osteon, which is formed by the concentric arrangement of collagen layers. It is often considered that these layers are associated, like those in plywood or composite laminates, with an angular shift of fibrils of adjacent layers. Recent conceptions suggest, however, that a continuous angular deviation exists between collagen layers, like in cholesteric structures (Giraud-Guille et al., 2004).

The mineral formation begins in the hole zone of the collagen fibrils and is then progressively extended to the inter- and intrafibrillar space. Vesicular mineralization, sometimes reported, seems to play a very limited role in the bone formation of vertebrates, as discussed in the review of Christoffersen and Landis (1991).

The study of mineralization of newly formed osteons reveals two distinct stages of mineralization (Meunier and Boivin, 1997):

- A primary mineralization corresponding to a fast process (a few days) and resulting in bone tissues containing about 60–70% of the mineral load of fully mineralized bone.
- A secondary mineralization, corresponding to a slow process lasting several months and possibly years, leading to fully mineralized bone tissue.

This aging process is associated with alterations to the composition and characteristics of the bone mineral and the organic matrix.

3.3.2 Hydrated domains and non-apatitic chemical environments in bone apatite crystals

Spectroscopic studies (FTIR spectroscopy and SSNMR) of bone mineral confirmed the formation of apatite phases with carbonate and phosphate ions in apatitic environments. However, they also enabled the existence of carbonate and hydrogen phosphate ions localized in non-apatitic environments to be evidenced (Rey et al., 1990; Wu et al., 1994). Combined studies on bone apatites and synthetic analogs using SSNMR spectroscopy have shown that these ions were close to water molecules in environments distinct from those of carbonate and phosphate ions in apatitic domains (Wu et al., 1994; Sfihi and Rey, 2002). Additional experiments involving ion-exchange studies have shown that non-apatitic CO_3^{2-} and HPO_4^{2-} were sharing the same hydrated domains and could easily substitute one another in rapid and reversible reactions (Cazalbou et al., 2004), suggesting, as proposed by Brown and Neuman, the existence of a hydrated layer containing these bivalent ions on the surface of the nanocrystals (Brown et al., 1984; Neuman et al., 1956). These findings imply a revision of bone composition, initially based solely on the chemical composition of relatively well-crystallized non-stoichiometric synthetic apatites, which do not take into account the presence of a hydrated layer. The new concept would be based on the formation of apatite domains that are less extended than those first considered, which contain fewer bivalent ionic species and a surface hydrated layer on the nanocrystals containing mainly bivalent ions. At this stage, however, it seems difficult to have a more precise description, as the evaluation of the amount of these non-apatitic environments by spectroscopic methods is rather inaccurate. Interesting data obtained by SSNMR on synthetic nanocrystalline apatites that are similar to bone crystals (Jäger et al., 2006) suggest that as much as 55% of the phosphate ions could belong to a disturbed surface layer on the synthetic nanocrystals. In the case of bone mineral, however, the amount of non-apatitic ionic species has not yet been accurately estimated, and only relative variations have been reported (Cazalbou et al., 2004).

From a structural point of view, little is known about this hydrated layer and most of its properties have been determined using synthetic analogs of bone crystals (see Section 3.4). It is a very sensitive surface structure that is very easily altered upon drying and as a result of changes in the surrounding solution (such as chemical composition or pH).

3.3.3 The question of possible precursor phases of bone apatite

Although most researchers now agree on the fact that other ionic environments to those found in regular apatite effectively exist in bone (Neuman and Bareham, 1975; Aue et al., 1984; Veis and Dorvee, 2013; Cazalbou et al., 2005; Rey et al., 1995b), the meaning of these environments is still a controversial topic. One of the most recurrent interpretations hypothesizes that bone mineral apatite does not form *de novo*, but results from the transformation of precursor phases (Weiner, 2006; Crane et al., 2006; Brown et al., 1987; Neuman and Bareham, 1975; Posner and Betts, 1975). These hypotheses are based on the frequent observation of intermediary phases during the synthesis of calcium orthophosphates, which can be compared to a more general process of transient phase formation known as Ostwald's step rule or the law of successive reactions, which states that, in many crystallization processes, the phases that generally form are not the most stable but the least stable, the free energies of which are the closest to the free energy of the original system. In the case of the calcium orthophosphate phase, under physiological conditions, the most stable phase should be hydroxyapatite; however, the growth rates of apatite crystals are very low (Onuma, 2006) and this phenomenon could promote the formation of unstable precursor phases in accordance with Ostwald's step rule. Additionally, the bone apatite crystals, which are not stoichiometric HA, show a variable composition and crystallinity, as established previously, and they exhibit an unusual morphology of irregular platelets elongated along the *c*-axis of the hexagonal apatite structure, which is far from the genuine needle-like shape of apatite crystals (Elliott, 1994; Weiner and Pryce, 1986). Three main precursor phases have been proposed historically, with some justifications as well as incompatibilities with the main characteristics of bone mineral crystals.

The formation of an amorphous calcium phosphate (ACP) phase is one of the oldest hypotheses. It was based on the improvement of the crystallinity of bone mineral apatites with time (XRD and FTIR data) (Termine and Posner, 1966), which appeared analogous to the amorphous-apatite conversion of synthetic samples. Several other data support the formation of an amorphous phase, with the question of possible artifacts not always mentioned or considered in the preparation of samples. Some arguments supporting the amorphous precursor phase theory have been discussed or revised. For example, the high background to noise ratio in the XRD pattern of bone samples is no longer assigned to a mixture of ACP and apatite, but is considered to correspond to lattice distortions and structural defects (vacancies and substituting ions) in the bone apatite crystals (Aoba et al., 1980; Posner and Betts, 1975). More detailed studies using the radial distribution function of bone mineral crystals, even at the earliest stage, did not allow the detection of an amorphous phase (Grynpas et al., 1984). The amorphous-apatite transition has been shown to be very rapid at 37 °C (just a few hours) (Somrani et al., 2005) and is described as autocatalytic (Combes and Rey, 2010), that is, when the apatite crystals have begun to crystallize, the transformation kinetics of the amorphous phase accelerates. Although this process can be slowed down by additives such as Mg, carbonate, pyrophosphate ions, and organic molecules, the detection of an unstable amorphous precursor phase would remain rather difficult. Recently, several reports have revived the ACP precursor theory, mainly based on the *in vitro* observations of apatite formation involving the formation of a dense liquid phase formed of amorphous mineral nuclei stabilized by polyanionic molecules (Olszta et al., 2007; Veis and Dorvee, 2013; Nudelman et al., 2010). These amorphous nuclei eventually crystallize into apatite nodules, but with amorphous hydrated domains at their periphery (Veis and Dorvee, 2013). If the presence of amorphous-like environments, corresponding to a hydrated layer at the surface of dried biological nanocrystals, seems to be well established, this observation does not mean that a separate amorphous phase preexists.

Another important theory involves the triclinic octacalcium phosphate (OCP: $Ca_8(HPO_4)_2(PO_4)_4{\cdot}5H_2O$) precursor. The OCP structure is formed by a sandwich of hydrated layers and apatite layers (Brown et al., 1962). This compound has the advantage of being able to progressively convert into the more stable apatite structure through a topotactic transformation, preserving the plate-like morphology of OCP crystals and involving interlayers of apatite and OCP domains (Brown et al., 1979). Such a sandwich structure has been proposed to explain the non-stoichiometry in HPO_4^{2-}-containing apatites (Brown, 1966). Several characteristics of bone mineral seem to justify the existence of this precursor phase, as evoked earlier, such as the composition of early mineral deposits in bone, which exhibits a very low Ca/P ratio that is close to that of OCP (Rey et al., 1995b), and the elongated platelet shape of mineral crystals, which are unusual for apatite crystals. In addition, there is some analogy between the FTIR spectra of nascent biological crystals (in bone, cartilage, or enamel) with that of OCP (Rey et al., 1995b). However, the triclinic OCP structure exhibits an XRD pattern that is distinct from that of apatite, with strong characteristic lines that are not observed in well-preserved bone. In addition, the FTIR spectrum of nascent crystals exhibits several anomalies when compared with that of OCP. Another strong argument against the formation of OCP is the existence of carbonate in all biological minerals, even at very early stages, which is incompatible with the formation of OCP crystals.

The third theory is that of a dicalcium phosphate dihydrate (DCPD or Brushite: $CaHPO_4{\cdot}2H_2O$) precursor, and this theory is supported by SSNMR data and other considerations, especially the fact that bone fluid is supersaturated with respect to DCPD (Neuman and Bareham, 1975; Roufosse et al., 1984). However, this DCPD precursor hypothesis has little support today. The question of the supersaturation of bone fluid is still open, as discussed later in this review. The conditions for the formation of DCPD generally necessitate a medium with a pH value that is much lower than physiological pH, although it has been shown that DCPD can form at physiological pH values (Francis and Webb, 1971). It should be kept in mind that the precipitation of DCPD (as well as other, more basic, calcium phosphates) from ions in solution at physiological pH (i.e., in solutions containing mainly HPO_4^{2-} and $H_2PO_4^{-}$) involves the release of protons, lowering the pH value in the vicinity of the precipitation and, eventually, in the surrounding solution; this could favor the formation of DCPD crystals.

Some conceptions integrate some or all of these phases in a time sequence, which are most often based on the *in vitro* observations (Habraken et al., 2013). The non-apatitic chemical environments in the hypothesis of precursor phases are considered to be the remnants of such precursor phases. The precursor-phase theory supposes, however, that foreign phases exist as separate entities before apatite formation. However, the existence of separate precursor phases has never been conclusively demonstrated. No diffraction data (using X-rays, electrons, or neutrons) have ever been reproducibly produced to support this hypothesis. In addition, careful examination of spectroscopic data obtained by different techniques and an exploration of all spectroscopic characteristics (and not only special features) do not support the existence of separated phases that are different to apatite. No spatially distinct foreign phase can reasonably account for the high amount of non-apatitic chemical environments observed in bone mineral without being detected by spectroscopic and/or diffraction techniques. The non-apatitic environments observed using spectroscopic techniques exhibit rather specific, although changing, characteristics (Rey et al., 2009), and cannot be considered to originate from distinct well-identified phases, but clearly represent a local surface organization of mineral ions forming a hydrated layer. Episodic reports are published on the identification of non-apatitic phases in bone; however, it must be recalled that apatite nanocrystals and bone mineral are rather unstable, reactive nanocrystals (especially freshly formed crystals like in embryonic bone) that, in the presence of moisture and under acidic pH conditions that are related to the maturation process, can readily transform into other phases such as OCP and DCPD, which have occasionally been identified and reported as separate entities (Muenzenberg and Gebhardt, 1973). Drying supersaturated extracellular bone fluid for microscopic observations, for example, or treating bone with solvents may result in artifact formation in the form of calcium phosphate phases as separated entities, including ACP nodules. These considerations do not exclude the possible formation of transient precursor phases, but such phases would be very rapidly hydrolyzed with no long-lasting existence and no determining role in the tissue's final properties.

Interestingly, the OCP structure studied by Brown has also been proposed as a surface model of the apatite–water interface; this author considered that the sandwich of apatite layers and hydrated layers found in the OCP structure could be a representation of the interface between apatite and water (Brown et al., 1984). This conception seems corroborated, in part, by spectroscopic data. However, unlike a set structure, the hydrated layer offers a variety of possible ionic substitutions, to a greater extent than that observed in the three-dimensional triclinic OCP structure, and this first model needs some amendments, as discussed later in this chapter.

At this stage, bone crystals are best described as a rather stable (non-stoichiometric) apatite cores surrounded by a structured non-apatitic hydrated layer, which probably minimizes the surface energy of the nanocrystals, thereby playing a stabilizing role. However, questions surrounding the origin of the plate-like shape of the crystals and their mode of formation remain open. The growth conditions in a collagen matrix or the role of biological molecules cannot be invoked to explain such a crystal morphology, which is also observed in synthetic preparations, even when only calcium and phosphate ions are present.

3.3.4 The extracellular fluid of bone

Bone crystals are immersed in a bone fluid, the composition of which is difficult to analyze, owing to the separation difficulties and to the constant evolution of bone mineral crystals, involving correlative changes in the fluids that are at equilibrium with these crystals. Although very few studies have been reported to answer this important question, some early reports (Neuman, 1980) did discuss the composition of BEF and, more specifically, the calcium and phosphate concentrations, as well as its pH value. Although the bone extracellular matrix is surrounded by cells or cell membranes, experimental results suggest that free-ion exchanges exist between bone organs and blood serum. Only the flow and free diffusion of macromolecules are hampered. Some studies, however, suggest that the high potassium content of bone, which can readily be mobilized, testify to the existence of a specific bone fluid compartment (Rubinacci et al., 2000). Nevertheless, it is generally considered that the ion concentrations in BEF are similar to those in blood, with the exception of mineral ions bound to specific bone macromolecules. In blood serum, the calcium content is rather constant in different species, including humans, and is close to 2.5 mM. However, ultra-filterable calcium (not bound to blood proteins) appears in much lower concentrations (1.5 mM) and the free calcium-ion content (determined with a specific electrode) is 1.25 mM. The calcium concentration is well regulated and this value does not depend on species, sex, diet, or circadian variations. On the contrary, the phosphate-ion content of blood serum varies largely; for example, it is higher in newborn humans (2 mM) compared to adults (1 mM), and shows diurnal variations. Inorganic phosphate ions do not seem to bind strongly to macromolecules; however, several protonation states exist that are linked to the pH value, which result in different ion pairs. The estimated pH of mature BEF is considered to be slightly more acidic than the physiological pH in adults (Sections 3.3 and 3.4); however, this estimation has to be taken with caution, owing to the constant release of protons from bone crystals related to their maturation (see Section 3.3.5 on the maturation process).

The supersaturation level of BEF can then be estimated for ACP and well-crystallized calcium phosphates with a constant solubility product (Table 3.3). DCPD and ACP appear under-saturated and should dissolve, whereas OCP and HA are supersaturated. If these observations have little value regarding the process of mineral formation, which occurs in unknown conditions, they indicate that only OCP and stoichiometric HA would be stable in BEF, even though none of them are observed. Regarding bone mineral and considering its heterogeneities at different levels, it never reaches a stable thermodynamic equilibrium, and the concept of metastable solubility equilibrium (Figure 3.2), considering an apparent equilibrium independent of the existence of precursor phases but taking into account the variable chemistry of bone nanocrystals, seems interesting (Baig et al., 1999).

3.3.5 Evolution of bone apatite on aging

Several data have shown changes in bone mineral composition and the amount of non-apatitic ionic species upon aging and during different disease states. The increase in the amount of carbonate in bone during aging was observed in early studies (Taylor and Sheard, 1929) and was later found to be associated with a concomitant decrease in the amount of HPO_4^{2-} (Legros et al., 1987; Rey et al., 1991a,b; Cazalbou et al.,

Table 3.3 Supersaturation levels of BEF, with regard to the main calcium phosphates of biological interest

Calcium phosphates	Solubility product (pK)	Ion product in bone extracellular fluid (pK_i)
DCPD (Brushite) $CaHPO_4{\cdot}2H_2O$	6.62	12.9
Triclinic OCP: $Ca_8H_2(PO_4)_6{\cdot}5H_2O$	96.6	93.9
ACP: $Ca_3(PO_4)_2$	24–26	27.5
HA: $Ca_{10}(PO_4)_6(OH)_2$	117.2	98.3

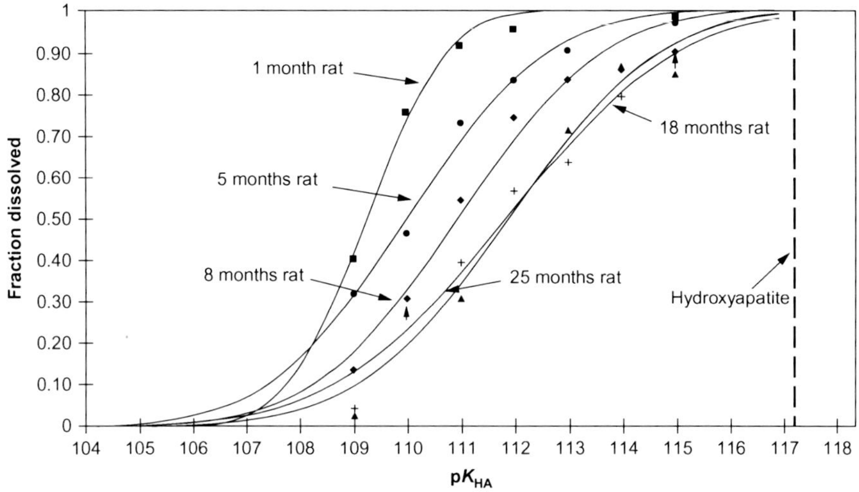

Figure 3.2 Example of metastable equilibrium solubility in rat bone at different ages, which is possibly related to bone heterogeneities. The solubility product varies to a large extent depending on the amount dissolved, in contrast to a standard dissolution curve of a well-crystallized apatite, showing a constant solubility product for any amount dissolved. Upon aging, the solubility curves are shifted toward the p*K*s of HA.
Adapted with permission from Barry et al. (2002).

2004; Pellegrino and Biltz, 1972). This is why aging does not seem to be associated with any change in the Ca/(P+C) molar ratio, which is in agreement with the global chemical composition of bone mineral (see Section 3.2.3). Thus, there is no significant evolution of bone mineral apatite toward stoichiometric hydroxyapatite with aging, as is frequently reported. The changes observed in the Ca/P ratio by chemical analyses or after firing bone samples are real; however, as mentioned earlier in this text, the Ca/P ratio is not representative of stoichiometry in carbonated apatites where a significant part of the orthophosphate is replaced by carbonate ions. On heating, the carbonate ions decompose and leave behind only calcium and orthophosphate ions, which recrystallize according to the Ca/P ratio (Rey et al., 2011) and these Ca/P variations appear to be related to the carbonate content.

The number of non-apatitic chemical environments decreases in bones with aging (Rey et al., 1991b; Wu et al., 1994; Cazalbou et al., 2004); see Figure 3.3. This can be assigned to the progressive development of stable apatite domains from the crystal core at the expense of the unstable surface hydrated layer. During this process, some ions from the hydrated layer can be irreversibly incorporated in the apatite domains, as is the case for carbonate ions. In addition to this purely physicochemical effect related to bone nanocrystal evolution, biological events should also be considered, such as the crystal formation rates during primary and secondary bone mineralization and the remodeling rate of bone. These biological processes result in heterogeneous crystal populations that represent an additional difficulty in view of precise characterization of the mineral as a whole. A way to progress the understanding of bone apatite properties

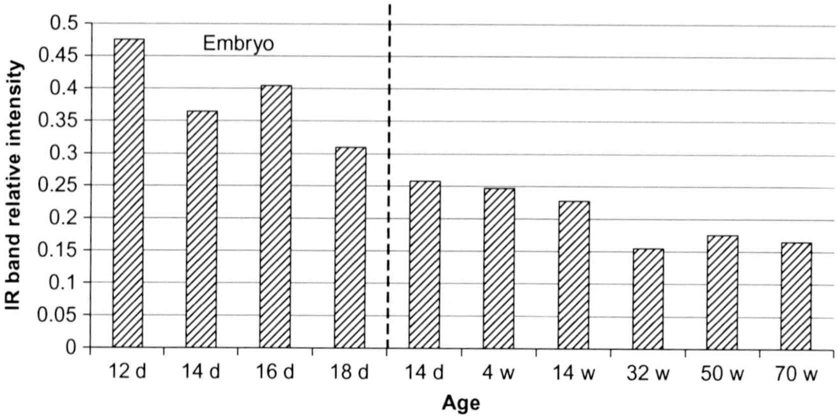

Figure 3.3 Evolution of the amount of non-apatitic HPO_4^{2-} in bone mineral with age in chicken bones. Evaluation of HPO_4^{2-} content by curve fitting of the spectra in the ν_4 PO_4 domain according to Bohic et al. (2000).

is to work on synthetic analogs. Advances in the knowledge of bone mineral structure, composition, and properties have already been made possible, by reference to apatite synthetic apatite chemistry and crystallography.

3.3.6 Synthetic analogs of bone apatite nanocrystals

3.3.6.1 Syntheses of biomimetic apatites

Several syntheses of bone mineral analogs have been proposed. Among these, our synthetic process is based on the precipitation by double decomposition between a soluble calcium salt solution and a soluble phosphate salt solution (Table 3.4) (Rey et al., 2011). Carbonate ions can be included in the phosphate solution and several other cationic or anionic additives can be incorporated in the calcium or phosphate solution, respectively. These ions can be included to evaluate their influence on nanocrystal formation and aging. The large phosphate (and carbonate) excess ensures pH buffering of the solution without an additional component.

Precipitation is obtained by rapidly pouring the calcium solution into the phosphate solution to ensure a homogeneous age of the crystal population. An amorphous phase forms first in this highly supersaturated solution, which transforms into nanocrystalline apatite within minutes at room temperature. The precipitate can be aged in the mother solution for variable periods of time. The precipitate is then filtered, washed, freeze-dried, and kept dry in a freezer to avoid any alterations. These formation conditions are far from those experienced by bone crystals, which still remain largely unknown; however, the most important parameters of physiological pH and low (possibly also physiologic) temperature are preserved, which produce plate-like apatite crystals with a size and irregular morphology analogous to that of bone apatite crystals (Figure 3.4). The chemical composition, XRD pattern, and FTIR spectra are also analogous to those given by bone mineral crystals (Kim et al., 1995, 1996).

Table 3.4 Synthesis conditions for the precipitation of bone mineral synthetic analogs

Cation solution (250 mL)	Calcium nitrate (or chloride): 0.3 M Magnesium, strontium, or other cations may be added to this solution
Anion solution (500 mL)	Diammonium (or disodium) phosphate: 0.3–0.6 M Ammonium or sodium bicarbonate: 0–1 M (0.5 M) pH adjustment with KOH or NH_4OH (no pH adjustment) Pyrophosphate or other anions may be added to this solution

Preferred synthesis conditions are in bold.

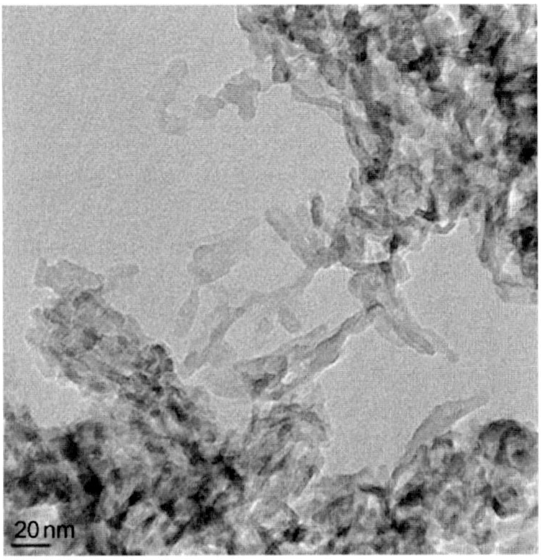

Figure 3.4 TEM picture of synthetic nanocrystalline apatite with a size and composition similar to bone mineral exhibiting a platelet habit. Characteristics: 31 days maturation, 5.6% carbonate (w/w), atomic Ca/(P + C) = 1.38.
Reproduced with permission from Pascaud et al. (2013).

3.3.6.2 Maturation of synthetic apatite nanocrystals and bone mineral evolution

The initial apatite nanocrystals obtained by the above-described method are very close to those obtained *in vivo* at the beginning of mineralization (Cazalbou et al., 2004). They show a low Ca/(P+C) ratio that is close to that of OCP (Table 3.5) and the carbonate content is small, even when they have been precipitated in a carbonate-rich solution, and these carbonate ions are mainly present in the hydrated surface layer. A similar situation is found in embryonic chicken bone, during calf cartilage calcification, and at the beginning of dental enamel formation in pigs (Rey et al., 1995b). In all cases, XRD data show a poorly crystalline apatite structure. FTIR spectroscopic analyses of wet synthetic samples demonstrated that these samples give narrow, specific bands that are different to those found in apatite (Eichert et al., 2004). A striking analogy with the triclinic

Table 3.5 Chemical composition of synthetic nanocrystalline carbonate–apatites at different maturation stages compared to that of chicken bones at different ages

Samples	Ca/P (±0.02)	Ca/(P+C) (±0.02)	C/P (±0.02)
Synthetic carbonated apatite maturation time			
0	1.44	1.41	0.030
6 h	1.51	1.42	0.060
24 h	1.51	1.40	0.089
3 days	1.54	1.39	0.145
10 days	1.62	1.40	0.162
4 months	1.62	1.38	0.200
Chicken bone age			
Embryonic (16 days)	1.51	1.27	0.186
4 weeks	1.62	1.31	0.234
10 weeks	1.64	1.35	0.218
70 weeks	1.88	1.45	0.293

Adapted with permission from Cazalbou et al. (2004).

OCP spectrum can be noticed for freshly formed biological crystals (Figure 3.5); however, this analogy was not confirmed by SSNMR (Eichert et al., 2004) and the comparison of FTIR spectra shows clear distinctive differences between nanocrystals and triclinic OCP (Figure 3.5). Upon drying, the spectral lines broaden considerably and the fine structure seen in the wet samples disappears (see Figure 3.9, spectra nos. 1 and 2). These spectral features have been assigned to the hydrated layer, which loses its fragile structure and transforms, partially irreversibly, into amorphous-like domains (Figure 3.5). Owing to this

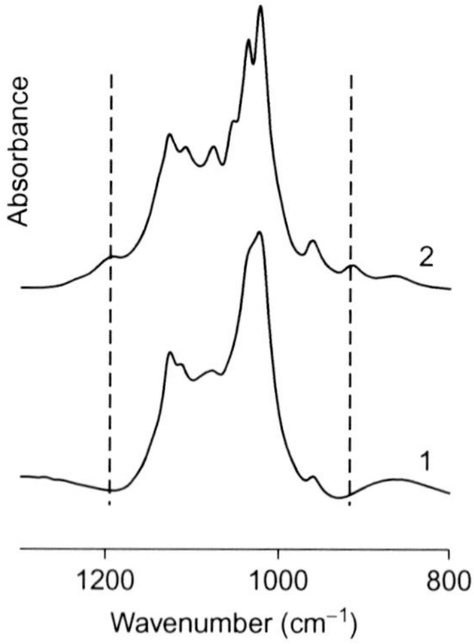

Figure 3.5 Comparison of the ν_3 PO_4 domain of FTIR spectra for freshly formed synthetic nanocrystals (1) and triclinic OCP (2). Several missing, but characteristic, bands of OCP are indicated by dotted lines.

artifact, the hydrated layer can be confused with an amorphous layer in dried samples. These considerations highlight the reactivity and fragility of nanocrystalline apatites as well as the need for physicochemical characterization, allowing the use of wet samples to preserve the original bone crystal environments and structures.

Upon aging, the evolution of synthetic apatites shows, in all cases, a decrease in the number of non-apatitic environments, as also seen in bone mineral aging. This purely physicochemical evolution indicates that non-apatitic environments are less stable than apatite domains, which grow at the expense of the hydrated layer. The apatite composition found strongly depends, however, on the composition of the solution and, more specifically, on the presence or absence of carbonates ions.

In the presence of carbonate ions, the Ca/(P+C) ratio remains almost constant, although the carbonate content increases through the replacement of HPO_4^{2-} ions with carbonate ions. This evolution is similar to that observed in bone mineral (Neuman et al., 1956; Cazalbou et al., 2004; Rey et al., 1991b). The use of labeled species allows a more precise follow-up of the maturation process. Experiments using ^{13}C carbonate (Cazalbou et al., 2005) indicate that:

- The uptake of carbonate in nanocrystalline apatites occurs during the maturation process and involves the hydrated layer.
- The distribution between the two carbonate sites of the apatite structure (referred to as type A carbonate ion on the OH^- site and type B carbonate ion on the PO_4^{3-} site) does not depend on the maturation time at a given pH, temperature, and solution composition.
- The uptake of carbonate ions from the surrounding solution is related to the development of the hydrated layer; there is a fast uptake in freshly formed apatites, which decreases considerably as the maturation time increases.
- Once the carbonate ions have entered the apatite structure, they cannot be removed.

A model of the apatite nanocrystals of bone can be proposed, illustrating its physicochemical evolution and surface reactivity (Figure 3.6).

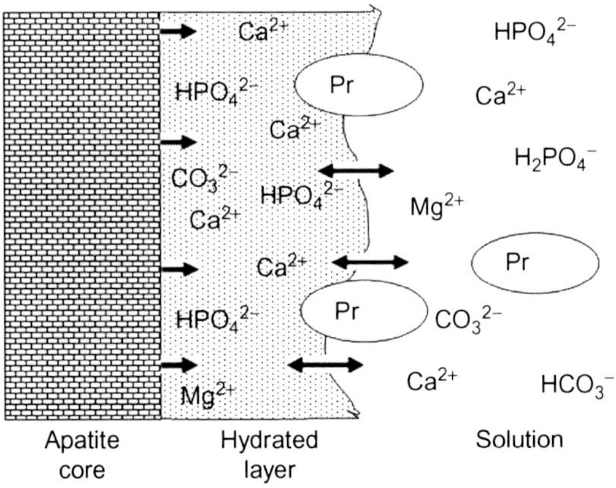

Figure 3.6 Model of bone nanocrystalline apatite: the structured hydrated layer mainly contains loosely bound bivalent ions, some of which, like carbonate, HPO_4^{2-}, and Sr^{2+}, can be incorporated in the growing apatitic domains and can be exchanged with mineral ions from the solution and/or charged groups of proteins (Pr) or other molecules.
Reproduced with permission from Rey et al. (2007).

In the absence of carbonate ions, apatite nanocrystal evolution is different. The Ca/P ratio increases, indicating an evolution toward stoichiometry, and OH^- formation is observed. In all cases, the development of apatite domains at the expense of the hydrated layer is associated with protons draining toward the solution, owing to the formation of PO_4^{3-}, that are needed to build the apatite structure from HPO_4^{2-} in the hydrated layer. In some cases, for example in cell cultures or gel-like suspensions, when a limited amount of solution is present, this evolution can lead to a significant acidification of the solution and even to the crystallization of acidic calcium phosphates, such as DCPD coexisting with apatite. This phenomenon may also exist in bone and could explain the sporadic detection of non-apatitic crystalline phases such as DCPD and OCP.

3.4 Properties of biological apatites and synthetic analogs

The existence of a hydrated layer with labile ions on the surface of apatite nanocrystals can help determine some of their properties, especially ion-exchange capabilities, which are susceptible to involvement in the mineral-ion reservoir function of bone as well as adsorption and molecule binding.

3.4.1 Ion exchanges

The existence of a structured hydrated layer with loosely bound bivalent ions suggests the possibility of performing ion-exchange reactions between the surface of the crystals and the solution. Both cationic and anionic exchanges have been investigated.

Anionic exchanges have been made using mainly carbonate and hydrogen phosphate ions. The obtained data show that the CO_3^{2-}–HPO_4^{2-} ion-exchange reactions (or the reverse exchanges: HPO_4^{2-}–CO_3^{2-}) could be performed within minutes; they were reversible and did not alter the main crystal characteristics (crystal size and structure). The amount exchanged was directly related to the solution concentration of the exchanging ion and to the amount of exchangeable ion in the hydrated layer. As the maturation time increased, the amount exchanged decreased considerably and became negligible after a few months of maturation (Figure 3.7). The use of labeled ^{13}C carbonate perfectly illustrated the exchange process and its reversibility (Figure 3.8).

Cationic exchanges on apatite nanocrystals have also been studied (Mg^{2+}–Ca^{2+}, Sr^{2+}–Ca^{2+}, and the reverse reactions) (Cazalbou, 2000; Cazalbou et al., 2005; Drouet et al., 2008). Unlike anionic exchanges, cationic exchanges are still observed after several months of maturation (Figure 3.7), suggesting the possible implication of apatite ions as well as non-apatitic cations from the hydrated layer. The amount exchanged was, once again, related to the cation concentration in solution and to the development of the hydrated layer (maturation time). These exchanges, like the anionic ones, are fast and reversible, and the amount exchanged was found to decrease in matured apatites. We did not succeed in cation exchanges involving Na^+ and K^+ (Cazalbou, 2000). The fragile structure of the hydrated layer is reversibly altered by these ion

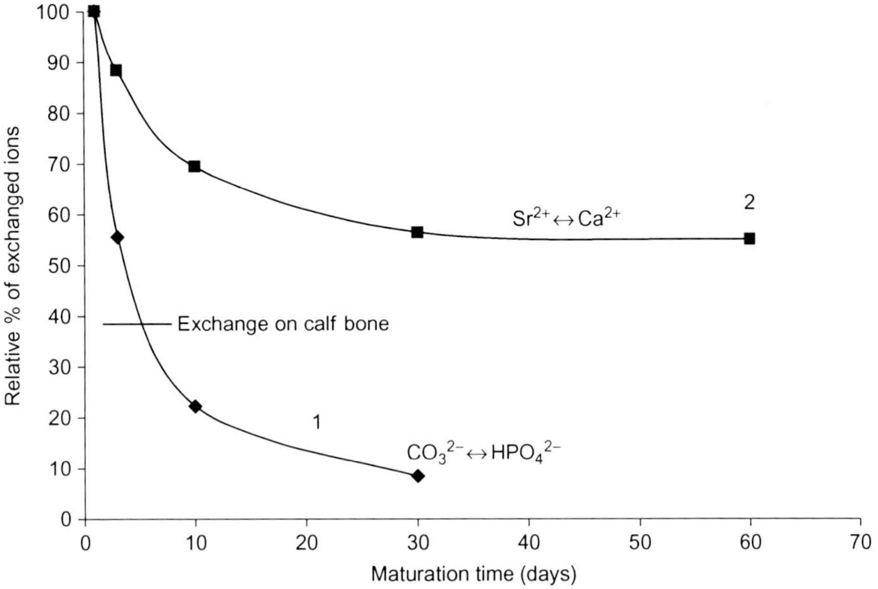

Figure 3.7 Example of the evolution of ion-exchange capacity of synthetic nanocrystalline apatites with the maturation time and calf bone expressed as the % of the exchange capability after 1 day maturation: (1) $CO_3^{2-} \leftrightarrow HPO_4^{2-}$ exchange capacity of carbonate nanocrystalline apatite determined as the difference in carbonate content between a carbonate-loaded sample (sample treated for 10 min in a 1 M ammonium bicarbonate solution) and HPO_4^{2-}-loaded sample (sample treated in 1 M diammonium hydrogen phosphate solution). The line represents this same experiment using a ground scapula bone of calf (6 months old); the exchange capacity appears surprisingly high. (2) $Sr^{2+} \leftrightarrow Ca^{2+}$ exchange.

exchanges, as illustrated in Figure 3.9, adding complexity to the study of real bone mineral crystals.

When the composition of the hydrated layer is changed by ion-exchange reactions, the maturation process proceeds, although it might be slowed down (e.g., as is the case with carbonate or magnesium). Two phenomena can be distinguished, that is, the incorporation of the foreign ions into the growing apatite domain, in which case they are no longer available for reverse exchange reactions, or the persistence of the ions in the hydrated layer, in which case the ion remains re-exchangeable (Cazalbou et al., 2005). Strontium and carbonate belong to the first category, whereas magnesium belongs to the second one and can be reversibly exchanged, even after prolonged maturation times. Although it has been shown that magnesium can be incorporated in limited quantities in synthetic calcium phosphate apatites, it appears that this element is predominantly stored in the hydrated layer. This finding appears to be in agreement with the conclusion that magnesium is at the surface of bone mineral crystals (Burnell et al., 1980).

The existence of an exchangeable ion pool in bone raises the question of its role in the ion-reservoir function of bone.

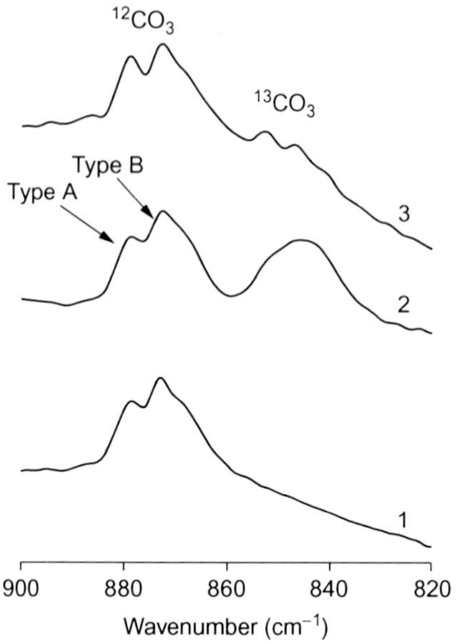

Figure 3.8 Fourier self-deconvoluted FTIR spectra in the ν_2 CO_3 domain: (1) original precipitated ^{12}C carbonate apatite matured for 1 day; (2) original apatite 1 treated in 1 M ^{13}C carbonate solution for 20 min. The initial ^{12}C carbonate species are not altered and a featureless broad line corresponding to the uptake of ^{13}C carbonate appears, which is associated with a decrease in the number of surface HPO_4^{2-} ions; (3) apatite 2 reverse exchange in 1 M HPO_4^{2-} solution for 20 min. Only the ^{13}C carbonate line is removed, owing to HPO_4^{2-} uptake from the solution; faint residual ^{13}C carbonate lines can be seen, corresponding to ^{13}C carbonate species that have entered the growing apatitic domains during the course of the experiment (maturation). All samples were freeze-dried after washing.
Reproduced with permission from Rey et al. (2014).

3.4.2 Mineral-ion reservoir function of bone

Bone is considered an ion reservoir for many essential elements: it contains 99% of the body's calcium in addition to 85% of P, 65% of Mg, 99% of F, and 30% of Zn. This function is believed to exist in very primitive unicellular organisms, in which calcium phosphate concentrations in mitochondria act as ion reservoirs for phosphorus. This primitive ion-reservoir function is still apparent in birds, for example, where the formation of egg shells mobilizes the calcium of bones, which occurs daily in the case of hens. Also, this calcium carbonate egg shell later serves as a calcium source for the embryo to build its own bones. Such an ion-reservoir function is also in use in mammals and one of the most important roles of bone mineral is to maintain homeostasis. Two main processes are considered: a passive physicochemical process based on the mineral solution equilibrium and an active process involving dissolution of bone by osteoclast (Talmage and Talmage, 2006). In both cases, trabecular bone favoring, by its structure, bone–blood exchange, plays a crucial role. In addition, physicochemical studies have shown that bone mineral crystals appeared less mature in trabecular than cortical bone,

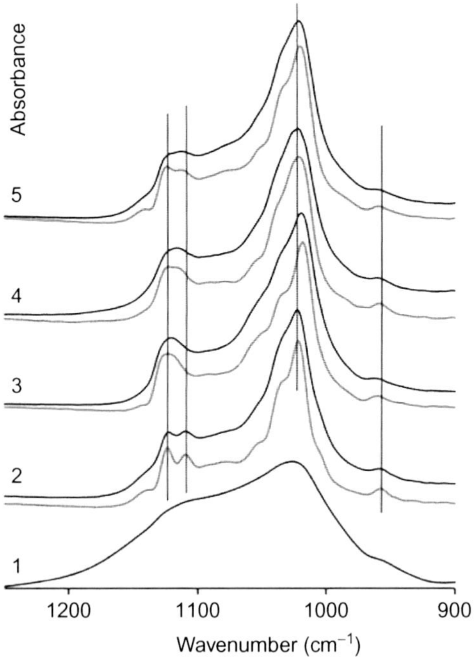

Figure 3.9 FTIR spectra (ν_3 PO_4 domain) showing the alterations of the hydrated layer of apatite nanocrystals upon drying and during the ion-exchange reaction. Raw (black lines) and resolution-enhanced spectra (gray lines) are shown: (1) freeze-dried carbonate-free apatite nanocrystals, 1 h maturation; (2) sample 1, but in wet state (not freeze-dried); (3) sample exchanged with a carbonate solution (1 M NH_4HCO_3) wet state; (4) sample exchange with a magnesium solution (1 M $Mg(NO_3)_2$), wet state; (5) sample exchanged with a strontium solution (1 M $Sr(NO_3)_2$), wet state. Faint, but significant, changes in the spectra can be observed, especially on the resolution-enhanced ones (Fourier self-deconvolution: OMNIC software, bandwidth 9.2–9.6 cm^{-1}; smoothing factor 1.2), characterized by a change in the number of lines or shoulders and different positions and intensities. With the exception of freeze-drying, all alterations of the hydrated layer are reversible.

with larger amounts of non-apatitic environments (Kuhn et al., 2008). Thus, bone mineral dissolution or ion-exchange processes would be considerably favored in trabecular bone. Maturation is believed to play a crucial role in the ion-reservoir function, and remodeling appears to be a necessity in order to preserve the properties of freshly formed apatite nanocrystals, which allow homeostasis.

Foreign ions (i.e., other than Ca and phosphates) in bones have several roles regarding bone mineral characteristics, essential element reservoir function, and toxic element storage.

One of the most striking observations is that bone mineral contains many more non-apatitic environments than synthetic apatites, showing similar maturation levels based on carbonate evaluation (Rey et al., 1991b). This observation suggests that the non-apatitic environments detected in bone crystals have some capacity for regeneration and/or contain stabilizing elements. The first hypothesis has not yet been analyzed. Several studies, however, support the second hypothesis, and it has been shown that mineral ions like Mg, pyrophosphates, and carbonate inhibit apatite formation and may thus prevent apatite growth and favor the stability of the hydrated layer. Organic molecules may also interfere with the maturation process.

From a physicochemical point of view, the mobilization of essential elements is, in some ways, comparable to their uptake. Ion exchanges can occur between body fluids and the surface hydrated layer of bone apatite nanocrystals. In addition, ion substitutions in the hydrated layer or in the apatite domains may change their properties considerably, even when rather small amounts are involved. Fluoride, for example, favors apatite formation at the expense of the hydrated layer and accelerates the maturation process. It significantly decreases the solubility of the apatite and slows down the turnover rate. The mobilization of this element is difficult because of its recycling during remodeling. Lead seems to have a similar behavior. Strontium, on the contrary, seems easily mobilized; it is taken up in bone mineral in Sr-rich diets, but can easily be removed from bone after returning to a normal diet (Farlay et al., 2005). This phenomenon may be related to a higher solubility of Sr-containing apatites (Christoffersen et al., 1997). For other elements, like Zn, mobilization from bone tissues has been evidenced to some extent in the case of deficiency (King et al., 2000), and about a third of the bone content can be removed, although no explanation of this limited mobilization has been proposed on a physicochemical basis. Upon aging, the increase in bone crystal size and the decrease in the number of surface non-apatitic environments reduce the ion-reservoir capacity of elements stored mainly in the hydrated layer, like Mg. This phenomenon could be responsible for the deficiency of such elements with age. On the contrary, in young persons, the small size of the crystals and their high surface reactivity increase their uptake capability, but could also play a part in lead poisoning of infants.

The storage of toxic elements in bone may also be seen as preservation against these elements, as the elemental activity is completely inhibited when they are included in apatite domains. It should be recalled that the formation of extracellular calcium carbonate and phosphate has been described as a detoxification of intracellular fluid (Simkiss, 1977). It is only during turnover or remodeling processes that potential toxic elements can interact with body fluids and possibly exert their deleterious activity in soluble form. Thus, Al and Pb, for example, which are potent neurotoxic elements, can be stored in bone with a low turnover rate. They do not show toxic effects on bone cells at low doses, although high doses have been shown to prevent bone mineralization and to initiate bone disorders (Monir et al., 2010; Rüster et al., 2002).

3.4.3 Adsorption of molecules on biological apatites and synthetic analogs

Several molecules of biological interest have been shown to adsorb on apatites; however, only a few studies have been made regarding their adsorption on nanocrystalline apatites or bone mineral crystals. Generally, the adsorption of different molecules of biological interest on apatites follows a Langmuir adsorption isotherm. This is the case for molecules with negative functional groups, which show the strongest affinity. Their adsorption on apatites, however, presents some unusual characteristics: (1) it often appears irreversible; once a molecule has been adsorbed it cannot easily be removed; in other words, adsorption equilibrium can be reached when the concentration of adsorbate in solution is increased; however, they cannot be removed when it is reduced by dilution. (2) The adsorption reaction of negatively charged molecules is associated with the release of phosphate ions. These phenomena can be explained if we consider that the adsorption of charged molecules corresponds to an ion-exchange process between the molecule in solution and phosphate ions of the surface of crystals:

$$M_{sol} + P_{surf} \leftrightarrow M_{ads} + P_{sol}$$

with the equilibrium constant:

$$K_{ex} = \left(\mathrm{M}_{ads} \cdot P_{sol}\right) / \left(M_{sol} \cdot P_{surf}\right)$$

where M_{ads} is the amount adsorbed at equilibrium, P_{sol} is the concentration of exchanged phosphate ions at equilibrium, M_{sol} is the concentration of M in solution at equilibrium, and P_{surf} is the amount of exchangeable phosphate ions still on the surface at equilibrium. This chemical equilibrium can lead to a Langmuir-type isotherm, assuming that the maximum number of adsorption sites for M on the surface is N. The exchangeable amount of P surface ions becomes $P_{surf}=N-M_{ads}$ and we can write:

$$M_{ads} = NK_{ex}\left(M_{sol} / P_{sol}\right) / \left[1 + K_{ex}\left(M_{sol} / P_{sol}\right)\right]$$

which is comparable to the original Langmuir isotherm equation:

$$M_{ads} = NKM_{sol} / \left(1 + KM_{sol}\right)$$

where K is the Langmuir affinity constant.

Then, it appears that it is no longer the concentration of molecules in equilibrium with the adsorbed species that determines the reaction, but the ratio M_{sol}/P_{sol}. Thus, when a dilution is made, this ratio remains almost constant, except for marginal re-equilibration of P_{sol} related to the apatite solubility equilibrium, and only very small changes are observed in the amount of molecules adsorbed on the surface. The release of the adsorbed molecules necessitates a change in the M_{sol}/P_{sol} ratio, for example an increase in the phosphate concentration in solution at constant M_{sol} values.

Similar reactions can be written for the adsorption of molecules with positive functional groups, which are exchangeable with surface calcium. Then, adsorption appears as a particular series of ion-exchange reactions, involving only surface ions. As expected, adsorption parameters on nanocrystalline apatites appear to be linked to the hydrated layer (Ouizat et al., 2000; Benaziz et al., 2001; Barroug et al., 2008). Desorption of the adsorbed molecules can be obtained by increasing the phosphate concentration or by displacement of the adsorbed molecules with other adsorbing molecular ions, like carbonate ions or organic molecules with stronger affinities. The observed equilibrium can appear more complex, depending on the number of adsorbing groups per molecule, for the availability of surface sites, owing to the steric hindrance, or for the mode of interaction of the adsorbing molecules with the surface ions (Pascaud et al., 2013; Mukherjee et al., 2009).

Although several reports have been produced on the adsorption of macromolecules of bone on apatites, the isotherms have rarely been obtained. Several functional groups have been found to attach strongly to apatites; thus, phosphoric esters or related phosphorus-containing molecules like bisphosphonates and phytic acid adsorb rather strongly on apatites. Silicate and sulfate esters also show good affinity. Carboxylates generally show poor binding ability on apatite surfaces; however, these are improved in polycarboxylates, and molecules such as citrates and polyaspartate can attach strongly to apatite surfaces.

Some proteins (bovine serum albumin (BSA), growth factor, and bone-specific molecules) (Cazalbou et al., 2000; Ouizat et al., 2000; Midy et al., 1998) are easily adsorbed on nanocrystalline apatites. Ouizat et al. (2000) demonstrated that the ability of nanocrystalline apatites to adsorb BSA was related to the amount of labile species, because the adsorption capacity diminished as the maturation time increased. Thus, it might be suggested that such proteins interact with labile ions of the hydrated layer at the surface of apatite crystals. Recently, Grunenwald et al. (2014) studied the adsorption of DNA on biomimetic apatites to shed some light on DNA preservation over time in skeletal remains. They demonstrated that the strong binding affinity of DNA for nanocrystalline apatite surfaces in addition to the irreversibility of DNA adsorption toward dilution and phosphate-promoted DNA desorption. These results suggest the involvement of an ion-exchange phenomenon between the phosphate anionic groups of DNA and the non-apatitic hydrogen phosphate ions potentially released from the hydrated layer of biomimetic apatite nanocrystals in the adsorption process.

A similar relationship has been observed in the adsorption of bisphosphonates (BP) on synthetic nanocrystalline apatites or bone crystals (Pascaud et al., 2013; Mukherjee et al., 2009). The number of phosphate ions released per molecule adsorbed has been shown to depend on the adsorbing BP molecules involved, as well as on the nature of the nanocrystalline apatites used as an adsorption substrate.

Several other molecules can adsorb on apatites. Heparin shows very specific adsorption behavior, owing to its calcium-ion chelating ability (Errassifi et al., 2010). Non-collagenous bone proteins and citrate–ion interactions with bone mineral are discussed in the following section.

3.4.4 Interactions of bone mineral crystals with bone tissue constituents

Several interactions of bone mineral crystals with the different components of bone tissue can be distinguished, that is, with the mineral crystals themselves, with collagen, with non-collagenous bone proteins, and with bone cells. Only the first three of these interactions will be reviewed here.

3.4.4.1 Intercrystalline bonding

Recently, the problem of intercrystalline bonding has been raised. The association of crystals in bone, dentine, or enamel was recognized in early studies of bone mineral (Weiner and Pryce, 1986). A number of papers have been published that observed a "crystal fusion" phenomenon during enamel formation (Ichijo et al., 1993; Chen et al., 2011) corresponding to the junction of close mineral crystals to form a single crystal with a "dark line" appearing in the middle as a residue of the mode of formation of these crystals. In bone mineral, crystal platelets have been shown to have parallel orientations in the tissue, which could be analogous to a crystal fusion process (Landis et al., 1993). These strong associations are supposed to prevent the disaggregation of bone mineral into individual crystals (Chen et al., 2011), although several processes exist to obtain at least some individual crystals (Kim et al., 1995). In recent years, the formation of crystal associations was considered to be related to the hydrated layer surface favoring crystal junctions (Wang et al., 2013). More recently, however, a new hypothesis has been presented in which citrate ions are proposed as the link between adjacent crystals (Davies et al., 2014), which is based on the existence of OCP–citrate crystals, in which the citrate ion, located in the hydrated layer of the OCP structure, is bonding two adjacent apatite layers. This single-crystal model applied to bone crystals recalls Brown's hypothesis concerning the surface structure of hydrated layers at the apatite surface. If the model of OCP–citrate is well established (Davies et al., 2014), the formation of such associations needs to be clarified, considering the mode of mineralization of bone (primary and secondary mineralization) and the fact that citrate has also been described as a dispersing agent of agglomerated apatite crystals (Tan et al., 2009).

3.4.4.2 Collagen–apatite interactions in bone

The question of the interactions between the two major constituents of bone, apatite mineral crystals and the collagen matrix, seems of uttermost importance. The mineralization of the collagen assembly gives some clues on collagen–mineral interactions. It has been shown that the first mineral crystals were formed in the hole zone of a collagen molecule assembly and were then spread in the inter- and intrafibrillar space (Glimcher, 2006). Interactions between amino acids of terminal groups and those of adjacent molecules were suggested, but without direct evidence. The parallel orientation of the *c*-axis of the apatite crystals with the axis of the collagen fibers is well documented, although with some divergence that is possibly related to the formation of cholesteric crystals (Weiner and Traub, 1992; Sasaki and Sudoh, 1997). This orientation corresponds to the faster growing crystal direction oriented in the free available space left by the collagen fiber assembly, and not necessarily to a templating effect of collagen molecules.

Recent studies on collagen–bone mineral interactions by rotational echo double resonance SSNMR (Nikel et al., 2012) indicate that the citrate ions and glucosaminoglycans (GAGs) are the closer organic bone molecules found in the vicinity of phosphate apatite. Among collagen amino acids, alanine and hydroxyproline carbon atoms were found to interact with bone mineral, although with some variability depending on age. These data did not succeed in demonstrating stable bonding between collagen and bone crystals, thus intermediate interacting molecules were considered, such as GAGs and citrate. Other hypotheses suggest that water molecules interacting with both the mineral crystals and the collagen could play a role in bridging the two main fractions of the bone composite (Wilson et al., 2006; Wang et al., 2013).

3.4.4.3 Interactions of apatites with non-collagenous molecules

Several non-collagenous molecules interact specifically with apatites. Very often, however, there is only partial information available on these interactions. Among those non-collagenous molecules that are bonding more specifically to apatite, we can distinguish osteocalcin (OC), osteopontin (OPN), BSP, albumin, GAGs, and citrate ions.

OC is characterized by the presence of gamma-carboxyglutamic acid (Gla) residues (Hauschka and Wians, 1989). It is a major bone protein secreted by osteoblasts, the blood plasma level of which is frequently used to assess bone remodeling activity. This protein binds calcium and other bivalent ions and adsorbs strongly on apatite according to a Langmuir-type isotherm (Hauschka and Wians, 1989). This adsorption process seems to be related to the presence of bound calcium ions. Divergent structural models have been proposed for OC binding to bone crystals, based on stoichiometric hydroxyapatite (Hoang et al., 2003; Hauschka and Wians, 1989). OC has been found to inhibit or stimulate the formation of apatite crystals, as found for other adsorbing molecules. However, its ability to bind to the mineral has suggested that OC could function as a matrix signal in bone remodeling. Its adsorption behavior on nanocrystalline apatites at different maturation stages would be interesting to investigate from this perspective.

OPN is a protein that is very rich in acidic amino acids (25%) (Hunter, 2013). In addition, it has the ability to be phosphorylated at about 30 potent sites, but it has a variable phosphorylation ratio, which has been shown to determine its capability to bind to apatite (Hunter, 2013). Like OC, OPN has been shown to inhibit or stimulate apatite formation, but with a strong modulation related to the phosphorylation ratio. OPN is involved in the differentiation and recruitment of osteoclasts, although its role in bone remodeling could also be a regulation of vascularization (Gericke et al., 2005). Interestingly, recent studies using SSNMR showed that the lack of these molecules (in genetically modified mice lacking OC and/or OPN) does not significantly modify the bone crystal characteristics (Nikel et al., 2013). It can be observed that OC, OPN, or other large molecules are unable to diffuse into the organic matrix and their effect on mineral crystals would be localized to the site at which they were deposited. In addition, their strong adsorption to the apatite crystals considerably limits their ability to regulate global mineral formation in bone via a physicochemical process. Even though they could block active growth sites, owing to their low concentration in bone mineral, their effect would be limited to very few mineral crystals. Most *in vitro* studies on the effects of these proteins on mineral formation do not take into account the strong limitation related to their small number and limited diffusing capabilities. The physicochemical role of these molecules regarding the mineralization process cannot be considered as a dominant phenomenon, although they could act on mineral crystal formation at specific strategic zones. The main role of these proteins could be matrix signaling in remodeling.

Among the adsorbing molecules in close contact with bone mineral, only citrate ions, which adsorb strongly on apatite and exist in rather high numbers in bone, could produce a physicochemical effect. As with other molecules binding strongly to the apatite surface, citrate-ion adsorption has been shown to correspond to ion exchange with surface phosphate ions (Skwarek et al., 2013). Citrate ions have also been found to inhibit the growth of apatite crystals like many other adsorbing molecules. The effect of this molecule on the stability of the hydrated layer and the aging process of the nanocrystals have not yet been investigated.

3.5 Conclusions and perspectives

Bone mineral can be described as an apatitic phase associated with mineral ions in non-apatitic environments belonging to hydrated domains. The non-apatitic ionic environments are localized within a hydrated layer on the crystal surface, that is, in close association with apatite domains forming the nanocrystal core. This structured surface layer has been found to be very fragile and can easily be altered upon drying or when being exposed to other bone tissue preparation techniques. Accurate characterization of these mineral domains appears challenging and some reports have been published to address this question (Farlay et al., 2010), distinguishing the characteristics of the apatite domains and those of the surface hydrated layer, although, in many cases, bone mineral is treated as a global entity. Progress must be made concerning the quantitative composition of the hydrated layer as well as the location of trace elements. The exact content of HPO_4^{2-} in bone apatite is still unknown, although these ions appear to be essential constituents of the mineral, determining its dissolution properties and evolution. Some determinations have been proposed, but they lack accuracy and often do not distinguish HPO_4^{2-} in apatitic domains from that in the hydrated layer (Combes et al., 2001; Jäger et al., 2006; Spevak et al., 2013). The distribution of these ions and that of carbonate ions must, however, be carefully determined, as they could be indicative of the very first crystals formed in bone tissue and indicate aging alterations. The structure of the hydrated layer is still unknown, although vibrational spectroscopies in the wet state indicate that it is very sensitive to mineral-ion composition and that it is altered during crystal aging. It is often described as amorphous; however, this term could be inadequate, as wet investigation techniques, which are the only ones to preserve the hydrated layer's integrity, are rarely used. It seems probable that several kinds of environment could be found in this hydrated layer, owing to the reorganization capabilities of the mineral ions, and *in vitro* investigations could lead to interesting results on this point. The exact role of the hydrated layer in crystal–crystal and crystal–matrix interactions also needs to be identified, keeping in mind that these interactions can also modify the original ionic organization of individual crystals. An interesting domain for bone science and, more broadly, materials science concerns the organization of nanocrystals at a supracrystalline level in order to form monoliths or composites. A few groups have worked on this problem from the bone formation point of view (Wang et al., 2013; Davies et al., 2014; Duer and Veis, 2013) using different approaches. The dynamic of bone tissue appears to be important, especially concerning bone remodeling with the two stages of mineralization (primary and secondary) and the role of the hydrated layer in these aspects. A hypothesis that has not yet been examined is that primary mineralization would correspond to a stage of crystallization involving nanocrystals with a well-developed hydrated layer, introducing some swelling into the crystal assembly, and the secondary mineralization would develop the apatite domains within these crystals, involving denser associations and the release of water. Another aspect to consider in bone dynamics is the blood–mineral relationships, including circadian variations of mineral–ion concentrations and their role in the hydrated layer in order to establish possible regeneration, as well as to determine the rate of mineral formation with age and its consequences on different aspects regarding the nascent bone crystal constitution and their interactions. Another characteristic of bone is a strong heterogeneity at different levels, which is not well known at the tissue level or at the molecular level. In some of these aspects, the use of *in vitro* experimental modeling can be carried out with the view of approaching all of the subtlety of bone tissue organization and evolution.

References

Agna, B.J.W., Knowles, H.C., Alverson, G., 1958. The mineral content of normal human bone. J. Clin. Invest. 37 (10), 1357–1361.

Aoba, T., et al., 1980. Diffuse X-ray scattering from apatite crystals and its relation to amorphous bone mineral. J. Osaka Univ. Dent. Sch. 20, 81–90.

Arends, J., Jongebloed, W.L., 1981. Apatite single crystals. Formation, dissolution and influence of CO_3^{2-} ions. Recl. Trav. Chim. Pays-Bas 100 (1), 3–9.

Aue, W.P., et al., 1984. Solid-state phosphorus-31 nuclear magnetic resonance studies of synthetic solid phases of calcium phosphate: potential models of bone mineral. Biochemistry 23, 6110–6114.

Baig, A.A., et al., 1999. Metastable equilibrium solubility behavior of bone mineral. Calcif. Tissue Int. 64, 329–339.

Barroug, A., Legrouri, A., Rey, C., 2008. Exchange reactions at calcium phosphate surface and applications to biomaterials. Key Eng. Mater. 361–363, 79–82.

Barry, A.B., Baig, A.A., Miller, S.C., Higuchi, W.I., 2002. Effect of age on rat bone solubility and crystallinity. Calcif. Tissue Int. 71, 167–171.

Benaziz, L., et al., 2001. Adsorption of O-phospho-L-serine and L-serine onto poorly crystalline apatite. J. Colloid Interface Sci. 238 (1), 48–53.

Biltz, R.M., Pellegrino, E.D., 1981. Skeletal carbonates and acid–base regulation. Miner. Electrolyte Metab. 5, 1–7.

Bohic, S., et al., 2000. Characterization of the trabecular rat bone mineral: effect of ovariectomy and bisphosphonate treatment. Bone 26 (4), 341–348.

Boskey, A.L., 2003. Biomineralization: an overview. Connect. Tissue Res. 44 (Suppl. 1), 5–9.

Brown, W.E., 1966. Crystal growth of bone mineral. Clin. Orthop. Relat. Res. 44, 205–220.

Brown, W.E., et al., 1962. Octacalcium phosphate and hydroxyapatite: crystallographic and chemical relations between octacalcium phosphate and hydroxyapatite. Nature 196, 1050–1055.

Brown, W.E., Schroeder, L.W., Ferris, J.S., 1979. Interlayering of crystalline octacalcium phosphate and hydroxyapatite. J. Phys. Chem. 83 (11), 1385–1388.
Brown, W.E., Mathew, M., Chow, L.C., 1984. Roles of octocalcium phosphate in surface chemistry of hydroxyapatite. In: Misra, D.N. (Ed.), Adsorption on and Surface Chemistry of Hydroxyapatite. Plenum Press, New York, pp. 13–28.
Brown, W.E., Eidelman, N., Tomazic, B., 1987. Octacalcium phosphate as a precursor in biomineral formation. Adv. Dent. Res. 1, 306–313.
Burnell, J.M., Teubner, E.J., Miller, A.G., 1980. Normal maturational changes in bone matrix, mineral, and crystal size in the rat. Calcif. Tissue Int. 31, 13–19.
Cardoso, L., et al., 2013. A review of recent advances in the assessment of bone porosity, permeability, and interstitial fluid flow. J. Biomech. 46 (2), 253–265.
Cazalbou, S., 2000. Echanges cationiques impliquant des apatites nanocristallines analogues au minéral osseux. PhD thesis, INPT, Toulouse, France.
Cazalbou, S., Hina, A., Rey, C., 1999. Interactions between trace elements and bone mineral matrix. In: Abdulla, M., et al. (Eds.), New Aspects of Trace Element Research. Smith-Gordon, London, pp. 58–62.
Cazalbou, S., Hina, A., Rey, C., 2000. Poorly crystalline apatites for drug delivery. In: Ravaglioli, A., Krajewski, A. (Eds.), Ceramics, Cells and Tissues. Drug Delivery Systems. IRTEC-CNR, Faenza, Italy, pp. 42–47.
Cazalbou, S., et al., 2004. Poorly crystalline apatites: evolution and maturation *in vitro* and *in vivo*. J. Bone Miner. Metab. 22 (4), 310–317.
Cazalbou, S., et al., 2005. Ion exchanges in apatites for biomedical application. J. Mater. Sci. Mater. Med. 16, 405–409.
Chen, P.-Y., et al., 2011. Minerals form a continuum phase in mature cancellous bone. Calcif. Tissue Int. 88 (5), 351–361.
Cho, G., Wu, Y., Ackerman, J.L., 2003. Detection of hydroxyl ions in bone mineral by solid-state NMR spectroscopy. Science 300, 1123–1127.
Christoffersen, J., Landis, W.J., 1991. A contribution with review to the description of mineralization of bone and other calcified tissues *in vivo*. Anat. Rec. 230 (4), 435–450, Available at: http://www.ncbi.nlm.nih.gov/pubmed/1928750.
Christoffersen, J., et al., 1997. Effects of strontium ions on growth and dissolution of hydroxyapatite and on bone mineral detection. Bone 20 (1), 47–54.
Combes, C., Rey, C., 2010. Amorphous calcium phosphates: synthesis, properties and uses in biomaterials. Acta Biomater. 6 (9), 3362–3378.
Combes, C., Rey, C., Mounic, S., 2001. Identification and evaluation of HPO_4 ions in biomimetic poorly crystalline apatites and bone mineral. Key Eng. Mater. 192–195, 143–146.
Crane, N.J., et al., 2006. Raman spectroscopic evidence for octacalcium phosphate and other transient mineral species deposited during intramembranous mineralization. Bone 39, 434–442.
Currey, J.D., 2002. Bone: Structure and Mechanics. Princeton University Press, Princeton, NJ.
Davies, E., et al., 2014. Citrate bridges between mineral platelets in bone. Proc. Natl. Acad. Sci. U. S. A. 111, E1354–E1363.
De Jong, W.F., 1926. The mineral substance in bones. Recl. Trav. Chim. Pays-Bas 45, 445–448.
Drouet, C., et al., 2008. Surface enrichment of biomimetic apatites with biologically-active ions Mg^{2+} and Sr^{2+}: a preamble to the activation of bone repair materials. Mater. Sci. Eng. C 28 (8), 1544–1550.
Duer, M., Veis, A., 2013. Bone mineralization: water brings order. Nat. Mater. 12 (12), 1081–1082.

Eichert, D., et al., 2004. Specific characteristics of wet nanocrystalline apatites. Consequences on biomaterials and bone tissue. Key Eng. Mater. 254–256, 927–930.

Elliott, J.C., 1994. Structure and Chemistry of the Apatites and other Calcium Orthophosphates. Elsevier Science, Amsterdam.

Errassifi, F., et al., 2010. Adsorption onto nanocrystalline apatitic calcium phosphates. Applications to growth factors and drugs delivery. In: Narayan, R., McKittrick, J., Singh, M. (Eds.), Advances in Bioceramics and Biotechnologies. Wiley, Hoboken, NJ.

Farlay, D., et al., 2005. Long-term strontium ranelate administration in monkeys preserves characteristics of bone mineral crystals and degree of mineralization of bone. J. Bone Miner. Res. 20 (9), 1569–1578.

Farlay, D., et al., 2010. Mineral maturity and crystallinity index are distinct characteristics of bone mineral. J. Bone Miner. Metab. 28 (4), 433–445.

Francis, M.D., Webb, N.C., 1971. Hydroxyapatite formation from a hydrated calcium monohydrogen phosphate precursor. Calcif. Tissue Res. 6 (4), 335–342.

Gericke, A., Qin, C., Spevak, L., 2005. Importance of phosphorylation for osteopontin regulation of biomineralization. Calcif. Tissue Int. 77 (1), 45–54.

Giraud-Guille, M.M., Belamie, E., Mosser, G., 2004. Organic and mineral networks in carapaces, bones and biomimetic materials. Comptes Rendus Palevol 3, 503–513.

Glimcher, M.J., 2006. Bone: nature of the calcium phosphate crystals and cellular, structural, and physical chemical mechanisms in their formation. Rev. Mineral. Geochem. 64, 223–282.

Grunenwald, A., et al., 2014. Adsorption of DNA on biomimetic apatites: toward the understanding of the role of bone and tooth mineral on the preservation of ancient DNA. Appl. Surf. Sci. 292, 867–875.

Grynpas, M.D., Bonar, L.C., Glimcher, M.J., 1984. Failure to detect an amorphous calcium-phosphate solid phase in bone mineral: a radial distribution function study. Calcif. Tissue Int. 36, 291–301.

Habraken, W.J.E.M., et al., 2013. Ion-association complexes unite classical and non classical theories for the biomimetic nucleation of calcium phosphates. Nat. Commun. 4, 1507–1512.

Hauschka, P.V., Wians, F.H., 1989. Osteocalcin-hydroxyapatite interaction in the extracellular organic matrix of bone. Anat. Rec. 224 (2), 180–188.

Hoang, Q., et al., 2003. Bone recognition mechanism of porcine osteocalcin from crystal structure. Nature 425 (6961), 977–980.

Hodge, A.J., 1989. Molecular models illustrating the possible distribution of holes in simple systematically staggered arrays of type I collagen molecules in native-type fibrils. Connect. Tissue Res. 21 (1–4), 137–147.

Hunter, G.K., 2013. Role of osteopontin in modulation of hydroxyapatite formation. Calcif. Tissue Int. 93 (4), 348–354.

Ichijo, T., Yamashita, Y., Terashima, T., 1993. Observations on structural features and characteristics of biological apatite crystals. 8. Observation on fusion of human enamel crystals. Bull. Tokyo Med. Dent. Univ. 40 (4), 207–216.

Iyengar, G.V., Tandon, L., 1999. Minor and Trace Elements in Human Bones and Teeth. International Atomic Energy Agency, Section of Nutritional and Health-Related Environmental Studies, Vienna, Austria.

Jäger, C., et al., 2006. A solid-state NMR investigation of the structure of nanocrystalline hydroxyapatite. Magn. Reson. Chem. 44, 573–580.

Kim, H.-M., Rey, C., Glimcher, M.J., 1995. Isolation of calcium-phosphate crystals of bone by non-aqueous methods at low temperature. J. Bone Miner. Res. 10 (10), 1589–1601.

Kim, H.-M., Rey, C., Glimcher, M.J., 1996. X-ray diffraction, electron microscopy, and Fourier transform infrared spectroscopy of apatite crystals isolated from chicken and bovine calcified cartilage. Calcif. Tissue Int. 59, 58–63.
King, J.C., Shames, D.M., Woodhouse, L.R., 2000. Zinc and health: current status and future directions zinc homeostasis in humans. J. Nutr. 130 (Suppl.), 1360–1366.
Kuhn, L.T., et al., 2008. A comparison of the physical and chemical differences between cancellous and cortical bovine bone mineral at two ages. Calcif. Tissue Int. 83 (2), 146–154.
Labarthe, J.C., Bonel, G., Montel, G., 1973. Structure and properties of B-type phospho calcium carbonate apatites. Ann. Chim. (Fr.) 8, 289–301.
Landis, W.J., et al., 1986. Detection and localization of silicon and associated elements in vertebrate bone tissue by imaging ion microscopy. Calcif. Tissue Int. 38, 52–59.
Landis, W.J., et al., 1993. Mineral and organic matrix interaction in normally calcifying tendon visualized in 3 dimensions by high-voltage electron-microscopic tomography and graphic image-reconstruction. J. Struct. Biol. 110 (1), 39–54.
Laurencin, D., et al., 2010. Probing the calcium and sodium local environment in bones and teeth using multinuclear solid state NMR and X-ray absorption spectroscopy. Phys. Chem. Chem. Phys. 12 (5), 1081–1091.
LeGeros, R.Z., 1991. Calcium phosphates in oral biology and medicine. Monogr. Oral Sci. 15, 1–201.
Legros, R., Balmain, N., Bonel, G., 1987. Age-related changes in mineral of rat and bovine cortical bone. Calcif. Tissue Int. 41, 137–144.
Loong, C.K., et al., 2000. Evidence of hydroxyl-ion deficiency in bone apatites: an inelastic neutron-scattering study. Bone 26, 599–602.
Meunier, P.J., Boivin, G., 1997. Bone mineral density reflects bone mass but also the degree of mineralization of bone: therapeutic implications. Bone 21 (5), 373–377.
Midy, V., et al., 1998. Basic fibroblast growth factor adsorption and release properties of calcium phosphate. J. Biomed. Mater. Res. 41 (3), 405–411.
Monir, A.U., et al., 2010. The effect of lead on bone mineral properties from female adult C57/BL6 mice. Bone 47 (5), 888–894.
Montel, G., et al., 1981. New concepts in the composition, crystallization and growth of the mineral components of calcified tissues. J. Cryst. Growth 53, 74–79.
Muenzenberg, K.J., Gebhardt, M., 1973. Brushite, octacalcium phosphate, and carbonate-containing apatite in bone. Clin. Orthop. Relat. Res. 90, 271–273.
Mukherjee, S., et al., 2009. Thermodynamics of bisphosphonates binding to human bone: a two-site model. J. Am. Chem. Soc. 131 (24), 8374–8375.
Neuman, W.F., 1980. Bone mineral and calcification mechanism. In: Urist, M.R. (Ed.), Fundamental and Clinical Bone Physiology. J.B. Lippincott, Philadelphia, PA, pp. 83–107.
Neuman, W.F., Bareham, B.J., 1975. Evidence for the presence of secondary calcium phosphate in bone and its stabilization by acid production. Calcif. Tissue Res. 18 (3), 161–172.
Neuman, W.F., Toribara, T.Y., Mulryan, B.J., 1956. The surface chemistry of bone. IX. Carbonate:Phosphate exchange. J. Am. Chem. Soc. 78 (17), 4263–4266.
Nikel, O., et al., 2012. Solid state NMR investigation of intact human bone quality: balancing issues and insight into the structure at the organic-mineral interface. J. Phys. Chem. C 116 (10), 6320–6331.
Nikel, O., et al., 2013. NMR investigation of the role of osteocalcin and osteopontin at the organic–inorganic interface in bone. Langmuir 29 (45), 13873–13882.
Nudelman, F., et al., 2010. The role of collagen in bone apatite formation in the presence of hydroxyapatite nucleation inhibitors. Nat. Mater. 9 (12), 1004–1009.

Olszta, M.J., et al., 2007. Bone structure and formation: a new perspective. Mater. Sci. Eng. R-Rep. 58 (3–5), 77–116.

Onuma, K., 2006. Recent research on pseudobiological hydroxyapatite crystal growth and phase transition mechanisms. Prog. Cryst. Growth Charact. Mater. 52 (3), 223–245.

Ouizat, S., et al., 2000. Adsorption of bovine serum albumin on poorly crystalline apatite: influence of maturation. Mater. Res. Bull. 34 (14/15), 2279–2289.

Pascaud, P., et al., 2013. Interaction between a bisphosphonate, tiludronate, and biomimetic nanocrystalline apatites. Langmuir 29 (7), 2224–2232.

Pasteris, J.D., et al., 2004. Lack of OH in nanocrystalline apatite as a function of degree of atomic order: implications for bone and biomaterials. Biomaterials 25 (2), 229–238.

Pellegrino, E.D., Biltz, R.M., 1972. Mineralization in the chick enbryo. I. Monohydrogen phosphate and carbonate relationships during maturation of the bone crystal complex. Calcif. Tissue Res. 10 (2), 128–135.

Posner, A., Betts, F., 1975. Synthetic amorphous calcium-phosphate and its relation to bone-mineral structure. Acc. Chem. Res. 8 (8), 273–281.

Quelch, K.J., Melick, R.A., Bingham, P.J., Mercuri, S.M., 1983. Chemical composition of human bone. Arch. Oral Biol. 28 (8), 665–674.

Rey, C., et al., 1989. The carbonate environment in bone mineral. A resolution-enhanced Fourier Transform infrared spectroscopy study. Calcif. Tissue Int. 45, 157–164.

Rey, C., et al., 1990. Resolution-enhanced Fourier transform infrared spectroscopy study of the environment of phosphate ions in the early deposits of a solid phase of calcium phosphate in bone and enamel, and their evolution with age. I: investigations in the n_4 PO_4 domain. Calcif. Tissue Int. 46, 384–394.

Rey, C., et al., 1991a. A resolution-enhanced Fourier Transform Infrared spectroscopic study of the environment of the CO_3^{2-} ion in the mineral phase of enamel during its formation and maturation. Calcif. Tissue Int. 49, 259–268.

Rey, C., et al., 1991b. Fourier Transform infrared spectroscopic study of the carbonate ions in bone mineral during aging. Calcif. Tissue Int. 49, 251–258.

Rey, C., et al., 1995a. Hydroxyl groups in bone mineral. Bone 16, 583–586.

Rey, C., et al., 1995b. Maturation of poorly crystalline apatites: chemical and structural aspects *in vivo* and *in vitro*. Cell Mater. 5 (4), 345–356.

Rey, C., et al., 2007. Physico-chemical properties of nanocrystalline apatites: implications for biominerals and biomaterials. Mater. Sci. Eng. C 27, 198–205.

Rey, C., et al., 2009. Bone mineral: update on chemical composition and structure. Osteoporos. Int. 20 (6), 1013–1021.

Rey, C., et al., 2011. Bioactive ceramics: physical chemistry. In: Ducheyne, P., et al. (Eds.), Comprehensive Biomaterials. Elsevier, Amsterdam, The Netherlands, pp. 187–221.

Rey, C., Marsan, O., Combes, C., Drouet, C., Grossin, D., Sarda, S., 2014. Characterization of calcium phosphates using vibrational spectroscopies. In: Ben Nissan, B. (Ed.), Advances in Calcium Phosphate Biomaterials. Springer, London, pp. 229–266.

Roufosse, A.H., et al., 1984. Investigation of the mineral phases of bone by solid-state phosphorus-31 magic angle sample spinning nuclear magnetic resonance. Biochemistry 23 (23), 6115–6120.

Rubinacci, A., et al., 2000. Bone as an ion exchange system: evidence for a pump-leak mechanism devoted to the maintenance of high bone K(+). Am. J. Physiol. Endocrinol. Metab. 278 (1), E15–E24.

Ruppel, M.E., Miller, L.M., Burr, D.B., 2008. The effect of the microscopic and nanoscale structure on bone fragility. Osteoporos. Int. 19 (9), 1251–1265.

Rüster, M., et al., 2002. Aluminum deposition in the bone of patients with chronic renal failure—detection of aluminum accumulation without signs of aluminum toxicity in bone using acid solochrome azurine. Clin. Nephrol. 58 (4), 305–312.
Sasaki, N., Sudoh, Y., 1997. X-ray pole figure analysis of apatite crystals and collagen molecules in bone. Calcif. Tissue Int. 60, 361–367.
Sfihi, H., Rey, C., 2002. 1-D and 2-D heteronuclearmagnetic resonance study of the local structure of type B carbonate fluorapatite. In: Fraissard, F., Lapina, B. (Eds.), Magnetic Resonance in Colloid and Interface Science. NATO Science Series, Kluwer Academic, Dordrecht, The Netherlands, pp. 409–422.
Simkiss, K., 1977. Biomineralization and detoxification. Calcif. Tissue Res. 24, 199–200.
Skwarek, E., Janusz, W., Sternik, D., 2013. Adsorption of citrate ions on hydroxyapatite synthesized by various methods. J. Radioanal. Nucl. Chem. 299 (3), 2027–2036.
Somrani, S., et al., 2005. Physico-chemical and thermochemical studies of the hydrolytic conversion of amorphous tricalcium phosphate into apatite. J. Solid State Chem. 178 (5), 1337–1348.
Spevak, L., et al., 2013. Fourier Transform infrared spectroscopic imaging parameters describing acid phosphate substitution in biologic hydroxyapatite. Calcif. Tissue Int. 92 (5), 418–428.
Talmage, R.V., Talmage, D.W., 2006. Calcium homeostasis: solving the solubility problem. J. Musculoskelet. Neuronal Interact. 6 (4), 402–407.
Tan, J., Chen, M., Xia, J., 2009. Water-dispersible hydroxyapatite nanorods synthesized by a facile method. Appl. Surf. Sci. 255 (21), 8774–8779.
Taylor, N.W., Sheard, C., 1929. Microscopic and X-ray investigations on the calcification of tissue. J. Biol. Chem. 81, 479–493.
Taylor, M.G., Parker, S.F., Mitchell, P.C.H., 2003. A study by high energy transfer inelastic neutron scattering spectroscopy of the mineral fraction of ox femur bone. J. Mol. Struct. 123–126, .
Termine, J.D., Posner, A.S., 1966. Infrared analysis of rat bone: age dependency of amorphous and crystalline mineral fractions. Science 153, 1523–1525.
Veis, A., Dorvee, J.R., 2013. Biomineralization mechanisms: a new paradigm for crystal nucleation in organic matrices. Calcif. Tissue Int. 93 (4), 307–315.
Wang, Y., et al., 2013. Water-mediated structuring of bone apatite. Nat. Mater. 12 (12), 1144–1153.
Weiner, S., 2006. Transient precursor strategy in mineral formation of bone. Bone 39, 431–433.
Weiner, S., Pryce, P.A., 1986. Disaggregation of bone into crystals. Calcif. Tissue Int. 39 (6), 365–375.
Weiner, S., Traub, W., 1992. Bone structure: from angstroms to microns. FASEB J. 6 (3), 879–885.
Wilson, R., Elliott, J., Dowker, S., 1999. Rietveld refinement of the crystallographic structure of human dental enamel apatites. Am. Mineral. 84, 1406–1414.
Wilson, R.M., et al., 2005. Rietveld refinements and spectroscopic studies of the structure of Ca-deficient apatite. Biomaterials 26, 1317–1327.
Wilson, E.E., et al., 2006. Three structural roles for water in bone observed by solid-state NMR. Biophys. J. 90 (10), 3722–3731.
Wopenka, B., Pasteris, J.D., 2005. A mineralogical perspective on the apatite in bone. Mater. Sci. Eng. C 25 (2), 131–143.
Wu, Y., et al., 1994. A unique protonated phosphate group in bone mineral not present in synthetic calcium phosphates. Identification by phosphorus-31 solid state NMR spectroscopy. J. Mol. Biol. 244, 423–435.
Yoshinaga, J., Suzuki, T., Morita, M., 1989. Sex- and age-related variations in elemental concentrations of contemporary Japanese ribs. Sci. Total Environ. 79, 209–221.

Mineralization processes in hard tissue: Bone

4

J. An, S. Leeuwenburgh, J. Wolke, J. Jansen
Radboud University Nijmegen Medical Center, Biomaterials, Nijmegen, The Netherlands

4.1 Introduction

Bone tissue consists of bone cells and an extracellular matrix (ECM), which is composed of collagen fibrils, noncollagenous proteins (NCPs), and inorganic minerals. Bone has a highly complex mineralized structure, which provides the support for a body to hold its weight. Bone remodeling is the net result; therefore, bone achieves its intricate balance between local formation and resorption. In this chapter, we address the function of bone, the relationship between its main constituents, and focus on the bone mineralization process. From a clinical perspective, we summarize the most common bone diseases, which are generally caused by bone mineralization disorders. Finally, the mineralization of therapeutic biomaterial implants is discussed in the last section. However, the precise mechanism of bone mineralization requires interdisciplinary knowledge, deeper understanding of bone biology, and extensive research in the future.

4.2 Composition of bone

Bone is comprised of a collagenous fibrous matrix which is enriched with bone mineral crystals (Figure 4.1; Hutmacher, 2000; Rho et al., 1998). This complex nanostructure endows bone with exceptional mechanical strength and provides it with the load-bearing capacity that is necessary to protect the internal organs. Moreover, bone also provides a unique microenvironment for a cooperative action among osteoblasts, osteocytes, osteoclasts, hematopoietic cells, mesenchymal cells, and immune cells (Compton and Lee, 2014). Specifically, osteoblasts and osteoclasts are responsible for the processes of bone formation and resorption, respectively (Weiner and Wagner, 1998; Suda et al., 1997).

4.2.1 Bone cells

Bone is a highly complex, metabolically active organ, which contains several cell types. Three of these are mainly involved in the formation and resorption of bone tissue, i.e. osteoblasts, osteoclasts, and osteocytes (Sommerfeldt and Rubin, 2001). Osteoblasts are bone-forming cells derived from mesenchymal stem cells which primarily deposit osteoid prior to bone mineralization (Compton and Lee, 2014). Functionally, osteoblasts are the cells within bone that lay down the ECM and regulate

Biomineralization and Biomaterials. http://dx.doi.org/10.1016/B978-1-78242-338-6.00005-3

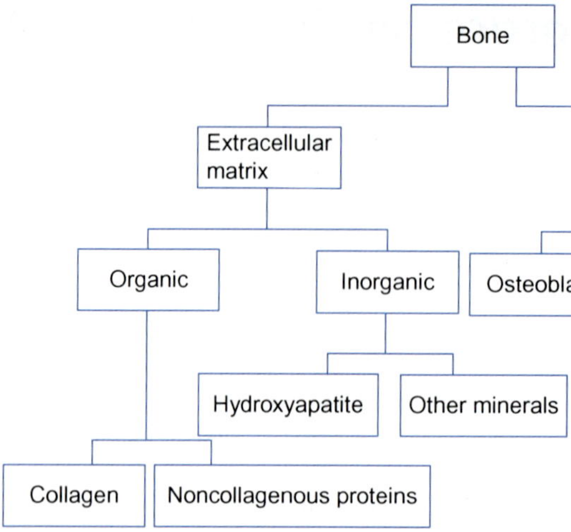

Figure 4.1 Composition of bone.

its mineralization. Morphologically, these cells are cuboidal in shape and located at the bone surface (Sommerfeldt and Rubin, 2001; Ferrari et al., 2000).

Bone resorption is carried out by osteoclasts. Their number and activity are determined by cell lineage allocation, proliferation and differentiation of osteoclast precursors, and the resorptive efficacy of mature osteoclasts (Harada and Rodan, 2003). Osteoclasts are derived from highly migratory, multinucleated and polarized cells, which carry an arsenal of lysosomal enzymes (Teitelbaum, 2000). Osteoclasts are highly specialized cells that contain several unique ultrastructural characteristics, such as pleomorphic mitochondria, vacuoles, and lysosomes (Walker, 1972). Bone resorption and formation are regulated locally by the combined action of osteoblasts and osteoclasts. Several factors, such as parathyroid hormone (PTH), prostaglandin E2 (PGE2), fibroblast growth factor, transforming growth factor beta 1 (TGF-β1), and receptor activator of nuclear factor kappa B (NF-κB) ligand (RANKL) (also known as osteoprotegerin ligand, OPGL; TNF-related activation-induced cytokine, TRANCE), have been shown to stimulate both resorption and formation. Biomechanical processes can also affect both bone resorption and formation, since weakening of bone resulting from, for example, resorption may induce corrective bone formation (Harada and Rodan, 2003).

Osteocytes are the most abundant cells in bone and comprise about 90–95% of all bone cells in the adult skeleton. These cells are matured osteoblasts trapped within the bone matrix. Osteocytes respond to mechanical strain and produce the biochemical signals to orchestrate the resorption and/or formation of bone (Lanyon, 1993). In mature bone, osteocytes are connected with each other via long cytoplasmic extensions that occupy tiny canals called lacunae/canaliculi, which are used for exchange of nutrients and waste. Osteocytes are regularly dispersed throughout the mineralized matrix and approach the bone surface and bone marrow through a dendritic process (Bonewald, 2007; Kamioka et al., 2001). These connections enable recruitment of osteoclast precursors, to stimulate bone resorption and to regulate mesenchymal stem cell differentiation (Baylink et al., 1973; Heino et al., 2004; Zhao et al., 2002a). Various roles have been proposed for both osteocytes and the lacunar/canalicular labyrinth they occupy. These include arrest of fatigue cracks, mineral exchange, osteocytic osteolysis, renewed remodeling activity after release by resorption, hormonal stimulation, and guidance of osteoclastic cutting cones involved in mineral exchange and repair of microdamage, strain detection, and control of mechanically related bone modeling/remodeling (Lanyon, 1993).

4.2.2 Extracellular matrix

The ECM of bone is a complex supramolecular structure composed of organic and inorganic components. The major constituents of the extracellular bone matrix are collagen, NCPs, an inorganic phase consisting of dispersed mineral nanocrystals, and water, thereby forming a template for cell attachment, proliferation, and differentiation (Labat-Robert et al., 1990; Frisch and Ruoslahti, 1997; Schönherr and Hausser, 2000; Velleman, 2000).

4.2.2.1 Collagen

The main constituent of the ECM of bone is collagen, which is present in bone tissue as a fibrillar network. Collagen is the most abundant protein in which three parallel polypeptide strands are arranged in a triple helix. The formation of this secondary structure is facilitated by the repeated sequence glycine–X–Y, where X and Y are most often proline or hydroxyproline (Shoulders and Raines, 2009). So far more than 27 different collagen types have been identified in body tissue. Collagen types I, II, III, V, and XI are found in tissues which are able to resist mechanical forces such as tendon, bone, cartilage, and skin (Velleman, 2000; Gelse et al., 2003; Viguet-Carrin et al., 2006). In bone tissue, collagen type I is abundant and comprises approximately 95% of the entire collagen content. Other collagen types (e.g. collagen types II, III, and V) are present at low levels in bone (Niyibizi and Eyre, 1989, 1994). Collagen gives bones their tensile strength and is spatially arranged to accommodate high stress levels. Collagen fibrils have been regarded as a passive template for mineral formation for decades (Stetlerstevenson and Veis, 1986; Fleisch and Bisaz, 1962). However, recent insights have shown that collagen is actively involved in controlling the mineralization of apatite (Nudelman et al., 2010; Kawska et al., 2008; Landis and Silver, 2008). Collagen type I not only initiates and directs the growth of carbonated apatite mineral but also affects the size, shape, orientation, and three-dimensional distribution of apatite crystals.

4.2.2.2 Noncollagenous proteins

NCPs are embedded in the ECM and function as signaling molecules during the bone mineralization process. The most abundant NCPs are summarized in Table 4.1 (Robey et al., 1993). Osteopontin, osteocalcin, bone sialoprotein (BSP), bone acidic glycoprotein, thrombospondin, biglycan, decorin, and osteonectin are the principal NCPs of

Table 4.1 Structural and functional characteristics of noncollagenous proteins present in bone

Noncollagenous protein	Structural features	Function
Osteopontin	RGD amino acid sequence, phosphorylation	Cell attachment, HA binding
Bone sialoprotein	RGD amino acid sequence, sulfation of tyrosines	Cell attachment
Bone acidic glycoprotein	Polyaspartic acid	Mineralization
Thrombospondin	RGD amino acid sequence, EGF homology	Cell attachment
Biglycan	Leucine repeat structure, two GAG chains near NH_2 terminus	Cell–cell or cell–protein interactions
Decorin	Leucine repeat structure, one GAG chain near NH_2 terminus	Binds collagen, regulates fibril formation
Osteonectin	EF hand motif, NH_2-terminal domain rich in asparagines and glutamate	Ca^{2+} and hydroxyapatite binding, cell spreading
Osteocalcin	Gamma carboxylation of glutamic acid	Bone turnover

bone (Young et al., 1992). Extensive investigations have focused on NCPs due to their potential influence on cell attachment, binding to the inorganic phase, and mineralization of bone (Delany et al., 2000; Ducy et al., 2000). Osteocalcin binds strongly to bone mineral, while osteopontin was shown to be essential for cell adhesion via an acid sequence in the center of the molecule (Oldberg et al., 1986). These striking features render osteopontin bone bridging, thereby being able to attach to both cells and hydroxyapatite (HA). Osteocalcin is a low-molecular-weight bone matrix protein (about 5.8 kDa) found exclusively in bone and teeth. This protein contains three γ-carboxyglutamic acid residues (Price, 1988). These acidic residues provide the protein with calcium-binding properties, thereby facilitating tight bonding to HA. The protein was suggested to have an important role in the assembly of mineralized bone, most likely by participating in the regulation of HA crystal growth (Heinegård and Oldberg, 1989). Osteopontin contains a stretch of negatively charged amino acids as well as a cell-binding RGD (Arg-Gly-Asp) tripeptide amino acid sequence near its N-terminal region (Oldberg et al., 1988). Poundarik et al. found that osteocalcin and osteopontin function in both independent and dependent ways to regulate various aspects of bone mineralization, including crystal size, morphology, organization, and composition (Poundarik, 2013).

BSP is a highly post translationally modified protein that is the most abundant NCP in human bone (Fisher et al., 1983). Unlike osteopontin, the main acidic amino acid sequence is located at the C-terminus in BSP, in the form of one continuous stretch of 10 glutamic acid residues and several shorter segments of this amino acid. This sequence provides calcium-binding properties to the protein (Ryden et al., 1987). In addition, almost half of the serine residues in the protein are phosphorylated, so the protein can be classified as a phosphoprotein (Heinegård and Oldberg, 1989). In addition, BSP has high concentrations of tyrosine residues, which is rich in O-glycosidically linked oligosaccharides, which match the sequence for sulfation (Ecarot-Charrier et al., 1989; Midura et al., 1990). Research has indicated that BSP is a late marker for bone differentiation. Furthermore, this NCP (i) binds with high affinity to certain strains of *Staphylococcus aureus* isolated from patients with infected bone tissue, (ii) plays a role in matrix attachment, and (iii) affects cell adhesion to bone tissue (Young et al., 1992; Bianco et al., 1990; Ryden et al., 1989).

4.2.2.3 Inorganic phase

The inorganic phase of bone, also called bone mineral, is composed of HA crystals and has the formula $Ca_{10}(PO_4)_6(OH)_2(OH)$ (Fujisawa and Tamura, 2011; Boskey, 1998). This apatitic mineral in bone tissue has a plate-like crystal morphology with dimensions ranging between 20 and 80 nm in length, 15–30 nm in width, and 2–10 nm in thickness (Nudelman et al., 2010; Farbod et al., 2013; Beniash, 2011; Palmer et al., 2008; Glimcher, 2006). This morphological characteristic renders HA reactive, soluble, and chemically sensitive. However, bone crystals are not formed by calcium and phosphate ions only, since bone mineral consists of 4–6% of carbonate together with magnesium, sodium, potassium, and other ions. The structure and composition of bone minerals can become more crystalline with increasing age and maturation.

4.3 Bone mineralization process

Bone mineralization is the process by which mineral crystals are deposited in an organized fashion onto the organic ECM. This matrix surrounding the cells provides a template for mineral deposition and defines the sites where mineralization will commence, as well as the final dimensions of the crystals. Even though a plethora of studies have been dedicated to elucidate the mechanisms that determine where, when, and how mineral crystals form in each of these organisms, the exact mechanism of bone mineralization remains to be elucidated (Beniash, 2011; Boskey, 1996, 1998).

According to the classical theories on biomineralization, NCPs are actively involved in the matrix mineralization process. In the early 1990s, Hunter and Goldberg (1994) hypothesized that the mineralization effects of BSP are exclusively related to acidic, aspartate- and glutamate-rich peptide sequences. Later studies considered whether other groups (such as phosphate, sulfate, and sialic acid groups) are also essential factors for mineralization (Stubbs et al., 1997). According to the classical theories, the nucleation of inorganic minerals is considered to occur in supersaturated solutions of their constituent ions. This results in the nucleation of the solid phase by stochastic solute clustering, where a cluster of critical size is regarded as the initial crystal precursor (Gebauer et al., 2008; Pouget et al., 2009).

Currently, two main theories account for the mineralization of bone tissue, that is, (i) collagen template-mediated mineralization and (ii) matrix vesicle-mediated matrix mineralization. It is generally accepted that physiologic mineral deposition, that is, the formation of mineral in an organized fashion in bone, does not occur without a matrix. The matrix provides a hierarchical template for mineral deposition, and may participate directly in the mineralization process by serving as a nucleator. It is also well accepted that the matrix is not the same in all mineralized tissues (Chai et al., 2012). Matrix vesicles, on the other hand, are extracellular membrane-derived particles, 100 nm in diameter, which are selectively located within the matrix of bone, cartilage, and predentin. They are generated by polarized budding and release from the surfaces of chondrocytes, osteoblasts, and odontoblasts, and serve as the initial site of calcification in all skeletal tissues (Anderson, 1995, 2003). In bone tissue, matrix vesicles are regarded as the site of initial mineral deposition (Landis, 2005; Aparicio et al., 2002). The generation of initial HA mineral crystals occurs within the matrix vesicle membrane during mineralization under cellular control (Boskey, 2007).

4.3.1 Collagen template-mediated mineralization

Collagen provides the template for mineral deposition in bone tissue, where the size and the organization of the collagen fibrils determine the dimensions that mineral crystals can attain. However, without the NCPs the mineralization process generally does not proceed. As mentioned in Section 4.2.2.2, NCPs function as signaling molecules during the bone matrix mineralization process (Palmer et al., 2008; Chai et al., 2012). Specifically, BSP was reported to act as a crystal nucleator (Stubbs et al., 1997; Hunter et al., 1994; He et al., 2005; Monfoulet et al., 2010), osteocalcin was proposed to affect remodeling and recognition of mineral surfaces (Farbod et al., 2013; Meyer et al., 1999; Hoang et al., 2003), while osteopontin and osteonectin were suggested to regulate the size, type, and growth of the formed crystals (Roach, 1994; Rosenthal et al., 2007). In the periodic 67 nm cross-striated pattern of the collagen fibril (Miller et al., 2005; Orgel et al., 2006), the less dense 40-nm-long gap zone was suggested as the site of crystal nucleation from an amorphous phase followed by a subsequent growth phase. As mentioned above, this process was suggested to be directed by highly acidic NCPs (Nudelman et al., 2010; Glimcher and Muir, 1984; Landis et al., 1993; Traub et al., 1992). The theory of collagen template-mediated mineralization emphasizes the activity of collagen fibrils in bone mineralization. Briefly, by applying high-resolution cryo-transmission electron microscopy in the observation of mineralizing collagen fibrils, soluble polyaspartic acid was shown to play an important role in the mineralization of collagen (Nudelman et al., 2010; Cölfen, 2010). This acidic polypeptide stabilized prenucleation clusters, which were found to be composed of calcium triphosphate ion-association complexes (Gebauer et al., 2008). The resulting negatively charged prenucleation clusters were attracted toward positively charged regions within collagen (Nudelman et al., 2010). Subsequently, these complexes diffused into the collagen fibril and solidified into an amorphous phase, and were subsequently altered into oriented apatite crystals directed by the collagen fibril arrangement (Chai et al., 2012). Figure 4.2a schematically depicts the above-described process for biomineralization of collagen fibril. Chapter 7 provides further discussion about this mechanism of bone mineralization.

4.3.2 Matrix vesicle-mediated mineralization

Bone mineralization proceeds according to a matrix vesicle-mediated mechanism *in vivo* (Stewart, 1969). In this process, the initial phase of the bone mineralization

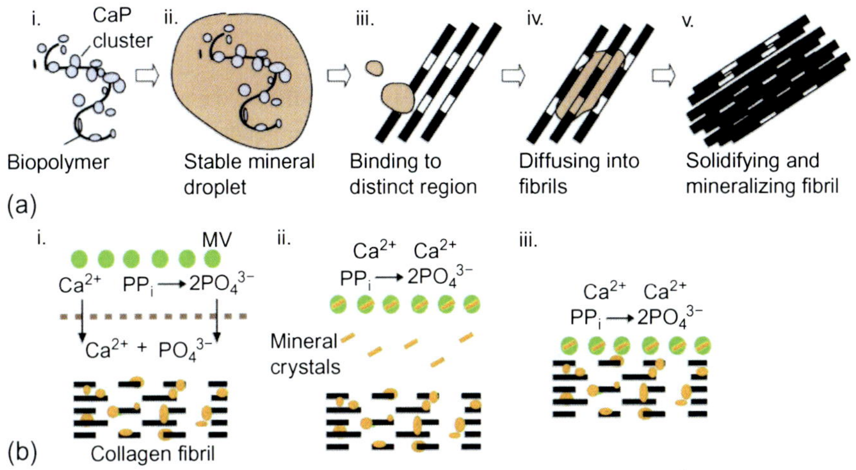

Figure 4.2 Mechanisms for mineralization of collagen fibrils. (a) Collagen templated-mediated mineralization and (b) matrix vesicle-mediated mineralization.
Reprinted from Chai et al. (2012) with permission.

concerns the formation of Ca^{2+} ions and inorganic phosphate (Pi) within mineral vesicles (Anderson, 1995; Wu et al., 1989). Calcium-binding molecules such as annexins, phosphatidylserine, and BSP attract and control the accumulation of calcium and phosphate ions prior to the formation of insoluble HA crystals (Wu et al., 1995; Hunter and Goldberg, 1993). In this phase, the intravesicular pH may be raised by the action of carbonic anhydrase that is concentrated in mineral vesicles (Stechschulte et al., 1992) and stabilizes the initial mineral crystals (Anderson, 2003). The second phase of matrix-mediated mineralization involves the breakdown of the mineral vesicle membranes, thereby exposing the preformed HA to the extracellular fluid (Stewart, 1969).

The extracellular fluid containing matrix vesicles contains homeostatically maintained levels of Ca^{2+} and PO_4^{3-} sufficient to support propagation of new HA crystals on preformed HA crystals. Perforation of the matrix vesicle membranes is facilitated by phospholipases (Wuthier, 1973) and proteases (Hirschman et al., 1983). Propagation of mineral into the perivesicular matrix is enabled by metalloproteinases of mineral vesicles, which are capable of degrading mineral-inhibiting proteoglycans (Dean et al., 1992).

Recent work indicates that collagen types II and X are bound to the outer surfaces of matrix vesicles and serve as a bridge for crystal propagation into the extravesicular matrix (Wu et al., 1991). Three possible functions of matrix vesicles have been suggested in the process of matrix vesicle-mediated mineralization (Figure 4.2b):

1. Matrix vesicles solely regulate ion concentrations in the matrix to initiate mineral formation in collagen fibrils.
2. Matrix vesicles directly regulate ionic composition and form intravesicular apatite crystals. The mineralization process is initiated upon their migration to collagen.
3. Mineral deposition into the fibrils due to the direct interaction of mineral vesicles with collagen.

4.4 Pathological mineralization

The bone mineralization process is essential for the hardness and strength of bone (Yeni et al., 1998). If this process is not properly regulated, the resulting mineralization will be either insufficient or excessive. As a consequence, the quality of bone tissue can be compromised. Such pathological bone mineralization can be the main cause of many bone diseases.

Osteomalacia is a metabolic bone disease caused by lack of bone mineral or overactive bone resorption in adults. Osteomalacia in children is known as rickets, therefore osteomalacia is referring to the adult form of the disease. As explained in Section 4.2, in healthy bone, osteoclasts remove bone, while osteoblasts deposit osteoid and are active in the mineralization process. In osteomalacia, there is a decrease in the rate of calcification, but an increase in bone surface area, which is covered with unmineralized osteoid. Symptoms of osteomalacia include body pains, muscle weaknesses, and bone fragility (Parfitt, 1990; Feng et al., 2006; Aaron et al., 1974).

Fibrous osteodystrophy is another common type of bone disease caused by pathological mineralization. Several *in vivo* studies have confirmed that bone becomes flexible and deformed due to the continuous and excessive exposure of PTH, thus reducing its load-bearing capacity (Lynch et al., 1999; Fincham et al., 1985; Long et al., 1975; Flom et al., 1978; Jaffe et al., 1931). Osteoclasia, which is defined as the destruction and resorption of bony tissue by osteoclasts, becomes a pathognomonic feature when the remodeling space formerly occupied by calcified bone is filled by fibrous connective tissue. With the progress of this disease, cortical bone is replaced by unmineralized woven bone. The mineralization of osteoid shows a distinctive feature, namely the remodeling space formerly occupied by osseous tissue (Fetter et al., 1985).

Osteitis deformans, which is widely known as Paget's disease, is a chronic and disorder of adult bone (Whyte, 2006; Siris, 1998). In general, Paget's disease typically manifests from middle age onward and is slightly more commonly seen in men than in women (Whyte, 2006; Kanis, 1998; Altman et al., 1973). Paget's disease accelerated bone resorption followed by the deposition of a thick and chaotically mineralized bone matrix. This process leads to weakening of the bone structure, which results in pain, misshapen bones, fractures, and arthritis in the joints in the vicinity of affected bones. Occasionally, Paget's disease develops into a primary bone cancer known as Paget's sarcoma. The pathological mineralization of Paget's disease is initiated by numerous multinucleated osteoclasts of large size and high resorptive activity (Rebel et al., 1976). In addition to this resorption process, osteoblasts deposit bone, ultimately resulting in disorganized bone remodeling (Meunier et al., 1980). This localized, disorganized bone remodeling results in irregular cortical thickening, trabecular coarsening, and vascular fibrous tissue, which is not able to withstand high stresses (Siris, 1998; Merkow and Lane, 1990; Ooi and Fraser, 1997).

The most common bone disorder is osteoporosis, which is characterized by a low mass, structural deterioration, and increased porosity of bone issue (Kanis et al., 2007). In contrast to Paget's disease, osteoporosis usually affects all bones in the body. Bone is known to become more brittle when the mineral content is reduced below a critical value, resulting in reduced load-bearing capacity and toughness of bone

(Gao et al., 2003). Bone mineral density (BMD) is directly related to mechanical strength, and a decreased BMD is associated with increased fracture risks for patients with osteoporosis (Faibish et al., 2006; Cefalu, 2004; McCalden and McGeough, 1997; Koh and Ng, 2002).

Three main mechanisms for the development of osteoporosis can be identified. First, osteoporosis may be caused by underdevelopment during the growth of individuals. Second, osteoporosis develops as a result of excessive bone resorption and, third, upon insufficient new bone formation during remodeling (Chiappelli, 2011). Osteoporosis was previously considered as the consequence of aging, but recent evidence suggests that osteoporosis is multidetermined with potential causes such as malnutrition, chronic acidosis, excessive use of hormones, or changes in biomechanical loading (NIH Consensus Development Panel on Osteoporosis Prevention, Diagnosis, and Therapy, 2001).

4.5 Mineralization of synthetic biomaterials

Biomaterials have been developed for several decades to substitute for diseased or damaged bone tissue. Since the unique properties of bone tissue are mainly related to its mineralized nature, understanding and controlling the mineralization behavior of synthetic bone substitutes is pivotal. Therefore, the mineralization behavior of synthetic biomaterials, which is described in detail in Part 2 of this book, is briefly reviewed in the following section.

4.5.1 *Inorganic materials*

Due to the similarity with the inorganic phase of bone mineral (Table 4.2; Sáenz et al., 1999), calcium phosphate-based materials (CaPs)—and in particular HA and bioactive glasses—have been extensively investigated as synthetic bone substitutes (Jarcho, 1981). In the late 1960s and early 1970s, Hench et al. (1971) showed that bioactive glasses bonded to bone tissue without the formation of surrounding fibrous tissue. Since then, several types of ceramics such as sintered HA (Jarcho et al., 1977), sintered tricalcium phosphate (Ducheyne and De Groot, 1981), apatite/β-tricalcium phosphate biphasic ceramics (LeGeros et al., 2003), and glass–ceramic composites (Kitsugi et al., 1986) were also shown to bond to living bone, and they are used clinically on a routine basis for reconstructive and regenerative surgery (Kokubo and Takadama, 2006).

4.5.1.1 *Calcium phosphates*

Bioactive CaP-based ceramics are often applied for bone tissue engineering purposes and widely used in orthopedics and dentistry. Bioactive CaPs are most frequently used as bulk bone substitute or as coatings on top of titanium implants. Exposure to fluids that mimic the composition of blood plasma is often regarded as a first step toward understanding the mineralization behavior of synthetic biomaterials. In this context, an

Table 4.2 **Composition of bone and synthetic hydroxyapatite**

Compound	Bone	Synthetic hydroxyapatite
Calcium	35.5	39.0
Phosphorus	17.1	18.5
Carbon dioxide	4.4	–
Magnesium	0.9	–
Sodium	1.1	–
Potassium	0.1	–
Chlorine	0.1	–
Fluorine	0.02	–
Sulfur	0.6	–
Zinc	–	–
Silicon	0.04	–
Ca/P ratio	1.71	1.667
Crystallinity	33–37	80–100

acellular fluid called simulated body fluid (SBF), with ion concentrations nearly equal to those in human blood plasma, has been extensively used to study mineralization for biomaterials *in vitro*. In 1991, Kokubo et al. proposed that the essential requirement for an artificial material to bond to living bone is the formation of bone-like apatite on its surface, when implanted in the living body, and that this *in vivo* apatite formation can be reproduced in SBF (Kokubo, 1991). Kokubo et al. postulated that the *in vivo* bone response of a synthetic biomaterial can be predicted from the apatite formation on its surface in SBF, but this assumption has now been challenged (Bohner and Lemaitre, 2009).

With respect to their tissue response, calcium phosphate bioceramics display an osteoconductive bone response characterized by guidance of bone tissue along the surface of synthetic biomaterials in orthotopic implantation sites, the possible mechanism of mineralization on the surface of HA surfaces was proposed by Bertazzo et al. (2010), which is shown in Figure 4.3. Recently, certain types of CaP ceramics were also shown to possess osteoinductive properties (i.e. capable of forming bone at ectopic implantation sites). The exact mechanism of osteoinduction by CaP remains, however, to be elucidated although several factors such as topography, geometry, composition, and porosity were proposed to be responsible for osteoinduction.

4.5.1.2 Bioactive glasses

In 1969, reactive silicate-based glasses were shown be bond to bone tissue for the first time (Hench and Wilson, 1984). During the following decades, several melt- and sol–gel-derived glasses such as Bioglass® 45S5 (composed of SiO_2, Na_2O, CaO, and P_2O_5) were translated toward clinical applications due to their favorable bone response. To understand how bioglasses interact with the local environment *in vivo*, bioglasses have been immersed in SBF followed by physicochemical analysis of the mineralization of the material surfaces. It appeared that carbonated apatite layers were deposited onto bioglass, which revealed chemical similarity to bone mineral. Similar to the mineral-

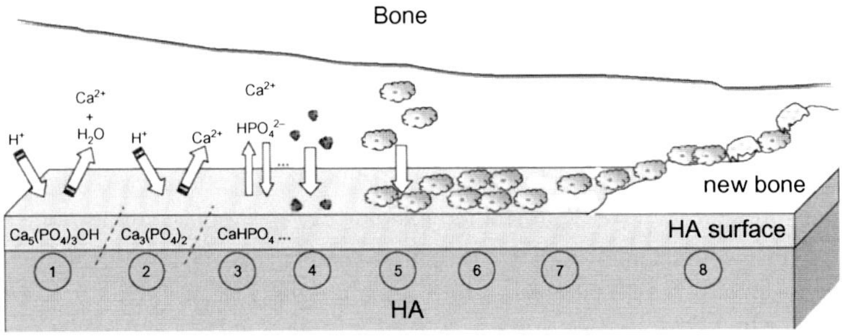

Figure 4.3 Proposed mechanism of hydroxyapatite mineralization after implantation: (1) Solubilization of the surface of hydroxyapatite (1 and 2). (2) Ion exchange (3 and 4). (3) Cell adhesion and protein adsorption (5 and 6). (4) New bone formation (7 and 8). Reprinted from Bertazzo et al. (2010) with permission.

ization behavior of calcium phosphate ceramics, it has also been suggested for bioactive glasses that this apatite layer is responsible for the formation of a chemical bond between bioglass and host bone (Gross et al., 1988).

4.5.2 Organic materials

In addition to inorganic materials, polymers such as collagen and hyaluronic acid, polylactic acid (PLA), polyglycolic acid (PGA), and copolymers made thereof are interesting candidates for tissue engineering. These polymers generally do not match the biological performance of calcium phosphates in terms of osteoconductive and inductive properties, but these polymers are often used as matrix component for the development of bioinspired polymer–ceramic nanocomposites for bone substitution.

4.5.2.1 Hydrated biomaterials

Hydrogels are polymer networks swollen with water. Due to their high water absorption capacity, hydrogels have potential applications in a wide variety of field such as tissue engineering, cell culture, and drug delivery. However, for bone regenerative applications, hydrogels are associated with a number of disadvantages such as their inherent mechanical weakness which limit their applicability to nonload-bearing sites (Hoffman, 2002; Griffith, 2000). Furthermore, many hydrogels are difficult to sterilize due to their high water content (Driessens et al., 1994). In addition, hydrogels are unable to mineralize and cannot form chemical bonds with hard tissue such as bone (Mano et al., 2004; Gkioni et al., 2010).

As such, current research focuses on the development of hydrogels (natural or synthetic) with the ability to mineralize. Besides simple incorporation of inorganic (nano) particles into hydrogel matrices, several biomimetic strategies have been adopted to mineralize hydrogels, including soaking of hydrogels in solutions that are saturated with respect to calcium phosphate, incorporation of enzymes that catalyze deposition of bone mineral, and incorporation of synthetic analogs to matrix vesicles that act as the initial sites of biomineralization. In addition, functionalization of the polymeric hydrogel backbone with negatively charged groups is a third mechanism to attract calcium cations and promote mineralization in otherwise inert hydrogels (Gkioni et al., 2010).

4.5.2.2 Nonhydrated synthetic polymers

Synthetic polymers such as polyfumarates, PLA, PGA, polylactic-*co*-glycolic acid, and polycaprolactone have been studied extensively for application in bone regenerative surgery. Scaffolds made of these polymers can be processed using techniques such as porogen leaching (Mikos et al., 1993), gas foaming (Ma and Choi, 2001), phase separation (Zhao et al., 2002b), fiber meshing (Langer and Vacanti, 1998), supercritical fluid processing (Ginty et al., 2006), microsphere sintering, and three-dimensional printing (Sherwood et al., 2002) to generate a range of three-dimensional scaffolds with different porosities and surface characteristics (Stevens, 2008). Mineralization of nonhydrated polymers can be achieved by several methods including soaking in SBF or anionic surface modification to stimulate accumulation of calcium cations in the vicinity of the biomaterial surface.

Similar to inorganic biomaterials, SBF is also used to allow for the evaluation of the mineralization behavior of solid polymers or to create mineral coatings on top of otherwise inert polymers. The type of SBF used for mineralization influences the chemical composition and morphology of the resultant mineral phase, i.e. the use of more concentrated SBFs accelerates the mineralization process. Anionic proteins were shown to play a role in the regulation of mineral phase deposition in natural bone and affect mineralization in SBF. Therefore, solid polymers have also been functionalized with anionic groups or bone-specific proteins to induce mineralization (Tanahashi and Matsuda, 1997; Stephansson et al., 2002; Thorwarth et al., 2005; Filmon et al., 2002; Song et al., 2005).

4.6 Conclusion

Bone is a highly complex and hierarchically structured tissue which provides mechanical support and protection for the internal organs and brain due to its mineralized nature. Mineralization of bone proceeds via both collagen template-mediated and matrix vesicle-mediated pathways in a complex interplay that is still not fully understood. Increased understanding of the mineralization of bone tissue will provide insight into various bone diseases that are caused by pathological mineralization as well as the bio-inspired development of a novel generation of synthetic bone substitutes that exhibit biological performance levels equal to native bone tissue.

References

Aaron, J., et al., 1974. Frequency of osteomalacia and osteoporosis in fractures of the proximal femur. Lancet 303 (7851), 229–233.

Altman, R.D., et al., 1973. Influence of disodium etidronate on clinical and laboratory manifestations of Paget's disease of bone (osteitis deformans). N. Eng. J. Med. 289 (26), 1379–1384.

Anderson, H.C., 1995. Molecular biology of matrix vesicles. Clin. Orthop. Relat. Res. 314, 266–280.
Anderson, H.C., 2003. Matrix vesicles and calcification. Curr. Rheumatol. Rep. 5 (3), 222–226.
Aparicio, S., et al., 2002. Optimal methods for processing mineralized tissues for Fourier transform infrared microspectroscopy. Calcif. Tissue Int. 70 (5), 422–429.
Baylink, D., et al., 1973. Vitamin D-enhanced osteocytic and osteoclastic bone resorption. Am. J. Physiol. 224 (6), 1345–1357.
Beniash, E., 2011. Biominerals—hierarchical nanocomposites: the example of bone. Wiley Interdiscip. Rev. Nanomed. Nanobiotechnol. 3 (1), 47–69.
Bertazzo, S., et al., 2010. Hydroxyapatite surface solubility and effect on cell adhesion. Colloids Surf. B Biointerfaces 78 (2), 177–184.
Bianco, P., et al., 1990. Expression and localization of the two small proteoglycans biglycan and decorin in developing human skeletal and non-skeletal tissues. J. Histochem. Cytochem. 38 (11), 1549–1563.
Bohner, M., Lemaitre, J., 2009. Can bioactivity be tested *in vitro* with SBF solution? Biomaterials 30 (12), 2175–2179.
Bonewald, L.F., 2007. Osteocytes as dynamic multifunctional cells. Ann. N. Y. Acad. Sci. 1116 (1), 281–290.
Boskey, A.L., 1996. Matrix proteins and mineralization: an overview. Connect. Tissue Res. 35 (1–4), 357–363.
Boskey, A.L., 1998. Biomineralization: Conflicts, challenges, and opportunities. J. Cell. Biochem. 72 (S30–31), 83–91.
Boskey, A.L., 2007. Mineralization of bones and teeth. Elements 3 (6), 385–391.
Cefalu, C.A., 2004. Is bone mineral density predictive of fracture risk reduction? Curr. Med. Res. Opin. 20 (3), 341–349.
Chai, Y.C., et al., 2012. Current views on calcium phosphate osteogenicity and the translation into effective bone regeneration strategies. Acta Biomater. 8 (11), 3876–3887.
Chiappelli, F., 2011. Osteoimmunopathology: Evidence-Based Perspectives from Molecular Biology to Systems Biology. Springer, Dordrecht.
Cölfen, H., 2010. Biomineralization: a crystal-clear view. Nat. Mater. 9 (12), 960–961.
Compton, J.T., Lee, F.Y., 2014. A review of osteocyte function and the emerging importance of sclerostin. J. Bone Joint Surg. Am. 96 (19), 1659–1668.
Dean, D.D., et al., 1992. Matrix vesicles are enriched in metalloproteinases that degrade proteoglycans. Calcif. Tissue Int. 50 (4), 342–349.
Delany, A., et al., 2000. Osteopenia and decreased bone formation in osteonectin-deficient mice. J. Clin. Invest. 105 (7), 915–923.
Driessens, F., et al., 1994. Effective formulations for the preparation of calcium phosphate bone cements. J. Mater. Sci. Mater. Med. 5 (3), 164–170.
Ducheyne, P., De Groot, K., 1981. *In vivo* surface activity of a hydroxyapatite alveolar bone substitute. J. Biomed. Mater. Res. 15 (3), 441–445.
Ducy, P., et al., 2000. Leptin inhibits bone formation through a hypothalamic relay: a central control of bone mass. Cell 100 (2), 197–207.
Ecarot-Charrier, B., Bouchard, F., Delloye, C., 1989. Bone sialoprotein II synthesized by cultured osteoblasts contains tyrosine sulfate. J. Biol. Chem. 264 (33), 20049–20053.
Faibish, D., Ott, S.M., Boskey, A.L., 2006. Mineral changes in osteoporosis: a review. Clin. Orthop. Relat. Res. 443, 28.
Farbod, K., et al., 2013. Interactions between inorganic and organic phases in bone tissue as a source of inspiration for design of novel nanocomposites. Tissue Eng. Part B Rev. 20 (2), 173–188.
Feng, J.Q., et al., 2006. Loss of DMP1 causes rickets and osteomalacia and identifies a role for osteocytes in mineral metabolism. Nat. Genet. 38 (11), 1310–1315.

Ferrari, S.L., et al., 2000. A role for N-cadherin in the development of the differentiated osteoblastic phenotype. J. Bone Miner. Res. 15 (2), 198–208.
Fetter, A., Siemering, G., Riser, W., 1985. Osteoporosis and osteopetrosis. In: Newton, C.D., Nunamaker, D.M. (Eds.), Textbook of Small Animal Orthopaedics. Lippincott, Philadelphia, PA, pp. 627–629.
Filmon, R., et al., 2002. Effects of negatively charged groups (carboxymethyl) on the calcification of poly(2-hydroxyethyl methacrylate). Biomaterials 23 (14), 3053–3059.
Fincham, J., et al., 1985. Mseleni joint disease. Part I. An animal model? S. Afr. Med. J. 67 (2), 51–57.
Fisher, L.W., et al., 1983. Matrix sialoprotein of developing bone. J. Biol. Chem. 258 (20), 12723–12727.
Fleisch, H., Bisaz, S., 1962. Mechanism of calcification: inhibitory role of pyrophosphate. Nature 1, 195, 911.
Flom, J.O., Brown, R.J., Jones, R.E., 1978. Fibrous osteodystrophy in a wild dolphin. J. Am. Vet. Med. Assoc. 173 (9), 1124–1126.
Frisch, S.M., Ruoslahti, E., 1997. Integrins and anoikis. Curr. Opin. Cell Biol. 9 (5), 701–706.
Fujisawa, R., Tamura, M., 2011. Acidic bone matrix proteins and their roles in calcification. Front. Biosci. 17, 1891–1903.
Gao, H., et al., 2003. Materials become insensitive to flaws at nanoscale: lessons from nature. Proc. Natl. Acad. Sci. U.S.A. 100 (10), 5597–5600.
Gebauer, D., Völkel, A., Cölfen, H., 2008. Stable prenucleation calcium carbonate clusters. Science 322 (5909), 1819–1822.
Gelse, K., Pöschl, E., Aigner, T., 2003. Collagens—structure, function, and biosynthesis. Adv. Drug Deliv. Rev. 55 (12), 1531–1546.
Ginty, P.J., et al., 2006. Mammalian cell survival and processing in supercritical CO_2. Proc. Natl. Acad. Sci. U.S.A. 103 (19), 7426–7431.
Gkioni, K., et al., 2010. Mineralization of hydrogels for bone regeneration. Tissue Eng. Part B Rev. 16 (6), 577–585.
Glimcher, M.J., 2006. Bone: Nature of the calcium phosphate crystals and cellular, structural, and physical chemical mechanisms in their formation. Rev. Mineral. Geochem. 64 (1), 223–282.
Glimcher, M., Muir, H., 1984. Recent studies of the mineral phase in bone and its possible linkage to the organic matrix by protein-bound phosphate bonds. Philos. Trans. R. Soc. B 304 (1121), 479–508.
Griffith, L., 2000. Polymeric biomaterials. Acta Mater. 48 (1), 263–277.
Gross, U., Schmitz, H.J., Strunz, V., 1988. Surface activities of bioactive glass, aluminum oxide, and titanium in a living environment. Ann. N. Y. Acad. Sci. 523 (1), 211–226.
Harada, S.-I., Rodan, G.A., 2003. Control of osteoblast function and regulation of bone mass. Nature 423 (6937), 349–355.
He, G., et al., 2005. Phosphorylation of phosphophoryn is crucial for its function as a mediator of biomineralization. J. Biol. Chem. 280 (39), 33109–33114.
Heinegård, D., Oldberg, A., 1989. Structure and biology of cartilage and bone matrix noncollagenous macromolecules. FASEB J. 3 (9), 2042–2051.
Heino, T.J., Hentunen, T.A., Väänänen, H.K., 2004. Conditioned medium from osteocytes stimulates the proliferation of bone marrow mesenchymal stem cells and their differentiation into osteoblasts. Exp. Cell Res. 294 (2), 458–468.
Hench, L.L., Wilson, J., 1984. Surface-active biomaterials. Science 226 (4675), 630–636.
Hench, L.L., et al., 1971. Bonding mechanisms at the interface of ceramic prosthetic materials. J. Biomed. Mater. Res. 5 (6), 117–141.

Hirschman, A., et al., 1983. Neutral peptidase activities in matrix vesicles from bovine fetal alveolar bone and dog osteosarcoma. Calcif. Tissue Int. 35 (1), 791–797.
Hoang, Q.Q., et al., 2003. Bone recognition mechanism of porcine osteocalcin from crystal structure. Nature 425 (6961), 977–980.
Hoffman, A.S., 2002. Hydrogels for biomedical applications. Adv. Drug Deliv. Rev. 54 (1), 3–12.
Hunter, G.K., Goldberg, H.A., 1993. Nucleation of hydroxyapatite by bone sialoprotein. Proc. Natl. Acad. Sci. U.S.A. 90 (18), 8562–8565.
Hunter, G.K., Goldberg, H.A., 1994. Modulation of crystal formation by bone phosphoproteins: role of glutamic acid-rich sequences in the nucleation of hydroxyapatite by bone sialoprotein. Biochem. J. 302, 175–179.
Hunter, G.K., Kyle, C.L., Goldberg, H.A., 1994. Modulation of crystal formation by bone phosphoproteins: structural specificity of the osteopontin-mediated inhibition of hydroxyapatite formation. Biochem. J. 300, 723–728.
Hutmacher, D.W., 2000. Scaffolds in tissue engineering bone and cartilage. Biomaterials 21 (24), 2529–2543.
Jaffe, H., Bodansky, A., Blair, J.E., 1931. Fibrous osteodystrophy (osteitis fibrosa) in experimental hyperparathyroidism of guinea-pigs. Arch. Pathol. 11, 207.
Jarcho, M., 1981. Calcium phosphate ceramics as hard tissue prosthetics. Clin. Orthop. Relat. Res. 157, 259–278.
Jarcho, M., et al., 1977. Tissue, cellular and subcellular events at a bone-ceramic hydroxylapatite interface. J. Bioeng. 1 (2), 79–92.
Kamioka, H., Honjo, T., Takano-Yamamoto, T., 2001. A three-dimensional distribution of osteocyte processes revealed by the combination of confocal laser scanning microscopy and differential interference contrast microscopy. Bone 28 (2), 145–149.
Kanis, J.A., 1998. Pathophysiology and Treatment of Paget's Disease of Bone. Taylor & Francis, Oxford.
Kanis, J., et al., 2007. The use of clinical risk factors enhances the performance of BMD in the prediction of hip and osteoporotic fractures in men and women. Osteoporos. Int. 18 (8), 1033–1046.
Kawska, A., et al., 2008. The nucleation mechanism of fluorapatite–collagen composites: Ion association and motif control by collagen proteins. Angew. Chem. Int. Ed. 47 (27), 4982–4985.
Kitsugi, T., et al., 1986. Bone bonding behavior of three kinds of apatite containing glass ceramics. J. Biomed. Mater. Res. 20 (9), 1295–1307.
Koh, L., Ng, D., 2002. Osteoporosis risk factor assessment and bone densitometry–current status and future trends. Ann. Acad. Med. Singapore 31 (1), 37–42.
Kokubo, T., 1991. Bioactive glass ceramics: properties and applications. Biomaterials 12 (2), 155–163.
Kokubo, T., Takadama, H., 2006. How useful is SBF in predicting *in vivo* bone bioactivity? Biomaterials 27 (15), 2907–2915.
Labat-Robert, J., Bihari-Varga, M., Robert, L., 1990. Extracellular matrix. FEBS Lett. 268 (2), 386–393.
Landis, W., 2005. Chemistry and Biology of Mineralized Tissues. University of Toronto Press, Toronto.
Landis, W.J., Silver, F.H., 2008. Mineral deposition in the extracellular matrices of vertebrate tissues: Identification of possible apatite nucleation sites on type I collagen. Cells Tissues Organs 189 (1–4), 20–24.
Landis, W., et al., 1993. Mineral and organic matrix interaction in normally calcifying tendon visualized in three dimensions by high-voltage electron microscopic tomography and graphic image reconstruction. J. Struct. Biol. 110 (1), 39–54.

Langer, R.S., J.P. Vacanti, 1998. Preparation of three-dimensional fibrous scaffold for attaching cells to produce vascularized tissue *in vivo*. Google Patents.

Lanyon, L., 1993. Osteocytes, strain detection, bone modeling and remodeling. Calcif. Tissue Int. 53 (1), S102–S107.

LeGeros, R., et al., 2003. Biphasic calcium phosphate bioceramics: preparation, properties and applications. J. Mater. Sci. Mater. Med. 14 (3), 201–209.

Long, G.G., et al., 1975. Fibrous osteodystrophy in an opossum. J. Wildl. Dis. 11 (2), 221–223.

Lynch, M.J., et al., 1999. Fibrous osteodystrophy in dromedary camels (Camelus dromedarius). J. Zoo Wildl. Med. 30, 577–583.

Ma, P.X., Choi, J.W., 2001. Biodegradable polymer scaffolds with well-defined interconnected spherical pore network. Tissue Eng. 7 (1), 23–33.

Mano, J.F., et al., 2004. Bioinert, biodegradable and injectable polymeric matrix composites for hard tissue replacement: state of the art and recent developments. Compos. Sci. Technol. 64 (6), 789–817.

McCalden, R.W., McGeough, J.A., 1997. Age-related changes in the compressive strength of cancellous bone. The relative importance of changes in density and trabecular architecture. J. Bone Joint Surg. Am. 79 (3), 421–427.

Merkow, R., Lane, J., 1990. Paget's disease of bone. Orthop. Clin. North Am. 21 (1), 171–189.

Meunier, P.J., et al., 1980. Bone histomorphometry in Paget's disease: quantitative and dynamic analysis of pagetic and nonpagetic bone tissue. Arthritis Rheum. 23 (10), 1095–1103.

Meyer, U., et al., 1999. Decreased expression of osteocalcin and osteonectin in relation to high strains and decreased mineralization in mandibular distraction osteogenesis. J. Cranio Maxill. Surg. 27 (4), 222–227.

Midura, R.J., et al., 1990. A rat osteogenic cell line (UMR 106–01) synthesizes a highly sulfated form of bone sialoprotein. J. Biol. Chem. 265 (9), 5285–5291.

Mikos, A.G., et al., 1993. Preparation of poly(glycolic acid) bonded fiber structures for cell attachment and transplantation. J. Biomed. Mater. Res. 27 (2), 183–189.

Miller, B.F., et al., 2005. Coordinated collagen and muscle protein synthesis in human patella tendon and quadriceps muscle after exercise. J. Physiol. 567 (3), 1021–1033.

Monfoulet, L., et al., 2010. Bone sialoprotein, but not osteopontin, deficiency impairs the mineralization of regenerating bone during cortical defect healing. Bone 46 (2), 447–452.

NIH Consensus Development Panel on Osteoporosis Prevention, Diagnosis, and Therapy, 2001. Osteoporosis prevention, diagnosis, and therapy. JAMA 285 (6).

Niyibizi, C., Eyre, D.R., 1989. Identification of the cartilage α 1 (XI) chain in type V collagen from bovine bone. FEBS Lett. 242 (2), 314–318.

Niyibizi, C., Eyre, D.R., 1994. Structural characteristics of cross-linking sites in type V collagen of bone. Eur. J. Biochem. 224 (3), 943–950.

Nudelman, F., et al., 2010. The role of collagen in bone apatite formation in the presence of hydroxyapatite nucleation inhibitors. Nat. Mater. 9 (12), 1004–1009.

Oldberg, A., Franzén, A., Heinegård, D., 1986. Cloning and sequence analysis of rat bone sialoprotein (osteopontin) cDNA reveals an Arg-Gly-Asp cell-binding sequence. Proc. Natl. Acad. Sci. U.S.A. 83 (23), 8819–8823.

Oldberg, A., et al., 1988. Identification of a bone sialoprotein receptor in osteosarcoma cells. J. Biol. Chem. 263 (36), 19433–19436.

Ooi, C., Fraser, W., 1997. Paget's disease of bone. Postgrad. Med. J. 73 (856), 69–74.

Orgel, J.P., et al., 2006. Microfibrillar structure of type I collagen *in situ*. Proc. Natl. Acad. Sci. U.S.A. 103 (24), 9001–9005.

Palmer, L.C., et al., 2008. Biomimetic systems for hydroxyapatite mineralization inspired by bone and enamel. Chem. Rev. 108 (11), 4754–4783.
Parfitt, A., 1990. Osteomalacia and related disorders. Metabolic Bone Disease and Clinically Related Disorders. second ed. WB Saunders, Philadelphia, PA, pp. 329–396.
Pouget, E.M., et al., 2009. The initial stages of template-controlled $CaCO_3$ formation revealed by cryo-TEM. Science 323 (5920), 1455–1458.
Poundarik, A., 2013. The role of non-collagenous proteins, in particular, osteocalcin and osteopontin, in the determination of bone matrix quality. Rensselaer Polytechnic Institute, New York.
Price, P.A., 1988. Role of vitamin-K-dependent proteins in bone metabolism. Annu. Rev. Nutr. 8 (1), 565–583.
Rebel, A., et al., 1976. Osteoclast ultrastructure in Paget's disease. Calcif. Tissue Res. 20 (1), 187–199.
Rho, J.-Y., Kuhn-Spearing, L., Zioupos, P., 1998. Mechanical properties and the hierarchical structure of bone. Med. Eng. Phys. 20 (2), 92–102.
Roach, H., 1994. Why does bone matrix contain non-collagenous proteins? The possible roles of osteocalcin, osteonectin, osteopontin and bone sialoprotein in bone mineralisation and resorption. Cell Biol. Int. 18, 617–628.
Robey, P.G., et al., 1993. Structure and molecular regulation of bone matrix proteins. J. Bone Miner. Res. 8 (S2), S483–S487.
Rosenthal, A.K., et al., 2007. Osteopontin promotes pathologic mineralization in articular cartilage. Matrix Biol. 26 (2), 96–105.
Ryden, C., et al., 1987. Selective binding of bone matrix sialoprotein to *Staphylococcus aureus* in osteomyelitis. Lancet 330 (8557), 515.
Ryden, C., et al., 1989. Specific binding of bone sialoprotein to *Staphylococcus aureus* isolated from patients with osteomyelitis. Eur. J. Biochem. 184 (2), 331–336.
Sáenz, A., et al., 1999. Ceramic biomaterials: an introductory overview. J. Mater. Educ. 21 (5/6), 267–276.
Schönherr, E., Hausser, H.-J., 2000. Extracellular matrix and cytokines: a functional unit. J. Immunol. Res. 7 (2–4), 89–101.
Sherwood, J.K., et al., 2002. A three-dimensional osteochondral composite scaffold for articular cartilage repair. Biomaterials 23 (24), 4739–4751.
Shoulders, M.D., Raines, R.T., 2009. Collagen structure and stability. Annu. Rev. Biochem. 78, 929.
Siris, E.S., 1998. Paget's disease of bone. J. Bone Miner. Res. 13 (7), 1061–1065.
Sommerfeldt, D., Rubin, C., 2001. Biology of bone and how it orchestrates the form and function of the skeleton. Eur. Spine J. 10 (2), S86–S95.
Song, J., Malathong, V., Bertozzi, C.R., 2005. Mineralization of synthetic polymer scaffolds: a bottom-up approach for the development of artificial bone. J. Am. Chem. Soc. 127 (10), 3366–3372.
Stechschulte Jr., D.J., et al., 1992. Presence and specific concentration of carbonic anhydrase II in matrix vesicles. Bone Miner. 17 (2), 187–191.
Stephansson, S.N., Byers, B.A., García, A.J., 2002. Enhanced expression of the osteoblastic phenotype on substrates that modulate fibronectin conformation and integrin receptor binding. Biomaterials 23 (12), 2527–2534.
Stetlerstevenson, W.G., Veis, A., 1986. Type-I collagen shows a specific binding-affinity for bovine dentin phosphophoryn. Calcif. Tissue Int. 38 (3), 135–141.
Stevens, M.M., 2008. Biomaterials for bone tissue engineering. Mater. Today 11 (5), 18–25.

Stewart, T., 1969. The presence of delayed hypersensitivity reactions in patients toward cellular extracts of their malignant tumors. 1. The role of tissue antigen, nonspecific reactions of nuclear material, and bacterial antigen as a cause for this phenomenon. Cancer 23 (6), 1368–1379.

Stubbs, J.T., et al., 1997. Characterization of native and recombinant bone sialoprotein: delineation of the mineral-binding and cell adhesion domains and structural analysis of the RGD domain. J. Bone Miner. Res. 12 (8), 1210–1222.

Suda, T., et al., 1997. Regulation of osteoclast function. J. Bone Miner. Res. 12 (6), 869–879.

Tanahashi, M., Matsuda, T., 1997. Surface functional group dependence on apatite formation on self-assembled monolayers in a simulated body fluid. J. Biomed. Mater. Res. 34 (3), 305–315.

Teitelbaum, S.L., 2000. Bone resorption by osteoclasts. Science 289 (5484), 1504–1508.

Thorwarth, M., et al., 2005. Bioactivation of an anorganic bone matrix by P-15 peptide for the promotion of early bone formation. Biomaterials 26 (28), 5648–5657.

Traub, W., Arad, T., Weiner, S., 1992. Origin of mineral crystal growth in collagen fibrils. Matrix 12 (4), 251–255.

Velleman, S., 2000. The role of the extracellular matrix in skeletal development. Poult. Sci. 79 (7), 985–989.

Viguet-Carrin, S., Garnero, P., Delmas, P., 2006. The role of collagen in bone strength. Osteoporos. Int. 17 (3), 319–336.

Walker, D.G., 1972. Enzymatic and electron microscopic analysis of isolated osteoclasts. Calcif. Tissue Res. 9 (1), 296–309.

Weiner, S., Wagner, H.D., 1998. The material bone: structure–mechanical function relations. Annu. Rev. Mater. Sci. 28 (1), 271–298.

Whyte, M.P., 2006. Paget's disease of bone. N. Eng. J. Med. 355 (6), 593–600.

Wu, L., et al., 1989. Induction of mineral deposition by primary cultures of chicken growth plate chondrocytes in ascorbate-containing media. Evidence of an association between matrix vesicles and collagen. J. Biol. Chem. 264 (35), 21346–21355.

Wu, L., et al., 1991. Collagen-binding proteins in collagenase-released matrix vesicles from cartilage. Interaction between matrix vesicle proteins and different types of collagen. J. Biol. Chem. 266 (2), 1195–1203.

Wu, L.N., et al., 1995. Morphological and biochemical characterization of mineralizing primary cultures of avian growth plate chondrocytes: Evidence for cellular processing of Ca^{2+} and Pi prior to matrix mineralization. J. Cell. Biochem. 57 (2), 218–237.

Wuthier, R.E., 1973. The role of phospholipids in biological calcification: Distribution of phospholipase activity in calcifying epiphyseal cartilage. Clin. Orthop. Relat. Res. 90, 191–200.

Yeni, Y., Brown, C., Norman, T., 1998. Influence of bone composition and apparent density on fracture toughness of the human femur and tibia. Bone 22 (1), 79–84.

Young, M.F., et al., 1992. Structure, expression, and regulation of the major noncollagenous matrix proteins of bone. Clin. Orthop. Relat. Res. 281, 275–294.

Zhao, S., et al., 2002a. MLO-Y4 osteocyte-like cells support osteoclast formation and activation. J. Bone Miner. Res. 17 (11), 2068–2079.

Zhao, F., et al., 2002b. Preparation and histological evaluation of biomimetic three-dimensional hydroxyapatite/chitosan-gelatin network composite scaffolds. Biomaterials 23 (15), 3227–3234.

Mineralization processes in hard tissues: Teeth

5

R.A. Giacaman[1], *V.A. Perez*[1,2], *C.A. Carrera*[3]
[1]University of Talca, Talca, Chile, [2]University of Melbourne, Melbourne, VIC, Australia,
[3]University of Minnesota, Minneapolis, MN, USA

5.1 Introduction

Besides their indisputable critical function for food trituration, teeth are also important in phonetics and play an important role in esthetic appearance and self-esteem. Created by extremely complex processes and with a unique combination of tissues of different origin, teeth are a series of organs located in the mouth. Figure 5.1 shows a schematic representation of a human mature tooth. Anatomically, a tooth can be divided into two major portions: one visible in the mouth, the crown; and the other covered by the gingiva (gums), the root. A tooth is a highly mineralized organ that comprises several different calcified tissues, including enamel, the hardest tissue of the body, dentin, cementum, and the noncalcified dental pulp. Under normal conditions, the crown is covered by enamel, whereas root is covered by cementum. Beneath enamel and cementum, a thick portion of dentin extends from the crown to the apical portion of the root. In turn, dentin surrounds a nonmineralized soft tissue residing in the central area of the tooth, called dental pulp, which contains the vascular and nerve component of the tooth.

A tooth is well adapted to the intense masticatory function due to its external covering of enamel. Its highly mineralized structure can resist the constant impact of food and the antagonist dentition. Despite the frequent mechanical challenge posed by a hard diet, a sound human tooth can last the entire life of a person. Hence, optimal mineralization of dental hard tissues is critical in maintaining oral health. In spite of its unrivaled hardness and high durability against long-lasting mechanical forces from mastication, enamel is susceptible to acid dissolution during dental caries, one of the most prevalent chronic conditions in the human being and the most prevalent of the oral diseases. In its current understanding, caries is defined as a complex disease caused by an imbalance in the physiologic equilibrium between the tooth mineral and the oral biofilm fluid (Fejerskov, 2004). Carious lesions are signs of the disease that result from a progressive and localized demineralization of the hard tissues of the tooth due to a pH drop induced by fermentation of sugars when they are exposed to the oral biofilm (Fejerskov et al., 2008). If the carious process is concomitantly associated with developmental dental defects (D3) (www.thed3group.org; Simmer and Hu, 2001), increased damage from mineral loss can be clinically detected. Therefore, better understanding of

Biomineralization and Biomaterials. http://dx.doi.org/10.1016/B978-1-78242-338-6.00006-5

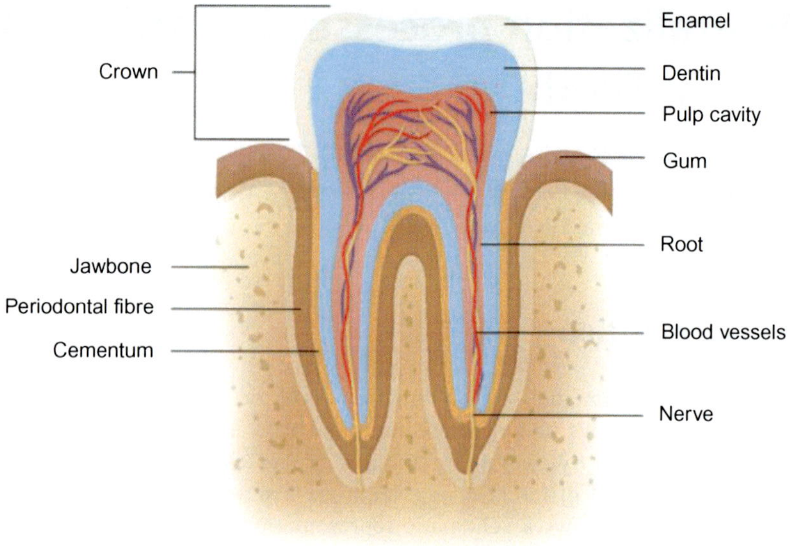

Figure 5.1 Human tooth structure. A mature tooth comprises two main anatomical regions. The crown is clinically visible and covered by enamel. The root in a healthy tooth is surrounded by the periodontal tissues and covered by dental cementum. The inner part of the tooth is composed of pulp and dentin.

the mechanisms involved in enamel and dentin mineralization will allow a more accurate diagnosis, prevention, and treatment of caries and D3s.

This chapter describes the structure of the different tissues of a human tooth and the complex processes involved in biomineralization. Besides its remarkable complexity, tooth formation and biomineralization require an exquisite biological synchronization of multiple players, most of which are still under investigation. The chapter will try to summarize classic and updated available knowledge, in a simple and didactic manner. The reader should understand that presenting all the research on the topic of biomineralization of a tooth largely surpasses the scope of this chapter, nonetheless. Section 5.2 describes the sequence and stages of tooth formation. Section 5.3 explains how enamel is formed and how its mineral content is built upon, before and after eruption, to obtain its mature mechanical and biological properties. Section 5.4 explores the current status in the understanding of the mineralization of the highly complex structure of dentin. Although dentin biomineralization is a growing field with novel research being frequently reported, this chapter has incorporated the most recent findings on the topic to portray the current state of knowledge. Since this subject is evolving and many aspects are still unknown, the chapter will cover only the most relevant topics of tooth biomineralization. The reader is referred to more specific publications for more in-depth knowledge of each of the issues covered here. A special section has been incorporated to provide directions to the reader for further information.

5.2 Tooth formation: Odontogenesis

5.2.1 Introduction

A tooth undergoes a series of highly complex processes to become the final organ in the mouth. With some variation in different individuals, the first temporary tooth usually erupts at the sixth month of age. The deciduous dentition is gradually replaced by the permanent teeth starting with the eruption of the first molar at about six years old. A mixed dentition occurs between 6 and approximately 12 years of age, when all deciduous teeth are exfoliated and replaced by the permanent teeth. This section describes the general aspects of tooth formation in the maxillary bone during embryogenesis and after birth in the case of some of the permanent teeth. For details of the embryogenesis of these structures, the reader is referred to embryology texts, for example *Orban's Oral Histology and Embryology*, 13th edition, Elsevier.

Teeth may vary from species to species in form, composition, and location, but there are certain common characteristics in the teeth of different animals. A tooth comprises a crown, which is usually exposed to the oral cavity and protected by enamel, and a root surrounded by cementum that is firmly attached to the maxillary bone by the periodontal ligament and covered by the gingival soft tissues (gums). The number of dentitions varies widely across the different vertebrates. While a monophyodont species has only a single set of teeth during their entire life, a diphyodont, like humans, develops two sets of teeth: deciduous or primary and permanent teeth. Although this chapter focuses mainly on human teeth, the embryogenesis, development, and mineralization processes are similar across the different vertebrate teeth.

5.2.2 Odontogenesis: The complex and dynamic process of tooth formation

The complex process of developing a tooth is called odontogenesis. Tooth formation originates from embryonic cells derived from the ectoderm of the first branchial arch and the ectomesenchyme from the neural crest. A forming tooth germ is organized into three structures: the enamel organ that leads to enamel in a mature tooth, the dental papilla that originates the pulp–dentin complex, and the dental follicle or sac, which is the precursor to cementum, alveolar bone, and periodontal ligament (Figure 5.1).

The tooth undergoes a continuous process of cell organization, mineralization, and eruption in the oral cavity, which can be divided into four stages: initiation, bud, cap, and bell (Figure 5.2).

Initiation stage. This is the earliest sign of tooth formation at the sixth week of embryonic development. A band of epithelial cells from the oral ectoderm, known as the dental lamina, extends internally to connect the developing tooth bud with the epithelium of the oral cavity. The invagination of the dental lamina into the mesenchymal tissue originates the ameloblast, which in turn will form the enamel. The dental

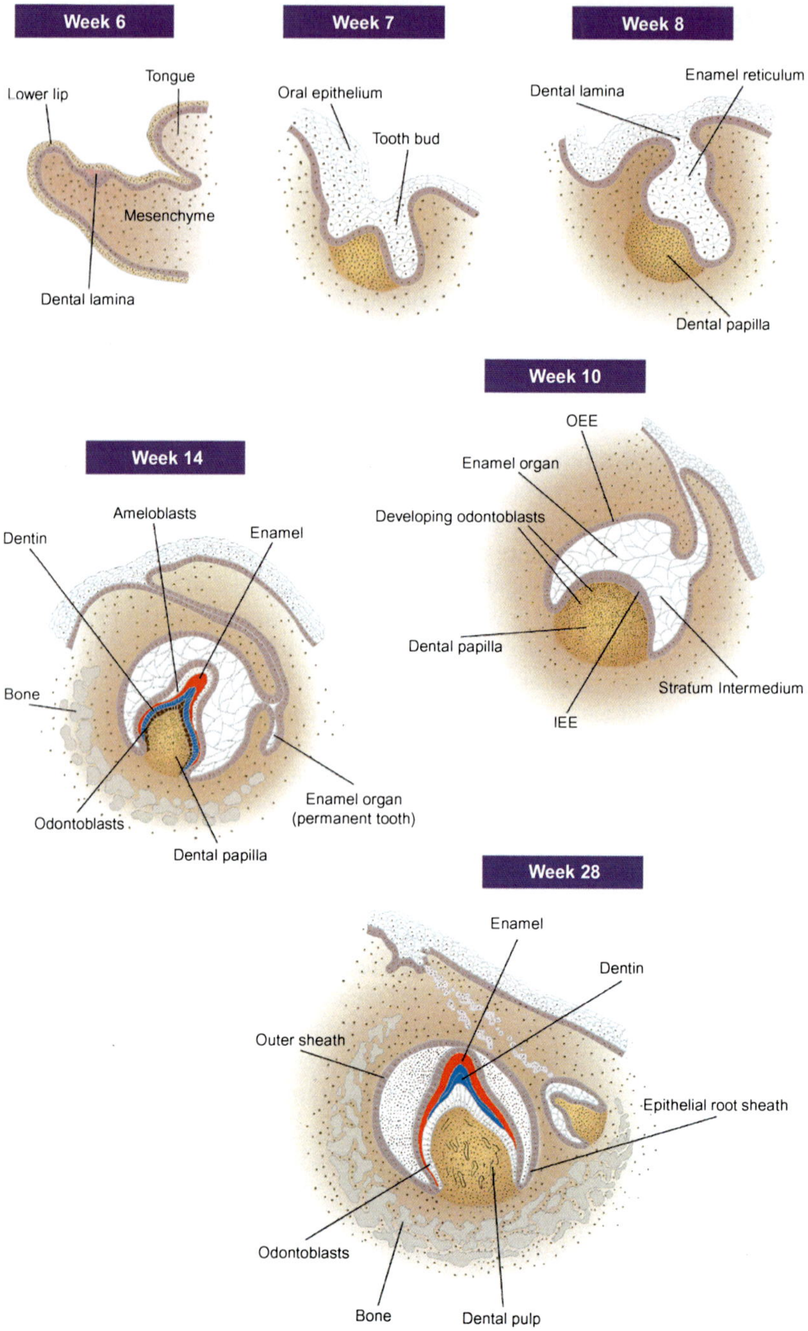

Figure 5.2 Schematic representation of odontogenesis. The figure shows a sequential representation of events from week 6 to 28 in the formation of a human incisor.

lamina extends in depth and covers the ectomesenchymal cells of the arch, which will form the dental papilla. The cells of the dental papilla will form the odontoblast, the dentin, and the pulp.

Bud stage. At the seventh week, cells from the dental lamina organize to form 10 round epithelial structures (buds) to the distal side of the lamina in each arch. Each bud is isolated from the ectomesenchymal cells by a basal membrane. Thus, a bud is formed by a proliferation of cells of the dental lamina into the mesenchyme. On the other hand, ectomesenchymal cells close to the in-growing epithelium start to coalesce to form the dental papilla.

Cap stage. At eight weeks, the tooth bud grows and expands to encircle a group of condensed ectomesenchymal cells forming the dental papilla. The epithelial cells resemble a cap, which will become the enamel organ covering the dental papilla. The cells of the cap can be seen as an outer and an inner layer of epithelium, leaving an intermediate region in between that forms the cap and is called the stellate or enamel reticulum. A condensation of ectomesenchymal cells called the dental sac or follicle surrounds the enamel organ and limits the dental papilla. The dental sac will give rise to the periodontium, the supporting tooth tissues (alveolar bone and periodontal ligament).

Bell stage. During this stage, the forming dental organ adopts a bell shape. The bell can be divided into the early and late stage at the 10th and 14th weeks, respectively. Overall, during this stage the dental germ differentiates the enamel-forming ameloblasts from the epithelium and the dentin-forming odontoblasts from the mesenchymal cells. The enamel organ differentiates distinctive layers: the outer enamel epithelium (OEE), the cells of the enamel organ adjacent to the enamel papilla, called inner enamel epithelium (IEE), and the cells between the IEE and the OEE, the stratum intermedium. In the rim of the enamel organ, the OEE and IEE join in a region known as the epithelial root sheath. This region invaginates, originating the typical bell shape of the forming tooth. As the stage progresses, a series of events will end with dentin and enamel formation.

The odontoblast, the dentin precursor cell, differentiates from mesenchymal cells of the dental papilla. This secretory cell produces organic matrix or predentin adjacent to the IEE, which will further calcify to form mature dentin. Incremental dentin formation occurs by the regression of the odontoblast toward the dental papilla. During its movement, the odontoblast leaves its cytoplasmic extension behind, called Tomes' dentinal fibers or processes, creating the unique tubular structure of dentin. This complex mixture of calcified matrix surrounding the cytoplasmic extensions of cells located in the pulp is described in more detail later in this chapter. Simultaneously with dentin formation, cells of the IEE modify their shape to become pre-ameloblasts. Nuclear polarization and lack of mitotic activity of the IEE precede secretory activity of these cells. Like the odontoblasts, ameloblasts regress externally, secreting organic matrix. This matrix immediately mineralizes and becomes the initial layer of enamel (revised in Section 5.3). Ameloblasts are formed in response to the presence of dentin. The continued interaction between dentin and enamel completes tooth's crown formation starting from the cusp. In the outer region of the germ, the dental lamina disintegrates. Thus, the tooth loses its connection with the oral epithelium until tooth eruption. The cervical loop, where the IEE and OEE gather, form the "neck" or cervical region of the tooth to start root formation. The internal cells of the dental follicle or sac form cementoblasts, which will produce cementum covering the root. As the root sheath grows internally, extending root formation, the odontoblasts form dentin constraining the size of the pulp. Externally to the germ, the dental follicle starts bone formation to cover the root with ossified tissue. The dental follicle will also form the structures that bind the bone to the root, the periodontal ligament.

5.3 Enamel mineralization

5.3.1 *Introduction*

Looking at the enamel of a tooth, it is difficult, if not impossible, to appreciate the extremely complex biological process behind its development (Hubbard, 2000). Enamel formation or amelogenesis is a very slow developmental process that takes longer to complete than the time that the entire embryo requires to be formed *in utero*. In fact, human enamel formation can take as long as 4–5 years to complete on the crown of some permanent teeth (Smith, 1998). Enamel is unique among mineralized tissues in vertebrates, as it is the only epithelium-derived calcified tissue. Furthermore, enamel is also the hardest substance in the body, matching carbon steel (Simmer and Fincham, 1995). Importantly, most of the understanding of amelogenesis comes from animal-model studies (e.g. rats and pigs). Although current evidence strongly supports conservation of this process across mammals, at least in its general aspects, variability in tooth formation and developmental timeframes among species should be taken into consideration when interpreting values of animal studies.

In this section, a general overview of the histological composition of the mature enamel will be provided. Also, the multifaceted process of enamel development will be covered, dividing it broadly into two main phases. *Pre-eruptive cell-mediated enamel mineralization*: enamel from the secretory stage of the odontogenesis is hardened through active mineral deposition by the ameloblasts. *Posteruptive enamel maturation*: not mediated by cells, this phase occurs when the outermost enamel matures via a residual mineralization from mineral ions from saliva.

5.3.2 *Histological organization of mature enamel*

Mature enamel has an intricate and organized mineral structure composed by closely packed crystallites, which comprise 87% of its volume and 95% of the enamel weight (Simmer and Fincham, 1995). In contrast to bone and cartilage, which contain around 20–30% of organic material by weight, mature enamel possesses less than 1% of organic material (Bartlett and Simmer, 1999; Deakins and Volker, 1941). Light microscope analysis has shown that enamel crystallites are organized into rods (prisms) separated by interrod enamel. Rods and interrod enamel (4–5 μm diameter) extend from the dentin–enamel junction (DEJ) to the outer crown surface. Both structures differentiate from each other by crystal orientation.

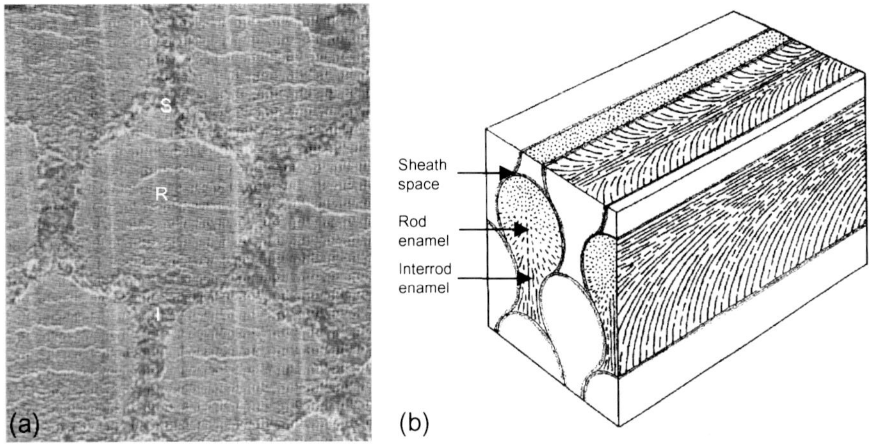

Figure 5.3 Microstructure of mature dental enamel. (a) The electron micrograph shows a perpendicular cross-section to the long axes of human primary molar enamel prisms. Sheath space, rod, and interrod enamel are indicated as S, R, and I, respectively. (b) The schematic representation of crystal arrangement shows different orientations in rod and interrod enamel. Additionally, the protein-rich sheath space (separating rod from interrod) can be visualized as indicated.
Modified from: Habelitz et al. (2001) and Meckel et al. (1965).

While rod crystallites run parallel to the long axis of the rods, interrod crystallite inclination deviates from the rods by 40° to 65° (Orams, 1966). A thin area deprived of crystals, called rod sheath or sheath space, is located between the rods and interrod enamel (Orams, 1966; Simmer and Fincham, 1995; Travis and Glimcher, 1964). It has been proposed that the sheath space contains proteins, which would confer enamel its unusual high elasticity by acting as a torsional cushion between prisms (Xie et al., 2009a) (Figure 5.3).

5.3.3 *Pre-eruptive enamel mineralization*

5.3.3.1 *Cellular modeling of the primordial tooth*

Control over enamel mineralization is made possible by the formation of an isolated chamber within the tooth organ during odontogenesis (Section 5.2). Ectomesenchymal cells trigger differentiation and multiplication of basal cells of the stratum basale of the overlying epithelium, forming two adjacent layers with separate basement membranes. Thus, enamel organ epithelium is divided into an outer epithelium that will give rise to the odontoblast layer from mesenchymal cells and an inner epithelium that will form the ameloblast layer from ectodermal cells. The space generated between both epithelial layers will contain the future enamel cellular matrix. Genetically programmed cells from the IEE undergo morphological and cytological changes, ultimately differentiating into specialized enamel-forming cells called ameloblasts (Nanci and Ten Cate, 2008).

Enamel development is under the direct control of ameloblasts. Through a series of sequential stages, ameloblasts control the synthesis of enamel matrix proteins and proteinases to regulate the patterning of the nascent enamel crystals, known as enamel ribbons, before transporting bulk calcium to promote enamel mineralization and hardening (Bartlett, 2013; Sasaki et al., 1997). These complex processes require a progressive change of the ameloblast morphology and function during two main developmental stages called secretion and maturation (Warshawsky and Smith, 1974). A third stage between secretion and maturation is sometimes described as a transition stage, but it will be considered here as an early step of the maturation stage.

5.3.3.2 Differentiation and secretion stage during amelogenesis

After collagenous predentin is deposited by the odontoblast during the establishment of the DEJ, ameloblasts start enamel deposition from the cusp tips and progress toward the cervical region of the tooth (Slavkin, 1979). The main function of the ameloblasts at this stage is to secrete and partially mineralize the enamel cellular matrix until the full thickness of the enamel is formed. Cytoplasmic projections of the ameloblast extend through the basement membrane until it is removed and destroyed. In addition, cuboidal presecretory ameloblasts differentiate to columnar cells over 60 μm in height and 2–3 μm in width. Ameloblasts connect to each other by specialized cell adhesion structures at the proximal and distal ends (Reith, 1970). At this stage, the distal portion of each ameloblast is shaped as an angulated cone named Tomes' process, near the forming enamel matrix (Warshawsky and Smith, 1974). Large amounts of enamel matrix proteins are primarily secreted from one side of the Tomes process (secretory face). Ameloblasts within a row secrete proteins from the same side of the Tomes processes as they move away from the dentin surface so that the nascent enamel becomes thicker (Bartlett, 2013; Rönnholm, 1962). Daily increments in secretion of matrix protein create appositional growth lines called Striae of Retzius, which can be observed as macroscopic structures on the surface of an erupted tooth, known as perikymata (Simmer et al., 2010).

Four major proteins are secreted by the ameloblasts at this stage: a proteinase and three putative structural proteins (Fincham et al., 1999). Amelogenin (AMEL), ameloblastin (AMBN), and enamelin (ENAM) participate in controlling the shape and spatial relationships between the hydroxyapatite (HAp) enamel mineral ribbons (Hu et al., 1997; Krebsbach et al., 1996; Simmer et al., 2012; Snead et al., 1985).

AMEL is the most abundant (80–90%) ameloblast-secreted organic product in the enamel cellular matrix during the secretion stage. This protein is critical for proper enamel formation as has been demonstrated by its contribution to amelogenesis imperfecta, when mutated in human genetic analyses as well as genetically engineered AMEL-null mice (Gibson et al., 2001; Hu et al., 2007). These studies support the hypothesis that AMEL plays a role in the organization and growth of enamel crystals during tissue patterning (Moradian-Oldak, 2012). In addition to its structural role, AMEL, particularly the small isoform leucine-rich amelogenin peptide, has been associated with signaling activities, as assessed in a variety of cell culture systems (Gibson, 2008; Veis, 2003; Warotayanont et al., 2008). It has also been reported that AMEL interacts with bone cells and, because of its potential to promote cell differentiation, the protein has been considered for periodontal regenerative therapies (Gibson, 2008). Possibly, the structural characteristics of AMEL may explain its signaling potential and multifunctionality. Recent studies have classified AMEL as part of the family of "intrinsically disordered" proteins or "unstructured proteins" where large segments of proteins lack well-structured three-dimensional folding and exist as a dynamic group of interconverting structures (Delak et al., 2009; Dunker et al., 2008). Despite their lack of a well-defined three-dimensional structure, such proteins are indeed functional. Studies in vertebrates have showed that AMEL contains 50–60% disorder-promoting residues, reporting that AMEL bulk is enriched with proline and glutamine amino acid residues, and its hydrophilic C-terminus also contains other disorder-promoting residues such as glutamic acid, arginine, and lysine (Lakshminarayanan et al., 2009; Moradian-Oldak, 2012).

A largely unfolded monomeric form of recombinant porcine AMEL isoform rP172 (at pH 3.8) has been reported, containing a local residual secondary structure. It was proposed that the structure of rP172 is in line with reported functions of AMEL during enamel mineralization, such as: self-assembly into oligomers, nanospheres, and nanochains; associations with other enamel matrix proteins and with calcium phosphate minerals; and interaction with cell surfaces (Gibson, 2008; Hoang et al., 2002). Inducing oligomeric interaction, it was found that the central domain of the native-like structure of recombinant porcine AMEL is resistant to folding. However, the N- and C-termini are conformationally responsive and represent potential interactive sites for AMEL self-association during enamel matrix mineralization (Ndao et al., 2011).

AMEL molecular interactions have been assessed *in vitro* and *in vivo*. Studies *in vitro* have reported that AMEL assembles into nanospheres, oligomers, nanochains, and other elongated assemblies (Bromley et al., 2011). It has been proposed that AMEL forms nanospheres via intermolecular hydrophobic interactions (Fincham et al., 1994; Moradian-Oldak, 2007). In addition, oligomeric interaction of AMEL nanospheres has also been reported (Fang et al., 2011). The interactions that cause monomers to assemble into oligomers are primarily N-terminal mediated (potentially N-terminal to N-terminal). However, those that cause oligomers to bind and form nanospheres are mainly regulated by hydrophobic interactions via the AMEL central portion (Bromley et al., 2011). Nanochains and other elongated assemblies of AMEL have been reported using different approaches (e.g. slow solvent evaporation, incubation in a calcium-based and neutral pH environment; Beniash et al., 2005; Du et al., 2005; Wiedemann-Bidlack et al., 2007). Moreover, AMEL can also form nanorods during proteolysis with matrix metalloproteinase 20 (MMP20) and as the result of its co-assembly with proteolytic products (Yang et al., 2011). It has been proposed that AMEL oligomers form nanospheres that break up into oligomers in a mildly acidic environment similar to that present during the later stage (maturation) of enamel development. This finding suggests that AMEL oligomeric structures might be functional components during maturation of enamel apatite (Bromley et al., 2011; Moradian-Oldak, 2012).

AMEL self-assembly *in vivo* has been described as electron-lucent structures by transmission electron microscopy images. Those interactions were primarily reported as "nanospheres" (Diekwisch et al., 1993; Moradian-Oldak and Goldberg, 2005; Robinson et al., 1981). However, subsequent analysis of the dimension of electron-lucent spherical particles in some sections of forming enamel suggests that these structures may be oligomers rather than nanospheres (Moradian-Oldak et al., 1994; Moradian-Oldak and Goldberg, 2005). The recent finding that the initial amorphous calcium phosphate (ACP) mineral in enamel has the form of defined ribbon-like shapes that are separated by chain-like structures of organic matrix supports the notion that AMEL self-assembly into nanochains guides mineral formation. This AMEL interaction may occur at the very early stage of mineral formation, separating the mineral and dictating the morphology of early apatite crystallites (Du et al., 2005). It is now largely accepted that the AMEL-based supramolecular assemblies exert control over the organization and directionality of HAp crystal growth (Moradian-Oldak, 2012).

The function of AMEL has been widely studied. *In vitro* research has provided clues of the detailed molecular mechanisms responsible for AMEL-guided mineralization. It is suggested that AMEL controls the formation and organization of HAp crystals (Beniash et al., 2005). Studies assessing crystal interaction have been reported that AMEL has the potential to control calcium phosphate polymorphism, promote parallel crystal organization, and modulate crystal morphology (Tarasevich et al., 2007; Wang et al., 2008). A nucleation model was suggested in which AMEL was proposed to interact with calcium phosphate in a cooperative manner to form intermediate complex clusters, which in turn aggregated on the sides of apatite crystals. In this model, phosphate-containing AMEL proteolytic cleavage products may suppress unwanted rapid mineral deposition within the enamel crystals during the secretory stage of amelogenesis (Yang et al., 2010) (Figure 5.4).

In vivo studies using AMEL-null mice have improved the understanding of the physiological function of AMEL in controlling enamel thickness and prism organization. Modified mice teeth presented chalky-white opacities with disorganized and hypoplastic (thin) enamel. No AMEL was expressed in these animals. However, a thin layer of mineralized enamel was secreted, suggesting that AMEL does not participate in the initiation of mineralization, but is critical for getting a proper enamel crystal elongation and thickness (Gibson et al., 2001).

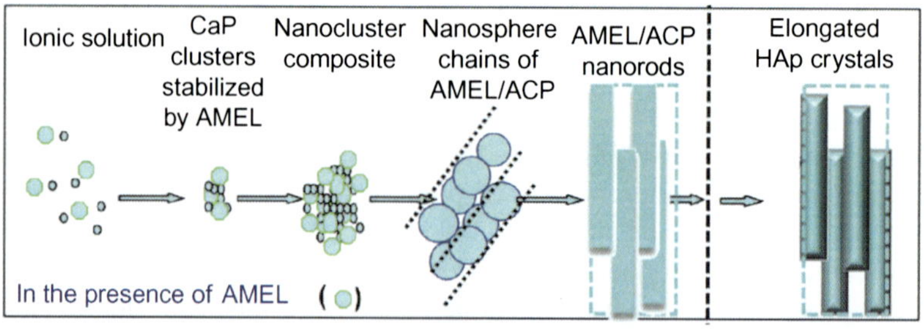

Figure 5.4 Suggested mechanism for *in vitro* hierarchically organized, elongated HAp microstructures formed by co-assembly of Amel–Ca–P nanoclusters. CaP, calcium phosphate; ACP, amorphous calcium phosphate.
Modified from Yang et al. (2010).

AMBN is the second most abundant protein in enamel, constituting about 5% of the total protein present in the organic matrix during the secretion stage (Hu et al., 2007). In contrast to AMEL, experimental data about AMBN's structure, assembly, and function are limited (due to limited expression and rapid proteolysis). However, AMBN is secreted together with AMEL and interaction between these proteins has been suggested *in vitro*. However, the significance of this warrants further investigation (Ravindranath et al., 2004). Studies using AMBN-null mice suggest that AMBN appears to be critical for enamel formation since a pathologically affected ameloblast layer and junctional epithelium were noted in these animals. Ameoblastin has showed cell adhesion properties and is believed to control ameloblast cell differentiation (Fukumoto et al., 2004).

On the other hand, ENAM is much less abundant (3–5%) than AMEL (Hu et al., 2007). Although it is expressed during the three main stages of enamel formation, ENAM expression terminates prior to the expression of AMEL (Hu et al., 2001). In ENAM null-mice, the final enamel layer is not formed, and disorganization and apoptosis of ameloblasts have been reported. Moreover, a progressive decrease in mineral is associated with reduced ENAM expression (Hu et al., 2008). *In vitro* studies have shown that ENAM has relevant participation in controlling crystal nucleation or growth. The 32-kDa ENAM appears to have a relatively high affinity for apatite crystals (Yamakoshi et al., 2001). It has been reported that 32-kDa ENAM isolated from developing porcine enamel in a calcium-rich environment undergoes spatial conformational changes that could facilitate its interaction with apatite crystals during enamel mineralization (Moradian-Oldak, 2012). When AMEL is added to gelatine, 32-kDa ENAM promotes nucleation of HAp crystals, suggesting a cooperative function (Bouropoulos and Moradian-Oldak, 2004). In fact, biophysical and spectroscopic studies have shown that interaction between the 32-kDa ENAM and AMEL occurs, suggesting that the two proteins might actually co-assemble (Fan et al., 2009). In this regard, when particles of AMEL were combined with 32-kDa ENAM, an increment in AMEL particle size was visualized, suggesting direct AMEL–ENAM interaction (Fan et al., 2011). However, *in vivo* approaches supporting the relationship between AMEL and ENAM in the enamel cellular matrix need to be investigated further if their cooperative participation in controlling enamel mineral formation is to be established.

These three enamel proteins (AMEL, AMBN, and ENAM) are almost immediately processed after secretion by a proteinase named MMP20, the predominant enamel-resident AMEL/AMBN/ENAM-processing enzyme at the secretory stage of enamel development. Intact (undegraded) proteins and their proteolytic fragments reside in different locations in the growing enamel cellular matrix, suggesting different functional characteristics (Bartlett and Simmer, 1999). Intact AMEL, AMBN, and ENAM are more abundant in the outer enamel layers (most recently formed enamel) (Fukae et al., 1993; Hu et al., 1997; Tanabe et al., 1992; Uchida et al., 1991, 1997). In the inner enamel, however, only AMEL/AMBN/ENAM cleavage products seem to be present (Smith, 1998).

The mineral phase formed in the enamel resembles HAp $[Ca_5OH(PO_4)_3]$, but it may contain variable amounts of carbonate (20–30%) and low quantities of sodium and magnesium (Simmer and Fincham, 1995). From the DEJ, enamel prisms grow through the entire thickness of the enamel layer, creating its characteristic structure.

Enamel proteins, mostly AMEL, cover immature prisms by the sides, while calcium and phosphate are deposited mainly on the tips of the elongating crystals (Miake et al., 1993). All these events are facilitated by the selective proteolytic processing of enamel proteins (e.g. AMEL, AMEBN, and ENAM) by MMP20. This metalloproteinase seems to be critical for the attachment of enamel to dentin as a thin line of hypermineralized enamel at the DEJ. Experiments in animals have shown that *Mmp20*-null mice fail to form these structures (Hu et al., 2011). Furthermore, the mineralization process may be interrupted by disturbances affecting ameloblasts at the secretion stage, which in turn induce failures in crystal elongation, resulting in thin and hypoplastic enamel (Suckling, 1989).

5.3.3.3 Maturation stage during amelogenesis

Maturation comprises a long stage during amelogenesis, whereby enamel acquires its final hardness through mineral intake. This phase may last as long as 3–4 years for some permanent teeth (Simmer and Hu, 2002). Once the final thickness of the enamel layer is reached in the secretion stage, ameloblasts switch function to mineralization (Lu et al., 2008). From a predominantly protein secretory function, ameloblasts rapidly reorganize to adopt calcium transportation functions (Reith, 1970; Smith and Nanci, 1995; Warshawsky and Smith, 1974). The Tomes process is lost and the tall, columnar enamel-secretory cells at the secretory stage change into shorter, squatter cells at the maturation stage (Smith and Nanci, 1995; Warshawsky and Smith, 1974). Around 25% of the ameloblasts die by apoptosis at an early post secretion or early maturation stage. The remaining 75% of the ameloblasts initiate the rhythmic modulation at the surface of maturing enamel, but 25% die as they modulate the tooth surface (Smith and Warshawsky, 1977). After producing a new basal lamina, ameloblasts begin modulating between two different functional states. Histologically, ameloblasts appear as ruffled and smooth-ended cells at the enamel surface (Smith, 1998). Ruffle-ended ameloblasts have unique morphological characteristics presenting tight junctions and membrane infolding at their apical surface (Josephsen and Fejerskov, 1977; Warshawsky and Smith, 1974). Cells acquire a smooth-ended shape when the apical invaginations are lost and the apical junctional complexes become disassociated and leaky (Crenshaw and Takano, 1982; Josephsen and Fejerskov, 1977; Takano, 1995). The ruffle/smooth cellular modulation is accompanied by a distinctive regulation of the composition of the enamel fluid in the enamel cellular matrix environment, which is largely attributed to ameloblasts. Critical for an optimal mineralization, these processes seem to be influenced by mineral concentration and pH. A summary of enamel-forming cell changes during enamel development (secretion and maturation) is presented in Figure 5.5.

The route taken by minerals (e.g. calcium and phosphate) varies during the maturation stage with the modulation of ameloblasts. It appears that during the "ruffled-ended ameloblast" state, calcium may be primarily transported transcellularly, leading to mineral accretion. On the other hand, the migrating wave of smooth-ended ameloblasts causes a transient breach of membrane permeability, allowing calcium and other interstitial fluid constituents to enter the enamel fluid (paracellularly).

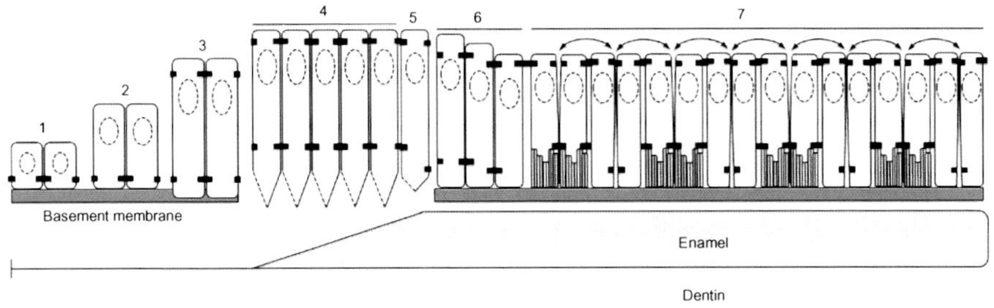

Figure 5.5 Changes of ameloblast during enamel development. The figure represents: (1–5) the differentiation and secretion stage ameloblast and (6 and 7) early maturation (transition) and maturation stage ameloblast, respectively. (1) Inner enamel–epithelial cells lie on a basement membrane. (2) They increase in length and (3) participate in basement membrane degradation along with the initial secretion of enamel proteins on the dentin surface. (4 and 5) At the secretion stage and after DEJ establishment, ameloblasts (from the secretory face of the Tomes process) increase the secretion of enamel proteins (including MMP20) modeling crystals grow in length. Once enamel has achieved its final thickness, (6) ameloblasts change function to mineralization. (7) At the maturation stage, KLK4 is secreted, degrading almost all the proteins in the enamel cellular matrix and a new basement membrane is established. At this stage, ameloblasts modulate as ruffled and smoothed states, hardening the enamel layer by facilitating lateral mineral deposition on the crystals laid down at the secretory stage. Modified from Hu et al. (2007).

During the "smooth-ended ameloblast" state little calcium is incorporated into the mineralizing enamel cellular matrix (Hubbard, 2000). The mechanism of phosphate transport is still unclear. It has been suggested that phosphate uptake might be linked with the passive smooth-ended ameloblast state. However, both transcellular and paracellular transport mechanisms may be involved (Lacruz et al., 2013).

The pH at the enamel fluid changes as ameloblasts modulate. While ruffled ameloblasts appear to associate with mildly acidic enamel fluid (~pH 6.2), favoring mineralization (higher calcium/phosphate ratio), the enamel fluid under smooth-ended ameloblasts appears more neutral (~pH 7.2), similar to that in the interstitial fluid with lower calcium phosphate ratio (Kawamoto and Shimizu, 1990; Sasaki et al., 1991). The dynamics of ameloblast modulation allow deposition of minerals into enamel fluid and lateral expansion of immature crystals until they contact adjacent crystals, eventually leading to hardening of the whole enamel layer (Lu et al., 2008). To facilitate this sideways enlargement, massive degradation and elimination of almost all enamel matrix proteins is needed (Robinson et al., 1997). Early in this stage, the secreted enamel protease kallikrein 4 (KLK4) almost completely degrades enamel matrix proteins, creating space for mineral crystal expansion (Lu et al., 2008; Simmer and Hu, 2002). After degradation, enamel protein fragments are absorbed from enamel cellular matrix via endocytosis and pynocitosis by ameloblasts, being finally degraded by lysosomes (Reith and Cotty, 1967; Sasaki et al., 1984). Despite proteolytic degradation, a residual 1% (by weight) of organic material has been reported to be retained within the mature enamel. Low abundance precludes accurate characterization. It has been suggested that AMEL/AMBN/ENAM peptides may be present (Porto et al., 2011).

These proteins may be involved in providing resilience to torsional deformation of enamel during mastication (Xie et al., 2009b). Disturbances of the above-mentioned complex processes of amelogenesis may result in pathologically soft enamel with normal thickness, termed hypomaturation or hypocalcification depending on the lesion protein content (Mangum et al., 2010).

At the end of the maturation stage, the pre-eruptive mineralization is complete and the enamel is ready for eruption. Concomitantly, a final component of apoptotic death apparently occurs in reduced ameloblasts during eruption. The epithelium of the reduced enamel organ fuses with the oral epithelium as part of the events leading to the formation of the junctional epithelium (Shibata et al., 1995).

5.3.4 Posteruptive enamel maturation

Following the maturation stage, where most of the enamel mineral content is acquired, the final mineralization is completed after eruption (Dirks, 1966; Robinson, 2014). In this nonameloblast-controlled stage, interaction with saliva contributes a posteruptive maturation of the enamel surface via diffusion of ions such as calcium, phosphorus, magnesium, fluoride, and potentially other components (Briner et al., 1971; Mandel, 1987). Almost immediately after eruption, enamel is covered by an acellular layer of protective HAp-binding salivary proteins (e.g. proline-rich proteins, mucinous proteins, and statherine) called acquired enamel pellicle (AEP) that directly contributes to the posteruptive maturation of enamel (Hannig et al., 2004). It has been reported that statherin may function in the transport of calcium and phosphate during secretion in the salivary glands (Bennick et al., 1981; Schlesinger and Hay, 1977). In saturated solutions of calcium phosphate minerals, statherin inhibits spontaneous precipitation and crystal growth, providing a protective and reparative but stable environment for the teeth (Hay et al., 1984; Johnsson et al., 1991). Acting together with the acidic proline-rich and mucinous proteins, statherin maintains the supersaturated state of saliva with respect to most calcium phosphate salts in the AEP, a condition needed for the recalcification and stabilization of the tooth enamel and for the inhibition of formation of mineral accretions on tooth surfaces (Raj et al., 1992). In this environment, enamel mineralization is completed by replacing more acid-soluble HAp (e.g. carbonate-substituted HAp) with more acid-resistant HAp (e.g. fluorapatite). After this process, the porosity and permeability of the 200 μm outer enamel layer is greatly reduced (García-Godoy and Hicks, 2008). It is important to take into consideration that AEP also contributes to bacterial adhesion, favoring demineralization, and subsequent caries development. For more details on the dynamics of the cariogenic bacterial biofilm, the reader is referred to specific caries literature (Nyvad et al., 2013).

A schematic representation of pre-eruptive enamel mineralization and posteruptive maturation is shown in Figure 5.6.

5.3.5 Future avenues for research on enamel mineralization

As mentioned at the beginning of this section, most amelogenesis-related research has been conducted using animal models. *In vitro* biochemical models, *in vivo*

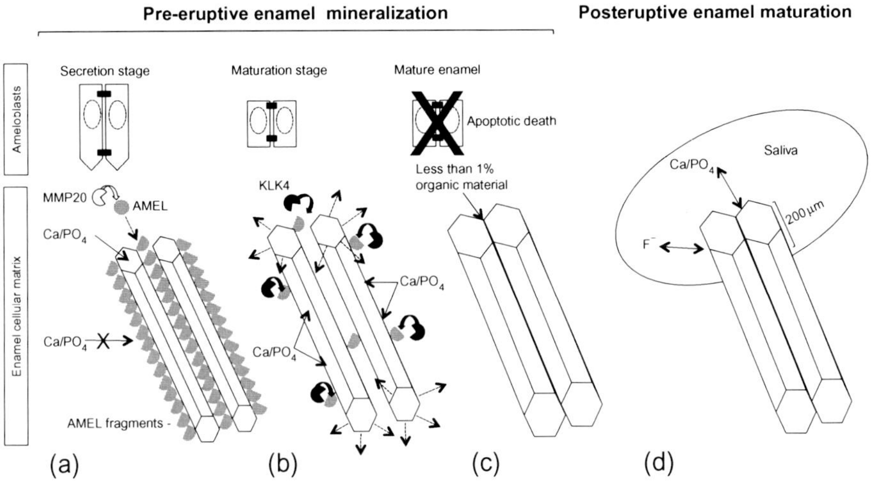

Figure 5.6 Phases of enamel development. Pre-eruptive mineralization: (a) At the secretion stage, ameloblast-secreted proteins (represented by AMEL) are selectively processed by MMP20 into stable fragments, guiding crystal elongation via mineral deposition (calcium/phosphate; Ca/PO_4) mainly on the crystal tips. (b) At the maturation stage, ameloblasts switch function to mineralization. In this regard, a massive degradation of AMEL by KLK4 occurs. Then, minerals in enamel fluid allow immature crystals to undergo lateral expansion until coming into contact with adjacent crystals. (c) After apoptotic death of ameloblasts, pre-eruptive enamel is characterized as a tightly packed and hard structure containing less than 1% organic material and is ready for eruption and functional occlusion. Posteruptive maturation: (d) Into the oral cavity, the outermost enamel (200 μm) continues to mature as it interacts with mineral ions from saliva (e.g. Ca/PO_4 and fluoride, F^-).

studies based on the molecular approaches, including identification of proteins and enamel proteinases and their degradation products, and histological observation of animal tissues have been the cornerstone of amelogenesis research. Although great advancements have been made, a more comprehensive understanding of enamel development is far from achieved. A better characterization of amelogenesis at the biochemical and genetic levels seem to be of increasing interest to researchers. Indeed, recent genetic identification of the main enamel proteins/proteinases has allowed the development of genetically modified animal models that have contributed to elucidate the functional roles of some novel molecular players during the process of enamel development (Simmer et al., 2009; Smith et al., 2011; Yamakoshi et al., 2011). Additionally, the complex relationship between enamel protein gene transcripts and proteins seems to be another promising path for further research in this field (Chen et al., 2000; Li et al., 1995).

A better understanding of normality may arise from studying abnormality. Hence, over the last decades, a particular type of developmental dental defect called molar hypomineralization (MH) has become extremely important among clinicians due to severe implications for child oral health worldwide (Mangum et al., 2010; Weerheijm et al., 2001). Recent translational approaches to understanding the pathogenesis of

MH have prompted molecular characterization of the abnormal enamel present in MH lesions (www.thed3group.org). The finding of extrinsic enamel proteins, such as albumin, instead of intrinsic proteins-like AMEL in MH-affected enamel has presented novel opportunities for better understanding of the enamel mineralization process and the pathogenesis of MH as well (Mangum et al., 2010).

Finally, as the routinely restorative materials used in dentistry cannot replicate enamel (mainly due to their different structure and composition), different research efforts have been made to synthesize a human-like enamel by using proteins-like AMEL to mimic the biomineralization process to restore the enamel layer (Moradian-Oldak, 2009; Tian et al., 2012).

5.3.6 Sources of further information

To get more information about oral histology (including tooth formation), please see: *Ten Cate's Oral Histology: Development, Structure, and Function* by Antonio Nanci, 8th edition, Mosby Elsevier (2012). For ameloblast cell function and its role in calcium transport during amelogenesis, the reader is referred to Mike Hubbard's work. For a deeper understanding of enamel development, please see in detail the work of Robinson, Smith, Simmers, Bartlett, and Moradian-Oldak. For more details about developmental dental defect (D3) and D3-related translational research, see Grace Suckling's work and The D3 Group website (www.thed3group.org), respectively.

5.3.7 Summary

Enamel contains an exquisitely intricate and ordered arrangement of mineral crystals packed so densely that the final product is harder than iron (Hubbard, 2000). Dental enamel formation is highly specialized and requires the concomitant activity of the ameloblasts and some specific proteins directly involved in enamel biomineralization (Simmer et al., 2010).

Enamel development or amelogenesis can be classified into two stages: secretion, where crystals grow in length; and maturation, in which crystals thicken leading to enamel hardening. Although mostly completed before tooth eruption, enamel mineralization is completed after the tooth erupts into the oral cavity by mineral ions from saliva. Ameloblast disruptions during the secretion and maturation stages may result in hypoplastic and hypomineralized enamel, respectively (Suckling, 1989). Understanding the pathologically different protein composition of D3-affected enamel may lead to new diagnostic and therapeutic approaches (Mangum et al., 2010).

5.4 Dentin mineralization

5.4.1 Introduction

As mentioned, dentin is a thick and mineralized connective tissue that comprises most of the tooth structure. Since dentin has many similarities with bone, in terms of the composition and mineralization process, an important amount of knowledge about dentin mineralization comes from *in vitro* and *in vivo* models of bone tissue. Also known as collagenous mineralization, dentin mineralization is an intricate process that entails highly orchestrated interactions of several cellular and matrix components. Despite the research on dentin and its biomineralization, there are many aspects that are still controversial. This section describes the general composition and histology of dentin, along with a broad overview of the mineralization process, with an emphasis on the role played by the different components of the tissue. Additionally, some mechanistic aspects of collagen mineralization will also be covered by the end of the section.

5.4.2 Dentin composition

The composition and structure of dentin vary considerably throughout the different parts of the tooth. As a whole, however, the mineral phase accounts for approximately 70% of the weight percentage (40–45% by volume); the organic matrix, mostly type I collagen, comprises about 20% (30% by volume); and the remaining 10% corresponds to water (20–25% by volume) (Goldberg et al., 2011).

Dentin formation begins with odontoblast differentiation. These cells undergo elongation, acquiring a cylindrical shape, synthesis of organelles, and nuclear polarization toward the proximal pole of the cell (Arana-Chavez and Katchburian, 1997). Differentiating odontoblasts start to secrete a mixture of unmineralized collagen ground and other matrix components, denominated predentin. This matrix becomes mineralized through the incorporation of noncollagenous proteins and mineral phase at the mineralization front. Predentin maintains a constant thickness of 10–40 μm between the odontoblast layer and the mineralization front. Thus, after mineralization of the outer portion, a new layer of predentin is deposited by the odontoblast. While secreting new unmineralized matrix, the odontoblasts start to extend a cytoplasmic process called the odontoblast process or Tomes fiber into the dentin matrix. At the same time, the cells begin to migrate to the center of the dental papilla (Nancy, 2013) (Figure 5.6). Tight junctions assemble between the odontoblasts and an apical domain is established on their plasma membrane as the cell becomes fully differentiated (Arana-Chavez and Katchburian, 1997). Cell migration generates a unique structural characteristic of dentin, the dentinal tubules, which are embedded within the mineralized collagen matrix through the entire length of dentin as the odontoblast recedes (Figure 5.7).

Because of the complexity of its formation, dentin is not a homogeneous tissue. Therefore, several types of dentin can be described at different stages of dentinogenesis and at different anatomical localizations. To simplify its description, dentin will be divided into three major types: mantle, circumpulpal, and root dentin, which are part of the physiological formation process. For further classifications not included in this section, supplementary sources of information will be given at the end of the section.

5.4.2.1 Mantle dentin

Measuring 5–30 μm in thickness in humans, the outermost layer in immediate contact with enamel or cementum is called mantle dentin (Linde and Goldberg, 1993).

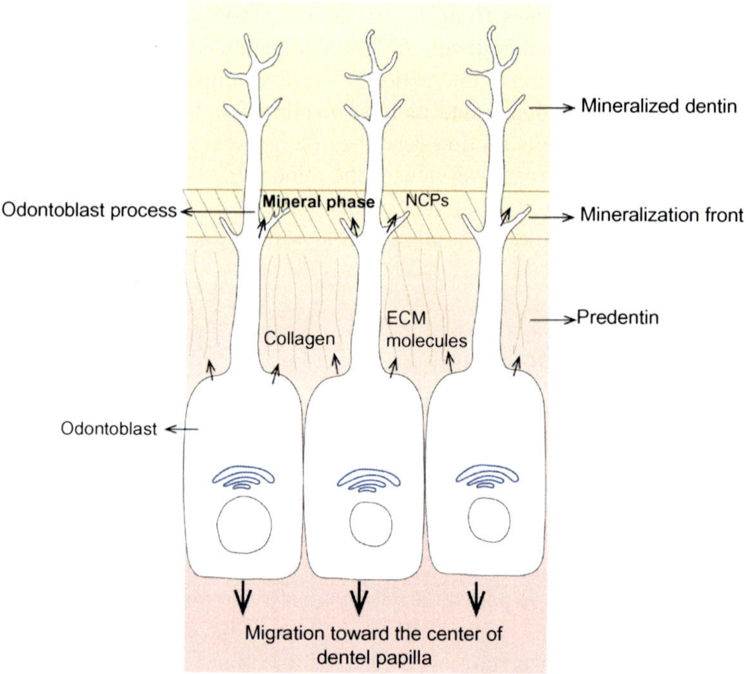

Figure 5.7 Dentin formation and mineralization. As the odontoblast starts to differentiate, various events take place such as nuclear polarization, extension of a cellular process into the dentin matrix, and migration toward the center of the dental papilla. The differentiating odontoblast starts to secrete an unmineralized matrix composed mostly of collagen I together with other extracellular matrix (ECM) molecules. The unmineralized matrix is called predentin and its thickness remains constant between the odontoblast layer and the mineralization front. Dentin mineralization takes place at the mineralization front, where the mineral phase is secreted as well as NCPs. Acidic NCPs have a dual role inhibiting and promoting mineralization in a site-specific manner. The complex interaction of these molecules results in a controlled process in which mineralization of the collagen fibers is induced.

This dentin is formed by deposition of collagen fibers, mostly type III (von Korff fibers), with a small proportion of type I collagen (Shroff and Thomas, 1992). Characteristically long and aligned perpendicular to the basal lamina, collagenous fibers originate at the odontoblast layer and extend up to the IEE (Nagata et al., 1992). In the crown, mantle dentin is mostly atubular, either depleted or with few thin curved tubules. Due to its lower mineralization than the rest of the tissue, mantle dentin is considered to be responsible for the elastic properties of the teeth, acting as a cushion to dissipate stress (Tesch et al., 2001). However, another study has shown that this resilience might not only be limited to mantle dentin, but rather extends 200 μm into the thickness of the circumpulpal dentin (Wang and Weiner, 1998). The mineralization process of mantle dentin is thought to occur through budding membrane-bound matrix vesicles at the distal end of the odontoblasts. Matrix vesicles contain crystals that are deposited to form a continuous layer of mineralized matrix (Takano et al., 2000). The matrix vesicle-mediated mechanism is only observed at this stage of dentin formation.

5.4.2.2 Circumpulpal dentin

After mantle dentin mineralization, the odontoblast layer starts the secretion of thinner and packed type I collagen fibers together with the production and secretion of noncollagenous proteins that participate in the regulation of the mineralization process. The newly deposited matrix is usually called circumpulpar dentin and it accounts for the bulk of dentin. Circumpulpar dentin is termed primary or secondary according to the stage of development. Primary dentin corresponds to the circumpulpar dentin secreted until completion of the root formation or until the tooth becomes mechanically functional. Dentin is continuously deposited throughout life, but that secreted after the tooth is completely formed is called *secondary* dentin. Although its composition and structure are similar to primary, the rate of deposition of secondary dentin is slower. The steady secretion of secondary dentin throughout life constricts the amplitude of the pulp chamber because it is mostly deposited at the roof and the floor of the chamber. Besides the slight differences observed between primary and secondary dentin, a more accentuated S shape of the dentinal tubules is found in secondary dentin (Nancy, 2013). A third type of dentin, often called tertiary dentin, is an irregular kind of collagen matrix deposited in response to external irritants to protect the pulp from injury, and will not be covered in this chapter.

5.4.2.3 Root dentin

Root dentin is formed by odontoblasts differentiated from epithelial cells from Hertwig's root sheath. In general, the formation of root dentin is very similar to the circumpulpal dentin described above. However, certain differences are of interest. For instance, collagen fibers present different orientation and organization in root mantle dentin compared to crown mantle dentin. The reason for this is that collagen fibers of the root connect with fibers extending from the cementum layer, causing the change in direction (Nancy, 2013).

5.4.3 Dentin histology

As mentioned before, the presence of dentin tubules is unique to dentin formation and relevant to describe the histological characteristics of dentin. The tubules contain the odontoblast process and dentinal fluid for most of their extension. They lengthen from the DEJ to the pulp, forming a network for nutrient supply. Their "S"-shaped configuration represents the migration track of the odontoblasts toward the pulp. The dentinal tubules display a tapered lumen of 1 μm in diameter in the mid portion of dentin, reaching a diameter of 2.5 μm near the pulp (Nancy, 2013). The tubular density in the outer and middle dentin ranges from 18,000 to almost 60,000 tubules per mm^2, whereas in deep dentin it varies between 21,000 and 82,000 tubules per mm^2 (Schilke et al., 2000). Due to these regional differences, significant variations in dentin structure can be detected from the DEJ to the pulp chamber. Moreover, increased branching and ramifications from the main tubule are observed in areas with lower tubule density (Mjor and Nordahl, 1996).

According to its histological characteristics, dentin can be divided into peritubular and intertubular dentin. Dentinal tubules are concentrically surrounded by peritubular (or intratubular) dentin. The spaces between the tubules are occupied by intertubular dentin. The latter is mostly composed of deposited type I collagen fibers reinforced with apatite crystals (Marshall et al., 1997). The mineral phase consists of two parallelepiped-shaped crystals 2–3 nm thick and 60 nm long, located either along or in between the fibers (Goldberg et al., 2011; Linde and Goldberg, 1993).

Peritubular dentin is hypermineralized when compared with intertubular dentin and forms the walls of the dentinal tubule. The model for peritubular dentin comprises thin plate-shaped crystals arranged in layers similar to the intertubular dentin, but porous in nature (Gotliv and Veis, 2007, 2009), and reduced collagenous content (Weiner et al., 1999). Indeed, recent data using SEM in combination with time-of-flight secondary ion mass spectroscopy suggested the absence of collagen fibers (Gotliv and Veis, 2007). Instead, the organic composition of this dentin is a mixture of glutamic acid-containing proteins and phospholipids, such as phosphatidylserine, phosphatidylinositol, and phosphatidylcholine, along with the presence of glycosaminoglycans. Peritubular mineralization is hypothesized to occur by adsorption along the lumen of the tubules of amorphous mineral matrix and not from the transition of predentin into dentin as occurs in intertubular dentin. The origin of the mineral phase is still unclear, but it is assumed to be secreted by odontoblasts or originated from serum-containing ECM molecules involved in matrix formation (Goldberg et al., 2011).

5.4.4 The process of dentin biomineralization

Dentin mineralization is considered to be a dynamic process, in which matrix secretion is sequentially followed by crystal deposition. Three different types of mineralization processes are described to participate in dentin mineralization. The first is based on matrix vesicles, as it occurs in bone biomineralization (Takano et al., 2000). The second involves transformation of predentin into dentin with collagen deposition, ECM molecule secretion, and the mineral phase playing a major role in mineralization (Nancy, 2013). The third potential mechanism involves passive deposition of serum-derived molecules along the wall of the dentinal tubules that would be responsible for peritubular mineralization (Goldberg et al., 2011). In this section, the role of several cellular and matrix components will be described to provide a deeper insight into current knowledge on dentin biomineralization.

5.4.4.1 Odontoblasts' role in dentin biomineralization

In order to achieve full mineralization of the collagenous ECM, a large influx of calcium and phosphate ions is needed to form HAp. After differentiation, odontoblasts are aligned at the surface of the pulp, forming palisade-like structures. Gap, desmosome-like, and tight junctions keep the plasma membrane together (Nancy, 2013). Despite the presence of these junction structures, odontoblast cells form a selective barrier that allows minerals and blood-serum-derived components to diffuse between the cells, though with minimal contribution to the final mineral volume. Most of the matrix, including collagenous or noncollagenous components involved in the mineralization process, are secreted and controlled by the odontoblasts (Goldberg et al., 1995a). Since some molecules are deposited at proximal predentin and others are secreted at the mineralization front, this process appears to be site-specific. In this scenario, odontoblasts play a critical role in the transport of Ca^{2+} ions to the mineral front (Linde and Lundgren, 1995). Early data suggested the presence of a transcellular pathway for Ca^{2+} transportation during dentinogenesis. The rationale for this mechanism is derived from studies that showed impaired ion transport when odontoblasts' microtubules were disturbed or when Ca^{2+} channels were blocked (Lundgren and Linde, 1992). Further investigations reported that odontoblasts are capable of Ca^{2+} uptake through Ca^{2+} voltage-gated channels (L type), through nonvoltage-gated, agonist-sensitive Ca^{2+} uptake pathways (Lundgren and Linde, 1997). These mechanisms would be responsible for calcium influx into the cell. Two calcium-extrusion mechanisms have been identified for calcium release into the ECM: Na^+/Ca^{2+} antiports (Tsumura et al., 2010) and a Ca^{2+}-ATPase (Borke et al., 1993; Lundgren and Linde, 1988). The precise mechanism for intracellular transport remains unknown.

Although little is known about inorganic phosphate ion (P_i) transport in odontoblasts, recent data have shown the expression of type II high-capacity $Na^+{-}P_i$ cotransporters in a rat odontoblast-like cell line. Also, the gene products displayed changes in the glycosylated state that were associated with an increase in P_i uptake by the cells (Lundquist et al., 2002). The mechanisms for P_i extrusion and transport still remain unknown.

5.4.4.2 *Role of the ECM in dentin biomineralization*

The ECM of dentin consists of mineral (HAp), collagen, noncollagenous matrix proteins (NCPs), extracellular organelles, and water. As in bone, HAp mineral crystals of dentin are spatially arranged, with their long axis parallel to the axis of collagen. The dynamic process of dentin mineralization is site specific and has varying rates of apatite deposition. Interactions among all the structural components of the ECM in the formation of dentin are thought to be crucial for biomineralization.

Collagen is the main constituent of the organic matrix. It accounts for 90% of the dentinal organic component. Type I collagen is the predominant type, found in the form of a heteropolymer with two α1(I) chains and one α2(I) chain (Hodge and Petruska, 1963). Traces of type III and type V collagen have been reported as well (Lukinmaa et al., 1993; Palosaari et al., 2001; Sodek and Mandell, 1982). The triple-helical polypeptide is arranged in a quasi-hexagonal form with parallel self-assembly of 300-nm-long collagen units. Their ends are separated by 40 nm, creating open spaces called holes/gaps. The fibrils are staggered with the adjacent fibril by 67 nm leaving an overlapped zone of 27 nm of densely packed fibers (Hodge and Petruska, 1963; Orgel et al., 2006).

For many years the role of collagen was underestimated to be only a passive organic scaffold, mostly because early studies showed that collagen alone was not capable of inducing a controlled mineralization process (Hunter et al., 2001; Saito et al., 1997). However, evidence suggests that both the structure and assembly of collagen are essential in order to act as an active template for mineralization. In the classical model of collagen organization, consecutive assembly of 2D arrays into 3D structures occurs with precise registration of the holes and overlap regions that leads to the formation of channels through the collagen structure (Landis et al., 1993, 1996; McEwen et al., 1991). Collagen spatial organization is used as a guide for crystal deposition in parallel arrays within type I collagen, with preferential growth in the axial direction of the fibrils (Weiner and Traub, 1986). Although mineral precipitates preferentially in the holes of the fibrils (Hodge, 1989; McEwen et al., 1991), crystals have been observed to expand beyond the gap zone and spread into the interstices of the fibrils (Landis et al., 1993; Traub et al., 1989). The higher mineral concentration at the gap leads to a banded periodic appearance. While mineral deposition within the hole zone is considered intrafibrillar, that occurring on the outer surface of the fibrils is called interfibrillar or extrafibrillar mineralization (Landis et al., 1996) (Figure 5.8). The exact mechanisms driving each mineralization mode are still being investigated.

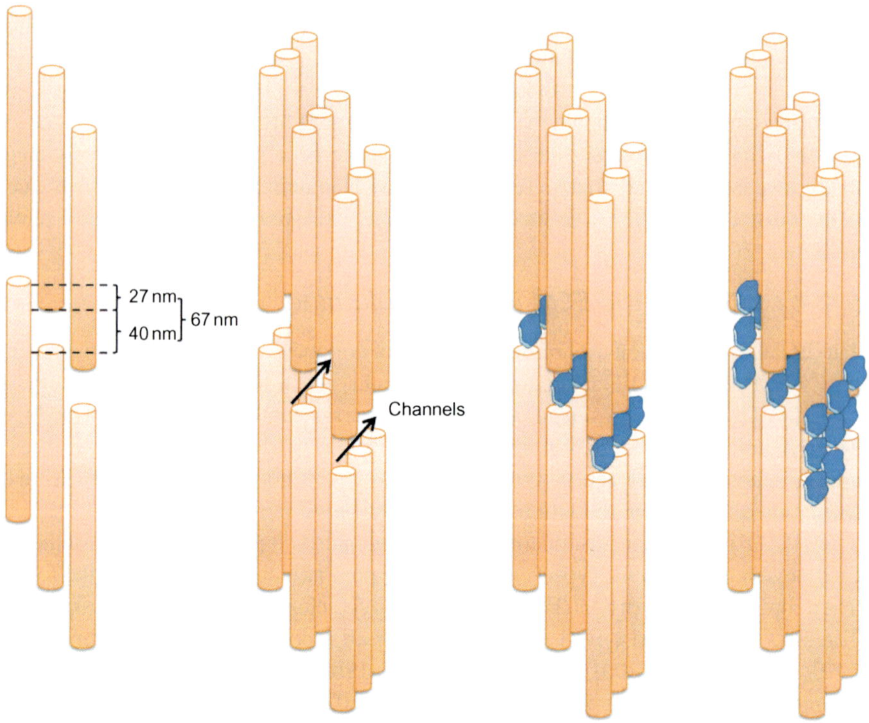

Figure 5.8 Collagen mineralization model proposed by Landis et al. (1993). The scheme shows how collagen molecules are staggered with their neighboring molecules by 67 nm, generating an open region called a hole/gap of 40 nm and an overlapped region of 27 nm. Assembly of collagen molecules from 2D to 3D organization leads to the formation of channels or gaps within collagen structures. The formed channels are the primary sites for nucleation of mineral crystals within collagen (shown in blue; intrafibrillar mineralization). After nucleation and mineralization within the gap zones, the crystals extend to the interstices between the molecules (interfibrillar mineralization).

Recent studies reported that collagen requires the participation of inhibitors of HAp nucleation to control mineralization. Indeed, an experimental approach used the negatively charged polyaspartic acid to mimic the function of NCPs in combination with ACP (Nudelman et al., 2010). The concomitant presence of both molecules led to the formation of a complex interacting with the positively charged C-terminal end of collagen. This interaction promoted the infiltration of the mineral phase into the fibrils. Furthermore, it was noted that charged amino acids, both in the gap and overlap regions, formed clusters that acted as nucleation sites, controlling the orientation of apatite crystals. Moreover, a recent *in vitro* study has also challenged the idea of collagen as a passive scaffold. The study showed that type I collagen-like matrix can promote the growth of carbonated apatite mineral, even in the absence of other ECM molecules. Besides the role in promoting mineral deposition, collagen organization has been shown to influence the size and 3D allocation of apatite crystals (Wang et al., 2012). Indeed, the analysis of the collagen structure revealed putative candidate sites for Ca^{2+} and PO^{3-} binding. It was also suggested that the structure and stereochemistry of the amino acidic residues of collagen may mediate crystal growth direction (Silver and Landis, 2011). Wang et al.'s study further supported the idea that charged groups in the collagen structure could deliver nucleation sites that induce nucleation of apatite. Taken together, all this novel information on the role of collagen have helped to unveil and expand the involvement of collagen in several steps of dentin biomineralization, such as controlling the infiltration of calcium phosphate into the collagen, nucleation, and crystal orientation.

5.4.4.3 Role of NCPs in dentin biomineralization (SIBLINGs)

Several acidic proteins have been recognized to be active in promoting or inhibiting mineral deposition. One group of proteins receiving major attention is the small integrin-binding ligand N-linked glycoprotein (SIBLING) family. This group of proteins constitutes the major group of NCPs in both bone and dentin, and includes: osteopontin (OPN), bone sialoprotein (BSP), dentin matrix protein 1 (DMP1), dentin sialophosphoprotein (DSPP), and matrix extracellular phosphoglycoprotein (MEPE) (Fisher et al., 2001). All of these proteins have shown the ability to bind to some specific ECM components or cells, and the capacity to interact and bind Ca^{2+} ions. They are intrinsically disordered, with relatively random structures and an open conformation that allows them to interact with a variety of other matrix components (Evans, 2003; George and Veis, 2008). Their importance in mineralization comes from studies in which the lack of individual SIBLINGs causes defective mineralization *in vivo* (Maciejewska and Chomik, 2012; Xiao et al., 2001; Zhang et al., 2001). Nonetheless, some level of redundancy in their function was suggested, as none of the proteins induced total suppression of mineralization.

Several studies have demonstrated that DMP1 is a multifunctional protein with a relevant role in odontoblast differentiation and mineral nucleation events (He et al., 2003a; He and George, 2004; Qin et al., 2007). Studies using recombinant DMP1 (rDMP1) have shown that the protein undergoes self-assembly into a β-sheet configuration only in the presence of calcium (He et al., 2003a,b). This finding led to the concept that DMP1 oligomerization temporarily stabilizes the newly formed calcium phosphate precursors by sequestering and preventing their further aggregation and precipitation (He et al., 2005). Moreover, peptide mapping revealed collagen-binding sites at the DMP1 C-terminal (He and George, 2004). Later experiments demonstrated that, in the presence of type I collagen, both the full-length rDMP1 and the phosphorylated native DMP1 (p-DMP1) induce HAp nucleation and growth, while the N-terminal domain inhibited HAp formation and stabilized the amorphous mineral phase (Gajjeraman et al., 2007). Interestingly, DMP1 has been localized in the peritubular dentin, which lacks collagen fibrils. This finding suggests that, *in vivo*, DMP1 might be involved in mineral organization outside the collagen fibril and in the mineralization of peritubular dentin (Beniash et al., 2011). Further studies will contribute more information to better comprehend the function, but it is likely that DMP1 function is controlled by its state of phosphorylation. Therefore, DMP1 might have a dual role involving crystal growth inhibition and promotion of mineral nucleation.

DSPP is highly expressed in odontoblasts and transiently expressed in ameloblasts (Begue-Kirn et al., 1998; D'Souza et al., 1997). This protein is cleaved into two main products: dentin sialoprotein (DSP) derived from the DSPP N-terminal and dentin phosphoprotein (DPP), or phosphorin, from the C-terminal region. Mutations in the DSPP gene have been associated with human dentinogenesis imperfecta type II/III, suggesting its involvement in the mineralization process (McKnight et al., 2008). In fact, knockout (KO) mice studies have demonstrated that deletion or modification of the protein affected dentin development (von Marschall et al., 2012) and mineralization, generating similar defects to human dentinogenesis imperfecta III (Sreenath et al., 2003). DPP, one of the cleavage products, was actually discovered much earlier than its precursor (Veis and Perry, 1967) and it is indeed the most abundant NCP in dentin ECM, accounting for 50% of the NCPs (MacDougall et al., 1985). DPP is highly expressed and secreted directly at the dentin mineralization front by polarized odontoblasts (D'Souza et al., 1997). This protein is considered as a phosphate carrier, as 85–90% of the Ser residues are phosphorylated (Butler et al., 1983; Fujisawa and Sasaki, 1983; Sabsay et al., 1991). Binding assays of DPP to collagen fibrils have shown that DPP attaches to a specific band in the hole region of collagen, which suggests a possible regulation of mineral deposition within the gap region (Traub et al., 1992). In addition, the high levels of Asp and phosphorylated serine make DPP a very polyanionic macromolecule that binds large amounts of calcium with relatively high affinity. Expressed by odontoblasts and secreted into the ECM, the DSP fragment is less abundant. Studies using conditional DPP-KO mice to isolate the role of DSP showed a partial rescue of the phenotype with significant dentin volume formation, but with lower mineral density. Based on these results, the authors suggested that DSP might be implicated in the initiation of dentin mineralization (Suzuki et al., 2009).

Other less investigated proteins may also play important roles. In dentin, however, a potential function during dentinogenesis is yet to be elucidated. For example, BSP, initially isolated from bone, exhibits strong Ca^{+2}-binding properties (Zurick et al., 2013). *In vitro*, BSP has been shown to promote HAp nucleation by interacting with collagen (Baht et al., 2008). Likewise, OPN is a negatively charged acidic protein that contains a collagen-binding moiety (Lee et al., 2007). Several *in vitro* studies revealed that

OPN has either an inhibitory or enhancing effect on HAp formation depending on its level of phosphorylation and concentration (Gericke et al., 2005; Hunter et al., 1994, 1996; Pampena et al., 2004). One of the mechanisms that would explain its inhibitory effect is based on the adsorption of the phosphate groups to the HAp crystal, preventing further crystal growth, but the specific interaction is not yet fully understood (George and Veis, 2008). With a putative role in phosphate homeostasis, MEPE is highly expressed in differentiated odontoblasts and it has been shown to inhibit mineralization (MacDougall et al., 2002). An acidic serine/aspartic acid-rich motif located at the C-terminal of MEPE has been identified as a strong mineralization inhibitor after enzymatic cleavage (Addison et al., 2008; Salmon et al., 2013). A recent study reported abnormal localization of MEPE and OSP in human dentin in patients with X-hypophosphatemia rickets (XLH), suggesting a role of both proteins in the impaired dentin mineralization observed in XLH (Salmon et al., 2014).

Overall, all these proteins are active players in the mineralization process, showing multifunctional roles that will affect mineralization. These proteins will either act to inhibit or promote mineralization depending on their concentration, their phosphorylated state, the degree of other posttranslational modifications, and if they are present in solution or bound to some ECM component.

5.4.4.4 Role of other noncollagenous proteins and ECM components

The role of other extracellular components is under intense investigation. *In vivo* studies have identified osteocalcin (OCN) and osteonectin in fully differentiated human odontoblasts (Papagerakis et al., 2002). A recent immunolocalization study in turkey tendon revealed that OCN is localized at the surface, outside and within the collagen fiber. In the presence of Ca^{2+}, OCN binds to specific bands in the collagen fibril, suggesting that OCN might mediate nucleation and crystal growth (Chen et al., 2014).

Evidence for other EMC components, such as biglycan (BGN) and decorin (DCN) (members of the small leucine-rich repeat proteoglycans family), comes from DSPP-KO mouse studies, in which increased levels of both components were found in the altered dentin phenotype. Using a double-KO mouse (DSPP/BGN or DCN) it was shown that the absence of DCN greatly rescues the DSPP −/− phenotype. In addition, the alteration of the mineralization process was even more accentuated in the absence of BGN compared to the DSPP-null background (Haruyama et al., 2009). Targeted disruption of BGN and DCN displays a hypomineralized dentin phenotype in the newborn mice, but the effect is transient and dentin is recovered at the adult state (Goldberg et al., 2005). These results suggest a complex interaction among different matrix components that must work in an orchestrated way in order to achieve proper dentin mineralization. The precise individual roles need to be further investigated.

Several phospholipids have been isolated from dentin: phosphatidylcholine, phosphatidylethanolamine, sphingomyelin, phosphatidylinositol, phosphatidylserine, and phosphatidic acid. Based on the fact that some of them can be only recovered after dentin demineralization, phospholipids might be associated with the mineral phase

(Goldberg and Septier, 2002). Other studies have detected phospholipids between collagen fibers in rat incisor predentin and in dentin as needle-like structures along mineralizing collagen fibers (Goldberg et al., 1995b).

5.4.4.5 Role of mineral amorphous phase

For many years, it was thought that mineralization followed the classical crystal growth theory in which ionic crystals precipitated from an aqueous physiological solution to form crystal clusters that initiated mineralization. Whereas many studies support this theory, others propose the existence of an ACP precursor (Posner, 1974; Termine and Posner, 1966). However, the lack of evidence of the presence of an ACP precursor in *in vivo* studies led the former idea to predominate. Recent *in vitro* studies, however, generated a new shift in this paradigm. Gower's group work revisited the crystallization process using an *in vitro* biomimetic approach. In their model, crystallization occurred via a mineral precursor. The authors pioneered the concept that the formation of liquid nanoprecursors surrounded by polyanionic polymers may play a fundamental role in biomineralization (Gower, 2008; Olszta et al., 2003a,b). In addition to these data, evidence of the presence of transient precursors has now been found in *in vivo* studies. Crane et al. (2006) showed the existence of an octacalcium phosphate-like mineral precursor in mouse calvarian tissues. Also, Beniash et al. (2009) showed that newly formed enamel mineral is ACP that later transforms into HAp crystals. Overall, these studies have changed the classical view of the biomineralization process, although many mechanistic questions remain to be clarified.

5.4.4.6 Process of intrafibrillar collagen mineralization

At least two main hypotheses have been proposed to explain the intrafibrillar mineralization in the presence of a precursor mineral phase. One model proposes that the polyelectrolytic nature of noncollagenous acidic proteins is critical to form complexes that stabilize the transient ACP. These NCP–ACP complexes bind to the hole zones of the collagen fibrils, which facilitates the transformation from an amorphous to a crystalline phase. Crystal growth would be controlled by collagen steric hindrance of neighboring molecules and preferential adsorption of NCPs to specific apatite mineral crystallographic planes (reviewed by Beniash, 2011). Alternatively, Gower et al. proposed a different mechanism. They suggested that the anionic charged polymer acts as a directing agent, via which the conventional solution crystallization is converted into a precursor. This process generates an amorphous liquid-phase mineral precursor prior to HAp; the fluidic character of the amorphous precursor facilitates intrafibrillar mineralization as the mineral phase can infiltrate nanoscopic gaps and grooves of collagen fibrils by capillary action. The precursor then solidifies and crystallizes after loss of hydration waters, forming HAp crystals in the collagen fibrils. The mineralization patterns of the highly mineralized collagen fibrils have been shown to be indistinguishable from those occurring in natural bone, with the HAp crystals oriented along the collagen fibril axes (Gower, 2008; Olsztaa et al., 2007).

5.4.5 Dentin biomineralization: A biomimetic approach

Several biomimetic model systems have been developed to study the collagen biomineralization process. In these systems highly negatively charged polyelectrolyte and polyacid molecules are used to mimic NCPs, such as poly-aspartic acid. Noncollagenous recombinant proteins and demineralized tissue containing matrix-bound NCPs are also used to better understand their individual roles. One of these models developed by Gower's group is based on the polymer-induced liquid precursor (PILP; Olszta et al., 2003a,b). The PILP process is carried out by adding small quantities of acidic polypeptides to the remineralizing solution. The anionic polymer sequesters calcium ions, which then attract oppositely charged ions, such as phosphate or carbonate, to induce liquid–liquid phase separation in the crystallizing medium (Olsztaa et al., 2007). As mentioned above, Gower et al. have been successful in mineralizing a range of organic matrices, including intrafibrillar mineralization of collagen matrices. Hence, the PILP model has been used recently to induce remineralization of dentin lesions. Recent data showed that PILP is capable of inducing remineralization of 140-μm depth dentin lesions, in which 91% recovery of the dentin modulus was achieved after 28 days of treatment (Burwell et al., 2012). In this same line of thought, Pashley and Tay's collaborative studies have used a dual biomimetic approach to induce remineralization of artificial caries lesions (Liu et al., 2011a,b; Qi et al., 2012). Their model comprises poly-acrylic acid as an apatite-nucleation inhibitor to stabilize ACP nanoprecursors and polyvinylphosphonic acid to template apatite nucleation and growth within the collagen matrix. The results showed that both partially demineralized dentin and mineral depleted dentin were completely remineralized (Liu et al., 2011b). Further examination confirmed that intrafibrillar and extrafibrillar remineralization of the collagen fibers were obtained. Similar results were achieved when sodium trimethaphosphate or sodium tripolyphophate were used to replace polyvinylphosphonic acid (Liu et al., 2011a; Qi et al., 2012). These data demonstrate that biomimetic approaches may be a promising alternative strategy for dentin remineralization. Nonetheless, further work is needed to translate these *in vitro* achievements into a clinically applicable therapy.

5.4.6 Summary

Three different types of mineralization processes have been identified in dentin formation. The first one is only seen in mantle dentin and it is based on crystal-containing matrix vesicles, similar to bone. The second involves transformation of predentin into dentin with a more complex mechanism and participation of several extracellular molecules. The third mechanism comprises a passive deposition of serum-derived molecules along the wall of the dentinal tubules that contributes to peritubular dentin mineralization (Figure 5.9).

In order to achieve proper dentin mineralization, numerous cellular and molecular interactions must be highly orchestrated. This process is initiated by odontoblast differentiation and unmineralized collagen matrix deposition. As the cells continue to differentiate, a site-specific secretion of ECM components and mineral phase begins in a sequential manner. The course of dentin formation obeys a continuous pattern of

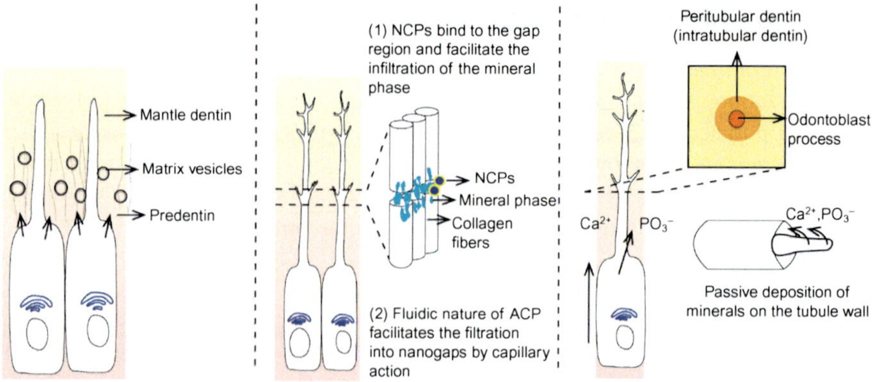

Figure 5.9 Dentin mineralization. The scheme shows a summary of the three different types of mineralization processes involved in dentin mineralization. The left panel shows a mineralization process mediated by mineral-containing matrix vesicles. This mechanism is only observed in mantle dentin mineralization. The middle panel shows the process of mineralization that involves transformation of predentin into dentin. Mineral phase and NCPs are sequentially secreted at the mineralization front. The right panel represents a third mechanism potentially responsible for peritubular mineralization with passive deposition of serum-derived molecules along the wall of the dentinal tubules.

collagen matrix deposition followed by mineralization. The 3D architecture of the collagen protein offers an active template for HAp nucleation and mineralization. NCPs and other ECM components are fundamental to achieve both stabilization of the mineral phase and induction of the mineralization process.

Current findings have proposed the presence of an ACP precursor that is later transformed into HAp mineral, which diverges from the classic mineralization theory. In the case of intrafibrillar mineralization, two main models are investigated. The first model suggests that NCP–ACP complexes bound to collagen drive the transformation of the amorphous phase to a crystalline state. The second model proposes that ACP infiltrates nanoscopic gaps and grooves of collagen fibrils by capillary action.

Finally, biomimetic models have been key to gaining valuable knowledge and comprehension of the already complex mineralization process. There is no doubt that research in this field continues to grow, revealing important concepts in order to fully understand the mechanism behind dentin mineralization.

5.4.7 Sources of further information

To get more information about oral histology (including tooth formation), please see Antonio Nanci, *Ten Cate's Oral Histology: Development, structure, and function*. 8th edition, Mosby, Elsevier, 2012.

For more information on dentin formation and structure, see Tjäderhaned, L., et al. 2011. Dentin basic structure and composition—an overview. *Endod. Top.* 20, 3–29.

For more information on odontoblasts, see Arana-Chavez, V., Massa, L., 2004. Odontoblasts: The cells forming and maintaining dentine. *Int. J. Biochem. Cell Biol.* 36, 1367–1373.
For more information on NCPs, see George, A., Veis A., 2008. Phosphorylated proteins and control over apatite nucleation, crystal growth, and inhibition. *Chem. Rev.* 108 (11), 4670–4693.
For an overview in biomineralization, see Beniash, E., 2011. Biominerals—hierarchical nanocomposites: the example of bone. *Wiley Interdiscip. Rev. Nanomed. Nanobiotechnol.* 3 (1), 47–69; Nudelman, F., et al., 2013. In vitro models of collagen biomineralization. *J. Struct. Biol.* 183, 258–269; Landis, W., Jacquet, R., 2013. Association of calcium and phosphate ions with collagen in the mineralization of vertebrate tissues. *Calcif. Tissue Int.* 93, 329–337.
For information on biomimetics and dentin remineralization, see Niu, L., et al., 2014. Biomimetic remineralization of dentin. *Dent. Mater.* 30, 77–96.

References

Addison, W.N., Nakano, Y., Loisel, T., Crine, P., McKee, M.D., 2008. MEPE-ASARM peptides control extracellular matrix mineralization by binding to hydroxyapatite: an inhibition regulated by PHEX cleavage of ASARM. J. Bone Miner. Res. 23 (10), 1638–1649.

Arana-Chavez, V.E., Katchburian, E., 1997. Development of tight junctions between odontoblasts in early dentinogenesis as revealed by freeze-fracture. Anat. Rec. 248 (3), 332–338.

Baht, G.S., Hunter, G.K., Goldberg, H.A., 2008. Bone sialoprotein-collagen interaction promotes hydroxyapatite nucleation. Matrix Biol. 27 (7), 600–608.

Bartlett, J.D., 2013. Dental enamel development: proteinases and their enamel matrix substrates. ISRN Dentistry 2013, 684607.

Bartlett, J.D., Simmer, J.P., 1999. Proteinases in developing dental enamel. Crit. Rev. Oral Biol. Med. 10 (4), 425–441.

Begue-Kirn, C., Krebsbach, P.H., Bartlett, J.D., Butler, W.T., 1998. Dentin sialoprotein, dentin phosphoprotein, enamelysin and ameloblastin: tooth-specific molecules that are distinctively expressed during murine dental differentiation. Eur. J. Oral Sci. 106 (5), 963–970.

Beniash, E., 2011. Biominerals–hierarchical nanocomposites: the example of bone. Wiley Interdiscip. Rev. Nanomed. Nanobiotechnol. 3 (1), 47–69.

Beniash, E., Simmer, J.P., Margolis, H.C., 2005. The effect of recombinant mouse amelogenins on the formation and organization of hydroxyapatite crystals in vitro. J. Struct. Biol. 149 (2), 182–190.

Beniash, E., Metzler, R.A., Lam, R.S., Gilbert, P.U., 2009. Transient amorphous calcium phosphate in forming enamel. J. Struct. Biol. 166 (2), 133–143.

Beniash, E., Deshpande, A.S., Fang, P.A., Lieb, N.S., Zhang, X., Sfeir, C.S., 2011. Possible role of DMP1 in dentin mineralization. J. Struct. Biol. 174 (1), 100–106.

Bennick, A., McLaughlin, A.C., Grey, A.A., Madapallimattam, G., 1981. The location and nature of calcium-binding sites in salivary acidic proline-rich phosphoproteins. J. Biol. Chem. 256 (10), 4741–4746.

Borke, J.L., Zaki, A.E., Eisenmann, D.R., Ashrafi, S.H., Ashrafi, S.S., Penniston, J.T., 1993. Expression of plasma membrane Ca^{++} pump epitopes parallels the progression of enamel and dentin mineralization in rat incisor. J. Histochem. Cytochem. 41 (2), 175–181.

Bouropoulos, N., Moradian-Oldak, J., 2004. Induction of apatite by the cooperative effect of amelogenin and the 32-kDa enamelin. J. Dent. Res. 83 (4), 278–282.

Briner, W., Francis, M., Widder, J., 1971. Factors affecting the rate of post-eruptive maturation of dental enamel. Calcif. Tissue Res. 7 (1), 249.

Bromley, K.M., Kiss, A.S., Lokappa, S.B., Lakshminarayanan, R., Fan, D.M., Ndao, M., et al., 2011. Dissecting amelogenin protein nanospheres: characterization of metastable oligomers. J. Biol. Chem. 286 (40), 34643–34653.

Burwell, A.K., Thula-Mata, T., Gower, L.B., Habelitz, S., Kurylo, M., Ho, S.P., et al., 2012. Functional remineralization of dentin lesions using polymer-induced liquid-precursor process. PLoS ONE 7 (6), e38852.

Butler, W.T., Bhown, M., DiMuzio, M.T., Cothran, W.C., Linde, A., 1983. Multiple forms of rat dentin phosphoproteins. Arch. Biochem. Biophys. 225 (1), 178–186.

Chen, W.Y., Bell, A.W., Simmer, J.P., Smith, C.E., 2000. Mass spectrometry of native rat amelogenins: primary transcripts, secretory isoforms, and C-terminal degradation. J. Dent. Res. 79 (3), 840–849.

Chen, L., Jacquet, R., Lowder, E., Landis, W.J., 2014. Refinement of collagen-mineral interaction: a possible role for osteocalcin in apatite crystal nucleation, growth and development. Bone 71C, 7–16.

Crane, N.J., Popescu, V., Morris, M.D., Steenhuis, P., Ignelzi Jr., M.A., 2006. Raman spectroscopic evidence for octacalcium phosphate and other transient mineral species deposited during intramembranous mineralization. Bone 39 (3), 434–442.

Crenshaw, M.A., Takano, Y., 1982. Mechanisms by which the enamel organ controls calcium entry into developing enamel. J. Dent. Res. 61, 1574–1579.

D3G, the D3 Group Website.

Deakins, M., Volker, J.F., 1941. Amount of organic matter in enamel from several types of human teeth. J. Dent. Res. 20 (2), 117.

Delak, K., Harcup, C., Lakshminarayanan, R., Sun, Z., Fan, Y.W., Moradian-Oldak, J., et al., 2009. The tooth enamel protein, porcine amelogenin, is an intrinsically disordered protein with an extended molecular configuration in the monomeric form. Biochemistry 48 (10), 2272–2281.

Diekwisch, T., David, S., Bringas Jr., P., Santos, V., Slavkin, H.C., 1993. Antisense inhibition of AMEL translation demonstrates supramolecular controls for enamel HAP crystal growth during embryonic mouse molar development. Development 117 (2), 471–482.

Dirks, O.B., 1966. Posteruptive changes in dental enamel. J. Dent. Res. 45 (3), 503.

D'Souza, R.N., Cavender, A., Sunavala, G., Alvarez, J., Ohshima, T., Kulkarni, A.B., et al., 1997. Gene expression patterns of murine dentin matrix protein 1 (Dmp1) and dentin sialophosphoprotein (DSPP) suggest distinct developmental functions in vivo. J. Bone Miner. Res. 12 (12), 2040–2049.

Du, C., Falini, G., Fermani, S., Abbott, C., Moradian-Oldak, J., 2005. Supramolecular assembly of amelogenin nanospheres into birefringent microribbons. Science 307 (5714), 1450–1454.

Dunker, A.K., Silman, I., Uversky, V.N., Sussman, J.L., 2008. Function and structure of inherently disordered proteins. Curr. Opin. Struct. Biol. 18, 756–764.

Evans, J.S., 2003. 'Apples' and 'oranges': comparing the structural aspects of biomineral- and ice-interaction proteins. Curr. Opin. Colloid Interface Sci. 8 (1), 48–54.

Fan, D., Du, C., Sun, Z., Lakshminarayanan, R., Moradian-Oldak, J., 2009. In vitro study on the interaction between the 32 kDa enamelin and amelogenin. J. Struct. Biol. 166, 88–94.

Fan, D.M., Iijima, M., Bromley, K.M., Yang, X.D., Mathew, S., Moradian-Oldak, J., 2011. The cooperation of enamelin and amelogenin in controlling octacalcium phosphate crystal morphology. Cells Tissues Organs 194 (2–4), 194–198.

Fang, P.A., Margolis, H.C., Conway, J.F., Simmer, J.P., Dickinson, G.H., Beniash, E., 2011. Cryogenic transmission electron microscopy study of amelogenin self-assembly at different pH. Cells Tissues Organs 194 (2–4), 166–170.

Fejerskov, O., 2004. Changing paradigms in concepts on dental caries: consequences for oral health care. Caries Res. 38 (3), 182–191.

Fejerskov, O., Kidd, E., Nyvad, B., Baelum, V., 2008. Dental Caries. The Disease and the Clinical Management, second ed. Blackwell Munksgaard, UK.

Fincham, A.G., Moradian-Oldak, J., Simmer, J.P., Sarte, P., Lau, E.C., Diekwisch, T., et al., 1994. Self-assembly of a recombinant amelogenin protein generates supramolecular structures. J. Struct. Biol. 112 (2), 103–109.

Fincham, A.G., Moradian-Oldak, J., Simmer, J.P., 1999. The structural biology of the developing dental enamel matrix. J. Struct. Biol. 126 (3), 270–299.

Fisher, L.W., Torchia, D.A., Fohr, B., Young, M.F., Fedarko, N.S., 2001. Flexible structures of SIBLING proteins, bone sialoprotein, and osteopontin. Biochem. Biophys. Res. Commun. 280 (2), 460–465.

Fujisawa, R., Sasaki, S., 1983. Phosphophoryns: highly phosphorylated Ca^{2+}-binding proteins in tooth. Tanpakushitsu Kakusan Koso 26, 333–344.

Fukae, M., Tanabe, T., Uchida, T., Yamakoshi, Y., Shimizu, M., 1993. Enamelins in the newly formed bovine enamel. Calcif. Tissue Int. 53 (4), 257–261.

Fukumoto, S., Kiba, T., Hall, B., Iehara, N., Nakamura, T., Longenecker, G., et al., 2004. Ameloblastin is a cell adhesion molecule required for maintaining the differentiation state of ameloblasts. J. Cell Biol. 167 (5), 973–983, Rockefeller University Press.

Gajjeraman, S., Narayanan, K., Hao, J., Qin, C., George, A., 2007. Matrix macromolecules in hard tissues control the nucleation and hierarchical assembly of hydroxyapatite. J. Biol. Chem. 282 (2), 1193–1204.

García-Godoy, F., Hicks, M.J., 2008. Maintaining the integrity of the enamel surface: the role of dental biofilm, saliva and preventive agents in enamel demineralization and remineralization. J. Am. Dent. Assoc. 139, 25S–34S.

George, A., Veis, A., 2008. Phosphorylated proteins and control over apatite nucleation, crystal growth, and inhibition. Chem. Rev. 108 (11), 4670–4693.

Gericke, A., Qin, C., Spevak, L., Fujimoto, Y., Butler, W.T., Sorensen, E.S., et al., 2005. Importance of phosphorylation for osteopontin regulation of biomineralization. Calcif. Tissue Int. 77 (1), 45–54.

Gibson, C.W., 2008. The amelogenin "enamel proteins" and cells in the periodontium. Crit. Rev. Eukaryot. Gene Expr. 18 (4), 345–360.

Gibson, C.W., Yuan, Z.A., Hall, B., Longenecker, G., Chen, E., Thyagarajan, T., et al., 2001. Amelogenin-deficient mice display an amelogenesis imperfecta phenotype. J. Biol. Chem. 276 (34), 31871–31875.

Goldberg, M., Septier, D., 2002. Phospholipids in amelogenesis and dentinogenesis. Crit. Rev. Oral Biol. Med. 13 (3), 276–290.

Goldberg, M., Septier, D., Lecolle, S., Chardin, H., Quintana, M.A., Acevedo, A.C., et al., 1995a. Dental mineralization. Int. J. Dev. Biol. 39 (1), 93–110.

Goldberg, M., Septier, D., Lecolle, S., Vermelin, L., Bissila-Mapahou, P., Carreau, J.P., et al., 1995b. Lipids in predentine and dentine. Connect. Tissue Res. 33 (1–3), 105–114.

Goldberg, M., Septier, D., Rapoport, O., Iozzo, R.V., Young, M.F., Ameye, L.G., 2005. Targeted disruption of two small leucine-rich proteoglycans, biglycan and decorin, excerpts divergent effects on enamel and dentin formation. Calcif. Tissue Int. 77 (5), 297–310.

Goldberg, M., Kulkarni, A.B., Young, M., Boskey, A., 2011. Dentin: structure, composition and mineralization. Front. Biosci. (Elite Ed). 3, 711–735.

Gotliv, B.A., Veis, A., 2007. Peritubular dentin, a vertebrate apatitic mineralized tissue without collagen: role of a phospholipid-proteolipid complex. Calcif. Tissue Int. 81 (3), 191–205.

Gotliv, B.A., Veis, A., 2009. The composition of bovine peritubular dentin: matching TOF-SIMS, scanning electron microscopy and biochemical component distributions. New light on peritubular dentin function. Cells Tissues Organs 189 (1–4), 12–19.

Gower, L.B., 2008. Biomimetic model systems for investigating the amorphous precursor pathway and its role in biomineralization. Chem. Rev. 108 (11), 4551–4627.

Habelitz, S., Marshall, S.J., Marshall, J.G.W., Balooch, M., 2001. Mechanical properties of human dental enamel on the nanometre scale. Arch. Oral Biol. 46, 173–183.

Hannig, M., Fiebiger, M., Güntzer, M., Döbert, A., Zimehl, R., Nekrashevych, Y., 2004. Protective effect of the in situ formed short-term salivary pellicle. Arch. Oral Biol. 49, 903–910.

Haruyama, N., Sreenath, T.L., Suzuki, S., Yao, X., Wang, Z., Wang, Y., et al., 2009. Genetic evidence for key roles of decorin and biglycan in dentin mineralization. Matrix Biol. 28 (3), 129–136.

Hay, D.I., Smith, D.J., Schluckebier, S.K., Moreno, E.C., 1984. Relationship between concentration of human salivary statherin and inhibition of calcium phosphate precipitation in stimulated human parotid saliva. J. Dent. Res. 63 (6), 857.

He, G., George, A., 2004. Dentin matrix protein 1 immobilized on type I collagen fibrils facilitates apatite deposition in vitro. J. Biol. Chem. 279 (12), 11649–11656.

He, G., Dahl, T., Veis, A., George, A., 2003a. Nucleation of apatite crystals in vitro by self-assembled dentin matrix protein 1. Nat. Mater. 2 (8), 552–558.

He, G., Dahl, T., Veis, A., George, A., 2003b. Dentin matrix protein 1 initiates hydroxyapatite formation in vitro. Connect. Tissue Res. 44 (1), 240–245.

He, G., Gajjeraman, S., Schultz, D., Cookson, D., Qin, C., Butler, W.T., et al., 2005. Spatially and temporally controlled biomineralization is facilitated by interaction between self-assembled dentin matrix protein 1 and calcium phosphate nuclei in solution. Biochemistry 44 (49), 16140–16148.

Hoang, A.M., Klebe, R.J., Steffensen, B., Ryu, O.H., Simmer, J.P., Cochran, D.L., 2002. Amelogenin is a cell adhesion protein. J. Dent. Res. 81 (7), 497–500.

Hodge, A.J., 1989. Molecular models illustrating the possible distributions of 'holes' in simple systematically staggered arrays of type I collagen molecules in native-type fibrils. Connect. Tissue Res. 21 (1–4), 137–147.

Hodge, A.J., Petruska, J.A., 1963. Recent studies with the electron microscope on ordered aggregates of the tropocollagen molecule. Aspects of Protein Structure. Academic Press, New York, 288–300.

Hu, C.C., Fukae, M., Uchida, T., Qian, Q., Zhang, C.H., Ryu, O.H., et al., 1997. Cloning and characterization of porcine enamelin mRNAs. J. Dent. Res. 76 (11), 1720.

Hu, J.C.C., Sun, X.L., Zhang, C.H., Simmer, J.P., 2001. A comparison of enamelin and amelogenin expression in developing mouse molars. Eur. J. Oral Sci. 109 (2), 125–132.

Hu, J.C.C., Chun, Y.-H.P., Al Hazzazzi, T., Simmer, J.P., 2007. Enamel formation and amelogenesis imperfecta. Cells Tissues Organs 186 (1), 78–85.

Hu, J.C.C., Hu, Y., Smith, C.E., McKee, M.D., Wright, J.T., Yamakoshi, Y., et al., 2008. Enamel defects and ameloblast-specific expression in Enam knock-out/lacz knock-in mice. J. Biol. Chem. 283 (16), 10858–10871.

Hu, Y.Y., Hu, J.C.C., Smith, C.E., Bartlett, J.D., Simmer, J.P., 2011. Kallikrein-related peptidase 4, matrix metalloproteinase 20, and the maturation of murine and porcine enamel. Eur. J. Oral Sci. 119, 217–225.

Hubbard, M.J., 2000. Calcium transport across the dental enamel epithelium. Crit. Rev. Oral Biol. Med. 11 (4), 437–466.

Hunter, G.K., Kyle, C.L., Goldberg, H.A., 1994. Modulation of crystal formation by bone phosphoproteins: structural specificity of the osteopontin-mediated inhibition of hydroxyapatite formation. Biochem. J. 300, 723–728.

Hunter, G.K., Hauschka, P.V., Poole, A.R., Rosenberg, L.C., Goldberg, H.A., 1996. Nucleation and inhibition of hydroxyapatite formation by mineralized tissue proteins. Biochem. J. 317, 59–64.

Hunter, G.K., Poitras, M.S., Underhill, T.M., Grynpas, M.D., Goldberg, H.A., 2001. Induction of collagen mineralization by a bone sialoprotein–decorin chimeric protein. J. Biomed. Mater. Res. 55 (4), 496–502.

Johnsson, M., Richardson, C.F., Bergey, E.J., Levine, M.J., Nancollas, G.H., 1991. The effects of human salivary cystatins and statherin on hydroxyapatite crystallization. Arch. Oral Biol. 36 (9), 631–636.

Josephsen, K., Fejerskov, O., 1977. Ameloblast modulation in the maturation zone of the rat incisor enamel organ. A light and electron microscopic study. J. Anat. 124 (Pt 1), 45–70.

Kawamoto, T., Shimizu, M., 1990. Changes in the mode of calcium and phosphate transport during rat incisal enamel formation. Calcif. Tissue Int. 46 (6), 406–414.

Krebsbach, P.H., Lee, S.K., Matsuki, Y., Kozak, C.A., Yamada, K.M., Yamada, Y., 1996. Full-length sequence, localization, and chromosomal mapping of ameloblastin. A novel tooth-specific gene. J. Biol. Chem. 271 (8), 4431–4435.

Lacruz, R.S., Smith, C.E., Kurtz, I., Hubbard, M.J., Paine, M.L., 2013. New paradigms on the transport functions of maturation-stage ameloblasts. J. Dent. Res 92 (2), 122–129.

Lakshminarayanan, R., Yoon, I., Hegde, B.G., Fan, D.M., Du, C., Moradian-Oldak, J., 2009. Analysis of secondary structure and self-assembly of amelogenin by variable temperature circular dichroism and isothermal titration calorimetry. Proteins 76 (3), 560–569.

Landis, W.J., Song, M.J., Leith, A., McEwen, L., McEwen, B.F., 1993. Mineral and organic matrix interaction in normally calcifying tendon visualized in three dimensions by high-voltage electron microscopic tomography and graphic image reconstruction. J. Struct. Biol. 110 (1), 39–54.

Landis, W.J., Hodgens, K.J., Song, M.J., Arena, J., Kiyonaga, S., Marko, M., et al., 1996. Mineralization of collagen may occur on fibril surfaces: evidence from conventional and high-voltage electron microscopy and three-dimensional imaging. J. Struct. Biol. 117 (1), 24–35.

Lee, J.Y., Choo, J.E., Choi, Y.S., Park, J.B., Min, D.S., Lee, S.J., et al., 2007. Assembly of collagen-binding peptide with collagen as a bioactive scaffold for osteogenesis in vitro and in vivo. Biomaterials 28 (29), 4257–4267.

Li, R., Li, W., DenBesten, P.K., 1995. Alternative splicing of amelogenin mRNA from rat incisor ameloblasts. J. Dent. Res. 74 (12), 1880–1885.

Linde, A., Goldberg, M., 1993. Dentinogenesis. Crit. Rev. Oral Biol. Med. 4 (5), 679–728.

Linde, A., Lundgren, T., 1995. From serum to the mineral phase. The role of the odontoblast in calcium transport and mineral formation. Int. J. Dev. Biol. 39 (1), 213–222.

Liu, Y., Li, N., Qi, Y., Niu, L.N., Elshafiy, S., Mao, J., et al., 2011a. The use of sodium trimetaphosphate as a biomimetic analog of matrix phosphoproteins for remineralization of artificial caries-like dentin. Dent. Mater. 27 (5), 465–477.

Liu, Y., Mai, S., Li, N., Yiu, C.K., Mao, J., Pashley, D.H., et al., 2011b. Differences between top-down and bottom-up approaches in mineralizing thick, partially demineralized collagen scaffolds. Acta Biomater. 7 (4), 1742–1751.

Lu, Y., Papagerakis, P., Yamakoshi, Y., Hu, J.C.C., Bartlett, J.D., Simmer, J.P., 2008. Functions of KLK4 and MMP-20 in dental enamel formation. Biol. Chem. 389 (6), 695–700.

Lukinmaa, P.L., Vaahtokari, A., Vainio, S., Sandberg, M., Waltimo, J., Thesleff, I., 1993. Transient expression of type III collagen by odontoblasts: developmental changes in the distribution of pro-alpha 1(III) and pro-alpha 1(I) collagen mRNAs in dental tissues. Matrix 13 (6), 503–515.

Lundgren, T., Linde, A., 1988. Na^+/Ca^{2+} antiports in membranes of rat incisor odontoblasts. J. Oral Pathol. 17 (9–10), 560–563.
Lundgren, T., Linde, A., 1992. Calcium ion transport kinetics during dentinogenesis: effects of disrupting odontoblast cellular transport systems. Bone Miner. 19 (1), 31–44.
Lundgren, T., Linde, A., 1997. Voltage-gated calcium channels and nonvoltage-gated calcium uptake pathways in the rat incisor odontoblast plasma membrane. Calcif. Tissue Int. 60 (1), 79–85.
Lundquist, P., Ritchie, H.H., Moore, K., Lundgren, T., Linde, A., 2002. Phosphate and calcium uptake by rat odontoblast-like MRPC-1 cells concomitant with mineralization. J. Bone Miner. Res. 17 (10), 1801–1813.
MacDougall, M., Zeichner-David, M., Slavkin, H.C., 1985. Production and characterization of antibodies against murine dentine phosphoprotein. Biochem. J. 232 (2), 493–500.
MacDougall, M., Simmons, D., Gu, T.T., Dong, J., 2002. MEPE/OF45, a new dentin/bone matrix protein and candidate gene for dentin diseases mapping to chromosome 4q21. Connect. Tissue Res. 43 (2–3), 320–330.
Maciejewska, I., Chomik, E., 2012. Hereditary dentine diseases resulting from mutations in DSPP gene. J. Dent. 40 (7), 542–548.
Mandel, I.D., 1987. The functions of saliva. J. Dent. Res. 66, 623–627.
Mangum, J.E., Crombie, F.A., Kilpatrick, N., Manton, D.J., Hubbard, M.J., 2010. Surface integrity governs the proteome of hypomineralized enamel. J. Dent. Res. 89 (10), 1160–1165.
Marshall Jr., G.W., Marshall, S.J., Kinney, J.H., Balooch, M., 1997. The dentin substrate: structure and properties related to bonding. J. Dent. 25 (6), 441–458.
McEwen, B.F., Song, M.J., Landis, W.J., 1991. Quantitative determination of the mineral distribution in different collagen zones of calcifying tendon using high voltage electron microscopic tomography. J. Comput.-Assist. Microsc. 3 (4), 201–210.
McKnight, D.A., Suzanne Hart, P., Hart, T.C., Hartsfield, J.K., Wilson, A., Wright, J.T., et al., 2008. A comprehensive analysis of normal variation and disease-causing mutations in the human DSPP gene. Hum. Mutat. 29 (12), 1392–1404.
Meckel, A.H., Griebstein, W.J., Neal, R.J., 1965. Structure of mature human dental enamel as observed by electron microscopy. Arch. Oral Biol. 10 (5), 775–783.
Miake, Y., Shimoda, S., Fukae, M., Aoba, T., 1993. Epitaxial overgrowth of apatite crystals on the thin-ribbon precursor at early stages of porcine enamel mineralization. Calcif. Tissue Int. 53 (4), 249–256.
Mjor, I.A., Nordahl, I., 1996. The density and branching of dentinal tubules in human teeth. Arch. Oral Biol. 41 (5), 401–412.
Moradian-Oldak, J., 2007. The emergence of "nanospheres" as basic structural components adopted by amelogenin. J. Dent. Res. 86 (6), 487–490.
Moradian-Oldak, J., 2009. The regeneration of tooth enamel: changing the face of restorative dental materials with synthetic enamel. Dimensions Dent. Hyg. 7 (8), 12–15.
Moradian-Oldak, J., 2012. Protein-mediated enamel mineralization. Front. Biosci. (Landmark Ed.) 17, 1996–2023.
Moradian-Oldak, J., Goldberg, M., 2005. Amelogenin supra-molecular assembly in vitro compared with the architecture of the forming enamel matrix. Cells Tissues Organs 181 (3–4), 202–218.
Moradian-Oldak, J., Simmer, J.P., Lau, E.C., Sarte, P.E., Slavkin, H.C., Fincham, A.G., 1994. Detection of monodisperse aggregates of a recombinant amelogenin by dynamic light scattering. Biopolymers 34 (10), 1339–1347.
Nagata, K., Huang, Y.H., Ohsaki, Y., Kukita, T., Nakata, M., Kurisu, K., 1992. Demonstration of type III collagen in the dentin of mice. Matrix 12 (6), 448–455.

Nanci, A., Ten Cate, A.R., 2008. Ten Cate's Oral Histology: Development, Structure, and Function/Antonio Nanci, seventh ed. Mosby Elsevier, St. Louis, MO.
Nancy, A., 2013. Ten Cate's Oral Histology. Elsevier, St. Louis, MO.
Ndao, M., Dutta, K., Bromley, K.M., Lakshminarayanan, R., Sun, Z., Rewari, G., et al., 2011. Probing the self-association, intermolecular contacts, and folding propensity of amelogenin. Protein Sci. 20 (4), 724–734.
Nudelman, F., Pieterse, K., George, A., Bomans, P.H., Friedrich, H., Brylka, L.J., et al., 2010. The role of collagen in bone apatite formation in the presence of hydroxyapatite nucleation inhibitors. Nat. Mater. 9 (12), 1004–1009.
Nyvad, B., Crielaard, W., Mira, A., Takahashi, N., Beighton, D., 2013. Dental caries from a molecular microbiological perspective. Caries Res. 47 (2), 89–102.
Olszta, M.J., Douglas, E.P., Gower, L.B., 2003a. Scanning electron microscopic analysis of the mineralization of type I collagen via a polymer-induced liquid-precursor (PILP) process. Calcif. Tissue Int. 72 (5), 583–591.
Olszta, M.J., Odom, D.J., Douglas, E.P., Gower, L.B., 2003b. A new paradigm for biomineral formation: mineralization via an amorphous liquid-phase precursor. Connect. Tissue Res. 44 (1), 326–334.
Olsztaa, M.J., Chenga, X., Jeea, S.S., Kumara, R., Kima, Y., Douglas, E.P., et al., 2007. Bone structure and formation: a new perspective. Mater. Sci. Eng. R. Rep. 58 (3–5), 77–116.
Orams, H.J., 1966. An examination of the prism core, prism sheath and interprismatic substance using the electron microscope. Aust. Dent. J. 11 (2), 93.
Orgel, J.P., Irving, T.C., Miller, A., Wess, T.J., 2006. Microfibrillar structure of type I collagen in situ. Proc. Natl. Acad. Sci. U.S.A. 103 (24), 9001–9005.
Palosaari, H., Tasanen, K., Risteli, J., Larmas, M., Salo, T., Tjaderhane, L., 2001. Baseline expression and effect of TGF-beta 1 on type I and III collagen mRNA and protein synthesis in human odontoblasts and pulp cells in vitro. Calcif. Tissue Int. 68 (2), 122–129.
Pampena, D.A., Robertson, K.A., Litvinova, O., Lajoie, G., Goldberg, H.A., Hunter, G.K., 2004. Inhibition of hydroxyapatite formation by osteopontin phosphopeptides. Biochem. J. 378 (Pt 3), 1083–1087.
Papagerakis, P., Berdal, A., Mesbah, M., Peuchmaur, M., Malaval, L., Nydegger, J., et al., 2002. Investigation of osteocalcin, osteonectin, and dentin sialophosphoprotein in developing human teeth. Bone 30 (2), 377–385.
Porto, I.M., Laure, H.J., Tykot, R.H., de Sousa, F.B., Rosa, J.C., Gerlach, R.F., 2011. Recovery and identification of mature enamel proteins in ancient teeth. Eur. J. Oral Sci. 119, 83–87.
Posner, A.S., 1974. Amorphous calcium phosphate-hydroxyapatite-bone-mineral formation analog system. J. Bone Joint Surg. Am. 4 (20).
Qi, Y.P., Li, N., Niu, L.N., Primus, C.M., Ling, J.Q., Pashley, D.H., et al., 2012. Remineralization of artificial dentinal caries lesions by biomimetically modified mineral trioxide aggregate. Acta Biomater. 8 (2), 836–842.
Qin, C., D'Souza, R., Feng, J.Q., 2007. Dentin matrix protein 1 (DMP1): new and important roles for biomineralization and phosphate homeostasis. J. Dent. Res. 86 (12), 1134–1141.
Raj, P.A., Johnsson, M., Levine, M.J., Nancollas, G.H., 1992. Salivary statherin. Dependence on sequence, charge, hydrogen bonding potency, and helical conformation for adsorption to hydroxyapatite and inhibition of mineralization. J. Biol. Chem. 267 (9), 5968–5976.
Ravindranath, H.H., Chen, L.-S., Zeichner-David, M., Ishima, R., Ravindranath, R.M.H., 2004. Interaction between the enamel matrix proteins amelogenin and ameloblastin. Biochem. Biophys. Res. Commun. 323, 1075–1083.
Reith, E.J., 1970. The stages of amelogenesis as observed in molar teeth of young rats. J. Ultrastruct. Res. 30 (1), 111–151.

Traub, W., Arad, T., Weiner, S., 1989. Three-dimensional ordered distribution of crystals in turkey tendon collagen fibers. Proc. Natl. Acad. Sci. U.S.A. 86 (24), 9822–9826.

Traub, W., Jodaikin, A., Arad, T., Veis, A., Sabsay, B., 1992. Dentin phosphophoryn binding to collagen fibrils. Matrix 12 (3), 197–201.

Travis, D.F., Glimcher, M.J., 1964. The structure and organization of, and the relationship between, the organic matrix and the inorganic crystals of embryonic bovine enamel. J. Cell Biol. 23, 447–497.

Tsumura, M., Okumura, R., Tatsuyama, S., Ichikawa, H., Muramatsu, T., Matsuda, T., et al., 2010. Ca^{2+} extrusion via $Na^{+}-Ca^{2+}$ exchangers in rat odontoblasts. J. Endod. 36 (4), 668–674.

Uchida, T., Tanabe, T., Fukae, M., Shimizu, M., Yamada, M., Miake, K., et al., 1991. Immunochemical and immunohistochemical studies, using antisera against porcine 25 kDa amelogenin, 89 kDa enamelin and the 13–17 kDa nonamelogenins, on immature enamel of the pig and rat. Histochemistry 96 (2), 129–138.

Uchida, T., Murakami, C., Dohi, N., Wakida, K., Satoda, T., Takahashi, O., 1997. Synthesis, secretion, degradation, and fate of ameloblastin during the matrix formation stage of the rat incisor as shown by immunocytochemistry and immunochemistry using region-specific antibodies. J. Histochem. Cytochem. 45 (10), 1329–1340.

Veis, A., 2003. Amelogenin gene splice products: potential signaling molecules. Cell. Mol. Life Sci. 60 (1), 38.

Veis, A., Perry, A., 1967. The phosphoprotein of the dentin matrix. Biochemistry 6 (8), 2409–2416.

von Marschall, Z., Mok, S., Phillips, M.D., McKnight, D.A., Fisher, L.W., 2012. Rough endoplasmic reticulum trafficking errors by different classes of mutant dentin sialophosphoprotein (DSPP) cause dominant negative effects in both dentinogenesis imperfecta and dentin dysplasia by entrapping normal DSPP. J. Bone Miner. Res. 27 (6), 1309–1321.

Wang, R.Z., Weiner, S., 1998. Strain-structure relations in human teeth using Moire fringes. J. Biomech. 31 (2), 135–141.

Wang, L.J., Guan, X.Y., Yin, H.Y., Moradian-Oldak, J., Nancollas, G.H., 2008. Mimicking the self-organized microstructure of tooth enamel. J. Phys. Chem. C 112 (15), 5892–5899.

Wang, Y., Azais, T., Robin, M., Vallee, A., Catania, C., Legriel, P., et al., 2012. The predominant role of collagen in the nucleation, growth, structure and orientation of bone apatite. Nat. Mater. 11 (8), 724–733.

Warotayanont, R., Zhu, D., Snead, M.L., Zhou, Y., 2008. Leucine-rich amelogenin peptide induces osteogenesis in mouse embryonic stem cells. Biochem. Biophys. Res. Commun. 367, 1–6.

Warshawsky, H., Smith, C.E., 1974. Morphological classification of rat incisor ameloblasts. Anat. Rec. 179 (4), 423–446.

Weerheijm, K.L., Jalevik, B., Alaluusua, S., 2001. Molar-incisor hypomineralisation. Caries Res. 35 (5), 390–391.

Weiner, S., Traub, W., 1986. Organization of hydroxyapatite crystals within collagen fibrils. FEBS Lett. 206 (2), 262–266.

Weiner, S., Veis, A., Beniash, E., Arad, T., Dillon, J.W., Sabsay, B., et al., 1999. Peritubular dentin formation: crystal organization and the macromolecular constituents in human teeth. J. Struct. Biol. 126 (1), 27–41.

Wiedemann-Bidlack, F.B., Beniash, E., Yamakoshi, Y., Simmer, J.P., Margolis, H.C., 2007. pH triggered self-assembly of native and recombinant amelogenins under physiological pH and temperature in vitro. J. Struct. Biol. 160, 57–69.

Xiao, S., Yu, C., Chou, X., Yuan, W., Wang, Y., Bu, L., et al., 2001. Dentinogenesis imperfecta 1 with or without progressive hearing loss is associated with distinct mutations in DSPP. Nat. Genet. 27 (2), 201–204.

Xie, Z., Swain, M.V., Hoffman, M.J., 2009a. Structural integrity of enamel: experimental and modeling. J. Dent. Res. 88 (6), 529–533.

Xie, Z.H., Swain, M.V., Swadener, G., Munroe, P., Hoffman, M., 2009b. Effect of microstructure upon elastic behaviour of human tooth enamel. J. Biomech. 8, 1075.

Yamakoshi, Y., Tanabe, T., Oida, S., Hu, C.C., Simmer, J.P., Fukae, M., 2001. Calcium binding of enamel proteins and their derivatives with emphasis on the calcium-binding domain of porcine sheathlin. Arch. Oral Biol. 46 (11), 1005–1014.

Yamakoshi, Y., Richardson, A.S., Nunez, S.M., Yamakoshi, F., Milkovich, R.N., Hu, J.C.C., et al., 2011. Enamel proteins and proteases in Mmp20 and Klk4 null and double-null mice. Eur. J. Oral Sci. 119, 206–216.

Yang, X.D., Wang, L.J., Qin, Y.L., Sun, Z., Henneman, Z.J., Moradian-Oldak, J., et al., 2010. How amelogenin orchestrates the organization of hierarchical elongated microstructures of apatite. J. Phys. Chem. B 114 (6), 2293–2300.

Yang, X., Sun, Z., Ma, R., Fan, D., Moradian-Oldak, J., 2011. Amelogenin "nanorods" formation during proteolysis by Mmp-20. J. Struct. Biol. 176, 220–228.

Zhang, X., Zhao, J., Li, C., Gao, S., Qiu, C., Liu, P., et al., 2001. DSPP mutation in dentinogenesis imperfecta Shields type II. Nat. Genet. 27 (2), 151–152.

Zurick, K.M., Qin, C., Bernards, M.T., 2013. Mineralization induction effects of osteopontin, bone sialoprotein, and dentin phosphoprotein on a biomimetic collagen substrate. J. Biomed. Mater. Res. Part A 101 (6), 1571–1581.

Biomimetic mineralization of collagen

6

L.B. Gower
University of Florida, Gainesville, FL, USA

6.1 Prolog

This chapter is not a traditional review, but rather a critical analysis of specific papers on collagen mineralization. I hope that the reader will not take offense to my rather informal and argumentative critique of the subject, and I apologize in advance to the authors of the papers that I have criticized. Often times the criticism is related to how things are worded, or where a particular bias may be clouding the analysis. I recognize that I am likely guilty of the same offenses, and may not always word things as carefully as I should (and deserve to be corrected for this); and I am certainly prone to having my own biases. I personally consider hypothesis-driven research to be inherently flawed because it is human nature to look for things that harmonize with our preconceived notions. To help amend for this, it is important to occasionally step back and look at things from alternative perspectives. Therefore, I have taken the liberty with this book chapter to bring forth some concerns and alternative perspectives pertinent to collagen mineralization, in the hope of keeping the discussion on such issues active, which is needed if we are ever to fully resolve the mechanisms of collagen infiltration, and the corresponding activities in bone formation.

6.2 *In vitro* model systems for studying mechanisms of mineralization

A primary focus of my group is the development of *in vitro* model systems for examining the physicochemical mechanisms involved in biomineralization (Gower, 2008), with a specific focus on the use of polymeric process-directing agents to modulate the mineralization reaction.

Over the years, I have found that many people in the biomineral community belittle the value of *in vitro* model systems. Although it is certainly desirable to study biological reactions *in situ*, this is simply not possible for many scenarios, particularly when deciphering molecular-level interactions within such complex media, and inside the organism. After all, one needs to first understand the basic crystallochemical reactions before one can then piece together why the cells are secreting various proteins and such. For many years, it was assumed that crystallization processes were well understood and could be readily modeled by classical nucleation theory for example, but in recent years this has been found not to be the case, and particularly in

Biomineralization and Biomaterials. http://dx.doi.org/10.1016/B978-1-78242-338-6.00007-7

the case of mineral salt reactions, for which there seems to be a variety of precursor species (or even phases) present before the actual crystallization reaction takes place (Habraken et al., 2013; Gebauer and Colfen, 2011; Weiner, 2008; Bewernitz et al., 2012). Therefore, it's not as simple as just searching for a potential "epitaxial" or stereochemical relationship between an organic matrix and crystal lattice, as was typical of the approach when I first entered into this field. *In vitro* model systems are needed because one can create a well-controlled environment in which individual parameters can be modulated independently or in logistical groupings, enabling one to decipher various signature features to suggest a role for a molecule's actions in the biological realm. Of course, this cannot prove that the same thing occurs in the biological system, but it can lead to clues that point to specific things for which to search. For example, in our case, when we found a way to achieve intrafibrillar mineralization of collagen in the beaker (as explained in detail below), we put forth the hypothesis that this might be how it occurs during bone formation, and suggested that the bone biophysicists should re-examine the issue of an amorphous precursor being utilized in bone formation (Olszta et al., 2007). An amorphous precursor had in fact already been suggested (and hotly debated) years ago (back in the 1960s–1970s), but for various reasons that concept had been dismissed (see discussion in our *Materials Science and Engineering R: Reports* paper (Olszta et al., 2007)). But in light of the fact that bone's nanostructure could now be duplicated using a simple *in vitro* model system, this revelation seemed to prompt the bone community to return to that issue, and being armed with modern characterization tools, evidence was finally found to prove that bone (at least from zebrafish fins (Mahamid et al., 2008)) is indeed formed from an amorphous precursor (as will be discussed in Section 6.3). But before we get to that, it will be useful to first see how the mineralization process works in the *in vitro* model systems that we and other groups have developed, where the reaction can be more clearly visualized at the fibrillar level (another advantage of model systems). This has enabled us to offer an explanation of the potential crystallochemical mechanism involved. However, an agreement has not even been reached for interpreting the results of the *in vitro* systems, as will be highlighted in this review.

6.2.1 How is intrafibrillar mineralization of collagen achieved with NCP mimics?

I will start with my group's model system, where acidic macromolecules are added to the crystallizing solution to investigate the role acidic proteins might play in modulating biomineralization. Acidic proteins have been known for years to be intimately associated with biominerals, both the calcium carbonate ($CaCO_3$) biominerals of invertebrate shells (Weiner and Addadi, 1991) and the calcium phosphate (CaP) biominerals of vertebrate bones and teeth (Qin et al., 2004). These proteins are often enriched with aspartic acid residues, so we chose to use polyaspartic acid-sodium salt (pAsp) in our initial model system. In actuality, most biomineral proteins do not contain as many aspartate residues, but they are further acidified with post translational modifications such as phosphorylation of serine/threonine residues, as well as glycosylation with side chains containing sulfates and carboxylates

(Qin et al., 2004) (the latter of which is a whole other level of complexity that most of us in the biomineral community avoid altogether). Although a simple poly(amino acid) seems like a rather crude mimic to proteins, we have found that the specific amino acid composition and sequences are not as important as one might think (in some aspects), because the key feature that pAsp provides is polyanionic character, which seems to promote a nonclassical crystallization pathway that we termed the polymer-induced liquid precursor (PILP) process. In fact, this degree of nonspecificity has been broadened even further by Meldrum and colleagues, who have shown that analogous reactions ($CaCO_3$ precursor droplets and films) can be induced with some polycations as well (Cantaert et al., 2012). It seems the generic reaction entails polymeric chains that are decorated with many charged groups that sequester ions and/or ion clusters, which then induces liquid–liquid phase separation of tiny nanodroplets of a hydrated amorphous mineral precursor. To be more accurate, recent studies on the $CaCO_3$ system find that a liquid condensed phase (LCP) may already be present in these mineral salt solutions (Bewernitz et al., 2012; Wolf et al., 2011b), but the polymer seems to stabilize and thereby promote much further accumulation of the liquid precursor (Bewernitz et al., 2012). With time, the liquid precursor, or what we refer to as the PILP phase, solidifies to an amorphous solid. Importantly, if the precursor is sufficiently stabilized, it can undergo a pseudomorphic transformation where the crystals retain the shape of the amorphous precursor (Gower and Odom, 2000), thereby leading to crystals with non equilibrium morphologies, the hallmark of biominerals. Indeed, our group and others have been able to emulate many biomineral morphologies with this one simple process, as described in overly abundant detail in my review of amorphous precursors (Gower, 2008). When referring to "sufficiently stabilized," I mean the phase must be stable enough to avoid undergoing dissolution and recrystallization (where it would then form a faceted morphology, which simply leads to an outcome analogous to the classical crystallization), but not too stable, so that it eventually undergoes a pseudomorphic transformation (which appears to be a pseudo-solid-state transformation), thereby retaining the morphology of the amorphous precursor.

Note that a key feature of the PILP process is that the precursor phase is so highly hydrated that it is in the form of nanodroplets, rather than nanoparticles, at least in the early stages, as seen in the $CaCO_3$ system (Gower, 1997; Gower and Odom, 2000). The liquid-like character of the CaP precursor, however, remains to be proven because the nanodroplets/particles are too small to observe on an optical microscope, as was seen in the original $CaCO_3$ PILP discovery. The liquid-like character of the $CaCO_3$ precursor was further confirmed with NMR relaxation times (Bewernitz et al., 2012), but the lower solubility of the CaP system creates sensitivity problems for NMR analysis. Nevertheless, given the analogous reaction conditions using pAsp additive, and remarkably similar results in achieving intrafibrillar mineralization for both $CaCO_3$ and CaP (Figure 6.1), we surmised that a similar process is most likely occurring (Olszta et al., 2007). Given that the phenomenon of capillary action seemed to be prevalent in the $CaCO_3$ PILP phase, where the precursor likes to seep into cracks and crevices, such as under a microscope coverslip or glass beads (see Figure 45 of our *Chem. Rev.* paper: Gower, 2008), it seemed reasonable to consider that capillary action might

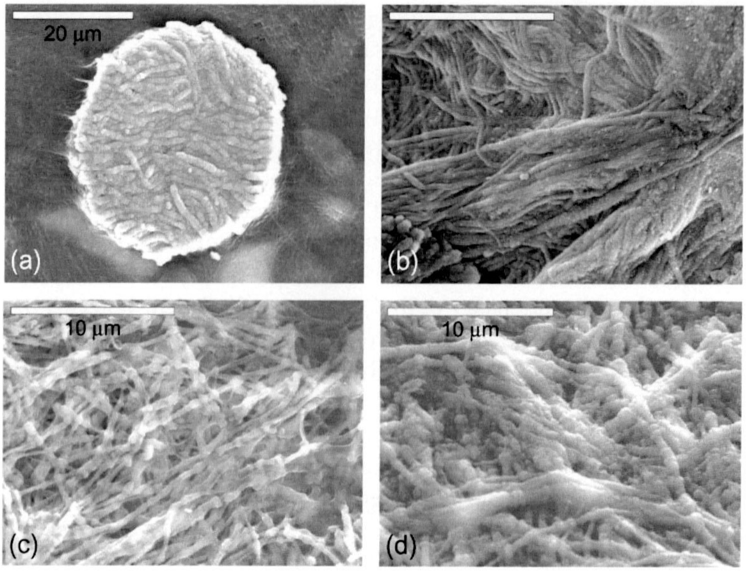

Figure 6.1 Comparison between $CaCO_3$ versus CaP mineralization of collagen sponges in the presence of pAsp additive, which are shown in (a) and (b) versus (c) and (d), respectively. (a) The heterogeneous mineralization in this sample shows a dramatic difference between the texture of the collagen that has been PILP-mineralized versus the surrounding nonmineralized collagen. In the central bright region, the mineral has replaced the water in the swollen fibrils so that they do not collapse upon drying, leaving a thicker patch with fibrillar texture as compared to the surrounding mat of collapsed collagen. Note that the fibrils also become brighter due to Z-contrast of intrafibrillar mineral, and the surface texture sometimes becomes rougher as compared to the smooth texture of purely organic collagen. Quite often, one can readily detect whether or not intrafibrillar mineralization has occurred simply by the emergence of a pronounced fibrillar texture. (b) A fully PILP-mineralized sponge shows thickened fibrils with a slightly rough surface texture, but no evidence of extrafibrillar crystals of calcite, which are readily seen in the control reaction (without pAsp) as large rhombs (see Olszta et al., 2003c). The texture is essentially identical to the CaP mineralized sponges, as seen in (d). (c) Fibrillar thickening can be seen in the CaP PILP mineralization as well, and in the early stages (day 3) the collagen fibrils sometimes have lumps and nodules if the mineral infiltration is not completely uniform. If one zooms in (on the electronic image) to examine the surface of the fibrils, they appear smooth because most of the mineral has penetrated into the interior at the early stage. (d) At 6 days, the fibrils are thicker and more uniformly extended throughout, and the surface appears rougher from deposition of some extrafibrillar mineral coating.

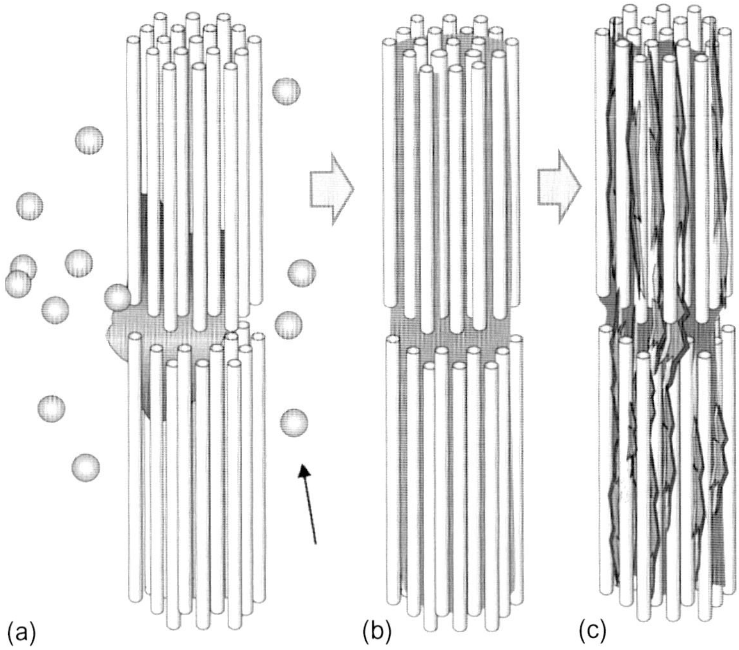

Figure 6.2 Schematic depicting our proposed PILP mechanism of intrafibrillar mineralization of collagen. Each of these pictures represents a collagen fibril in the region of a hole zone, surrounded by the aqueous mineralizing solution that contains polymer additive (e.g. pAsp). (a) The negatively charged polymer sequesters ions/clusters and induces liquid–liquid-phase separation within the crystallizing solution, forming nanoscopic droplets of a highly hydrated, amorphous calcium phosphate phase. The nanoscopic droplets of this polymer-induced liquid precursor (PILP) phase adsorb to the collagen fibril, and due to their fluidic character become wicked inside the hole zones and interstices of the collagen fibril by capillary action. (b) The collagen fibril becomes fully imbibed with the amorphous mineral precursor, which eventually solidifies to ACP as the hydration waters are excluded. (c) The ACP precursor crystallizes, leaving the collagen fibril embedded with nanoscopic crystals of hydroxyapatite.
Reprinted with permission from Olszta et al. (2007), © 2007 from the American Chemical Society.

also occur with a CaP PILP phase. This led to our hypothesis that a fluidic mineral precursor could be drawn inside the gaps and grooves of collagen fibrils by capillary action, thus providing a physical means for achieving intrafibrillar mineralization (as schematically illustrated in Figure 6.2). Because the mineral precursor seemed to infiltrate far beyond the hole zones, into the intermolecular spaces between tropocollagen molecules to generate an interpenetrating composite (Olszta et al., 2007), capillarity seemed (to us) to be the most probable mechanism for achieving that nanostructure.

6.2.1.1 Liquid-displace-liquid capillary action

Given the substantial deviation of this hypothesis from the conventional perspective of collagen mineralization reactions, it is perhaps prudent to start off with a definition of capillarity. Capillary action can be defined as the ascension of a liquid through a slim tube, cylinder, or permeable substance, as a result of the adhesive forces between the liquid and the surface. The cohesive forces within the liquid create surface tension, which acts to hold the surface intact, so instead of just the edges moving upward, the whole liquid is dragged upward, until being balanced by the weight of the liquid column. An important thing to keep in mind is the inverse relationship between transport of the liquid and radii of the capillary, where greater heights (forces) occur for narrower channels.

I have found that when I present this hypothesis of capillary infiltration of mineral precursor into collagen, people often seem befuddled, and this is probably because definitions of capillarity usually assume that the capillary tube is initially filled with air, before it is replaced by the infiltrating fluid. In the collagen system, the channels within the collagen fibril are already saturated with water. But given that capillary action is caused by surface tension and favorable adhesive forces, it seems reasonable to predict that liquid–liquid interfaces could lead to a similar action, with one liquid displacing another liquid within a narrow tube, rather than air. In the collagen system, I am referring to one of the liquids being the fluidic PILP phase, which appears to have phase boundaries (as seen in the $CaCO_3$ system), thereby creating capillary forces that could displace the aqueous-filled channels within a collagen fibril. Thus, if the PILP phase has an interface with favorable adhesive interactions with the interior of the collagen fibril, it could conceivably be wicked up into its exceedingly narrow channels. Theoretically, this seemed feasible to me, but I haven't been able to find any literature that shows liquid-displacing-liquid capillary action, so we conducted a simple set of experiments to prove to ourselves that it works, which is now posted as three videos on YouTube (narrated: http://youtu.be/XG5OGl5mgAM; clearer image of capillary infiltration: http://youtu.be/uWeswp2kHng; wet collagen wicking of dye: http://youtu.be/gVCfSwfI880).

The experiments were successful in demonstrating displacement of a nonpolar organic solvent in a glass capillary tube by a droplet of aqueous dye, as well as wicking of the polar dye droplet into a solvent-soaked collagen sponge. The reverse situation of liquids was also effective, where a nonpolar capillary tube was infiltrated when a nonpolar solvent droplet displaced the water-filled interior of thin plastic tubing (not shown here due to opacity of the plastic tubing). Of course, neither of these systems is very similar to the collagen system, which does not involve a nonpolar liquid, but instead has two aqueous-based phases (droplets of ion-enriched PILP phase sequestered by polymer, which is within a less concentrated mother solution). Nevertheless, these simple experiments serve to demonstrate the concept of liquid-displacing-liquid capillarity. Admittedly, I am not a physical chemist, so I keep hoping that someone with computational expertise will examine the physicochemical requirements for this possibility with collagen and PILP phase. The reason any of this matters is because it likely impacts the depth of penetration that can be achieved, as discussed later in Section 6.5.1.

Based on the correlations between bone nanostructure and that produced by our *in vitro* model system, we proposed the following hypothesis regarding the physicochemical processes involved in bone formation: The intrafibrillar mineralization that occurs in collagenous hard tissues (i.e. bone and dentin) may be caused by a nonclassical crystallization process that is modulated by some of the acidic noncollagenous proteins (NCPs) associated with bone, where their polyanionic character leads to an ion-sequestering ability that not only inhibits crystal nucleation but also induces a liquid-phase amorphous precursor that can be drawn into the interstices of a collagen matrix and its fibrils by capillary action, and upon sustained contact leads to solidification and crystallization of the precursor, generating collagen fibrils embedded with nanoscopic platelets of hydroxyapatite (HA).

In studying the influence of polymer molecular weight (MW; Jee et al., 2010a), we found that the higher MW pAsp led to an enhanced ability to generate purely intrafibrillar mineral (Figure 6.3). Amazingly, under these optimal conditions, the mineral

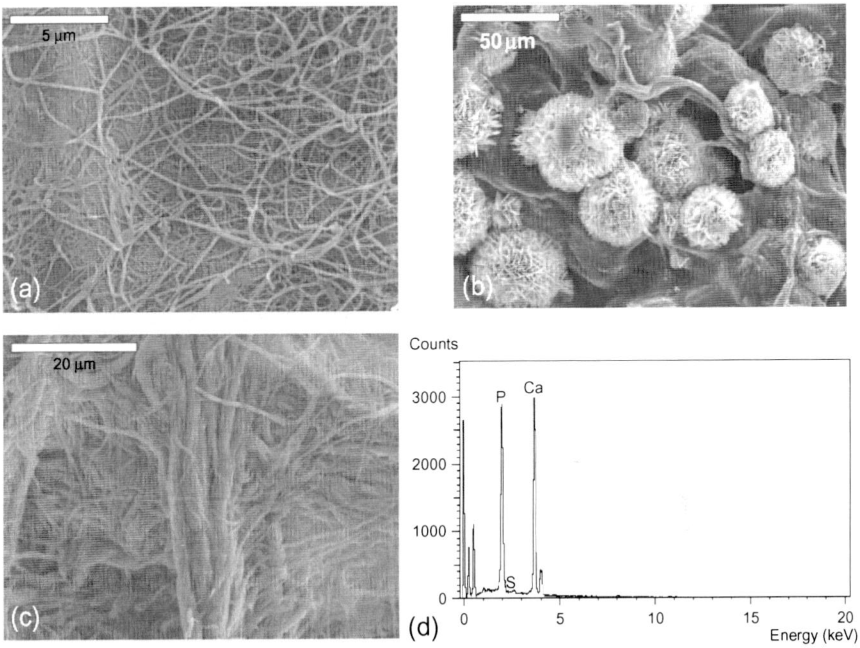

Figure 6.3 Purely intrafibrillar mineralization achieved using poly-L-aspartic acid of MW = 32,200 Da. (a) SEM of a type I collagen sponge prior to mineralization (Cellagen™). Note that this brand of sponge had a more fibrous texture than many of the commercially available collagen sponges. (b) Mineralization via the conventional reaction (without polymer additive) leads to spherulitic clusters of hydroxyapatite on the surface of the collagen sponge. Although frequently described as being "biomimetic" simply because an SBF reaction medium was employed, the mineral is clearly not intimately associated with the collagen as it is in bone. Heterogeneous nucleation of hydroxyapatite on most substrates, even dust, has this same appearance. Thus, I do not consider the structure, and therefore the reaction, to be biomimetic. (c) PILP mineralization using pAsp of relatively high MW (32 kDa) enhances the degree of intrafibrillar mineralization relative to lower MW pAsp (see Jee et al., 2010a,b). No extrafibrillar mineral is apparent in this sample (which was scanned all over); thus, the presence of large quantities of mineral, as confirmed by the following techniques, indicates that the mineral must be intrafibrillar. (d) Energy dispersive spectroscopy (EDS) shows large Ca and P peaks, demonstrating that the sponge is highly mineralized in this local region, even though no mineral can be seen on the surface.

(Continued)

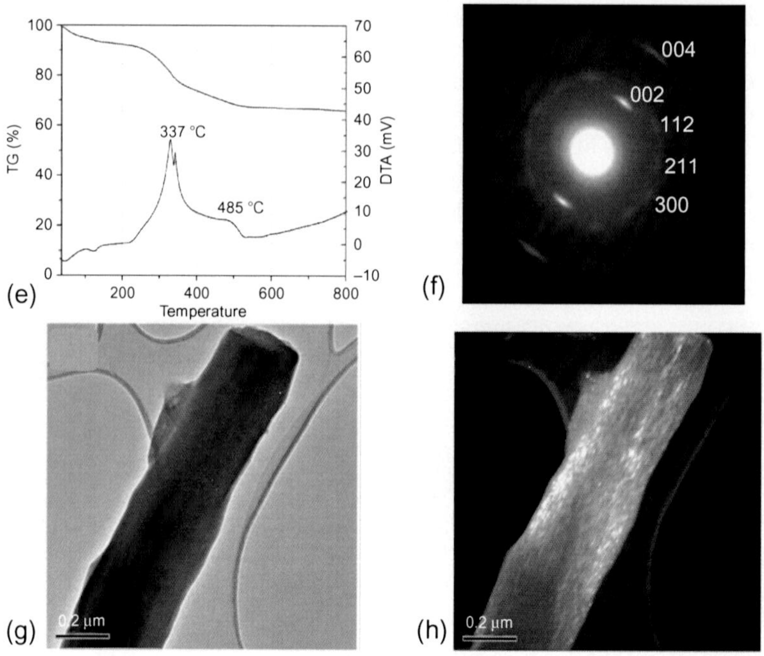

Figure 6.3 Continued. (e) Thermogravimetric analysis (TGA) definitively proves that the sponge is well mineralized throughout, with ~60 wt% mineral in this sample. In addition, the DTA curve shows a thermal decomposition peak at 330 °C, which matches that of bone and dentin. Pure collagen, without intrafibrillar mineral, has a peak around 500 °C. (f) Selected area electron diffraction (SAED) of the isolated fibril shown in (g) and (h) shows *d*-spacings and angles between reflections (see Olszta et al. (2007)) that match hydroxyapatite, and demonstrate that the crystals have preferential orientation with the crystallographic [001] direction parallel to the long axis of the collagen fibril. (g) Bright-field TEM micrograph of a PILP-mineralized collagen fibril which is not stained, but is dark due to mineral incorporation. Although not seen here, the crystals are often seen as dark streaks within the fibril. (h) Dark-field TEM micrograph (selected from [002] beam) shows that the same fibril is well laden with nanocrystals. Note that even in the high magnification TEM micrographs, none of the crystals appear to be on the surface of the fibrils. Scale bars = 0.2 µm.
(a) Reprinted with permission from Jee et al. (2010b). (b and c) Reprinted with permission from Gower (2008), © 2008 American Chemical Society. (f–h) Reprinted with permission from Jee et al. (2010a), © 2010 Acta Materialia Inc., published by Elsevier Ltd.

seemed to selectively go inside the fibrils (Figures 6.1c and 6.3c), and did not coat the surface of the fibril until the interior was fully infiltrated with mineral (Figure 6.1d). Therefore, we assumed that the precursor droplets had some preferential affinity to the hole zones, as illustrated in Figure 6.4 (Jee, 2008). In our prior studies of the $CaCO_3$ PILP system, we had frequently observed that the PILP droplets preferentially adsorbed at step edges of films and crystal defects (Kim and Gower, 2003; Kim et al., 2007), so it seemed reasonable to speculate that the proposed CaP PILP droplets might also selectively bind to high-energy defects, which in this case would be the hole zones of colla-

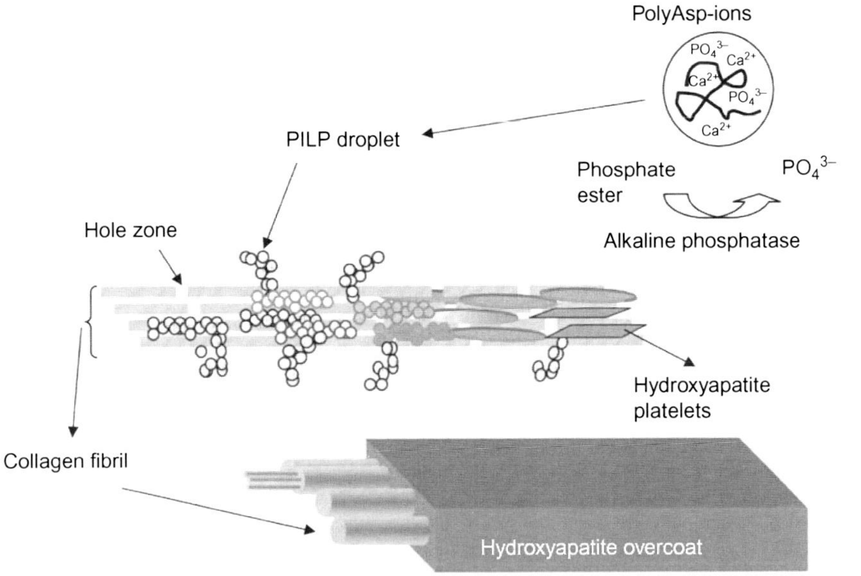

Figure 6.4 Schematic illustrating the stages that would be required for the proposed PILP mineralization mechanism to sequentially yield intra- and then extrafibrillar mineral. In this study, alkaline phosphatase was used to convert an organic phosphoester into inorganic phosphate ions to generate the mineralization reaction, mimicking the temporal control provided by enzymes in biological reactions. This graphic is a bit too simplistic in showing the extrafibrillar mineral overcoat as being a parallelepiped (it actually forms a conformal coating surrounding the fibrils, as seen in Figure 6.1d); nevertheless, the resemblance between our predicted scheme, showing nanodroplets lining up at the hole zones, with Nudelman et al.'s cryo-EM images, showing nanocomplexes lining up at the hole zones (Figure 6.6), is quite remarkable.
Reprinted with permission from Jee et al. (2010b), © 2009 Elsevier B.V.

gen. The high-energy "defects" could simply be from surface energy of chain ends, or more specific chemical domains. Alternatively, if one considers that the nanodroplets are bouncing around from Brownian motion and yet not readily adsorbing to the collagen surface (presumably due to less favorable surface energetics since the fibril surface is not coated), they may finally enter the gaps and get trapped inside the hole zones, where restraints on their Brownian motion may force them to finally settle down and attach, where they then would get rapidly pulled inside the interstices by capillary forces.

6.2.1.2 Other hypotheses on mineral infiltration into collagen

Our hypothesis of capillary infiltration was not readily accepted, which is understandable given its significant departure from classical theory on crystal nucleation and growth; and multistep nucleation mechanisms were not well recognized at that time. Deshpande and Beniash (2008) followed up on our work using a nice technique they developed to assemble the collagen fibrils directly on a TEM grid, enabling them to more readily examine the isolated fibrils *ex situ* at different stages of mineralization

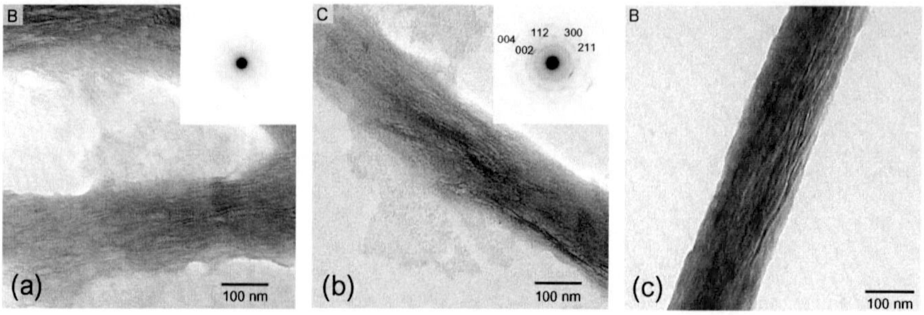

Figure 6.5 Deshpande et al. examined different stages of the mineralization process *ex situ* by assembling isolated collagen fibrils directly onto TEM grids. In the earliest stage, amorphous mineral was seen in the fibrils (a), which in the later stages became streaked with dark ribbon-like crystals (b), and which were highly oriented in the [001] direction along the fibril axis. (c) I note, in the earlier stages of (a) and (b), there also appears to be amorphous matter outside the fibrils, but for some reason this was not considered in their discussion.
Reprinted with permission from Deshpande and Beniash (2008), Copyright © 2008 American Chemical Society.

(Figure 6.5). Their reaction conditions were similar to ours, using polyaspartate additive, but with a phosphate buffer of very high concentration (9.5 mM PO_4^{3-}) rather than the Tris buffer we used in our reaction (2.1 mM PO_4^{3-}). For some reason, they considered this system to be closer to physiological conditions, based on comparison to the enamel compartment, which reportedly has a phosphate concentration of ~4 mM. In any case, we have since compared the two buffer systems, and found they give similar results, at least in the bulk collagen scaffolds we use. The most critical factor of the buffer system (aside from pH) was found to be the NaCl content (unpublished results). Regardless of the buffer, it seems the results of our two systems were similar in yielding intrafibrillar mineral, yet the authors proposed a substantially different mechanism: "our data suggest that polyAsp can induce mineralization of collagen fibrils, probably by binding to the collagen and creating local Ca^{2+} supersaturation around the fibrils, while simultaneously inhibiting mineral formation in solution." I agree that both of these things can occur; pAsp is certainly inhibitory to crystal nucleation, and indeed, the mineralization solution stays clear with no *obvious* precipitates for prolonged times when pAsp is added. But this explanation was not satisfying to me because it seems like this scenario would promote mineral nucleation on the surface of the fibrils, which we had found was not the case under optimal conditions, where the mineral is strictly intrafibrillar. One could ignore that aspect and just consider pAsp as a total inhibitor of mineral, both in solution and on the surface of the collagen, thereby only allowing crystals to nucleate and grow inside the fibrils. This seems a reasonable hypothesis, and is analogous to the hypothesis of "mineralization by inhibitor exclusion" proposed by others (Toroian et al., 2007; Price et al., 2009). But the work that followed by Sommerdijk's group seemed to prove otherwise, as will be discussed next.

Nudelman et al. (2010) utilized cryo-electron microscopy to study this same collagen mineralization system with polyaspartate additive, as well as with some acidic proteins. One of their goals was to use tomographic imaging to more definitively prove that the mineral was intrafibrillar, as we had claimed, given that the TEM images in our work (and Desphande's) are only two-dimensional (2D) projections. Actually, we felt this was already proven, not by TEM alone, but by using a combination of techniques (as described in Figure 6.3), which I recommend to those who do not have easy access to tomographic imaging. Nevertheless, and more importantly, the Nudelman paper provided valuable information on the mechanism of mineralization. Their beautiful freeze-captured images show that there actually *are* some types of colloids in solution (Figure 6.6), countering the inhibited solution hypothesis (Deshpande and Beniash, 2008; Price et al., 2009). They were careful with their wording, and referred to the colloids as "calcium phosphate–pAsp complexes," presumably because it is impossible to determine by TEM if they are liquid or solid nanoparticles. Remarkably, their freeze-captured images showed that the "complexes" seem to selectively adsorb at the gap regions of the fibrils (Figure 6.6a–c). In fact, I was amazed to see how closely their images, which show little

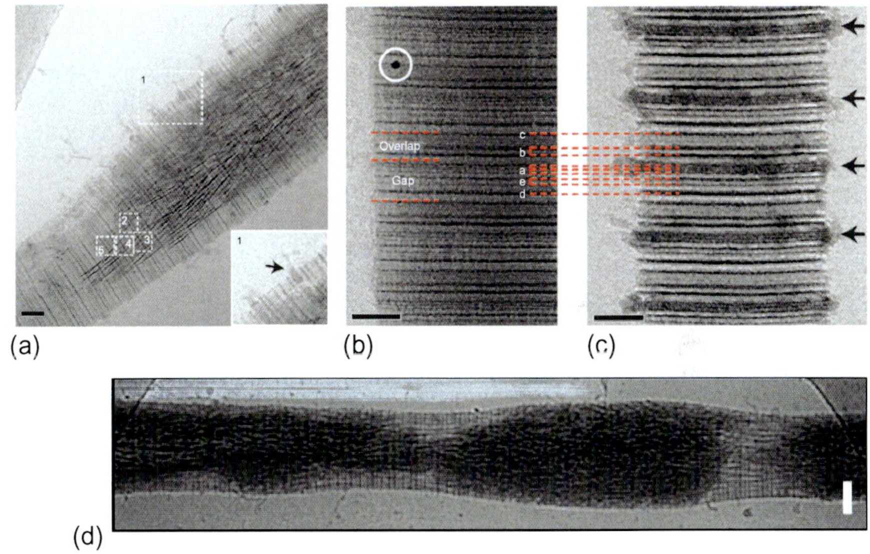

Figure 6.6 Cryo-TEM imaging study of *in situ* collagen mineralization from Nudelman et al. (2010). (a) Freeze-captured image of a collagen fibril mineralized for 48 h in the presence of 10 μg/ml pAsp. Apatite crystals are found within an amorphous calcium phosphate bed within the fibril, while on its surface there is a collection of amorphous "complexes" that seem to be lining up as they selectively infiltrate the fibril at the hole zones. Note that mineral is seen more in the interior than the exterior of the fibril. (b) Collagen fibril stained with uranyl acetate to map various staining bands, for comparison to (c), which shows amorphous nanoparticles selectively binding/entering the collagen at the *a*-bands when mineralized in the presence of 10 μg/ml of pAsp for 24 h. (d) Freeze-capture of a fibril mineralized for 48 h shows a nonuniform mineral infiltration which seems analogous to what we had observed as nodules in SEM (see Figure 6.1c). In this case, however, the collagen has not collapsed from dehydration because it is cryo-EM; apparently the collagen can become deformed from excessive mineralization. Again, the mineral appears to start at the interior and work its way outward.

(Continued)

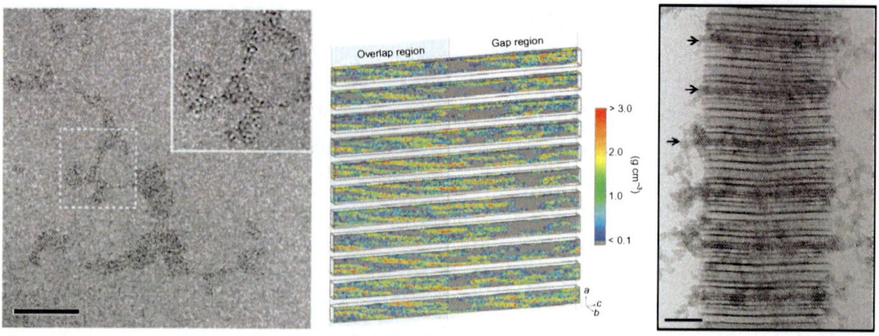

(e) (f) (g)

Figure 6.6 Continued. (e) Freeze-capture image of calcium phosphate aggregates formed after 10 min of reaction in the presence of pAsp. Inset: Higher magnification of the marked area. The authors describe these as "pAsp stabilized pre-nucleation clusters, forming loosely packed, diffuse structures that slowly aggregate and densify." Scale bars = 50 nm.
(f) Simulation of the mass density of a microfibril demonstrating the space within the microfibril through which the "complexes" could potentially "diffuse." (g) Cryo-TEM image of a collagen fibril mineralized in the presence of 15 μg/ml of C-DMP1 and 10 μg/ml pAsp for 8 h. Note the large size of the accumulated amorphous phase at this relatively early stage. It is difficult to conceive of anything this large diffusing into the hole zones and beyond, particularly between tropocollagens or microfibrils. All scale bars = 50 nm.
Reprinted with permission from Nudelman et al. (2010),

"nanocomplexes" lining up at the hole zones (Figure 6.6a), resemble the schematic we had made prior to their work (Figure 6.4), which showed little "nanodroplets" lining up at the hole zones. Our schematic was simply based on the frequent observations that the mineral precursor preferentially infiltrates into the interior of the fibril first (Figures 6.1 and 6.3), rather than forming a coating on the surface. Through staining of the collagen bands, Nudelman et al. further surmised that selective binding is occurring specifically at the *a*-bands (Figure 6.6, compare b and c), which are described as being ~9-nm-wide bands at the C-terminal region of the collagen molecules. These bands, which span the interface between the hole and overlap zones, apparently provide the entry points of the mineral precursor into the interior of the fibril. Their data not only refutes the inhibitor exclusion hypothesis, but seems to nicely support our capillary infiltration hypothesis, yet they offered a different interpretation, as will be discussed.

First, they observed the following: "at the early stages of mineralization, pAsp binds to and stabilizes the pre-nucleation clusters, forming loosely packed, diffuse structures that slowly aggregate and densify." I note that this phase has very low electron density in the early stages, lower than the amorphous calcium phosphate (ACP) that is formed later, which suggests it must contain more water than ACP. This would be consistent with a hydrated PILP phase, and the coalescence of the early phase also appears indicative of its liquid-like character (Figure 6.6e). Second, they proposed that "the positive net charge close to the C-terminal end of the collagen molecules promotes the infiltration of the fibrils with amorphous calcium phosphate (ACP)," which they seemed to consider as being nanoparticles that diffuse into and throughout the fibril. This interpretation seems (to me) difficult to rationalize, because if diffusion were the operative mechanism of entry of the colloids, there should be a concentration gradient that goes from highest to lowest, from exterior to interior of the fibrils, as these so-called complexes diffuse into the fibril. That does not appear to be the case, where mineral seems to consistently be found more prevalently at the interior of the fibril in the earlier stages of the reaction (see Figure 6.6a and d, as well as all of the other images in their paper and supplement). This can be seen even better in the 3D tomographic imaging provided in both of their Supplementary Movies. Third, if it was simply electrostatic interactions, which they propose is what attracts the particles to the *a*-bands in the first place, then it is not clear why the particles would not stick to the surface of the fibril and build up an external coating, given that the fibril surface is presumably composed of the same molecules and contains the same positively charged amino acids. Ignoring this concern and assuming the complexes can somehow selectively enter the fibril, they would then encounter the positive domains inside the collagen along their "diffusion" path, where again one would expect the electrostatic interactions to cause adsorption to those domains as they encounter them, thereby limiting their progress into the interior of the fibril and making that concentration gradient even more pronounced. Again, a declining mineral gradient is not seen in <u>any</u> of their images. In addition, it seems likely that electrostatic attractions would cause the nanoparticles to adsorb to those domains and block further particle entry as they accumulate. Instead, one sees bulges in fibrils as the mineral seems to fill them up from inside-out, as shown in their fascinating TEM image of Figure 6.6d. It is also not clear (to me) how diffusion of complexes could provide a driving force to push aside large tropocollagen molecules and cause the fibrils to deform, yet I can readily envision deformation of a fibril as it becomes imbibed with a distinctly different liquid phase via capillary action.

Fourth, the Nudelman paper provided a simulation of the mass density of a microfibril, which provides a nice visualization of the small spaces within a fibrillar compartment (Figure 6.6f). It is not clear how such large "complexes" could diffuse into these limited spaces if they were solid-like, and especially beyond the hole zones and into the intermolecular spaces between tropocollagens or microfibrils. This is particularly conspicuous for their results using the protein C-DMP1 and 10 μg/ml pAsp, where the agglomerates seem far too large to infiltrate into the gap zones if one were to assume they were solid particles (Figure 6.6g), much less the interstices of the fibril (such as between tropocollagen molecules or microfibrils); yet upon crystallization, it is clear the mineral precursor traversed well beyond the hole zones because the crystals appear continuous down the length of the fibril (see dark streaks in Figure 6.6a and d shown here, as well as numerous other images in their paper, as well as in reports on bone nanostructure). In comparison of hypotheses, a fluid precursor would be expected to infiltrate more readily into the smaller channels (capillary transport varies inversely with the radius of the capillary), thereby enabling transport of mineral in between tropocollagen molecules and/or microfibrils.

In actuality, it has been known for some time that the crystals in bone are also much larger than can possibly fit in the hole zones (Traub et al., 1989). In the pioneering tomographic imaging studies by Landis et al. (1993), which were performed on naturally mineralizing turkey tendon, not only did they observe in 3D that the single crystals grew beyond the hole zones, they also found that the crystals forming in adjacent zones appeared to "fuse together to create larger units." This fusion description would be consistent with the capabilities of an amorphous precursor phase, and particularly one with fluidic character. Although fusion can occur with amorphous solids, capillary action of a liquid precursor could enable the mineral to penetrate throughout all the interstices of the fibril, including the narrower spaces between tropocollagens or microfibrils, thereby enhancing the interconnectivity. This is skillfully demonstrated by Li and Aparicio (2013), whose high-resolution SEM and TEM images of collagen mineralization with pAsp show a subfibrillar texture (Figure 6.7), with mineral intercalated in between and surrounding the microfibrils (microfibrils, the basic subunit of collagen, are composed of five tropocollagen molecules that intertwine within a fibril). Notably, a similar subfibrillar texture was also discerned in bone collagen (Figure 6.7b). It is also worth noting that mineralized collagen in the PILP system (Olszta et al., 2007), as well as in bone (Rosen et al., 2002; Chen et al., 2011), behaves as an interpenetrating composite, and not as an assortment of isolated crystals. Thus, we argued that a liquid precursor could more readily infiltrate throughout the interstices of a fibril, as well as fuse, to ultimately provide this interconnectivity of the final mineral phase.

Overall, it seems the mechanism we proposed of capillary infiltration could help overcome these ostensibly insurmountable hurdles to selective mineral entry and penetration throughout, as long as the precursor phase is fluidic, and has phase boundaries that provide the favorable surface-adhesive forces to create capillary forces. The first aspect of this statement (as long as the precursor phase is fluidic) is an issue that deserves further discussion, and will be considered in Section 6.5.1. To summarize this section, I hope to have shown from the following four observations that the data from the Nudelman paper actually provides stronger support for our proposed mechanism of capillary infiltration of a liquid-like amorphous phase than for their <u>proposed</u> mechanisms of diffusion of solid complexes: (1) the polymer stabilizes some type of colloidal amorphous mineral precursor in the solution, which is less electron dense and thus more hydrated than ACP, and appears to coalesce like a liquid; (2) the mineral seems to fill the interior of the fibrils from inside-out (opposite of a diffusion gradient); (3) the amorphous phase selectively infiltrates into gaps in the fibrils, without forming a coating, even though the complexes have a negative zeta potential and should purportedly be attracted to the positively charged domains of collagen on the surface as well as interior; and (4) the amorphous precursor complexes and agglomerates that are visualized in the freeze-captured cryo images appear to be too large to be able to infiltrate into the gap zones, and then diffuse all throughout the interstices of a collagen fibril.

The word "proposed" is underlined above, but in reality the wording in their review article implied more (Nudelman et al., 2013). They stated that "It has been <u>demonstrated</u> that the interaction of the polymer with the incipient mineral results in a negatively charged complex that interacts with a positively charged region at the C-terminus end of the gap zone, thus mediating the infiltration of the ACP into the fibril." While I agree with the first part of the statement, where they measured zeta

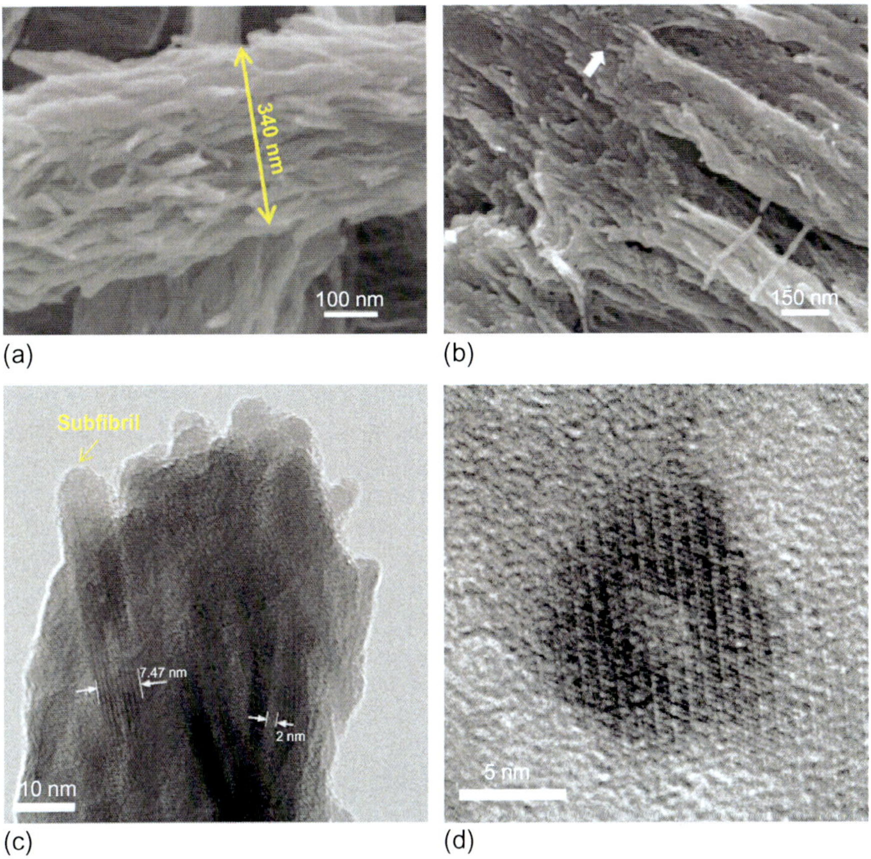

Figure 6.7 Subfibrillar texture of mineralized collagen, as discerned by the high-resolution imaging of Li and Aparicio (2013). (a) SEM of one collagen fibril, which appears to be comprised of many mineral-coated and interdigitated microfibrils. (b) SEM of the fracture surface of bovine cortical bone shows a similar texture (albeit not as pronounced), comprised of similarly sized (7–10 nm) subfibrillar constituents. One would expect that in bone collagen, the microfibrils would assemble and be crosslinked differently, so the lesser degree of microfibril splay is not surprising. Likewise, the flatter profile may arise from the lamellar organization of collagen in bone. (c) TEM lattice image of the subfibrils from within one PILP-mineralized collagen fibril. Based on the low mass contrast in the interior of the subfibrils here, and in the transverse section of image (d), the authors proposed that these entities might have a core-shell structure from mineral encapsulated microfibrils.

potential of the complexes, the rest of the statement is only their interpretation. Such statements can mislead the community to assume things were proven, which in my opinion, they were not (see further discussion on the colloidal aspects in Section 6.5.1).

Before concluding this section, the fascinating work of Niu et al. (2011, 2013) deserves mention because it shows that intrafibrillar mineralization of collagen can also be achieved with silica, and this is done using what appears to be a similar strategy of stabilizing a precursor phase that may have fluidic character (Figure 6.8). Silicic acid precursors were stabilized with choline chloride (to prevent polymerization rather than crystallization), but the polyallylamine hydrochloride (PAH) seemed to contribute by promoting sequestration of the nanoprecursors into supramolecular complexes that enabled "complete infiltration of the water-filled spaces by the liquid-like silica precursor phase," which results in "replication of the crossbanding patterns and microfibrillar architecture of the collagen fibril." In addition to the PILP-infiltration hypothesis, the authors also suggested a hypothesis analogous to that of Nudelman et al., that "PAH covers the collagen sponge surface, with its positive charge attracting the negatively charged silica species to facilitate silicification of the PAH-coated sponges." However, their TEM image in the supplement (shown here as Figure 6.8), which captures the early stage silicification of a collagen fibril, appears to show a **seeping** precursor phase with a sharply defined phase boundary, rather than a mineral concentration gradient that one would expect from diffusion (Figure 6.8c). In either case, it is certainly likely that charge groups on or within the fibrils will influence the precursor penetration, because capillary forces are strongly influenced by the surface energy of the capillary and its matching to the fluid being transported. Although further studies are needed to verify this concept of capillary infiltration in both systems, it is exciting to think that a similar infiltration strategy could be used on even nonbiological mineralization systems to create interpenetrating organic–inorganic composites.

6.2.2 How is intrafibrillar mineralization of collagen achieved without NCP mimics?

A new model system has been developed by Wang et al. (2012) which does not utilize any type of polymer additive or NCP, yet is still able to achieve intrafibrillar mineralization of collagen (Figure 6.9). They claim that their work "contradicts the mechanism proposed for the *in vitro* polymer-induced liquid-precursor collagen mineralization process." It is not clear what this is referring to, such as whether they think the mechanism we proposed for intrafibrillar mineralization is disproven, or the relevance to bone formation is not valid, so both issues will be addressed. In the latter case, it comes across throughout the paper that this group considers their *in vitro* model system to more closely match the physiological processes of bone formation. With respect to the collagen assembly, I agree that the work originally developed by Giraud-Guille et al. (2005, 2008) proficiently supports the hypothesis that the cholesteric liquid-crystalline properties of collagen molecules may be responsible for the rotating plywood organization of collagen within bone lamellae. But with respect to the mineralization aspects, I strongly disagree that their process is more physiological. One might be tempted to believe that their reaction conditions are close to physiological because the collagen matrix is continuously bathed in simulated body fluid (1.5×

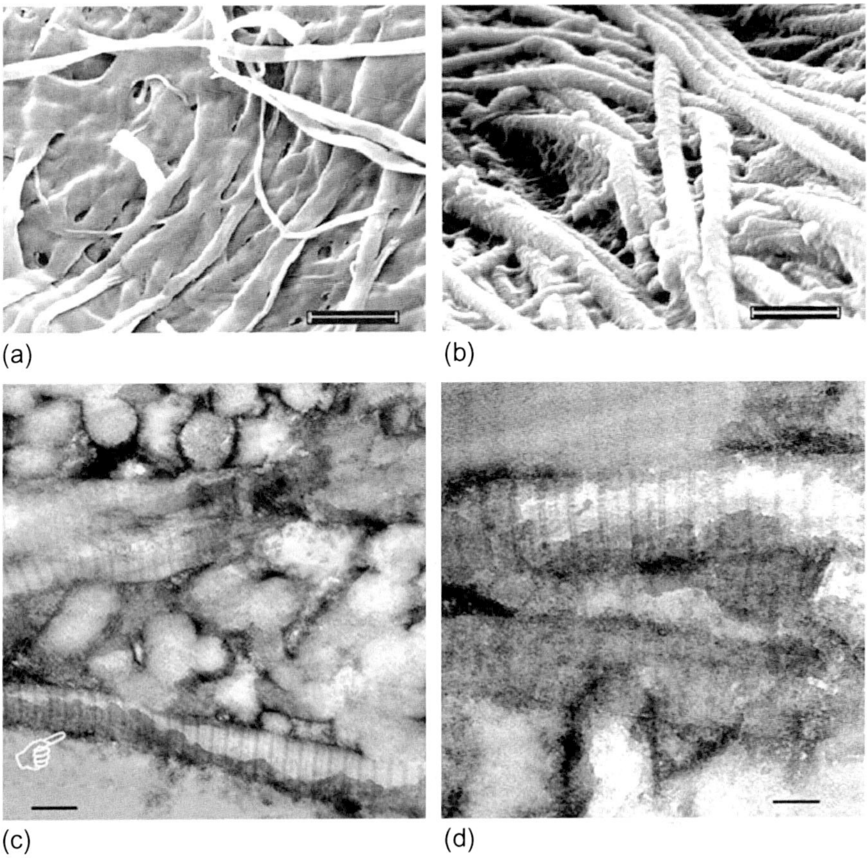

Figure 6.8 Silicification of collagen fibrils using a PAH-stabilized precursor, as presented in Niu et al. (2011). (a) SEM image of the "as received" collagen sponge. (b) SEM of the collagen sponge after treatment with PAH and silicification in choline-stabilized silicic acid shows fibrils that have plumped up after silicification, and have a rougher surface texture, similar to what we have observed with CaP mineralization via the PILP process (a and b: bar = 2 µm). (c) TEM shows a nongradient entry of mineral precursor with a sharply defined interface that appears to be seeping across the fibril. One cannot say from this image if the mineral is coating the surface or is intrafibrillar, but the overall high degree of intrafibrillar mineral was shown through a variety of techniques, and extrafibrillar mineral does not appear to be present in the SEM image of (b). (d) Above and below the mineralizing fibril (that is still bright), one can see that the more uniformly gray areas also correspond to mineralized fibrils because the banding pattern can still be discerned (c and d: bar = 200 nm and 100 nm, respectively).
Reprinted with permission for Niu et al. (2011), © 2011 Wiley-VCH Verlag GmbH & Co. KGaA, Weinheim.

SBF), but in reality the SBF alone failed to produce intrafibrillar mineral, and only produced superficial spherulitic forms of HA (Wang et al., 2012), as seen in many studies using SBF. It was only when the collagen was preloaded with a very high concentration of ions (110 mM $CaCl_2$, 33 mM NaH_2PO_4, and 33 mM $NaHCO_3$) during fibrillogenesis that subsequent intrafibrillar mineralization occurred. The concentrations

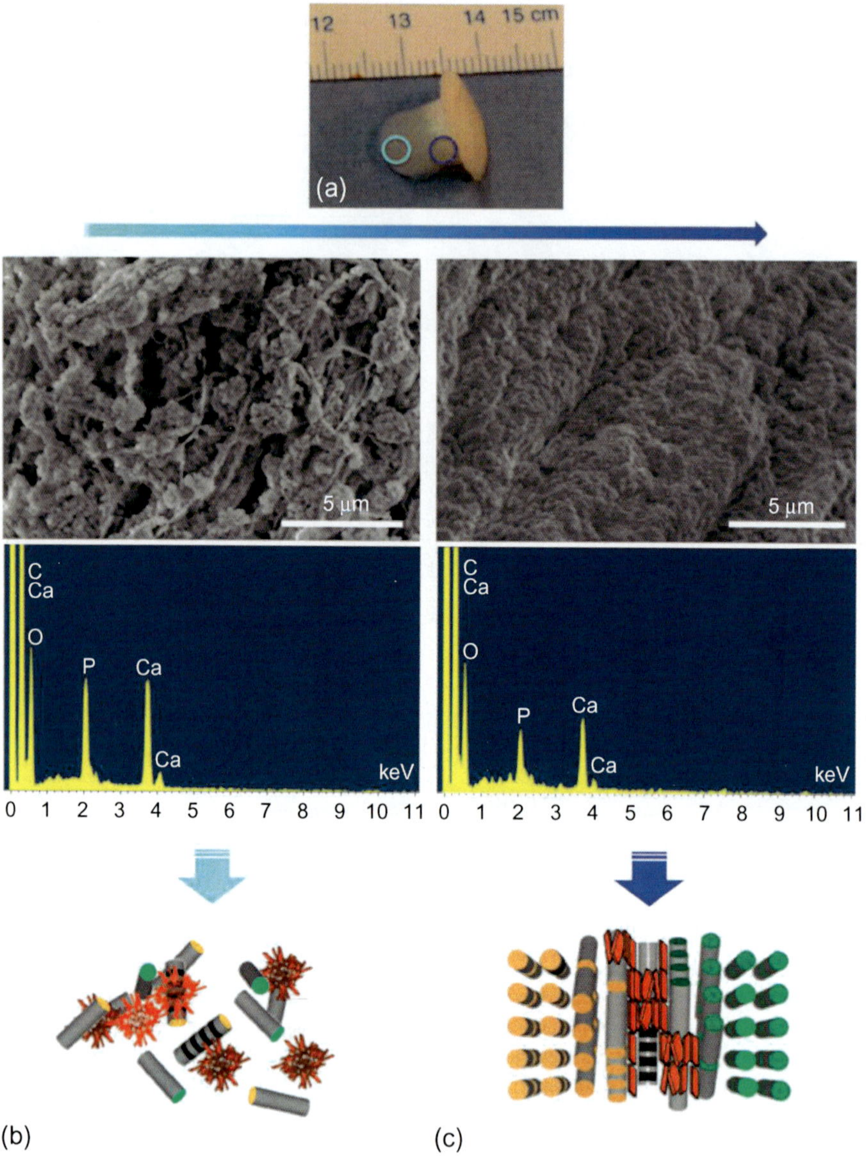

Figure 6.9 Mineralization of dense collagen matrices produced without polymer additive, as demonstrated by Wang et al. (2012). The schematics below illustrate the different organization of the two regions. (a) The dense collagen matrix is produced by liquid crystal processing, where collagen molecules self-assemble when the solution is evaporated to a concentration of at least 80 mg/ml. The blue color gradation indicates increasing collagen concentration across the matrix, from about 20 to 100 mg/ml (marked with blue circles). (b) The low-density collagen region did mineralize, as seen by the Ca and P peaks in the corresponding EDS spectrum below, but the collagen was poorly organized, and the mineral appeared to be mainly spherulitic clusters.

Figure 6.9 Continued. (c) The dense collagen region appears relatively smooth, yet the Ca and P peaks in EDS indicate that it also contains mineral, but in a more organized fashion (see TEM in their paper). I note that there seems to be somewhat less mineral in sample (c) than sample (b), as judging by a comparison of the CaP peaks relative to the oxygen peak, which correlates with the discrepancy I discovered in their TGA compositional analysis. The PILP system (and bone) shows Ca/P peaks that are larger than the C peak, so there is far less mineral here.

were only provided in the Supplement, so some readers may not have caught this supposed inconsequential information. In fact, at these levels of supersaturation (44× that of SBF), there will most surely be amorphous precipitates within the collagen as they are co-assembled, yet this issue was not discussed. I tried to bring this issue up in a letter to the editor of *Nature Materials*, but the editor rejected publishing my letter based on the authors' private rebuttal, which said that the paper did not preclude the existence of an amorphous phase, and did not contradict the implication of NCPs in mineralization *in vivo*. Upon searching the paper, one can find that ACP is indeed mentioned in the NMR section: "The Gaussian line shape of the broad signal is characteristic of a distribution of ^{31}P chemical shifts due to a variety of chemical environments in the hydrated domain, described as a putative amorphous calcium phosphate layer around synthetic and biological apatite platelets." From what I can discern, their mention of a putative ACP layer around the platelets was considered sufficient discussion of ACP by the editor. In my discussions with colleagues at conferences, and in subsequent literature reports (Nudelman et al., 2013; Cantaert et al., 2013a), it seemed as though the bone community did not realize that the collagen was preloaded with an amorphous precursor, so I feel it is important to bring this issue forward here. In addition, the paper claims that "a mineral content close to that published for mature bone was found after the immersion in SBF (~50%, Coll/CHA/SBF)" (Supplementary Fig. S3, Supplementary data associated with this article can be found, in the online version, at doi: http://dx.doi.org/10.1016/B978-1-78242-338-6.00007-7), but when one examines the thermogravimetric analysis data in Fig. S3, there is only 20 wt% mineral (residual mass). Also found in the Supplement is the NMR data showing a big broad peak, which to those who know ^{31}P NMR apparently shows there was an amorphous precursor. Their rebuttal to the editor also claimed that their paper did not suggest any contradiction to the implications of NCPs *in vivo*, yet the abstract of their paper states the following: "These results call into question recent consensus in the literature on the need for Ca-rich noncollagenous proteins for collagen mineralization to occur *in vivo*." Given that physiological conditions do not likely use a CaP concentration that is 44× that of SBF to preload the collagen with ACP (see Section 6.3), then it seems probable that the NCPs must serve this important role *in vivo* of providing a means for infiltrating the collagen with ACP, as we had originally proposed. Thus, I don't see how their reaction conditions, which are not even remotely close to being physiological, "call into question" the proposed role of NCPs, or "contradict the PILP mechanism" we had proposed.

Basically, it seems that this group has simply developed an alternative approach for getting an ACP precursor inside the fibrils, which in this case is done during fibril assembly. On the one hand, this is a nice accomplishment because it offers an alternative method for producing highly mineralized collagen which could offer certain advantages over other routes toward commercialization of bone substitutes, such as scale-up with large depths of penetration (which is a limitation of our postmineralization method, as discussed in Section 6.5.1). It also definitively proves that NCPs are not needed to control the orientation of the HA crystals (as long as an amorphous phase is within the fibrils!). On the other hand, I feel this paper is misleading in implying that their system is more physiological, and that collagen alone is sufficient for causing intrafibrillar mineralization during bone formation. The paper thus caused some of those in the bone community to return to their favorite hypothesis that collagen, all by itself, directs intrafibrillar mineralization through epitaxy or stereospecific templating. Some groups have promoted this hypothesis for many years. For example, Silver and Landis (2011) stated from their collagen simulation work that "the interaction and binding between specific charged amino acid residues of collagen and Ca^{2+} and PO_3^{-4} ions supersaturated in the fluid bathing collagen throughout its inter- and intrafibrillar spaces offer a conceptually attractive mechanism by which collagen may mineralize. The events of this nature preclude other recent suggestions based on work *in vitro* and *in situ* that collagen mineralization may be induced by the presence of exogenous polyaspartic acid (they cite our work here) and that collagen internal spaces may be mineralized." I guess the authors consider *in silico* predictions to preclude *in vitro* evidence of a viable mechanism. In fact, *ex vivo* observations of bone formation do not support their view either, as discussed in the following section. Computational simulations can be a powerful approach for predicting material properties, but one really should have supportive experimental data to validate the models. I have noticed over the years that this is often lacking in the biomineral community at large, many of whom seem to rely on faith and devotion when it comes to supporting their favorite "conceptually attractive" hypothesis with computational simulations.

To conclude this argument, the Wang data does not "contradict the PILP hypothesis," but circumvents the PILP method. Perhaps it was simply a language barrier issue with a poor choice of wording, where the authors could have more accurately stated that their method provides an alternative route toward infiltrating collagen with an amorphous precursor. Furthermore, this pre-embedding method is not physiological (as discussed in the next section), so it does not refute the proposed role of the NCPs in the biological realm. I maintain that it still seems likely that the NCPs serve a critical role in sequestering ions into some colloidal phase, which then somehow selectively enters into the fibrils, as judging by the observations of bone formation, as described in the next section.

Before leaving this section, it should be pointed out that an analogous co-precipitation/fibrillogenesis reaction had already been demonstrated by Kikuchi et al. back in 2001. They, however, did not try to present their reaction conditions as being biomimetic because they used a titration reaction between $Ca(OH)_2$ with H_3PO_4; and at lower collagen concentrations than the Wang method (Kikuchi et al., 2001, 2004), they did not produce the cholesteric collagen organization. In the Kikuchi system, it was suggested that the oriented HA–collagen nanocomposites were formed by self-organization during the concurrent crystallization and fibrillogenesis, where both are driven to occur as the pH becomes neutralized. What I wish to point out here is that in both of these co-precipitation reactions (Wang and Kikuchi), it seems that the mineral reactant species can promote fibrillogenesis of collagen, which in turn leads to an alternative pathway for achieving intrafibrillar HA that is oriented along the [001] direction. I therefore wonder if the reports that I occasionally hear about, suggesting that intrafibrillar mineralization of collagen can occur without polymeric additives, might in fact be inadvertently formed via such a co-precipitation process. In other words, I think that sometimes the collagen fibrils are not well formed to begin with, so even though the experimenter is adding the mineralization solution to what they think is a fibrillar scaffold, if some amount of gelatinous collagen is still present, it may further assemble into fibrils as the mineral reactants are added. This would then lead to some amount of intrafibrillar crystals as the poorly organized collagen further assembled and mineralized simultaneously.

6.3 *Ex situ* studies of biological mineralization of collagenous tissue

The biomineralization research group from the Weizmann Institute has been pivotal in demonstrating that many biominerals are formed by an amorphous precursor, both for $CaCO_3$ and now for CaP in bone. In the latter case, they examined the continuously forming fin bony ray of the zebrafish, where different stages of mineralization could be readily isolated and characterized by advanced techniques (Mahamid et al., 2008, 2010). Although there are many interesting facets of this landmark work, this chapter is focused on biomimetic model systems of collagen mineralization, so my discussion here will mainly point out some features from their data that correlate well with our hypotheses. Their first paper involving the zebrafish system (Mahamid et al., 2008) was impactful because it demonstrated that the early stage mineral associated with collagen was amorphous, which has been instrumental in amending the long-held view that bone formation occurs from crystallization of HA *de novo* (from solution) through a classical nucleation and growth mechanism. But it is their second paper (Mahamid et al., 2010), which mapped the mineralizing activities within the zebrafish tissue, that has some intriguing images that could be viewed from an alternative perspective. In the growth zone, they observed (with cryo-SEM) that the extracellular matrix contained spherical entities, which when viewed in backscattered mode can be seen to have bright contrast, suggesting they are mineral rich (Figure 6.10a and b). They noted that "the nonmineralized matrix contains mineral-bearing globules that exist as separate entities within the collagen matrix." This can be seen by the low contrast of the matrix, which clearly is not preloaded with ions or amorphous mineral, as was seemingly suggested by the Wang et al. (2012) paper. These "mineral-bearing globular entities" were around 200–500 nm in diameter, with Ca/P ratios (determined by EDS) of around 1.3 or more, and appeared to be comprised of a "nanogranulated material"

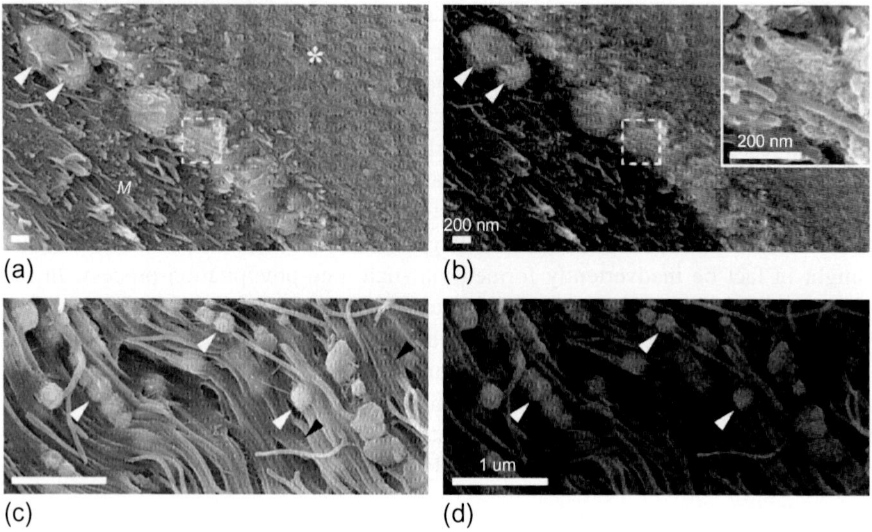

Figure 6.10 Parts of Figure 6.4 from Mahamid et al. (2010), demonstrating features of bone formation that I consider supportive of the PILP mechanism. (a and b) Cryo-SEM micrographs of the bone growth zone in secondary electron and backscatter modes respectively, to highlight where mineral is present. Neighboring the mineralized bone matrix (asterisk) is newly deposited, nonmineralized matrix (M), which contains large, "mineral-bearing globular entities" (white arrowheads) that appear to "fuse into the collagenous matrix." (Inset in b) Higher magnification of the delineated area from (a), showing the "nanosphere subparticles composing the large mineral globules." (c and d) Higher magnification of the growth zone region shows mineral-bearing globules adsorbing to the collagen fibrils and infusing them with mineral. Note that some of the fibrils shown in backscatter mode (d) become bright all along their length, even when there is some distance between them and the nearest mineral-bearing globule, yet no nanoparticles can be seen to be diffusing along the surface in these freeze-captured images. Reprinted with permission from Mahamid et al. (2010), © copyright 2008 National Academy of Sciences, USA.

(Figure 6.10b, inset). A particularly intriguing set of images is shown in Figure 6.10c and d, where they stated: "We also noted that the mineral-bearing globules appear to be *fusing* into the collagen fibrils within the growth zone." Taking this apparent "fusion" a step further, I note that where each globule had adsorbed onto a fibril, that fibril seems to have become brighter all along its length (Figure 6.10d), which indicates it is becoming mineralized, yet there does not appear to be any external release of the so-called nanoparticles from the globules (i.e. nanoparticles are not seen diffusing on or around the surface of the collagen). This would seem to indicate that the "released nanoparticles" must be traveling down the length of the fibril via some interior pathway within the fibril. It is hard to envision that the mineral could be transported that far by diffusion of solid nanoparticles. Thus, I believe that these observations lend support to the idea that these globules may have liquid-like character, because they appear to just meld into the matrix and the fibrils, and then capillary forces could be responsible for drawing their contents into and down the length of the fibril.

Nudelman et al. (2013) also seemed to question their description of nanoparticle release, and suggested that "one possibility is that infiltration of calcium phosphate into the collagen could occur through a dissolution-re-precipitation mechanism, where once deposited in the extracellular medium, the ACP could dissolve and the ions diffuse into the collagen. This hypothesis reconciles the idea that mineralization occurs through the diffusion of ions into the collagen and their subsequent nucleation into apatite." I personally do not have the goal of trying to reconcile the literature, because there is no evidence that a sufficient amount of ions can diffuse into collagen to promote nucleation within (unless the collagen is presoaked in 44× SBF, as in the Wang et al. method). Veis and Dorvee (2013) suggested something similar in their review, where they discuss matrix vesicles, saying "the questions which remain unresolved deal with the mechanisms by which the mineral is transferred from the mitochondria to the mineralizing matrix and then reordered, *in essence dissolved and recrystallized*, on the collagen fibril matrix on the fibril surfaces or within the fibrils themselves." This seems a rather broad assumption, which ignores the possibility of the capillary mechanism we had proposed, or the nanoparticle diffusion mechanism proposed by others (Nudelman et al., 2010; Mahamid et al., 2010). If the globules/vesicles did dissolve and produce a local high supersaturation, why would not the ions simply precipitate on the surface of the fibrils, as occurs in the conventional mineralization reactions performed at high supersaturation? Instead, it appears (to me) that the globules are delivering some phase that directly enters into the fibrils upon contact.

Although the "nanogranulated" texture described by Mahamid et al. does not sound like something that would be liquid-like, such nanoscale features seen in these freeze-captured SEM micrographs may be related to the nanoscale entities seen in the freeze-captured TEM micrographs of Nudelman et al. (Figure 6.6e), which were attributed to aggregates of prenucleation clusters (Nudelman et al., 2010; Dey et al., 2010). Habraken et al. (2013) have also observed nanometric clusters, polymeric aggregates, and spherical entities in their cryo-TEM studies of the various stages leading to HA formation (without polymer), so it would not be surprising that such entities are present and even stabilized by polymer in the more complex biological system containing proteins, lipids, etc. In our $CaCO_3$ PILP studies, we find a nanocolloidal texture in the final products (Amos et al., 2007), so even though the PILP phase coalesces into dense single-crystalline structures, there is some remnant memory of the colloidal precursor that one would not expect if it were a pure liquid. I think this is an issue that bothers people when considering the PILP process, which is understandable because clearly it is not clear-cut chemistry that is easy to define, particularly when the phase rapidly transforms into a gel or solid upon contact with substrates.

To conclude this section, more studies are needed to determine the actual composition and structure/consistency of these mineral-bearing globules, as well as their mechanism of infiltration into the interstices of the collagen fibril. Nevertheless, these *ex vivo* freeze-captured observations support the basic concepts of bone formation occurring through some type of sequestered and stabilized amorphous mineral precursor that is formed outside of the fibrils (presumably within the cells), which then fuses and infiltrates into a preformed collagen matrix (refuting the Wang et al. method, along with the traditional hypothesis of *de novo* nucleation via epitaxy on collagen domains). In my opinion, the globular fusion and infiltration also appear to provide support to our hypothesis of capillary infiltration.

6.4 How does collagen control crystal nucleation and orientation?

Although we had predicted in our first proposal in this area (thankfully funded by the NSF) that the PILP process might enable a mineral precursor to infiltrate the collagen fibrils, we were surprised to find that the HA crystals were oriented with their crystallographic *c*-axes parallel to the long axis of the fibrils (Figure 6.3f), just as they are in bone (Olszta et al., 2007). Even we had assumed that the NCPs would be needed to provide that level of control. The electron diffraction pattern showed spots with some arcing, as in bone, which at the time we assumed was from tilting and rotational disorder. More recent studies of the structure of collagen fibrils have shown that they consist of supertwisted microfibrils (five tropocollagen molecules) that interdigitate with one another (Orgel et al., 2006), so this may be what leads to the off-axis tilting of the crystals as they may follow the path that meanders between the microfibrils, as demonstrated by Li and Aparicio (2013), shown in Figure 6.7.

As to the oriented crystal alignment, this is clearly directed by the collagen itself, but with the important caveat that this seems to only occur when an amorphous mineral precursor is within the fibrils (if my understanding of the bemusing Wang et al. (2012) paper is correct). Many groups have shown that when HA nucleates via conventional crystallization (without polymer additive), the crystals heterogeneously nucleate to form spherulitic clusters on the surface of the fibril (Figure 6.3b), with radial or randomly oriented platelets within the cluster. Therefore, we suggested two possibilities for how the crystal orientation might be templated by the collagen (Olszta et al., 2007): (1) specific domains within the collagen may template or provide specific interactions (such as an epitaxial type of relationship) with the mineral or (2) the narrow confines within a collagen fibril may lead to constrained growth, where the naturally favored [001] rapid growth direction of HA outcompetes other directions and becomes the dominant product. Both mechanisms have since found support in the literature, as follows.

6.4.1 *Scenario 1: Specific templating interactions of collagen with mineral*

For scenario 1, it was suggested by Nudelman et al. (2010) that preferential nucleation of HA occurred at the staining bands, where "clusters of charged amino acids, both in hole and overlap regions, form nucleation sites controlling the conversion of ACP into a parallel array of oriented apatite crystals." This proposition was based on observations that the dark streaks appeared to initiate in those regions. Although the data was not entirely convincing (to me), it nevertheless seems reasonable that an amorphous phase would be triggered to transform (dehydrate and crystallize) in regions with more favorable interactions. This would not necessarily require any specific molecular recognition or epitaxy. For example, we have observed that PILP-formed amorphous $CaCO_3$ films crystallized much faster on negatively charged surfaces than on polar or uncharged surfaces (Kim et al., 2007). Even crystal orientation was influenced by surface charge/polarity alone, where simply the method of cleaning the glass coverslips with a rinse of ethanol versus acetone led to more [001] oriented calcite. Therefore, I am hesitant to refer to domains in collagen as providing epitaxy or specific control, as many groups strive to predict with computational simulations (Silver and Landis, 2011; Kawska et al., 2008; Kniep and Simon, 2007). And because collagen does not nucleate oriented HA on its surface, the simulations have advanced to now include a 3D microenvironment from a set of domains associated from different neighboring tropocollagens (Silver and Landis, 2011; Xu et al., 2015). It's not clear (to me) how epitaxy would work in a pocket of surrounding molecules, and presumably with a fair amount of flexibility and motion. It seems more likely that a pocket would just provide regions of high charge density that collect ions, or channels might foster dehydration of the amorphous precursor, both of which work to promote some form of nucleation (but not necessarily with any specific interactions, or in any particular orientation, as discussed below). Indeed, in the most recent simulations of collagen's internal structure (Xu et al., 2015), which focused specifically on the e1 and e2 bands of the hole zones (which purportedly promote nucleation of the HA), it was found that clustering of ions occurred, not as an organized lattice, but as a disordered cluster. The authors argue that the charged side groups of the collagen "templated" the formation of an amorphous CaP phase. Although I am happy that the computational community is now considering the presence of an amorphous phase, I am concerned about the context of how they use the word "templated." After all, the *in vitro* and *ex vivo* experiments find that an amorphous phase does not nucleate within preformed collagen fibrils (unless co-precipitated), but rather an amorphous phase is already formed outside the fibril, which is then somehow transported into the interior (such as by capillarity or diffusion). So any templating aspects would be more involved during transformation of the ACP phase, rather than induction of the original ACP phase. I was actually most intrigued by their data which showed a lower water density in the hole zones, which they suggest could allow desolvation of ions to more readily occur at this band. From a transformation perspective, this would likely promote the densification and ultimate crystallization of the PILP or ACP precursor, thereby fitting with Nudelman's observations.

6.4.2 Scenario 2: Orientation via confined growth

For scenario 2, Meldrum's group (Cantaert et al., 2013a) elegantly demonstrated the concept of nanoscale confinement that we had proposed in our MSE-Report (Olszta et al., 2007), by growing CaP crystals in the narrow pores of track-etch anodized alumina membranes (Figure 6.11). With pore diameters ranging from 25 to 200 nm, they found that the spatial constraints of the smaller pores not only stabilized an amorphous precursor, but the crystals that then formed were [001] oriented, as in bone. In fact, the smallest pore size of 25 nm led to diffraction arcs with even narrower widths than those of bone, even though the pore size is still considerably larger than the 2–6 nm gap regions of intermolecular spaces between tropocollagen molecules. It is interesting that nanoscale confinement from the porous membranes alone (no polymer) induced ACP to form within, yet this does not seem to happen with collagen alone (no polymer). Perhaps the hydration waters associated with collagen are held more strongly, thus requiring some other force to displace the water (maybe capillary forces).

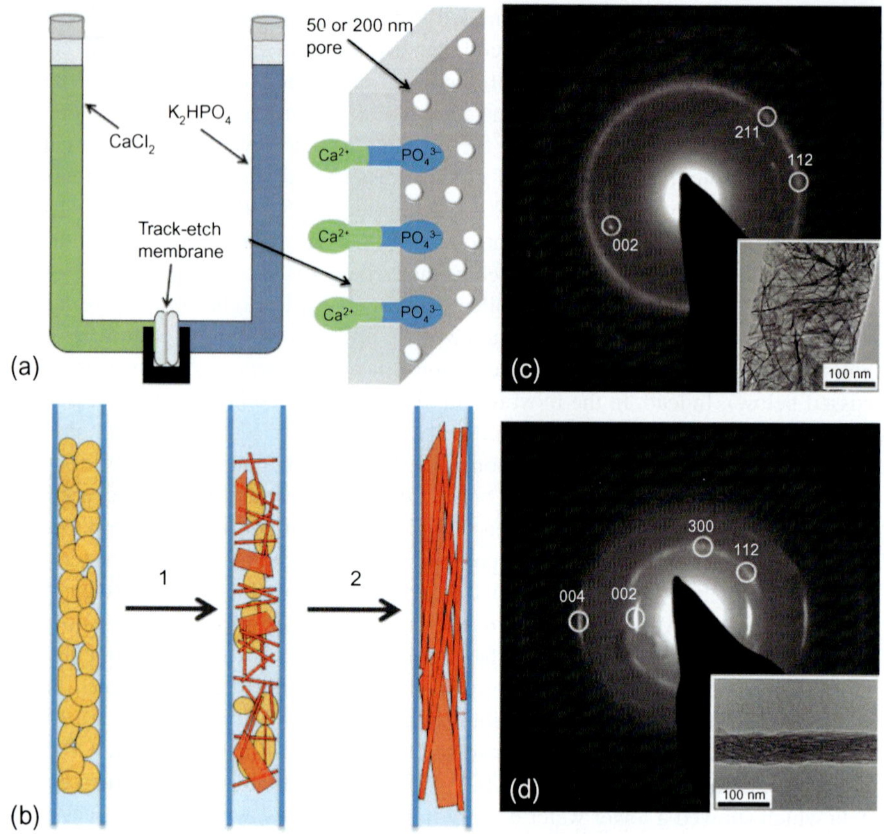

Figure 6.11 Effect of nanoscale confinement on crystallization of hydroxyapatite (HA), as demonstrated by Cantaert et al. (2013a). Interestingly, nanoscale confinement itself caused precipitation of an intermediate amorphous phase, and polymer additive was not needed. (a) Schematic illustration of their method of growing CaP within the pores of track-etch membrane, and in (b), how competitive growth leads to alignment of the crystals in their preferred growth direction. The SAED patterns and corresponding TEM images show that the HA crystals were randomly oriented when formed in 300 nm pores (c), but became [001] oriented when formed in 50 nm pores (d). The SAED pattern of (d) is very similar to bone. Reprinted with permission from Cantaert et al. (2013a), © 2013 Wiley-VCH Verlag GmbH&Co. KGaA, Weinheim.

This group did a subsequent study that included pAsp in the reaction (Cantaert et al., 2013b), this time using polycarbonate track-etch filtration membranes. They found that pAsp reduced the yield of rods and failed to increase their lengths. They argue that because the pAsp reduced the effectiveness of rod formation in the pores of track-etch membranes (unlike the $CaCO_3$ system), this "challenges the idea that a PILP phase forms in the CaP/pAsp system." While I readily admit that a CaP PILP phase has not been as definitively proven, I think one should not extrapolate to this conclusion. After all, an amorphous phase does not form within the pores of collagen without polymer, as it does in their track-etch membranes, so clearly something about these two scaffolds differs. Certainly the surface chemistry differs, and surface energetics are an important contributor to capillary forces (as well as nucleation of amorphous phases and their transformation). In addition, I believe their reaction conditions were inadequate to produce a stable PILP phase because the MW of the pAsp they used (2–11 kDa) was probably too low. This is evidenced by the large ACP solid-like particles that formed in the control reaction (without the membrane), whereas our optimal CaP conditions using 27 kDa pAsp leads to no precipitates in the control solution for at least 2 weeks in the presence of collagen. We have found that MWs below 10,300 Da do not effectively mineralize collagen, where mainly agglomerates of ACP form (Jee et al., 2010a). Regrettably, our results for pAsp of MW = 5500 Da were only included in the supplement of that Jee paper, so perhaps the Meldrum team was unaware of this limitation in the CaP PILP system, which seems to differ from the $CaCO_3$ PILP system that works well in the MW range of 10–15 kDa. It should be noted, though, that Nudelman's experiments used a pAsp of MW from 2 to 11 kDa (Nudelman et al., 2010), and they found nice mineralization. I believe the discrepancy here may be caused by the volume of the reactions. It is difficult to compare reactions performed in tiny volumes on a TEM grid, because transport distances and times are very short, thereby avoiding accumulation and solidification of less stable precursor phases, versus droplet diffusion in large-volume solutions for bulk scaffolds, which may require better stabilization of the liquid-like PILP state. In fact, my group has not been able to reproduce some of these microscale mineralizations in our bulk solution reactions, so one should be cautious when comparing these differing set ups.

In the Nudelman et al. (2013) review paper, the authors state: "the most important conclusion from these *in vitro* model studies is that collagen is not a passive scaffold as has been previously thought; rather, it actively controls and templates apatite formation during mineralization by directing ACP infiltration and mediating its nucleation into the crystalline phase." I argue that, given that the porous aluminum scaffolds create essentially identical diffraction patterns, showing the same crystal orientation is can be produced in confinement, that this aspect of the process could certainly be more passive than credit is given. After all, collagen fibrils were already present in soft tissue organisms long before bony tissues evolved, so why would specific collagen domains need to further evolve to provide a high degree of specificity when the crystals would likely have already been oriented from confinement alone? So even though one might be tempted to conclude that because the crystals appear to initiate in specific regions of the fibril, their nucleation is directed by specific interactions, the crystallographic orientation of those first formed nuclei is not known, and a dominant orientation could still have been modulated by growth in confinement. If confinement is breached, as we had found in some excessively mineralized fibrils where the HA crystals seemed to escape from the fibrils, their orientation then becomes randomized as they splay outward (Figure 6.12), seemingly following the common spherulitic growth pattern (Olszta et al., 2007). Perhaps, the NCPs that were shown by Beniash's group to lead to clusters of oriented HA crystals *outside* of the collagen could play a role in maintaining the crystal orientation of extrafibrillar mineral (Deshpande et al., 2011), since confinement is not available.

(a) (b)

Figure 6.12 An atypical but illustrative example of mineral crystals growing within and emerging from the constraints of the collagen fibrils. (a) In these relatively high magnification SEM images, the edges of HA crystals can be seen protruding from the surfaces of overly mineralized collagen fibrils. The crystal edges appear to be oriented roughly parallel to the fibrils, as one would expect from the diffraction patterns that are produced. (b) In this region, the bulges along the fibril appear to arise from mineral that is no longer constrained by the fibril, forming outgrowths of crystal clusters that are starting to resemble the common spherulitic growth pattern of hydroxyapatite.
Reprinted with permission from Olszta et al. (2007), © 2007 from the American Chemical Society.

6.4.3 Other functions of the NCPs

In comparison of the two scenarios, given the perspective that simpler is more easily achieved through evolution, I am still inclined to favor the confinement hypothesis of scenario 2. That's not to say I do not think NCPs have evolved to have more specific functionality. The extrafibrillar mineral may need further control of crystal orientation (as mentioned above), as well as size and morphology. Moreover, the NCPs are also thought to contribute to the mechanical toughness and self-healing in bone (Thurner et al., 2009; Fantner et al., 2005). NCPs are also critically important with respect to cell signaling functions. This is demonstrated in our recent work with osteopontin (OPN), which was found to be multifunctional in that it can both promote intrafibrillar mineralization of collagen, as well as stimulate osteoclast activation (Rodriguez et al., 2014) (see Figure 7.16, which is presented later). The NCPs might also template the organization of the intrafibrillar mineral, such as in the hierarchical banding pattern that is often seen in the biological tissues. This may be regulated by the phosphate groups of the proteins, as suggested by the perceptive work of Liu et al. (2011), who were able to recreate the banded mineralization pattern sometimes found in biological hard tissues by using a combination of "sequestering" and "templating" analogs of the NCPs (Figure 6.13). Although they interpret the layered bands of mineral as being caused by the phosphate groups acting to promote nucleation of apatite, I can offer alternative interpretation to consider. Given that the sequestering analog (polyacrylic acid) is used in both cases (Figure 6.13a and b), one would think that there would be

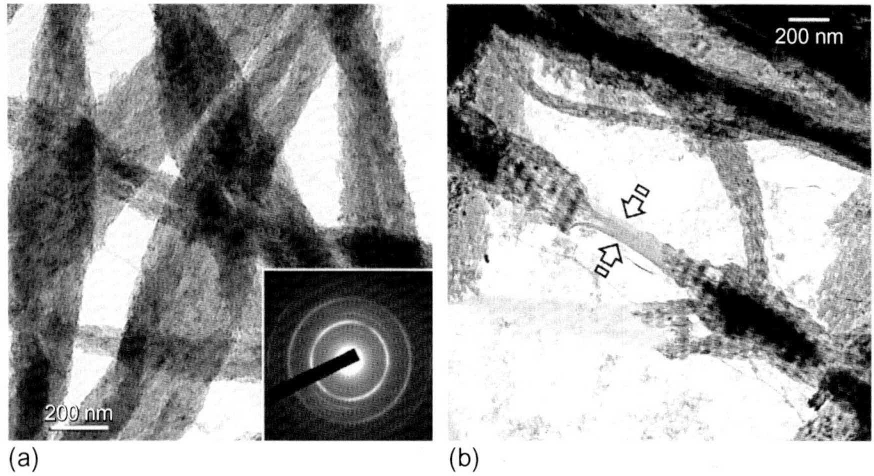

Figure 6.13 TEM images of mineralized collagen prepared by Liu et al. (2011), comparing non-hierarchical versus hierarchical intrafibrillar mineralization of collagen. The collagen was prepared via fibrillogenesis on TEM grids. (a) Collagen fibrils with dark contrast from intrafibrillar mineral (no stain), generated by mineralization in SBF containing the sequestration analog (polyacrylic acid). The mineral appears to have a continuous texture throughout the entirety of the fibrils. (b) Collagen fibrils with pronounced cross-banding from organized layers of crystals, generated by pretreating the collagen with a templating analog of 2.5 wt% sodium tripolyphosphate, and then mineralizing with the sequestration analog of polyacrylic acid-containing SBF. This example also shows the extensive shrinkage of collagen that occurs upon dehydration in regions that do not contain mineral (arrows), as well as the contrast variation seen in the partly mineralized fibrils below.
Reprinted with permission from Liu et al. (2011), © 2011 Elsevier Ltd.

mineral throughout the fibril, regardless of where it nucleates once inside. So perhaps the "templating" analog is blocking the precursor from penetrating the entirety of the fibril. It may be reducing the infiltration via favorable charge interactions with the precursor, or it could be promoting rapid nucleation, which if it causes the precursor to solidify/crystallize too quickly, might block further precursor penetration as well. In this case, I do not actually favor any particular hypothesis, but am just offering an alternative perspective.

6.4.4 Molecular recognition—Must it always be involved?

Although these other aspects such as hierarchical structures are intriguing and deserve further study, the excitement we felt in our first reports (Olszta et al., 2003a,b, 2007) was that we were able to show that intrafibrillar mineralization of collagen does not require highly specific, "epitaxial" types of interactions between an NCP and HA crystal, which at the time was the cherished belief. Indeed, the Wang et al. (2012) paper further illustrates this fact, given that they do not add any polymer or protein additives to their collagen system at all. Thus, given that the NCPs apparently do not provide the epitaxy, the focus then shifted back to the hypothesis that collagen is the organic template that must surely provide epitaxy, the next best cherished belief. I should clarify (given that people seem to be touchy about this issue), I am not suggesting that collagen is not important in bone formation. Of course it is! The inimitable structure that stems from the quarter stagger assembly leads to unique nanocompartments that template an interpenetrating mineral phase; it's just open to debate whether or not that templating is passive or requires specific molecular recognition.

Of course, one can understand why the biological community has always tended to fancy the theory of epitaxy, given that a basic tenet of biochemistry is that protein interactions with ligands and substrates involve "molecular recognition." In fact, when we first tried to publish our findings in *Science*, one reviewer's comments aptly illustrates this bias: "Less than a year ago, the crystal structure of osteocalcin was published by *Nature*, suggesting very strongly that epitaxial growth in bone is likely to be one of the mechanisms by which the magnificent structure of bone is formed." For some reason, the computational simulation suggesting a potential epitaxial relationship (Hoang et al., 2003), which had not even been physically demonstrated, was considered a reason to negate our hypothesis, which was based on real experimental data that showed that an amorphous mineral was first infiltrated into the collagen fibrils. What was troubling to the reviewer (and apparently many others in the field) was that we considered a simple polypeptide, polyaspartic acid, to be able to effectively mimic the actions of a protein. Given the traditional perspective on proteins, with well-defined tertiary structures and molecular recognition, clearly this model system seemed much too simplistic to have any validity. I agree that pAsp is a simple mimic to a protein; however, given the non-specificity of the interaction in this particular reaction, where the main function of the polymer is to sequester ions or ion clusters into some colloidal precursor, it becomes understandable that it does not require a specific protein conformation, and therefore sequence. Indeed, studies have now shown that many of the acidic proteins associated with biominerals, both $CaCO_3$ and CaP, are intrinsically disordered proteins (Delak et al., 2009a,b; Ndao et al., 2010; Evans, 2012; Fisher et al., 2001; Huq et al., 2004; Cross et al., 2005; Uversky, 2002). Therefore, it becomes less surprising that such a simple model system could yield such nice results with respect to emulating bone's nanostructure (i.e. aligned HA nanocrystals embedded throughout the interstices of collagen fibrils to form an interpenetrating composite).

Even with all this evidence, many groups prefer to hold onto their favorite hypothesis of protein binding to crystal faces. Of course, this can and does occur, but it is not clear how relevant this is to modulating bone formation. For example, in light of the revelation that many of the biomineral proteins are intrinsically disordered, Hunter et al. (2010) have modified their perspective (slightly) to now include "the flexible polyelectrolyte hypothesis," suggesting that "the molecular disorder assists these proteins in forming multiple bonds with cations of the crystal face." Yes, adsorbed macromolecules can certainly modify crystal size and shape, but given the prevalence of biominerals that seem to form from an amorphous phase, it is not clear to me how relevant this is given that no crystal faces are even present when the biomineral is formed into its primary shape. Even the originators of this concept (the group from the Weizmann Institute) seemed to have changed their focus once they found that an amorphous precursor was present in nearly all the biominerals they examined. The influence of protein binding to crystals is likely more pertinent to the extrafibrillar crystals in the bone and dentin system. Or maybe in kidney stones, which is the system Hunter et al. were primarily discussing in that particular paper. However, there is evidence to suggest that some types of stones may form from an amorphous precursor as well, as suggested by mineralogical signatures identified in our *in vitro* model systems of stone formation (Amos et al., 2009; Gower et al., 2010).

Veis and Dorvee (2013) have provided a recent review paper on "Biomineralization mechanisms: a new paradigm for crystal nucleation in organic matrices." It is not clear (to me) what the authors are considering the new paradigm—the role of structured water that has become a recent focus for them, or the work of others in demonstrating prenucleation clusters (PNCs) and LCPs? In any case, a word of caution about this paper—they use terminology which is not consistent with much of the biomineral literature, where considerable attention has been focused in recent years on PNCs (Demichelis et al., 2011; Gebauer and Colfen, 2011; Gebauer et al., 2008, 2014), which are clusters and/or strands of ions. It is an accumulation of these clusters that leads to a liquid-condensed or amorphous phase described by Nudelman et al. (2010), Dey et al. (2010), and Habraken et al. (2013), and it is this accumulated phase that Nudelman et al. observed attaching to collagen fibrils, not PNCs (as incorrectly described in the Veis review). It was also a bit disconcerting to me personally that the authors state that "Although the nature of the PILP process is clear, the mechanism of enhancement of the nucleation by the participation of the inducing polymer, for example, the polyAsp in the heterogeneous nucleation process described above (Nudelman paper), has not been described in equal depth." Apparently the point that we have been trying to make, that a "nucleating protein" is not needed to achieve intrafibrillar mineralization, or even oriented nucleation, was not grasped. The pAsp is inhibitory to nucleation of HA, which thereby promotes intrafibrillar mineralization of collagen through accumulation of the amorphous precursor (described by Nudelman et al. (2010) as an accumulation of PNCs). We have tried to make this point in several papers, that it is the very proteins that people have long considered to be "nucleation inhibitors" that lead to this novel mineralization process, not nucleation promoters.

In Veis and Dorvee's discussion on nucleation, they simulate the hydration shells of structured water surrounding the calcium and phosphate ions, and suggest that the "enhanced stability of the dense liquid water allows time for the final steps of densification to establish order and crystallinity from the nucleation clusters." Simply simulating waters around ions is not sufficient, because even in our control reaction (without polymer), ACP still forms in solution at this supersaturation, yet it does not infiltrate the collagen fibrils! Therefore, the amorphous phase is somehow different when combined with polymer. We surmised that it might be more liquid-like (Olszta et al., 2007). Also, it is not clear why their simulations are focused on the phosphate ion, when the predominant species at physiological pH are hydrogen phosphates, and thus are likely the predominant species in the LCP. Indeed, in the $CaCO_3$ system, we found (through NMR) that the LCP was dominated by bicarbonate species (Bewernitz et al., 2012). Thus, if one considers this, not only is the release of structured water important, but also the deprotonation of the phosphate species would contribute to retarding nucleation in the condensed phases. Nevertheless, even though I do not agree with their oversimplified computational model, I do agree with their conclusion, that "from this perspective it is easy to see why and how the introduction of a stable but compatible polymeric polyion interface also having a restricted water layer can be so important in modulating and/or guiding the final location and rate of formation of the final stable crystals." Indeed, this is what we have proposed for the PILP mechanism, that the polymer helps to accumulate and stabilize a highly hydrated LCP (Bewernitz et al., 2012). Moreover, we proposed a mechanism to explain how the fluidic character of the phase enables infiltration into the collagen fibril, which is a point that doesn't seem to be addressed by this group. However, Veis and Dorvee (2013) do consider the NCPs in stating that "Though polyAsp can do the *in vitro* job of nucleating carbonated apatite (cAp), the evidence is strong that, although the sequence of DPP contains approximately 40% Asp residues, phosphorylation of the Ser residues of DPP (50% of the amino acids) is crucial for its function in mediating mineralization (He et al., 2005)." I think this was meant to imply that pAsp does not effectively serve as a model for proteins (at least that was what I was told at the last Gordon Research Conference on Biomineralization), but I would argue this the other way around—the fully carboxylated pAsp is needed in our model system because it doesn't have the more acidic phosphate groups that are in the NCPs. I would agree that a protein with only 40% Asp is not likely to be sufficient in mediating the mineralization if it was not phosphorylated, but I would argue that this is because it is lacking sufficient charge density to effectively induce the PILP process until it is phosphorylated. In some of our unpublished work, we found that phosphopeptides were highly potent at forming $CaCO_3$ PILP films, even when using peptides with MWs much lower than that needed with pAsp.

6.5 Influence of various parameters on collagen mineralization

6.5.1 Polymer process-directing agent and mineral infiltration

Capillary infiltration should, in theory, lead to continuous mineral infiltration as long as the precursor phase remains liquid-like and the collagen matrix maintains its sorptivity. As alluded to earlier, I believe the fluidity of the PILP could be a limiting factor to achieving large depths of penetration into dense collagen scaffolds, where I assume solidification of the PILP would eventually block the entry of further mineral precursor. In the case of the $CaCO_3$ PILP system, we have observed that the PILP droplets grow in solution and solidify with time (DiMasi et al., 2005), which seems to happen more rapidly upon contacting the glass surface (unpublished results). In contrast, the CaP precursor phase seems to be relatively stable when a higher MW polymer is used, remaining as nanodroplets/particles within the solution, which stays clear for extended times (up to a week or more). Although it is not known precisely how long it would take for solidification of the presumed CaP PILP droplets upon interacting with an organic and soft surface such as collagen, it is clear that the precursor phase does ultimately solidify and crystallize after entering into the fibrils. Solidification to ACP would likely block further entry of more precursor, ultimately causing any later stage deposition to occur on the surface of the fibrils, as we have observed. Thus, our working hypothesis is that certain attributes of the polymer additive, such as MW, charge density/distribution, and concentration, play a vital role in regulating the metastability of the precursor phase. Note that the metastability of the precursor can refer to two stages of the process: (1) retaining its liquid-like PILP state in the first stage, which should impact its ability to infiltrate the collagen and then (2) crystallization of the PILP/ACP into HA crystals in the second stage. These same attributes of the polymer are also important with respect to influencing the colloidal properties of the precursor phase, which is discussed later.

We have found that polymer MW can strongly influence the PILP mechanism, where higher MW pAsp led to faster kinetics of infiltration. MWs in the range of 27–32 kDa seemed ideal for generating purely intrafibrillar mineral (Jee et al., 2010a; Thula et al., 2011b), as illustrated earlier in Figure 6.3. It may seem counterintuitive that higher MW pAsp would yield the most optimal intrafibrillar mineralization given that the polymer coils will be larger. We believe this is because the higher MW polymer stabilizes smaller nanodroplets by preventing growth via coalescence (see light scattering study in the supplement of Jee et al. (2010a)). Smaller droplets could increase the diffusion rate across the solution and throughout porous sponge scaffolds, as well as possible entry and entrapment of tiny nanodroplets in the hole zones. Of course, smaller nanodroplets/particles could favor diffusion into the collagen hole zones, which would be consistent with Nudelman et al.'s hypothesis that I argued against earlier (and is discussed further below). Certainly, the issue of particle size of colloidal precursors deserves further study. In any case, the rate of mineralization becomes an important consideration when trying to avoid solidification and achieve large depths of penetration, and particularly for dense collagen scaffolds. We have observed mineral penetration depths of the order of 3 mm for spongy scaffolds, whereas mineral was limited to about 25–100 μm in PILP-remineralized bone specimens (Thula et al., 2011a), and to up to 200 μm in dense collagen scaffolds prepared by liquid-crystalline processing (unpublished results).

The polymer also seems to play a central role in influencing the colloidal aspects of the precursor phase. For example, in the $CaCO_3$ PILP system, Wolf et al. (2011a) describe the role of the polymer as providing depletion stabilization and de-emulsification of the liquid amorphous calcium carbonate precursor. In our light scattering study, the growth kinetics of the $CaCO_3$ PILP droplets seem to correlate with growth by accumulation of droplets, not ion by ion (DiMasi et al., 2005). The $CaCO_3$ PILP droplets seem to simultaneously grow and solidify, ultimately leading to their precipitation, so short deposition distances are needed to get coalescence of the liquid-like phase into smooth continuous $CaCO_3$ films. In contrast, in the CaP system, the nanodroplets (if they are indeed PILP) seem to remain relatively stable (with higher MW pAsp) until they enter the collagen fibril. As described by Holt et al. (2009, 2014) and Holt (2013), milk OPN and casein phosphopeptides (CPP) can sequester ACP to form thermodynamically stable complexes. Although this work assumed solid-like CPP–ACP complexes, we have found that OPN peptides are highly effective at achieving intrafibrillar mineralization of collagen (Rodriguez et al., 2014), and therefore believe these CPP–ACP complexes are one and the same as PILP. Once the precursor has infiltrated into the collagen, the stability limit may be overcome by interactions with the collagen that promote solidification/nucleation, or perhaps the stabilizing polymer is left at the surface, no longer able to provide the stabilizing influence to the infiltrated mineral (see Section 6.5.2). This excluded polymer could then be "recycled" to further sequester more ions, or it may simply be left at the surface of the fibril or along intrafibrillar zones where transport has ceased.

With respect to the CaP colloidal aspects, Nudelman et al. also considered the charged features of the colloids to be important, in this case for directing their migration into the fibrils. Although I don't fully agree with their electrostatic attraction argument (which is stated in their subsequent review paper (Nudelman et al., 2013) as having been demonstrated), I do think the surface energetics of the precursor colloids are important with respect to their stability and selective infiltration into the fibril via capillarity. It should be noted that the buffers that we all use (Tris or PBS) are all quite high in ionic strength, which would presumably help screen these surface charges, yet it is the NaCl in the buffer that seems critical for the reaction to work in our system (although Nudelman et al. (2010) did not use NaCl in their buffer, but again the microscale reaction droplet may be less restrictive with respect to stability of precursor). In fact, if one considers the hierarchical templating work by Dai et al. (2011), where functionalization of the collagen with phosphate-containing compounds provides negatively charged domains that seemingly attract the precursor phase (Figure 6.13), this would apparently oppose this hypothesis of infiltration via attraction to positive domains. Likewise, Li et al. (2012) found that increased crosslinking of collagen increased its degree of mineralization, even though the crosslinking reaction presumably consumes the amine groups that are what provide the positive charges on collagen. So I am not convinced that this aspect of positive domain interactions has been demonstrated.

Nudelman et al. also considered the colloidal characteristics in their follow-up study, where the concentration of pAsp was examined (Nudelman et al., 2012), and they found that a lower concentration of pAsp increased both the infiltration rate of the amorphous phase as well as its transformation into crystalline HA. They therefore suggested that the smaller size of stabilized particles allowed for faster diffusion. While I agree that diffusion of colloids to the collagen fibrils would be increased, which would then enable mineralization (and subsequent transformation) to occur more quickly, I disagree with their argument that "diffusion" into the fibrils was enhanced. This is because in their prior experiment using a combination of pAsp with the protein C-DMP1 (Nudelman et al., 2010), the kinetics were said to be further accelerated relative to pAsp alone, yet the size of the agglomerates seemed very large, even at the early stage of 8 h, as seen in Figure 6.6g. Thus, this argument that presumes increased intrafibrillar diffusion does not seem consistent with the data when comparing "particle" sizes for the pAsp versus C-DMP1 systems. Strangely, even though the authors discuss the mineralization in terms of diffusion, their promotional movie in the R&D magazine, entitled "Making bone in the laboratory," seems to illustrate the system exactly as we had described for a PILP system, with liquid-like "droplets" that merge and morph as they seep into and throughout the fibril (http://www.youtube.com/watch?v=nEOyF0C24xA).

As an aside, it should be noted that C-DMP1 by itself was not able to produce intrafibrillar mineral, but its ability to promote HA nucleation did seem to transfer over to increasing nucleation from the amorphous phase as well. One can imagine that the NCPs have been optimized to provide the proper stability for enabling full infiltration of a mineral precursor, but with protein domains or combinations to augment the crystallization stage as well. Another interesting aspect of Nudelman's follow-up paper was that copper, which was found to leach from the copper TEM grids, disrupted the process by overly stabilizing the amorphous CaP phase (when pAsp was present). Thus, impurity ions, in combination with the polymer additive and/or collagen, can further influence the process. We have tried adding barium and strontium in the hope of increasing z-contrast of the precursor, but found that these ions also strongly disrupted the process (unpublished results).

Although our PILP hypothesis considers capillary infiltration as the means for achieving intrafibrillar mineral, it should be noted that *diffusion* is still relevant to the necessity of nanodroplet transport across the reaction solution to reach the collagen scaffold. In bone formation, the distances of transport of matrix vesicles or mineral-bearing globules between cells and osteoid matrix are small, but in the beaker this becomes an important consideration for scale-up. We have found that dynamic conditions (stirring) lead to a pronounced increase in the mineralization rate. Dynamic flow conditions were also examined by Antebi et al. (2013), who developed a perfusion-flow system to improve the efficiency and homogeneity of the PILP mineralization system in bulk scaffolds (Figure 6.14). Homogeneity can be a significant issue with the PILP system, where there can be regions of well-mineralized collagen alongside nonmineralized regions. In the $CaCO_3$ PILP system, where the droplets become large enough to scatter light, I have even observed cloudy patches and strands of connected PILP droplets, where they seem to be connected by a gel (the particles all move together). The cryo TEM images of the deformed fibril captured by Nudelman et al. (2010) show that the mineralization reaction can be heterogeneous even at the nanoscale (Figure 6.6d). This correlates well with our less exquisite SEM images showing heterogeneous infiltration leading to bulges and nodules in the fibrils at the early stages of the reaction (see Figure 6.1c). The bulges are more pronounced in our SEM images due to the drying procedure needed for conventional SEM, which causes collagen to collapse upon dehydration except for where the mineral has displaced the water. In both their and our observations, the mineral infiltration seems to not be perfectly uniform throughout the fibrils in the early stages, but eventually becomes more uniform upon completion of the reaction. We have noticed (tentative observation) that this heterogeneous infiltration occurs more commonly in the reconstituted collagens than in the biogenic collagens (e.g. tendon and demineralized bone scaffolds). This may be due to differences in fibrillogenesis or crosslinking, both of which differ substantially in the biological setting. We have also observed that aging of the collagen scaffolds can impact the mineralization process. For example, when I visited the Sommerdijk lab 2 years after their seminal paper in *Nature Materials*, we were no longer able to mineralize the exact same collagen, and under the exact same conditions. Although the collagen appeared in the TEM to be well-formed fibrils, the staining bands were slightly disrupted, indicating the organization had changed ever so slightly upon aging, which had a surprisingly pronounced impact in totally nullifying its ability to mineralize. I say

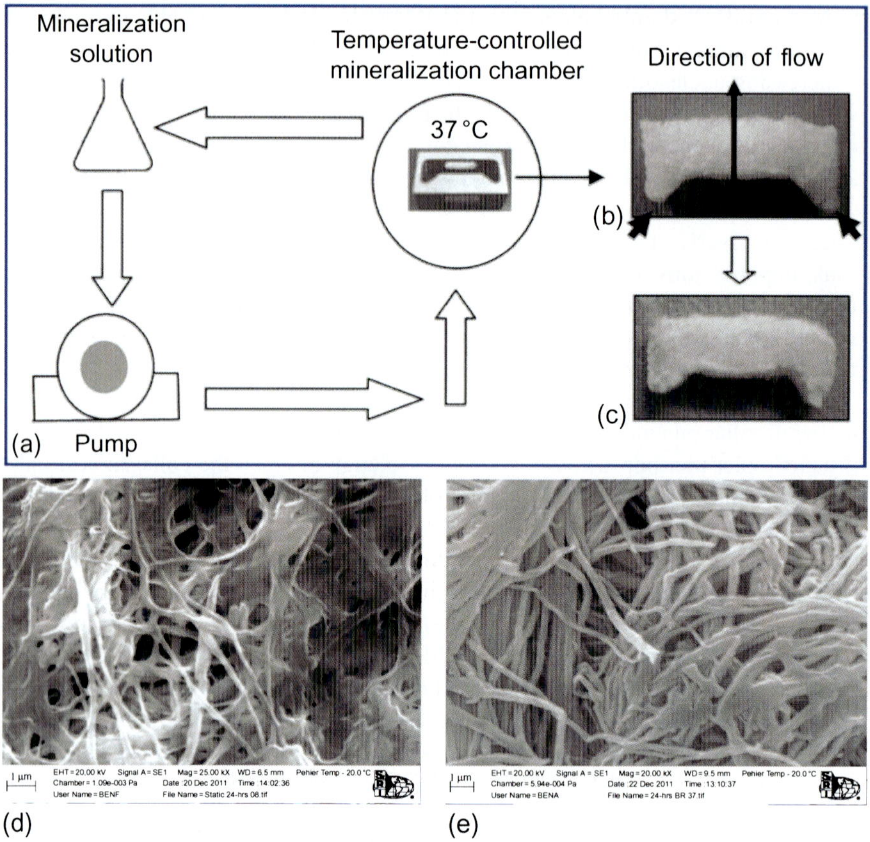

Figure 6.14 A flow perfusion system developed by Antebi et al. (2013) for enhancing the PILP mineralization of bulk scaffolds. (a) Schematic illustration shows how the stainless steel mold was designed to prevent fluid from flowing around the scaffold, maximizing the transport of PILP droplets into the collagen sponge. Photographs of the collagen sponge are shown before (b) and after (c) mineralization. SEM micrographs show the mineralized fibrils after 24 h of static conditions (d) and 24 h of dynamic perfusion (e). Note the web-like texture of the collagen which has not yet mineralized in (d), as compared to the nice fibrillar texture of the uniformly expanded fibrils in (e).
Reprinted with permission from Antebi et al. (2013), © 2013 Mary Ann Liebert, Inc.

surprising because we have been able to mineralize so many different types of collagen scaffolds (including various reconstituted sponges prepared in-house or commercially; tendon from turkey, bovine, and rat tail; and manatee and bovine bone), that I had come to think that the collagen fine structure was not that critical (as long as it was well-formed fibrils). But something about aging seems to be important, which we presume to be some type of degradation, even though it is not obvious at the micro- or nanoscale. We have occasionally observed brushite formation in our older commercial collagen scaffolds as well, which suggests bacterial contamination could be involved in some cases.

6.5.2 Collagen mineralization with NCPs

Several proteins have been found to promote intrafibrillar mineralization, including C-DMP1 (Nudelman et al., 2010), which was discussed earlier; dentin matrix protein 1 (DMP1) and dentin phosphophoryn (DPP), with comparisons between the proteins with and without phosphorylation (Deshpande et al., 2011); milk OPN (Rodriguez et al., 2014); fetuin (Toroian et al., 2007; Price et al., 2009; Nudelman et al., 2010); and even amelogenin (Deshpande et al., 2010). In the latter case, even though amelogenin is associated with enamel formation, which is not a collagen-based hard tissue, it was argued that amelogenin may associate with collagen at the dentin–enamel junction to modulate the mineralization at this border region.

For fetuin protein, the Nudelman et al. (2010) paper had perplexing results. Fetuin is an acidic blood protein which is considered to be an inhibitory protein that helps prevent ectopic calcifications in soft tissues, plaques, and stones. Interestingly, they found that fetuin promoted intrafibrillar mineralization of collagen, yet they stated that "As opposed to pAsp, we did not observe any amorphous calcium phosphate entering the fibril, suggesting that infiltration of mineral occurs through a different mechanism." They argue that "this protein induces collagen mineralization by inhibiting calcium phosphate precipitation in solution and thus allowing the mineral phase to penetrate into the fibril," and further state that "the protein itself cannot diffuse into the fibril owing to its large molecular weight (48 kDa)." However, even though they seem to be arguing for the "mineralization by inhibitor exclusion" hypothesis first put forth by Price et al. (2009) and Toroian et al. (2007), they also show with light scattering that there are 50–100 nm particles/complexes in the solution, so it seems to me that the solution is not really fully inhibited. It also was not clear how they can claim there was no amorphous CaP, at least not given the images provided (see their Supplement Fig. S10). The fibril they show already contains a lot of crystals, suggesting the image was captured at a late stage; perhaps the fetuin gave faster mineralization kinetics, so it may have been too late to capture particle adsorption on the exterior. In addition, the interior of the fibril is quite dark with some material that does not appear striated (like crystals do), so it is not clear if that is stain or amorphous mineral. Amorphous mineral is not particularly easy to see, and especially if it happens to be highly hydrated, giving it lower contrast (or blurry dark contrast with stain).

Regarding the size of fetuin, Price and colleagues had previously designed a clever experiment using collagen beads in a size exclusion column, where they determined that molecules larger than a 40 kDa protein should be excluded from the water/pore space of type I collagen. Thus, Price et al. (and Nudelman et al.) concluded that fetuin's MW of 48 kDa would be too large to fit into type I collagen, thereby leading to the "mineralization by inhibitor exclusion" hypothesis. The fact that fetuin forms 50–100 nm colloidal complexes with mineral ions (light scattering data) leads me to offer another suggestion—perhaps the fetuin is filtered out and remains at the surface of the fibril as the rest of the precursor phase (if it is liquid-like) is separately wicked inside. Alternatively (and I think more likely), given the difficulty in separating the protein and its complexed ions, it may be that the entire PILP is wicked inside the fibril by capillary forces (which are strong and fast), which could cause the protein to be sheared into a thin molecular strand, which would have a diameter much thinner than theoretically predicted for a globular or random coil protein. According to 2D electrophoresis methods (Csizmok et al., 2006), fetuin ran closer to the intrinsically unstructured proteins than the globular proteins, so even though fetuin is commonly considered a globular protein, perhaps it is "soft" enough for shear deformation to denature it into a thin strand. As the capillary transport slows, the protein will eventually recoil due to entropy (as in most polyelectrolytes), and at this point the protein transport will likely cease. This is of course speculation, but it would be interesting to see if relatively large unstructured proteins can be conveyed into fibrils by this mode of transport.

I will now address the question I posed earlier: Why does it matter how the amorphous precursor gets inside the collagen fibril? One of my favorite experiments that illustrates the difference between polymer diffusion versus capillary infiltration of a (supposed) PILP phase was in an experiment by Jee et al. (2010b), a doctoral student in my group who examined the transport of fluorescently labeled pAsp additive into a dense collagen scaffold. Turkey tendon was chosen because the dense molecular packing of collagen within tendons should reduce the amount of diffusion of polymer in extrafibrillar space around the fibrils (as occurs in a porous collagen sponge). The pAsp plus calcium showed limited transport via diffusion, while the addition of phosphate, which presumably generated the PILP phase, showed pronounced penetration of pAsp into the tendon to a depth of about 500 μm (Figure 6.15). Admittedly, it could be argued that the pAsp + Ca created positively charged complexes, while the addition of pAsp + Ca + P created negatively charged complexes, and according to the Nudelman hypothesis, the latter would favor infiltration due to electrostatic attraction. But again, it's hard for me to imagine that those complexes could have diffused 500 μm passing all along the way many positive domains, while the calcium–pAsp complexes alone did not. In either case, there clearly is a dramatic difference in the depth of penetration of the mineral precursor, depending on the precursor's form, and this could be important for future design and scale-up of biomimetic bone composites.

Interestingly, there seemed to be a cutoff point where the majority of the polymer became trapped at around 300 μm, while some continued further. The PILP process directing agent used at that time was poly-(α,β)-DL-aspartic acid with MW = 5500 Da, which was not monodisperse, and may have been trapped at different depths (although a size distribution of polymer MW wouldn't be expected to give a sharp cutoff). If one considers the elastic recoil mechanism mentioned earlier, this might be expected to yield a sharper cutoff because the sheared chains would all recoil (regardless of size) at about the same time as the transport velocity is reduced to a critical value as the sorptivity is reduced in the scaffold (along with chains encountering favorable domains to interact with). We have since learned that this low an MW is not optimal for stabilizing PILP droplets and achieving intrafibrillar mineralization, so the precursor likely solidified and partially blocked further transport at a certain distance (contact time) as well. This penetration depth

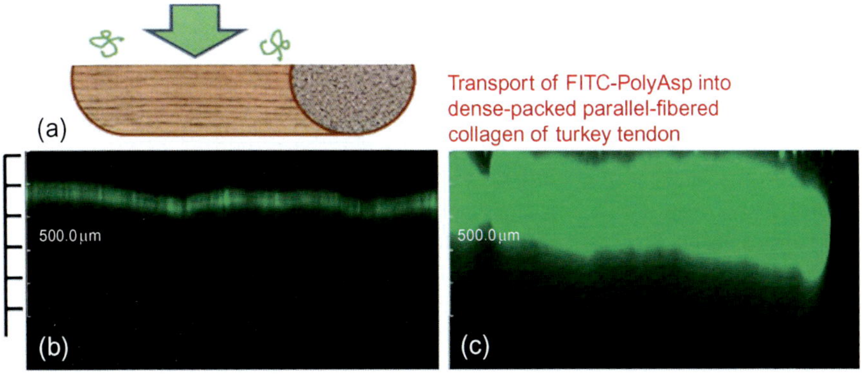

Figure 6.15 Confocal microscopy study examining transport of pAsp into a dense-packed collagen scaffold composed of turkey tendon. (a) As schematically illustrated, the tendon was sliced open longitudinally to expose the collagen fibrils because tendons are surrounded by a sheath which was found to block mineral penetration. Each half was placed in one of two solutions, face up for 3 days. The pAsp was labeled with fluorescein isothiocyanate (FITC) to monitor its transport into the depth of the tendon (indicated by the arrow), with each bar on the image's left scale representing 100 μm in depth into the tendon. (b) When the FITC-polyAsp was only provided with calcium ions, confocal imaging showed a depth of penetration less than 100 μm. Note that the surface of this sample was about 100 μm below the start of the scale bar. (c) With addition of phosphate, FITC-polyAsp was presumably incorporated into a calcium phosphate PILP, which dramatically enhanced its infiltration into the turkey tendon to around 500 μm. Different depths of pAsp transport seemed to occur.
Reprinted with permission from Jee et al. (2010b), © 2009 Elsevier B.V.

could likely be improved with further optimization of conditions. In the biological realm, bone's NCPs have probably been optimized to promote stability of the precursor. However, when considering secondary bone formation, this limitation in penetration depth that we encountered *in vitro* would not be a problem for bone formation *in vivo* because bone is laid down in the form of lamellae, which are only a few microns in thickness. Or perhaps this is the reason for the limited lamellar thickness. But given the similar size scale of cells and lamellae, it seems reasonable to assume that the amount of matrix they can secrete and organize is a limiting factor. Of course, evolutionary controls could have also led to a particular lamellar thickness that optimized mechanical properties. Much more work is needed to understand the intricacies of the NCPs, but I believe that these studies point to at least one central role they play, which is in modulating the degree of intra- versus extrafibrillar mineralization of collagen. Given that both types of crystals are found in biological hard tissues, and there is some debate as to how much mineral is actually intrafibrillar (McNally et al., 2013), a point of current interest to our group is in determining the impact that these different types of mineral crystals have on the biocomposite properties.

6.6 Concluding remarks

As has hopefully become apparent in this review, *in vitro* model systems can be of value from both the science and engineering perspectives. From the science perspective, I believe these model systems may be providing new insights into the crystallochemical mechanisms involved in bone formation. Of course, they cannot <u>prove</u> how bone is formed, but let's be frank, the concept of a fluidic amorphous mineral precursor infiltrating a collagen fibril via capillary action was not likely to have ever occurred to anyone examining biomineral progress *ex situ*. So even though further work is needed to resolve the issue of how the amorphous precursor infiltrates the collagen, now at least there are new hypotheses to consider. I would be particularly pleased to find a research group with computational expertise that would be willing to steer away from the epitaxy simulations and instead examine the influence of surface chemistries, topologies, energetics, etc., of collagen–PILP interactions, to determine how this system can be further regulated, such as maintaining stability, optimizing capillary forces, etc. Of course, it will be difficult to simulate a system containing water, ions, polymer coils, collagen fibrils, and precursor phases, so this may be a grand challenge for the future.

From the engineering perspective, *in vitro* model systems may ultimately lead to the development of the next generation bone substitutes. By fabricating bone in the beaker from a bottom-up approach, one has the potential to tailor properties, both biocompatibility and mechanical properties. With respect to mechanical properties, the hierarchical structure of bone provides it with a unique set of properties, consisting of a load-bearing modulus that falls somewhere between ceramics and plastics, as well as high strength and toughness. Therefore, our current efforts are directed at mimicking both the nano- and microstructure of bone, with a microstructure comprised of densely packed collagen lamellae, which are then PILP mineralized to create an interpenetrating organic–inorganic composite that emulates bone's nanostructure. From the biocompatibility perspective, we hope that if the cells "see" the biomimetic bone as being bone-like, they will be able to remodel it through the native processes of the bone remodeling unit (BRU). In the BRU, the osteoclasts dig out a small canal, which is then filled in with bone by the osteoblasts that follow closely behind (Figure 6.16). In other words, we hope to surmount the problems associated with the common approach of using biodegradable composites, which require careful matching of the degradation rate (due to simple hydrolysis) to the rate of new bone infiltration (which may be highly variable amongst patients and locations in the body). We hypothesize that through the coordinated activity of the cells within the BRU, a biomimetic bone substitute could be replaced a little at a time such that it maintains its load-bearing capacity, as occurs during the natural processes of bone remodeling. Thankfully, our bones do not fall apart as they are remodeled because the activities of the cells within the BRU are carefully synchronized. Thus, our long-range goal is to build a biomimetic bone substitute with bone-like structure and cell modulating signals, such that it has mechanical properties like bone, and can maintain those load-bearing properties as it is remodeled through the BRU into native bone.

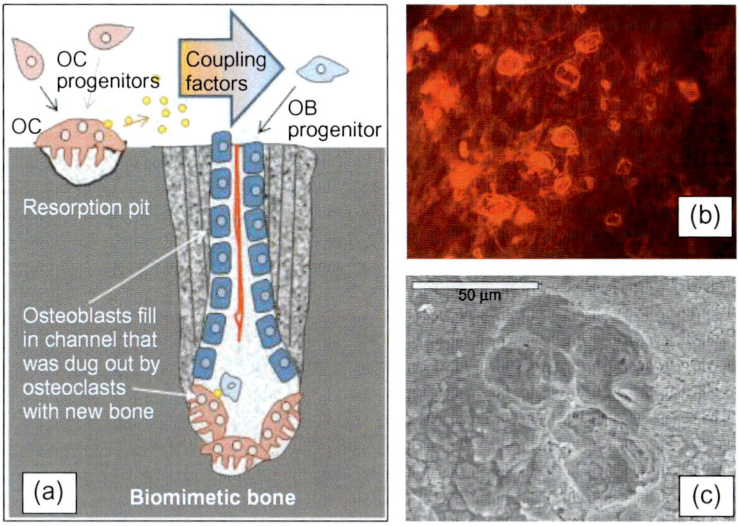

Figure 6.16 Biomimetic approach to the next generation bone substitutes. (a) Schematic representation of the remodeling of a biomimetic bone substitute that we propose could follow the natural pathway of the bone remodeling unit (BRU). If the biomimetic bone closely resembles native bone, the osteoclasts may be activated to resorb the biomimetic bone. Through osteoanabolic factors, the osteoclasts may recruit osteoblast precursors and stimulate them to remodel the synthetic bone in synchrony with its resorption. Through this coordinated cellular activity, a small canal is dug out, yet the entirety of the material still retains its load-bearing capacity, as occurs during natural bone remodeling. (b) Fluorescence image showing activated osteoclasts that are resorbing bone that had been PILP-remineralized using osteopontin as the process-directing agent. In addition to enabling intrafibrillar mineralization of the native collagen matrix, the osteopontin had a strong influence on activating the osteoclasts. This suggests a means for tailoring the resorption rate of bone implants. Specimens were stained with Texas Red-tagged phalloidin to show the presence of actin rings which form the sealing zone. Scale bars = 20 μm. (c) SEM showing the resorption pit created by osteoclasts on the OPN PILP-remineralized bone samples. Reprinted with permission from Rodriguez et al. (2014), © 2013 Acta Materialia Inc. Published by Elsevier Ltd. All rights reserved.

References

Amos, F.F., Sharbaugh, D.M., Talham, D.R., Gower, L.B., 2007. Formation of single-crystalline aragonite tablets/films via an amorphous precursor. Langmuir 23 (4), 1988–1994.

Amos, F.F., Dai, L., Kumar, R., Khan, S.R., Gower, L.B., 2009. Mechanism of formation of concentrically laminated spherules: implication to Randall's plaque and stone formation. Urol. Res. 37 (1), 11–17.

Antebi, B., Cheng, X., Harris, J.N., Gower, L.B., Chen, X.-D., Ling, J., 2013. Biomimetic collagen-hydroxyapatite composite fabricated via a novel perfusion-flow mineralization technique. Tissue Eng. Part C Methods 19 (7), 487–496.

Bewernitz, M.A., Gebauer, D., Long, J., Cölfen, H., Gower, L.B., 2012. A metastable liquid precursor phase of calcium carbonate and its interactions with polyaspartate. Faraday Discuss. 159, 291–312.

Cantaert, B., Kim, Y.-Y., Ludwig, H., Nudelman, F., Sommerdijk, N.A.J.M., Meldrum, F.C., 2012. Think positive: phase separation enables a positively charged additive to induce dramatic changes in calcium carbonate morphology. Adv. Funct. Mater. 22, 907–915.

Cantaert, B., Beniash, E., Meldrum, F.C., 2013a. Nanoscale confinement controls the crystallization of calcium phosphate: relevance to bone formation. Chem. Eur. J. 19, 14918–14924.

Cantaert, B., Beniash, E., Meldrum, F.C., 2013b. The role of poly(aspartic acid) in the precipitation of calcium phosphate in confinement. J. Mater. Chem. B 1 (48), 6586–6595.

Chen, P.-Y., Toroian, D., Price, P.A., McKittrick, J., 2011. Minerals form a continuum phase in mature cancellous bone. Calcif. Tissue Int. 88 (5), 351–361.

Cross, K.J., Huq, N.L., Reynolds, E.C., 2005. Protein dynamics of bovine dentin phosphophoryn. J. Pept. Res. 66, 59–67.

Csizmok, V., Szollosi, E., Friedrich, P., Tompa, P., 2006. A novel two-dimensional electrophoresis technique for the identification of intrinsically unstructured proteins. Mol. Cell. Proteomics 5, 265–273.

Dai, L., Qi, Y.-P., Niu, L.-N., Liu, Y., Pucci, C.R., Looney, S.W., Ling, J.-Q., Pashley, D.H., Tay, F.R., 2011. Inorganic–organic nanocomposite assembly using collagen as a template and sodium tripolyphosphate as a biomimetic analog of matrix phosphoprotein. Cryst. Growth Des. 11, 3504–3511.

Delak, K., Collino, S., Evans, J.S., 2009a. Polyelectrolyte domains and intrinsic disorder within the prismatic asprich protein family. Biochemistry 48, 3669–3677.

Delak, K., Harcup, C., Lakshminarayanan, R., Sun, Z., Fan, Y., Moradian-Oldak, J., Evans, J.S., 2009b. The tooth enamel protein, porcine amelogenin, is an intrinsically disordered protein with an extended molecular configuration in the monomeric form. Biochemistry 48, 2272–2281.

Demichelis, R., Raiteri, P., Gale, J.D., Quigley, D., Gebauer, D., 2011. Stable prenucleation mineral clusters are liquid-like ionic polymers. Nat. Commun. 2, 590. http://dx.doi.org/10.1038/ncomms1604.

Deshpande, A.S., Beniash, E., 2008. Bioinspired synthesis of mineralized collagen fibrils. Cryst. Growth Des. 8, 3084–3090.

Deshpande, A.S., Fang, P.-A., Simmer, J.P., Margolis, H.C., Beniash, E., 2010. Amelogenin-collagen interactions regulate calcium phosphate mineralization in vitro. J. Biol. Chem. 285 (25), 19277–19287.

Deshpande, A.S., Fang, P.-A., Zhang, X., Jayaraman, T., Sfeir, C., Beniash, E., 2011. Primary structure and phosphorylation of dentin matrix protein 1 (DMP1) and dentin phosphophoryn (DPP) uniquely determine their role in biomineralization. Biomacromolecules 12, 2933–2945.

Dey, A., Bomans, P.H.H., Mãller, F.A., Will, J., Frederik, P.M., De With, G., Sommerdijk, N.A.J.M., 2010. The role of prenucleation clusters in surface-induced calcium phosphate crystallization. Nat. Mater. 9, 1010–1014.

Dimasi, E., Liu, T., Olszta, M.J., Gower, L.B., 2005. Laser light scattering studies of a polymer-induced liquid-precursor (PILP) process for mineralization. In: Sandhage, K.H., Yang, S., Douglas, T., Parker, A.R., Dimasi, E. (Eds.), Biological and Bio-Inspired Materials and Devices. Materials Research Society Proceedings, Warrendale, PA.

Evans, J.S., 2012. Aragonite-associated biomineralization proteins are disordered and contain interactive motifs. Bioinformatics 28, 3182–3185.

Fantner, G.E., Hassenkam, T., Kindt, J.H., Weaver, J.C., Birkedal, H., Pechenik, L., Cutroni, J.A., Cidade, G.A.G., Stucky, G.D., Morse, D.E., Hansma, P.K., 2005. Sacrificial bonds and hidden length dissipate energy as mineralized fibrils separate during bone fracture. Nat. Mater. 4, 612–616.

Fisher, L.W., Torchia, D.A., Fohr, B., Young, M.F., Fedarko, N.S., 2001. Flexible structures of SIBLING proteins, bone sialoprotein, and osteopontin. Biochem. Biophys. Res. Commun. 280, 460–465.

Gebauer, D., Colfen, H., 2011. Prenucleation clusters and non-classical nucleation. Nano Today 6, 564–584.

Gebauer, D., Völkel, A., Cölfen, H., 2008. Stable prenucleation calcium carbonate clusters. Science 322, 1819–1822.

Gebauer, D., Kellermeier, M., Gale, J.D., Bergstro, M.L., Cölfen, H., 2014. Pre-nucleation clusters as solute precursors in crystallisation. Chem. Soc. Rev. 43, 2348–2371.

Giraud-Guille, M.M., Mosser, G., Helary, C., Eglin, D., 2005. Bone matrix like assemblies of collagen: from liquid crystals to gels and biomimetic materials. Micron 36, 602–608.

Giraud-Guille, M.M., Belamie, E., Mosser, G., Helary, C., Gobeaux, F., Vigier, S., 2008. Liquid crystalline properties of type I collagen: perspectives in tissue morphogenesis. C. R. Chim. 11, 245–252.

Gower, L.A., 1997. The influence of polyaspartate additive on the growth and morphology of calcium carbonate crystals. PhD dissertation, University of Massachusetts, Amherst.

Gower, L.B., 2008. Biomimetic model systems for investigating the amorphous precursor pathway and its role in biomineralization. Chem. Rev. 108, 4551–4627.

Gower, L.B., Odom, D.J., 2000. Deposition of calcium carbonate films by a polymer-induced liquid-precursor (PILP) process. J. Cryst. Growth 210 (4), 719–734.

Gower, L.B., Amos, F.F., Khan, S.R., 2010. Mineralogical signatures of stone formation mechanisms. Urol. Res. 38 (4), 281–292.

Habraken, W.J.E.M., Tao, J., Brylka, L.J., Friedrich, H., Bertinetti, L., Schenk, A.S., Verch, A., Dmitrovic, V., Bomans, P.H.H., Frederik, P.M., Laven, J., Van Der Schoot, P., Aichmayer, B., De With, G., Deyoreo, J.J., Sommerdijk, N.A.J.M., 2013. Ion-association complexes unite classical and non-classical theories for the biomimetic nucleation of calcium phosphate. Nat. Commun. 4, 1507.

He, G., Ramachandran, A., Dahl, T., George, S., Schultz, D., Cookson, D., Veis, A., George, A., 2005. Phosphorylation of phosphophoryn is crucial for its function as a mediator of biomineralization. J. Biol. Chem. 280, 33109–33114.

Hoang, Q.Q., Sicheri, F., Howard, A.J., Yang, D.S.C., 2003. Bone recognition mechanism of porcine osteocalcin from crystal structure. Nature 425, 977–980.

Holt, C., 2013. Unfolded phosphopolypeptides enable soft and hard tissues to coexist in the same organism with relative ease. Curr. Opin. Struct. Biol. 23, 420–425.

Holt, C., Sorensen, E.S., Clegg, R.A., 2009. Role of calcium phosphate nanoclusters in the control of calcification. FEBS J. 276, 2308–2323.

Holt, C., Lenton, S., Nylander, T., Sørensen, E.S., Teixeira, S.C.M., 2014. Mineralisation of soft and hard tissues and the stability of biofluids. J. Struct. Biol. 185, 383–396.

Hunter, G.K., O'Young, J., Grohe, B., Karttunen, M., Goldberg, H.A., 2010. The flexible polyelectrolyte hypothesis of protein-biomineral interaction. Langmuir 26 (24), 18639–18646.

Huq, N.L., Cross, K.J., Ung, M., Reynolds, E.C., 2004. A review of protein structure and gene organisation for proteins associated with mineralised tissue and calcium phosphate stabilisation encoded on human chromosome 4. Arch. Oral Biol. 50 (7), 599–609.

Jee, S.S. 2008. The development of collagen-hydroxyapatite nanostructured composites via a calcium phosphate precursor mechanism. Dissertation, University of Florida.

Jee, S.-S., Thula, T.T., Gower, L.B., 2010a. Development of bone-like composites via the polymer-induced liquid-precursor (PILP) process. Part 1: influence of polymer molecular weight. Acta Biomater. 6, 3676–3686.

Jee, S.S., Culver, L., Li, Y., Douglas, E.P., Gower, L.B., 2010b. Biomimetic mineralization of collagen via an enzyme-aided PILP process. J. Cryst. Growth 312, 1249–1256.

Kawska, A., Hochrein, O., Brickmann, A., Kniep, R., Zahn, D., 2008. The nucleation mechanism of fluorapatite-collagen composites: ion association and motif control by collagen proteins. Angew. Chem. Int. Ed. 47, 4982–4985.

Kikuchi, M., Itoh, S., Ichinose, S., Shinomiya, K., Tanaka, J., 2001. Self-organization mechanism in a bone-like hydroxyapatite/collagen nanocomposite synthesized in vitro and its biological reaction in vivo. Biomaterials 22, 1705–1711.

Kikuchi, M., Ikoma, T., Itoh, S., Matsumoto, H.N., Koyama, Y., Takakuda, K., Shinomiya, K., Tanaka, J., 2004. Biomimetic synthesis of bone-like nanocomposites using the self-organization mechanism of hydroxyapatite and collagen. Compos. Sci. Technol. 64, 819–825.

Kim, Y.-Y., Gower, L.B., 2003. Formation of complex non-equilibrium morphologies of calcite via biomimetic processing. In: Materials Inspired by Biology; Thomas, J., Kiick, K., Gower, L. (Eds.), Materials Inspired by Biology. Materials Research Society, San Francisco, CA, vol. 774, O6.7.1–O6.7.8.

Kim, Y.-Y., Douglas, E.P., Gower, L.B., 2007. Patterning inorganic ($CaCO_3$) thin films via a polymer-induced liquid-precursor process. Langmuir 23, 4862–4870.

Kniep, R., Simon, P., 2007. Fluorapatite-gelatine-nanocomposites: self-organized morphogenesis, real structure and relations to natural hard materials. Top. Curr. Chem. 270, 73–125.

Landis, W.J., Song, M.J., Leith, A., McEwen, L., McEwen, B.F., 1993. Mineral and organic matrix interaction in normally calcifying tendon visualized in three dimensions by high-voltage electron microscopic tomography and graphic image reconstruction. J. Struct. Biol. 110, 39–54.

Li, Y., Aparicio, C., 2013. Discerning the subfibrillar structure of mineralized collagen fibrils: a model for the ultrastructure of bone. PLoS One 8, e76782.

Li, Y., Thula, T.T., Jee, S., Perkins, S.L., Aparicio, C., Douglas, E.P., Gower, L.B., 2012. Biomimetic mineralization of woven bone-like nanocomposites: role of collagen cross-links. Biomacromolecules 13, 49–59.

Liu, Y., Kim, Y.-K., Dai, L., Li, N., Khan, S.O., Pashley, D.H., Tay, F.R., 2011. Hierarchical and non-hierarchical mineralisation of collagen. Biomaterials 32, 1291–1300.

Mahamid, J., Sharir, A., Addadi, L., Weiner, S., 2008. Amorphous calcium phosphate is a major component of the forming fin bones of zebrafish: indications for an amorphous precursor phase. Proc. Natl. Acad. Sci. U.S.A. 105, 12748–12753.

Mahamid, J., Aichmayer, B., Shimoni, E., Ziblat, R., Li, C.H., Siegel, S., Paris, O., Fratzl, P., Weiner, S., Addadi, L., 2010. Mapping amorphous calcium phosphate transformation into crystalline mineral from the cell to the bone in zebrafish fin rays. Proc. Natl. Acad. Sci. U.S.A. 107, 6316–6321.

McNally, E., Nan, F., Botton, G.A., Schwarcz, H.P., 2013. Scanning transmission electron microscopic tomography of cortical bone using Z-contrast imaging. Micron 49, 46–53.

Ndao, M., Keene, E., Amos, F.F., Rewari, G., Ponce, C.B., Estroff, L., Evans, J.S., 2010. Intrinsically disordered mollusk shell prismatic protein that modulates calcium carbonate crystal growth. Biomacromolecules 11, 2539–2544.

Niu, L.-N., Jiao, K., Qi, Y.-P., Yiu, C.K.Y., Ryou, H., Arola, D.D., Chen, J.-H., Breschi, L., Pashley, D.H., Tay, F.R., 2011. Infiltration of silica inside fibrillar collagen. Angew. Chem. Int. Ed. 50 (49), 11688–11691.

Niu, L.-N., Jiao, K., Ryou, H., Diogenes, A., Yiu, C.K.Y., Mazzoni, A., Chen, J.-H., Arola, D.D., Hargreaves, K.M., Pashley, D.H., Tay, F.R., 2013. Biomimetic silicification of demineralized hierarchical collagenous tissues. Biomacromolecules 14, 1661–1668.

Nudelman, F., Pieterse, K., George, A., Bomans, P.H.H., Friedrich, H., Brylka, L.J., Hilbers, P.A.J., De With, G., Sommerdijk, N.A.J.M., 2010. The role of collagen in bone apatite formation in the presence of hydroxyapatite nucleation inhibitors. Nat. Mater. 9, 1004–1009.

Nudelman, F., Bomans, P.H.H., George, A., With, G.D., Sommerdijk, N.A.J.M., 2012. The role of the amorphous phase on the biomimetic mineralization of collagen. Faraday Discuss. 159, 357–370.

Nudelman, F., Lausch, A.J., Sommerdijk, N.A.J.M., Sone, E.D., 2013. In vitro models of collagen biomineralization. J. Struct. Biol. 183 (2), 258–269.

Olszta, M.J., Douglas, E.P., Gower, L.B., 2003a. Scanning electron microscopic analysis of the mineralization of type I collagen via a polymer-induced liquid-precursor (PILP) process. Calcif. Tissue Int. 72, 583–591.

Olszta, M.J., Odom, D.J., Douglas, E.P., Gower, L.B., 2003b. A new paradigm for biomineral formation: mineralization via an amorphous liquid-phase precursor. Connect. Tissue Res. 44, 326–334.

Olszta, M.J., Douglas, E.P., Gower, L.B., 2003c. Intrafibrillar mineralization of collagen using a liquid-phase mineral precursor. In: Thomas, J., Kiick, K., Gower, L. (Eds.), In: Mat. Res. Soc. Symp. Proc.—Materials Inspired by Biology, vol. 774. MRS, San Francisco, pp. 127–134.

Olszta, M.J., Cheng, X., Jee, S.S., Kumar, R., Kim, Y.-Y., Kaufman, M.J., Douglas, E.P., Gower, L.B., 2007. Bone structure and formation: a new perspective. Mater. Sci. Eng. R. Rep. 58, 77–116.

Orgel, J., Irving, T.C., Miller, A., Wess, T.J., 2006. Microfibrillar structure of type I collagen in situ. Proc. Natl. Acad. Sci. U.S.A. 103, 9001–9005.

Price, P.A., Toroian, D., Lim, J.E., 2009. Mineralization by inhibitor exclusion—the calcification of collagen with fetuin. J. Biol. Chem. 284, 17092–17101.

Qin, C., Baba, O., Butler, W.T., 2004. Post-translational modifications of SIBLING proteins and their roles in osteogenesis and dentinogenesis. Crit. Rev. Oral Biol. Med. 15, 126–136.

Rodriguez, D.E., Thula-Mata, T., Toro, E.J., Yeh, Y.-W., Holt, C., Holliday, L.S., Gower, L.B., 2014. Multifunctional role of osteopontin in directing intrafibrillar mineralization of collagen and activation of osteoclasts. Acta Biomater. 10 (1), 494–507.

Rosen, V.B., Hobbs, L.W., Spector, M., 2002. The ultrastructure of anorganic bovine bone and selected synthetic hydroxyapatites used as bone graft substitute materials. Biomaterials 23, 921–928.

Silver, F.H., Landis, W.J., 2011. Deposition of apatite in mineralizing vertebrate extracellular matrices: a model of possible nucleation sites on type I collagen. Connect. Tissue Res. 52 (3), 242–254.

Thula, T.T., Rodriguez, D.E., Lee, M.H., Pendi, L., Podschun, J., Gower, L.B., 2011a. In vitro mineralization of dense collagen substrates: a biomimetic approach toward the development of bone-graft materials. Acta Biomater. 7 (8), 3158–3169.

Thula, T.T., Svedlund, F., Rodriguez, D.E., Podschun, J., Pendi, L., Gower, L.B., 2011b. Mimicking the nanostructure of bone: comparison of polymeric process-directing agents. Polymers 3 (1), 10–35.

Thurner, P.J., Lam, S., Weaver, J.C., Morse, D.E., Hansma, P.K., 2009. Localization of phosphorylated serine, osteopontin, and bone sialoprotein on bone fracture surfaces. J. Adhes. 85, 526–545.

Toroian, D., Lim, J.E., Price, P.A., 2007. The size exclusion characteristics of type I collagen—implications for the role of noncollagenous bone constituents in mineralization. J. Biol. Chem. 282, 22437–22447.

Traub, W., Arad, T., Weiner, S., 1989. Three-dimensional ordered distribution of crystals in turkey tendon collagen fibers. Proc. Natl. Acad. Sci. U.S.A. 86, 9822–9826.

Uversky, V.N., 2002. What docs it mcan to be natively unfolded? Eur. J. Biochem. 269, 2–12.

Veis, A., Dorvee, J., 2013. Biomineralization mechanisms: a new paradigm for crystal nucleation in organic matrices. Calcif. Tissue Int. 93, 307–315.

Wang, Y., Azais, T., Robin, M., Vallee, A., Catania, C., Legriel, P., Pehau-Arnaudet, G., Babonneau, F., Giraud-Guille, M.-M., Nassif, N., 2012. The predominant role of collagen in the nucleation, growth, structure and orientation of bone apatite. Nat. Mater. 11, 724–733.

Weiner, S., 2008. Biomineralization: a structural perspective. J. Struct. Biol. 163, 229–234.

Weiner, S., Addadi, L., 1991. Acidic macromolecules of mineralized tissues: the controllers of crystal formation. Trends Biochem. Sci. 16 (7), 252–256.

Wolf, S.E., Leiterer, J., Pipich, V., Barrea, R., Emmerling, F., Tremel, W., 2011a. Strong stabilization of amorphous calcium carbonate emulsion by ovalbumin: gaining insight into the mechanism of 'polymer-induced liquid precursor' processes. J. Am. Chem. Soc. 133 (32), 12642–12649.

Wolf, S.E., Muller, L., Barrea, R., Kampf, C.J., Leiterer, J., Panne, U., Hoffmann, T., Emmerling, F., Tremel, W., 2011b. Carbonate-coordinated metal complexes precede the formation of liquid amorphous mineral emulsions of divalent metal carbonates. Nanoscale 3, 1158–1165.

Xu, Z., Yang, Y., Zhao, W., Wang, Z., Landis, W.J., Cui, Q., Sahai, N., 2015. Molecular mechanisms for intrafibrillar collagen mineralization in skeletal tissues. Biomaterials 39, 59–66.

Part Two

Applications

Ion substitution in biological and synthetic apatites

7

A. Bigi[1], E. Boanini[1], M. Gazzano[2]
[1]University of Bologna, Bologna, Italy, [2]ISOF-CNR, Bologna, Italy

7.1 Introduction

Biological apatites are characterized by the presence of a number of foreign ions, most of which play an important metabolic role. On this basis, hydroxyapatites (HAs) substituted with biologically relevant ions are regarded as attractive biomaterials for hard tissue substitution/repair. This chapter begins with a few general remarks on ionic substitutions and with a brief mention of the main available methods to study the structural modifications induced by ion replacement. Then, Section 7.3 reports a detailed description of the structure of synthetic HA, which is highly flexible and allows a variety of cationic and anionic substitutions. Section 7.4 discusses the peculiar features of biological apatites, their main differences with respect to synthetic HA, and the biological role of the associated foreign ions. The section on ionic substitution in synthetic HA is focused on some specific ions: the anions include carbonate, fluoride, and silicate, whereas the cations are magnesium, strontium, zinc, iron, and silver. For each ion, Section 7.5 reviews the structural and chemical modifications induced by its incorporation into the HA structure, and discusses some relevant results on the employment of substituted/doped HA for the preparation of biomaterials.

7.2 Ionic substitutions

The number of papers which abuse the term "ion substituted" without proper experimental proof of the regular incorporation of the foreign ions into the structure of HA, as well as of other calcium phosphates, is substantial. Although there can be objective experimental difficulties in determining foreign ion incorporation into a structure, such as low degree of crystallinity of the inorganic phase, very low amounts of the substituting ion and nonstoichiometry, it is worthwhile to consider briefly what the meaning of ionic substitution is and how it can be verified.

Two main factors, namely the size of the substituting ion and the ability of the structure to tolerate a certain degree of distortion, influence the possibility of a structure to accommodate different compositions. According to Goldschmit's rules, the ions of one element can extensively replace those of another in ionic crystals when their radii differ by no more than 15% and their charges by no more than ± 1, provided that electrical neutrality is maintained. When the substitution involves heterovalent ions, charge balance can be maintained by coupling a further parallel substitution of

Biomineralization and Biomaterials. http://dx.doi.org/10.1016/B978-1-78242-338-6.00008-9

heterovalent ions, such as CO_3^{2-} for PO_4^{3-} coupled to Na^+ for Ca^{2+} in HA, or by vacant sites in nonstoichiometric compounds, such as calcium-deficient apatite. Even when the above criteria are satisfied, substitution can be limited by poor structure flexibility and by the different electronegativities of the competing ions, which form bonds of different ionic character (Giacovazzo, 2002). Isostructural end members are a necessary, although not sufficient, condition for the occurrence of solid solutions over the whole range of compositions.

The presence of ionic substitution can be easily established by X-ray diffraction (XRD) analysis (Figure 7.1). The powder XRD pattern of a solid solution is consistent with the presence of just one crystalline phase, and the accurate evaluation of the lattice parameters of the unit cell allows one to verify variations related to composition. The dimensions of the unit cell generally increase when a smaller ion is substituted by a larger one, but the variation of the single axes also depends on occupancy factors, as

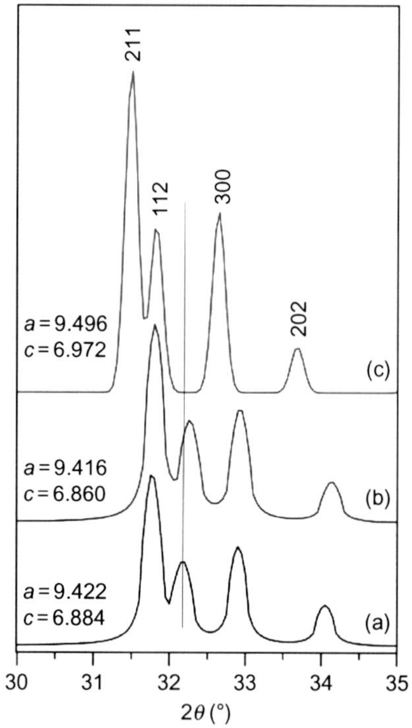

Figure 7.1 X-ray diffraction patterns of stoichiometric HA (a), Zn-substituted HA (15 Zn atom%) (b), and Sr-substituted HA (23 Sr atom%) (c); the unit *a*- and *c*-cell parameters are reported in Å; the vertical line is an eye guide to evidence peak shifts; the Miller indexes of the reflections are reported. The peak shift toward lower angles corresponds to an increase of interlayer distances. The partial replacement of Ca^{2+} ions (radius 1.00 Å) with Sr^{2+} (radius 1.18 Å) causes an expansion of the crystal planes and unit cell, while Zn^{2+} (radius 0.74 Å) induces a contraction of the distances.

shown by the discontinuity at about 50–60 Pb atom% (with respect to the total number of metal atoms) in the c-axis curve of lead–calcium HA solid solutions (Bigi et al., 1991) (Figure 7.2).

Ion substitution also provokes modification of the infrared absorption spectrum. In particular, band splitting and shifts of the frequencies of the PO_4^{3-} and OH^- modes in the spectra of substituted HA have been related to ion dimensions and cation–anion bond character (Fowler, 1974a,b; Bigi et al., 2007; Badraoui et al., 2009).

Further structural information can be obtained through solid-state NMR investigation (Marchat et al., 2013; Laurencin et al., 2011). The single P site of HA accounts for the main narrow peak in the ^{31}P MAS NMR spectrum of pure HA, whereas the ^{1}H MAS NMR spectrum of pure HA shows two main signals, a narrow one at 0 ppm, corresponding to the OH group, and the other at 5.3 ppm, which has been assigned to water molecules mainly located at the surface of HA crystallites (Yesinowski and Eckert, 1987). Shifts and changes in the ^{1}H or ^{31}P signals can be used as indicators of the preferential location of substitutional cations (Pizzala et al., 2009; Badraoui et al., 2001). The recent resolution of the signals corresponding to the two crystallographic Ca sites in a ^{43}Ca NMR study of a nonsubstituted HA provides a further tool to

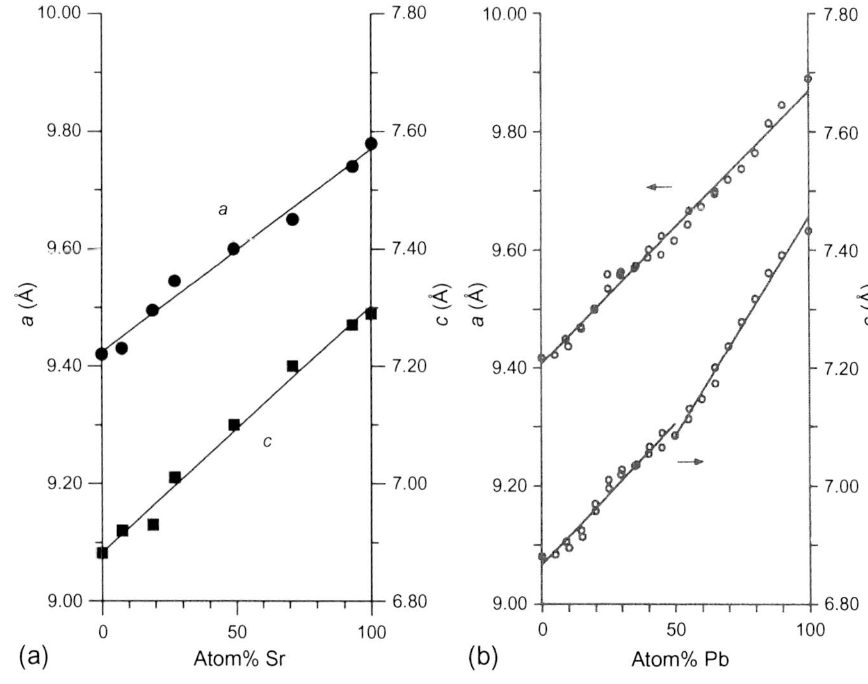

Figure 7.2 Plots of the a- and c-cell parameters of Sr-substituted HA (a) and Pb-substituted HA (b) as a function of composition expressed as atom%. Both ions are bigger than calcium and provoke an increase of the a- and c-axes; the variation of the c-axis in Pb-substituted HA is not linear.

(a) Data from Bigi et al. (2007), (b) reproduced from Bigi et al. (1991) with permission of RSC.

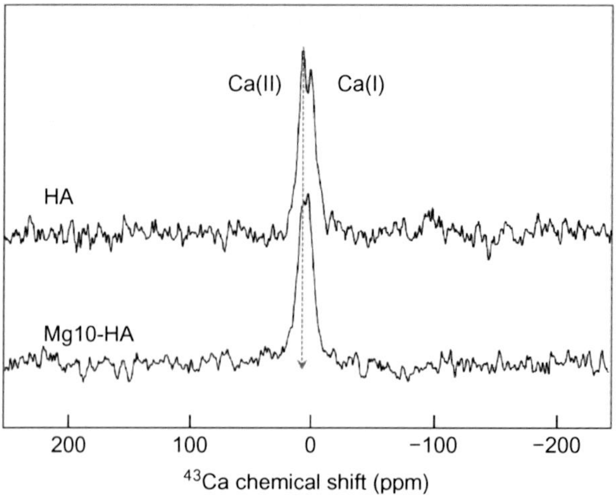

Figure 7.3 Natural abundance ^{43}Ca MAS NMR spectra of HA (top), and of a Mg-substituted derivative containing 10 Mg atom% (bottom). The signals of the two independent metal environments are observed in a magnetic field of 18.8 T (top). The signal of Ca(II) is decreased in Mg-substituted HA, suggesting that Mg is incorporated at this position (bottom). Reproduced with permission from Laurencin and Smith (2013).

investigate structural modifications due to the ionic substitution in both synthetic and biological apatites (Figure 7.3) (Laurencin et al., 2011; Laurencin and Smith, 2013).

The development of instrumentations opens new possibilities for the study of ionic substitutions, as shown by the direct atomic-resolution imaging of every individual atomic column in shark tooth enameloid recently obtained using aberration corrected transmission electron microscopy (TEM) with a low-dose electron beam (Chen et al., 2014). As shown in Figure 7.4, the approach allows one to directly resolve all atoms in fluoroapatite.

7.3 Hydroxyapatite

The formula $M_{10}(XO_4)_6Y_2$ is generally used to indicate a family of inorganic compounds known as apatites. M usually is a bivalent cation, although mono- and trivalent cations can be hosted as well. XO_4 can be PO_4^{3-}, VO_4^{3-}, or AsO_4^{3-}, but also SiO_4^{4-}, CO_3^{2-}, and SO_4^{2-}. Y is usually a monovalent anion, typically OH^-, F^-, Cl^-, or Br^- (Elliott, 1994). Within this family, synthetic HA, $Ca_{10}(PO_4)_6(OH)_2$, is the most similar to biological apatites. As a consequence, it is widely employed for the preparation of biomaterials for hard tissue substitution and/or repair. Moreover, HA displays a variety of further potential applications, including as chemical sensor, lightning material, catalyst and catalyst carrier, liquid chromatographic columns, fertilizer, corrosion-resistant material, and toxic ion immobilizer (Dorozhkin, 2011; Mossaad et al., 2010). HA can be easily prepared in aqueous solution by direct synthesis, as well as by other methods,

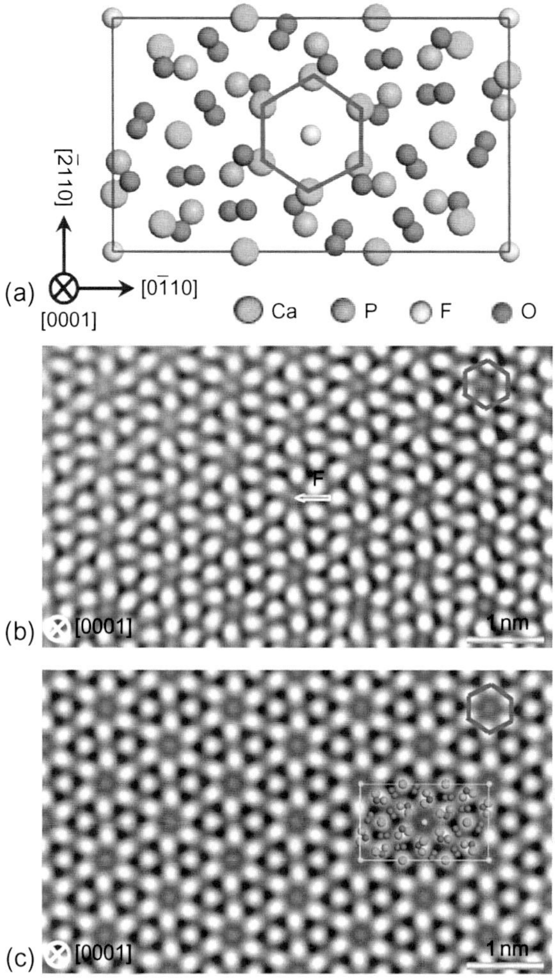

Figure 7.4 Atomic-resolution HRTEM image of a shark tooth. (a) Atomic model of fluorapatite viewed along the [0001] direction. (b) Aberration corrected HRTEM image taken along the [0001] zone axis. The image shows periodically aligned hexagonal units (marked) with a clear bright spot at their center. (c) Simulated HRTEM image along the [0001] direction. An atomic model of fluorapatite is overlapped on the image. Bright spots in a hexagon represent the very closely spaced Ca, P, and O atomic columns. The bright spot at the center of each hexagon represents an F atom.
Reproduced with permission from Chen et al. (2014).

including mechanochemical, hydrothermal, sol–gel, phase transition, and crystallization from simulated body fluid (Boanini et al., 2010; Dorozhkin, 2011; Mossaad et al., 2010). Direct syntheses of crystalline HA in aqueous solution are usually performed at a basic pH and at relatively high temperatures in order to avoid incorporation of carbonate ions. pH, temperature, reaction times, and a controlled atmosphere are the main factors employed to control the degree of crystallinity, crystal dimensions, morphology, size of the coherently scattering domains, thermal stability, and stoichiometry (Sadat-Shojai et al., 2013).

The presence of two formula units in the unit cell of HA accounts for the frequently used double chemical formula ($Ca_{10}(PO_4)_6(OH)_2$ vs. $Ca_5(PO_4)_3(OH)$). Two polymorphs of HA are known: the monoclinic form, which occurs in the monoclinic space group $P2_1/b$, and the hexagonal one, space group $P6_3/m$. The monoclinic form is easily destabilized by the presence of impurities, such as even very small quantities of foreign ions, which is the reason why the hexagonal form is generally found in biological apatites (Wang and Nancollas, 2008). Moreover, the monoclinic form is transformed into the hexagonal phase at temperatures above 250 °C (Dorozhkin, 2011). The hexagonal structure exhibits lattice parameters: $a = 9418$ Å, $c = 6884$ Å (ICDD PDF No. 9-432), and contains one crystallographically independent phosphate site and two different cation sites, M(I) and M(II). The six Ca(II) atoms present in the unit cell are arranged at the apexes of staggered equilateral triangles, that form the walls of the OH^- channels running parallel to the *c*-axis (Figure 7.5). Each Ca(II) is surrounded by seven oxygen atoms. The oxygen atoms of the OH groups are situated either above or below the Ca(II) planes, pointing either upward or downward in the structure. The four Ca(I) atoms are aligned in columns parallel to the OH channels, and are tightly bonded to six oxygen atoms arranged in a distorted trigonal prism, and less strongly to three further oxygen atoms (Kay et al., 1964; Elliott, 1994).

This "open" and flexible structure justifies the great variety of cationic and anionic substitutions that HA can host. In particular, the shorter distances between Ca(I) atoms, and their strict alignment in columns, makes the Ca(I) site more suitable for smaller cations or for modest amounts of cations slightly larger than Ca. On the other hand, Ca(II)–Ca(II) distances are relatively large and Ca(II) atoms in consecutive layers are staggered, so that this site can accommodate larger cations. Cation substitutions in the whole range of composition are known for Sr^{2+}, Cd^{2+}, and Pb^{2+}, whereas

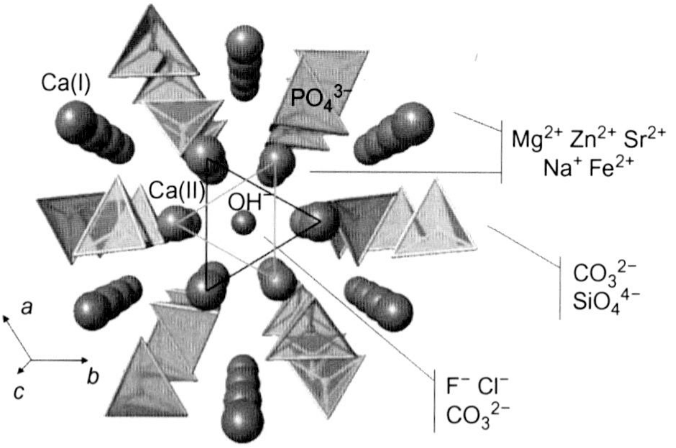

Figure 7.5 Perspective view along the *c*-axis of the OH channel environment. Some of the possible substituting ions are reported. The two triangles join Ca(II) ions at the same level along the *c*-axis.

bivalent cations smaller than Ca^{2+} usually inhibit the crystallization of HA and yield just partial substitutions. Among the possible anionic substitutions, F^- and Cl^- can substitute for OH^- in the whole range, whereas carbonate ion displays a peculiar behavior since it can substitute for OH^- (type A) or for PO_4^{3-} (type B). The two different types of substitutions have opposite effects on the dimensions of HA lattice parameters (LeGeros, 1965), and are characterized by different positions of the carbonate infrared absorption bands (Ren et al., 2014).

7.4 Biological apatites

Although less abundant than calcium carbonates and silicates among biomineralized tissues, calcium phosphates are of great relevance because they constitute the mineral components of the hard tissues of vertebrates. Type I collagen is the main component of the organic matrix of bone, dentine, and cementum, the thin layer of calcified tissue that covers the roots of teeth and anchors them to the jaw. At variance enamel contains no collagen, but phosphorylated proteins, including amelogenins, ameloblastins, and enamelins (Bartlett et al., 2006). The mineral phase, which accounts for about 70 wt.% of bone and 95–97% of enamel (Palmer et al., 2008), is indicated as biological apatite, even if it would be more correct to talk about "biological apatites." In fact, synthetic HA is just a model to describe the variety of compositions and features which characterize the basic calcium phosphates deposited in biological tissues. In contrast to HA, biological apatites are nonstoichiometric and composed of crystals with small dimensions and low degree of structural order. The crystals in bone have a mean length of ~20–50 nm and a width of 12–20 nm, depending on age and species (Glimcher, 2006). The crystals in dentin are of similar size, but enamel crystals are about 10 times larger in all dimensions (Kirkham et al., 1998).

A number of ions not present in stoichiometric HA are associated with biological apatites (Table 7.1), which is not surprising when considering the rich ionic composition of the biological environment where mineralization occurs (Wang and Nancollas, 2008). Ionic composition and, as a consequence, crystallinity and solubility, depend on type and location of the biological tissue. For instance, carbonate content decreases from about 5% in bone and dentin to about 3% in enamel. Moreover, the amount of carbonate varies in the different types of bone and within the tooth crown (Bigi et al., 1997; Asscher et al., 2011). Increase of carbonate content from outer to inner enamel and the associated decreased crystallinity of the apatitic phase have been related to the decreased enamel mechanical properties (Xu et al., 2012). Carbonate is the most abundant foreign ion in biological apatites, often referred to as "carbonated apatites," where it mostly replaces PO_4^{3-} (Elliott, 1994; LeGeros, 1965). Identification of carbonate location is relatively easy, even when it is located at the two different HA sites in the same tissue, as in enamel where it replaces hydroxyl (type A) and phosphate (type B), in a ratio of 1:9 (Elliott et al., 1985). However, in most cases the simultaneous presence of a number of different foreign ions makes it difficult to state if a particular ion is incorporated into the apatite structure through a regular ionic substitution, or if it is just adsorbed onto the crystal surface. Anyway, whatever the location

Table 7.1 Compositional data of the most significant elements in enamel, dentin, and bone

	Enamel	Dentine	Bone
Ca (wt.%)[a]	37.6	40.3	36.6
P (wt.%)[a]	18.3	18.6	17.1
CO_2 (wt.%)[a]	3.0	4.8	4.8
Na (wt.%)[a]	0.70	0.1	1.0
K (wt.%)[a]	0.05	0.07	0.07
Mg (wt.%)[a]	0.2	1.1	0.6
Sr (wt.%)[a]	0.03	0.04	0.05
Cl (wt.%)[a]	0.4	0.27	0.1
F (wt.%)[a]	0.01	0.07	0.1
Zn (ppm)[b]	263	173	126–217[c]
Ba (ppm)[b]	125	129	
Fe (ppm)[b]	118	93	
Ag (ppm)[b]	0.6	2	
Si (ppm)[d]			80
Ca/P[a]	1.59	1.67	1.65

[a] The composition is relative to the ash content of whole enamel (96%) and of the whole dentin (73 wt.%). Bone is from bovine cortical bone. The percentage is relative to the mineral fraction (73 wt.%) of dry fat-free bone (Elliott, 1994).
[b] McConnell (1973).
[c] Li et al. (2009).
[d] Yamada et al. (2003).

of the foreign ions, the rich composition of biological apatites does play an essential, dose-dependent, role on the properties of mineralized tissues. Fluoride stimulates bone formation by increasing osteoblast functions *in vivo* (Chavassieux et al., 1993) and it has been successfully used in the treatment of osteoporosis (Boivin et al., 1988; Guanabens et al., 2000). Moreover, it is considered the most important caries-preventive agent in dentistry. On the other hand, excessive fluoride intake with respect to its narrow therapeutic window causes deterioration of the biomechanical properties of bone and enamel (Aaseth et al., 2012; Aoba and Fejerskov, 2002). Since the primary source of dietary fluoride is water, it is not surprising that fluorosis is endemic in regions with high drinking water fluoride levels (Boivin et al., 1988).

A further element which has been suggested to play a role in postmenopausal osteoporosis is silicon (Hench, 1998). The relative bone content of Si, which is known as an essential trace element in connective tissue metabolism, varies with age and sex of the individuals (Carlisle, 1988). Silicon provides stability for biological apatites, and it has been found to stimulate osteoblast production of type I collagen, as well as of other biochemical markers of osteoblast differentiation, and to promote bone repair at the wounded sites (Waked and Grauer, 2008). Regular dietary silicon intake has been reported to display a beneficial effect on the density of bone mineral (Jugdaohsingh et al., 2004; Macdonald et al., 2012).

Widespread in biomineralized tissues, magnesium can be considered a natural calcium antagonist and it is one of the most important cations associated with biological apatites.

It has been verified that, in calcified tissues, the amount of magnesium associated with the apatitic phase is higher at the beginning of the calcification process and decreases on increasing calcification (Bigi et al., 1992). Moreover, it plays a key role in bone metabolism, since it influences osteoblast and osteoclast activity, and therein bone growth (Percival, 1999). Accordingly, its depletion has been associated with decreased osteoblast activity, bone fragility, and bone loss (Rude and Gruber, 2004).

The trace element strontium has received increasing attention since the development and introduction of strontium ranelate in the treatment osteoporosis (Marie, 2005; Ammann, 2005). Sr is reported to influence bone turnover by promotion of bone formation and simultaneous reduction of bone resorption (Gallacher and Dixon, 2010). In agreement with this, Sr presence in bone is greater at the regions of high metabolic turnover, and its content in a new compact bone is three- to fourfold higher than in an old compact bone, and ~2.5-fold higher in new than in old cancellous bone (Marie et al., 2001).

Also, the involvement of zinc, an essential trace element relatively abundant in bone, is implied in biomineralization, since it inhibits osteoclast differentiation and promotes osteoblast activity (Moonga and Dempster, 1995). This ion is present in all human tissues and it is involved in a variety of biological functions, including enzyme activity, nucleic acid metabolism, and protein synthesis (Prasad, 1995). A relevant proportion of the total body content of zinc is in bone, and a correlation between zinc content and bone strength in both men and women has been reported (Alhava et al., 1977). Its deficiency is commonly linked to reduced bone density and retardation of bone growth in animals (Lynch, 2011), and to risk of osteoporosis in humans (Ilich and Kerstetter, 2000; Gür et al., 2002).

Iron concentration within hard tissues is low and it does not significantly affect bone remodeling. However, Fe deficiency has been reported to lead to decreased mechanical strength and decreased bone mass density, whereas Fe excess has been correlated with decreased osteoblast cell number and activity (Kyriacou et al., 2013).

Generally, most of the foreign ions associated with the bone mineral phase influence osteogenesis and/or angiogenesis (Bose et al., 2013) in a dose-dependent way and, as a consequence, play a vital role on the mineralization process.

7.5 Synthetic apatites

Due to the high flexibility of its structure, HA represents an ideal material to study the chemical, structural, and morphological modifications induced by ionic substitutions. Moreover, enrichment of HA for biomedical applications with biologically relevant ions can improve osteointegration and even provide a tool for local administration of ions with specific therapeutic properties. HA is highly insoluble and it resorbs very slowly *in vivo*. Resorption is a cell-mediated process, which is influenced not only by the solubility of the material but also by a series of parameters, such as chemistry, morphology, crystallinity, and superficial area, which are affected by ionic substitution. The high interest in substituted apatites and their use for the preparation of biomaterials is demonstrated by the large number of studies in this field.

7.5.1 Carbonate

In view of its abundant and ubiquitous presence in biological apatites, carbonate is of great interest as a substituent of HA for biomedical applications. As already mentioned, carbonate (CO_3^{2-}) is a peculiar anion since it can substitute both the OH^- group (type A) and the PO_4^{3-} ion (type B) (Figure 7.6). Both A and B substitutions imply charge imbalance, which can be compensated by additional charged defects, such as OH^- vacancies for type A and coupled OH^- and Ca^{2+} vacancies, as well as Na^+ substitution for Ca^{2+}, for type B (Astala and Stott, 2005; Ulian et al., 2014). Self-compensation, with simultaneous A and B site carbonate substitutions, is possible (Elliott 1994).

Carbonate as a complete replacement for hydroxyl ions in type A carbonated hydroxyapatite (CHA) was reported to yield monoclinic symmetry, with space group $P2_1/b$ and lattice constants $a=9.557$ Å, $b=2a$, $c=6.872$ Å, and $\gamma=120.36°$ (Elliott, 1980). Slightly different values were reported for lower CO_3^{2-} contents (Fleet and Liu, 2003, 2005). The monoclinic $P2_1/b$ structure was recently confirmed through Rietveld refinement diffraction data, which indicated that CO_3^{2-} occupies two independent sites and exhibits a disordered configuration along the *c*-axis (Tonegawa et al., 2010).

In contrast, replacement of the larger tetrahedral phosphate ion in type B hexagonal CHA provokes a decrease in the *a*-axis and an increase in the *c*-axis dimensions. Mean parameters of $a=9.4047(47)$ Å and $c=6.8960(17)$ Å have been reported for Na-free CHA containing about 11 wt.% CO_3^{2-}, where XRD powder pattern refinements confirmed that carbonate ion occupies the sloping face of the PO_4^{3-} site in twofold disorder about the mirror plane (Wilson et al., 2006).

A number of theoretical studies have been performed on different types of CHA in order to clarify the role of CO_3^{2-} substitution in the structural modifications of HA. Calculations based on density functional theory indicated type B with charge compensation by a calcium vacancy together with a hydrogen atom which bonds

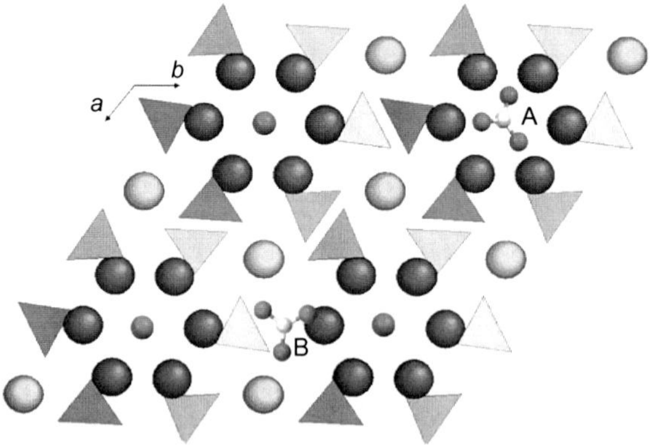

Figure 7.6 A projection of the apatite structure along the *c*-axis. The sites of CO_3^{2-} substitution in A- or B-type CHA are indicated. Tetrahedra are PO_4^{3-}, big dark spheres are Ca(II), big bright spheres are Ca(I), small spheres are OH^- ions; CO_3^{2-} ion is indicated in stick and ball mode.

to a neighboring phosphate as the most stable substitution (Astala and Stott, 2005). However, simulation methods often led to contradictory results: Peroos et al. (2006) found that the energetically favored order was type A, type A–B, type B, whereas Ren et al. (2013) indicated A–B as the most stable substitution, followed by type A and then type B. In particular, the most stable model for type B implies Na^+ substitution to Ca^{2+}, as further supported by the work of Ulian et al. (2014), who concluded that in spite of the lattice parameter variation, accomodation of CO_3^{2-} ion provokes just a slight rotation of the PO_4^{3-} tetrahedra.

Carbonate substitution in HA is easily detected by FTIR spectroscopy. The characteristic bands of carbonate are the ν_3 symmetric stretching in the 1400–1600 cm^{-1} region and the 873–880 cm^{-1} ν_2 out-of-plane banding vibration. The detailed positions of these bands are commonly used to distinguish A- and B-type CHA (Rey et al., 1991; Paschalis et al., 1996; Antonakos et al., 2007; Fleet, 2009). In particular, 1546, 1456, and 880 cm^{-1} are indicated as characteristic of type A substitution, and 1465, 1413, and 873 cm^{-1} of type B. A recent experimental and computational study indicates that the carbonate ν_3 band at about 1546 cm^{-1} can be considered diagnostic of type A substitution, whereas the ν_3 band at 1465 cm^{-1} is characteristic of type B. In contrast, the further bands cannot establish the type of substitution since they can also be present when carbonate is just adsorbed onto HA or present in a separate phase (Ren et al., 2014).

Type A CHA is usually synthesized by ion exchange, whereas types B and A–B are obtained using a precipitation method (Shepherd et al., 2012). Type B CHA has also been prepared by hydrolysis of monetite (Pieters et al., 2010).

On considering that biological apatites are indeed carbonated apatites where CO_3^{2-} mostly replaces PO_4^{3-}, the preparations of HA for biomedical applications generally aim to obtain carbonated, preferably type B, HA. However, most of the studies do not provide sufficient data to establish the anion position and/or to discriminate between substituted or just adsorbed carbonate, which is at least partially due to the decreased crystallinity and crystal size of the synthesized product with respect to pure HA (Yao and LeGeros, 2010). In agreement with the reduced crystallinity, CHA displays increased solubility (Wang and Nancollas, 2008) and reduced thermal stability (Barralet et al., 2002). The increased solubility of type B CHA was exploited to increase the efficiency of gene transfer of a DNA–fibronectin–apatite composite layer (Yazaki et al., 2011). Moreover, carbonate-containing HA has been shown to increase osteoclastogenesis and mineral phase resorbability in a dose-dependent manner and, as a consequence, to enhance collagen synthesis by osteoblasts (Spence et al., 2009, 2010). However, the results reported on *in vitro* osteoblast response to CHA are different and sometimes contradictory, varying from no appreciable differences in cell behavior onto apatites at different carbonate contents (Melville et al., 2006; Landi et al., 2006) to enhanced cell adhesion and proliferation (Pieters et al., 2010), to increased osteogenic gene expression levels (Hesaraki et al., 2014). The variety of results depends not only on the different carbonate content and location but also on the different morphology, crystallinity, crystal size, and surface roughness, all of which can affect cell differentiation alone (Okada et al., 2010). This problem is not limited to carbonate: ion doping and/or substitution produce a series of chemical and structural modifications with respect to stoichiometric HA, which often hinder the efforts to find a clear relationship between the biological response and the single properties of the material (Boanini et al., 2010).

7.5.2 Silicon

Silicon has been shown to stimulate calcium phosphate precipitation even in the presence of inhibitors of HA precipitation (Damen and Cate, 1992). Since it has been demonstrated that the bioactivity of HA is considerably improved by the inclusion of silicate ions into its lattice (Patel et al., 2002), silico-susbstituted apatites (SiHAs) are promising materials for bone repair applications (Khan et al., 2014). The adhesion of human osteoblasts on SiHA increases as a function of Si substitution degree (Botelho et al., 2006), but the effect is reversed when the substitution is over 1 wt.%.

Incorporation of silicon into the HA structure implies the replacement of a phosphate ion, PO_4^{3-}, with a silicate one, SiO_4^{4-}, which requires a concomitant charge compensation mechanism like the HPO_4^{2-} formation from PO_4^{3-}, cationic replacement or the creation of anionic vacancies at OH sites (Arcos et al., 2004; Gibson et al., 1999). As a consequence, the maximum incorporation of Si into the HA structure is limited by the vacancy that the lattice can tolerate (about 4 wt.%), that is more than enough for biological effect. The data about the maximum incorporation of silicon in the synthetic HA structure are spread up to 5 wt.%. Silicon substitution provokes small *a*-axis contraction or fluctuations, moderate *c*-axis and unit cell volume expansion, and a reduction of the mean crystallite size (Gibson et al., 1999; Marchat et al., 2013). Lattice parameters values of $a=9.416$, $c=6.920$ Å are reported by Marchat et al. (2013) for a Si content of about 3 wt.%. Silicon promotes precipitation of a biomimetic apatite by increasing the solubility of the material through the creation of crystalline defects with substitution for PO_4^{3-} ions and associated charge compensation mechanism. The result is a nanocrystalline material with a more electronegative surface due to exchange of SiO_4^{4-} for PO_4^{3-} (Pietak et al., 2007).

Several different methods have been used to prepare SiHAs, but the aqueous precipitation methods are the most widespread. Recently Marchat et al. (2013) reported a new route which avoids the contamination of secondary phases or undesired ions. The original procedure consisted of the hydrolysis, condensation, and depolymerization reaction of the tetraethylorthosilicate used as the source of silicate ions.

The simultaneous substitution of SiO_4^{4-} with other bioactive ions like Mg^{2+} and CO_3^{2-} was tested in order to obtain materials with features closer to biological apatites (Kim et al., 2003; Sprio et al., 2008). At low concentration, CO_3^{2-} ions balance the charge excess induced by SiO_4^{4-} ions, but since both can compete for PO_4^{3-} sites, the structure can host only a limited amount before collapse. Apatite coatings containing Sr and Si were studied as a vehicle for delivery of biologically active ions with the purpose of further optimizing bone response to metal implants (Lindahl et al., 2012). The poorly crystalline apatite dissolves and releases Sr and Si ions in phosphate-buffered saline solution during the re-precipitation of apatite. The surface charge of apatites containing Fe^{3+} and SiO_4^{4-} ions showed a good bovine serum albumin absorption and controlled release, which suggests that FeSiHA particles could be used for the sustained release of proteins or other biomolecules (Pon-on et al., 2013).

7.5.3 Fluorine

Fluorine is essential for the formation of teeth and bones: about 99% of the fluorine in the body is localized in the mineralized tissues. Free fluoride ions in solution are very reactive and can interact with HA crystals so that F–OH exchange at the apatite–solution interface readily occurs even at low fluoride levels. *In vitro* F^- ions released from F-substituted apatite significantly stimulate cell attachment and proliferation and increase alkaline phosphatase activity of osteoblastic cells (Qu and Wei, 2006). However, it is necessary to optimize the F^- content in fluoride-substituted HA so as to achieve the maximum bioactivity of the material (Qu and Wei, 2006), and avoid severe adverse effects, such as anomalous enamel formation (Aoba, 1997).

Fluoride can be substituted in the apatite structure over the whole range of composition. If F^- ions substitute completely for hydroxyl ions on the hexagonal axis, fluorapatite (FA) is formed, with the formula $Ca_{10}(PO_4)_6F_2$, where fluoride ion occupies the center of the Ca(II) triangles. As a result, the substitution of smaller F^- (ionic radius, 1.32 Å) for OH^- leads to a contraction of cell parameters in the *a*-axis direction ($a = 9.3684$ Å, $c = 6.8841$ Å, and ICDD PDF No. 15-876) and of the unit cell volume, the lattice becomes more dense, and its chemical stability is greatly enhanced (Elliott, 1994). When fluoride partially replaces hydroxyl groups, it occupies a position that brings it close to the adjacent OH^- ions, which promotes hydrogen bonding between the hydroxide and fluoride within the 6_3 channel (Rodriguez-Lorenzo et al., 2003). A very small amount of fluoride substituted for hydroxyl ions drives the building of quite long rows of equiversed O–H groups. This promotes a regular arrangement of the molecular packing into the crystal lattice, which fosters the crystalline stability of apatite (Young, 1975).

Most of the studies about F-substitution in the apatitic structure use precipitation (Jha et al., 1997; Stanić et al., 2014) or sol–gel preparation methods (Kim et al., 2004; Tredwin et al., 2014). Partially substituted fluoro-HAs are homogeneous polycrystalline solid solutions. Calculations of cell parameters are often reported as evidence of F^- ion incorporation into the apatitic structure.

By careful selection of the precipitation conditions and fluoride contents, the composition and morphology of fluoride-substituted apatite may be controlled and this has interesting implications for the development of these materials for biomedical issues (Jha et al., 1997). Their thermal stability in the whole range of composition has been studied up to 1200 °C, where no secondary phases are formed and crystals grow in dimensions (Gross and Rodriguez-Lorenzo, 2004; Wei et al., 2003). In the case of sol–gel preparations, the as-prepared samples display a very low degree of crystallinity that definitely increases after heat treatment. However, Kim et al. (2004) report that the phase stability and crystallinity can vary as a function of the level of fluoride substitution: formation of tricalcium phosphate (TCP) occurs at increasing temperatures (always above 400 °C) on increasing fluoride content in the crystals.

Sol–gel methods have also been used for producing fluoro-HA and fluoroapatite coatings on metallic substrates by spin-coating (Tredwin et al., 2013) or dip-coating techniques (Cavalli et al., 2001), with good results from the points of view of purity, crystallinity, and surface homogeneity. In particular, Tredwin et al. (2013) report that increasing fluoride substitution in the apatite structure significantly increased the bond strength with titanium support, so that the mean bond strength of FA was ~40 MPa, which is higher than or comparable to that usually reported for HA plasma-sprayed coatings.

Substitution of F^- ions at OH^- ion sites in the HA lattice can be obtained so easily that many studies have been addressed to develop new strategies for obtaining shape-controlled crystals. The addition of EDTA and citric acid produced hexagonal disks with predominant (001) faces, hexagonal prisms, icosahedrons, and hexagonal microrods with tunable aspect ratios (Jiang et al., 2010), whereas a surfactant-free biomimetic method produced a highly organized and tightly packed needle-like fluoro-HA layer with a tailored F^- ion content on pretreated titanium surfaces (Xia et al., 2012).

7.5.4 Magnesium

The destabilizing role of magnesium on the HA structure is well known. The Mg^{2+} ion is indeed quite smaller than calcium ion and it usually prefers structures with available octahedral coordination sites, such as β-TCP. In fact, biological β-TCP is always partially magnesium substituted and the thermal conversion of biological apatites into β-TCP has been quantitatively related to magnesium content (Bigi et al., 1992). As far as synthetic HA is concerned, a number of studies have been addressed to establish the maximum extent of Mg^{2+} substitution for Ca^{2+} in the HA structure. The possibility to discriminate between substitution and simple adsorption depends on the degree of crystallinity of the material and therefore on the synthetic method (Bigi et al., 1993; Yasukawa et al., 1996; Laurencin et al., 2011; Bertoni et al., 1998; Landi et al., 2008; Ren et al., 2010). However, it is generally accepted that, although a much greater amount can be associated to HA most likely adsorbed onto the crystal surface, the extent of Mg substitution for Ca does not exceed 10 atom% (Bigi et al., 1996; Yasukawa et al., 1996; Laurencin et al., 2011). In agreement with the different ionic dimensions, substitution causes a reduction of the lattice parameters (Bigi et al., 1996; Ren et al., 2010). The shorter mean metal–oxygen distances and the more suitable coordination geometry of M(II) (due to the presence of six short M–O distances) with respect to the M(I) site suggest preferential location of the small Mg^{2+} ion into M(II). Indeed, except for a study which indicates M(I) as the energetically favored site on the basis of simulation results (Ren et al. 2010), both theoretical and experimental data confirm the preference of Mg for M(II) (Matsunaga, 2008; Bigi et al., 1996; Laurencin et al., 2011). In particular, M(II) preferential occupancy was demonstrated by the results of structural refinements performed with the Rietveld method on HA synthesized at increasing Mg concentration (Bigi et al. 1996) and by ^{43}Ca NMR and Ca-K-edge XAS data obtained on an about 10 atom% Mg-substituted HA (Figure 7.3) (Laurencin et al., 2011). These data are further supported by the results of density functional theory and interatomic potential computation, which also show a preference for an ordered pattern of Mg ions in alternating positions of the Ca(II) triangles along a single OH channel. The presence of Mg at M(II) sites, which is close to OH groups, has been invoked to justify the shift of the $3572\,cm^{-1}$ infrared band, due to the OH stretching mode, to lower wavenumbers on increasing Mg concentration (Diallo-Garcia et al., 2011). Both infrared and XRD data indicate that Mg substitution and/or adsorption induces a decrease of crystallinity (Bigi et al., 1996; Gibson and Bonfield, 2002; Laurencin et al., 2011; Diallo-Garcia et al., 2011; Ren et al., 2010). Reduced crystallinity, which might be coupled to increased surface hydration (Bertinetti et al., 2009), results in increased solubility and decreased thermal stability (Bigi et al., 1993; Landi et al., 2008; Suchanek et al., 2004). A synergic effect on crystallinity reduction and solubility increase has been observed in apatites also containing CO_3^{2-} (LeGeros et al., 1995; Mayer et al., 1997; Dorozhkin, 2011; Kolmas et al., 2011; Sader et al., 2013).

HA morphology depends on the preparation method (Boanini et al., 2010; Han et al., 2013; Sadat-Shojai et al., 2013). However, crystals synthesized in the presence of Mg show reduced dimensions with respect to pure HA (Ren et al., 2010; Diallo-Garcia et al., 2011). Moreover, high resolution transmission electron microscopy (HRTEM), images revealed the presence of an amorphous/poorly crystalline phase at the grain boundaries, which increases with Mg content (Figure 7.7) and is likely related to increased solubility (Ren et al., 2010). A similar effect on particle dimensions have been observed on samples prepared through a step reaction and ion-exchange process at different Ca:Mg molar ratios (Yuan et al. 2013).

Mg has been proposed as an additive in the composition of calcium phosphate bone cements in order to control degradability, mechanical properties, and *in vitro* cell response (Lilley et al., 2005; Julien et al., 2007; Lu et al. 2011). In particular, addition of Mg to an amorphous calcium phosphate (ACP)-based cement was recently reported to influence the hydration reaction, yielding a dense cement matrix and improved compressive strength (Yu et al., 2013). The improved mechanical properties can be related to the Mg stabilizing effect on ACP that blocks its transition to HA and delays HA growth (Ding et al., 2014). The presence of Mg was also reported to influence crystallinity, morphology, and crystal size of HA coatings deposited by means of different methods (Barrere et al., 2002; Bracci et al., 2009; Mroz et al., 2010; Zhao et al., 2013). Moreover, the presence of Mg in apatitic-based materials has been shown to display a beneficial role in cell proliferation and differentiation (Lu et al., 2011; Bracci et al., 2009; Mroz et al., 2010; Zhao et al., 2013), as well as *in vivo* osteoconductivity and resorption (Landi et al., 2008).

7.5.5 Strontium

Among the bivalent cations that can replace calcium in HA, strontium is a trace element in the human body that has attracted remarkable interest regarding its possible biological role. There is no significant structural difference between physiological and pathological calcifications regarding the content of Sr^{2+} cations (Bazin et al., 2014). Sr^{2+} has been shown to be taken up, like many other bone seeking elements, preferably by recent mineral deposits (Marie et al., 2001). As a drug, strontium ranelate increases

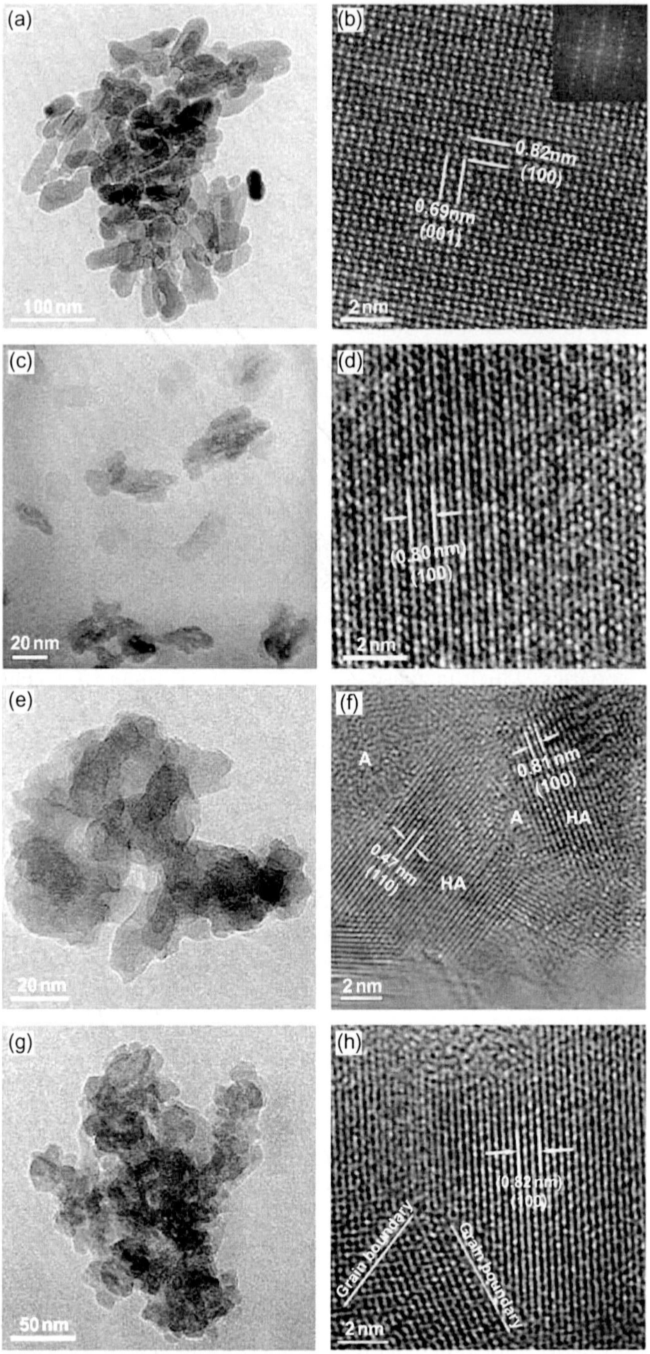

Figure 7.7 TEM images of HA samples at various Mg contents. HA: (a) bright field and (b) HRTEM with FFT. Mg-substituted HA containing 1.5 Mg atom%: (c) bright field and (d) HRTEM. Mg-substituted HA containing 3.1 Mg atom%: (e) bright field and (f) HRTEM. Mg-substituted HA containing 4.0 Mg atom%: (g) bright field and (h) HRTEM. The individual crystal sizes decrease with increasing Mg content. Some of the crystal planes are highlighted; some amorphous regions visible in (f) are marked as "A" and along the grain boundary in (h). Reproduced with permission from Ren et al. (2010).

osteoblast proliferation and differentiation, and inhibits osteoclast formation and activity *in vitro* (Bonnelye et al., 2008), and it is currently prescribed to treat postmenopausal osteoporosis. In fact, it is able to play a unique dual role in bone metabolism by stimulating bone formation and inhibiting bone resorption, hence improving bone strength and resistance. It has been shown that strontium ranelate treatment promotes an increase in the formation of mineralized nodules in osteoblast cell cultures and preserves the overall bone-like nature of the matrix. Moreover, Sr^{2+} is incorporated into the intact mineralized nodules in a dose-dependent way (Querido and Farina, 2013). Biological apatites can also exchange ions with the surrounding fluids and the exchange of calcium ions with strontium ions in solution readily occurs at the surface hydrated layer of nanocrystals. High reversibility of these surface exchanges and a nearly 1:1 substituting ratio have been observed (Drouet et al., 2008). It has been shown that Sr can influence cells' response even when incorporated into the HA structure. In particular, the results of the investigations carried out on Sr-substituted HA using normal and osteopenic osteoblast-like cells and human osteoclasts indicated that strontium influences bone cell behavior in a dose-dependent manner (Capuccini et al., 2009; Boanini et al., 2011).

From a chemical point of view, calcium and strontium share a similar chemistry, since they are close in the second group of the periodic table of elements. As a result, strontium can be substituted into the apatite structure over the whole range of composition. Complete substitution yields strontium-HA, $Sr_{10}(PO_4)_6(OH)_2$ (SrHA). The substitution of the larger Sr^{2+} (ionic radius, 1.20 Å) for Ca^{2+} leads to an expansion of *a* and *c* cell parameters ($a=9.7660$ Å, $c=7.2760$ Å, and ICDD PDF No. 33-1348) and, as a consequence, of the unit cell volume. Partial replacement results in homogeneous solid solutions with linear increase of cell parameters as a function of Sr content. In contrast, the effect of strontium on crystallinity and morphology changes with composition: relatively low Sr content induces a reduction in crystal size and disturbs the shape of the crystals, whereas crystallinity, as well as the mean dimensions of the crystals, increase at relatively high strontium levels (Figure 7.8) (O'Donnell et al., 2008; Bigi et al., 2007).

Since calcium can occupy two different crystallographic positions, many experimental and theoretical studies have been devoted to describe the specific site preference for strontium isomorphous substitution. In the literature, many data are apparently conflicting, some studies concluding a preferential occupancy of the strontium ions at the M(I) site (Kikuchi et al., 1994) and some others at the M(II) site (O'Donnell et al., 2008; Zhu et al., 2006). Recent findings have concluded that a slight Sr enrichment at the M(II) site prevails in most of the range of compositions, whereas a modest preference of Sr for the smaller M(I) site can be appreciated only at a very low strontium content (Bigi et al., 2007; Terra et al., 2009).

Most of the studies on strontium-substituted HAs were performed on samples prepared in aqueous solutions by coprecipitation methods. In an attempt to produce samples with improved crystallinity, many other preparation methods were investigated, such as hydrothermal (Zhu et al., 2006), electrolytic (Montalbert-Smith and Montero, 2013), and solid-state methods (Aina et al., 2013). Sr substitution for Ca always affects HA solubility, which significantly and steadily increases with increasing Sr content

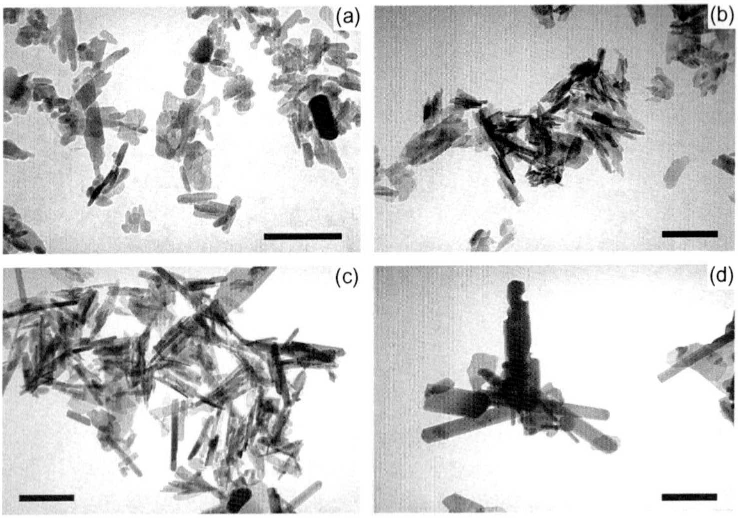

Figure 7.8 TEM micrographs of HA (a), Sr-substituted HA containing: 20 Sr atom% (b), 70 Sr atom% (c), and fully substituted SrHA (d). Scale bars = 200 nm.
Reproduced with permission from Bigi et al. (2007).

(Pan et al., 2009a; Ni et al., 2012), thus affecting ionic release (both for Sr and Ca) and finally biological response. Also for this reason, Sr-substituted HA coatings of metallic substrates are expected to enhance bone formation and osteointegration. Coatings have been produced using different techniques such as pulsed laser deposition (Capuccini et al., 2008), plasma spray (Xue et al., 2007), pulsed electrodeposition (Drevet and Benhayoune, 2013), sol–gel dip method (Li et al., 2010), and biomimetic deposition (Oliveira et al., 2007; Bracci et al., 2009; Pan et al., 2009b). When compared to HA coatings, substituted coatings show similar morphology, surface roughness, and adhesion strength. A slight broadening of the diffraction peaks are consistent with a shortening of the perfect crystalline domains when strontium concentration is increased. However, even when the crystallinity of the inorganic phase is high (Capuccini et al., 2008; Li et al., 2010), it is difficult to find direct evidence of ionic substitution, also because the degree of replacement is always relatively low (maximum value around 10 atom%). In fact, Sr content in the coatings is always within its biological range of activity. Indirect evidence of strontium substitution can be found by analyzing the powders precipitated in the absence of the substrates (Bracci et al., 2009). These data indicate that the presence of Sr^{2+} in solution reduces the extent of precipitation as a function of its concentration and leads to a shift of HA diffraction peaks to larger interatomic distances, which is actually consistent with the ionic substitution hypothesis.

7.5.6 Zinc

Zinc is an essential trace element and Zn-substituted HA has been shown to possess enhanced bioactivity, since it promotes the growth of human adipose-derived mesenchymal stem cells and the production of bone cell differentiation markers (Thian et al., 2013).

The antibacterial properties of Zn^{2+} ions released from calcium phosphate compounds are exploited in the preparation of apatitic biomaterials and in the coating of metal prostheses (Sutha et al., 2013).

Zn^{2+} ion inhibits HA crystal growth (Fuierer et al., 1994). Nonetheless, it can be quantitatively incorporated into the HA lattice by direct synthesis under mild conditions. The substitution for calcium occurs up to about 20 atom%, whereas the apatite structure does not sustain higher Zn^{2+} incorporation. The substitution of zinc into HA resulted in a decrease in both the *a*- and *c*-axes of the unit cell, consistent with the smaller ionic radius of Zn^{2+} with respect to Ca^{2+} (Bigi et al., 1995). A correct evaluation of cell parameters at relatively high zinc content are hindered by the progressive broadening of the XRD peaks, due to the reduction of crystal sizes and/or the increase of crystal strain with increasing Zn^{2+} concentration. Some authors used thermal annealing or autoclave treatment to recover good crystal features and improve characterization data, but these procedures might be inappropriate for the preparation of biomaterials where low crystal size and crystal irregularity are desirable to increase dissolution and osteointegration. Owing to the smaller ionic sizes of Zn^{2+}, its substitution for Ca brings about considerable inward atomic relaxation of the surrounding oxygen ions, which results in smaller coordination numbers with oxygen as compared with that of Ca in HA. It is also found that the substitutional Zn^{2+} ions tend to form covalent bonds with oxygen in the fourfold coordination state (Matsunaga, 2008).

Zinc is often used in association with other ions to prepare several biomimetic materials. As an example, it is used in association with Sr^{2+} to improve the protein adsorption efficiency onto HA (Mavropoulos et al., 2011). The copresence of Mg^{2+} and Zn^{2+} as doping ions was found to enhance the mechanical and biological properties of nanocrystalline HA ceramics for tissue engineering applications (Kalita and Bhatt, 2007).

7.5.7 Iron

Fe is present in small amounts in biological apatites and scarcely affects bone remodeling (Morrissey et al., 2005). The interest concerning Fe-doped apatites (Fe-HA) started about 25 years ago, when hyperthermia treatment was demonstrated as a promising technique for tumor eradication in oncological patients. The most common approach to access difficult locations in the body has been to manufacture a thermoseed that can be heated remotely. Paramagnetic materials are suitable thermoseeds that produce hysteresis heating by application of an alternating magnetic field (Gross et al., 2002). Moreover, paramagnetic nanoparticles have been progressively explored for their potential as contrast agents for diagnostic imaging, for magnetic drug delivery, and recently for tissue engineering applications. To avoid the use of poorly tolerated magnetite-based nanoparticles, Fe-HA has been proposed as an innovative biocompatible and bioresorbable superparamagnetic-like phase (Panseri et al., 2012). To obtain the desired properties, the copresence of Fe^{2+} and Fe^{3+} ions is necessary. Some authors introduced only Fe^{2+} ions at the beginning of the synthetic procedure, necessitating partial oxidation during the process (Hou et al., 2009). The products are often not fully characterized and it is not clear if the resulting material is a single Fe-HA phase or a mixture of HA/Fe-HA and magnetite, but good performances have been reported (Birsan et al., 2007; Wu et al., 2007). Reduced crystal size and unit cell modifications, *a*-axis expansion and *c*-axis contraction, suggesting the presence of oxyhydroxyapatite, have been observed in Fe-HA. In particular, $a=9.464$ and $c=6.879$ Å have been reported at Fe content of 5 atom% (Low et al., 2008). Atomistic simulations revealed that Fe^{3+} ions are energetically favored at the Ca(I) site and Fe^{2+} ions at the Ca(II) site (Jiang et al., 2002). Even HA-coated magnetic Fe_3O_4 nanoparticles have been reported to enhance osteoblast proliferation and have been suggested as potential candidates to increase bone growth at desired bone defect sites (Tran and Webster, 2011). A controlled method implying the simultaneous addition of Fe^{2+}/Fe^{3+} ions during apatite nucleation has been reported to produce Fe-HA containing about 20 mol.% of Fe ions and a small amount of magnetite as secondary phase (Tampieri et al., 2012). The magnetic property of the new Fe-HA phase together with its biocompatibility open the door of regenerative medicine to a conceptually new family of biomimetic scaffolds able to be biologically manipulated or activated *in situ* by means of an external magnetic field (Panseri et al., 2012). As an example the use of a magnetic scaffold in conjunction with magnetically functionalized cells would allow an increase in the efficiency of both cell seeding and growth by simply employing a static magnetic field (Figure 7.9).

7.5.8 Silver

Ag is used in a wide range of applications, from water and air purification to cosmetics, clothing, and numerous consumer products, because of its well-known antibacterial properties (Marambio-Jones and Hoek, 2010). Recently its use against many

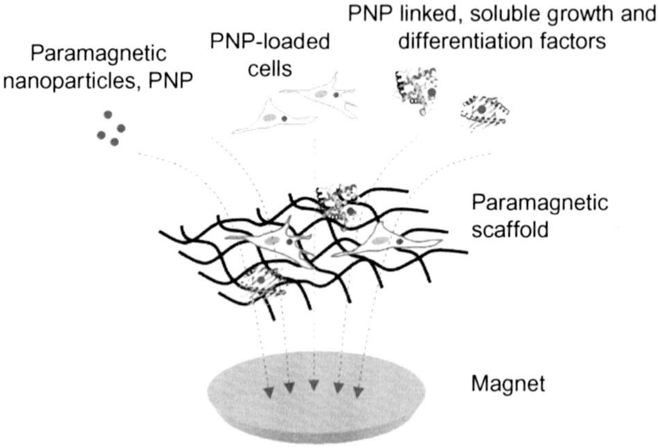

Figure 7.9 Sketch showing a possible use of a paramagnetic scaffold in regenerative medicine. The application of an external magnetic field induces magnetic activity in the paramagnetic materials. Cells and soluble molecules previously linked with paramagnetic nanoparticles (PNP) are attracted and located in the scaffold to promote tissue growth. When the thickness of the new material in the scaffold is sufficient, the removal of the magnet will stop the driving force.

bacterial strains that show resistance toward antibiotics has attracted a lot of interest also in the medical and biomedical fields for the production of medical implants and instruments for the prevention of infections. A possible mechanism of action involve disruption of the mitochondrial respiratory chain by Ag, leading to production of reactive oxygen species and interruption of ATP synthesis, which in turn causes DNA damage (Asharani et al., 2009).

Many efforts have been made to investigate the properties of silver-doped HA, which should combine the biocompatibility of HA with the antimicrobial properties of silver. Often Ag is associated with calcium phosphate crystals in the form of metallic nanoparticles (Honda et al., 2013; Boanini et al., 2014), which allows one to achieve a high silver content in the composite material. Indeed, the structure of HA is not changed by the presence of metallic silver, which is usually visible with electron microscopy and XRD as a separate phase.

Actually, silver ion incorporation into the structure of HA is hindered by its ionic charge (Ag ions are only monovalent) and by a steric factor (the large ionic radius of Ag, 1.26 Å). As a result, silver-substituted HAs have been obtained with very low degrees of replacement (<5 atom%). Successful preparation of Ag-substituted HA powders has been obtained using silver oxide as a precursor (Stanić et al., 2011), as demonstrated by a slight increase of cell parameters due to the presence of the larger Ag^+ ions. In many papers, Ag-substituted HA is indicated as the unique crystalline phase obtained via wet preparation methods (Ciobanu et al., 2012, 2013; Costescu et al., 2013). However, the reported values of lattice parameters are generally in good agreement with those of pure HA and are not affected by the presence of the larger Ag ions, which might happen either because of the very low Ag content or indeed because silver could not be incorporated into the lattice. The latter possibility could lead to mixed phases in the case of a subsequent sintering. The charge deficiency caused by Ag^+ substitution in the HA lattice led to the presence of a minor secondary phase of TCP when using a wet coprecipitation method followed by sintering (Singh et al., 2011; Iqbal et al., 2012). Similar sintering approaches have also resulted in mixed phases containing silver oxide, likely due to poor incorporation of Ag (Dubnika et al., 2013; Baştan and Özbek, 2013).

The association of silver with HA to produce new biomaterials has been studied for coatings and cements in an attempt to meet the current needs for the treatment of implant-based bacterial infections. Coatings have been produced using different techniques such as pulsed laser deposition (Eraković et al., 2014), thermal substrate method (Yanovska et al., 2014), plasma spray (Fielding et al. 2012; Roy et al. 2012), electrophoretic deposition (Eraković et al., 2013), and biomimetic deposition (Saidin et al., 2013). There is no evidence that these techniques provide coatings where silver is able to replace the calcium ions in HA. Most of the coatings show XRD patterns of HA as the unique crystalline phase, but the degree of crystallinity is not sufficient to calculate the variation of cell parameters (Eraković et al., 2013, 2014; Yanovska et al., 2014). A layered structure where Ag nanoparticles alternate with HA has been obtained via a biomimetic method (Saidin et al. 2013), and also a plasma spray technique produces phase separation where Ag is present as metallic Ag or silver oxide (Fielding et al., 2012; Roy et al., 2012). Nevertheless, all these coatings have proven to be effective against bacteria *in vitro*, regardless of the form of silver present. Good efficacy has also been demonstrated with silver-doped calcium apatitic cements (Ewald et al., 2011), but the very low amount of silver present (about 1 atom%) does not allow any structural evaluation of this material.

7.6 Concluding remarks

Isomorphous substitutions in synthetic apatites have been investigated for a long time. However, the increased awareness of the potentially improved performance of substituted HA-based biomaterials has recently produced an exponential growth of scientific papers in this field. On one hand, the continuous upgrading of the characterization techniques has stimulated the revisiting of already widely studied ions, with the aim to gain deeper information on the chemical and structural changes brought about by the presence of the foreign ion. On the other hand, an increasing part of the studies, often carried out via a multidisciplinary approach, has focused on the modifications induced by the replacement of ions that can positively influence the biological response. On this basis, ion-substituted HA is now often utilized as a tool to tailor the properties of biomaterials and/or to provide local delivery of therapeutically useful ions. Moreover, besides the long-standing interest in the most studied ions, such as carbonate, fluoride, strontium, magnesium, and zinc, a great deal of attention has recently turned to less common ions, such as silicon, iron, and silver, as well as to multiple substitutions. In spite of the increasing number of studies on the use of ion-substituted HA in biomedical applications, the amount of data available on the *in vitro* and *in vivo* performance of these biomaterials is still relatively scarce. It can be concluded that this is still an open and exciting research field, which needs the combined efforts of basic and applied studies to yield successful outcomes.

References

Aaseth, J., Boivin, G., Andersen, O., 2012. Osteoporosis and trace elements – an overview. J. Trace Elem. Med. Biol. 26, 149–152.

Aina, V., Bergandi, L., Lusvardi, G., Malavasi, G., Imrie, F.E., Gibson, I.R., Cerrato, G., Ghigo, D., 2013. Sr-containing hydroxyapatite: morphologies of HA crystals and bioactivity on osteoblast cells. Mater. Sci. Eng. C 33, 1132–1142.

Alhava, E.M., Olkkonen, H., Puittinen, J., Nokso-Koivisto, V.M., 1977. Zinc content of human cancellous bone. Acta Orthop. Scand. 48, 1–4.

Ammann, P., 2005. Strontium ranelate: a novel mode of action leading to renewed bone quality. Osteoporos. Int. 16, S11–S15.

Antonakos, A., Liarokapis, E., Leventouri, T., 2007. Micro-Raman and FTIR studies of synthetic and natural apatites. Biomaterials 28, 3043–3054.

Aoba, T., 1997. The effect of fluoride on apatite structure and growth. Crit. Rev. Oral Biol. Med. 8, 136–153.

Aoba, T., Fejerskov, O., 2002. Dental fluorosis: chemistry and biology. Crit. Rev. Oral Biol. Med. 13, 155–170.

Arcos, D., Rodríguez-Carvajal, J., Vallet-Regí, M., 2004. The effect of silicon incorporation on the hydroxylapatite structure. A neutron diffraction study. Solid State Sci. 6, 987–994.

Asharani, P.V., Low Kah Mun, G., Hande, M.P., Valiyaveettil, S., 2009. Cytotoxicity and genotoxicity of silver nanoparticles in human cells. ACS Nano 3, 279–290.

Asscher, Y., Weiner, S., Boaretto, E., 2011. Variations in atomic disorder in biogenic carbonate hydroxyapatite using the infrared spectrum grinding curve method. Adv. Funct. Mater. 21, 3308–3313.

Astala, R., Stott, M.J., 2005. First principles investigation of mineral component of bone: CO_3 substitutions in hydroxyapatite. Chem. Mater. 17, 4125–4133.

Badraoui, B., Bigi, A., Debbabi, M., Gazzano, M., Roveri, N., Thouvenot, R., 2001. X-ray powder diffraction and solid-state NMR investigations in cadmium-lead hydroxyapatites. Eur. J. Inorg. Chem. 5, 1261–1267.

Badraoui, B., Aissa, A., Bigi, A., Debbabi, M., Gazzano, M., 2009. Synthesis and characterization of $Sr_{(10-x)}Cd_x(PO_4)_6Y_2$ (Y=OH and F): a comparison of apatites containing two divalent cations. Mater Res. Bull. 44, 522–530.

Barralet, J., Knowles, J.C., Best, S., Bonfield, W., 2002. Thermal decomposition of synthesised carbonate hydroxyapatite. J. Mater. Sci. Mater. Med. 13, 529–533.

Barrere, F., van Blitterswijk, C.A., de Groot, K., Layrolle, P., 2002. Nucleation of biomimetic Ca–P coatings on Ti_6Al_4V from a $SBFx_5$ solution: influence of magnesium. Biomaterials 23, 2211–2220.

Bartlett, J.D., Ganss, B., Goldberg, M., Moradian-Oldak, J., Paine, M.L., Snead, M.L., Wen, X., White, S.N., Zhou, Y.L., 2006. Protein–protein interactions of the developing enamel matrix. Curr. Top. Dev. Biol. 74, 57–115.

Baştan, F.E., Özbek, Y.Y., 2013. Producing antibacterial silver-doped hydroxyapatite powders with chemical precipitation and reshaping in a spray dryer. Mater. Technol. 47, 431–434.

Bazin, D., Dessombz, A., Nguyen, C., Ea, H.K., Lioté, F., Rehr, J., Chappard, C., Rouziere, S., Thiaudiére, D., Reguerh, S., Daudon, M., 2014. The status of strontium in biological apatites: an XANES/EXAFS investigation. J. Synchrotron Radiat. 21, 136–142.

Bertinetti, L., Drouet, C., Combes, C., Rey, C., Tampicri, A., Coluccia, S., Martra, G., 2009. Surface characteristics of nanocrystalline apatites: effect of Mg surface enrichment on morphology, surface hydration species, and cationic environments. Langmuir 25, 5647–5654.

Bertoni, E., Bigi, A., Cojazzi, G., Gandolfi, M., Panzavolta, S., Roveri, N., 1998. Nanocrystals of magnesium and fluoride substituted hydroxyapatite. J. Inorg. Biochem. 72, 29–35.

Bigi, A., Gandolfi, M., Gazzano, M., Ripamonti, A., Roveri, N., Thomas, S.A., 1991. Structural modifications of hydroxyapatite induced by lead substitution for calcium. J. Chem. Soc. Dalton Trans. 11, 2883–2886.

Bigi, A., Foresti, E., Gregorini, R., Ripamonti, A., Roveri, N., Shah, J.S., 1992. The role of magnesium on the structure of biological apatites. Calcif. Tissue Int. 50, 439–444.

Bigi, A., Falini, G., Foresti, E., Gazzano, M., Ripamonti, A., Roveri, N., 1993. Magnesium influence on hydroxyapatite crystallization. J. Inorg. Biochem. 49, 69–78.

Bigi, A., Foresti, E., Gandolfi, M., Gazzano, M., Roveri, N., 1995. Inhibiting effect of zinc on hydroxylapatite crystallization. J. Inorg. Biochem. 58, 49–58.

Bigi, A., Falini, G., Foresti, E., Gazzano, M., Ripamonti, A., Roveri, N., 1996. Rietveld structure refinements of calcium hydroxylapatite containing magnesium. Acta Crystallogr. B52, 87–92.

Bigi, A., Cojazzi, G., Panzavolta, S., Ripamonti, A., Roveri, N., Romanello, M., Suarez, K.N., Moro, L., 1997. Chemical and structural characterization of the mineral phase from cortical and trabecular bone. J. Inorg. Biochem. 68, 45–51.

Bigi, A., Boanini, E., Capuccini, C., Gazzano, M., 2007. Strontium-substituted hydroxyapatite nanocrystals. Inorg. Chim. Acta 360, 1009–1016.

Birsan, M., Predoi, D., Birsan, C., Secu, C.E., Andronescu, E., 2007. Characterisation of magnetic iron oxide in hydroxyapatite. J. Optoelectron. Adv. Mater. 9, 1829–1832.

Boanini, E., Gazzano, M., Bigi, A., 2010. Ionic substitutions in calcium phosphates synthesized at low temperature. Acta Biomater. 6, 1882–1894.

Boanini, E., Torricelli, P., Fini, M., Bigi, A., 2011. Osteopenic bone cell response to strontium-substituted hydroxyapatite. J. Mater. Sci. Mater. Med. 22, 2079–2088.

Boanini, E., Torricelli, P., Cassani, M.C., Gentilomi, G.A., Ballarin, B., Rubini, K., Bonvicini, F., Bigi, A., 2014. Cationic-anionic polyelectrolyte interaction as a tool to graft silver nanoparticles on hydroxyapatite crystals and prevent cytotoxicity. RSC Adv. 4, 645–652.

Boivin, G., Chapuy, M.C., Baud, C.A., Meunier, P.J., 1988. Fluoride content in human iliac bone: results in controls, patients with fluorosis, and osteoporotic patients treated with fluoride. J. Bone Miner. Res. 3, 497–501.

Bonnelye, E., Chabadel, A., Saltel, F., Jurdic, P., 2008. Dual effect of strontium ranelate: stimulation of osteoblast differentiation and inhibition of osteoclast formation and resorption in vitro. Bone 42, 129–138.

Bose, S., Fielding, G., Tarafder, S., Bandyopadhyay, A., 2013. Understanding of dopant-induced osteogenesis and angiogenesis in calcium phosphate ceramics. Trends Biotechnol. 31, 594–605.

Botelho, C.M., Brooks, R.A., Best, S.M., Lopes, M.A., Santos, J.D., Rushton, N., Bonfield, W., 2006. Human osteoblast response to silicon-substituted hydroxyapatite. J. Biomed. Mater. Res. A 79A, 723–730.

Bracci, B., Torricelli, P., Panzavolta, S., Boanini, E., Giardino, R., Bigi, A., 2009. Effect of Mg^{2+}, Sr^{2+} and Mn^{2+} on the chemico-physical and in vitro biological properties of calcium phosphate biomimetic coatings. J. Inorg. Biochem. 103, 1666–1674.

Capuccini, C., Torricelli, P., Sima, F., Boanini, E., Ristoscu, C., Bracci, B., Socol, G., Fini, M., Mihailescu, I.N., Bigi, A., 2008. Strontium-substituted hydroxyapatite coatings synthesized by pulsed-laser deposition: *in vitro* osteoblast and osteoclast response. Acta Biomater. 4, 1885–1893.

Capuccini, C., Torricelli, P., Boanini, E., Gazzano, M., Giardino, R., Bigi, A., 2009. Interaction of Sr-doped hydroxyapatite nanocrystals with osteoclast and osteoblast-like cells. J. Biomed. Mater. Res. A 89A, 594–600.

Carlisle, E.M., 1988. Silicon as a trace nutrient. Sci. Total Environ. 73, 95–106.

Cavalli, M., Gnappi, G., Montenero, A., Bersani, D., Lottici, P.P., Kaciulis, S., Mattogno, G., Fini, M., 2001. Hydroxy- and fluorapatite films on Ti alloy substrates: sol-gel preparation and characterization. J. Mater. Sci. 36, 3253–3260.

Chavassieux, P., Boivin, G., Serre, M., Meunier, P.J., 1993. Fluoride increases osteoblastic function and population after in vivo administration, but not after in vitro exposure. Bone 14, 721–725.

Chen, C., Wang, Z., Saito, M., Tohei, T., Takano, Y., Ikuhara, Y., 2014. Fluorine in shark teeth: its direct atomic-resolution imaging and strengthening function. Angew. Chem. Int. Ed. 53, 1543–1547.

Ciobanu, C.S., Iconaru, S.L., Le Coustumer, P., Constantin, L.V., Predoi, D., 2012. Antibacterial activity of silver-doped hydroxyapatite nanoparticles against gram-positive and gram-negative bacteria. Nanoscale Res. Lett. 7, 324–333.

Ciobanu, C.S., Iconaru, S.L., Pasuk, I., Vasile, B.S., Lupu, A.R., Hermenean, A., Dinischiotu, A., Predoi, D., 2013. Structural properties of silver doped hydroxyapatite and their biocompatibility. Mater. Sci. Eng. C 33, 1395–1402.

Costescu, A., Ciobanu, C.S., Iconaru, S.L., Ghita, R.V., Chifiriuc, C.M., Marutescu, L.G., Predoi, D., 2013. Fabrication, characterization, and antimicrobial activity, evaluation of low silver concentrations in silver-doped hydroxyapatite nanoparticles. J. Nanomater. 2013, 9, Article ID 194854.

Damen, J.J.M., Cate, T.J.M., 1992. Silica-induced precipitation of calcium phosphate in the presence of inhibitors of hydroxyapatite formation. J. Dent. Res. 71, 453–457.

Diallo-Garcia, S., Laurencin, D., Krafft, J.M., Casale, S., Smith, M.E., Lauron-Pernot, H., Costentin, G., 2011. Influence of magnesium substitution on the basic properties of hydroxyapatites. J. Phys. Chem. C 115, 24317–24327.

Ding, H., Pan, H., Xu, X., Tang, R., 2014. Towards a detailed understanding of magnesium ions on hydroxyapatite crystallization inhibition. Cryst. Growth Des. 14, 763–769.

Dorozhkin, S.V., 2011. Calcium orthophosphates: occurrence, properties, biomineralization, pathological calcification and biomimetic applications. Biomatter 1, 121–164.

Drevet, R., Benhayoune, H., 2013. Pulsed electrodeposition for the synthesis of strontium-substituted calcium phosphate coatings with improved dissolution properties. Mater. Sci. Eng. C 33, 4260–4265.

Drouet, C., Carayon, M.T., Combes, C., Rey, C., 2008. Surface enrichment of biomimetic apatites with biologically-active ions Mg^{2+} and Sr^{2+}: a preamble to the activation of bone repair materials. Mater. Sci. Eng. C 28, 1544–1550.

Dubnika, A., Loca, D., Reinis, A., Kodols, M., Berzina-Cimdina, L., 2013. Impact of sintering temperature on the phase composition and antibacterial properties of silver-doped hydroxyapatite. Pure Appl. Chem. 85, 453–462.

Elliott, J.C., 1980. Space group and lattice constants of $Ca_{10}(PO_4)_6CO_3$. J. Appl. Crystallogr. 13, 618–621.

Elliott, J.C., 1994. Structure and Chemistry of the Apatites and Other Calcium Orthophosphates. Elsevier, Amsterdam.

Elliott, J.C., Holcomb, D.W., Young, R.A., 1985. Infrared determination of the degree of substitution of hydroxyl by carbonate ions in human dental enamel. Calcif. Tissue Int. 37, 372–375.

Eraković, S., Janković, A., Veljović, D., Palcevskis, E., Mitrić, M., Stevanović, T., Janaćković, D., Misković-Stanković, V.J., 2013. Corrosion stability and bioactivity in simulated body fluid of silver/hydroxyapatite and silver/hydroxyapatite/lignin coatings on titanium obtained by electrophoretic deposition. J. Phys. Chem. B 117, 1633–1643.

Eraković, S., Janković, A., Ristoscu, C., Duta, L., Serban, N., Visan, A., Mihailescu, I.N., Stan, G.E., Socol, M., Iordache, O., Dumitrescu, I., Luculescu, C.R., Janaćković, D.J., Miškovic-Stanković, V., 2014. Antifungal activity of Ag:hydroxyapatite thin films synthesized by pulsed laser deposition on Ti and Ti modified by TiO_2 nanotubes substrates. Appl. Surf. Sci. 293, 37–45.

Ewald, A., Hösel, D., Patel, S., Grover, L.M., Barralet, J.E., Gbureck, U., 2011. Silver-doped calcium phosphate cements with antimicrobial activity. Acta Biomater. 7, 4064–4070.

Fielding, G.A., Roy, M., Bandyopadhyay, A., Bose, S., 2012. Antibacterial and biological characteristics of silver containing and strontium doped plasma sprayed hydroxyapatite coatings. Acta Biomater. 8, 3144–3152.

Fleet, M.E., 2009. Infrared spectra of carbonate apatites: ν2-Region bands. Biomaterials 30, 1473–1481.

Fleet, M.E., Liu, X., 2003. Carbonate apatite type A synthesized at high pressure: new space group (*P*3) and orientation of channel carbonate ion. J. Solid State Chem. 174, 412–417.

Fleet, M.E., Liu, X., 2005. Local structure of channel ions in carbonate apatite. Biomaterials 26, 7548–7554.

Fowler, B.O., 1974a. Infrared studies of apatites. I. Vibrational assignments for calcium, strontium, and barium hydroxyapatites utilizing isotopic substitution. Inorg. Chem. 13, 194–207.

Fowler, B.O., 1974b. Infrared studies of apatites. II. Preparation of normal and isotopically substituted calcium, strontium, and barium hydroxyapatites and spectra-structure-composition correlations. Inorg. Chem. 13, 207–214.

Fuierer, T.A., LoRe, M., Puckett, S.A., Nancollas, G.H., 1994. A mineralization adsorption and mobility study of hydroxyapatite surfaces in the presence of zinc and magnesium ions. Langmuir 10, 4721–4725.

Gallacher, S.J., Dixon, T., 2010. Impact of treatments for postmenopausal osteoporosis (bisphosphonates, parathyroid hormone, strontium ranelate and denosumab) on bone quality: a systematic review. Calcif. Tissue Int. 87, 469–484.

Giacovazzo, C., 2002. Fundamentals of Crystallography. Oxford University Press, Oxford.

Gibson, I.R., Bonfield, W., 2002. Preparation and characterization of magnesium/carbonate co-substituted hydroxyapatites. J. Mater. Sci. Mater. Med. 13, 685–693.

Gibson, I.R., Best, S.M., Bonfield, W., 1999. Chemical characterization of silicon-substituted hydroxyapatite. J. Biomed. Mater. Res. 44, 422–428.

Glimcher, M.J., 2006. Bone: nature of the calcium phosphate crystals and cellular, structural, and physical chemical mechanisms in their formation. In: Sahai, N., Schoonen, M.A.A. (Eds.), Medical Mineralogy and Geochemistry. Series: Reviews in Mineralogy & Geochemistry, vol. 64. Mineralogical Society of America, Chantilly, VA, pp. 223–282.

Gross, K.A., Rodriguez-Lorenzo, L.M., 2004. Sintered hydroxyfluorapatites. Part I: sintering ability of precipitated solid solution powders. Biomaterials 25, 1375–1384.

Gross, K.A., Jackson, R., Cashion, J.D., Rodriguez-Lorenzo, L.M., 2002. Iron substituted apatites: a resorbable biomaterial with potential magnetic properties. Eur. Cell. Mater. 3, 114–117.

Guanabens, N., Farrerons, J., Perez-Edo, L., Monegal, A., Renau, A., Carbonell, J., Roca, M., Torra, M., Pavesi, M., 2000. Cyclical etidronate versus sodium fluoride in established postmenopausal osteoporosis: a randomized 3 year trial. Bone 27, 123–128.

Gür, A., Çolpan, L., Nas, K., Çevik, R., Saraç, J., Erdoğan, F., Düz, M.Z., 2002. The role of trace minerals in the pathogenesis of postmenopausal osteoporosis and a new effect of calcitonin. J. Bone Miner. Metab. 20, 39–43.

Han, G.S., Lee, S., Kim, D.W., Kim, D.H., Noh, J.H., Park, J.H., Roy, S., Ahn, T.K., Jung, H.S., 2013. A simple method to control morphology of hydroxyapatite nano and microcrystals by altering phase transition route. Cryst. Growth Des. 13, 3414–3418.

Hench, L.L., 1998. Sol–Gel Silica: Properties, Processing and Technology Transfer. Noyes Publications, Westwood, NJ, pp. 117–132.

Hesaraki, S., Nazarian, H., Pourbaghi-Masouleh, M., Borhan, S., 2014. Comparative study of mesenchymal stem cells osteogenic differentiation on low-temperature biomineralized nanocrystalline carbonated hydroxyapatite and sintered hydroxyapatite. J. Biomed. Mater. Res. B 102B, 108–118.

Honda, M., Kawanobe, Y., Ishii, K., Konishi, T., Mizumoto, M., Kanzawa, N., Matsumoto, M., Aizawa, M., 2013. In vitro and in vivo antimicrobial properties of silver-containing hydroxyapatite prepared via ultrasonic spray pyrolysis route. Mater. Sci. Eng. C 33, 5008–5018.

Hou, C.H., Hou, S.M., Hsueh, Y.S., Lin, J., Wu, H.C., Lin, F.H., 2009. The in vivo performance of biomagnetic hydroxyapatite nanoparticles in cancer hyperthermia therapy. Biomaterials 30, 3956–3960.

Ilich, J.Z., Kerstetter, J.E., 2000. Nutrition in bone health revisited: a story beyond calcium. J. Am. Col. Nutr. 19, 715–737.

Iqbal, N., Kadir, M.R.A., Malek, N.A.N.N., Mahmood, N.H., Murali, M.R., Kamarul, T., 2012. Rapid microwave assisted synthesis and characterization of nanosized silver-doped hydroxyapatite with antibacterial properties. Mater. Lett. 89, 118–122.

Jha, L.J., Best, S.M., Knowles, J.C., Rehman, I., Santosk, J.D., Bonfield, W., 1997. Preparation and characterization of fluoride-substituted apatites. J. Mater. Sci. Mater. Med. 8, 185–191.

Jiang, M., Terra, J., Rossi, A.M., Morales, M.A., Saitovitch, E.M.B., Ellis, D.E., 2002. Fe^{2+}/Fe^{3+} substitution in hydroxyapatite: theory and experiment. Phys. Rev. B: Condens. Mater. Phys. 66, 224107–224115.

Jiang, D., Li, D., Xie, J., Zhu, J., Chen, M., Lü, X., Dang, S., 2010. Shape-controlled synthesis of F-substituted hydroxyapatite microcrystals in the presence of Na_2EDTA and citric acid. J. Colloid Interface Sci. 350, 30–38.

Jugdaohsingh, R., Tucker, K.L., Qiao, N., Cupples, L.A., Kiel, D.P., Powell, J.J., 2004. Dietary silicon intake is positively associated with bone mineral density in men and premenopausal women of the Framingham Offspring cohort. J. Bone Miner. Res. 19, 297–307.

Julien, M., Khairoun, I., LeGeros, R.Z., Delplace, S., Pilet, P., Weiss, P., Daculsi, G., Bouler, J.M., Guicheux, J., 2007. Physicochemical-mechanical and in vitro biological properties of calcium phosphate cements with doped amorphous calcium phosphates. Biomaterials 28, 956–965.

Kalita, S.J., Bhatt, H.A., 2007. Nanocrystalline hydroxyapatite doped with magnesium and zinc: Synthesis and characterization. Mater. Sci. Eng. C 27, 837–848.

Kay, M.I., Young, R.A., Posner, A.S., 1964. Crystal structure of hydroxyapatite. Nature 204, 1050–1052.

Khan, A.F., Saleem, M., Afzal, A., Ali, A., Khan, A., Khan, A.R., 2014. Bioactive behavior of silicon substituted calcium phosphate based bioceramics for bone regeneration. Mater. Sci. Eng. C35, 245–252.

Kikuchi, M., Yamazaki, A., Otsuka, R., Akao, M., Aoki, H., 1994. Crystal structure of Sr-substituted hydroxyapatite synthesized by hydrothermal method. J. Solid State Chem. 113, 373–378.

Kim, S.R., Lee, J.H., Kim, Y.T., Riu, D.H., Jung, S.J., Lee, Y.J., Chung, S.C., Kim, Y.H., 2003. Synthesis of Si, Mg substituted hydroxyapatites and their sintering behaviors. Biomaterials 24, 1389–1398.

Kim, H.W., Li, L.H., Koh, Y.H., Knowles, J.C., Kim, H.E., 2004. Sol–gel preparation and properties of fluoride-substituted hydroxyapatite powders. J. Am. Ceram. Soc. 87, 1939–1944.

Kirkham, J., Brookes, S.J., Shore, R.C., Bonass, W.A., Smith, D.A., Wallwork, M.L., Robinson, C., 1998. Atomic force microscopy studies of crystal surface topology during enamel development. Connect. Tissue Res. 38, 91–100.

Kolmas, J., Jaklewicz, A., Zima, A., Bućko, M., Paszkiewicz, Z., Lis, J., Slosarczyk, A., Kolodziejski, W., 2011. Incorporation of carbonate and magnesium ions into synthetic hydroxyapatite: the effect on physicochemical properties. J. Mol. Struct. 987, 40–50.

Kyriacou, A., Leventouri, Th., Chakoumakos, B.C., Garlea, V.O., dela Cruz, C.B., Rondinone, A.J., Sorge, K.D., 2013. Combined X-ray and neutron diffraction Rietveld refinement in iron-substituted nano-hydroxyapatite. J. Mater. Sci. 48, 3535–3545.

Landi, E., Tampieri, A., Mattioli-Belmonte, M., Celotti, G., Sandri, M., Gigante, A., Fava, P., Biagini, G., 2006. Biomimetic Mg- and Mg, CO_3-substituted hydroxyapatites: synthesis, characterization and in vitro behaviour. J. Eur. Ceram. Soc. 26, 2593–2601.

Landi, E., Logroscino, G., Proietti, L., Tampieri, A., Sandri, M., Sprio, S., 2008. Biomimetic Mg-substituted hydroxyapatite: from synthesis to in vivo behaviour. J. Mater. Sci. Mater. Med. 19, 239–247.

Laurencin, D., Smith, M.E., 2013. Development of ^{43}Ca solid state NMR spectroscopy as a probe of local structure in inorganic and molecular materials. Prog. Nucl. Magn. Reson. Spectrosc. 68, 1–40.

Laurencin, D., Almora-Barrios, N., de Leeuw, N.H., Gervais, C., Bonhomme, C., Mauri, F., Chrzanowski, W., Knowles, J.C., Newport, R.J., Wong, A., Gan, Z., Smith, M.E., 2011. Magnesium incorporation into hydroxyapatite. Biomaterials 32, 1826–1837.

Laurencin, D., Smith, M.E., 2013. Development of ^{43}Ca solid state NMR spectroscopy as a probe of local structure in inorganic and molecular materials. Prog. Nucl. Magn. Reson. Spectrosc. 68, 1–40.

LeGeros, R.Z., 1965. Effect of carbonate on the lattice parameters of apatite. Nature 206, 403–404.

LeGeros, R.Z., Kijkowska, R., Bautista, C., LeGeros, J.P., 1995. Synergistic effects of magnesium and carbonate on properties of biological and synthetic apatites. Connect. Tissue Res. 33, 203–209.

Li, X., Sogo, Y., Ito, A., Mutsuzaki, H., Ochiai, N., Kobayashi, T., Nakamura, S., Yamashita, K., LeGeros, R.Z., 2009. The optimum zinc content in set calcium phosphate cement for promoting bone formation in vivo. Mater. Sci. Eng. C 29, 969–975.

Li, Y., Li, Q., Zhu, S., Luo, E., Li, J., Feng, G., Liao, Y., Hu, J., 2010. The effect of strontium-substituted hydroxyapatite coating on implant fixation in ovariectomized rats. Biomaterials 31, 9006–9014.

Lilley, K.J., Gbureck, U., Knowles, J.C., Farrar, D.F., Barralet, J.E., 2005. Cement from magnesium substituted hydroxyapatite. J. Mater. Sci. Mater. Med. 16, 455–460.

Lindahl, C., Wei, X., Jukka, L., Per, B., Haakan, E., 2012. Strontium and silicon co-doped apatite coating: preparation and function as vehicles for ion delivery. J. Biomater. Nanobiotechnol. 3, 335–341.

Low, H.R., Phonthammachai, N., Maignan, A., Stewart, G.A., Bastow, T.J., Ma, L.L., White, T.J., 2008. The crystal chemistry of ferric oxyhydroxyapatite. Inorg. Chem. 47, 11774–11782.

Lu, J., Wei, J., Yan, Y., Li, H., Jia, J., Wei, S., Guo, H., Xiao, T., Liu, C., 2011. Preparation and preliminary cytocompatibility of magnesium doped apatite cement with degradability for bone regeneration. J. Mater. Sci. Mater. Med. 22, 607–615.

Lynch, R.J.M., 2011. Zinc in the mouth, its interactions with dental enamel and possible effects on caries; a review of the literature. Int. Dent. J. 61 (Suppl. 3), 46–54.

Macdonald, H.M., Hardcastle, A.C., Jugdaohsingh, R., Fraser, W.D., Reid, D.M., Powell, J.J., 2012. Dietary silicon interacts with oestrogen to influence bone health: evidence from the Aberdeen Prospective Osteoporosis Screening Study. Bone 50, 681–687.

Marambio-Jones, C., Hoek, E.M.V., 2010. A review of the antibacterial effects of silver nanomaterials and potential implications for human health and the environment. J. Nanopart. Res. 12, 1531–1551.

Marchat, D., Zymelka, M., Coelho, C., Gremillard, L., Joly-Pottuz, L., Babonneau, F., Esnouf, C., Chevalier, J., Bernache-Assolant, D., 2013. Accurate characterization of pure silicon-substituted hydroxyapatite powders synthesized by a new precipitation route. Acta Biomater. 9, 6992–7004.

Marie, P.J., 2005. Strontium ranelate: a novel mode of action optimizing bone formation and resorption. Osteoporos. Int. 16, S7–S10.

Marie, P.J., Ammann, P., Boivin, G., Rey, C., 2001. Mechanisms of action and therapeutic potential of strontium in bone. Calcif. Tissue Int. 69, 121–129.

Matsunaga, K., 2008. First-principles study of substitutional magnesium and zinc in hydroxyapatite and octacalcium phosphate. J. Chem. Phys. 128, 245101–245110.

Mavropoulos, E., Kede, M.L.F.M., da Rocha, N.C.C., Machado Costa, A., Tosi, A., Prado da Silva, M.H., Rossi, A.M., 2011. BSA attachment on apatite surface modified with Zn^{2+} and Sr^{2+}. Bioceram. Dev. Appl. 1, D101115.

Mayer, I., Schlam, R., Featherstone, J.D.B., 1997. Magnesium-containing carbonate apatites. J. Inorg. Biochem. 66, 1–6.

McConnell, D., 1973. Apatite. Its Crystal Chemistry, Mineralogy, Utilization, and Geologic and Biologic Occurrences. Springer-Verlag, Wien.

Melville, A.J., Harrison, J., Gross, K.A., Forsythe, J.S., Trounson, A.O., Mollard, R., 2006. Mouse embryonic stem cell colonization of carbonated apatite surfaces. Biomaterials 27, 615–622.

Montalbert-Smith Echeverría, R., Montero, M.L., 2013. Electrolytic one-pot synthesis of group II nanohydroxyapatites. J. Mater. Sci. 48, 5253–5260.

Moonga, B.S., Dempster, D.W., 1995. Zinc is a potent inhibitor of osteoclastic bone resorption in vitro. J. Bone Miner. Res. 10, 453–457.

Morrissey, R., Rodriguez-Lorenzo, L.M., Gross, K.A., 2005. Influence of ferrous iron incorporation on the structure of hydroxyapatite. J. Mater. Sci. Mater. Med. 16, 387–392.

Mossaad, C., Starr, M., Patil, S., Riman, R.E., 2010. Thermodynamic modeling of hydroxyapatite crystallization with biomimetic precursor design considerations. Chem. Mater. 22, 36–46.

Mroz, W., Bombalska, A., Burdynska, S., Jedynski, M., Prokopiuk, A., Budner, B., Slosarczyk, A., Zima, A., Menaszek, E., Scisłowska-Czarnecka, A., Niedzielski, K., 2010. Structural studies of magnesium doped hydroxyapatite coatings after osteoblast culture. J. Mol. Struct. 977, 145–152.

Ni, G.X., Shu, B., Huang, G., Lu, W.W., Pan, H.B., 2012. The effect of strontium incorporation into hydroxyapatites on their physical and biological properties. J. Biomed. Mater. Res. B 100B, 562–568.

O'Donnell, M.D., Fredholm, Y., de Rouffignac, A., Hill, R.G., 2008. Structural analysis of a series of strontium-substituted apatites. Acta Biomater. 4, 1455–1464.

Okada, S., Ito, H., Nagai, A., Komotori, J., Imai, H., 2010. Adhesion of osteoblast-like cells on nanostructured hydroxyapatite. Acta Biomater. 6, 591–597.

Oliveira, A.L., Reis, R.L., Li, P., 2007. Strontium-substituted apatite coating grown on Ti_6Al_4V substrate through biomimetic synthesis. J. Biomed. Mater. Res. B 83B, 258–265.

Palmer, L.C., Newcomb, C.J., Kaltz, S.R., Spoerke, E.D., Stupp, S.I., 2008. Biomimetic systems for hydroxyapatite mineralization inspired by bone and enamel. Chem. Rev. 108, 4754–4783.

Pan, H.B., Li, Z.Y., Lam, W.M., Wong, J.C., Darvell, B.W., Luk, K.D.K., Lu, W.W., 2009a. Solubility of strontium-substituted apatite by solid titration. Acta Biomater. 5, 1678–1685.

Pan, H.B., Li, Z.Y., Wang, T., Lam, W.M., Wong, C.T., Darvell, B.W., Luk, K.D.K., Hu, Y., Lu, W.W., 2009b. Nucleation of strontium-substituted apatite. Cryst. Growth Des. 9, 3342–3345.

Panseri, S., Cunha, C., D'Alessandro, T., Sandri, M., Giavaresi, G., Marcacci, M., Hung, C.T., Tampieri, A., 2012. Intrinsically superparamagnetic Fe-hydroxyapatite nanoparticles positively influence osteoblast-like cell behaviour. J. Nanobiotechnol. 10, 32–42.

Paschalis, E.P., DiCarlo, E., Betts, F., Sherman, P., Mendelsohn, R., Boskey, A.L., 1996. FTIR microspectroscopic analysis of human osteonal bone. Calcif. Tissue Int. 59, 480–487.

Patel, N., Best, S.M., Bonfield, W., Gibson, I.R., Hing, K.A., Damien, E., Revell, P.A., 2002. A comparative study on the in vivo behavior of hydroxyapatite and silicon substituted hydroxyapatite granules. J. Mater. Sci. Mater. Med. 13, 1199–1206.

Percival, M., 1999. Bone health and osteoporosis. Appl. Nutr. Sci. Rep. 5, 1–5.

Peroos, S., Du, Z., de Leeuw, N.H., 2006. A computer modelling study of the uptake, structure and distribution of carbonate defects in hydroxy-apatite. Biomaterials 27, 2150–2161.

Pietak, A.M., Reid, J.W., Stott, M.J., Sayer, M., 2007. Silicon substitution in the calcium phosphate bioceramics. Biomaterials 28, 4023–4032.

Pieters, I.Y., Van den Vreken, N.M.F., Declercq, H.A., Cornelissen, M.J., Verbeeck, R.M.H., 2010. Carbonated apatites obtained by the hydrolysis of monetite: influence of carbonate content on adhesion and proliferation of MC3T3-E1 osteoblastic cells. Acta Biomater. 6, 1561–1568.

Pizzala, H., Caldarelli, S., Eon, J.G., Rossi, A.M., Laurencin, D., Smith, M.E., 2009. A solid-state NMR study of lead and vanadium substitution into hydroxyapatite. J. Am. Chem. Soc. 131, 5145–5152.

Pon-On, W., Charoenphandhu, N., Teerapornpuntakit, J., Thongbunchoo, J., Krishnamra, N., Tang, I., 2013. Physicochemical and biochemical properties of iron-loaded silicon substituted hydroxyapatite. Mater. Chem. Phys. 141, 850–860.

Prasad, A.S., 1995. Zinc: an overview. Nutrition 11, 93–99.

Qu, H., Wei, M., 2006. The effect of fluoride contents in fluoridated hydroxyapatite on osteoblast behavior. Acta Biomater. 2, 113–119.

Querido, W., Farina, M., 2013. Strontium ranelate increases the formation of bone-like mineralized nodules in osteoblast cell cultures and leads to Sr incorporation into the intact nodules. Cell Tissue Res. 354, 573–580.

Ren, F., Leng, Y., Xin, R., Ge, X., 2010. Synthesis, characterization and ab initio simulation of magnesium-substituted hydroxyapatite. Acta Biomater. 6, 2787–2796.

Ren, F., Lu, X., Leng, Y., 2013. Ab initio simulation on the crystal structure and elastic properties of carbonated apatite. J. Mech. Behav. Biomed. Mater. 26, 59–67.

Ren, F., Ding, Y., Leng, Y., 2014. Infrared spectroscopic characterization of carbonated apatite: a combined experimental and computational study. J. Biomed. Mater. Res. A 102A, 496–505.

Rey, C., Renugopalakrishman, V., Collins, B., Glimcher, M., 1991. Fourier transform infrared spectroscopic study of the carbonate ions in bone mineral during aging. Calcif. Tissue Int. 49, 251–258.

Rodriguez-Lorenzo, L.M., Hart, J.N., Gross, K.A., 2003. Structural and chemical analysis of well-crystallized hydroxyfluorapatites. J. Phys. Chem. B 107, 8316–8320.

Roy, M., Fielding, G.A., Beyenal, H., Bandyopadhyay, A., Bose, S., 2012. Mechanical, in vitro antimicrobial, and biological properties of plasma-sprayed silver-doped hydroxyapatite coating. ACS Appl. Mater. Interfaces 4, 1341–1349.

Rude, R.K., Gruber, H.E., 2004. Magnesium deficiency and osteoporosis: animal and human observations. J. Nutr. Biochem. 15, 710–716.

Sadat-Shojai, M., Khorasani, M.T., Dinpanah-Khoshdargi, E., Jamshidi, A., 2013. Synthesis methods for nanosized hydroxyapatite with diverse structures. Acta Biomater. 9, 7591–7621.

Sader, M.S., Lewis, K., Soares, G.A., LeGeros, R.Z., 2013. Simultaneous incorporation of magnesium and carbonate in apatite: effect on physico-chemical properties. Mater. Res. 16, 779–784.

Saidin, S., Chevallier, P., Kadir, M.R.A., Hermawan, H., Mantovani, D., 2013. Polydopamine as an intermediate layer for silver and hydroxyapatite immobilisation on metallic biomaterials surface. Mater. Sci. Eng. C 33, 4715–4724.

Shepherd, J.H., Shepherd, D.V., Best, S.M., 2012. Substituted hydroxyapatites for bone repair. J. Mater. Sci. Mater. Med. 23, 2335–2347.

Singh, B., Dubey, A.K., Kumar, S., Saha, N., Basu, B., Gupta, R., 2011. In vitro biocompatibility and antimicrobial activity of wet chemically prepared $Ca_{10-x}Ag_x(PO_4)_6(OH)_2$ $(0.0 \leq x \leq 0.5)$ hydroxyapatites. Mater. Sci. Eng. C 31, 1320–1329.

Spence, G., Patel, N., Brooks, R., Rushton, N., 2009. Carbonate substituted hydroxyapatite: resorption by osteoclasts modifies the osteoblastic response. J. Biomed. Mater. Res. A 90A, 217–224.

Spence, G., Patel, N., Brooks, R., Bonfield, W., Rushton, N., 2010. Osteoclastogenesis on hydroxyapatite ceramics: the effect of carbonate substitution. J. Biomed. Mater. Res. A 92A, 1292–1300.

Sprio, S., Tampieri, A., Landi, E., Sandri, M., Martorana, S., Celotti, G., Logroscino, G., 2008. Physicochemical properties and solubility behaviour of multi-substituted hydroxyapatite powders containing silicon. Mater. Sci. Eng. C 28, 179–187.

Stanić, V., Janaćković, D., Dimitrijević, S., Tanasković, S.B., Mitrić, M., Pavlović, M.S., Krstić, A., Jovanović, D., Raičević, S., 2011. Synthesis of antimicrobial monophase silver-doped hydroxyapatite nanopowders for bone tissue engineering. Appl. Surf. Sci. 257, 4510–4518.

Stanić, V., Dimitrijević, S., Antonović, D.G., Jokić, B.M., Zec, S.P., Tanasković, S.T., Raičević, S., 2014. Synthesis of fluorine substituted hydroxyapatite nanopowders and application of the central composite design for determination of its antimicrobial effects. Appl. Surf. Sci. 290, 346–352.

Suchanek, W.L., Byrappa, K., Shuk, P., Riman, R.E., Janas, V.F., Ten-Huisen, K.S., 2004. Preparation of magnesium-substituted hydroxyapatite powders by the mechanochemical–hydrothermal method. Biomaterials 25, 4647–4657.

Sutha, S., Karunakaran, G., Rajendran, V., 2013. Enhancement of antimicrobial and long-term biostability of the zinc-incorporated hydroxyapatite coated 316L stainless steel implant for biomedical application. Ceram. Int. 39, 5205–5212.

Tampieri, A., D'Alessandro, T., Sandri, M., Sprio, S., Landi, E., Bertinetti, L., Panseri, S., Pepponi, G., Goettlicher, J., Bañobre-López, M., Rivas, J., 2012. Intrinsic magnetism and hyperthermia in bioactive Fe-doped hydroxyapatite. Acta Biomater. 8, 843–851.

Terra, J., Dourado, E.R., Eon, J.G., Ellis, D.E., Gonzalez, G., Rossi, A.M., 2009. The structure of strontium-doped hydroxyapatite: an experimental and theoretical study. Phys. Chem. Chem. Phys. 11, 568–577.

Thian, E.S., Konishi, T., Kawanobe, Y., Lim, P.N., Choong, C., Ho, B., Aizawa, M., 2013. Zinc-substituted hydroxyapatite: a biomaterial with enhanced bioactivity and antibacterial properties. J. Mater. Sci. Mater. Med. 24, 437–445.

Tonegawa, T., Ikoma, T., Yoshioka, T., Hanagata, N., Tanaka, J., 2010. Crystal structure refinement of A-type carbonate apatite by X-ray powder diffraction. J. Mater. Sci. 45, 2419–2426.

Tran, N., Webster, T.J., 2011. Increased osteoblast functions in the presence of hydroxyapatite-coated iron oxide nanoparticles. Acta Biomater. 7, 1298–1306.

Tredwin, C.J., Georgiou, G., Kim, H.W., Knowles, J.C., 2013. Hydroxyapatite, fluoro-hydroxyapatite and fluorapatite produced via the sol–gel method: bonding to titanium and scanning electron microscopy. Dent. Mater. 29, 521–529.

Tredwin, C.J., Young, A.M., Abou Neel, E.A., Georgiou, G., Knowles, J.C., 2014. Hydroxyapatite, fluor-hydroxyapatite and fluorapatite produced via the sol–gel method: dissolution behaviour and biological properties after crystallisation. J. Mater. Sci. Mater. Med. 25, 47–53.

Ulian, G., Valdrè, G., Corno, M., Ugliengo, P., 2014. DFT investigation of structural and vibrational properties of type B and mixed A-B carbonated hydroxylapatite. Am. Mineral. 99, 117–127.

Waked, W., Grauer, J., 2008. Silicates and bone fusion. Orthopedics 31, 591–597.

Wang, L., Nancollas, G.H., 2008. Calcium orthophosphates: crystallization and dissolution. Chem. Rev. 108, 4628–4669.

Wei, M., Evans, J.H., Bostrom, T., Grøndahl, L., 2003. Synthesis and characterization of hydroxyapatite, fluoride-substituted hydroxyapatite and fluorapatite. J. Mater. Sci. Mater. Med. 14, 311–320.

Wilson, R.M., Dowker, S.E.P., Elliott, J.C., 2006. Rietveld refinements and spectroscopic structural studies of a Na-free carbonate apatite made by hydrolysis of monetite. Biomaterials 27, 4682–4692.

Wu, H.C., Wang, T.W., Sun, J.S., Wang, W.H., Lin, F.H., 2007. A novel biomagnetic nanoparticle based on hydroxyapatite. Nanotechnology 18, 165601–165610.

Xia, W., Lausmaa, J., Thomsen, P., Engqvist, H., 2012. Highly packed and aligned fluoride substituted hydroxyapatite via a surfactant-free process. J. Biomed. Mater. Res. B 100B, 75–81.

Xu, C., Reed, R., Gorski, J.P., Wang, Y., Walker, M.P., 2012. The distribution of carbonate in enamel and its correlation with structure and mechanical properties. J. Mater. Sci. 47, 8035–8043.

Xue, W., Hosick, H.L., Bandyopadhyay, A., Bose, S., Ding, C., Luk, K.D.K., Cheung, K.M.C., Lu, W.W., 2007. Preparation and cell–materials interactions of plasma sprayed strontium-containing hydroxyapatite coating. Surf. Coat. Technol. 201, 4685–4693.

Yamada, M.-O., Tohno, Y., Tohno, S., Utsumi, M., Moriwake, Y., Yamada, G., 2003. Silicon compatible with the height of human vertebral column. Biol. Trace Elem. Res. 95, 113–121.

Yanovska, A.A., Stanislavov, A.S., Sukhodub, L.B., Kuznetsov, V.N., Illiashenko, V.Yu., Danilchenko, S.N., Sukhodub, L.F., 2014. Silver-doped hydroxyapatite coatings formed on Ti–6Al–4V substrates and their characterization. Mater. Sci. Eng. C 36, 215–220.

Yao, F., LeGeros, R.Z., 2010. Carbonate and fluoride incorporation in synthetic apatites: comparative effect on physico-chemical properties and in vitro bioactivity in fetal bovine serum. Mater. Sci. Eng. C 30, 423–430.

Yasukawa, A., Ouchi, S., Kandori, K., Ishikawa, T., 1996. Preparation and characterization of magnesium–calcium hydroxyapatites. J. Mater. Chem. 6, 1401–1405.

Yazaki, Y., Oyane, A., Sogo, Y., Ito, A., Yamazaki, A., Tsurushima, H., 2011. Control of gene transfer on a DNA-fibronectin-apatite composite layer by the incorporation of carbonate and fluoride ions. Biomaterials 32, 4896–4902.

Yesinowski, J.P., Eckert, H., 1987. Hydrogen environments in calcium phosphates: proton MAS NMR at high spinning speeds. J. Am. Chem. Soc. 109, 6274–6282.

Young, R.A., 1975. Some aspects of crystal structural modeling of biological apatites. Colloques Internationaux CNRS 230, 21–39.

Yu, T., Ye, J., Zhang, M., 2013. Effect of magnesium doping on hydration morphology and mechanical property of calcium phosphate cement under non-calcined synthesis condition. J. Am. Ceram. Soc. 96, 1944–1950.

Yuan, X., Zhu, B., Tong, G., Su, Y., Zhu, X., 2013. Wet-chemical synthesis of Mg-doped hydroxyapatite nanoparticles by step reaction and ion exchange processes. J. Mater. Chem. B 1, 6551–6559.

Zhao, S.F., Jiang, Q.H., Peel, S., Wang, X.X., He, F.M., 2013. Effects of magnesium-substituted nanohydroxyapatite coating on implant osseointegration. Clin. Oral Implants Res. 24 (Suppl. A100), 34–41.

Zhu, K., Yanagisawa, K., Shimanouchi, R., Onda, A., Kajiyoshi, K., 2006. Preferential occupancy of metal ions in the hydroxyapatite solid solutions synthesized by hydrothermal method. J. Eur. Ceram. Soc. 26, 509–513.

Accelerating mineralization of biomimetic surfaces

8

F.J. Gil Mur
Technical University of Catalonia, Barcelona, Spain

8.1 Introduction

Bone is the toughest tissue in the human body but can be damaged in both nontraumatic and traumatic ways. As is well known, autografts, allografts, and xenografts are the conventional choices for treating bone defects (Farbod et al., 2014). However, complications such as restricted availability, new surgeries and treatments, donor-site morbidity, and thus decreased quality of life for patients still remain problematic (Farbod et al., 2014; Porter et al., 2009; Spicer et al., 2012). These problems can be alleviated by the use of a number of synthetic biomaterials (alloplasts): metals, ceramics, polymers, or composite materials.

These materials have many clinical applications in response to many different diseases in the human body. However, synthetic materials do not reproduce the function of natural tissues, especially with regard to lack of adequate tissue formation around the biomaterials after implantation. Moreover, the use of metals can lead to corrosion, they do not have the potential for drug delivery, and in spite of their excellent mechanical properties (toughness, strength, etc.) they have an elastic modulus higher than the bone, which leads to undesirable load transfer from the biomaterial to the natural bone, and can cause bone resorption. This phenomenon is called stress shielding. Ceramics are very brittle and present serious problems with their high deformation rates. Polymer biomaterials are generally not osteoconductive and can cause stress relaxation behavior and creep. These properties can be improved with the use of composite materials, as these properties can be controlled via the properties and contents of the matrix and dispersed phase, as well as the shape, orientation, and interfacial interactions between the matrix and the reinforcement.

The development of a bioactive bond *in vivo* is a complex process that involves several interactions at the interface between the implant and the surrounding bone. These interactions take place at the atomic, molecular, and cellular scales. A simple overview of the most relevant stages occurring during bioactive bond formation is summarized in the diagram presented in Figure 8.1.

After implantation, the material immediately comes into contact with blood. Depending on its surface energy the implant can be wetted differently, i.e. a hydrophilic surface promotes interaction with the components of blood better than a hydrophobic one. Subsequently, blood proteins start a dynamic adsorption process that depends on surface chemistry, electric charge, and surface topography at the macro-, micro-, and nanoscales. At this stage, the main difference between bioactive and biostable

Biomineralization and Biomaterials. http://dx.doi.org/10.1016/B978-1-78242-338-6.00009-0

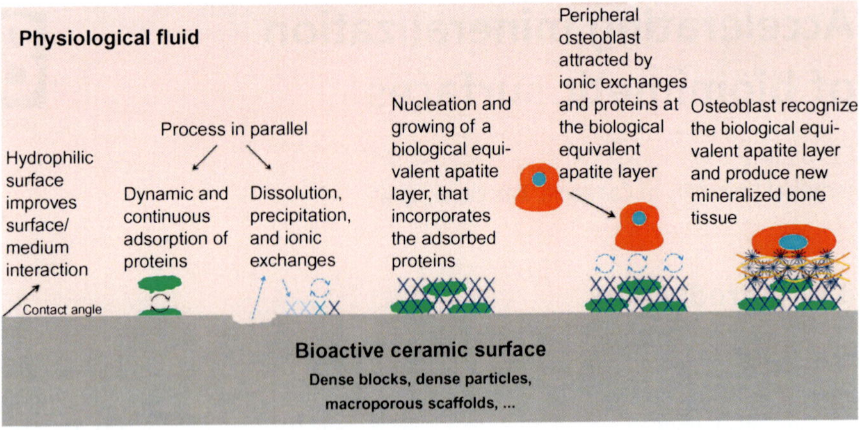

Figure 8.1 Simplified overview of the most relevant stages occurring *in vivo* during bone-bioactive material bonding. All these stages take place at the interface between the bioactive ceramic surface and the physiological fluid. The sequence of events is presented from left to right; however, it does not include a time scale.

materials is that, in parallel to the dynamic adsorption of proteins, bioactive surfaces interact with the ionic environment in several ways, whereas biostable surfaces do not. Three different phenomena can take place between the surface and the surrounding medium: dissolution, precipitation, and ion exchange. As a result, an inorganic layer is formed on the bioactive surface, which corresponds to an apatitic calcium phosphate, with very low crystallinity, which mimics the biological carbonated-calcium-deficient hydroxyapatite (HAp). Although physiological fluids like blood are ionically supersaturated with respect to HAp, the formation of a biologically equivalent apatite layer is not a spontaneous heterogenic nucleation process occurring on every surface. On the contrary, *in vitro* physicochemical analyses have shown that the more reactive a surface is, the faster precipitation of this layer takes place.

The objective of this chapter is to give an overview of the strategies used to accelerate the mineralization of some surfaces of metals, ceramics, polymers, and composites.

8.2 Metals

Stainless steel, cobalt–chromium alloys, titanium and its alloys including NiTi shape memory alloys have been used in orthopedics and dentistry for bone replacement for more than 50 years now. Excellent mechanical properties with high fracture toughness combined with a good corrosion resistance and biocompatibility make these materials suitable for bone substitution applications. Titanium and titanium alloys ($E = 110\,\text{GPa}$) have lower elastic moduli than stainless steel ($E = 210\,\text{GPa}$) and cobalt–chromium alloys ($E = 240\,\text{GPa}$) with a high specific strength. These properties prevent significant stress shielding to bone and thus facilitate healthier and faster bone formation (Brunski, 1996; Williams, 2001; Ratner, 2001; Aparicio et al., 2002, 2003; Placko et al., 2000; Kokubo et al., 1990). For these reasons, Ti and Ti alloys are now the usual choice for producing dental implants, joint replacements, and spine implants (Placko et al., 2000; Aparicio et al., 2003; Kokubo et al., 1990; Gil et al., 2002; Takadama et al., 2001).

A requisite for the clinical success of titanium, especially in dental implants, is the formation of a strong and long-lasting connection between the implant and bone (Buser, 2001). The improvement of the implant–bone interface in compromised clinical scenarios (Ronold et al., 2003; van Steenberghe et al., 2002), including procedures following immediate loading of the dental implant, is an ongoing problem (Schliephake and Scharnweber, 2008; Shrikhanzadeh, 1991). The modification of the surface of dental implants can overcome this limitation (Le Guehennec et al., 2007). Implant surfaces with improved topographical, physical, and chemical properties will provide a suitable environment where the natural biological potential for bone functional regeneration can be stimulated and maximized. These therapeutic strategies should ultimately enhance the osseointegration process of dental implants for their immediate loading and long-term success (Aparicio et al., 2011).

Surface topography is known to be a very important factor in bone fixation (Buser, 2001; Ronold et al., 2003; Giavaresi et al., 2003; Thomsen et al., 1997; Albrektsson et al., 1981; Bagno and Di Bello, 2004) and can be modified by blasting with abrasives of different natures and sizes, plasma-sprayed or laser ablation coatings, by acid etching and electrochemical processes with different solutions, photoluminescence, or a combination of some of these (Schliephake and Scharnweber, 2008; Shrikhanzadeh, 1991; Le Guehennec et al., 2007).

The changes in topography also correlate with selective protein adsorption and collagen synthesis, as well as with the maturation of chondrocytes, which all have a significant influence on implant osseointegration (Pegueroles et al., 2010, 2011; Boyan et al., 1996; Boyan and Schwartz, 1999; Martin et al., 1995). However, the bioinert nature of titanium surfaces cannot stimulate more rapid bone regeneration and mechanical fixation of dental implants, so they are not appropriate for compromised immediate-loading applications.

The deposition of apatites—the mineral phase in bone—on titanium surfaces can enhance implant performance at an early stage after implantation (Ronold et al., 2003). This inorganic coating has the potential to actively signal the cells that form on the surface after implantation so they can rapidly regenerate bone in contact with the mineral coating. A first generation of thick coatings applied on titanium dental implants by plasma spraying or electrodeposition processes showed that response was good in the first week. The bone grew around the apatite and increased the osseointegration in the dental implant (Shrikhanzadeh, 1991; Geesink et al., 1987).

The main disadvantage of these conventional methods (plasma spraying) arises from the high temperature used. Due to the low thermal properties of titanium alloys local heating of the surface occurs, thus changing the structure and the mechanical properties of the surface material. Another main problem concerns the control of the chemical composition and crystallinity of the apatite coating, as well as its physical and mechanical properties during and after deposition (Hulshoff et al., 1997; Cleries, 1999; Hero et al., 1994; De Groot et al., 1990). HAp is thermically unstable during cooling, thus producing amorphous calcium phosphate (ACP) phases. The presence of amorphous phases in the coating negatively affects the stability of the coating in the long term after implantation. As is known, ACPs and even phases such as tetracalcium and tricalcium phosphates, which are also formed during cooling, dissolve faster than HAp, thus leading to complete mechanical disintegration of the coating and loosening of the implant (Hulshoff et al., 1997; Posner et al., 1980; Park et al., 1998).

It has also been reported that adhesion of the coating to the implant is very limited, leading to early decohesion of the coating even in the short term after implantation. This is related to the absence of any chemical link between the titanium surface and the HAp coating. The new methods alleviate this as they improve adhesion by modifying the surface of the metal with different chemical agents so that apatite crystals can be deposited chemically. As an example, commercially pure titanium can be made bioactive by treating its oxide layer so that a gel of sodium titanate is produced, which is able to induce nucleation of apatite crystals.

New treatments have been developed to obtain coatings of crystalline calcium phosphate minerals with a chemical bond with titanium that overcome the aforementioned problems. Of these, the chemical modification of titanium by alkaline etching and further thermal treatment (Kokubo et al., 1996) has the potential to grow an apatitic layer under biomimetic conditions at 37 °C and is considered the most important. The treated titanium is potentially bioactive and so it might develop a layer of mineral *in vivo* after implantation.

Figure 8.2 illustrates the appearance of a microporous layer made up of an alkaline titanate hydrogel formed during alkaline/heat treatment. This surface layer is an amorphous sodium titanate layer containing small amounts of a mixture of crystalline sodium titanates ($Na_2Ti_5O_{11}$) and rutile (TiO_2). During the alkaline treatment, the surface-passive TiO_2 layer partially dissolves into alkaline solution because of the corrosive attack of hydroxyl groups (Kokubo et al., 1996; Hurlen and Wihelmsen, 1986; Healy and Ducheyne, 1992, 1993):

$$TiO_2 + OH^- \rightarrow HTiO^{3-}$$

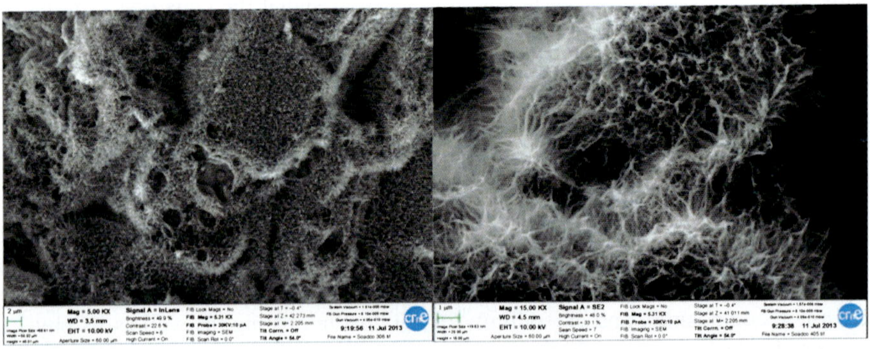

Figure 8.2 Sodium titanate formed on titanium surfaces at different magnifications.

This reaction is assumed to proceed simultaneously with the subsequent hydration of Ti metal (Arsov et al., 1991; Kim et al., 1996):

$$\mathrm{Ti} + 3\mathrm{OH}^- \rightarrow \mathrm{Ti}(\mathrm{OH})^{3+} + 4\mathrm{e}^-$$

$$\mathrm{Ti}(\mathrm{OH})^{3+} + \mathrm{e}^- \rightarrow \mathrm{TiO}_2 \cdot \mathrm{H}_2\mathrm{O} + 0.5\mathrm{H}_2(\mathrm{g})$$

$$\mathrm{Ti}(\mathrm{OH})^{3+} + \mathrm{OH}^- \Leftrightarrow \mathrm{Ti}(\mathrm{OH})_4$$

A further reaction of hydroxyl with hydrated TiO_2 produces negatively charged hydrates on the surfaces of the substrates as follows:

$$\mathrm{TiO}_2 \cdot n\mathrm{H}_2\mathrm{O} + \mathrm{OH}^- \Leftrightarrow \mathrm{HTiO}^{3-} \cdot n\mathrm{H}_2\mathrm{O}$$

These negatively charged species combine with alkali ions in aqueous solution, resulting in the formation of an alkali titanate layer. During heat treatment, the hydrogel layer is dehydrated and densified by the titanate layer.

When exposed to simulated body fluid (SBF), the alkali titanate layer is again hydrated to form TiO_2 hydrogel via the release of alkali ions from the alkali titanate layer into SBF. The alkali release and ion exchange with H_3O^+ ions in SBF result in a pH increase in the surrounding fluid. This pH increase gives rise to an increase in the ionic activity of apatite according to the following equilibrium in SBF (Gamble, 1967; Neuman and Neuman, 1958):

$$10\mathrm{Ca}^{2+} + 6\mathrm{PO}_4{}^{3-} + 2\mathrm{OH}^- \Leftrightarrow \mathrm{Ca}_{10}(\mathrm{PO}_4)_6(\mathrm{OH})_2$$

Figure 8.3a and b shows how the first apatite nuclei appeared after 3 days. Figure 8.3c shows that once the first nuclei has appeared the growth of the apatite layer is very rapid; in the figure it can be observed that the surface is completely coated with apatite. It is possible to design the thickness of the apatite layer by controlling the time of the treatment in SBF. Therefore, it is clear that the stage that controls the kinetics of the process is the nucleation stage.

Human osteoblast cultures have revealed that bioactive titanium surfaces show better cell responses than positive and negative controls (p-value < 0.05). The apatite layer has a significant effect on the adhesion (cell count) and differentiation (osteocalcine concentration) of osteoblast-like cells (Table 8.1).

The results of *in vivo* tests are shown in Figure 8.4; the percentage of early osseointegration in the bioactive implants was significantly improved as compared to all the other surface treatments between day 3 and week 3 (Fisher's test; $p < 0.05$). The value of osseointegration attained at week 2 was maintained until the end of the experiment without any significant changes (% direct contact $\approx 85\%$). Grit-blasted implants showed statistically significant differences in percentage direct contact between weeks 1, 2, 3, and 10, with the same level of direct contact as bioactive implants at the end of the study. These results indicate an acceleration of bone ingrowth in direct contact with the implant surface from the very early stages in the bioactive implants, which

Figure 8.3 (a) Nucleus of apatite formed on the titanate structure. (b) Detail at increased magnification of 2a. (c) Apatite is observed on the whole titanate structure.

Table 8.1 **Mean cell count and osteocalcin concentration values (± standard deviation)**

	Cell count (n/0.01 cm^2)	**Osteocalcine concentration (ng ml^{-1})**
Negative control	23.18 ± 2.80	1.87 ± 0.05
Positive control	21.00 ± 2.52	2.59 ± 0.12
Bioactive cp Ti	29.67 ± 0.93	4.39 ± 0.31

was significant just 3 days after implantation. This provides good primary stabilization for all implants, even before the hard tissues start to significantly regenerate.

These results show an improved short-term *in vivo* performance of the bioactive surfaces in comparison to the other types of implants tested. Osseointegration was accelerated and, consequently, implant failure decreased. Figure 8.5 clearly shows the osteoconductive behavior of the thermochemically treated surface (2Step), with the bone growing from the surface of the implant (Aparicio et al., 2004, 2007; Gil et al., 2014).

Other metals and alloys can be treated using this method, such as tantalum, niobium, and NiTi among others, and also porous materials. Of these, titanium is

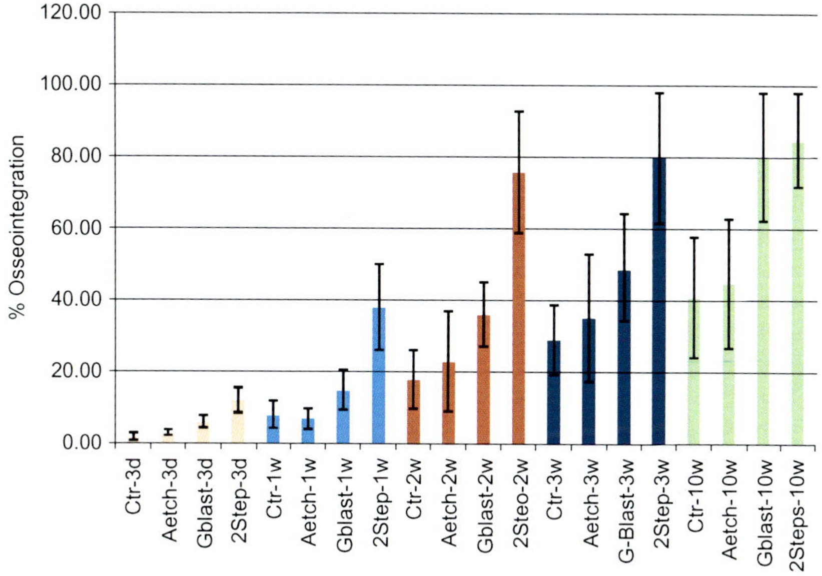

Figure 8.4 Percentage of osseointegration (BIC) for four different surfaces: Ctr, control (electropolished); Aetch, acid etching; Gblast, grit blasted; and 2Steps, bioactive. The study was carried out in minipigs at different implantation times: 3 days, 1, 2, 3, and 10 weeks.

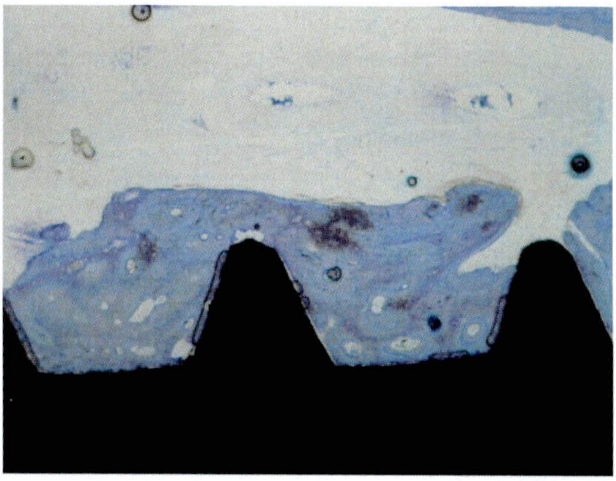

Figure 8.5 Bone growth from the bioactive surface of the dental implant. The surface has osteoconductive character.

the most widely used material for bone substitution, due to its biomechanical properties. However, it is necessary to reduce its elastic modulus for some special applications such as intervertebral substitution and/or reinforcement purposes. Although the most commonly used strategy is to construct alternative alloys (Arciniegas et al., 2007), the generation of titanium foams seems promising for modification of mechanical properties (Alvarez and Nakajima, 2009). Recently, different technologies, such as combustion synthesis and vapor deposition among others, have been used for the production of implants (Eck et al., 2000; Krishna et al., 2007; Karageorgiou and Kaplan, 2005; Kujalaa et al., 2003). Although these techniques allow one to attain high porosity rates, powder metallurgy allows straightforward modification and study of the structural properties of metallic foams with no compromise of the mechanical properties (Li et al., 2007; Niinomi, 2002). Intervertebral implants present lower load requirements than other implants, such as knee and hip implants, and must have low elastic moduli to avoid stress shielding effects on the bone (Arciniegas et al., 2007; Ryan et al., 2006). A high rate of interconnected porosity is important because it facilitates blood and nutrient transportation through the implant material, leading to bone ingrowth and fixation (Alvarez and Nakajima, 2009; Pilliar et al., 1979; Pilliar, 1998; Vasconcellos et al., 2010).

The incubation of the cells on metallic foams for 1, 7, and 14 days can be observed in Figure 8.6. The cultured cells colonized the external microspheres near the pore spaces for both microsphere sizes, occupying almost the total area of microspheres. Initially, cells covered the surfaces of the spheres. When final cell culture time was reached, cells colonized the interstices, more markedly on the fine foam.

Some macroporous metals with highly interconnected bioactivity and mechanical properties adequate for intervertebral applications have been obtained and characterized (Caparros et al., 2014). These materials have been thermochemically treated and apatite forms on the surface. The apatite-forming ability of the bioactive foams was predicted *in vitro* by immersion of the bioactivated metallic foams in SBF to elucidate the structural dependence of apatite formation on sodium titanate, as can be observed in Figure 8.7. In the same way the *in vivo* tests show the growth of bone tissue in the center of the prosthesis. Figure 8.8 shows the bone colonization inside the porous structure.

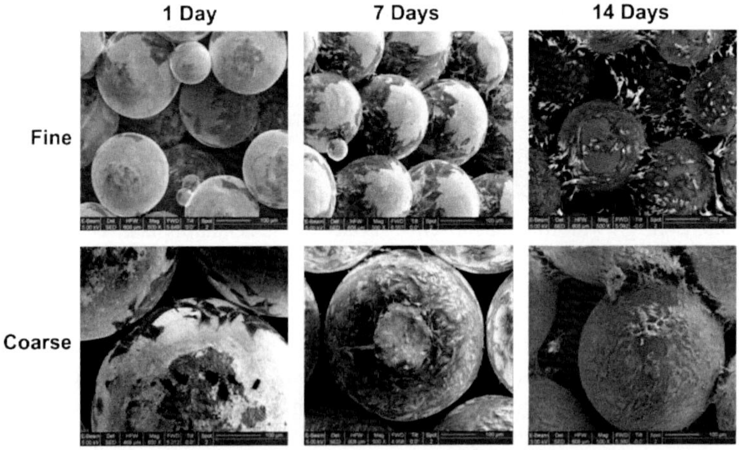

Figure 8.6 SEM images of SaoS-2 colonization on a metallic foam surface after 1, 7, and 14 days of incubation at 310.15 K (37 °C).

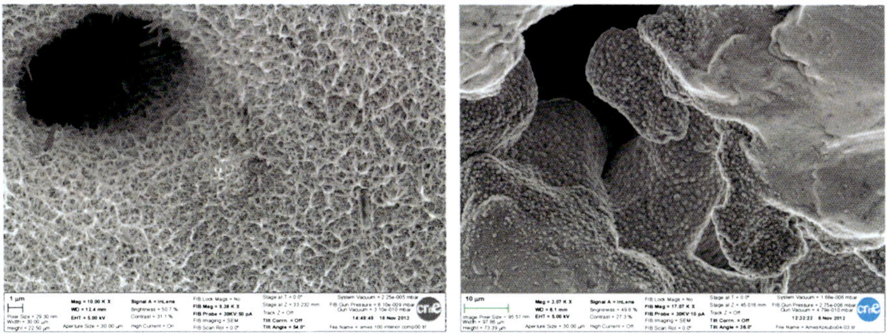

Figure 8.7 SEM micrographs of surface and pore after bioactive treatment.

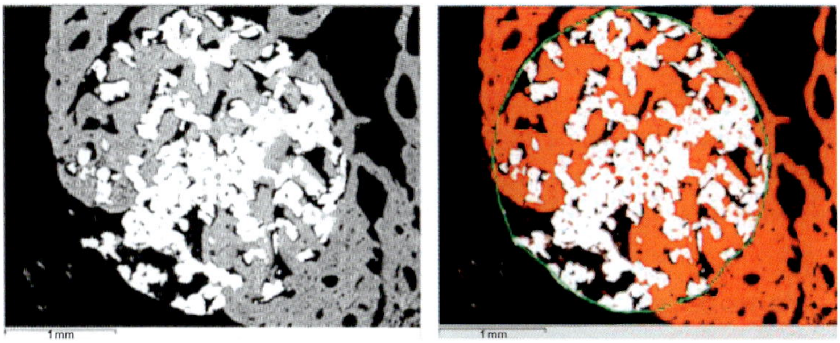

Figure 8.8 Transmission electron microscopy images; mature laminar bone can be seen inside the pores of the metallic structure.

8.3 Biomolecules and polymers

This section considers the use of polymers for calcium phosphate precipitation under physiological conditions. Liquid precursors have been studied in calcium carbonate and phosphate mineralization, the key concept being the polymer-induced liquid phases (PILPs). These liquids have high polymer and salt concentrations which, by way of their fluidity, can infiltrate templates with irregular shapes. The process is more complicated because many efficient additives are available, with an often very complex influence on the mineralization reaction (Bleek and Taubert, 2013).

Yang et al. (2011) studied the influence of asparagine (Asp), glycine (Gly), and lysine (Lys) on HAp crystallization in the presence and absence of magnesium. All amino acids accelerate HAp nucleation, while magnesium ions delay the nucleation process. Palazzo et al. (2009) investigated the role of alanine (Ala), arginine (Arg), and Asp on the nucleation, size, and morphology of calcium-deficient HAp (CDHAp) with a Ca/P ratio of 1.53 synthesized at pH 10. The results show that Ala, which has low polarity, only reduces the crystal domains with respect to the control samples. In contrast, the samples prepared with Ala, positively charged Arg, and negatively charged Asp contain needle-shaped particles.

Carboxylate binding to calcium phosphate has also been studied by Nancollas (Xie and Nancollas, 2010), and Schmidt-Rohr et al. (Hu et al., 2010) developed a citrate binding model for HAp crystallization on collagen fibrils in bone. Citrate molecules affect the thickening of HAp crystals and stabilize small crystals in the nanometer range. The distances of carboxylate groups in the citrate molecules (0.3 nm) fit well with the positions of the Ca^{2+} ions of the HAp to optimize the organic/inorganic interactions. According to the model, ACP clusters with only a few citrate molecules bound to the surface form in the supersaturated solution. The citrate molecules prevent further aggregation and increase the induction time of the crystallization.

Mann et al. (Schnepp et al., 2006) reported that dephosphorylated fluorenyl methyloxycarbonyl (Fmoc) tyrosine phosphate forms fibers in aqueous solution through π-stacking of the Fmoc moieties. The fibers have a helical orientation due to the hydrogen bonding-induced arrangement of the tyrosine residues. Depending on the mineralization time and salt concentration, viscoelastic hybrid gels, or hybrid monoliths were obtained with good mechanical properties (Figure 8.8).

Mineralized gelatin–beta-tricalcium phosphate (β-TCP) blends containing 5% dicalcium phosphate dihydrate (DPCD) for 7 days in SBF was achieved by Panzavolta et al. (2008). Transition rates from α-TCP to CDHAp decreased with increasing amounts of alendronate in the biomimetic cements.

8.3.1 Homopolymers

Synthetic polymers can carry functional groups that are identical to natural polymers or scaffolds or nonnatural groups such as biphosphonates or amino acids. These materials offer the possibility of different effects due to a large number of parameters on mineralization: topology, molecular weight, self-organization, chemistry, additives, etc.

Polyelectrolyte multilayers can be efficient nucleators or inhibitors of calcium phosphate, depending on their chemical composition (Ball et al., 2006; Ngankam et al., 2000; Kl et al., 2003; Hartgerink et al., 2001). Poly(allylamine, hydrochloride, lysine) coatings increase the calcium phosphate mineralizing ability of respective surfaces. At 4 weeks all samples were coated with a homogeneous layer of spheroidal mineral aggregates and the osteoblast-like cells show enhanced proliferation and viability.

Spraying or electrospinning have attracted much interest from the biomedical materials community; in the context of calcium phosphate-based composites it is preferable to co-electrospin preformed calcium phosphate nanoparticles with different polymers (Venugopal et al., 2010; Wu and Clark, 2008). Fletcher et al. prepared electrospun mats of ACP using polyethylene oxide and polyvinyl pyrrolidone (PVP). These ACP/PVP hybrid mats stimulated the *in vitro* remineralization of enamel in the presence of fluoride in artificial saliva. The ACP was transformed to a complete FAP layer on the enamel surface. Chen et al. (2010) found that electrospun poly(DL-lactide) (PDLLA) fibers promote the *in situ* mineralization of HAp within the PDLLA fibers. The same authors grafted chitosan onto PDLLA fibers by aminolysis. Chitosan attachment increased the hydrophilicity of the mats, with the carboxyl and amino groups providing active binding or chelating sites for calcium and phosphate ions.

Polylactic fibers, polyester, and poly(caprolactone), coupled with ethyldichlorophosphate and crosslinked poly(bis methacrylate) phosphazene, have been mineralized by Puska et al. (2009). High-density polyethylene (HDPE) showed mineralization achieved by surface modification of the HDPE with carboxylic acid groups (Huang et al., 2006).

Loebbicke et al. (2011) grafted polymethacrylic acid (PMAA) and poly(2-dimethyl aminoethyl methacrylate) (PDMAEMA) onto thiol-modified gold surfaces. Mineralization was done at pH 8.5 (PMAA) and 5.5 (PDMAEMA). At these pH values the respective polymer was charged and the ion addition was such that the oppositely charged ion was added first (i.e., calcium was added first onto the PMAA surfaces and phosphate was added first onto PDMARMA brushes).

Mineralization of collagen depends on the type of synthetic anionic polymer used for the PILPs. Polyacrylic acid (PAA) and polyvinyl phosphonic acid (PVPA) decelerated the calcium phosphate mineralization in solution, and combinations of these polymers with polyaspartic acid completely inhibited the mineralization (Thula et al., 2011). Grower et al. used alkaline phosphatase in the PILPs to release phosphate from rac-glycerol-1-phosphate, leading to calcium phosphate deposition on collagen fibers (Jee et al., 2010). The crystals were uniformly distributed throughout the collagen fibers.

Crosslinking of the collagen fibers enhanced mineralization. Crosslinking opens up intermolecular spaces and minimizes the excluded volume of the collagen fibers. A capillary-force-driven mechanism for PILP droplets has also been proposed for the interfibrillar mineralization of the collagen fibrils (Jee et al., 2010; Li et al., 2011). These studies demonstrate that PILPs are interesting precursors for controlled calcium phosphate deposition, in particular in complex templates or scaffolds such as gels and collagen.

Zhang et al. (2008) used bile acid groups on poly(3α-methacrylic cholic acid methyl ester) (PME-CAME), which formed fibrils with diameters of 1 nm in aqueous solution. Mineralization in SBF resulted in plate-like HAp crystals. The relatively short mineralization time indicates that these polymeric fibrils efficiently promoted local supersaturation, nucleation, or crystal growth.

Linear and branched poly(ethylene imine) (PEI) are efficient templates for synthesis of calcium phosphate hybrid particles. At pH 8, composite nanoparticles with up to 20 wt% polymer formed. The polymer was wrapped around one or more HAp nanoparticles with diameters of around 6 nm. The precipitation at pH 5 led to the formation of micrometer-sized DCPD platelets. The difference in the morphology and morphometry of the calcium phosphate is due to significantly different degrees of protonation and charge of PEI at pH 5 or 8 (Shkilnyy et al., 2008). Time-resolved mineralization experiments using the same synthesis protocol showed rapid precipitation, a slow decrease in the pH, and a very rapid decrease in the calcium phosphate concentrations (Shkilnyy et al., 2011). The initial precipitate was DCPD, which was transformed into HAp within 2 h. The redissolution–reprecipitation mechanism is the key reaction pathway. Because the DCPD–HAp transition was much faster than in studies using acidic additives (Bigi et al., 2003, 2011). This has been proposed to accelerate the DCPD–HAp transformation.

Natural polymers such as proteins and carbohydrates can be chemically modified. Falini et al. (2001) precipitated octacalcium phosphate (OCP) and HAp into a β-chitin scaffold; this offers carboxyl, hydroxyl, and amide groups as possible nucleation sites, and OCP grew oriented plate-like crystals. Chitosan improved processability compared to chitin. It is a polycation and is therefore also of interest with regard to calcium phosphate mineralization. Fratz et al. (Rusu et al., 2005) designed chitosan/HAp materials via stepwise coprecipitation from an aqueous medium by slowly increasing the pH of a chitosan/CaCl/NaH_2PO_4 solution from 3 to 11. Chiono et al. (2010) synthesized amphiphilic and water-soluble fluorescein-grafted chitosan-cast films to stimulate HAp deposition under ultraviolet/visible irradiation in SBF.

The analysis of the efficiency of mineralization inhibition, studied by Dogan et al. (2010), depended on the carboxylmethyl inulin (CMI) concentration and the degree of carboxylation. The growth rates and the coherence length of the crystals in the composite decreased with increasing CMI concentration and with negative charges in the inulin.

Carboxymethyl cellulose (CMC) has been used for mineralization in dental repair. Combining chlorhexidine with calcium phosphate nanoparticles resulted in a bifunctional antibacterial system which remineralized enamel and closed dentin tubules when coated with CMC.

Mineralization in the presence of proteins is difficult because they have a complex chemical composition and a three-dimensional structure. In spite of this, a number of proteins have been investigated as biomimetic scaffolds specially applied for bone formation in vertebral tissues.

Yang et al. (2010) found that amelogenin accelerates HAp nucleation. The growth of HAp proceeded from prenucleation clusters of hierarchical amelogenin and ACP over spherical nanoparticles and chain-like structures to enamel-like nano- and microrods. Water-based micelles from multiprotein casein isolated from skimmed milk was obtained by Mann et al. (Thachepan et al., 2010). Mineralization at 60 °C for several months in calcium and phosphate solution at pH 7 produced 12 nm ACP/protein particles. HAp nanofilaments also formed within intact micelles.

The protein matrix was suggested to be responsible for controlling the rate of structure development because of the close association of HAp nanofilaments and casein micelles. In contrast, synthetic β-casein alone accelerated the formation of uniaxially oriented HAp nanofilament bundles within the protein micelles.

Elastin-like polymers (ELPs) containing the SN_A 15 domain of statherin (a protein that inhibits calcium phosphate mineralization in saliva) can generate recombinant ELPs, which can interact with calcium phosphate (Prieto et al., 2011). Osteopontin (OPN) inhibits the *in vitro* mineralization of HAp and other biomaterials; OPN was also suggested to inhibit the *in vivo* calcification of soft tissues (Azzopardi et al., 2010).

Aparicio et al. (Li et al., 2014) studied the system of ELPs–human salivary statherin (HSS) tethered to nanorough surfaces and mineralized using an enzyme-directed process. A homogeneous layer of ACP minerals forms on HSS-coated surfaces retaining the original nanotopography of the substrate. The statherin-inspired and elastin-like recombinamers (ELRs) have the ability to induce and control growth of the minerals on the biofunctional surfaces. HSS–ELR coatings have similar bioactivity to that of statherin in human saliva. The hybrid nanorough surfaces improve adhesion and differentiation of preosteoblasts, and show potential for dental and orthopedic implant integration. This method enables the combination and tailoring of nanotopographical and biochemical cues to design functionalized surfaces to investigate and potentially direct the fate of stem cells. Mata et al. (Tejada-Montes et al., 2014) searched for alternative therapies to improve bone regeneration in order to enhance mineralization, osteogenesis, and *in vivo* bone regeneration properties of a bioactive ELR membrane. The ELR materials were then used to fabricate membranes comprising either a smooth surface or channel microtopographies. Smooth mineralization membranes were produced using static and dynamic (uniaxial strain of 8% at 1 Hz frequency) conditions. Smooth mineralization membranes in the static condition exhibited the highest quantity of calcium phosphate (Ca/P of 1.78) deposition with and without the presence of cells, the highest Young's modulus, and the highest production of alkaline phosphatase on day 10 in the presence of cells growing in nonosteogenic differentiation medium. The study demonstrated the possibility of using bioactive ELR membranes for bone regeneration applications.

However, bone morphogenetic protein 2 is the most powerful osteogenic factor for collagen-mediated calcium phosphate mineralization. This has potential for use in biomimetic mineralization or as a component in scaffolds for bone tissue engineering. Bone sialoprotein is also important in bone mineralization; it is an acidic phosphoprotein abundantly present in mineralized tissue. It can bind to type I collagen and is an efficient nucleator for HAp (Prasad et al., 2010).

8.3.2 Copolymers

The dynamics of the polymer surface have a strong influence on mineralization (Junginger et al., 2010). Polymer monolayer oscillation affects mineral deposition. Time-resolved studies showed that mineralization at pH 8 led to uniform deposits at the relatively short mineralization time of 4 h. This is in contrast to mineralization at pH 5, which only yielded uniform, dense features after 8–10 h. Templates dynamics must therefore also be considered in mineralization processes. There have been few studies on copolymer-controlled calcium phosphate mineralization from bulk aqueous solution. The deposition of HAp nanocrystals in the presence of random poly(vinylphosphonic acid)-*co*-poly(vinylimidazole) (PVPA-*co*-PVIm) copolymers is one such study. These copolymers promoted calcium phosphate precipitation. This crystallization rate was accelerated by a factor of four at higher polymer concentrations. Murphy et al. (Jongpaiboonkit et al., 2009) prepared poly(D,L,-lactide)-*co*-poly(glycolide) microspheres and coated them with a carbonated apatite layer by immersion in SBF. Mineralization occurred after the first 3 days, and after 7 days 45% of the microspheres were already aggregated. Aggregation of up to 90% of the particles occurred during mineralization. At higher aggregation levels, mineralization was faster and more efficient. The addition of a surfactant (Tween 20), which inhibits aggregation, also led to mineralization.

8.3.3 Synthetic hydrogels

Synthetic polymer hydrogels are interesting for mimicking biological scaffolds because they are highly flexible in terms of available or accessible monomers, polymer compositions, molecular weights, and topologies, among others.

One of the most important studies is the mineralization of poly(vinyl alcohol) (PVA) hydrogels containing calcium and phosphate ions via double diffusion (Suetsugu et al., 2009). Mineralizations were achieved by raising the pH of the hydrogel via sodium hydroxide diffusion from one edge. Crosslinked copolymer networks of poly(vinyl alcohol), poly(acrylic acid), and poly(acrylonitrile) (PVA-*co*-PAA-*co*-PAN) were obtained via free radical polymerization. The swelling ability decreased with increasing PVA fraction (Bajpai and Singh, 2007). Increasing the temperature to 35 °C also promoted mineralization. Apatite formation is a function of hydrogel swelling and temperature, parameters that can be easily adjusted for the development of implant materials.

The hydrogels showed significant mineralization in SBF when the hydrogel was impregnated with phosphoric acid. The resulting composites were a mixture of HAp and DCPD and had sponge-like assemblies near the surface (Figure 8.9). Kim and Lee (2009) prepared injectable poly(*N*-isopropylacrylamide-*co*-vinyl phosphonic acid) hydrogels and used a urea-based process for mineralization. Increasing PVA fractions resulted in enhanced calcium enrichment and higher calcium phosphate nucleation and growth rates.

Figure 8.10 shows scanning electron mircoscopy (SEM) and fluorescence microscopy data of mineralized PEI gels and dictyostelium seeding on the gels.

Figure 8.9 (a and b) Lightly mineralized Fmoc-tyrosine fibers. (c and d) Extensively mineralized monoliths after disassembly of the self-assembled matrix.
Images adapted from Schnepp et al. (2006). Reprinted with permission from Wiley-VCH, copyright 2006.

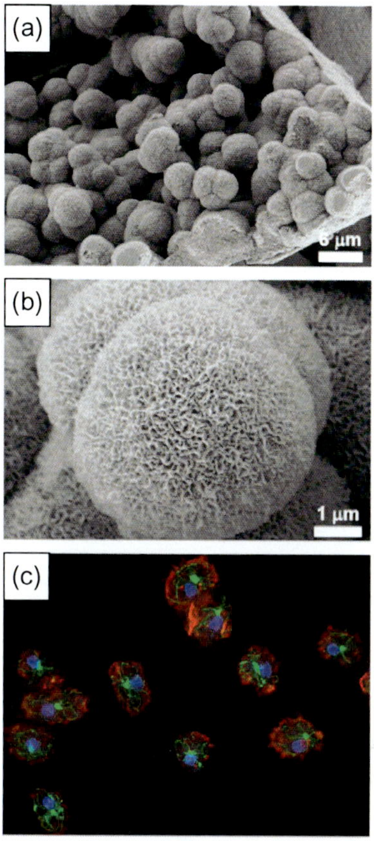

Figure 8.10 SEM and fluorescence microscopy data of mineralized PEI gels and dictyostelium seeding on the gels. (a and b) SEM images of mineralized hydrogel. (c) Fluorescence images of osteoblastic cells.

8.4 Ceramics

Calcium phosphates (CaPs) have become the most important candidates to promote osteoconductivity and osteogenesis (Termine et al., 1973; Le Geros, 2002; Ooms et al., 2003).

The specific chemical reactions, the ionic equilibria that occur, and the rate of these reactions depend on the ceramic composition and are described in the next paragraphs for most studied bioactive ceramics. In the last stages of this process, the osteoblasts located in the periphery of the implant recognize the biologically equivalent apatite layer as bone tissue and produce new bone, forming a biological union with the surrounding tissue. The attraction of osteoblast to the surface and the formation of new bone are additionally stimulated by the ionic modification of the microenvironment and the nature of the adsorbed proteins.

Hench proposed a sequence of surface reactions leading to the formation of a biologically equivalent apatite layer when bioactive glasses (Bioglass®) come into contact with body fluids. These glasses are based on silica as a network former, and can incorporate different ions as network modifiers, the most representative formulation (Bioglass® 45S5) containing 45% SiO_2, 24.5% Na_2O, 24.4% CaO, and 6% P_2O_5 in weight percent. The different steps in apatite layer formation are summarized in Figure 8.11 (Cao and Hench, 1996; Hench et al., 2002). First, immediately after coming into contact with the aqueous solution, a rapid exchange of Na^+ and Ca^{2+} ions from the Bioglass® surface with the H_3O^+ ions of the solution begins. This stage is controlled by diffusion in a $t^{1/2}$ dependence. As a result of the substitution of protons by metallic cations, the pH of the solution increases. Second, the increase in pH promotes network dissolution by breaking the Si–O–Si bonds, which is followed by formation of silanol groups (SiOH) at the glass–solution interface. Third, the continuous formation of silanol groups results in their condensation and the polymerization of a silica gel layer depleted of alkali and alkaline-earth cations. Fourth, Ca^{2+} and PO_4^{3-} groups migrate through the SiO_2-rich layer forming a $CaO-P_2O_5$-rich film on top of the SiO_2-rich layer. This layer grows by the incorporation of Ca^{2+} and PO_4^{3-} ions from the solution. Fifth, this $CaO-P_2O_5$-rich layer, initially amorphous, crystallizes, incorporating OH^- and CO_3^{2-} anions from the solution, giving rise to a crystalline hydroxyl carbonate apatite layer.

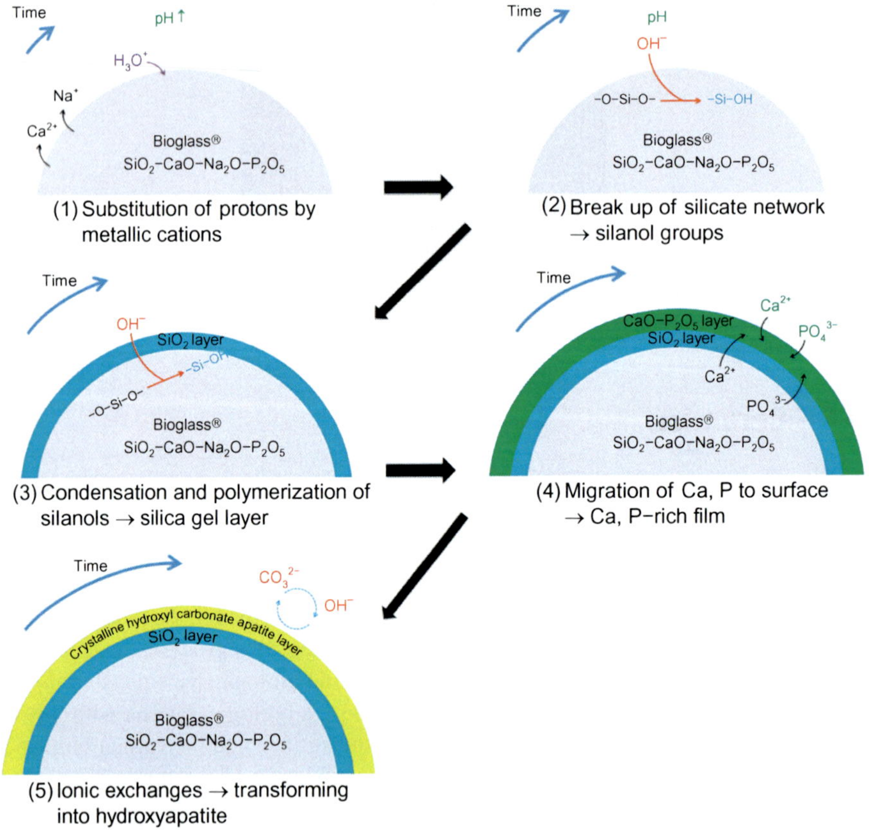

Figure 8.11 Summary of the sequence of reactions at the surface of bioactive glass that results in the formation of a biologically equivalent apatite layer.

Although this mechanism was first described for bioactive glasses, HAp and other calcium phosphate bioceramics are also bioactive. Nevertheless, the formation of a biologically equivalent apatite layer on HAp does not follow the process described above. Kokubo et al. (2003) proposed a mechanism based on the electrical charge variations on the HAp surface when it is immersed in SBF. The HAp surface has an initial negative zeta potential due to the presence of OH^- and PO_4^{3-} anions. According to Kokubo et al. (see Figure 8.12), calcium ions in solution are attracted by the negatively charged surface, forming a calcium-rich layer. As a consequence of calcium

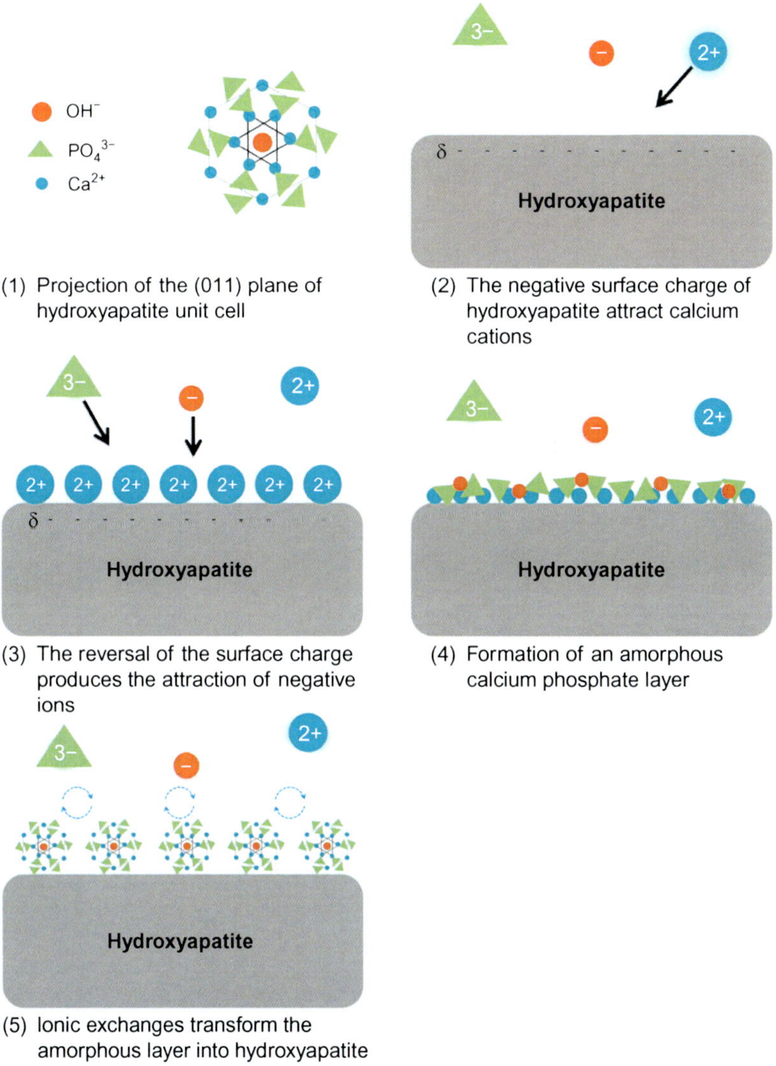

Figure 8.12 Summary of the sequence of reactions at the surface of HAp that results in the formation of a biologically equivalent apatite layer. Note that the size of the ions is only schematic and not proportional to reality.

accumulation, the zeta potential increases. Then, due to reversal of the surface charge, the surface interacts with the negatively charged phosphate ions from the solution, forming an ACP with a low Ca/P ratio. Throughout this period, the metastable ACP undergoes ionic exchanges and is transformed into a crystalline hydroxyl carbonate apatite layer.

β-TCP ($Ca_3(PO_4)_2$) and biphasic calcium phosphates (BCPs; composed of β-TCP and HA) also form a direct bonding with bone *in vivo*. Nevertheless, for these ceramics the mechanism of formation of the biologically equivalent apatite layer has not been described, and in fact the situation is more complex, especially for β-TCP, due to the higher solubility of this ceramic compared to HAp, which implies the lack of a stable surface. Thus, whereas in the case of BCPs it has been proposed that probably the higher solubility of β-TCP improves the ionic exchanges with the microenvironment, decreasing the time for the formation of the biologically equivalent apatite layer on the surface of HAp, in the case of β-TCP a dynamic process is established between bone bonding and material resorption, which has not yet been clarified.

References

Albrektsson, T., et al., 1981. Osseointegrated titanium implants. Requirements for ensuring a long-lasting direct bone-to-implant anchorage in man. Acta Orthop. Scand. 52, 155–170.

Alvarez, K., Nakajima, H., 2009. Metallic scaffolds for bone regeneration. Materials 2, 790–832.

Aparicio, C., Engel, E., Gil, F.J., Planell, J.A., 2002. Human-osteoblast proliferation and differentiation on grit-blasted and bioactive titanium for dental applications. J. Mater. Sci. Mater. Med. 13, 1105–1111.

Aparicio, C., Gil, F.J., Fonseca, C., Barbosa, M., Planell, J.A., 2003. Corrosion behaviour of commercially pure titanium shot blasted with different materials and sizes of shot particles for dental implant applications. Biomaterials 24, 263–273.

Aparicio, C., Gil, F.J., Thams, U., Munoz, F., Padros, A., Planell, J.A., 2004. Osseointegration of grit-blasted and bioactive titanium implants: histomorphometry in minipigs. Key Eng. Mater. 254–256, 737–740.

Aparicio, C., Manero, J.M., Conde, F., Pegueroles, M., Planell, J.A., Vallet-Regí, M., Gil, F.J., 2007. Acceleration of apatite nucleation on microrough bioactive titanium for bone-replacing implants. J. Biomed. Mater. Res. 82A, 521–529.

Aparicio, C., Padros, A., Gil, F.J., 2011. In vivo evaluation of micro-rough and bioactive titanium dental implants using histometry and pull-out tests. J. Mech. Behav. Biomed. Mater. 4, 1672–1682.

Arciniegas, M., Aparicio, C., Manero, J.M., et al., 2007. Low elastic modulus metals for joint prosthesis: tantalum and nickel–titanium foams. J. Eur. Ceram. Soc. 27, 3391–3398.

Arsov, L.D., Kormman, C., Plieth, W., 1991. In situ Raman spectra of anodically formed titanium dioxide layer in solution of H_2SO_4, KOH, and HNO_3. J. Electrochem. Soc. 138, 2964–2970.

Azzopardi, P.V., O'Young, J., Lajoie, G., Karttunen, M., Goldberg, H.A., Hunter, G.K., 2010. Roles of electrostatics and conformation in protein-crystal interactions. PLoS One 5.

Bagno, A., Di Bello, C., 2004. Surface treatments and roughness properties of Ti-based biomaterials. J. Mater. Sci. Mater. Med. 15, 939–945.

Bajpai, A.K., Singh, R., 2007. Study of biomineralization of poly(vinyl alcohol)-based scaffolds using an alternate soaking approach. Polym. Int. 56, 557–568.

Ball, V., Michel, M., Boulmedais, F., Hemmerle, J., Haikel, Y., Schaaf, P., 2006. Nucleation kinetics of calcium phosphates on polyelectrolyte multilayers displaying internal secondary structure. Cryst. Growth Des. 6, 327–334.

Bigi, A., Boanini, E., Bracci, B., Falini, G., Rubini, K., 2003. Interaction of acidic poly-amino acids with octacalcium phosphate. J. Inorg. Biochem. 95, 291–296.

Bigi, A., Boanini, E., Cojazzi, G., Falini, G., Panzavolta, S., 2011. Morphological and structural investigation of octacalcium phosphate hydrolysis in the presence of polyacrylic acids: effect of relative molecular weights. Cryst. Growth Des. 1, 239–244.

Bleek, K., Taubert, A., 2013. New developments in polymer-controlled, bioinspired calcium phosphate mineralization from aqueous solution. Acta Biomater. 9, 6283–6321. http://dx.doi.org/10.1016/j.actbio.2012.12.027.

Boyan, B.D., Schwartz, Z., 1999. Modulation of osteogenesis via implant surface design. In: Davies, J.E. (Ed.), Bone Engineering. EM Squared, Toronto, pp. 232–239.

Boyan, B.D., et al., 1996. Role of material surfaces in regulating bone and cartilage cell response. Biomaterials 17, 137–146.

Brunski, J.B., 1996. Classes of materials used in medicine. Metals. In: Rutner, B., Hoffman, A., Schoen, F., Lemons, J. (Eds.), Biomaterials Science, An Introduction to Materials in Medicine. Academic Press, San Diego, CA, pp. 137–152.

Buser, D., 2001. Titanium for dental applications (I). In: Brunette, D.M., Tengvall, P., Textor, M., Thomsen, P. (Eds.), Titanium in Medicine. Springer-Verlag, Berlin/NewYork/Heidelberg, pp. 875–888.

Cao, W., Hench, L.L., 1996. Bioactive materials. Ceram. Int. 22, 493–507.

Caparros, C., Guillem-Martí, J., Punset, M., Molmeneu, M., Calero, J., Gil, F.J., 2014. Mechanical properties and in vitro biological response to porous titanium alloys for intervertebral implants. J. Mech. Behav. Biomed. Mater. 39, 79–86. http://dx.doi.org/10.1016/j.jmbbm.2014.05.029.

Chen, J., Li, X., Xie, C., Chen, J., Zou, J., Zou, S., 2010. In situ grown fibrous composites of poly(D-lactide) and hydroxyapatite as potential tissue engineering scaffolds. Polymer 51, 6268–6277.

Chiono, V., Gentile, P., Boccafoschi, F., Carmagnola, I., Ninov, M., Georgieva, V., et al., 2010. Photoactive chitosan switching on bone-like apatite deposition. Biomacromolecules 11, 309–315.

Cleries, L., 1999. In vitro studies of calcium phosphate coatings obtained by laser ablation. PhD thesis, University of Barcelona, Spain.

De Groot, K., Klein, C.P., Wolke, J.G., Blieck-Hogervorst, J.M., 1990. Plasma sprayed coatings of calcium phosphate. In: Yamamuro, T., Hench, L., Wilson, J. (Eds.), CRC Handbook of Bioactive Ceramics. In: Calcium Phosphate and Hydroxyapatite Ceramics, vol. II. CRC Press, Boca Raton, FL, pp. 133–142.

Dogan, O., Öner, M., Cinel, O., 2010. The inhibitory effects of inulin biopolymers on the seeded growth of hydroxyapatite. J. Ceram. Soc. Jpn. 1118, 579–586.

Eck, K.R., Bridwell, K.H., Ungacta, F.F., et al., 2000. Analysis of titanium mesh cages in adults with minimum two-year follow-up. Spine 25, 2407–2415.

Falini, G., Fermani, S., Ripamonti, A., 2001. Oriented crystallization of octacalcium phosphate into beta-chitin scaffold. J. Inorg. Biochem. 84, 255–258.

Farbod, K., Nejadnik, M.R., Jansen, J.A., Leeuwenburgh, S.C.G., 2014. Interactions between inorganic and organic phases in bone tissue as a source of inspirations for design of novel nanocomposites. Tissue Eng. Part B Rev. 20, 173–188.

Gamble, J., 1967. Chemical Anatomy Physiology and Pathology of Extracellular Fluid, sixth ed. Harvard University Press, Cambridge, MA.

Geesink, R.G.T., de Groot, K., Klein, C.P.A.T., 1987. Chemical implant fixation using hydroxyl-apatite coatings. Clin. Orthop. Relat. Res. 225, 147–170.

Giavaresi, G., et al., 2003. Mechanical and histomorphometric evaluations of titanium implants with different surface treatments inserted in sheep cortical bone. Biomaterials 24, 1583–1594.

Gil, F.J., Padros, A., Manero, J.M., Aparicio, C., Nilsson, M., Planell, J.A., 2002. Growth of bioactive surfaces on titanium and its alloys for orthopaedic and dental implants. Mater. Sci. Eng. C 22, 53–60.

Gil, F.J., Manzanares, N., Badet, A., Aparicio, C., Ginebra, M.P., 2014. Biomimetic treatment on dental implants for short-term bone regeneration. Clin. Oral Invest. 18 (1), 59–66.

Hartgerink, J.D., Beniash, E., Stupp, S.I., 2001. Self-assembly and mineralization of peptide-amphiphile nanofibers. Science 294, 1684–1688.

Healy, K.E., Ducheyne, P., 1992. The mechanism of passive dissolution kinetics of titanium in a model physiological environment. J. Biomed. Mater. Res. 36, 319–338.

Healy, K.E., Ducheyne, P., 1993. Passive dissolution kinetics of titanium in vitro. J. Mater. Sci. Mater. Med. 4, 117–126.

Hench, L.L., Jones, J.R., Sepulveda, P., 2002. Bioactive materials for tissue engineering scaffolds. In: Polak, J.M., Hench, L.L., Kemp, P. (Eds.), Future Strategies for Tissue and Organ Replacement. Imperial College Press, London, pp. 3–24.

Hero, H., Wie, H., Jorgensen, R.B., Ruyter, I.E., 1994. Hydroxyapatite coating on titanium produced by isostatic pressing. J. Biomed. Mater. Res. 28, 344–348.

Hu, Y.Y., Rawal, A., Schmidt-Rohr, K., 2010. Strongly bound citrate stabilizes the apatite nanocrystals in bone. Proc. Natl. Acad. Sci. U.S.A. 107, 22425–22429.

Huang, S.P., Zhou, K.C., Zhu, W., Huang, B.Y., Li, Z.Y., 2006. Effects of in situ biomineralization on microstructural and mechanical properties of hydroxyapatite/polyethylene composites. J. Appl. Polym. Sci. 101, 1842–1847.

Hulshoff, J.E.G., et al., 1997. Mechanical and histologic evaluation of Ca-P plasma spray and magnetron sputter-coated implants in trabecular bone of the goat. J. Biomed. Mater. Res. 36, 75–83.

Hurlen, T., Wihelmsen, W., 1986. Passive behaviour of titanium. Electrochim. Acta 31, 1139–1146.

Jee, S.S., Culver, L., Li, Y., Douglas, E.P., Grower, L.B., 2010. Biomimetic mineralization of collagen via an enzyme-aided PILP process. J. Cryst. Growth 312, 1249–1256.

Jongpaiboonkit, L., Franklin-Ford, T., Murphy, W.L., 2009. Growth of hydroxyapatite coatings on biodegradable polymer microspheres. ACS Appl. Mater. Interfaces 1, 1504–1511.

Junginger, M., Bleek, K., Kita-Tokarczyk, K., Reiche, J., Shkilnyy, A., Schacher, F., et al., 2010. Calcium phosphate growth beneath a polycationic monolayer at the air water interface: effects of oscillating surface pressure on mineralization. Nanoscale 2, 2440–2446.

Karageorgiou, V., Kaplan, D., 2005. Porosity of 3D biomaterial scaffolds and osteogenesis. Biomaterials 26, 5474–5491.

Kim, S.Y., Lee, S.C., 2009. Thermo-responsive injectable hydrogel system based on poly(N-isopropylacrylamide-co-vinyl phosphonic acid). I. Biomineralization and protein delivery. J. Appl. Polym. Sci. 113, 3460–3469.

Kim, H.M., Miyaji, F., Kokubo, T., Nakamura, T., 1996. Preparation of bioactive Ti and its alloys via simple surface treatment. J. Biomed. Mater. Res. 32, 409–417.

Kokubo, T., Kushitani, H., Sakka, S., Kitsugi, T., Yamamuro, T., 1990. Solutions able to reproduce in vivo surface structure changes in bioactive glass-ceramic A-W. J. Biomed. Mater. Res. 24, 721–734.

Kokubo, T., Miyaji, F., Kim, H.M., 1996. Preparation of bioactive Ti and its alloys via simple chemical surface treatment. J. Am. Ceram. Soc. 79, 1127–1129.

Kokubo, T., Kim, H.-M., Kawashita, M., 2003. Novel bioactive materials with different mechanical properties. Biomaterials 24, 2161–2175.
Krishna, B.V., Bose, S., Bandyopadhyay, A., 2007. Low stiffness porous Ti structures for load-bearing implants. Acta Biomater. 3, 997–1006.
Kujalaa, S., Ryhanen, J., Danilovb, A., et al., 2003. Effect of porosity on the osteointegration and bone ingrowth of a weight-bearing nickel–titanium bone graft substitute. Biomaterials 24, 4691–4697.
Le Geros, R.Z., 2002. Properties of osteoconductive biomaterials: calcium phosphates. Clin. Orthop. Relat. Res. 395, 81–98.
Le Guehennec, L., et al., 2007. Surface treatments of titanium dental implants for rapid osseointegration. Dent. Mater. 23, 844–854.
Li, J.P., Habibovic, P., Doel, M.V.D., et al., 2007. Bone ingrowth in porous titanium implants produced by 3D fiber deposition. Biomaterials 28, 2810–2820.
Li, Y., Thula, T.T., Jee, S., Perkins, S.L., Aparicio, C., Douglas, E.P., et al., 2011. Biomimetic mineralization of woven bone-like nanocomposites: role of collagen crosslinks. Biomacromolecules 13, 49–59.
Li, Y., Chen, X., Ribeiro, A., Kyle, V., Rodriguez-Cabello, J.C., Aparicio, C., 2014. Hybrid nanotopographical surfaces obtained by biomimetic mineralization of statherin-inspired elastin-like recombinamers. Adv. Healthc. Mater. 3, 1638–1647. http://dx.doi.org/10.1002/adhm.201400015.
Loebbicke, R., Chanana, M., Sclaad, H., Pilz-Allen, C., Guenter, C., Moehwald, H., 2011. Polymer brush controlled bioinspired calcium phosphate mineralization and bone cell growth. Biomacromolecules 12, 3753–3760.
Martin, J.Y., et al., 1995. Effect of titanium surface roughness on proliferation, differentiation and protein synthesis of human osteoblast-like cells (MG63). J. Biomed. Mater. Res. 29, 389–401.
Neuman, W., Neuman, M., 1958. The Chemical Dynamics of Bone Mineral. University of Chicago Press, Chicago, IL.
Ngankam, P.A., Lavalle, P., Voegel, J.C., Szyk, L., Decher, G., Schaaf, P., 2000. Influence of polyelectrolyte multilayers film on calcium phosphate nucleation. J. Am. Chem. Soc. 122, 8998–9004.
Niece, K.L., Hartgerink, J.D., Donners, J., Stupp, S.I., 2003. Self-assembly combining two bioactive peptide-amiphiphile molecules into nanofibers by electrostatic attraction. J. Am. Chem. Soc. 125, 7146–7147.
Niinomi, M., 2002. Recent metallic materials for biomedical applications. Metall. Mater. Trans. A 33, 477–486.
Ooms, E.M., Egglezos, E.A., Wolke, J.G., Jansen, J.A., 2003. Soft tissue response to injectable calcium phosphate cements. Biomaterials 24, 749–757.
Palazzo, B., Walsh, D., Iafisco, M., Foresti, E., Bertinetti, L., Marta, G., 2009. Amino acid synergic effect on structure, morphology and surface properties of biomimetic apatite nanocrystals. Acta Biomater. 5, 1241–1252.
Panzavolta, S., Torricelli, P., Sturba, L., Bracci, B., Giardino, R., Bigi, A., 2008. Setting properties and in vitro activity of strontium-enriched gelatin-calcium phosphate bone cements. J. Biomed. Mater. Res. A 84A, 965–972.
Park, E., Condrate, S.R., Hoelzer, D.T., Fischman, G.S., 1998. Interfacial characterisation of plasma-spray coated calcium phosphate on Ti6Al4V. J. Mater. Sci. Mater. Med. 9, 643–649.
Pegueroles, M., et al., 2010. Spatial organization of osteoblast fibronectin-matrix on titanium surface – effects of roughness, chemical heterogeneity, and surface free energy. Acta Biomater. 6, 291–301.

Pegueroles, M., et al., 2011. Effect of blasting treatment and Fn coating on MG63 adhesion and differentiation on titanium: a gene expression study using real-time RT-PCR. J. Mater. Sci. Mater. Med. 22 (3), 617–627. http://dx.doi.org/10.1007/s10856-011-4229-3.

Pilliar, R.M., 1998. Overview of surface variability of metallic endosseous dental implants: textured and porous surface-structured designs. Implant. Dent. 7, 305–314.

Pilliar, R.M., Cameron, H.U., Macnab, I., 1979. Bone ingrowth and stress shielding with a porous surface coated fracture fixation plate. Biomed. Eng. 10, 126–131.

Placko, H.E., Mishra, S.H., Weimer, J.J., Lucas, L.C., 2000. Surface characterization of titanium-based implant materials. Int. J. Oral Maxillofac. Implants 15, 355–363.

Porter, J.R., Ruckh, T.T., Popat, K.C., 2009. Bone tissue engineering: a review in bone biomimetics and drug delivery strategies. Biotechnol. Prog. 25, 1539.

Posner, A.S., Blumenthal, N.C., Betts, F., 1980. In: Eon, C. (Ed.), Proceedings of the 2nd International Conference on Phosphorus Compounds. Imphos, Paris, p. 25.

Prasad, M., Butler, W.T., Qin, C., 2010. Dentin sialophosphoprotein in biomineralization. Connect. Tissue Res. 51, 404–417.

Prieto, S., Shkilnyy, A., Rumplasch, C., Ribeiro, A., Javier Arias, F., Carlos Rodriguez-Cabello, J., et al., 2011. Biomimetic calcium phosphate mineralization with multifunctional elastin-like recombinamers. Biomacromolecules 12, 1480–1486.

Puska, M., Nykanen, V.R.S., Korventausta, J., Nykanen, A., Narhi, T., Ruokolainen, J., 2009. Calcium phosphate formation on electrosphoester of poly(ε caprolactone) and poly(bis(methacrylate)) phosphazene in vitro. Key Eng. Mater. 396–398, 171–174.

Ratner, B.D., 2001. A perspective on titanium biocompatibility. In: Brunette, D.M., Tengvall, P., Textor, M., Thomsen, P. (Eds.), Titanium in Medicine: Material Science, Surface Science, Engineering, Biological Responses and Medical Applications. Springer-Verlag, Berlin/Heidelberg, pp. 1–12.

Ronold, H.J., Lyngstadaas, S.P., Ellingsen, J.E., 2003. Analysing the optimal value for titanium implant roughness in bone attachment using a tensile test. Biomaterials 24, 4559–4564.

Rusu, V.M., Ng, C.H., Wilke, M., Tiersch, B., Fratzl, P., Peter, M.G., 2005. Size-controlled hydroxyapatite nanoparticles as self-organized organic-inorganic composite materials. Biomaterials 26, 5414–5426.

Ryan, G., Pandit, A., Apatsidis, D.P., 2006. Fabrication methods of porous metals for use in orthopaedic applications. Biomaterials 27, 2651–2670.

Schliephake, H., Scharnweber, D., 2008. Chemical and biological functionalization of titanium for dental implants. J. Mater. Chem. 18, 2404–2414.

Schnepp, Z.A.C., Gonzalez McQuire, R., Mann, S., 2006. Hybrid biocomposites based on calcium phosphate mineralization of self-assembled supramolecular hydrogels. Adv. Mater. 18, 1869.

Shkilnyy, A., Friedrich, A., Tiersch, B., Schöne, S., Fechner, M., Koetz, J., et al., 2008. Poly(ethylene imine)-controlled calcium phosphate mineralization. Langmuir 24, 2102–2109.

Shkilnyy, A., Schöne, S., Rumplasch, C., Uhlmann, A., Hedderich, A., Guenter, C., et al., 2011. Calcium phosphate mineralization with linear poly(ethylene imine): a time resolved study. Colloid Polym. Sci. 289, 881–888.

Shrikhanzadeh, M., 1991. Bioactive calcium phosphate coating prepared by electrodeposition. J. Mater. Sci. Lett. 10, 1415–1417.

Spicer, P.P., Kretlow, J.D., Young, S., Jansen, J.A., Kasper, F.K., Mikos, A.G., 2012. Evaluation of bone regeneration using the rat critical size calvarial defect. Nat. Protoc. 7, 1918.

Suetsugu, Y., Walsh, D., Tanaka, J., Mann, S., 2009. Hydroxyapatite pattern formation in PVA gels. J. Mater. Sci. 44, 5806–5814.

Takadama, H., Kim, H.M., Kokubo, T., Nakamura, T., 2001. TEM-EDX study of mechanism of bonelike apatite formation on bioactive titanium metal in simulated body fluid. J. Biomed. Mater. Res. 57, 441–448.

Tejada-Montes, E., Klymov, A., Nejadnik, M., Alonso, M., Rodriguez-Cabello, J.C., Walboomers, X., Mata, A., 2014. Mineralization and bone regeneration using a bioactive elastin-like recombinamer membrane. Biomaterials 35, 8339–8347.

Termine, J.D., Eanes, E.D., Greenfield, D.J., Nylen, M.U., Harper, R.A., 1973. Hydrazine-deproteinated bone mineral. Physical and chemical properties. Calcif. Tissue Res. 12, 73–90.

Thachepan, S., Li, M., Mann, S., 2010. Mesoscale crystallization of calcium phosphate nanostructures in protein (casein) micelles. Nanoscale 2, 2400–2405.

Thomsen, P., et al., 1997. Bone response to machined cast titanium implants. J. Mater. Sci. Mater. Med. 8, 653–665.

Thula, T.T., Svedlund, F., Rodriguez, D.E., Poschun, J., Pendi, L., Grower, L.P., 2011. Mimicking the nanostructure of bone: composition of polymeric process-directing agents. Polymers (Basel) 3, 10–35.

van Steenberghe, D., et al., 2002. The relative impact of local and endogenous patient-related factors on implant failure up to the abutment stage. Clin. Oral Implants Res. 13, 617–622.

Vasconcellos, L.M.R., Leite, D.O., Oliveira, F.N., et al., 2010. Novel production method of porous surface Ti samples for biomedical application. Oral Res. 24, 399–405.

Venugopal, J., Prabhakaran, M.P., Zhang, Y.Z., Low, S., Choon, A.T., Ramakrishna, S., 2010. Biomimetic hydroxyapatite-containing composite nanofibers substrates for bone tissue engineering. Philos. Trans. R. Soc. A Math. Phys. Eng. Sci. 368, 2065–2081.

Williams, D.F., 2001. Titanium for medical applications. In: Brunette, D.M., Tengvall, P., Textor, M., Thomsen, P. (Eds.), Titanium in Medicine: Material Science, Surface Science, Engineering, Biological Responses and Medical Applications. Springer-Verlag, Berlin/Heidelberg, pp. 13–24.

Wu, Y.Q., Clark, R.L., 2008. Electrohydrodynamic atomization: a versatile process for preparing materials for biomedical applications. J. Biomater. Sci. Polym. Ed. 19, 573–601.

Xie, B., Nancollas, G.H., 2010. How to control the size and morphology of apatite nanocrystals in bone. Proc. Natl. Acad. Sci. U.S.A. 107, 22369–22370.

Yang, X., Wang, L., Qin, Y., Sun, Z., Henneman, Z.J., Moradian-Oldak, J., et al., 2010. How amelogenin orchestrates the organization of hierarchical elongated microstructures of apatite. J. Phys. Chem. B 114, 2293–2300.

Yang, X., Xie, B., Wang, L., Qin, Y., Henneman, Z., Nancollas, G.H., 2011. Influence of magnesium ions and amino acids on the nucleation and growth hydroxyapatite. CrystEngComm 13, 1153–1158.

Zhang, X., Li, Z., Zhu, X.X., 2008. Biomimetic mineralization induced by fibrils of polymers derived from a bile acid. Biomacromolecules 9, 2309–2314.

Biomimetic mineralization of hydrogels

9

Timothy E.L. Douglas
Ghent University, Ghent, Belgium

9.1 Introduction

9.1.1 Hydrogel biomaterials

Hydrogels, or highly hydrated polymer networks, have been applied in the areas of drug delivery and tissue engineering. Hydrogels are formed when a polymer solution undergoes a sol–gel transition to form a crosslinked polymer network. These crosslinks can be formed by covalent bonding or hydrophobic and/or ionic interactions between polymer chains.

These materials are thus injectable if they are injected into an implant site as a sol and the gelation process takes place after injection. Injectability is a minimally invasive implantation method and causes less scarring and patient discomfort and shorter hospitalization times than the more invasive surgical procedures often required for implantation of noninjectable materials.

The sol–gel transition can also be exploited to achieve superior fitting of the material to a defect site. Defect sites are generally irregular in shape, but an injected sol can assume the shape of the defect and then gel without any gap with surrounding tissue.

Furthermore, it is straightforward to add cells and soluble biomolecules which support tissue regeneration, such as drugs and growth factors, to the sol before gelation. After gelation, cells and biomolecules remain encapsulated in the crosslinked polymer network. It is relatively easy to distribute cells and biomolecules homogeneously in the sol. As a consequence, it is also relatively straightforward to attain homogeneous distribution of cells and biomolecules in the hydrogel.

Hydrogels consist predominantly of water, which facilitates exchange of dissolved substances with the surrounding environment by diffusion. This is advantageous for the maintenance of viability of cells incorporated in the hydrogel, as dissolved oxygen, nutrients, and biochemical cues can reach the cells by diffusion. Furthermore, toxic and undesirable products of cell metabolism can exit the hydrogel by diffusion. Biochemical signals produced by the cells incorporated in the hydrogel polymer network can also diffuse out and influence behavior of cells in the surrounding environment. The rate of diffusion into and out of the hydrogel polymer network is influenced by several factors. One factor is the ease by which a biomolecule diffuses through the pore network. Diffusion is hindered by decreasing the pore size, for example as a result of increasing the concentration of polymer and/or the degree of crosslinking between polymer chains. Diffusion of biomolecules is decelerated by steric hindrance, which increases with increasing biomolecule molecular weight and the bulkiness of

Biomineralization and Biomaterials. http://dx.doi.org/10.1016/B978-1-78242-338-6.00010-7

the three-dimensional (3D) conformation of the biomolecule. Diffusion can also be hampered by interactions (e.g. ionic and hydrophobic) between the diffusing substance and the polymer chains of the hydrogel network.

Since the main component of hydrogels is water, they exhibit low mechanical strength compared to solid polymeric, ceramic, and metallic biomaterials. Hence, they are non-load-bearing. Furthermore, cells entrapped in hydrogels exert tensile forces on the polymer network, which may lead to compaction of the network and shrinkage of the hydrogel.

Hydrogels have been produced from polymers of natural origin, e.g. chitosan, gelatin, fibrin, collagen, alginate, and gellan gum, as well as from polymers of synthetic origin, e.g. poly(ethylene glycol) (PEG) and poly(vinyl alcohol) (PVA). Natural polymers contain binding sites for cells and thus support cellular adhesion. In general, they are also biodegradable, which facilitates their replacement with newly formed tissue. However, they show batch-to-batch variation, which hinders reproducibility of results. Furthermore, in the case of animal-derived polymers, there are concerns related to possible disease transmission. Synthetic polymers are often not degradable. The majority do not contain binding sites for cells, which need to be incorporated by chemical means. On the other hand, synthetic polymers show less batch-to-batch variation than natural polymers, which improves the reproducibility of results.

9.1.2 Hydrogel–inorganic composite materials for bone tissue regeneration

Within the field of tissue engineering, hydrogels have predominantly been used as biomaterials to aid the regeneration of cartilage, intervertebral disks, and soft tissues. As a recent review demonstrates (Gkioni et al., 2010), composites consisting of a hydrogel phase and an inorganic phase are coming under increasing investigation as biomaterials for bone regeneration.

The most popular inorganic materials currently applied as biomaterials for bone tissue regeneration include various forms of calcium phosphate (CaP), including different types of apatite, the form of CaP found in natural bone, as well as tricalcium phosphates (TCPs) and biphasic apatite–TCP composites. CaP materials are generally produced in block or granulate form. Calcium carbonate ($CaCO_3$) materials derived from coral consist of the $CaCO_3$ polymorph aragonite and have also been used as bone regeneration materials, and have outperformed CaP materials in some cases (Viateau et al., 2013). Bioactive glasses are another popular class of inorganic materials.

The main strength of the aforementioned pure inorganic materials is their bioactivity, in other words their ability to form a direct bond with surrounding bone tissue (LeGeros, 1991). However, these pure inorganic materials also display disadvantages. They are brittle and not easy to handle in granulate form. When implanted, they do not optimally fit defect sites, and certain types of CaP such as hydroxyapatite (HA) are poorly resorbable.

The development of hydrogel–inorganic composites aims to combine the positive advantages of both hydrogels and inorganic materials. Bone itself can be considered as a mineralized hydrogel made up of a hydrogel organic phase consisting mainly of collagen in the form of fibrils, which is reinforced by an inorganic phase consisting predominantly of calcium-deficient hydroxyapatite. The production of a hydrogel–inorganic composite by artificial mechanisms mimics the formation, structure, and function of natural bone tissue to a certain extent and hence the general approach can be considered to be biomimetic.

The enrichment of a hydrogel with an inorganic component is seen as advantageous for bone regeneration applications for several reasons. One reason is increased bioactivity, as mentioned above. Another reason is mechanical reinforcement. Stiffer substrates are known to promote osteogenic differentiation of bone-forming cells (Engler et al., 2006; Evans et al., 2009; Rowlands et al., 2008). It can also be speculated that the presence of deposits of an inorganic phase in hydrogels would lead to increased surface roughness, although there is a lack of evidence in the literature confirming this hypothesis. Increased surface roughness has been reported to stimulate osteogenic differentiation of bone-forming cells (Boyan et al., 2003). In addition, CaP has been shown to exhibit an affinity for growth factors which stimulate bone regeneration (Ruhe et al., 2005). Hence mineralization of hydrogels may increase their ability to bind such growth factors. Lastly, the presence of an inorganic component in a hydrogel has been reported to enhance the adhesion, proliferation, and osteogenic differentiation of bone-forming cells and formation of new mineral within the hydrogel by cells (Douglas et al., 2014a; Bongio et al., 2011; Patel et al., 2010; Phadke et al., 2012).

Regarding the type of inorganic phase used to enrich hydrogels, most attention has been focused on incorporation of CaP or bioactive glasses (Douglas et al., 2013b; Gkioni et al., 2010). However, some authors have concentrated on hydrogel enrichment with $CaCO_3$. Certain marine invertebrates mineralize their exoskeletons with $CaCO_3$. $CaCO_3$ can occur as amorphous calcium carbonate and also as three polymorphs, or crystalline forms, namely: (i) calcite, the thermodynamically most stable polymorph, which is found in some sponges and crustaceans; (ii) aragonite, which occurs in corals and mollusks; and (iii) vaterite, the thermodynamically least stable polymorph (Sommerdijk and de With, 2008; Muller et al., 2013; Morse et al., 2007). Calcareous sponges, whose exoskeletons are mineralized with calcite, first appeared in the Cambrian period (540–485 million years ago) (Wood et al., 1992). The first bony fishes appeared in the late Silurian period (approximately 420 million years ago). Hence, from an evolutionary point of view, mineralization with $CaCO_3$ represents biomimicry of a far older process than mineralization with CaP. Other authors have concentrated on mineralization of hydrogels with silica. Silica-based materials have been widely applied as materials for bone regeneration (for a review, see Arcos and Vallet-Regi, 2010). Silica biomineralization is found in diatoms and certain marine sponges (for a review, see Andre et al., 2012) and is also a much older process than mineralization with CaP.

9.2 Overview of biomimetic strategies to promote hydrogel mineralization

Numerous biomimetic strategies to promote hydrogels have been described in the literature. Roughly speaking, the strategies to mineralize hydrogels with CaP, $CaCO_3$, and silica can be subdivided into two general categories. In the first category,

Table 9.1 Overview of biomimetic strategies to mineralize hydrogels

Hydrogel mineralization strategy	
1	Incorporation of preformed inorganic particles
2	Promotion of precipitation of the inorganic phase from constituent ions by: (a) delivery and reaction of ions to form the inorganic phase during hydrogel formation (b) enhanced binding of ions by hydrogel functionalization with biomolecules or chemical groups with a high affinity for such ions (c) external delivery of ions by incubation in ion-rich mineralization solutions (d) increasing intra-hydrogel ion concentrations

methods are involving preformation of inorganic particles, which are then added to the sol before gelation and entrapped in the hydrogel after gelation. In the second category, methods are where precipitation of the inorganic phase from its constituent ions is promoted, either during or after gelation. This second category can be subdivided into further subcategories, namely: (a) delivery and reaction of ions during gelation, leading to simultaneous formation of the hydrogel and the inorganic phase; (b) functionalization of hydrogels with biomolecules and/or chemical groups with a high affinity for the ions, leading to ion binding and an increase in local ion concentration, leading in turn to precipitation; (c) external delivery of an excess of ions, e.g. by pre-, post-, or alternate soaking of hydrogels in highly concentrated ion solutions; and (d) increasing intra-hydrogel ion concentrations, e.g. through the action of cells and enzymes.

An overview of the aforementioned categories and subcategories is found in Table 9.1. It should be stressed that some authors have used more than one strategy simultaneously. All strategies can be applied to mineralize hydrogels with CaP, $CaCO_3$, or silica.

Hydrogel mineralization with $CaCO_3$ and silica has been studied much less than hydrogel mineralization with CaP. There are publications on mineralization of hydrogels with CaP in the literature in all of the categories and subcategories in Table 9.1. In contrast, publications on mineralization of hydrogels with $CaCO_3$ and silica are not found in all of these categories and subcategories.

9.3 Biomimetic mineralization with CaP

9.3.1 Incorporation of preformed CaP particles

One of the most straightforward and commonly used methods to enrich hydrogels with CaP is simple addition of preformed CaP micro- or nanoparticles. Apatite particles dispersed in oligo(poly(ethylene glycol) fumarate) (OPF) and alginate hydrogels have served as nucleation sites for further CaP mineral growth (Nejadnik et al., 2012; Cardoso et al., 2014) and have mechanically reinforced a range of chemically cross-linked hydrogels (Sanginario et al., 2006; Sarvestani et al., 2007; Tolga Demirtas et al., 2008; Paxton et al., 2009; Gyawali et al., 2013; Lin et al., 2012; Song et al., 2009; Bongio et al., 2011; Anderson et al., 2011). They have also been added to alginate solutions together with glycerol and the acidifier glucono-delta-lactone (GDL) (Cardoso et al., 2014). Slow release of calcium ions from the apatite due to the decrease in pH caused by GDL leads to crosslinking of alginate and formation of a homogeneous alginate hydrogel.

Hydrogel enrichment with CaP particles has also positively influenced behavior of bone-forming cells *in vitro* and in bone healing *in vivo*. Incorporation of apatite particles in OPF hydrogels containing rat bone marrow-derived mesenchymal stem cells (MSCs) led to higher cell numbers after 8, 12, 16, and 24 days, and increased cell-mediated mineralization after 16 and 24 days (Bongio et al., 2011). Incorporation of nanograms of apatite particles into cyclic acetal hydrogels synthesized from poly(ethylene glycol) diacrylate (PEGDA) and 5-ethyl-5-(hydroxymethyl)-b,b-di-methyl-1,3-dioxane-2-ethanol diacrylate loaded with rat bone marrow-derived MSCs led to significantly increased gene expression of osteogenic differentiation markers after 1, 4, and 8 days of culture (Patel et al., 2010). Addition of apatite particles to peptide amphiphile hydrogels containing RGDS (Arg–Gly–Asp–Ser) ligands containing human bone marrow MSCs promoted gene expression of osteogenic markers after 14 days (Anderson et al., 2011). Implantation of the apatite-enriched hydrogels into critical-size rat femoral defects showed that they supported formation of new bone to a larger extent, resulting in defect bridging and closure. Addition of apatite nano- and microparticles to aldehyde-modified hyaluronic acid hydrogels containing bone morphogenetic protein 2 supported bone formation upon intra-muscular implantation in rat quadriceps muscle (Hulsart-Billstrom et al., 2011). The density of bone was dependent on particle size. Nanoparticles proved more effective in supporting denser bone formation than microparticles.

An alternative to direct incorporation of preformed apatite particles is incorporation of other inorganic particles which serve as templates for apatite growth in a suitable mineralization solution, e.g. simulated body fluid (SBF), which contains mineral ions at concentrations similar to those found in blood plasma. Addition of bioglass particles to collagen hydrogels and chitosan/β-glycerophosphate hydrogels resulted in apatite formation after incubation in SBF (Eglin et al., 2006; Couto et al., 2009; Marelli et al., 2011).

It has been reported that preformed CaP particles have aggregated when added to hydrogels, resulting in a nonhomogeneous distribution. Leeuwenburgh et al. (2010) alleviated this problem by dispersing particles with citrate, leading to superior distribution of apatite nanoparticles in gelatin hydrogels.

9.3.2 Enhanced binding of ions by hydrogel functionalization with biomolecules or chemical groups

A wide range of biomacromolecules have been incorporated into hydrogels to induce mineralization (for a review, see Silverman and Boskey, 2004). In the majority of these studies, the hydrogel serves as a matrix for immobilization of the biomolecules with the aim of testing the ability of the biomolecule to support mineralization. In most studies, the aim is the biomimicry of the function of the noncollagenous, mineralization-promoting biomolecules occurring in natural bone and enamel. These direct mineralization of the predominantly collagenous extracullar matrix (for reviews, see George and Veis, 2008; Nudelman et al., 2013). Hence, hydrogels have mostly been functionalized with mineral-nucleating proteins, or the anionic glutamate, aspartate, and phosphoserine groups which these proteins contain. One such mineral-nucleating bone protein is bone sialoprotein (BSP), which has been added to collagen and agarose hydrogels (Hunter et al., 2001; Baht et al., 2008). BSP is rich in glutamate anionic groups, and peptide sequences rich in glutamic acid have an affinity for apatite and have promoted CaP formation in hydrogels.

Besides BSP, biopolymers which have been particularly successful in supporting mineralization with CaP are alginate, which has been incorporated into gelatin hydrogels (Stancu et al., 2011) and grafted onto PEG-monomethacrylate and poly(propylene glycol)-monomethacrylate hydrogels (Cha et al., 2011), and casein, which has promoted mineralization of polyhydroxyethylmethacrylate (polyHEMA) hydrogels upon incubation in SBF (Dragusin et al., 2011).

Artificial proteins containing motifs from nacrein have been added to polyHEMA hydrogels to enhance CaP formation (Chirila et al., 2009). Dense collagen hydrogels produced by collagen hydrogel gelation followed by plastic compression to drive out fluid have shown higher mineralizability after incorporation of peptides derived from silk fibroin (Marelli et al., 2012). Silk fibroin has also promoted CaP formation in poly(acrylamide) (PAA) hydrogels (Zaharia et al., 2012). Increasing fibroin content led to increasing mineralization after incubation in SBF.

Several studies have focused on hydrogel functionalization with calcium-binding anionic groups. Song et al. (2005) methacrylated the anionic amino acids glycine, glutamic acid, serine and aspartic acid, and incorporated them into polyHEMA hydrogels by copolymerization, resulting in covalent binding. Hydrogels were incubated in an acidic solution of HA containing urea and urease. The digestion of urea by urease led to an increase in pH, which in turn led to a reduction in HA solubility. HA precipitated on the surfaces of the hydrogels was functionalized with the aforementioned amino acids. In another study, high formation of apatite was observed on crosslinked polyglutamic acid (PGA) hydrogels upon incubation in SBF (Miyazaki et al., 2013). The authors ascribed this high mineralizability to the high number of carboxyl groups, especially as mineralizability rose with increasing PGA molecular weight due to the higher availability of carboxyl groups after crosslinking, in the opinion of the authors. Pre-soaking of PGA hydrogels in $CaCl_2$ before incubation in SBF led to increased formation of apatite instead of amorphous CaP (Sugino et al., 2008).

To close this section, it should be mentioned that recombinant protein technology has been applied to develop biomolecules with high mineralizability with CaP. Examples are elastin-like recombinamers containing fragments of the salivary protein statherin, which show a high affinity for HA (Prieto et al., 2011), as well as recombinant mouse amelogenins, which are believed to play a role in enamel crystal growth (Beniash et al., 2005), and recombinant amelotin, which is a recently discovered protein expressed during the maturation stage of enamel (Abbarin et al., 2015). The promotion of hydrogel mineralization by incorporation of such recombinant proteins appears to be a promising strategy which has remained hitherto relatively unexplored.

9.3.3 Delivery of calcium and phosphate ions

Strategies to promote mineralization with CaP by enhanced delivery of calcium and phosphate ions from the environment can be roughly subdivided into the following categories: incubation in physiological mineralization solutions, pre-soaking, alternate soaking and post-soaking, single diffusion, double diffusion, assisted diffusion, and simultaneous apatite and hydrogel formation.

The most popular physiological mineralization solution used in the literature is SBF, which is generally considered to simulate the *in vivo* mineralization behavior of a biomaterial. Generally speaking, apatite forms on the immersed biomaterial. Indeed, apatite has formed in several hydrogel types after immersion in SBF (Ichibouji et al., 2009; Shkilnyy et al., 2009; Chesnutt et al., 2009; Ho et al., 2007). It should be noted that dried hydrogels have also been reswelled and simultaneously mineralized by incubation in SBF (Phadke et al., 2012).

Certain groups have combined pre-soaking of hydrogels in calcium solutions with incubation in SBF to increase apatite formation in PGA hydrogels (Koh et al., 2011; Sugino et al., 2008). The binding of calcium ions during the pre-soaking step increased the number of nucleation sites for further CaP growth.

Other groups have mineralized hydrogels with CaP by alternate soaking, whereby a hydrogel is immersed in a solution containing one type of ion (calcium or phosphate) and subsequently immersed in a solution of the other type of ion. Upon repetition of this cycle the amount of CaP formed increases. CaP has been formed in cellulose, PVA, and chitosan hydrogels by alternate soaking (Taguchi et al., 1999a,b; Madhumathi et al., 2009; Hutchens et al., 2006). Agarose hydrogels mineralized with apatite by an alternative soaking method have been implanted into rat cranial defects, leading to promotion of new bone formation and defect closure 8 weeks post-implantation due to apatite enrichment (Suzawa et al., 2010).

It is also possible to alter the type of CaP formed in a hydrogel by post-treatment. Chitosan hydrogels mineralized with amorphous CaP by alternate soaking were post-soaked in an alkali solution, namely 5% (w/v) NaOH, resulting in transformation of amorphous CaP into apatite (Li et al., 2009a, 2011). It is also possible to convert the CaP phase in a hydrogel by thermal post-treatment. Particles of α- and β-TCP dispersed in PVA hydrogels were converted into apatite after thermal treatment with superheated water vapor at 160 °C (Goto et al., 2012).

Yokoi et al. (2010a,b) formed PAA hydrogels which were preloaded with phosphate ions, which were subsequently brought into contact with $Ca(NO_3)_2$ solution. It was found that the type of CaP formed was dependent on the phosphate and calcium ion concentrations. At higher phosphate ion concentrations, HA was formed at the interface between the hydrogel and mineralization solution, while octacalcium phosphate was formed at lower phosphate concentrations.

In contrast to the single-diffusion system mentioned above, where one type of ion diffuses from a reservoir into a hydrogel, some authors have employed double-diffusion systems (for a review, see Dorvee et al., 2012), where two different ions enter the hydrogel network by diffusion from two separate reservoirs, leading to CaP formation within the hydrogel. By placing a hydrogel between reservoirs of calcium and phosphate ions, brushite was formed in hydrogels of gelatin (Azami et al., 2012). The authors later converted this brushite to apatite by soaking in SBF. Techniques similar to electrophoresis involving the use of electric current to assist diffusion have been applied to mineralize hydrogels of agarose and PAA with apatite (Watanabe and Akashi, 2008; Liu et al., 2009; Kamitakahara et al., 2012).

The inorganic reagents for CaP formation can be added during hydrogel formation, and react during the gelation process to form CaP particles within the hydrogel. This approach, called "chemical mixing" by Leeuwenburgh et al. (2007), was reported to result in more uniform distribution of apatite particles in OPF hydrogels than by simple addition of particles. CaP in the form of apatite particles has been generated in several hydrogel types using "chemical mixing" (Rajkumar et al., 2011; Xie et al., 2010; Chen et al., 2010; Luo et al., 2011).

9.3.4 Elevation of calcium and phosphate concentrations through enzymatic or cellular action

The fact that hydrogels consist of a polymer network formed from a polymer solution facilitates straightforward incorporation of mineralization-promoting enzymes. Enzymes are added to the polymer solution prior to gelation and entrapped within the hydrogel polymer network as a result of gelation.

The enzyme most commonly used to promote mineralization of hydrogels with CaP is alkaline phosphatase (ALP). ALP is an enzyme involved in mineralization of bone tissue *in vivo*, where it splits phosphate groups from organophosphate, which are then free to react with calcium ions, forming insoluble CaP.

ALP has been immobilized in hydrogels by simple physical entrapment, making the use of potentially toxic crosslinkers unnecessary. Depending on the hydrogel used, release of total ALP has ranged from 30% (w/v) to negligible amounts (Douglas et al., 2012b, 2014a). ALP has also been encapsulated in sodium alginate/hydroxypropylcellulose microparticles (Karewicz et al., 2014).

ALP-loaded hydrogels have been mineralized with CaP by incubation of the hydrogel in a solution of calcium ions and glycerophosphate, an organophosphate which is a substrate for ALP (Douglas et al., 2012a,b, 2013a, 2014a,b; Filmon et al., 2000; Spoerke et al., 2009; Gassling et al., 2013). Ions and glycerophosphate enter the hydrogel by diffusion. Insoluble CaP forms and precipitates and remains entrapped in the hydrogel. Reaction by-products, including glycerol, the organic part of glycerophosphate, diffuse out of the hydrogel.

The type of CaP formed in hydrogels depends greatly on the type of hydrogel, composition of mineralization medium (especially glycerophosphate and calcium concentrations), and ALP concentration. Some authors have used superphysiological concentrations of glycerophosphate, calcium, and ALP to maximize and accelerate mineral formation (Douglas et al., 2012a,b, 2013a, 2014a; Olderoy et al., 2012). At superphysiological concentrations (calcium and glycerophosphate: 0.1 M; ALP: 2.5 mg/ml), CaP but not apatite was detected in platelet-rich fibrin (PRF) containing entrapped ALP (Douglas et al., 2012a; Gassling et al., 2013). Apatite formed predominantly in ALP-loaded catechol/poly(ethylene glycol) (cPEG) and gellan gum hydrogels (Douglas et al., 2012b, 2014a,b), while a mixture of amorphous CaP and apatite formed in collagen, OPF, and chitosan hydrogels (Douglas et al., 2012b, 2013a). Images of mineral formed by enzymatic mineralization of PRF can be seen in Figure 9.1. Images of mineral formed by the action of ALP in gellan gum and cPEG hydrogels are displayed in Figure 9.2.

Other authors have focused on closely mimicking *in vivo* mineralization conditions and have chosen concentrations far closer to physiological values (Filmon et al., 2000; Gungormus et al., 2010; Spoerke et al., 2009). Apatite formed predominantly in poly-HEMA and hydrogels based on amphiphile polymers (Filmon et al., 2000; Spoerke et al., 2009). Two hydrogel types based on fibrils formed from artificial peptides differed in their ability to promote apatite formation (Gungormus et al., 2010).

Other metal ions, such as magnesium, have been incorporated into CaP formed in hydrogels by enzymatic mineralization via addition of magnesium ions to the mineralization medium (Douglas et al., 2014; Gassling et al., 2013). The type of CaP formed depends on the concentration of magnesium in the medium. At low concentrations of magnesium, apatite was formed in gellan gum hydrogels; however, the higher the concentration of magnesium, the more the formation of amorphous CaP was favored (Douglas et al., 2014). The presence of magnesium in both apatite and amorphous CaP promoted adhesion and proliferation of osteoblast-like MC3T3-E1 cells.

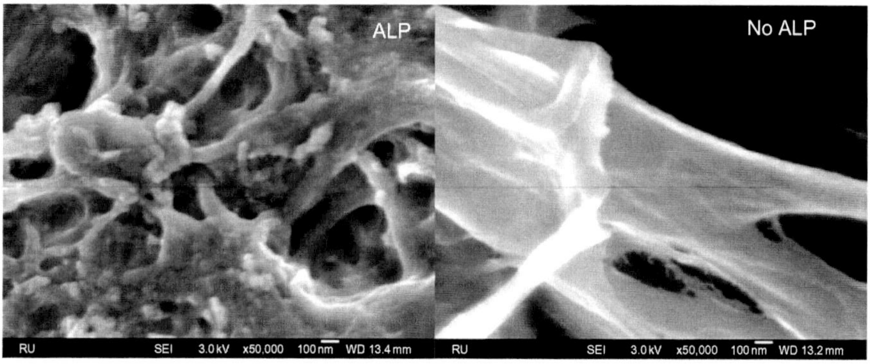

Figure 9.1 Enzymatic mineralization of platelet-rich-fibrin (PRF) using alkaline phosphatase (ALP). ALP was incorporated into patient blood at a concentration of 2 mg/ml and blood was centrifuged immediately to produce fibrin clots, which were subsequently compressed to form PRF membranes. PRF membranes were incubated in 0.1 M calcium glycerophosphate for 3 days, rinsed in double-distilled water for 24 h, and freeze–dried. SEM images of PRF membranes containing ALP (left) and without ALP (right) are shown. Mineral deposits are clearly visible in the PRF membrane containing ALP and clearly absent in the PRF membrane without ALP. Scale bar = 10 nm in both images.

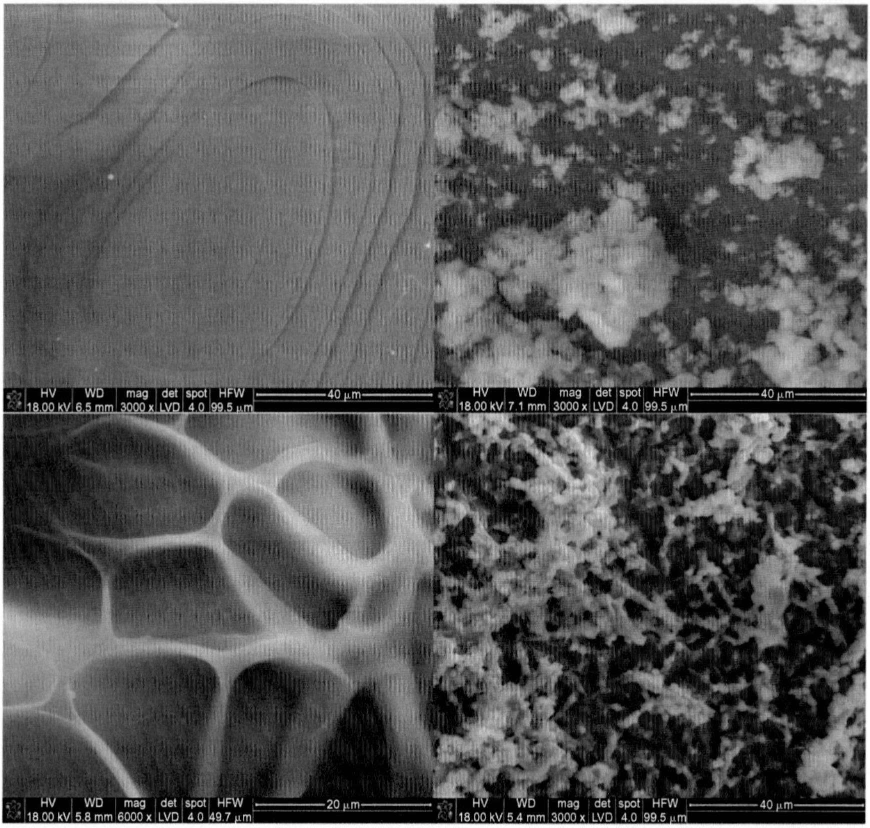

Figure 9.2 Enzymatic mineralization of gellan gum and catechol/poly(ethylene glycol) (cPEG) hydrogels using alkaline phosphatase (ALP). ALP was incorporated into both hydrogels at a concentration of 2.5 mg/ml. Hydrogels were incubated in 0.1 M calcium glycerophosphate for 7 days (gellan gum) or 6 days (cPEG), rinsed in double-distilled water for 24 h, and freeze–dried. SEM images are shown. Top row: gellan gum; bottom row: cPEG. Left column: without ALP; and right column: with ALP. Mineral deposits are clearly visible in the hydrogels containing ALP and clearly absent in hydrogels without ALP. Scale bar = 40 μm in both images.

Another enzyme which has been used to induce hydrogel mineralization with CaP is urease. Nijhuis et al. (2014) developed a system to mineralize a variety of biomaterials, including OPF hydrogels and OPF hydrogels functionalized with biphosphonates. Hydrogels were incubated in mineralization solutions containing $NaHPO_4$, $CaCl_2$, and NaCl at 37 °C and pH 6.0 enriched with urea. Addition of urease resulted in the formation of ammonia through decomposition of urea, leading to raised pH and precipitation, and deposition of CaP. It was found that hydrogel opacity and uptake of calcium was markedly higher for biphosphonate-functionalized hydrogels, demonstrating superior mineralization.

Regarding the incorporation of cells in hydrogels to generate CaP in hydrogels, poly(lactic-co-glycolic acid) (PLGA) microparticles were added to collagen hydrogels, leading to a rise in stiffness, leading in turn to increased mineral formation by mouse-derived MSCs, which was demonstrated by histological analysis and von Kossa staining for phosphate deposits (DeVolder et al., 2012). Incorporation of dental pulp cells in dense collagen hydrogels produced by a plastic compression method led to production of apatite, as evidenced by infrared spectroscopy, scanning electron microscopy (SEM), transmission electron microscopy (TEM), and TEM-assisted selected area electron diffraction (Coyac et al., 2013).

9.4 Biomimetic mineralization with $CaCO_3$

9.4.1 Incorporation of preformed $CaCO_3$ particles

There are far fewer publications on hydrogel enrichment with preformed $CaCO_3$ particles than with preformed CaP particles. Furthermore, the majority of articles dealing with addition of $CaCO_3$ particles focus on their incorporation into alginate solutions in combination with the acidifier GDL to serve as a slow release system for calcium ions to crosslink alginate and induce alginate hydrogel formation (Kuo and Ma, 2001, 2008; Jang et al., 2014; Bidarra et al., 2011; Fonseca et al., 2011, 2014; Alexander et al., 2012). This process has been referred to as "internal gelation" (Chan et al., 2002, 2006). Kuo and Ma (2001) demonstrated cytocompatibility for osteoblast-like MC3T3-E1 cells, while Jang et al. showed that the concentration of $CaCO_3$ particles used affected proliferation, production of extracellular matrix, and gene expression of chondrogenic differentiation markers by primary human chondrocytes encapsulated in the hydrogels. Alginate hydrogels formed by internal gelation have also served as carriers for endothelial cells and MSCs (Bidarra et al., 2011; Fonseca et al., 2011, 2014). Moreira et al. (2014) used $CaCO_3$ particles to neutralize acidic solutions of pectin to form hydrogels by internal gelation. This approach avoided the use of acidifiers such as GDL. Eluates from the resulting hydrogels were cytocompatible for L-929 fibroblasts.

The aforementioned publications did not aim specifically to produce materials for bone regeneration. However, Flausse et al. (2013) added nacre powder to alginate bead hydrogels containing human MSCs from bone marrow, resulting in enhanced osteogenic differentiation. Nacre consists of aragonite crystals embedded in an organic matrix (Corni et al., 2012). Nacre in powdered form has stimulated bone formation *in vivo* (Lamghari et al., 1999, 2001a,b; Liao et al., 2000) and organic components of nacre have stimulated mineral formation by osteoblast-like MC3T3-E1 cells (Rousseau et al., 2008). Coral scaffolds consisting of aragonite have supported growth and osteogenic differentiation of osteoblasts and MSCs (Tran et al., 2011; Foo et al., 2008; Abramovitch-Gottlib et al., 2006).

9.4.2 Delivery of calcium and carbonate ions during hydrogel formation

Xie et al. (2010) formed $CaCO_3$ in alginate hydrogels by addition of carbonate ions to alginate solution, which was added dropwise to $CaCl_2$. $CaCl_2$ caused crosslinking of alginate chains to enable hydrogel formation and provided calcium ions for $CaCO_3$ formation.

9.4.3 External delivery of calcium and carbonate ions by incubation in mineralization solutions

Some groups have incubated hydrogels in mineralization solutions to induce formation of $CaCO_3$. Generally speaking, four approaches have been reported, namely alternate soaking of hydrogels in solutions of calcium and carbonate ions, incubation in solutions of calcium and bicarbonate ions, diffusion of CO_2 into hydrogels containing calcium ions, and double-diffusion systems.

Ogomi et al. (2003) developed an alternate soaking approach to mineralize materials with $CaCO_3$ by immersion in solutions of 200 mM $CaCl_2$ and 200 mM Na_2CO_3. The same group later used this approach to mineralize PVA hydrogels with a layer of $CaCO_3$ which consisted of a mixture of calcite and aragonite, and served as a delivery vehicle for model drugs (Ogomi et al., 2005).

Suzawa et al. (2010) also used an alternate soaking mechanism involving 12 cycles of alternate immersion in 200 mM $CaCl_2$ and 200 mM Na_2CO_3 to mineralize agarose hydrogels with $CaCO_3$ in the form of a mixture of calcite and vaterite. Implantation of the agarose composites in rat calvarial defects led to superior bone regeneration after 8 weeks in comparison to pure agarose hydrogels.

Kitano (1962) developed a mineralization solution consisting of calcium bicarbonate saturated with carbon dioxide, which has been used to induce mineralization of biomolecules and biomaterials with $CaCO_3$ (Neira-Carrillo et al., 2008; Liang et al., 2004). Munro and McGrath (2011) prepared chitosan hydrogels which were subsequently mineralized with $CaCO_3$ in the form of calcite by combining alternate soaking in solutions of $CaCO_3$ precursors with incubation in Kitano solution. This combined approach increased the extent of mineralization and induced uniform mineralization throughout the hydrogel. The authors reported that higher concentrations of $CaCO_3$ precursors (680 mM) in pre-soaking solutions led to more nucleation events, more overall crystal growth, and more well-defined crystals in comparison to low precursor concentrations (10 mM). The authors attributed the differences in crystal morphology at high ionic strength to disruption of the polymer network. The authors also studied the effect of static versus dynamic incubation conditions (a rocking table was used to gently displace the mineralization solution to ensure refreshment of the solution). At high concentrations of $CaCO_3$ precursors, dynamic incubation conditions led to higher mineral formation. Addition of the biomolecule PAA to the Kitano solution resulted in changes in the direction of growth in calcite deposits. The same group also mineralized chitosan hydrogels with $CaCO_3$ containing Mg, or magnesian calcite, by using a mineralization medium similar to Kitano solution containing Mg ions (Munro and McGrath, 2011).

The same group also later converted chitosan hydrogels to chitin structures by treatment with acetic anhydride (Munro and McGrath, 2012). Mineralization by the combined mineralization approach involving alternate soaking in $CaCO_3$ precursor solutions and subsequent incubation in Kitano solution led to formation of calcite on both chitosan and chitin. However, addition of PAA to the mineralization solutions led to predominantly aragonite formation on chitin and formation of approximately equal amounts of calcite and vaterite on chitosan. In later work, the same group added not only PAA but also chitosan oligomers to the mineralization solutions, resulting in materials with a nacre-like structure (Munro et al., 2013).

Li and Estroff (2007) and Li et al. (2009b) used a method whereby CO_2 gas released by desiccation of ammonium carbonate diffused into agarose hydrogels containing calcium ions, resulting in formation of $CaCO_3$ inside the hydrogel. A carboxylate-terminated self-assembled monolayer of a gold-coated silicon wafer was incorporated into the hydrogels to direct crystal growth.

Grassmann and Lobmann (2003) used a double-diffusion method to induce mineralization of acrylamide- and sulfonic acid-based hydrogels. Hydrogels were placed between reservoirs of aqueous solutions of buffered 0.1 M $CaCl_2$ and 0.1 M $NaHCO_3$ for 1 week, leading to formation of calcite deposits, whose morphology changed with increasing sulfonic acid content in the hydrogel. The same group later used the same technique to mineralize hydrogels consisting of PAA copolymerized with acrylic acid (Grassmann and Lobmann, 2004). A multitude of small crystallites of calcite and vaterite formed within the hydrogel, which the authors attributed to the carboxylate functionalities. Helbig (2008) used the same double-diffusion method to mineralize PAA hydrogels and found that the concentration of polymer in the hydrogel affected $CaCO_3$ deposit morphology.

9.4.4 Increasing intra-hydrogel ion concentrations

Enzymatic approaches have been used to mineralize structures with $CaCO_3$. Sponges exploit an enzymatically catalyzed mineralization strategy to form $CaCO_3$ from calcium and carbon dioxide (CO_2) dissolved in sea water, which is a CO_2-poor environment, especially at depths inhabited by sponges. The enzyme carbonic anhydrase (CA) catalyzes the reaction of CO_2 with H_2O to form bicarbonate, which subsequently reacts with calcium ions to form $CaCO_3$ (Lee et al., 2010; Kim et al., 2014). CA is proposed to be responsible for mineralization of calcareous sponges (Muller et al., 2013), which first appeared in the Cambrian period (540–485 million years ago) (Wood et al., 1992). Hence, CA-catalyzed enzymatic mineralization evolved long before ALP-catalyzed mineralization in bone. Addition of CA to mineralization solutions containing CO_2 has accelerated $CaCO_3$ deposition on sponge spicules (Muller et al., 2013). The use of CA to mineralize hydrogels with $CaCO_3$ remains unexplored.

Rauner et al. (2014) immobilized urease in hydrogels based on hydroxyethyl acrylate and triethylene glycol dimethacrylate or poly(2-ethyl oxazoline). Urease catalyzes the conversion of urea and water to HCO_3^-, NH_4^+, and NH_3. A schematic illustration of enzymatic mineralization of a hydrogel with $CaCO_3$ by urease incorporation is shown in Figure 9.3. Upon incubation of the hydrogels in a mineralization solution containing urea and $CaCl_2$, deposition of aragonite and calcite occurred within the hydrogels.

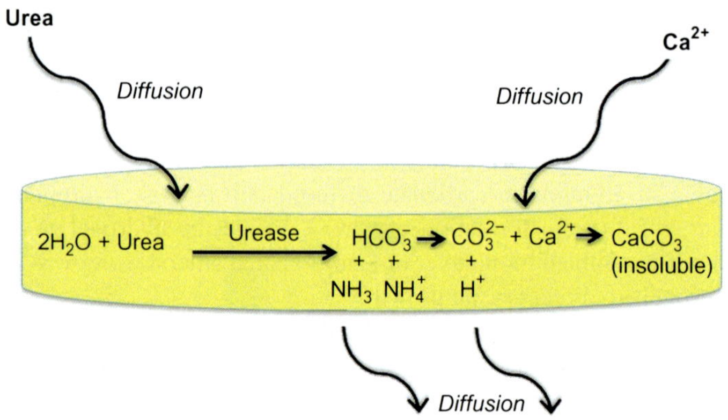

Figure 9.3 Schematic illustration of hydrogel mineralization with $CaCO_3$ using urease. Urease is immobilized in a hydrogel during hydrogel formation. Upon incubation in a solution containing urea, urea diffuses into the hydrogel. Urease catalyzes the reaction of urea with water to form ammonia, an ammonium ion, and a bicarbonate ion. The bicarbonate ion dissociates to form a hydrogen ion and a carbonate ion. A calcium ion in the mineralization solution enters the hydrogel by diffusion and reacts with the carbonate ion to form insoluble calcium carbonate, which precipitates and remains entrapped in the hydrogel. Reaction byproducts are free to exit the hydrogel by diffusion.

9.5 Biomimetic mineralization with silica

9.5.1 Incorporation of preformed silica particles

The most popular method of creating silica–hydrogel composites is the addition of preformed silica particles. Eglin et al. (2006) added preformed silica particles to collagen hydrogels, which promoted mineralization with apatite in SBF.

Some authors have used preformed silica as a crosslinker to enable hydrogel formation. Wu et al. (2013) incorporated silicate particles to promote physical crosslinking of PEG hydrogels formed from photopolymerization of PEGDA and methoxy-PEG-acrylate. Gaharwar et al. (2011) also used silicate particles to physically crosslink PEG hydrogels.

Incorporation of preformed silica particles to hydrogels has also been reported to influence the behavior of encapsulated cells. Desimone et al. (2010) added preformed silica nanoparticles to collagen hydrogels containing encapsulated fibroblasts. The presence of nanoparticles reduced cell-mediated hydrogel shrinkage and enhanced fibroblast metabolic activity. In a later publication, Desimone et al. (2011) added preformed silica nanoparticles or sodium silicate (as silica precursor) to collagen hydrogels. Fibroblasts seeded on the composites and cultured in a 2D environment proliferated better on composites with added sodium silicate. However, encapsulated fibroblasts showed superior proliferation in composites containing preformed nanoparticles. Subcutaneous implantation of silica–hydrogel composites in mice showed that silica did not cause a deleterious inflammatory response and that invasion by host cells took place, leading to vascularization.

Costantini et al. (2006) used preformed silica nanoparticles to enrich polyHEMA hydrogels. Silica nanoparticles were added to reach a final weight percentage of 30% (w/w). Enrichment with silica nanoparticles endowed the composite with the ability to support apatite formation upon immersion in SBF.

9.5.2 Delivery of silica precursors to form the inorganic phase during hydrogel formation

Certain authors have achieved formation of silica–hydrogel composites by simultaneous sol–gel transition of both silica and hydrogel phases. Costa et al. (2005) mixed silica precursors and different concentrations of polyHEMA hydrogel precursors to form a range of composites, on which apatite formed upon immersion in SBF. Costantini et al. (2008) mixed silica precursors and polyHEMA hydrogel precursors to achieve a final silica weight percentage of 30% (w/w in relation to the polymer) in the resulting composites. The authors reported separation of silica and organic phases. The same group later reported the production of samples with the same silica content by a similar method with reduced phase separation (Luciani et al., 2008). In a later publication, the same group generated composites with silica contents ranging from 10% to 30% (w/w in relation to the polymer) (Silvestri et al., 2009). All hybrid hydrogels supported apatite formation upon immersion in SBF and did not exhibit phase separation. Postnova et al. (2013) added silica precursors to a solution of hyperbranched polyglycerol in order to induce hydrogel formation as well as mineralization. The hydrogel formed consisted of crosslinked, interconnected nanoparticles.

9.5.3 External delivery of silica precursors by incubation in mineralization solutions

Another approach is the formation of a hydrogel followed by its incubation in a mineralization solution. Bonilla et al. (2013) formed a silica network by a sol–gel process involving immersion of a preformed poly(hydroxyethyl acrylate) hydrogel into a solution of silica precursors. A final silica content of 60% (w/v) was achieved. The presence of silica was observed in the pores of the hydrogel and the composite retained its integrity after destruction of the organic phase by pyrolysis, demonstrating the formation of a silica network within the hydrogel. The presence of silica promoted apatite formation after soaking in SBF. Jan et al. (2012) induced silicification of polypeptide hydrogels by incubation in orthosilicic acid solutions. Silicification did not lead to an increase in cytotoxicity towards a 3F3 fibroblast cell line.

9.6 Conclusion

Various strategies have been used to biomimetically mineralize hydrogels with CaP, $CaCO_3$, and silica. Mineralization with CaP has attracted much more attention than mineralization with $CaCO_3$ and silica. The most widely used method has been incorporation of preformed CaP, $CaCO_3$, and silica particles during hydrogel formation. Further studies on mineralization with $CaCO_3$ and silica are required, especially on the cell biological and *in vivo* effects of mineralization.

Acknowledgments

The author thanks the Research Foundation Flanders (FWO), Belgium for financial support in the framework of a postdoctoral fellowship. Prof. Elżbieta Pamuła, Department of Biomaterials, Faculty of Materials Science and Ceramics, AGH University of Science and Technology, Kraków, Poland is thanked for the SEM images of hydrogels.

References

Abbarin, N., Miguel, S.S., Holcroft, J., Iwasaki, K., Ganss, B., 2015. The enamel protein amelotin is a promoter of hydroxyapatite mineralization. J. Bone Miner. Res. 30 (5), 775–785.

Abramovitch-Gottlib, L., Geresh, S., Vago, R., 2006. Biofabricated marine hydrozoan: a bioactive crystalline material promoting ossification of mesenchymal stem cells. Tissue Eng. 12, 729–739.

Alexander, B.R., Murphy, K.E., Gallagher, J., Farrell, G.F., Taggart, G., 2012. Gelation time, homogeneity, and rupture testing of alginate-calcium carbonate-hydrogen peroxide gels for use as wound dressings. J. Biomed. Mater. Res. B Appl. Biomater. 100B, 425–431.

Anderson, J.M., Patterson, J.L., Vines, J.B., Javed, A., Gilbert, S.R., Jun, H.W., 2011. Biphasic peptide amphiphile nanomatrix embedded with hydroxyapatite nanoparticles for stimulated osteoinductive response. ACS Nano 5, 9463–9479.

Andre, R., Tahir, M.N., Natalio, F., Tremel, W., 2012. Bioinspired synthesis of multifunctional inorganic and bio-organic hybrid materials. FEBS J. 279, 1737–1749.

Arcos, D., Vallet-Regi, M., 2010. Sol–gel silica-based biomaterials and bone tissue regeneration. Acta Biomater. 6, 2874–2888.

Azami, M., Moosavifar, M.J., Baheiraei, N., Moztarzadeh, F., Ai, J., 2012. Preparation of a biomimetic nanocomposite scaffold for bone tissue engineering via mineralization of gelatin hydrogel and study of mineral transformation in simulated body fluid. J. Biomed. Mater. Res. A 100, 1347–1355.

Baht, G.S., Hunter, G.K., Goldberg, H.A., 2008. Bone sialoprotein-collagen interaction promotes hydroxyapatite nucleation. Matrix Biol. 27, 600–608.

Beniash, E., Simmer, J.P., Margolis, H.C., 2005. The effect of recombinant mouse amelogenins on the formation and organization of hydroxyapatite crystals in vitro. J. Struct. Biol. 149, 182–190.

Bidarra, S.J., Barrias, C.C., Fonseca, K.B., Barbosa, M.A., Soares, R.A., Granja, P.L., 2011. Injectable in situ crosslinkable RGD-modified alginate matrix for endothelial cells delivery. Biomaterials 32, 7897–7904.

Bongio, M., van den Beucken, J.J., Nejadnik, M.R., Leeuwenburgh, S.C., Kinard, L.A., Kasper, F.K., Mikos, A.G., Jansen, J.A., 2011. Biomimetic modification of synthetic hydrogels by incorporation of adhesive peptides and calcium phosphate nanoparticles: in vitro evaluation of cell behavior. Eur. Cell. Mater. 22, 359–376.

Bonilla, C.E.P., Gomez-Tejedor, J.A., Perilla, J.E., Ribelles, J.L.G., 2013. Silica phase formed by sol–gel reaction in the nano- and micro-pores of a polymer hydrogel. J. Non Cryst. Solids 379, 12–20.

Boyan, B.D., Lossdorfer, S., Wang, L., Zhao, G., Lohmann, C.H., Cochran, D.L., Schwartz, Z., 2003. Osteoblasts generate an osteogenic microenvironment when grown on surfaces with rough microtopographies. Eur. Cell. Mater. 6, 22–27.

Cardoso, D.A., van den Beucken, J.J., Both, L.L., Bender, J., Jansen, J.A., Leeuwenburgh, S.C., 2014. Gelation and biocompatibility of injectable alginate-calcium phosphate gels for bone regeneration. J. Biomed. Mater. Res. A 102, 808–817.

Cha, C., Kim, E.S., Kim, I.W., Kong, H., 2011. Integrative design of a poly(ethylene glycol)-poly(propylene glycol)-alginate hydrogel to control three dimensional biomineralization. Biomaterials 32, 2695–2703.

Chan, L.W., Lee, H.Y., Heng, P.W.S., 2002. Production of alginate microspheres by internal gelation using an emulsification method. Int. J. Pharm. 242, 259–262.

Chan, L.W., Lee, H.Y., Heng, P.W.S., 2006. Mechanisms of external and internal gelation and their impact on the functions of alginate as a coat and delivery system. Carbohydr. Polym. 63, 176–187.

Chen, J., Nan, K., Yin, S., Wang, Y., Wu, T., Zhang, Q., 2010. Characterization and biocompatibility of nanohybrid scaffold prepared via in situ crystallization of hydroxyapatite in chitosan matrix. Colloids Surf. B Biointerfaces 81, 640–647.

Chesnutt, B.M., Viano, A.M., Yuan, Y., Yang, Y., Guda, T., Appleford, M.R., Ong, J.L., Haggard, W.O., Bumgardner, J.D., 2009. Design and characterization of a novel chitosan/nanocrystalline calcium phosphate composite scaffold for bone regeneration. J. Biomed. Mater. Res. A 88, 491–502.

Chirila, T.V., Minamisawa, T., Keen, I., Shiba, K., 2009. Effect of motif-programmed artificial proteins on the calcium uptake in a synthetic hydrogel. Macromol. Biosci. 9, 959–967.

Corni, I., Harvey, T.J., Wharton, J.A., Stokes, K.R., Walsh, F.C., Wood, R.J.K., 2012. A review of experimental techniques to produce a nacre-like structure. Bioinspir. Biomim. 7, 031001.

Costa, R.O., Pereira, M.M., Lameiras, F.S., Vasconcelos, W.L., 2005. Apatite formation on poly(2-hydroxyethyl methacrylate)-silica hybrids prepared by sol–gel process. J. Mater. Sci. Mater. Med. 16, 927–932.

Costantini, A., Luciani, G., Annunziata, G., Silvestri, B., Branda, F., 2006. Swelling properties and bioactivity of silica gel/pHEMA nanocomposites. J. Mater. Sci. Mater. Med. 17, 319–325.

Costantini, A., Luciani, G., Silvestri, B., Tescione, F., Branda, F., 2008. Bioactive poly(2-hydroxyethylmethacrylate)/silica gel hybrid nanocomposites prepared by sol–gel process. J. Biomed. Mater. Res. B Appl. Biomater. 86B, 98–104.

Couto, D.S., Hong, Z., Mano, J.F., 2009. Development of bioactive and biodegradable chitosan-based injectable systems containing bioactive glass nanoparticles. Acta Biomater. 5, 115–123.

Coyac, B.R., Chicatun, F., Hoac, B., Nelea, V., Chaussain, C., Nazhat, S.N., McKee, M.D., 2013. Mineralization of dense collagen hydrogel scaffolds by human pulp cells. J. Dent. Res. 92, 648–654.

Desimone, M.F., Helary, C., Rietveld, I.B., Bataille, I., Mosser, G., Giraud-Guille, M.M., Livage, J., Coradin, T., 2010. Silica-collagen bionanocomposites as three-dimensional scaffolds for fibroblast immobilization. Acta Biomater. 6, 3998–4004.

Desimone, M.F., Helary, C., Quignard, S., Rietveld, I.B., Bataille, I., Copello, G.J., Mosser, G., Giraud-Guille, M.M., Livage, J., Meddahi-Pelle, A., Coradin, T., 2011. In vitro studies and preliminary in vivo evaluation of silicified concentrated collagen hydrogels. ACS Appl. Mater. Interfaces 3, 3831–3838.

Devolder, R.J., Kim, I.W., Kim, E.S., Kong, H., 2012. Modulating the rigidity and mineralization of collagen gels using poly(lactic-co-glycolic acid) microparticles. Tissue Eng. Part A 18, 1642–1651.

Dorvee, J.R., Boskey, A.L., Estroff, L.A., 2012. Rediscovering hydrogel-based double-diffusion systems for studying biomineralization. CrystEngComm 14, 5681–5700.

Douglas, T.E., Gassling, V., Declercq, H.A., Purcz, N., Pamula, E., Haugen, H.J., Chasan, S., De Mulder, E.L., Jansen, J.A., Leeuwenburgh, S.C., 2012a. Enzymatically induced mineralization of platelet-rich fibrin. J. Biomed. Mater. Res. A 100, 1335–1346.

Douglas, T.E., Messersmith, P.B., Chasan, S., Mikos, A.G., De Mulder, E.L., Dickson, G., Schaubroeck, D., Balcaen, L., Vanhaecke, F., Dubruel, P., Jansen, J.A., Leeuwenburgh, S.C., 2012b. Enzymatic mineralization of hydrogels for bone tissue engineering by incorporation of alkaline phosphatase. Macromol. Biosci. 12, 1077–1089.

Douglas, T.E., Skwarczynska, A., Modrzejewska, Z., Balcaen, L., Schaubroeck, D., Lycke, S., Vanhaecke, F., Vandenabeele, P., Dubruel, P., Jansen, J.A., Leeuwenburgh, S.C., 2013a. Acceleration of gelation and promotion of mineralization of chitosan hydrogels by alkaline phosphatase. Int. J. Biol. Macromol. 56C, 122–132.

Douglas, T.E.L., Pamula, E., Leeuwenburgh, S.C.G., 2013b. Biomimetic mineralization of hydrogel biomaterials for bone tissue engineering. In: Ramalingam, M., Wang, X., Chen, G., Ma, P., Cui, F.-Z. (Eds.), Biomimetics: Advancing Nanobiomaterials and Tissue Engineering. John Wiley, Hoboken, NJ.

Douglas, T., Wlodarczyk, M., Pamula, E., Declercq, H., De Mulder, E., Bucko, M., Balcaen, L., Vanhaecke, F., Cornelissen, R., Dubruel, P., Jansen, J., Leeuwenburgh, S., 2014a. Enzymatic mineralization of gellan gum hydrogel for bone tissue-engineering applications and its enhancement by polydopamine. J. Tissue Eng. Regen. Med. 8 (11), 906–918.

Douglas, T.E.L., Krawczyk, G., Pamula, E., Declercq, H., Schaubroeck, D., Bucko, M.M., Balcaen, L., van der Voort, P., Bliznuk, V., van der Vreken, N.M., Vanhaecke, F., Dash, M., Detsch, R., Boccaccini, A.R., Cornelissen, R., Dubruel, P., 2014b. Generation of composites for bone tissue engineering applications consisting of gellan gum hydrogels mineralized with calcium and magnesium phosphate phases by enzymatic means. J. Tissue Eng. Regen. Med.

Dragusin, D.M., Giol, D.E., Vasile, E., Zecheru, T., Stancu, I.C., 2011. Caesin—PHema: in vitro formation of nanometric Ca-P nuclei. Dig. J. Nanomater. Bios. 6, 1909–1918.

Eglin, D., Maalheem, S., Livage, J., Coradin, T., 2006. In vitro apatite forming ability of type I collagen hydrogels containing bioactive glass and silica sol–gel particles. J. Mater. Sci. Mater. Med. 17, 161–167.

Engler, A.J., Sen, S., Sweeney, H.L., Discher, D.E., 2006. Matrix elasticity directs stem cell lineage specification. Cell 126, 677–689.

Evans, N.D., Minelli, C., Gentleman, E., Lapointe, V., Patankar, S.N., Kallivretaki, M., Chen, X., Roberts, C.J., Stevens, M.M., 2009. Substrate stiffness affects early differentiation events in embryonic stem cells. Eur. Cell. Mater. 18, 1–13, discussion 13–14.

Filmon, R., Basle, M.F., Barbier, A., Chappard, D., 2000. Poly(2-hydroxy ethyl methacrylate)-alkaline phosphatase: a composite biomaterial allowing in vitro studies of bisphosphonates on the mineralization process. J. Biomater. Sci. Polym. Ed. 11, 849–868.

Flausse, A., Henrionnet, C., Dossot, M., Dumas, D., Hupont, S., Pinzano, A., Mainard, D., Galois, L., Magdalou, J., Lopez, E., Gillet, P., Rousseau, M., 2013. Osteogenic differentiation of human bone marrow mesenchymal stem cells in hydrogel containing nacre powder. J. Biomed. Mater. Res. A 101, 3211–3218.

Fonseca, K.B., Bidarra, S.J., Oliveira, M.J., Granja, P.L., Barrias, C.C., 2011. Molecularly designed alginate hydrogels susceptible to local proteolysis as three-dimensional cellular microenvironments. Acta Biomater. 7, 1674–1682.

Fonseca, K.B., Gomes, D.B., Lee, K., Santos, S.G., Sousa, A., Silva, E.A., Mooney, D.J., Granja, P.L., Barrias, C.C., 2014. Injectable MMP-sensitive alginate hydrogels as hMSC delivery systems. Biomacromolecules 15, 380–390.

Foo, L.H., Suzina, A.H., Azlina, A., Kannan, T.P., 2008. Gene expression analysis of osteoblasts seeded in coral scaffold. J. Biomed. Mater. Res. A 87, 215–221.
Gaharwar, A.K., Dammu, S.A., Canter, J.M., Wu, C.J., Schmidt, G., 2011. Highly extensible, tough, and elastomeric nanocomposite hydrogels from poly(ethylene glycol) and hydroxyapatite nanoparticles. Biomacromolecules 12, 1641–1650.
Gassling, V., Douglas, T.E., Purcz, N., Schaubroeck, D., Balcaen, L., Bliznuk, V., Declercq, H.A., Vanhaecke, F., Dubruel, P., 2013. Magnesium-enhanced enzymatically mineralized platelet-rich fibrin for bone regeneration applications. Biomed. Mater. 8, 055001.
George, A., Veis, A., 2008. Phosphorylated proteins and control over apatite nucleation, crystal growth, and inhibition. Chem. Rev. 108, 4670–4693.
Gkioni, K., Leeuwenburgh, S.C., Douglas, T.E., Mikos, A.G., Jansen, J.A., 2010. Mineralization of hydrogels for bone regeneration. Tissue Eng. Part B Rev. 16, 577–585.
Goto, T., Kim, I.Y., Kikuta, K., Ohtsuki, C., 2012. Hydrothermal synthesis of composites of well-crystallized hydroxyapatite and poly(vinyl alcohol) hydrogel. Mater. Sci. Eng. C Mater. Biol. Appl. 32, 397–403.
Grassmann, O., Lobmann, P., 2003. Morphogenetic control of calcite crystal growth in sulfonic acid based hydrogels. Chem. – Eur. J. 9, 1310–1316.
Grassmann, O., Lobmann, P., 2004. Biomimetic nucleation and growth of $CaCO_3$ in hydrogels incorporating carboxylate groups. Biomaterials 25, 277–282.
Gungormus, M., Branco, M., Fong, H., Schneider, J.P., Tamerler, C., Sarikaya, M., 2010. Self assembled bi-functional peptide hydrogels with biomineralization-directing peptides. Biomaterials 31, 7266–7274.
Gyawali, D., Nair, P., Kim, H.K., Yang, J., 2013. Citrate-based biodegradable injectable hydrogel composites for orthopedic applications. Biomater. Sci. 1, 52–64.
Helbig, U., 2008. Growth of calcium carbonate in polyacrylamide hydrogel: investigation of the influence of polymer content. J. Cryst. Growth 310, 2863–2870.
Ho, E., Lowman, A., Marcolongo, M., 2007. In situ apatite forming injectable hydrogel. J. Biomed. Mater. Res. A 83, 249–256.
Hulsart-Billstrom, G., Hu, Q., Bergman, K., Jonsson, K.B., Aberg, J., Tang, R., Larsson, S., Hilborn, J., 2011. Calcium phosphates compounds in conjunction with hydrogel as carrier for BMP-2: a study on ectopic bone formation in rats. Acta Biomater. 7, 3042–3049.
Hunter, G.K., Poitras, M.S., Underhill, T.M., Grynpas, M.D., Goldberg, H.A., 2001. Induction of collagen mineralization by a bone sialoprotein–decorin chimeric protein. J. Biomed. Mater. Res. 55, 496–502.
Hutchens, S.A., Benson, R.S., Evans, B.R., O'Neill, H.M., Rawn, C.J., 2006. Biomimetic synthesis of calcium-deficient hydroxyapatite in a natural hydrogel. Biomaterials 27, 4661–4670.
Ichibouji, T., Miyazaki, T., Ishida, E., Sugino, A., Ohtsuki, C., 2009. Apatite mineralization abilities and mechanical properties of covalently cross-linked pectin hydrogels. Mater. Sci. Eng. C 29, 1765–1769.
Jan, J.S., Chen, P.S., Hsieh, P.L., Chen, B.Y., 2012. Silicification of genipin-cross-linked polypeptide hydrogels toward biohybrid materials and mesoporous oxides. ACS Appl. Mater. Interfaces 4, 6864–6873.
Jang, J., Seol, Y.J., Kim, H.J., Kundu, J., Kim, S.W., Cho, D.W., 2014. Effects of alginate hydrogel cross-linking density on mechanical and biological behaviors for tissue engineering. J. Mech. Behav. Biomed. Mater. 37, 69–77.
Kamitakahara, M., Kimura, K., Ioku, K., 2012. Synthesis of nanosized porous hydroxyapatite granules in hydrogel by electrophoresis. Colloids Surf. B: Biointerfaces 97, 236–239.

Karewicz, A., Zasada, K., Bielska, D., Douglas, T.E., Jansen, J.A., Leeuwenburgh, S.C., Nowakowska, M., 2014. Alginate-hydroxypropylcellulose hydrogel microbeads for alkaline phosphatase encapsulation. J. Microencapsul. 31, 68–76.

Kim, I.G., Jo, B.H., Kang, D.G., Kim, C.S., Choi, Y.S., Cha, H.J., 2014. Biomineralization-based conversion of carbon dioxide to calcium carbonate using recombinant carbonic anhydrase. Chemosphere 87, 1091–1096.

Kitano, Y., 1962. The behavior of various inorganic ions in the separation of calcium carbonate from a bicarbonate solution. Bull. Chem. Soc. Jpn. 35, 1973–1980.

Koh, M.Y., Ohtsuki, C., Miyazaki, T., 2011. Modification of polyglutamic acid with silanol groups and calcium salts to induce calcification in a simulated body fluid. J. Biomater. Appl. 25, 581–594.

Kuo, C.K., Ma, P.X., 2001. Ionically crosslinked alginate hydrogels as scaffolds for tissue engineering: part 1. Structure, gelation rate and mechanical properties. Biomaterials 22, 511–521.

Kuo, C.K., Ma, P.X., 2008. Maintaining dimensions and mechanical properties of ionically crosslinked alginate hydrogel scaffolds in vitro. J. Biomed. Mater. Res. A 84A, 899–907.

Lamghari, M., Almeida, M.J., Berland, S., Huet, H., Laurent, A., Milet, C., Lopez, E., 1999. Stimulation of bone marrow cells and bone formation by nacre: in vivo and in vitro studies. Bone 25, 91S–94S.

Lamghari, M., Antonietti, P., Berland, S., Laurent, A., Lopez, E., 2001a. Arthrodesis of lumbar spine transverse processes using nacre in rabbit. J. Bone Miner. Res. 16, 2232–2237.

Lamghari, M., Berland, S., Laurent, A., Huet, H., Lopez, E., 2001b. Bone reactions to nacre injected percutaneously into the vertebrae of sheep. Biomaterials 22, 555–562.

Lee, S.W., Park, S.B., Jeong, S.K., Lim, K.S., Lee, S.H., Trachtenberg, M.C., 2010. On carbon dioxide storage based on biomineralization strategies. Micron 41, 273–282.

Leeuwenburgh, S.C., Jansen, J.A., Mikos, A.G., 2007. Functionalization of oligo(poly(ethylene glycol)fumarate) hydrogels with finely dispersed calcium phosphate nanocrystals for bone-substituting purposes. J. Biomater. Sci. Polym. Ed. 18, 1547–1564.

Leeuwenburgh, S.C., Ana, I.D., Jansen, J.A., 2010. Sodium citrate as an effective dispersant for the synthesis of inorganic–organic composites with a nanodispersed mineral phase. Acta Biomater. 6, 836–844.

Legeros, R.Z., 1991. Calcium Phosphates in Oral Biology and Medicine. Karger, San Francisco.

Li, H., Estroff, L.A., 2007. Hydrogels coupled with self-assembled monolayers: an in vitro matrix to study calcite biomineralization. J. Am. Chem. Soc. 129, 5480–5483.

Li, B., Wang, Y., Jia, D., Zhou, Y., Cai, W., 2009a. Mineralization of chitosan rods with concentric layered structure induced by chitosan hydrogel. Biomed. Mater. 4, 015011.

Li, H., Xin, H.L., Muller, D.A., Estroff, L.A., 2009b. Visualizing the 3D internal structure of calcite single crystals grown in agarose hydrogels. Science 326, 1244–1247.

Li, B., Wang, Y., Jia, D., Zhou, Y., 2011. Gradient structural bone-like apatite induced by chitosan hydrogel via ion assembly. J. Biomater. Sci. Polym. Ed. 22, 505–517.

Liang, P., Shen, Q., Zhao, Y., Zhou, Y., Wei, H., Lieberwirth, I., Huang, Y., Wang, D., Xu, D., 2004. Petunia-shaped superstructures of $CaCO_3$ aggregates modulated by modified chitosan. Langmuir 20, 10444–10448.

Liao, H.H., Mutvei, H., Sjostrom, M., Hammarstrom, L., Li, J.G., 2000. Tissue responses to natural aragonite (Margaritifera shell) implants in vivo. Biomaterials 21, 457–468.

Lin, G., Cosimbescu, L., Karin, N.J., Tarasevich, B.J., 2012. Injectable and thermosensitive PLGA-g-PEG hydrogels containing hydroxyapatite: preparation, characterization and in vitro release behavior. Biomed. Mater. 7, 024107.

Liu, G., Zhao, D.C., Tomsia, A.P., Minor, A.M., Song, X.Y., Saiz, E., 2009. Three-dimensional biomimetic mineralization of dense hydrogel templates. J. Am. Chem. Soc. 131, 9937–9939.

Luciani, G., Costantini, A., Silvestri, B., Tescione, F., Branda, F., Pezzella, A., 2008. Synthesis, structure and bioactivity of pHEMA/SiO(2) hybrids derived through in situ sol–gel process. J. Sol-Gel Sci. Technol. 46, 166–175.

Luo, D., Sang, L., Wang, X., Xu, S., Li, X., 2011. Low temperature, pH-triggered synthesis of collagen–chitosan–hydroxyapatite nanocomposites as potential bone grafting substitutes. Mater. Lett. 65, 2395–2397.

Madhumathi, K., Shalumon, K.T., Rani, V.V., Tamura, H., Furuike, T., Selvamurugan, N., Nair, S.V., Jayakumar, R., 2009. Wet chemical synthesis of chitosan hydrogel-hydroxyapatite composite membranes for tissue engineering applications. Int. J. Biol. Macromol. 45, 12–15.

Marelli, B., Ghezzi, C.E., Mohn, D., Stark, W.J., Barralet, J.E., Boccaccini, A.R., Nazhat, S.N., 2011. Accelerated mineralization of dense collagen-nano bioactive glass hybrid gels increases scaffold stiffness and regulates osteoblastic function. Biomaterials 32, 8915–8926.

Marelli, B., Ghezzi, C.E., Alessandrino, A., Barralet, J.E., Freddi, G., Nazhat, S.N., 2012. Silk fibroin derived polypeptide-induced biomineralization of collagen. Biomaterials 33, 102–108.

Miyazaki, T., Mukai, J., Ishida, E., Ohtsuki, C., 2013. Apatite mineralization behavior on polyglutamic acid hydrogels in aqueous condition: effects of molecular weight. Biomed. Mater. Eng. 23, 339–347.

Moreira, H.R., Munarin, F., Gentilini, R., Visai, L., Granja, P.L., Tanzi, M.C., Petrini, P., 2014. Injectable pectin hydrogels produced by internal gelation: pH dependence of gelling and rheological properties. Carbohydr. Polym. 103, 339–347.

Morse, J.W., Arvidson, R.S., Luttge, A., 2007. Calcium carbonate formation and dissolution. Chem. Rev. 107, 342–381.

Muller, W.E., Schroder, H.C., Schlossmacher, U., Neufurth, M., Geurtsen, W., Korzhev, M., Wang, X., 2013. The enzyme carbonic anhydrase as an integral component of biogenic Ca-carbonate formation in sponge spicules. FEBS Open Bio. 3, 357–362.

Munro, N.H., McGrath, K.M., 2011. Biomimetic mineralisation of polymeric scaffolds using a combined soaking approach: adaptation with various mineral salts. Dalton Trans. 40, 9269–9275.

Munro, N.H., McGrath, K.M., 2012. Biomimetic approach to forming chitin/aragonite composites. Chem. Commun. (Camb.) 48, 4716–4718.

Munro, N.H., Green, D.W., McGrath, K.M., 2013. In situ continuous growth formation of synthetic biominerals. Chem. Commun. (Camb.) 49, 3407–3409.

Neira-Carrillo, A., Acevedo, D.F., Miras, M.C., Barbero, C.A., Gebauer, D., Colfen, H., Arias, J.L., 2008. Influence of conducting polymers based on carboxylated polyaniline on in vitro $CaCO_3$ crystallization. Langmuir 24, 12496–12507.

Nejadnik, M.R., Mikos, A.G., Jansen, J.A., Leeuwenburgh, S.C., 2012. Facilitating the mineralization of oligo(poly(ethylene glycol) fumarate) hydrogel by incorporation of hydroxyapatite nanoparticles. J. Biomed. Mater. Res. A 100, 1316–1323.

Nijhuis, A.W., Takemoto, S., Nejadnik, M.R., Li, Y., Yang, X., Ossipov, D.A., Hilborn, J., Mikos, A.G., Yoshinari, M., Jansen, J.A., Leeuwenburgh, S.C., 2014. Rapid screening of mineralization capacity of biomaterials by means of quantification of enzymatically deposited calcium phosphate. Tissue Eng. Part C Methods 20 (10), 838–850.

Nudelman, F., Lausch, A.J., Sommerdijk, N.A., Sone, E.D., 2013. In vitro models of collagen biomineralization. J. Struct. Biol. 183, 258–269.

Ogomi, D., Serizawa, T., Akashi, M., 2003. Bioinspired organic–inorganic composite materials prepared by an alternate soaking process as a tissue reconstitution matrix. J. Biomed. Mater. Res. A 67, 1360–1366.

Ogomi, D., Serizawa, T., Akashi, M., 2005. Controlled release based on the dissolution of a calcium carbonate layer deposited on hydrogels. J. Control. Release 103, 315–323.

Olderoy, M.O., Xie, M., Andreassen, J.P., Strand, B.L., Zhang, Z., Sikorski, P., 2012. Viscoelastic properties of mineralized alginate hydrogel beads. J. Mater. Sci. Mater. Med. 23, 1619–1627.

Patel, M., Patel, K.J., Caccamese, J.F., Coletti, D.P., Sauk, J.J., Fisher, J.P., 2010. Characterization of cyclic acetal hydroxyapatite nanocomposites for craniofacial tissue engineering. J. Biomed. Mater. Res. A 94, 408–418.

Paxton, J.Z., Donnelly, K., Keatch, R.P., Baar, K., 2009. Engineering the bone-ligament interface using polyethylene glycol diacrylate incorporated with hydroxyapatite. Tissue Eng. Part A 15, 1201–1209.

Phadke, A., Shih, Y.R., Varghese, S., 2012. Mineralized synthetic matrices as an instructive microenvironment for osteogenic differentiation of human mesenchymal stem cells. Macromol. Biosci. 12, 1022–1032.

Postnova, I., Silant'ev, V., Kim, M.H., Song, G.Y., Kim, I., Ha, C.S., Shchipunov, Y., 2013. Hyperbranched polyglycerol hydrogels prepared through biomimetic mineralization. Colloids Surf. B: Biointerfaces 103, 31–37.

Prieto, S., Shkilnyy, A., Rumplasch, C., Ribeiro, A., Arias, F.J., Rodriguez-Cabello, J.C., Taubert, A., 2011. Biomimetic calcium phosphate mineralization with multifunctional elastin-like recombinamers. Biomacromolecules 12, 1480–1486.

Rajkumar, M., Meenakshisundaram, N., Rajendran, V., 2011. Development of nanocomposites based on hydroxyapatite/sodium alginate: synthesis and characterisation. Mater. Charact. 62, 469–479.

Rauner, N., Meuris, M., Dech, S., Godde, J., Tiller, J.C., 2014. Urease-induced calcification of segmented polymer hydrogels—a step towards artificial biomineralization. Acta Biomater. 10 (9), 3942–3951.

Rousseau, M., Boulzaguet, H., Biagianti, J., Duplat, D., Milet, C., Lopez, E., Bedouet, L., 2008. Low molecular weight molecules of oyster nacre induce mineralization of the MC3T3-E1 cells. J. Biomed. Mater. Res. A 85A, 487–497.

Rowlands, A.S., George, P.A., Cooper-White, J.J., 2008. Directing osteogenic and myogenic differentiation of MSCs: interplay of stiffness and adhesive ligand presentation. Am. J. Physiol. Cell Physiol. 295, C1037–C1044.

Ruhe, P.Q., Boerman, O.C., Russel, F.G., Spauwen, P.H., Mikos, A.G., Jansen, J.A., 2005. Controlled release of rhBMP-2 loaded poly(dl-lactic-co-glycolic acid)/calcium phosphate cement composites in vivo. J. Control. Release 106, 162–171.

Sanginario, V., Ginebra, M.P., Tanner, K.E., Planell, J.A., Ambrosio, L., 2006. Biodegradable and semi-biodegradable composite hydrogels as bone substitutes: morphology and mechanical characterization. J. Mater. Sci. Mater. Med. 17, 447–454.

Sarvestani, A.S., He, X., Jabbari, E., 2007. Effect of osteonectin-derived peptide on the viscoelasticity of hydrogel/apatite nanocomposite scaffolds. Biopolymers 85, 370–378.

Shkilnyy, A., Graf, R., Hiebl, B., Neffe, A.T., Friedrich, A., Hartmann, J., Taubert, A., 2009. Unprecedented, low cytotoxicity of spongelike calcium phosphate/poly(ethylene imine) hydrogel composites. Macromol. Biosci. 9, 179–186.

Silverman, L., Boskey, A.L., 2004. Diffusion systems for evaluation of biomineralization. Calcif. Tissue Int. 75, 494–501.

Silvestri, B., Luciani, G., Costantini, A., Tescione, F., Branda, F., Pezzella, A., 2009. In-situ sol–gel synthesis and characterization of bioactive pHEMA/SiO_2 blend hybrids. J. Biomed. Mater. Res. B Appl. Biomater. 89, 369–378.

Sommerdijk, N.A., De With, G., 2008. Biomimetic $CaCO_3$ mineralization using designer molecules and interfaces. Chem. Rev. 108, 4499–4550.

Song, J., Malathong, V., Bertozzi, C.R., 2005. Mineralization of synthetic polymer scaffolds: a bottom-up approach for the development of artificial bone. J. Am. Chem. Soc. 127, 3366–3372.
Song, J., Xu, J., Filion, T., Saiz, E., Tomsia, A.P., Lian, J.B., Stein, G.S., Ayers, D.C., Bertozzi, C.R., 2009. Elastomeric high-mineral content hydrogel-hydroxyapatite composites for orthopedic applications. J. Biomed. Mater. Res. A 89, 1098–1107.
Spoerke, E.D., Anthony, S.G., Stupp, S.I., 2009. Enzyme directed templating of artificial bone mineral. Adv. Mater. 21, 425–430.
Stancu, I.C., Dragusin, D.M., Vasile, E., Trusca, R., Antoniac, I., Vasilescu, D.S., 2011. Porous calcium alginate-gelatin interpenetrated matrix and its biomineralization potential. J. Mater. Sci. Mater. Med. 22, 451–460.
Sugino, A., Miyazaki, T., Ohtsuki, C., 2008. Apatite-forming ability of polyglutamic acid hydrogels in a body-simulating environment. J. Mater. Sci. Mater. Med. 19, 2269–2274.
Suzawa, Y., Funaki, T., Watanabe, J., Iwai, S., Yura, Y., Nakano, T., Umakoshi, Y., Akashi, M., 2010. Regenerative behavior of biomineral/agarose composite gels as bone grafting materials in rat cranial defects. J. Biomed. Mater. Res. A 93, 965–975.
Taguchi, T., Kishida, A., Akashi, M., 1999a. Apatite formation on/in hydrogel matrices using an alternate soaking process (III): effect of physico-chemical factors on apatite formation on/in poly(vinyl alcohol) hydrogel matrices. J. Biomater. Sci. Polym. Ed. 10, 795–804.
Taguchi, T., Kishida, A., Akashi, M., 1999b. Apatite formation on/in hydrogel matrices using an alternate soaking process: II. Effect of swelling ratios of poly(vinyl alcohol) hydrogel matrices on apatite formation. J. Biomater. Sci. Polym. Ed. 10, 331–339.
Tolga Demirtas, T., Karakecili, A.G., Gumusderelioglu, M., 2008. Hydroxyapatite containing superporous hydrogel composites: synthesis and in-vitro characterization. J. Mater. Sci. Mater. Med. 19, 729–735.
Tran, C.T., Gargiulo, C., Thao, H.D., Tuan, H.M., Filgueira, L., Michael Strong, D., 2011. Culture and differentiation of osteoblasts on coral scaffold from human bone marrow mesenchymal stem cells. Cell Tissue Bank. 12, 247–261.
Viateau, V., Manassero, M., Sensebe, L., Langonne, A., Marchat, D., Logeart-Avramoglou, D., Petite, H., Bensidhoum, M., 2013. Comparative study of the osteogenic ability of four different ceramic constructs in an ectopic large animal model. J. Tissue Eng. Regen. Med.
Watanabe, J., Akashi, M., 2008. Integration approach for developing a high-performance biointerface: sequential formation of hydroxyapatite and calcium carbonate by an improved alternate soaking process. Appl. Surf. Sci. 255, 344–349.
Wood, R., Zhuravlev, A., Debrenne, F., 1992. Functional biology and ecology of Archaeocyatha. Palaios 7, 131–156.
Wu, C.J., Wilker, J.J., Schmidt, G., 2013. Robust and adhesive hydrogels from cross-linked poly(ethylene glycol) and silicate for biomedical use. Macromol. Biosci. 13, 59–66.
Xie, M., Olderoy, M.O., Andreassen, J.P., Selbach, S.M., Strand, B.L., Sikorski, P., 2010. Alginate-controlled formation of nanoscale calcium carbonate and hydroxyapatite mineral phase within hydrogel networks. Acta Biomater. 6, 3665–3675.
Yokoi, T., Kawashita, M., Kikuta, K., Ohtsuki, C., 2010a. Biomimetic mineralization of calcium phosphate crystals in polyacrylamide hydrogel: effect of concentrations of calcium and phosphate ions on crystalline phases and morphology. Mater. Sci. Eng. C Mater. Biol. Appl. 30, 154–159.
Yokoi, T., Kawashita, M., Kikuta, K., Ohtsuki, C., 2010b. Crystallization of calcium phosphate in polyacrylamide hydrogels containing phosphate ions. J. Cryst. Growth 312, 2376–2382.
Zaharia, C., Tudora, M.-R., Stancu, I.-C., Galateanu, B., Lungu, A., Cincu, C., 2012. Characterization and deposition behavior of silk hydrogels soaked in simulated body fluid. Mater. Sci. Eng. C 32, 945–952.

Biomimetic mineralization of ceramics and glasses

10

M. Navarro[1], *T. Serra*[2]
[1]International Center for Numerical Methods in Engineering (CIMNE), Barcelona, Spain,
[2]University College London (UCL), London, UK

10.1 Introduction: The biomimetic approach

Mineralized tissues and biomineralization processes have been considered ideal examples to study in the biomaterials field, particularly the development and fabrication of materials and implants for bone-related applications. Natural mineralized tissues are thus excellent sources of inspiration for learning and creating biomimetic processes and materials mimicking living tissues.

Biomimetics has been described as a new field of science that studies how nature designs, processes, and builds various types of tissues such as organic/inorganic composites like the ones found in sea shells, teeth dentin and bone, among others, by orchestrating the assemblage of molecules in the right way, and subsequently applying these designs and processes to develop new molecules and materials with unique properties.

The biomimetic approach is currently being applied in the creation of new biomaterials and methodologies. Biomaterials that stimulate the tissues for eliciting specific responses at the interface with bone, in order to create a continuous transition from tissue to implant, are an example of the application of biomimetics. Nevertheless, there is still much to be understood, and the potential of biomimetics has not yet been properly fulfilled in the engineering of better biomedical systems.

When talking about biological hard tissues (such as bones, shells, and teeth), it is known that hierarchical structures with complex architecture are formed, ranging from the nanometer to the millimeter scales; these structures are a combination of minerals, proteins, polysaccharides, and lipids (Addadi and Weiner, 1992; Mann, 2001), which are combined following their specific genetic program. Among such natural biomineralized materials, calcium is present in some 50% of them (Addadi and Weiner, 1992). This reflects both the natural abundance of this metal and its importance (Addadi and Weiner, 1992).

The two major calcium compounds that are commonly found are (a) calcium carbonate, mainly found in shells of marine crustaceous and eggshells and (b) calcium phosphate, mainly found in bone and dentin in vertebrates. In the former case, three main polymorphs are found (calcite, aragonite, and vaterite), and of these, different shapes may be distinguished (Addadi and Weiner, 1992; Weiner and Addadi, 1997). It has been observed that, in nature, there are precise calcium-binding sites that are dictated by the structure of the nucleating proteins (negatively charged, interacting

Biomineralization and Biomaterials. http://dx.doi.org/10.1016/B978-1-78242-338-6.00011-9

electrostatically with the calcium ions), which are glycoproteins rich in aspartic acid, possibly with a β-sheet structure (Addadi and Weiner, 1992; Weiner and Addadi, 1997). Such stereochemical self-assembled templates induce and control both the nucleation and the growth of inorganic precipitation reactions, determining the final size, shape, and orientation of the crystalline structure formed (Addadi and Weiner, 1985).

These templates are usually the minor macromolecular component of a biological material, being intimately attached to a mineral phase. Minerals formed in organisms require either a 3D matrix for crystal formation and a substrate for the interaction between the control macromolecules and the mineral phase (Weiner and Addadi, 1997), or confinement from cell membranes or vesicles (Addadi and Weiner, 1992; Mann, 2001). This evidence suggests that analogous synthetic templates could be used in the production of tailor-made mineral structures. As in other biomimetic strategies, the conventional engineering and chemical approach must gradually be substituted by activities in other complementary fields, namely genetics, proteomics, and nanotechnologies, to reveal how to use the same tools as living organisms. This would include the understanding of molecular and supramolecular organization (also known as self-assembly).

10.1.1 Bone as a biomaterial

Bone is a complex living tissue which has a sophisticated structure on a range of different hierarchical scales. It is basically a composite comprising an organic phase (based on collagen fibrils) in which calcium-containing inorganic crystals (hydroxyapatite, HA) are embedded. Collagen is the framework macromolecule in bone, in contrast to the glycine- and alanine-rich proteins in sea shells. Although there are more important proteins in bone, collagen is the main component of the 3D matrix into which the HA is deposited in the form of thin nanocrystals in the form of plates with $50 \times 25\,nm^2$ of length and width and with 2–3 nm thickness (Nakahara, 1991).

Mimicking the exact calcification process occurring in bone is rather difficult. This is further complicated as all mineralization processes are ultimately controlled through cells directly associated with tissue formation (Simkiss and Wilbur, 1989). Nevertheless, the thorough understanding of bone biology that currently exists may provide key concepts that may be adapted within a synthetic context. A step further would be to control the precipitation of the minerals using protein templates that trigger biomineralization and control crystals' organization and properties. This would allow enhancement of the biological activity of the bone/cartilage mineral structure in terms of cell adhesion, spreading, and differentiation.

Besides the surface composition, which will provide the chemical substrate, the osteogenic process of bone cells will also depend on the roughness and topography of the surface. An important issue in the case of 3D porous structures is also their meso/macro organization. As stated by Green et al. (2002) pore diameters between 15 and 50 μm stimulate fibrovascular growth, between 50 and 150 μm stimulate osteoid formation, and diameters in the range of 150–500 μm lead directly to mineralized bone. Other authors (Yeni and Norman, 2000; Ayers et al., 1999; Kujala et al., 2003) have given different values for these porosities. Despite such insights, the ideal porous structure and pore size distribution expected to maximize bone growth are still unknown.

10.2 CaP coatings: Bioactivating nonbioactive materials

When considering an ideal material to replace and mimic bone, synthetic calcium phosphates (CaPs) could be suitable candidates, since they can replicate the structure and composition of bone mineral HA in a reproducible way. As expected, CaPs are biocompatible with most cell types found in calcified tissues, such as osteoblasts, osteoclasts, fibroblasts, and periodontal ligament cells (Midy et al., 2001; Heymann et al., 2001; Suzuki et al., 1999; Shirakata et al., 2002). Furthermore, CaPs have osteoconductive properties, allowing for the formation of bone on its surface by attachment, migration, proliferation, and differentiation of bone-forming cells (Ducheyne and Qiu, 1999; Kanazawa, 1989).

Therefore, these types of coatings have great potential for bone fixation applications and for scaffolding in tissue engineering applications. In fact, these coatings can be tailored in terms of chemical composition, crystallinity, and resorbability (Tanahashi and Matsuda, 1997; ter Brugge et al., 2002). Also, they can be loaded with osteogenic biological molecules such as growth factors or even anti-inflammatory molecules (Liu et al., 2003; Wen and Moradian-Oldak, 2003; Leonor et al., 2003), or serve as templates for the seeding of living cells that will stimulate bone formation (de Bruijn et al., 1999; Oliveira et al., 2002).

Most metals and polymers lack the ability to create this chemical bonding with bone. CaPs in the form of coatings have been shown to be a good option for providing a better tissue/material interaction.

Currently used techniques for CaP coating can be basically classified into two categories: physical deposition and wet-chemical techniques. Physical deposition methods are mainly based on plasma-spraying (PS) techniques to generate HA films on implants for bone and dental applications (Sumitomo Chemical Company Limited (Osaka Japan), 1976; Cao and Liu, 2013).

In the 1980s and 1990s, pioneering experiments with PS to prepare biocompatible HA coatings were undertaken by Newesly and Osborn (1978), Soballe et al. (1991), Geesink et al. (1987, 1988), Geesink and Manley (1993), Zyman et al. (1993a,b), Cook et al. (1988), McPherson et al. (1995), and Ji and Marquis (1993). The typical setup for the preparation of the films is shown in Figure 10.1a. A number of thermal spraying processes such as vacuum plasma spraying, atmospheric plasma spraying, liquid plasma spraying, suspension plasma spraying, high-velocity suspension flame spraying, high-velocity oxy-fuel, detonation gun spraying, and gas tunnel-type plasma spraying, were later developed in order to produce bioactive CaP-based films (Khor et al., 2004; Morks et al., 2008; Hasan and Stokes, 2011; Fauchais and Vardelle, 2011; Gledhill et al., 1999; Huang et al., 2010).

Pulsed laser deposition (PLD) is a physical vapor deposition technique to deposit thin films of materials on a particular surface. The PLD system is shown schematically in Figure 101c. As can be seen, the main components typically include a KrF excimer

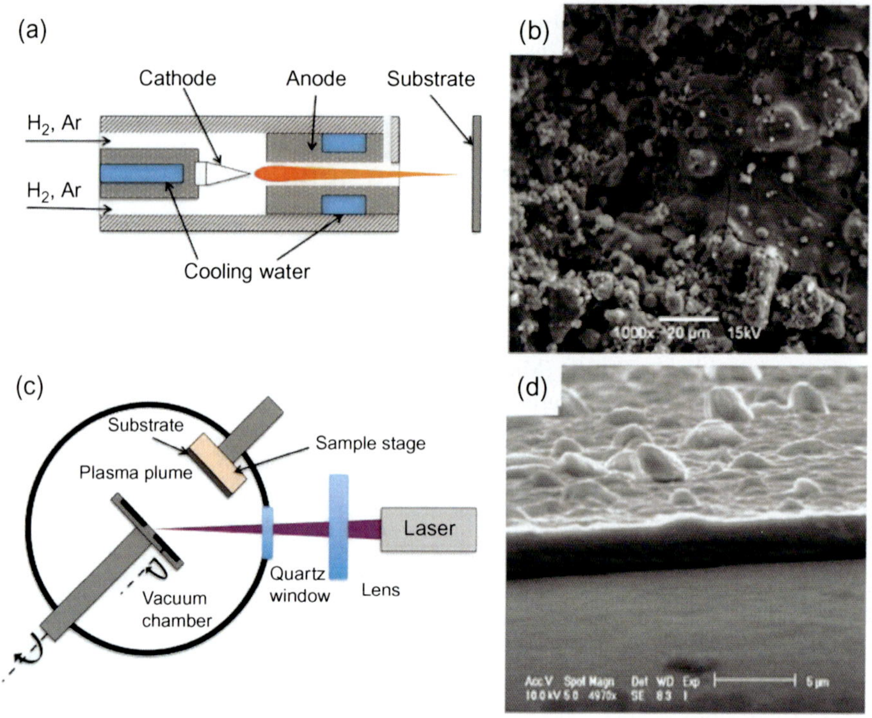

Figure 10.1 Plasma spraying (PS). (a) A schematic diagram of the gas plasma spraying of HA powders and (b) PS morphology. (c) Schematic setup of the pulsed laser deposition system (PLD) used to deposit HA coatings and (d) PLD HA coatings.
Reprinted from (a and b) Tang et al. (2010) and (c and d) Dinda et al. (2009).

laser source and an ultrahigh vacuum deposition chamber furnished with a rotating target and a fixed substrate holder connected to pumping systems. When PLD is used to prepare biocompatible CaP thin films, depositions are mostly performed using a UV KrF excimer laser ($\lambda = 248$ nm) (Surmenev, 2012).

In addition, there are other physical deposition techniques such as ion-beam-assisted deposition (IBAD; Choi et al., 2000; Cui et al., 1997; Cui and Luo, 1999; Granato et al., 2009; Yoon et al., 2009; Coelho et al., 2009), matrix-assisted pulsed laser evaporation (Negroiu et al., 2008), and radio-frequency (RF) magnetron sputtering (Mattox, 2010; Yamashita et al., 1994; van Dijk et al., 1996), which are currently available for creating CaP thin bioactive films. The typical setup for the RF coating technology is illustrated in Figure 10.2a.

Biomimetic coating processes, mostly in the family of wet-chemical techniques, were developed during 1990s and 2000s (Oyane et al., 2012). These processes are based on the surface modification of the base material with CaP seeds and/or functional groups that are effective for the induction of CaP nucleation, such as Si–OH (Kokubo et al., 2003), Ti–OH (Kokubo et al., 2003), Zr–OH (Kokubo et al., 2003), Nb–OH (Kokubo et al., 2003), COOH (Kokubo et al., 2003), Ta–OH (Kokubo et al., 2003), and SO_3H (Kawai et al., 2004).

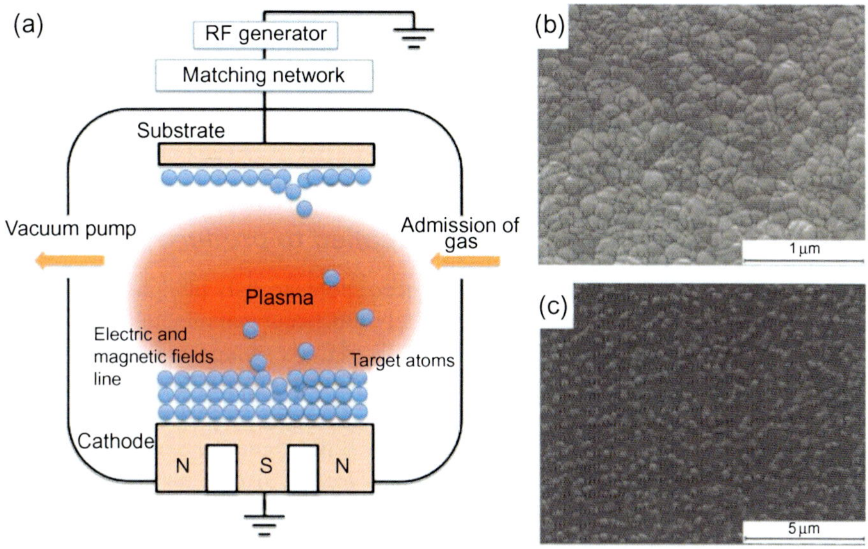

Figure 10.2 (a) Schematic representation of the typical RF magnetron sputtering facility. (b and c) Typical patterns of the surface morphology of RF magnetron sputter deposited coatings: RF power—290 W, deposition time 180 min, working atmosphere—argon. The substrate holder was grounded (b) and DC bias of −100 V was applied (c).
Surmenev (2012).

Biomimetic processes allow immobilization within the CaP layer of biofunctional substances, such as therapeutic agents (Zhao et al., 2011), trace elements (Xu et al., 2011), proteins (Haegi et al., 2010), bisphosphonates (Forsgren et al., 2011), bone morphogenetic protein (Zhao et al., 2011), vitamins (Ciobanu and Ciobanu, 2013), antibacterial agents (Zhao et al., 2009), and DNA (Luong et al., 2009). Due to the mild coating preparation conditions, organic substances are not deactivated completely in the resulting composite layer. This allows preparation of biocomposites that more closely resemble native bone tissue (Oyane et al., 2012).

The most frequently used wet-chemical techniques are the following: sol–gel (Azem et al., 2011), micro-arc oxidation, or plasma electrolytic oxidation/anodic spark oxidation (Legostaeva et al., 2013), electrophoretic (Bosco et al., 2012; Boccaccini et al., 2010), electrochemical deposition (Bosco et al., 2012; Qiu et al., 2011), electrospray deposition (Bosco et al., 2012), alternate soaking deposition (Bosco et al., 2012; Yoshioka et al., 2009), and drop-on-demand microdispensing (Thian et al., 2012). Wet-chemical deposition techniques are used to prepare various types of layers, such as HA (Liu et al., 2002), Si–HA (Ballo et al., 2012), Ag–HA (Qu et al., 2011), Sr–HA (Ballo et al., 2012), Sr- and Si-doped CaP (Lindahl et al., 2012), fluorine-doped hydroxyapatite (FHA) (Ge et al., 2011), and multifunctional, co-substituted HA thin films (Ag, Si–HA) (Thian et al., 2012), Zn–HA (Yang et al., 2012), and Zn-releasing FHA coatings (Miao et al., 2011).

Recently, there have been a number of reports on the preparation of thin bioactive coatings on biomedical substrates. Si–HA films have been prepared preferentially via sol–gel methods (Hijon et al., 2006). Studies have also focused on developing fluoride-substituted HA coatings on metallic surfaces using a sol–gel method. Results obtained from these studies demonstrated an interesting stimulatory effect on cell proliferation and differentiation *in vitro* (Wang et al., 2007).

10.2.1 The biomimetic approach applied to produce CaP coatings

CaPs found in natural hard tissues are produced spontaneously in a physiological environment at low temperatures from moderately supersaturated mineralizing solutions (Rey, 1990). The so-called biomimetic synthesis of CaP coatings on implant materials has since emerged as a new concept and several methodologies have been proposed. This type of approach is particularly suitable for coating polymeric materials (Abe et al., 1990), as it can be carried out in low-temperature reaction conditions. There is more than one approach for inducing the formation of a CaP on a surface. In fact, each coating methodology should be designed with consideration of the desired substrate in terms of its chemistry, structure, and surface morphology.

Kokubo et al. in the 1990s first reported the growth of a CaP layer in the form of apatite on a substrate via a biomimetic process (Abe et al., 1990). For this purpose, a bioactive $CaO–SiO_2$-based glass was used in the form of particles that were brought into contact with substrates to induce apatite nucleation on their surface in a solution that they called simulated body fluid (SBF), with ion concentrations nearly equal to those of the human blood plasma and at body temperature (Kokubo et al., 1990). The mechanism of apatite formation on Si-based glasses is now well known: the silanol groups (triple bond Si–OH) contained in the silicate ions are released from the bioactive glass and adsorbed on the substrate surface to induce the formation of a CaP layer (Abe et al., 1990). Similarly, other mechanisms for CaP formation on different ceramic surfaces have been reported. In general, the nucleation of apatite in the presence of a bioactive oxide is due to an electrostatically driven adsorption of Ca^{2+} ions at the ionized surface hydroxyl groups, onto which phosphate is subsequently absorbed. The deposited precipitated apatite layer acquires a characteristic cauliflower-like morphology, as shown in Figure 10.3.

Another fundamental aspect of this methodology is that it is highly dependent on the conditions of SBF immersion. Indeed, in order to tailor the apatite formation, there are many crucial parameters that should be taken into account, such as time, agitation, temperature, and ion concentration of the SBF solution.

It has also been reported that, by increasing the carbonate ion content, apatite with composition and structure nearly identical to those of bone-carbonated apatite can be produced. In fact, SBF solution is known to be deficient in HCO_3^- content, compared to human blood plasma (Kokubo et al., 1990; Kim et al., 2000). Therefore, a revised SBF with an ion composition closer to that of human blood plasma (higher amounts of HCO_3^-) was proposed to replace conventional SBF (Oyane et al., 2003). Though this new version of SBF solution is more like human plasma in composition, the original SBF is still being widely used in *in vitro* research (Lin et al., 2002).

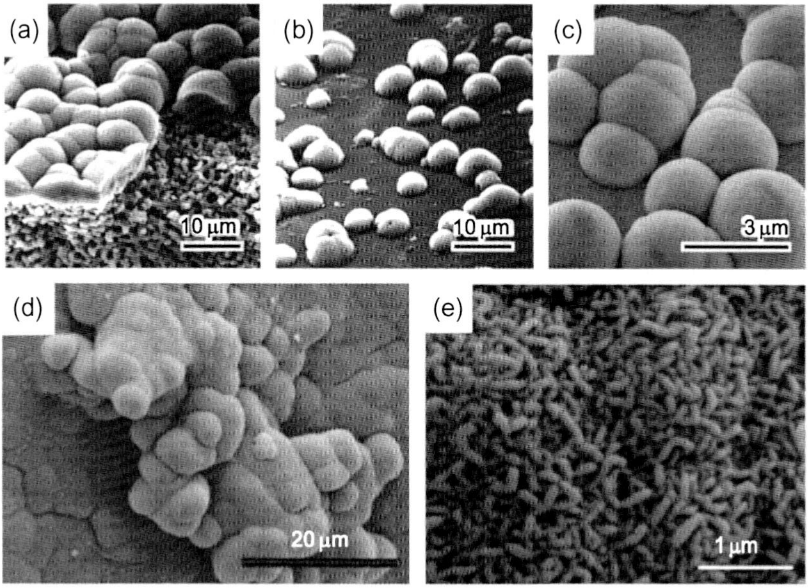

Figure 10.3 SEM photographs after immersion in SBF. Calcium phosphate formation induced on: (a) hydrogels of silica, (b) titania, and (c) zirconia (Kim, 2003). (d) HA scaffolds after 7 days of immersion in SBF (Oliveira et al., 2009). (e) High-magnification image showing apatite formation using SBF (Juhasz and Best, 2012).

In general, biomimetic methodology is based on the immersion of substrates in a supersaturated or metastable solution with ion concentrations and pH value similar to human blood plasma at a body temperature of 37 °C for several days, to induce the formation and growth of a several-micrometer-thick "bone-like" apatite layer. Biomimetic coatings are generally produced using simple, low-cost, and low-temperature processes. However, long induction and growth periods are required to form coatings with sufficient adhesion strength between the coating and substrate; thus, this technique still requires optimization.

10.2.2 Future challenges in CaP coating design

A significant number of the studies cited in this chapter have concluded that the biological performance of implants can be significantly improved by modifying their surfaces through the application of biologically active CaP-based coatings (Goodman et al., 2013).

Although it is generally accepted that these coatings improve the bone-bonding strength and the initial osteointegration rate, the ideal coating properties required to achieve the maximal bone response and best clinical performance are yet to be identified.

Moreover, the exact mechanism of osteoinduction by CaP is currently unknown (Chai et al., 2012). A comprehensive understanding of the structure–property

10.3.1 Calcium phosphates

Since bone consists of large amounts of HA, CaPs have been considered to develop scaffold materials for bone regeneration. Due to their close chemical and crystal resemblance to the mineral phase of bone HA and CaP exhibit excellent biocompatibility. CaPs are available in a number of chemical compositions and crystalline phases, among which HA, β-TCP, and biphasic HA/TCP ceramics are the most common for bone replacement. The bioactivity of these CaPs is strongly related to their solubility in the physiological environment, which increases with increasing amount of β-TCP (i.e. β-TCP > biphasic CaP > HA) (Dorozhkin, 2010).

The properties of the CaP family of compounds depend on the ratio of calcium to phosphorus ions in their structure. HA with a chemical formula of $Ca_{10}(PO_4)_6(OH)_2$ has a theoretical composition of 39.68 wt% Ca, 18.45 wt% P, a Ca/P weight ratio of 2.151, and a Ca/P molar ratio of 1.667. It has higher stability in aqueous media than other CaP ceramics within a pH range of 4.2–8.0. The close similarity of HA to the mineral component of bone, which is stable in the body, results however in the lack of biodegradation of HA in the body, which is generally an undesirable feature for tissue engineering scaffold materials.

In addition to sintered CaP-based ceramics, low-temperature CaPs such as calcium phosphate cements (CPCs) are often used clinically. CPCs' capacity to adapt to specific geometries is a valuable advantage. CPCs react with the surrounding aqueous media, producing preferably precipitated HA or brushite (dicalcium phosphate dihydrate). Moreover, these materials provide excellent biocompatibility, bioactivity, and osteconductivity (Ginebra et al., 2010).

10.3.2 Bioactive glasses

Bioactive glasses are a subset of inorganic bioactive ceramics, which are able to react with physiological fluids to form strong bonds to bone via the formation of bone-like HA layers and the biological interaction of collagen with the material surface (Hench, 1998b). It has been found that reactions on bioactive glass surfaces lead to the release of critical concentrations of soluble Si, Ca, P, and Na ions, which induce favorable intracellular and extracellular responses leading to rapid bone formation (Xynos et al., 2001; Chen et al., 2008).

There are now several types of bioactive glass: conventional silicates, such as Bioglass 45S5, phosphate-based glasses, and borate-based glasses. Recently, interest has increased in borate glasses (Chen et al., 2008), largely due to the very encouraging clinical results of healing of chronic wounds, such as diabetic ulcers, that would not heal using conventional treatments (Xynos et al., 2001). The soft tissue response may be due to their fast dissolution, which is quicker than for silica-based glasses. The benefits of phosphate glasses are also likely to be related to their very rapid solubility rather than their bioactivity (Chen et al., 2008; Navarro et al., 2003). Though phosphate glasses are also very interesting, they do not present bioactivity in the sense of the ability to form a CaP layer that bonds with bone.

Bioactive silicate glasses (e.g. 45S5 Bioglass) with compositions in the system SiO_2–Na_2O–CaO–P_2O_5 having <55% SiO_2 were discovered by Hench in 1969 (Hench, 1998b). They offer remarkable advantages as the inorganic components of composite scaffolds due to their high bioactivity index, and their ability to bond to both soft and hard connective tissues (Hench and Polak, 2002). These types of bioactive glasses are osteogenetic and osteoconductive materials while there is another type of bioactive materials (such as HA) that exhibit only osteoconductivity. It has also been found that reactions on bioactive glass surfaces release critical concentrations of soluble Si, Ca, P, and Na ions, which induce intracellular and extracellular responses (Hench, 1998b) For example, a synchronized sequence of genes is activated in osteoblasts that undergo cell division and synthesize an ECM, which mineralizes to become bone (Hench, 1998b; Hench and Polak, 2002). In addition, 45S5 Bioglass has been shown to increase the secretion of vascular endothelial growth factor *in vitro* and to enhance vascularization *in vivo*, suggesting that scaffolds containing controlled concentrations of Bioglass might stimulate neovascularization, which is beneficial to large tissue engineered constructs (Day et al., 2004).

10.3.2.1 Bioglass®

The name "Bioglass®" was trademarked by the University of Florida as a name for the original 45S5 composition. It should therefore only be used in reference to the 45S5 composition and not as a general term for bioactive glasses. *In vitro* tests showed that the 45S5 Bioglass® composition undergoes a surface reaction which occurs very rapidly. The surface reaction is a complex, multistage process, which results in the formation of a biologically active hydroxy-carbonate apatite (HCA) layer. This HCA phase is chemically and structurally similar to the mineral phase in bone and thus it provides a direct bonding by bridging host tissue with implants (Pantano et al., 1974; Ogino and Hench, 1980).

In order to assess this new class of highly bioactive materials, in 1991 Hench defined an *in vivo* bioactivity index IB (12.5), which is defined as IB = 100/*t*50bb, where *t*50bb is the time required for more than 50% of the interface to be bonded (Hench, 1998b). The rate of bone bonding and the strength and stability of the bond vary with the composition and microstructure of the bioactive materials. Hench et al. reported that, for their particular formulation of bioactive glass, bone formed a rapid bond when the silica levels were in the range 42–53%; glasses with 54–60% silica required 2–4 weeks for bone to bond; and bone did not form a direct bond with glasses containing more than 60% silica. In 1981, Dr. June Wilson reported that in addition to its excellent bone-bonding properties, soft connective tissues form a bond with Bioglass® (Best et al., 2008). Many glasses have been developed and reported in the literature. Most of them are based on the same system as Hench's original formulation.

10.3.3 A–W glass–ceramic

At around the same time that the original work on Bioglass® was being undertaken, Kokubo et al. were developing a new glass–ceramic material in Japan, and they first reported the production and behavior of A–W glass–ceramic in 1982. Apatite–wollastonite (A–W) glass–ceramic (also known by the trade name Cerabone®) became one of the most extensively studied glass–ceramics for use as a bone substitute. A dense and homogeneous composite was obtained after heat treatment of the parent glass, which comprised 38 wt% oxyfluorapatite ($Ca_{10}(PO_4)6(O,F)_2$) and 34 wt% wollastonite ($CaO \cdot SiO_2$) crystals, 50–100 nm in size, in a residual MgO–CaO–SiO_2 glassy matrix. Apatite–wollastonite glass–ceramic is an assembly of small apatite particles effectively reinforced by wollastonite. The bending strength, fracture toughness, and Young's modulus of A–W glass–ceramic are the highest among bioactive glass and glass–ceramics, enabling it to be used in bulk, granular, and porous forms in some major compression load-bearing applications, such as vertebral prosthesis replacement and other spinal surgery. It combines high bioactivity with suitable mechanical properties (Ikeda et al., 1999; Kokubo, 1993; Kokubo et al., 2000; Yamamuro, 1993).

10.4 The biomineralization process in CaPs and bioactive glasses

10.4.1 Surface chemistry and biomineralization

Interstitial blood plasma is the first phase that a bioactive ceramic comes into contact with once being implanted in the body. It is hypothesized that the ceramic surface is directed through mineralization of CaP via interaction with the blood plasma. The blood plasma contains many proteins that are significantly implicated in biomineralization; the protein and events cascade are too complex to be mimicked *ex vivo* for designing new bioactive materials. In terms of inorganic ion species, the blood plasma is a solution highly supersaturated with respect to CaPs, and accordingly has great potential to precipitate CaP (Neuman and Neuman, 1958). It is masterly controlled in a metastable condition to deposit the CaP by means of surface-induced heterogeneous nucleation and growth. Kokubo et al. (1990) earlier proposed a protein-free acellular SBF with pH (7.40) and ionic composition (Na^+ 142.0, K^+ 5.0, Ca^{2+} 2.5, Mg^{2+} 1.5, Cl^- 147.8, HCO_3^- 4.2, HPO_2^- 41.0, and SO_4^{2-} 0.5 mM) nearly equal to those of human blood plasma. The SBF has been extensively confirmed to produce CaP on different bioactive ceramics, as was found on their surfaces *in vivo* after implantation (Kokubo et al., 1990).

In vitro assessments using SBF (Kim et al., 1996) have provided various biomimetic models of biomineralization of CaP on bioactive ceramic surfaces with important information on the compositional and structural dependences of ceramic bioactivity. Glasses with different compositions in the systems CaO–P_2O_5–SiO_2, Na_2O–CaO–SiO_2, M_2O–SiO_2–TiO_2 ($M_{¼}$Na or/and K), and Na_2O–CaO–Al_2O_3, among which the former two systems are the basic compositions of glass–ceramic A–W and Bioglass®, have been subjected to immersion in SBF (Kim et al., 1996). Basic compositions able to induce CaP formation are based upon the binary systems CaO–SiO_2, M_2O–SiO_2 ($M_{¼}$Na or K), and M_2O–TiO_2, but interestingly not upon the systems CaO–P_2O_5 or M_2O–Al_2O_3. Surface characterization of the glasses as a function of immersion time in SBF indicates that the formation of crystalline CaP is preceded by formation of hydrated silica abundant in Si–OH groups on their surfaces. SBF characterizations indicate that the glasses form the surface silica by releasing Na^+, K^+, and/or Ca^{2+} ions via ion exchange with H_3O^+ ions in SBF. In view of the glass surface chemistry, the ion exchanges lead to an increase in the ionic activity product (IAP) of HA in SBF (Neuman and Neuman, 1958). This implies that a bioactive ceramic induces mineralization of CaP by forming specific surface functional groups such as Si–OH, which serve as effective sites for heterogeneous nucleation of CaP. An IAP increase in the surrounding fluid could thereby promote CaP nucleation and growth. Oxide hydrogels in SBF have provided simple models to predict effective compositions and structures of functional groups for CaP nucleation (Li et al., 1992; Miyazaki et al., 2001). As previously shown in Figure 10.3, CaP formation was induced on hydrogels of silica, titania, and zirconia, but not on alumina, implying the efficiencies of functional sites composed of groups such as Si–OH, Ti–OH, and Zr–OH in CaP nucleation in an *in vivo* environment.

Self-assembled monolayers (SAMs) and functionalized polymers in SBF indicated that COOH, PO_4H_2, and SO_3H groups could also serve as functional groups for CaP nucleation, but not NH_2, $CONH_2$, OH, and CH_3 groups (Tanahashi and Matsuda, 1997). Established techniques such as electron spectroscopies for chemical analysis (ESCA) and transmission electron microscopy (TEM) and energy dispersive X-ray analysis (EDX) were applied to these functional groups as a function of *in vitro* exposure time (Takadama et al., 2001), indicating that they induce the nucleation of CaP via formation of an amorphous calcium compound, such as calcium silicate and calcium titanate, and an amorphous CaP, which eventually stabilizes into a crystalline apatite.

Electrophoresis revealed that the surface potential at the functional sites is initially negative (Himeno et al., 2001), indicating that the initialization of CaP nucleation involves an electrostatic interaction between the functional sites and calcium ions. Therefore, functional groups able to develop negative charge in the blood plasma pH (7.4) are assumed to be potentially effective for CaP nucleation in an *in vivo* environment. There have been, however, strong implications concerning structural dependences of the functional groups with the same nominal compositions, which determine induction efficiencies of CaP nucleation (Cho et al., 1995; Uchida et al., 2003). For example, thermal annealing of sol–gel titania produces titanates with different structures, such as amorphous, anatase, or rutile. On immersion in SBF, CaP formation is much more pronounced on anatase gels than on amorphous and rutile gels (Uchida et al., 2003).

This structural dependence has been correlated with tissue-integration capability in animal models (Fujibayashi et al., 2003). At the time being, the nucleation of CaP appears to be processed by a certain type of nanotexture of the functional groups that may provide an epitaxial site arrangement for apatite crystals (Uchida et al., 2003). In fact, searching for surface functional textures likely to induce spontaneous biomineralization is one of the central challenges in bioactive ceramics and related biomimetic strategy.

The current knowledge on mechanisms of ceramic bioactivity is mostly *in vitro* and is poor in terms of biochemistry, but is inspiring simple effective biomimetic strategies for novel bio-interactive materials. On one hand, there may be discrepancies between *in vitro* and *in vivo* bioactive behaviors, which would imply a criticism of bioactivity predictions using SBF. On the other hand, SBF is a powerful tool for biomimetic design and processing for new bio-interactive surfaces and bulks. Examples of this have been shown in recent studies on bioactive surface functionalization on tough metallic and ceramic materials, sol–gel derivation of bioactive ceramic–polymer nano-hybrids, and biomimetic deposition of nano-CaP on polymer textures.

10.4.1.1 Biomineralization of bioactive glasses

As already mentioned, Bioglass 45S5 bonds rapidly with bone and it has also been shown that it stimulates bone growth away from the bone–implant interface. The mechanism for bone bonding is attributed to an HCA layer similar to the bone mineral on the surface of the glass, following initial glass dissolution (Hench et al., 1971). This HCA is thought to interact with collagen fibrils to bind the host bone.

The HCA layer forms following solution-mediated dissolution of the glass. Accumulation of dissolution products causes both a chemical composition change and a solution pH change, providing surface sites and a pH appropriate for HCA nucleation. There are five proposed stages for HCA formation in body fluid; this mechanism seems to be valid both *in vivo* and *in vitro* (Hench, 1998b; Clark et al., 1976).

1. Silanol groups are formed on the glass surface as a result of rapid cation exchange of Na^+ and/or Ca^{2+} with H^+ from solution. The pH of the solution increases and a silica-rich (cation-depleted) region form near the glass surface. If phosphate groups are present in the glass network, this is the stage where they are released from the glass to the medium.
2. pH values increase significantly locally and lead to attack of the silica glass network by OH, thus Si–O–Si bonds are broken. Soluble silica is released in the form of $Si(OH)_4$ to the solution, leaving more silanol groups at the glass–solution interface.
3. A repolymerization process forming a silica-rich layer takes place as Si–OH groups condense near the glass surface.
4. A film rich in amorphous CaO–P_2O_5 on the silica-rich layer is created due to the migration of Ca^{2+} and PO_4^{3-} groups to the surface through the silica-rich layer and from the solution.
5. Finally, there is incorporation of hydroxyls and carbonate from solution and crystallization of the CaO–P_2O_5 film to HCA.

Glass chemical composition is the variable that has the strongest influence on rate of HCA layer formation and bone bonding. In this sense, lower percentages of silica lead to a less connected silica network, which is more susceptible to dissolution, and therefore the five stages listed above occur more rapidly. Bioactivity has been shown to be directly related to the activation energy of silica dissolution in the glass (Arcos et al., 2003).

However, the connectivity of the network is key and depends on silica content and what other cations modify the glass. For example, adding sodium at the expense of silicon increases dissolution rate, but replacing cations such as sodium and calcium with multivalent ions such as Al^{3+}, and Ti^{4+} reduces bioactivity by reducing solubility (Greenspan and Hench, 1976; Gross and Strunz, 1980). As a rule of thumb, melt-derived glasses with compositions containing more than 60% SiO_2 do not bond and are bioinert. Sol–gel-derived glasses can, however, be bioactive with up to 90 mol% SiO_2 (Vallet-Regi et al., 2003).

Understanding glass structure and network connectivity in terms of the properties of bioactive glasses is most important. The atomic structure of bioactive glass and its relation to dissolution and HCA formation are a direct result of atomic structure. Understanding of the atomic structure is important when it comes to designing alternative glass compositions. The connectivity of the silica network is dictated by the composition and method of glass synthesis. High silica content results in a highly connected network containing a large proportion of bridging oxygen bonds and low dissolution and low bioactivity. Connectivity is lowered by adding more network-modifying cations such as sodium and calcium. As phosphate content increases, the network connectivity increases as the cations charge-balance the orthophosphate as monophosphate with or without diphosphate complexes (Mercier et al., 2011).

Osteogenesis is related to the action of dissolution products of the glasses on osteoprogenitor cells, stimulating new bone growth (Hench and Polak, 2002). However, the HCA layer also provides a surface suitable for osteogenic cell attachment and proliferation. The ideal surface chemistry and topography of a surface are yet to be identified.

10.4.2 Effect of geometry in CaP formation

In addition to compositional effects, surface morphology has been shown to affect cell response (Tampieri et al., 2005). It is now clear that micro- and nanoscale morphologies of the CaP surface can substantially affect cell behavior, including cell adhesion, proliferation, and differentiation (Li et al., 2012). Indeed, it has been demonstrated that surface concavities may play a major role in the osteogenic effect of CaP implants (Ripamonti et al., 2012). Ripamonti et al. initiated these studies on the effect of surface concavities on bone formation. The rationale for that was that concavities, by resembling the hemi-osteon channel, were able to activate bone induction and morphogenesis cascade (Ripamonti, 2009). In their investigations, Ripamonti et al. (2011) found that concavities with a diameter between 800 and 1600 μm, and depth between 400 and 800 μm, significantly increased the surface osteoinductive potential without adding exogenous osteogenic soluble molecules. Moreover, Ripamonti et al. (2011) also reported on the induction of bone formation in ectopic implantation due to the presence of concavities.

This mineralization process on concavities containing surfaces has been also studied under *in vitro* conditions (Bianchi et al., 2014). To this end, mineralization at the surface of HA and β-TCP disks prepared with hemispherical concavities of different diameters and sintered at 1100 or 1200 °C was evaluated upon immersion in SBF for up to 28 days. It was found that there is an important effect of both disk chemical composition (HA > β-TCP) and sintering temperature (1200 > 1100 °C) on CaP deposition from SBF. More interestingly, it was shown that CaP deposition was initiated exclusively within concavities and not on the planar surface of the disks. Finally, concavity dimensions had a significant effect on surface mineralization, with a fourfold reduction of concavity dimensions resulting in a 123-fold increase in calcium uptake. It was also observed that only HA disks sintered at 1200 °C enhanced rich CaP deposition. Since it is known that sintering has a direct influence on the grain size, porosity, crystalline phases, and surface charge of ceramics (Kim et al., 2005), it is suggested that larger size grains with higher crystallinity and lower porosity are more susceptible to trigger CaP deposition (Figure 10.5).

It seems that the faster dissolution of the β-TCP surface compared to the HA rather prevents the nucleation of stable CaP nuclei, despite locally increased ionic concentrations of calcium and phosphate ions at the material surface. The results of the present study suggested a two-step mineralization process on the HA disks heat treated at 1200 °C: (1) there was an initial aggregation of spherical-like CaP particles exclusively within the concavities and (2) these particles were subsequently replaced by a globular CaP phase consisting of crystals arranged in a reticular macroporous structure, with a chemical composition resembling that of the mineral phase in bone (Landi et al., 2005). The mechanism for the formation of this CaP phase has been described as follows in terms of surface charge and medium pH: in pH 7.4 SBF solution, HA possesses a negative surface charge; this negatively charged surface attracts cationic calcium ions from the SBF solution, forming a Ca-rich ACP. Consecutive calcium ion accumulation makes the Ca-rich ACP on the surface of the HA acquire a positive charge. At that stage, the Ca-rich ACP attracts negative phosphate ions in the SBF to form a Ca-poor ACP, which eventually crystallizes into a bone-like apatite by also incorporating other ions from SBF, such as sodium, magnesium, carbonate, and chloride (Kim et al., 2005).

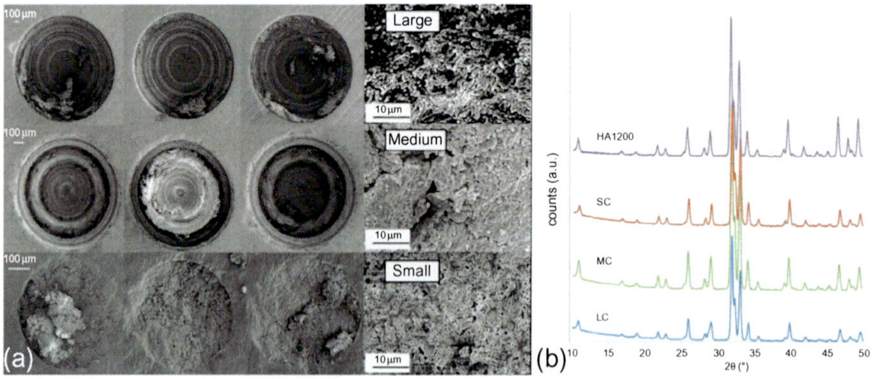

Figure 10.5 SEM images of large, medium, and small concavities at the surface of HA12 disks after 14 days immersion in SBF. (a) Top-view SEM images of large cavities (top), medium cavities (middle), and small cavities (bottom) on HA12_L, HA12_M, and HA12_S, respectively, after 14 days of immersion in SBF at 37 °C. Representative concavities for each dimension are displayed. The panels on the right show magnified images of the mineralization inside the concavities. (b) XRD spectra of planar surfaces of HA12_L, HA12_M, and HA12_S (and a reference HA1200 sample).
Bianchi et al. (2014).

Furthermore, the effect of geometrical features in CaP formation was also observed on titanium. Aparicio et al. (2007) demonstrated the influence of grit blasting on thermochemically treated titanium surfaces. The concave parts of the microroughness obtained during grit blasting provide the rough and bioactive surfaces with a chemically and electrostatically favored situation for apatite nucleation. This earlier nucleation significantly enhances the short-term integration in the surrounding hard tissues because the *in vivo* apatite-layer formation will be accelerated and, consequently, the osseointegration process will also be accelerated.

10.5 Conclusions

This chapter summarizes some of the latest achievements in the field of CaP-based coatings. Dense, homogeneous, highly adherent, biocompatible, pure inorganic CaP, or hybrid organic/inorganic CaP coatings with tailored properties can be deposited. A majority of the reported studies show that CaP coatings exert a significantly beneficial effect both *in vitro* and *in vivo* when comparing coated and uncoated implants; however, there are conflicting reports indicating no significant or even negative effects. Therefore, although CaP coatings generally show good promise for clinical use, specifically in bone repair and regeneration therapies, caution must be taken for the effective clinical use and development of these materials. A highly promising concept is the development of bioactive hybrid composite CaP-based coatings.

Some of the extensively studied plasma-assisted techniques (e.g. PLD, IBAD, and RF magnetron sputtering) are line-of-sight, with relatively low deposition rates that can result in expensive implants. Therefore, other effective routes for the fabrication of CaP coatings, particularly hybrid coatings, for clinical use still need to be considered for reasons of both economy and flexibility in materials design.

Moreover, methods that use hybrid materials that are similar to natural bone tissue, where a CaP inorganic phase is combined with either an organic ECM component such as collagen, or other proteins and biomolecules such as growth factors and peptides, represent a very promising option in the improvement and development of new bioactive coatings for bone regeneration.

In conclusion, further investigation is necessary to reveal the relative influences of implant design, surgical procedure, and coating characteristics (thickness, structure, topography, porosity, wettability, etc.) on long-term clinical effects of CaP coatings.

References

Abe, Y., Kokubo, T., Yamamuro, T., 1990. Apatite coating on ceramics, metals and polymers utilizing a biological process. J. Mater. Sci. Mater. Med. 1, 233–238.

Addadi, L., Weiner, S., 1985. Interactions between acidic proteins and crystals stereochemical requirements in biomineralization. Proc. Natl. Acad. Sci. U.S.A. 82, 4110–4114.

Addadi, L., Weiner, S., 1992. Control and design principles in biological mineralization. Angew. Chem. Int. Ed. 31, 153–169.

Aparicio, C., Manero, J.M., Conde, F., Pegueroles, M., Planell, J.A., Vallet-Regí, M., Gil, F.J., 2007. Acceleration of apatite nucleation on microrough bioactive titanium for bone-replacing implants. J. Biomed. Mater. Res. A 82 (3), 521–529.

Arcos, D., Greenspan, D.C., Vallet-Regi, M., 2003. A new quantitative method to evaluate the in vitro bioactivity of melt and sol–gel-derived silicate glasses. J. Biomed. Mater. Res. A 65A, 344–351.

Ayers, R.A., Simske, S.J., Bateman, T.A., Petkus, A., Sachdeva, R.L.C., Gyunter, V.E., 1999. Effect of nitinol implant porosity on cranial bone ingrowth and apposition after 6 weeks. J. Biomed. Mater. Res. 45, 42–47.

Azem, A.F., Eroglu, E.O., Cakir, A., 2011. Synthesis and structural properties of sol–gel derived Si-substituted hydroxyapatite coatings. J. Biomech. 44, 13.

Ballo, M., Xia, W., Palmquist, A., Lindahl, C., Emanuelsson, L., Lausmaa, J., et al., 2012. Bone tissue reactions to biomimetic ion-substituted apatite surfaces on titanium implants. J. R. Soc. Interface 9, 1615–1624.

Best, S.M., Porter, A.E., Thian, E.S., Huang, J., 2008. Bioceramics: past, present and for the future. J. Eur. Ceram. Soc. 28 (7), 1319–1327.

Bianchi, M., Urquia Edreira, E.R., Wolke, J.G., Birgani, Z.T., Habibovic, P., Jansen, J.A., Tampieri, A., Marcacci, M., Leeuwenburgh, S.C., van den Beucken, J.J., 2014. Substrate geometry directs the in vitro mineralization of calcium phosphate ceramics. Acta Biomater. 10 (2), 661–669.

Boccaccini, A.R., Keim, S., Ma, R., Li, Y., Zhitomirsky, I., 2010. Electrophoretic deposition of biomaterials. J. R. Soc. Interface 7, S581–S613.

Bosco, R., Van Den Beucken, J., Leeuwenburgh, S., Jansen, J., 2012. Surface engineering for bone implants: a trend from passive to active surfaces. Coatings 2, 95–119.

Cao, H., Liu, X., 2013. Plasma-sprayed ceramic coatings for osseointegration. Int. J. Appl. Ceram. Technol. 10, 1–10.

Chai, Y.C., Carlier, A., Bolander, J., Roberts, S.J., Geris, L., Schrooten, J., et al., 2012. Current views on calcium phosphate osteogenicity and the translation into effective bone regeneration strategies. Acta Biomater. 8, 3876–3887.

Chen, Q., Roether, J.A., Boccaccini, A.R., 2008. Tissue engineering scaffolds from bioactive glass and composite materials. In: Ashammakhi, N., Reis, R., Chiellini, F. (Eds.), Topics in Tissue Engineering, vol. 4. Biomaterials and Tissue Engineering Group, University of Oulu, Finland.

Cho, S.B., Nakanishi, K., Kokubo, T., Soga, N., Ohtsuki, C., Nakamura, T., et al., 1995. Dependence of apatite formation on silica gel on its structure: effect of heat treatment. J. Am. Ceram. Soc. 78, 1769–1774.

Choi, J.M., Kim, H.E., Lee, I.S., 2000. Ion-beam-assisted deposition (IBAD) of hydroxyapatite coating layer on Ti-based metal substrate. Biomaterials 21, 469–473.

Ciobanu, G., Ciobanu, O., 2013. Investigation on the effect of collagen and vitamins on biomimetic hydroxyapatite coating formation on titanium surfaces. Mater. Sci. Eng. C 33, 1683–1688.

Clark, A.E., Pantano, C.G., Hench, L.L., 1976. Auger spectroscopic analysis of bioglass corrosion films. J. Am. Ceram. Soc. 59, 37–39.

Coelho, P.G., Cardaropoli, G., Suzuki, M., Lemons, J.E., 2009. Early healing of nanothickness bioceramic coatings on dental implants. An experimental study in dogs. J. Biomed. Mater. Res. B: Appl. Biomater. 88, 387–393.

Cook, S.D., Thomas, K.A., Kay, J.F., Jarcho, M., 1988. Hydroxyapatite-coated titanium for orthopedic implant applications. Clin. Orthop. Relat. Res. 232, 225.

Cui, F.Z., Luo, Z.S., 1999. Biomaterials modification by ion-beam processing. Surf. Coat. Technol. 112, 278–285.

Cui, F.Z., Luo, Z.S., Feng, Q.L., 1997. Highly adhesive hydroxyapatite coatings on titanium alloy formed by ion beam assisted deposition. J. Mater. Sci. Mater. Med. 8, 403–405.
Day, R.M., Boccaccini, A.R., Shurey, S., Roether, J.A., Forbes, A., Hench, L.L., Gabe, S.M., 2004. Assessment of polyglycolic acid mesh and bioactive glass for soft-tissue engineering scaffolds. Biomaterials 25, 5857–5866.
de Bruijn, J.D., van den Brink, I., Mendes, S., Dekker, R., Bovell, Y.P., van Blitterswijk, C.A., 1999. Bone induction by implants coated with cultures osteogenic bone marrow cells. Adv. Dent. Res. 13, 74.
de Jonge, L.T., Leeuwenburgh, S.C.G., Wolke, J.G.C., Jansen, J.A., 2008. Organic–inorganic surface modifications for titanium implant surfaces. Pharm. Res. 25, 2357–2369.
Dinda, G.P., Shin, J., Mazumder, J., 2009 Jun. Pulsed laser deposition of hydroxyapatite thin films on Ti-6Al-4V: effect of heat treatment on structure and properties. Acta Biomater. 5 (5), 1821–1830.
Dorozhkin, S.V., 2010. Bioceramics of calcium orthophosphates. Biomaterials 31, 1465–1485.
Ducheyne, P., Qiu, Q., 1999. Bioactive ceramics: the effect of surface reactivity on bone formation and bone cell function. Biomaterials 20, 2287–2303.
Fauchais, P., Vardelle, A., 2011. Innovative and emerging processes in plasma spraying: from micro- to nano-structured coatings. J. Phys. D: Appl. Phys. 44, 194011.
Forsgren, J., Brohede, U., Piskounova, S., Mihranyan, A., Larsson, S., Stromme, M., et al., 2011. In vivo evaluation of functionalized biomimetic hydroxyapatite for local delivery of active agents. J. Biomater. Nanobiotechnol. 2, 150–155.
Förster, Y., Rentsch, C., Schneiders, W., Bernhardt, R., Simon, J.C., Worch, H., et al., 2012. Surface modification of implants in long bone. Biomatter 2, 149–157.
Fujibayashi, S., Neo, M., Kim, H.M., Kokubo, T., Nakamura, T., 2003. A comparative study between in vivo bone ingrowth and in vitro apatite formation on Na_2O–CaO–SiO_2 glasses. Biomaterials 24, 1335–1592.
Ge, X., Leng, Y., Ren, F., Lu, X., 2011. Integrity and zeta potential of fluoridated hydroxyapatite nanothick coatings for biomedical applications. J. Mech. Behav. Biomed. Mater. 4, 1046–1056.
Geesink, R.G.T., Manley, M.T., 1993. Hydroxyapatite Coatings in Orthopaedic Surgery. Raven Press, New York.
Geesink, R.G.T., de Groot, K., Klein, C.P.A.T., 1987. Chemical implant fixation using hydroxylapatite coatings. Clin. Orthop. Relat. Res. 225.
Geesink, R., de Groot, K., Klein, C.P.A.T., 1988. Bone bonding to apatite coated stems. J. Bone Joint Surg. 70-B, 17.
Ginebra, M.P., Espanol, M., Montufar, E.B., Perez, R.A., Mestres, G., 2010. New processing approaches in calcium phosphate cements and their applications in regenerative medicine. Acta Biomater. 6, 2863–2873.
Gledhill, H.C., Turner, I.G., Doyle, C., 1999. Direct morphological comparison of vacuum plasma sprayed and detonation gun sprayed hydroxyapatite coatings for orthopaedic applications. Biomaterials 20, 315–322.
Goodman, S.B., Yao, Z., Keeney, M., Yang, F., 2013. The future of biologic coatings for orthopaedic implants. Biomaterials 34, 3174–3183.
Granato, R., Marin, C., Suzuki, M., Gil, J.N., Janal, M.N., Coelho, P.G., 2009. Biomechanical and histomorphometric evaluation of a thin ion beam bioceramic deposition on plateau root form implants: an experimental study in dogs. J. Biomed. Mater. Res. B: Appl. Biomater. 90, 396–403.
Green, D., Walsh, D., Mann, S., Oreffo, R.O.C., 2002. The potential of biomimesis in bone tissue engineering: lessons from the design and synthesis of invertebrate skeletons. Bone 30, 810–815.

Greenspan, D.C., Hench, L.L., 1976. Chemical and mechanical behavior of bioglass-coated alumina. J. Biomed. Mater. Res. 10, 503–509.

Gross, U.M., Strunz, V., 1980. The anchoring of glass-ceramics of different solubility in the femur of the rat. J. Biomed. Mater. Res. 14, 607–618.

Haegi, T.T., Enggist, L., Michel, D., Ferguson, S.J., Liu, Y.L., Hunziker, E.B., 2010. Mechanical insertion properties of calcium-phosphate implant coatings. Clin. Oral Implants Res. 21, 1214–1222.

Hasan, S., Stokes, J., 2011. Design of experiment analysis of the Sulzer Metco DJ high velocity oxy-fuel coating of hydroxyapatite for orthopedic applications. J. Therm. Spray Technol. 20, 186–194.

Hench, L.L., 1991. Bioceramics: from concept to clinic. J. Am. Ceram. Soc. 74, 1487–1510.

Hench, L.L., 1998a. Bioactive materials: the potential for tissue regeneration. J. Biomed. Mater. Res. 15, 511–518.

Hench, L.L., 1998b. Bioceramics. J. Am. Ceram. Soc. 81, 1705–1728.

Hench, L.L., Polak, J.M., 2002. Third-generation biomedical materials. Science 295, 1014–1017.

Hench, L.L., Splinter, R.J., Allen, W.C., Greenlee, T.K., 1971. Bonding mechanisms at the interface of ceramic prosthetic materials. J. Biomed. Mater. Res. Symp. 334, 117–141.

Heymann, D., Guicheux, J., Rousselle, A.V., 2001. Ultrastructural evidence in vitro of osteoclast-induced degradation of calcium phosphate ceramic by simultaneous resorption and phagocytosis mechanisms. Histol. Histopathol. 16, 37–44.

Hijon, N., Cabanas, M.V., Pena, J., Vallet-Regi, M., 2006. Dip coated silicon substituted hydroxyapatite films. Acta Biomater. 2, 567–574.

Himeno, T., Kawashita, M., Kim, H.M., Kokubo, T., Nakamura, T., 2001. Zeta-potential variation of bioactive titanium metal during apatite formation on its surface in simulated body fluid. In: Brown, S., Clarke, I.R., Williams, P. (Eds.), Bioceramics, vol. 14. Trans Tech, Switzerland, pp. 641–644.

Huang, Y., Song, L., Liu, X., Xiao, Y., Wu, Y., Chen, J., et al., 2010. Hydroxyapatite coatings deposited by liquid precursor plasma spraying: controlled dense and porous microstructures and osteoblastic cell responses. Biofabrication 2, 045003.

Ikeda, N., Kawanabe, K., Nakamura, T., 1999. Quantitative comparison of osteoconduction of porous, dense A-W glass-ceramic and hydroxyapatite granules (effects of granule and pore sizes). Biomaterials 20, 1087–1095.

Ji, H., Marquis, P.M., 1993. Effect of heat treatment on the microstructure of plasma-sprayed hydroxyapatite coating. Biomaterials 14, 64.

Juhasz, J.A., Best, S.M., 2012. Bioactive ceramics: processing, structures and properties. J. Mater. Sci. 47, 610–624.

Kanazawa, T., 1989. General Background on Phosphate Materials. Elsevier, Amsterdam.

Kawai, T., Ohtsuki, C., Kamitakahara, M., Miyazaki, T., Tanihara, M., Sakaguchi, Y., et al., 2004. Coating of an apatite layer on polyamide films containing sulfonic groups by a biomimetic process. Biomaterials 25, 4529–4534.

Khor, K.A., Li, H., Cheang, P., 2004. Significance of melt-fraction in HVOF sprayed hydroxyapatite particles, splats and coatings. Biomaterials 25, 1177–1186.

Kim, H.M., 2001. Bioactive ceramics: challenges and perspectives. J. Ceram. Soc. Jpn. 109, S49–S57.

Kim, H.M., 2003. Ceramic bioactivity and related biomimetic strategy. Curr. Opinion Solid State Mater. Sci. 7 (4–5), 289–299.

Kim, H.M., Miyaji, F., Kokubo, T., Kobayashi, M., Nakamura, T., 1996. Bioactivity of M_2O–TiO_2–SiO_2 (M.Na, K) glasses: an in vitro evaluation. Bull. Chem. Soc. Jpn. 69, 2387–2394.

Kim, H.M., Kishimoto, K., Miyaji, F., Kokubo, T., Yao, T., Suetsugu, Y., et al., 2000. Composition and structure of apatite formed on organic polymer in simulated body fluid with a high content of carbonate ion. J. Mater. Sci. Mater. Med. 11, 421–426.

Kim, H.M., Himeno, T., Kokubo, T., Nakamura, T., 2005. Process and kinetics of bone like apatite formation on sintered hydroxyapatite in a simulated body fluid. Biomaterials 26, 4366–4373.

Kitsugi, T., Nakamura, T., Yamamuro, T., Kokubo, T., Shibuya, T., Takagi, M., 1987. SEM-EPMA observation of three types of apatite containing glass-ceramics implanted in bone: the variance of a Ca-P-rich layer. J. Biomed. Mater. Res. 21, 1255–1271.

Kokubo, T., 1991. Recent progress in glass-based materials for biomedical applications. J. Ceram. Soc. Jpn. (Seramikkusu Ronbunshi) 99, 965–973.

Kokubo, T., 1993. A/W glass-ceramic: processing and properties. In: Hench, L.L., Wilson, J. (Eds.), An Introduction to Bioceramics. World Scientific, Singapore, pp. 75–88.

Kokubo, T., Takadama, H., 2006 May. How useful is SBF in predicting in vivo bone bioactivity? Biomaterials 27 (15), 2907–2915.

Kokubo, T., Kushitani, H., Sakka, S., Kitsugi, T., Yamamuro, T., 1990. Solutions able to reproduce in vivo surface-structure changes in bioactive glass–ceramic A-W. J. Biomed. Mater. Res. 24, 721–734.

Kokubo, T., Kim, H.M., Kawashita, M., Nakamura, T., 2000. What kinds of materials exhibit bone-bonding? In: Davies, J.E. (Ed.), Bone Engineering. Em Squared, Toronto, pp. 190–194.

Kokubo, T., Kim, H.M., Kawashita, M., 2003. Novel bioactive materials with different mechanical properties. Biomaterials 24, 2161–2175.

Kujala, S., Ryhanen, J., Danilov, A., Tuukkanen, J., 2003. Effect of porosity on the osteointegration and bone ingrowth of a weight-bearing nickel–titanium bone graft substitute. Biomaterials 24, 4691–4697.

Landi, E., Tampieri, A., Celotti, G., Langenati, R., Sandri, M., Sprio, S., 2005. Nucleation of biomimetic apatite in synthetic body fluids: dense and porous scaffold development. Biomaterials 26, 2835–2845.

Legostaeva, E.V., Kulyashova, K.S., Komarova, E.G., Epple, M., Sharkeev, Y.P., Khlusov, I.A., 2013. Physical, chemical and biological properties of micro-arc deposited calcium phosphate coatings on titanium and zirconium-niobium alloy. Mat-wiss u Werkstofftech 44, 188–197.

Leonor, I.B., Azevedo, H.S., Alves, C.M., Reis, R.L., 2003. Effects of the incorporation of proteins and active enzymes on biomimetic calcium-phosphate coatings. Key Eng. Mater. 240–242, 97–100.

Li, P., Ohtsuki, C., Kokubo, T., Nakanishi, K., Soga, N., Nakamura, T., et al., 1992. Apatite formation induced on silica gel in a simulated body fluid. J. Am. Ceram. Soc. 75, 2094–2097.

Li, B., Liao, X.L., Zheng, L., Zhu, X.D., Wang, Z., Fan, H.S., et al., 2012. Effect of nanostructure on osteoinduction of porous biphasic calcium phosphate ceramics. Acta Biomater. 8, 3794–3804.

Lin, F.H., Hsu, Y.S., Lin, S.H., Sun, J.S., 2002. The effect of Ca/P concentration and temperature of simulated body fluid on the growth of hydroxyapatite coating on alkali-treated 316L stainless steel. Biomaterials 23, 4029–4038.

Lindahl, C., Xia, W., Lausmaa, J., Borchardt, P., Engqvist, H., 2012. Strontium and silicon co-doped apatite coating: preparation and function as vehicles for ion delivery. J. Biomater. Nanobiotechnol. 3, 335–341.

Liu, D., Yang, Q., Troczynski, T., 2002. Sol–gel hydroxyapatite coatings on stainless steel substrates. Biomaterials 23, 691–698.

Liu, Y., Hunziker, E.B., Randall, N.X., de Groot, K., Layrolle, P., 2003. Proteins incorporated into biomimetically prepared calcium phosphate coatings modulate their mechanical strength and dissolution rate. Biomaterials 24, 65–70.

Luong, L.N., McFalls, K.M., Kohn, D.H., 2009. Gene delivery via DNA incorporation within a biomimetic apatite coating. Biomaterials 30, 6996–7004.

Mann, S., 2001. Biomineralization, Principles and Concepts in Bioinorganic Materials Chemistry. Oxford University Press, Oxford.

Mattox, D.M., 2010. Handbook of Physical Vapor Deposition (PVD) Processing, second ed. Elsevier, Burlington, MA.

McPherson, R., Gane, N., Bastow, T.J., 1995. Structural characterization of plasma-sprayed hydroxyapatite coatings. J. Mater. Sci. Mater. Med. 6, 327–334.

Mercier, C., Follet-Houttemane, C., Pardini, A., Revel, B., 2011. Influence of P_2O_5 content on the structure of SiO_2–Na_2O–CaO–P_2O_5 bioglasses by ^{29}Si and ^{31}P MAS-NMR. J. Non-Cryst. Solids 357, 3901–3909.

Miao, S., Lin, N., Cheng, K., Yang, D., Huang, X., Han, G., et al., 2011. Zn-releasing FHA coating and its enhanced osseointegration ability. J. Am. Ceram. Soc. 94, 255–260.

Midy, V., Dard, M., Hollande, E., 2001. Evaluation of the effect of three calcium phosphate powders on osteoblast cells. J. Mater. Sci. Mater. Med. 12, 259–265.

Miyazaki, T., Kim, H.M., Kokubo, T., Kato, H., Nakamura, T., 2001. Induction and acceleration of bonelike apatite formation on tantalum oxide gel in simulated body fluid. J. Sol-Gel Sci. Tech. 21, 83–88.

Morks, M.F., Fahim, N.F., Kobayashi, A., 2008. Structure, mechanical performance and electrochemical characterization of plasma sprayed SiO_2/Ti-reinforced hydroxyapatite biomedical coatings. Appl. Surf. Sci. 255, 3426–3433.

Nakahara, H., 1991. Nacre formation in bivalve and gastropod mollusks. In: Suga, S., Nakahara, H. (Eds.), Mechanisms and Phylogeny of Mineralization in Biological Systems. Springer, New York, (Chap. 4.2) pp. 343–350.

Navarro, M., Ginebra, M.P., Clément, J., Martínez, S., Avila, G., Planell, J.A., 2003. Physicochemical degradation of resorbable phosphate glasses stabilized with TiO_2. J. Am. Ceram. Soc. 86 (8), 1345–1352.

Negroiu, G., Piticescu, R.M., Chitanu, G.C., Mihailescu, I.N., Zdrentu, L., Miroiu, M., 2008. Biocompatibility evaluation of a novel hydroxyapatite–polymer coating for medical implants (in vitro tests). J. Mater. Sci. Mater. Med. 19, 1537–1544.

Neo, M., Kotani, S., Nakamura, T., Yamamuro, T., Ohtsuki, C., Kokubo, T., et al., 1992. A comparative study of ultrastructure of the interface between four kinds of surface-active ceramic and bone. J. Biomed. Mater. Res. 26, 1419–1432.

Neo, M., Nakamura, T., Ohtsuki, C., Kokubo, T., Yamamuro, T., 1993. Apatite formation on three kinds of bioactive materials at an early stage in vivo: a comparative study by transmission electron microscopy. J. Biomed. Mater. Res. 27, 999–1006.

Neuman, W., Neuman, M., 1958. The Chemical Dynamics of Bone Mineral. University of Chicago Press, Chicago, IL.

Newesly, H., Osborn, J.F., 1978. Structure and texture of calcium phosphate in ceramics. In: 3rd Conference on Materials, for Use in Medicine and Biology. Keele University, Manchester.

Ogino, M., Hench, L.L., 1980. Formation of calcium phosphate films on silicate glasses. J. Non-Cryst. Solids 38–39, 673–678.

Ohura, K., Yamamuro, T., Nakamura, T., Kokubo, T., Ebisawa, Y., Kotoura, Y., et al., 1991. Bone-bonding ability of P_2O_5-free CaO SiO_2 glasses. J. Biomed. Mater. Res. 25, 357–365.

Oliveira, A.L., Alves, C.M., Reis, R.L., 2002. Cell adhesion and proliferation on biomimetic calcium-phosphate coatings produced by a sodium silicate gel methodology. J. Mater. Sci. Mater. Med. 13, 1181–1188.

Oliveira, J.M., Silva, S.S., Malafaya, P.B., Rodrigues, M.T., Kotobuki, N., Hirose, M., Gomes, M.E., Mano, J.F., Ohgushi, H., Reis, R.L., 2009. Macroporous hydroxyapatite scaffolds for bone tissue engineering applications: physicochemical characterization and assessment of rat bone marrow stromal cell viability. J. Biomed. Mater. Res. A 91 (1), 175–186.

Oyane, A., Kim, H.M., Furuya, T., Kokubo, T., Miyazaki, T., Nakamura, T., 2003. Preparation and assessment of revised simulated body fluids. J. Biomed. Mater. Res. 65A, 188–195.

Oyane, A., Wang, X., Sogo, Y., Ito, A., Tsurushima, H., 2012. Calcium phosphate composite layers for surface-mediated gene transfer. Acta Biomater. 8, 2034–2046.

Pantano Jr., C.G., Clark Jr., A.E., Hench, L.L., 1974. Multilayer corrosion films on glass surfaces. J. Am. Ceram. Soc. 57 (9), 412–413.

Qiu, D., Yang, L., Yin, Y., Wang, A., 2011. Preparation and characterization of hydroxyapatite/titania composite coating on NiTi alloy by electrochemical deposition. Surf. Coat. Technol. 205, 3280–3284.

Qu, J., Lu, X., Li, D., Ding, Y., Leng, Y., Weng, J., et al., 2011. Silver/hydroxyapatite composite coatings on porous titanium surfaces by sol–gel method. J. Biomed. Mater. Res. B: Appl. Biomater. 97, 40–48.

Rey, C., 1990. Calcium phosphate biomaterials and bone mineral. Differences in composition, structures and properties. Biomaterials 11, 13.

Ripamonti, U., 2009. Biomimetism, biomimetic matrices and the induction of bone formation. J. Cell. Mol. Med. 13, 2953–2972.

Ripamonti, U., Roden, L.C., Ferretti, C., Klar, R.M., 2011. Biomimetic matrices self-initiating the induction of bone formation. J. Craniofac. Surg. 22, 1859–1870.

Ripamonti, U., Roden, L.C., Renton, L.F., 2012. Osteoinductive hydroxyapatite-coated titanium implants. Biomaterials 33, 3813–3823.

Shirakata, Y., Oda, S., Kinoshita, A., Kikuchi, S., Tsuchioka, H., Ishikawa, I., 2002. Histocompatible healing of periodontal defects after application of an injectable calcium phosphate bone cement. A preliminary study in dogs. J. Periodontol. 73, 1043–1053.

Simkiss, K., Wilbur, K.M., 1989. Biomineralization. Cell Biology and Mineral Deposition. Academic Press, San Diego.

Soballe, K., Hansen, E.S., Brockstedt-Rasmussen, H., Hjortdal, V.E., Juhl, G.I., Møger Pedersen, K., Hvid, I., Bünger, C., 1991. Fixation of titanium and hydroxyapatite-coated implants in arthritic osteopenic bone. J. Arthroplasty 6 (4):307.

Sumitomo Chemical Company Limited (Osaka Japan), (1976). Implants for bones, joints and tooth roots. US Patent No. 4146936.

Surmenev, R.A., 2012. A review of plasma-assisted methods for calcium phosphate-based coatings fabrication. Surf. Coat. Technol. 206, 2035–2056.

Suzuki, T., Ohashi, R., Yokogawa, Y., Nishizawa, K., Nagata, F., Kawamoto, Y., et al., 1999. Initial anchoring and proliferation of fibroblast L-929 cells on unstable surface of calcium phosphate ceramics. J. Biosci. Bioeng. 87, 320–327.

Takadama, H., Kim, H.M., Kokubo, T., Nakamura, T., 2001. Mechanism of biomineralization of apatite on a sodium silicate glass: TEMEDX study in vitro. Chem. Mater. 13, 1108–1113.

Tampieri, A., Celotti, G., Landi, E., 2005. From biomimetic apatites to biologically inspired composites. Anal. Bioanal. Chem. 381, 568–576.

Tanahashi, M., Matsuda, T., 1997. Surface functional group dependence on apatite formation on self-assembled monolayers in a simulated body fluid. J. Biomed. Mater. Res. 34, 305–315.

Tang, Q., Brooks, R., Rushton, N., Best, S., 2010. Production and characterization of HA and SiHA coatings. J. Mater. Sci. Mater. Med. 21 (1), 173–181.

ter Brugge, P.J., Wolke, J.G.C., Jansen, J.A., 2002. Effect of calcium phosphate coating crystallinity and implant surface roughness on differentiation of rat bone marrow cells. J. Biomed. Mater. Res. 60, 70–78.

11.3 Various approaches for biomineralization on metals

Surface modifications of Ti metal and Ti-based alloys for the purpose of inducing apatite formation on their surfaces in SBF have been attempted by various methods, including ion implantation (Armitage et al., 2007; Nayab et al., 2007; Rautray et al., 2010), hydrothermal treatment (Dong et al., 2007; Park et al., 2007, 2010; Ueda et al., 2009; Chen et al., 2009; Zhang et al., 2010), and electrochemical reactions (Suh, 2003; Shibata et al., 2010; Bjursten et al., 2010; Zhao et al., 2010; Diefenback et al., 2011; Xie et al., 2011; Minagar et al., 2013; Whiteside et al., 2010; Zhang et al., 2013; Zhou et al., 2014). However, these techniques require special equipment and are not readily applicable to large-scale devices of complicated shape or those comprised of porous materials. In contrast with these techniques, simple chemical and heat treatments do not require any special equipment and are applicable to a broad range of shapes and materials.

Simple chemical and heat treatments for inducing apatite formation on Ti metal and Ti-based alloys in SBF have been attempted by various researchers (Wang et al., 2002; Wu et al., 2004; Takeuchi et al., 2005; Zhao et al., 2005, 2008; Cooper et al., 2006; Zhou et al., 2007; Liu et al., 2007; Lee et al., 2007; Sugino et al., 2009; Turkan and Guden, 2010; Li et al., 2010; Karthega and Rajendran, 2010; Ferraris et al., 2011). However, the general principle governing apatite formation has not been revealed by these studies. The present authors have systematically studied the chemical and heat treatments effective for inducing apatite formation for obtaining the general principle of apatite formation. In the following, this work is reviewed.

11.4 Simple chemical and heat treatments for biomineralization on metals

11.4.1 NaOH and heat treatments of Ti metal and its alloys

11.4.1.1 Surface structural change induced by NaOH and heat treatments

A typical procedure of this method is soaking of Ti metal in a 5 M NaOH solution at 60 °C for 24 h and its subsequent heat treatment at 600 °C for 1 h in an ambient atmosphere. Sodium and oxygen enter into the surface of the Ti metal to 1 μm in thickness by the first NaOH treatment, and only the oxygen among them penetrates into a slightly deeper region upon subsequent heat treatment, as shown in Figure 11.1 (Kim et al., 1999). According to thin-film X-ray diffraction, it was shown that sodium hydrogen titanate ($Na_xH_{2-x}Ti_yO_{2y+1}$; $0<x<2$ and $y=2, 3$, or 4) appears in fibrous form on the surface of the Ti metal within a level of 1 μm in thickness by the first NaOH treatment, and is transformed into sodium titanate ($Na_2Ti_yO_{2y+1}$; $y=5, 6$, etc.) and rutile TiO_2 without appreciable morphological change by the subsequent heat treatment, as shown in Figure 11.2 (Yamaguchi et al., 2009).

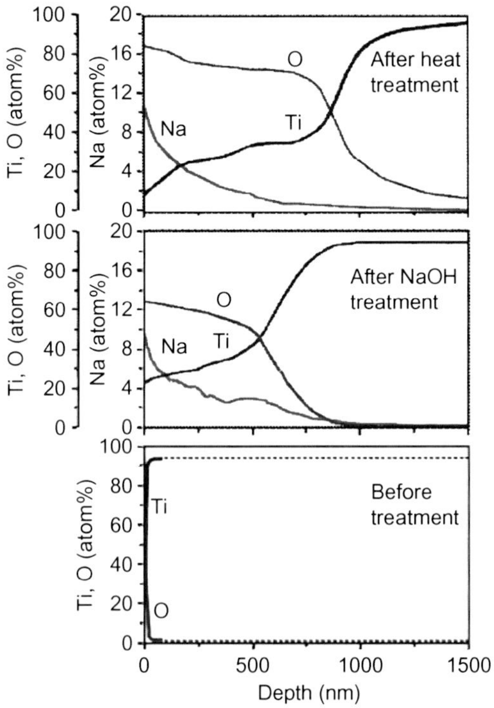

Figure 11.1 Depth profiles of Auger electron spectroscopy of Ti metal untreated and subjected to NaOH and heat treatments.
Reproduced from Kim et al. (1999), with permission John Wiley and Sons.

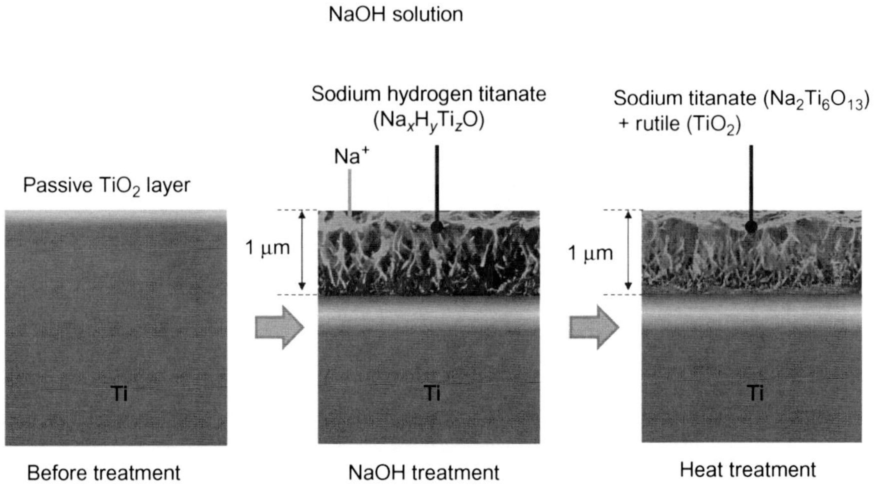

Figure 11.2 Structural change of the surfacc of Ti metal after NaOH and heat treatments.
Reproduced from Yamaguchi et al. (2009), with permission Ceramics Society of Japan.

Conventional Ti-based alloys such as Ti–6Al–4 V (Kim et al., 1996; Ueno et al., 2011), Ti–6Al–2Nb–Ta (Kim et al., 1996), and Ti–15Mo–5Zr–3Al (Kim et al., 1996) also form sodium titanate and rutile, by NaOH and heat treatments, since the alloying elements such as Al, V, and Mo are selectively released from the surface of the alloys during the first NaOH treatment (Kim et al., 2000a,b).

11.4.1.2 Apatite formation in SBF on metal surfaces subjected to NaOH and heat treatments

Ti metal subjected to NaOH and heat treatments described above fully forms apatite on its surface within 1 day in SBF, as shown in Figure 11.3 (Yamaguchi et al., 2009).

Based on X-ray photoelectron spectroscopy (Takadama et al., 2001a), transmission electron microscopy combined with energy dispersive X-ray analysis (Takadama et al., 2001b), and zeta potential measurements (Kim et al., 2003) on the Ti surface as a function of the soaking time in SBF, it was revealed that the apatite is formed during the process, as shown in Figure 11.4 (Kim et al., 2003). Sodium ions are first released from the surface of the sodium titanate layer via exchange with H_3O^+ ions in SBF. As a result, the pH of the surrounding environment on the surface increases and the surface becomes negatively charged (Gold et al., 1989). The positively charged calcium ions are adsorbed on the negatively charged surface. As the calcium ions accumulate, the surface becomes positively charged and adsorbs the negatively charged phosphate ions, forming a calcium phosphate. This phase is metastable and is eventually transformed into stable crystalline apatite.

Conventional Ti-based alloys such as Ti–6Al–4V (Kim et al., 1996; Ueno et al., 2011), Ti–6Al–2Nb–Ta (Kim et al., 1996), and Ti–15Mo–5Zr–3Al (Kim et al., 1996) with similar treatment also form an apatite layer on their surfaces in SBF.

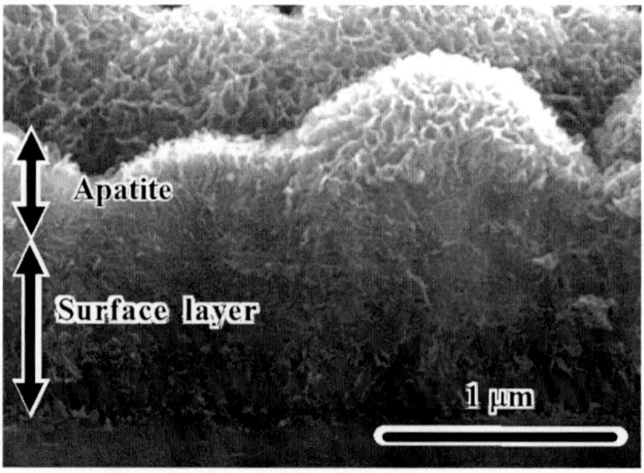

Figure 11.3 SEM photographs of a cross-section of the surface of Ti metal soaked in SBF for 1 day, after NaOH and heat treatments.
Reproduced from Yamaguchi et al. (2009), with permission Ceramics Society of Japan.

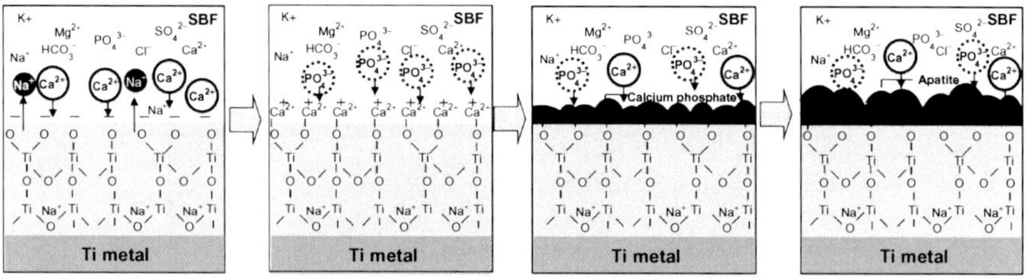

Figure 11.4 Process of apatite formation in SBF on Ti metal subjected to NaOH and heat treatments.
Reproduced from Kim et al. (2003), with permission John Wiley and Sons.

11.4.1.3 Bone-bonding of Ti metal and its alloys subjected to NaOH and heat treatments

The Ti metal subjected to NaOH and heat treatments described above forms an apatite layer on its surface in the living body of rabbit tibia, as in SBF, and so tightly bonds to the living bone, as shown in Figure 11.5 (Yan et al., 1997), such that fracture does not occur at the interface between the Ti metal and the bone under tensile stress. A Ti metal rod similarly treated cannot be removed from a medullary canal of rabbit femur without being accompanied by bone fragment after implantation for 12 weeks, as shown in Figure 11.6 (Nishiguchi et al., 2003).

Conventional Ti-based alloys such as Ti–6Al–4V, Ti–6Al–2Nb–Ta, and Ti–15Mo–5Zr–3Al after similar treatment also tightly bond to the surrounding bone without the intervention of fibrous tissue at the interface between the alloys and the bone (Nishiguchi et al., 1999).

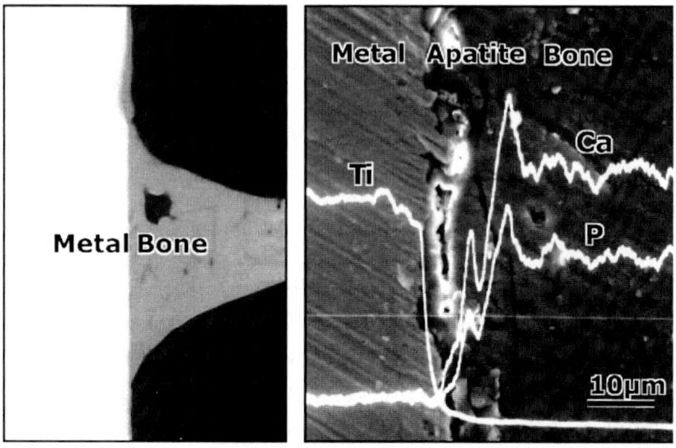

Figure 11.5 Contact microradiograph (left-hand side) and SEM picture with energy dispersive X-ray analysis (right-hand side) at the interface of Ti metal subjected to NaOH and heat treatments with neighboring bone, 8 weeks after implantation into rabbit tibia.
Reproduced from Yan et al. (1997), with permission John Wiley and Sons.

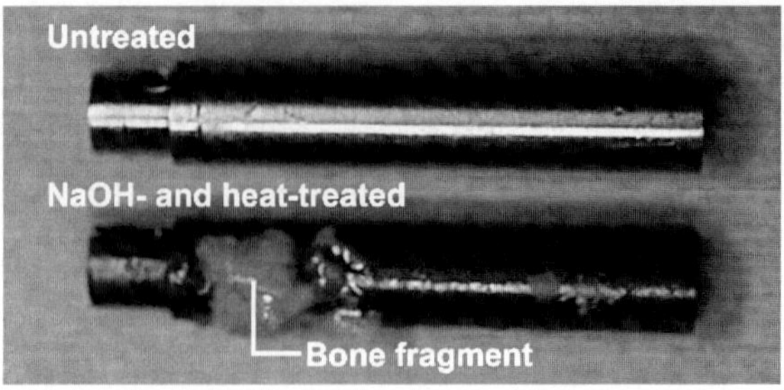

Figure 11.6 Untreated and NaOH- and heat-treated Ti metal rods removed from medullary canal of rabbit femur 12 weeks after implantation.
After Nishiguchi et al. (2003).

11.4.1.4 Some problems with NaOH and heat treatments

The NaOH and heat treatments are simple, but they also lead to certain problems that need to be solved.

Ca contamination

When even small amounts of calcium ions are contained in the NaOH solution, they preferentially enter into the sodium hydrogen titanate in the place of sodium ions during the first chemical treatment, and remain in the sodium titanate formed by the subsequent heat treatment so as to inhibit apatite formation by suppressing the release of sodium ions into SBF. Even a high-purity NaOH reagent often contains calcium contamination (Kizuki et al., 2013). Therefore, the reagent for the chemical treatment must be carefully chosen.

Humidity

The sodium ions of the sodium titanate are easily released even in a humid environment via exchange with the H_3O^+ ions in the humid moisture. As a result, the sodium concentration of the sodium titanate decreases before the soaking in SBF or implantation so as to suppress the apatite formation in SBF and the living body. Therefore, the treated specimens must be carefully stored in dry conditions (Kawai et al., 2010).

Ti–Zr–Nb–Ta alloys

NaOH and heat treatments are effective not only for Ti metal, but also for conventional Ti-based alloys in inducing surface apatite formation in SBF as well as enhancing bone-bonding ability, as described above. However, these treatments are not effective for a new kind of Ti-based alloy in the system Ti–Zr–Nb–Ta, since alloying elements such as Zr, Nb, and Ta are sparingly released during the NaOH treatment and remain in the surface layer, inhibiting apatite formation by suppressing the release of the Na^+ ions (Yamaguchi et al., 2011).

These problems are overcome by the following treatments.

11.4.2 NaOH, $CaCl_2$, heat, and water treatments of Ti and its alloys

11.4.2.1 Surface structural change induced by NaOH, $CaCl_2$, heat, and water treatments

A typical procedure of this method is soaking the Ti metal in NaOH solution under the same conditions as described above, successive soaking in 100 mM $CaCl_2$ solution at 40 °C for 24 h, subsequent heat treatment under the same conditions as described above, and additional soaking in hot water at 80 °C for 24 h (Kizuki et al., 2010).

The sodium hydrogen titanate formed by the first NaOH treatment has a layered structure in which the Na^+ ions are located at the interspace of the titanium oxide layers. These Na^+ ions are completely replaced with Ca^{2+} ions during the subsequent $CaCl_2$ treatment so as to form a calcium hydrogen titanate. This phase is transformed into calcium titanate and rutile by the subsequent heat treatment, as shown in Figure 11.7. The calcium ions in the calcium titanate near the surface are partly replaced with H_3O^+ ions so as to form a calcium-deficient calcium titanate by the additional hot water treatment, as shown in Figure 11.8 (Kizuki et al., 2010).

Ti-based alloys, including the new kind of alloys in the system Ti–Zr–Nb–Ta, also form essentially the same crystalline phases on their surfaces via the same treatments (Yamaguchi et al., 2010, 2012).

11.4.2.2 Apatite formation in SBF on metal surfaces subjected to NaOH, $CaCl_2$, heat, and water treatments

Ti metal subjected to NaOH, $CaCl_2$, and heat treatment does not form apatite on its surface in SBF within 3 days, but does form apatite after subsequent hot water treatment, as shown in Figure 11.9 (Kizuki et al., 2010). This is because the calcium ions

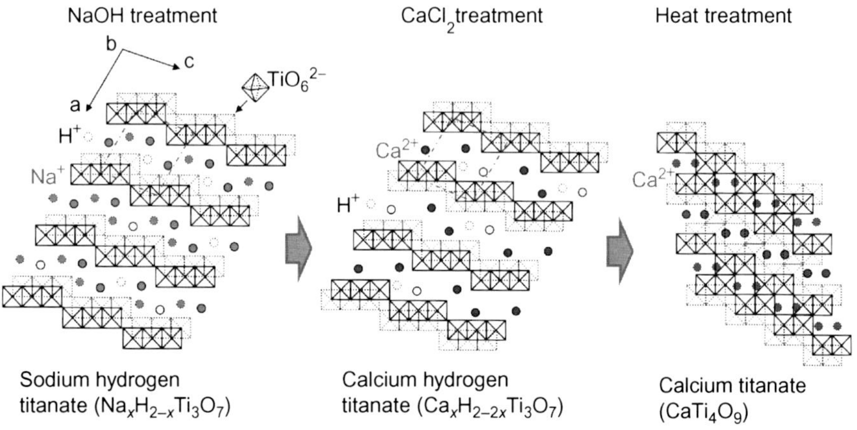

Figure 11.7 Structures of sodium hydrogen titanate, calcium hydrogen titanate, and calcium titanate perpendicular to the crystallographic b axis.

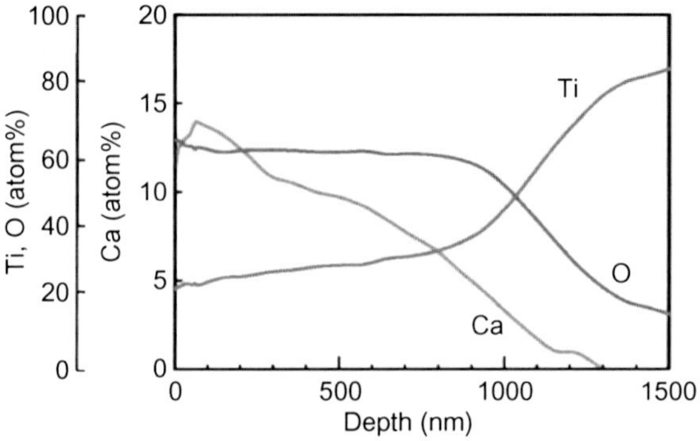

Figure 11.8 Depth profiles of Auger electron spectroscopy of Ti metal subjected to NaOH, $CaCl_2$, heat, and water treatments.
Reproduced from Kizuki et al. (2010), with permission Elsevier.

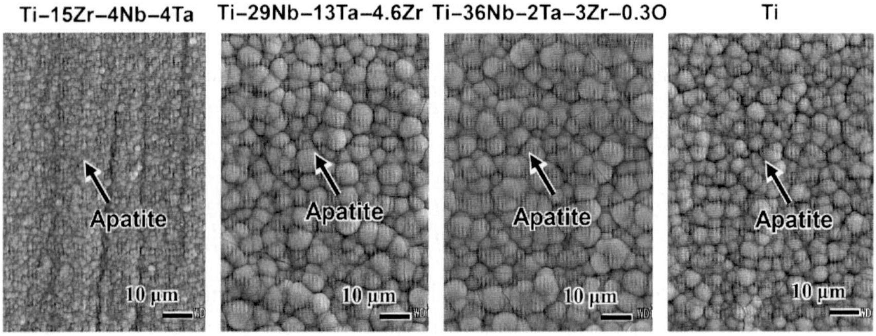

Figure 11.9 SEM photographs of Ti metal and its alloys subjected to NaOH, $CaCl_2$, heat, and water treatments after soaking in SBF for 3 days.
Reproduced from: Kizuki et al. (2010), with permission Elsevier and Yamaguchi et al. (2010, 2012), with permission Springer.

in the calcium titanate formed by the heat treatment are sparingly released in SBF, but those in the calcium-deficient calcium titanate after the hot water treatment are easily released.

In contrast to the NaOH and heat treatments, the NaOH, $CaCl_2$, heat, and water treatments are effective not only for Ti metal, but also for Ti-based alloys including the new generation of Ti–Zr–Nb–Ta alloys in inducing apatite formation on their surfaces in SBF, as shown in Figure 11.9 (Yamaguchi et al., 2010, 2012). This is because the released calcium ions are more effective than the released sodium ions for apatite formation.

In contrast to the NaOH and heat treatments, the apatite-forming capacity conferred by the NaOH, $CaCl_2$, heat, and water treatments is maintained even after storage in a humid environment over a long period of time (Yamaguchi et al., 2010). This is because the calcium ions in the calcium-deficient calcium titanate are not so easily released in the humid environment as the sodium ions of the sodium titanate.

11.4.2.3 Bone-bonding of the Ti metal and its alloys subjected to NaOH, $CaCl_2$, heat, and water treatments

Ti metal as well as Ti-based alloys including the new generation of Ti–Zr–Nb–Ta alloys bond to living bone without any intervention of fibrous tissue at the interface between the metals and the bone, as shown in Figure 11.10 (Fukuda et al., 2011a; Tanaka et al., 2014), when they are implanted into rabbit tibia after NaOH, $CaCl_2$, heat, and water treatments. Under tensile stress, fracture does not occur at their interfaces, but in the bone.

11.4.2.4 The evolution of NaOH, $CaCl_2$, heat, and water treatments

Various kinds of functional ions can be incorporated into the calcium-deficient calcium titanate, by adding them to the $CaCl_2$ solution, since these ions are able to enter into the interspace of the titanium oxide layers of the calcium hydrogen titanate and transform it into calcium titanate containing the functional ions via the subsequent heat treatment. If these ions are also added to the water in the final treatment, the functional ions are only sparingly released during the water treatment. These incorporated functional ions are slowly released in the body environment to produce various effects such as the promotion of bone growth and antibacterial activity.

For example, Mg (Yamaguchi et al., 2013a), Sr (Yamaguchi et al., 2014), and Zn (Yamaguchi et al., 2013b) ions, which are known to promote bone growth, are incorporated into the surface calcium-deficient titanate layer by the modified NaOH, $CaCl_2$, heat, and water treatments, as shown in Figure 11.11 (Yamaguchi et al., 2014). The resultant products not only form apatite on their surfaces in SBF, but also slowly release the functional ions into phosphate-buffered saline at 36.5 °C, as shown in Figure 11.12 (Yamaguchi et al., 2014). These ions would be expected to be released from the Ti metal also in the living body to promote growth of the surrounding bone. Consequently, early bonding of the Ti metal with the surrounding bone takes place.

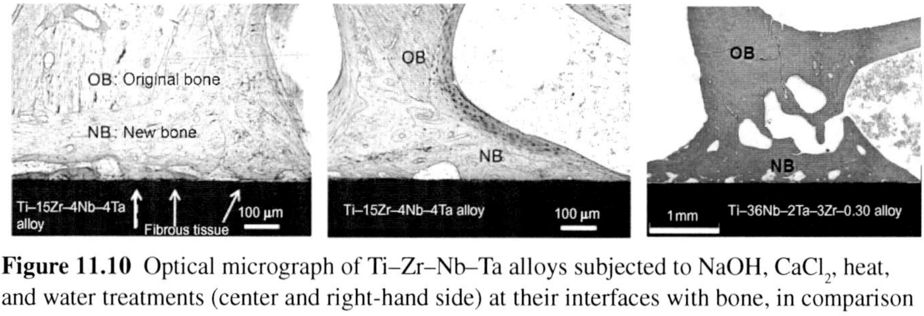

Figure 11.10 Optical micrograph of Ti–Zr–Nb–Ta alloys subjected to NaOH, $CaCl_2$, heat, and water treatments (center and right-hand side) at their interfaces with bone, in comparison with that untreated (left-hand side), 16 weeks after impanation into rabbit tibia for left-hand side and center, and 26 weeks after impanation for the right-hand side.
Reproduced from: Fukuda et al. (2011a), with permission Elsevier and Tanaka et al. (2014), with permission Springer.

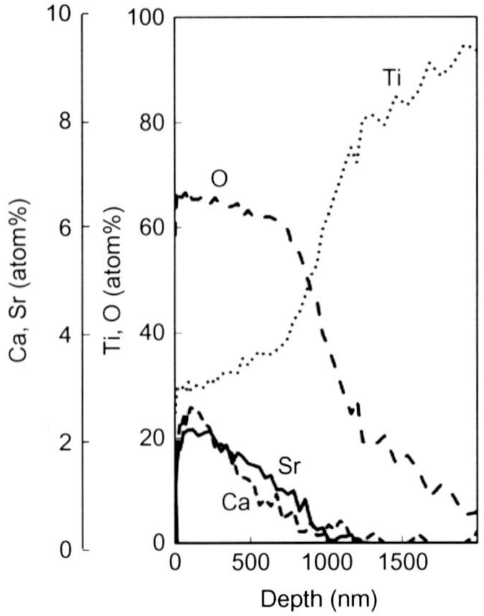

Figure 11.11 Depth profile of XPS of Ti metal subjected to NaOH, $CaCl_2/SrCl_2$, heat, and $SrCl_2$ treatments.
Reproduced from Yamaguchi et al. (2014), with permission Elsevier.

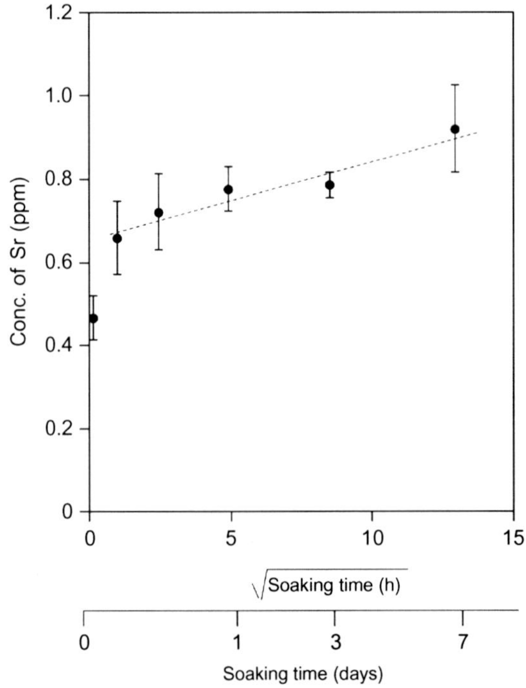

Figure 11.12 Sr ion release from Ti metal subjected to NaOH, $CaCl_2/SrCl_2$, heat, and $SrCl_2$ treatments in phosphate-buffered saline at 36.5 °C.
Reproduced from Yamaguchi et al. (2014), with permission Elsevier.

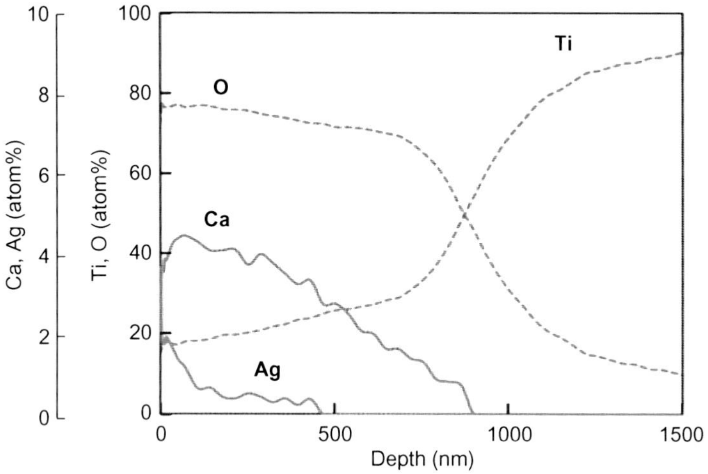

Figure 11.13 Depth profiles of Auger electron spectroscopy of Ti metal subjected to NaOH, $CaCl_2$, heat, and $AgNO_3$ treatments.
Reproduced from Kizuki et al. (2014), with permission Springer.

Silver ions, which are known to exhibit antibacterial activity, may also be incorporated into the surface calcium-deficient calcium titanate layer by adding $AgNO_3$ to the water of the final treatment, as shown in Figure 11.13 (Kizuki et al., 2014). The resultant product not only forms apatite on its surface in SBF, but also slowly releases Ag^+ ions into fetal bovine serum so as to exhibit antibacterial activity. These Ag^+ ions are expected to be released from the Ti metal in the living body to prevent infection.

There still remain certain problems with both the NaOH and heat treatments as well as the NaOH, $CaCl_2$, heat, and water treatments. The surface layer formed by these treatments, which is approximately 1 μm thick with nanometer-scale roughness, is liable to be removed by shear stress, for example when the treated metals are screwed into bone. In addition, the Na^+ and Ca^{2+} ions released from the treated metals are liable to exert an adverse effect on living cell activity, especially in the narrow space of a porous metal. These problems are overcome by the following treatments.

11.4.3 Acid and heat treatments of Ti metal

11.4.3.1 Surface structural change induced by acid and heat treatments

A typical procedure of this method is soaking the Ti metal into 66.3% H_2SO_4–10.6% HCl = 1:1 mixed acid solution at 70 °C for 1 h with a subsequent heat treatment at a temperature ranging from 500 to 650 °C for 1 h in an ambient atmosphere (Kokubo et al., 2010).

A layer of titanium hydride (TiH_x, where $0 < x < 2$) adsorbed with SO_4^{2-} and/or Cl^- ions about 1 μm in thickness with micrometer-scale roughness is produced on the surface of the Ti metal by the acid treatment. This layer is converted into rutile TiO_2

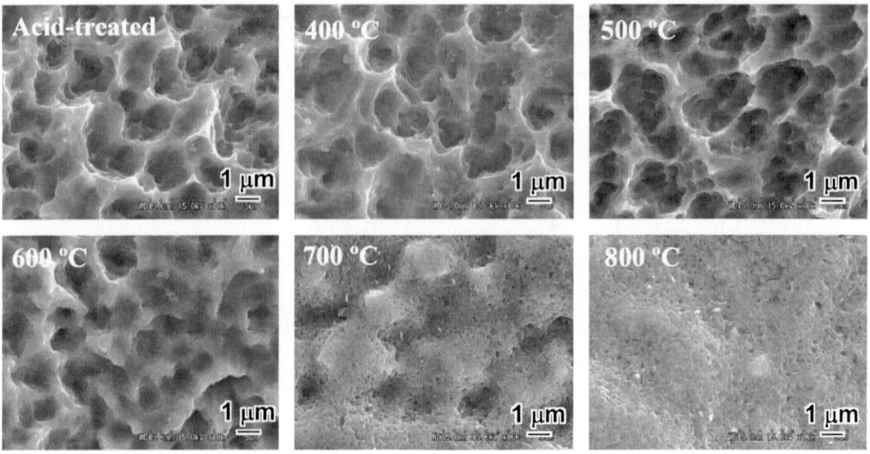

Figure 11.14 SEM photographs of surfaces of Ti metals subjected to acid and heat treatments. Reproduced from Kokubo et al. (2010), with permission The Royal Society.

adsorbed with SO_4^{2-} and/or Cl^- ions without appreciable morphological change by the subsequent heat treatment, as shown in Figure 11.14 (Kokubo et al., 2010).

11.4.3.2 Apatite formation in SBF on the Ti surfaces subjected to acid and heat treatments

Ti metal subjected to the acid and heat treatments described above forms apatite fully on its surface in SBF within 1 day, in contrast with that which was treated with only the acid, as shown in Figure 11.15 (Kokubo et al., 2010).

According to X-ray photoelectron spectroscopy and zeta potential measurements of the Ti surface as a function of the soaking time in SBF, it was shown that apatite is formed by the process shown in Figure 11.16 (Kokubo et al., 2010). The titanium oxide on the surface of the Ti metal dissociates the adsorbed SO_4^{2-} and/or Cl^- ions into SBF to

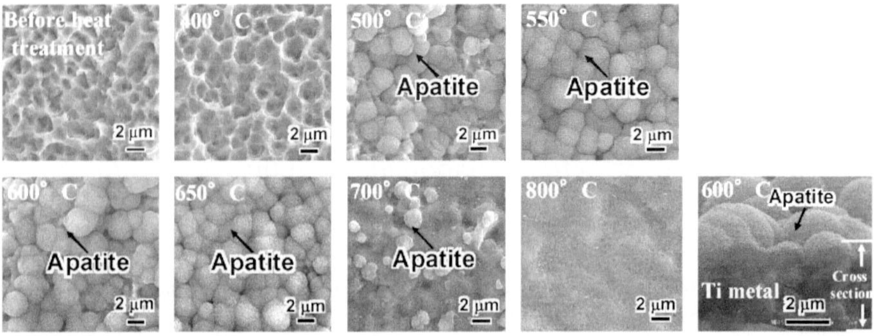

Figure 11.15 SEM photographs of surfaces of Ti metal soaked in SBF for 1 day after acid and heat treatments at various temperatures.
Reproduced from Kokubo et al. (2010), with permission The Royal Society.

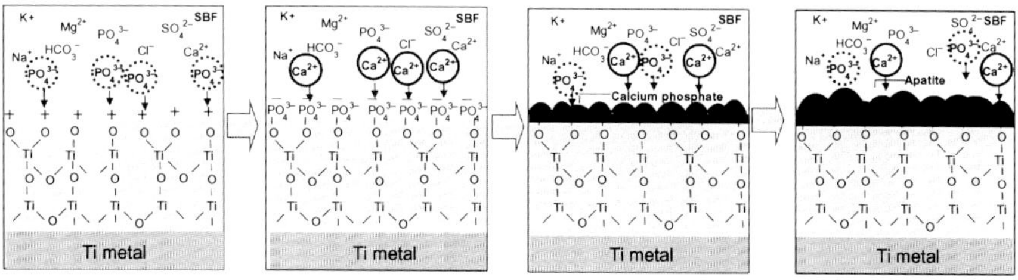

Figure 11.16 Process of apatite formation in SBF on Ti metal subjected to acid and heat treatments.
Reproduced from Kokubo et al. (2010), with permission The Royal Society.

produce an acidic environment. It is reported that titanium oxide is positively charged in an acidic environment (Gold et al., 1989). Therefore, the surface of the titanium oxide is positively charged to preferentially adsorb the negatively charged phosphate ions in SBF. As the phosphate ions accumulate, the surface becomes negatively charged so as to adsorb the positively charged calcium ions, forming a calcium phosphate. This phase is metastable and is eventually transformed into stable crystalline apatite.

The surface of the acid-treated Ti metal is not charged, and hence does not induce apatite formation (Kokubo et al., 2010).

The apatite-forming capacity conferred by the acid and heat treatments is also maintained even after storage in a humid environment over a long period of time (Kokubo et al., 2010).

11.4.3.3 Bone-bonding of Ti metal subjected to acid and heat treatments

Ti metal subjected to the acid and heat treatments described above tightly bonds to the living bone without the intervention of fibrous tissue, in contrast to that of the acid-treated and heat-treated material when it is implanted into a rabbit tibia, as shown in Figure 11.17 (Kawai et al., 2012). The Ti metal subjected to the acid and heat treatments does not produce fracture at the interface between the metal and the bone under tensile stress, but in the bone.

11.4.3.4 Osteoconduction of Ti metal subjected to acid and heat treatments

When porous Ti metal subjected to the acid and heat treatments described above is implanted into a rabbit femur, the surrounding bone deeply penetrates into the porous structure within 3 weeks, in contrast to that subjected to only acid or heat treatment, as shown in Figure 11.18 (Kawai et al., 2013).

11.4.3.5 Osteoinduction of Ti metal subjected to acid and heat treatments

When the same porous Ti metal subjected to the acid and heat treatments is implanted into a muscle of dog, bone tissue is ectopically formed in the porous

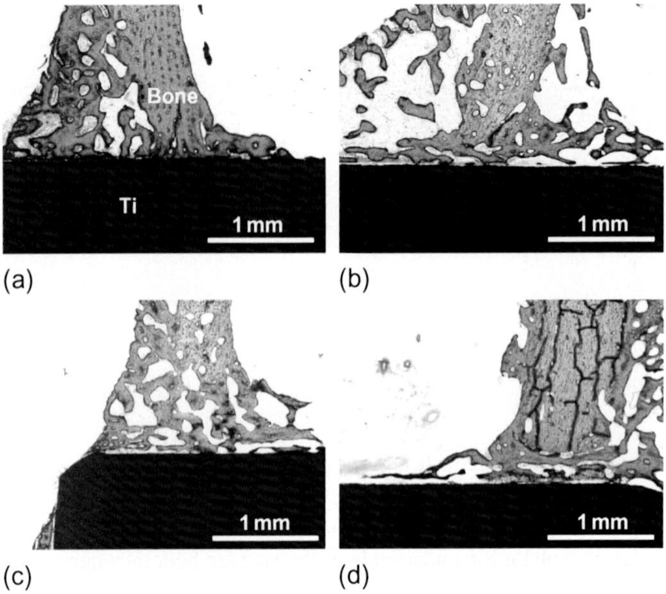

Figure 11.17 Optical micrographs of nondecalcified sections of Ti metals subjected to acid and heat treatments (a), in comparison with those subjected to no treatment (b), only heat treatment (c), and only acid treatment (d), 4 weeks after implantation into rabbit tibia. Reproduced from Kawai et al. (2012), with permission Springer.

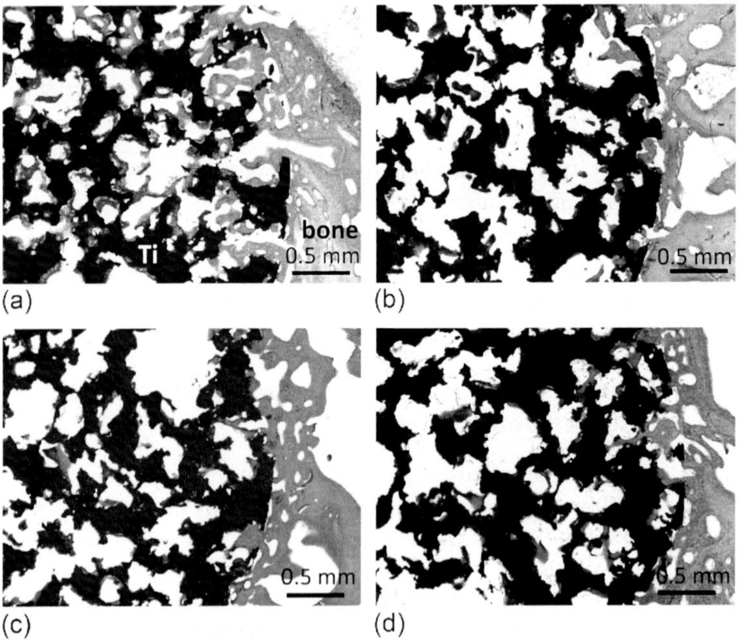

Figure 11.18 Optical micrographs of nondecalcified sections of porous Ti metals subjected to acid and heat treatments (a), in comparison with those subjected to no treatment (b), only heat treatment (c), and only acid treatment (d), 3 weeks after implantation into femoral condyle of rabbit. Reproduced from Kawai et al. (2013), with permission Springer.

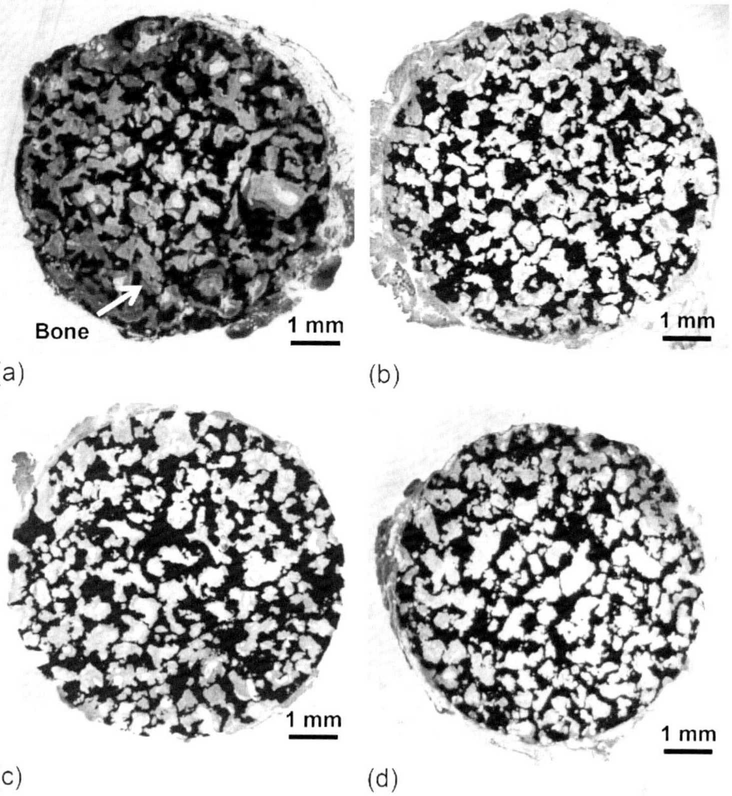

Figure 11.19 Optical micrographs of nondecalcified sections of porous Ti metal subjected to acid and heat treatment (a), in comparison with those subjected to no treatment (b), only heat treatment (c), and only acid treatment (d), 12 months after implantation into muscle of dog. Reproduced from Kawai et al. (2014), with permission PLOS.

structure within 12 months, in contrast to that subjected to only acid or heat treatment, as shown in Figure 11.19 (Kawai et al., 2014).

11.4.4 The dependence of apatite formation on Ti metal upon pH of the exposed solution

It has been shown by the experiments described above that various chemical treatments followed by heat treatment are effective in inducing apatite formation on Ti metal in SBF. In order to investigate the dependence of apatite formation on the Ti metal in SBF upon the pH of the exposed solution, Ti metal was soaked in a solution with pH systematically changed from approximately 0 to 14 using HCl and NaOH at 60 °C for 24 h, and then heat-treated at 600 °C for 1 h. As a result, it was found that the Ti metal forms apatite on its surface in SBF within 3 days, when it was exposed to solutions of pH values lower than 1.1 or higher than 13.6, as shown in Figure 11.20 (Pattanayak et al., 2012).

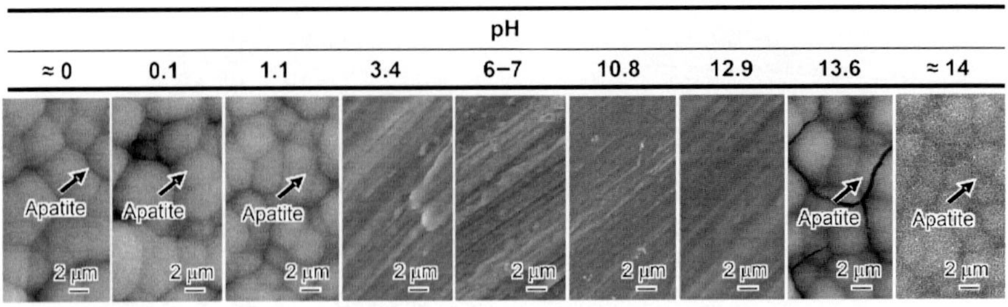

Figure 11.20 SEM photographs of surfaces of Ti metal soaked in SBF for 3 days after heat treatment following exposure to solutions with different pH values.
Reproduced form Pattanayak et al. (2012), with permission The Royal Society.

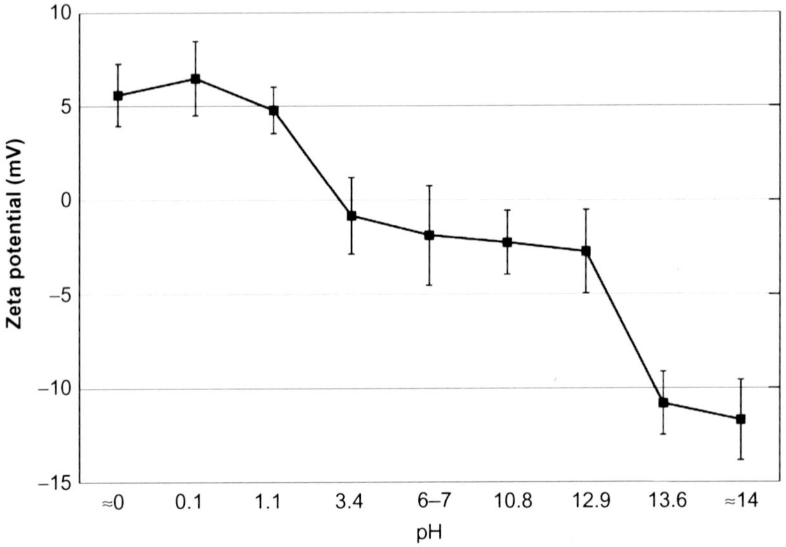

Figure 11.21 Zeta potentials of Ti metal heat-treated after exposure to solutions with different pH values.
Reproduced from Pattanayak et al. (2012), with permission The Royal Society.

Apatite formation depends not on surface roughness or specific crystalline phase, but upon surface charge. The Ti metal is positively charged, when heat-treated after exposure to strong acid solutions, and negatively charged when heat-treated after exposure to strong alkali solutions, as shown in Figure 11.21 (Pattanayak et al., 2012). The reason why the Ti metal heat-treated after exposure to acid solution is positively charged in SBF was already discussed in the case of the acid and heat treatments. The reason why the Ti metal heat-treated after exposure to the alkali solution is negatively charged was already discussed in the case of the NaOH and heat treatments. The reason why the positively or negatively charged Ti metal forms apatite on its surface in SBF was already discussed and illustrated in Figures 11.4 and 11.16.

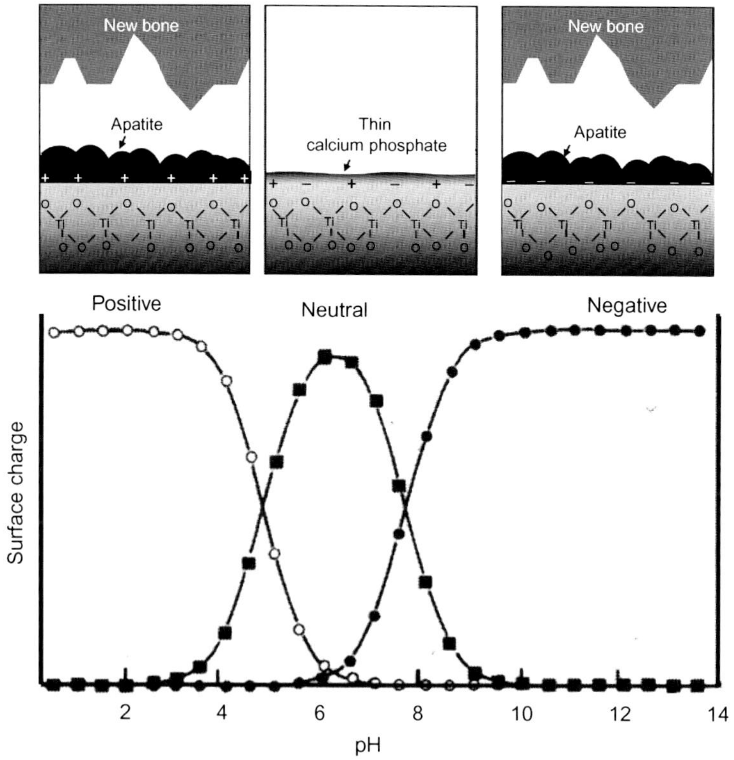

Figure 11.22 Surface charge of titanium oxide and apatite formation on its surface as a function of pH of exposed solution.
Reproduced from Pattanayak et al. (2012), with permission The Royal Society.

It is clear from Figures 11.20 and 11.21 that the Ti metal forms apatite on its surface in SBF, when it is heat-treated after exposure to strong acid or alkali solutions, because it becomes positively or negatively charged on its surface, as shown in Figure 11.22 (Pattanayak et al., 2012). The natural Ti metal does not form apatite on its surface in SBF even after heat treatment, since it is neither positively nor negatively charged. This indicates that the Ti metal is able to bond to living bone, if it is heat-treated after exposure to a strong acid or alkali solution, even though the natural Ti metal is unable to do so.

11.4.5 Biomineralization on metals other than Ti metal and its alloys

A zirconia gel prepared by a sol–gel method also forms apatite on its surface in SBF within 3 days, when it is heat-treated at 600 °C (Uchida et al., 2002b). This indicates the possibility for zirconium metal to form apatite on its surface in SBF, when its surface is slightly chemically modified. However, in fact the zirconium metal formed apatite only sparingly on its surface in SBF even after 28 days, when it was treated with

5–15 M NaOH solution at 95 °C for 24 h (Uchida et al., 2002a). Chen et al. (2008) reported that zirconium metal plate produced by sintering with Zr metal powders forms apatite on its surface in SBF within 7 days, when it was soaked in 10 M NaOH solution at 60 °C for 24 h and then heat-treated at 600 °C for 1 h in a vacuum.

A niobium oxide gel as prepared by a sol–gel method also forms apatite on its surface in SBF within 7 days (Miyazaki et al., 2001a). This indicates that niobium metal might be able to form apatite on its surface in SBF, if the surface is chemically modified. However, the niobium metal does not form apatite on its surface within 14 days, when treated with 0.5–2.0 M NaOH solution at 60 °C for 24 h (Miyazaki et al., 2001a). Treatment with NaOH solution at concentrations greater than 2 M produced cracks on the surface of the metal. Wang et al. (2008) reported that the niobium metal formed apatite on its surface in SBF within 7 days, when soaked in 0.5 M NaOH solution at 80 °C for 24 h, and then heat-treated at 600 °C for 1 h in a vacuum.

A tantalum oxide gel prepared by a sol–gel method also forms apatite on its surface in SBF within 7 days (Miyazaki et al., 2001b). This indicates that tantalum metal forms apatite on its surface in SBF when its surface is chemically modified. The tantalum metal forms apatite on its surface in SBF without the need of any surface chemical modification (Miyazaki et al., 2000). However, it takes an induction period of as long as 4 weeks for apatite to form. It forms apatite within 7 days when treated with 0.2–0.5 M NaOH solution at 60 °C for 24 h so as to form a sodium tantalate hydrogel layer on its surface. When it was subsequently heat-treated at 300 °C for 1 h after NaOH treatment, the adhesive strength of the surface layer to the substrate increased without any decrease in its apatite-forming ability (Miyazaki et al., 2001c, 2002). Tantalum metal subjected to 0.5 M NaOH treatment and heat treatment at 300 °C was confirmed to be strongly bonded to the living bone without any intervention of fibrous tissue at the bone interface within 16 weeks in an animal experiment using rabbit tibia (Kato et al., 2000).

An alumina gel prepared by a sol–gel method does not form apatite on its surface in SBF (Li et al., 1994). This indicates there is little possibility for aluminum metal to form apatite on its surface in SBF, even if its surface is chemically modified.

11.5 Clinical applications of metallic biomaterials inducing biomineralization

11.5.1 Applications to total hip joint

A porous Ti metal layer formed on an acetabular shell and femoral stem of a total artificial hip joint made of Ti–6Al–2Nb–Ta alloy was subjected to NaOH and heat treatments to induce apatite formation on its surface in the living body. The resultant joint has been clinically used since 2007 in Japan, as shown in Figure 11.23 (Kawanabe et al., 2009). It was confirmed by two implants retrieved by treatment for bone fracture and infection that the Ti metal was tightly integrated with newly grown bone as early as 2 weeks after implantation (So et al., 2013).

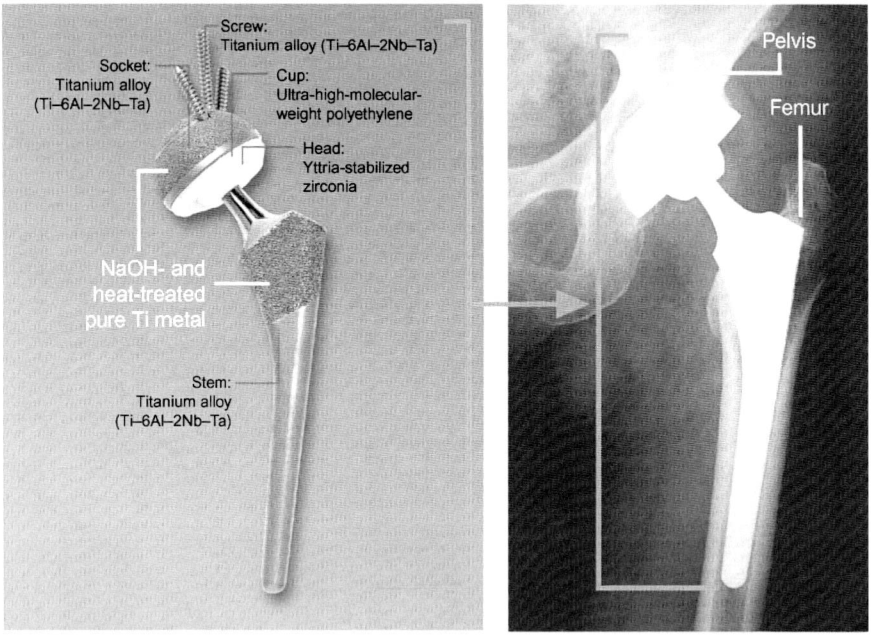

Figure 11.23 Application of NaOH and heat treatments to porous Ti metal layer on an artificial hip joint.
After Kawanabe et al. (2009).

11.5.2 Application to spinal fusion device

The porous Ti metal subjected to NaOH and heat treatments to form sodium titanate and rutile TiO_2 on its surface exhibits only low osteoinduction. However, if the same porous Ti metal is subjected to HCl treatment after the NaOH treatment to remove the sodium ions from the sodium hydrogen titanate (Pattanayak et al., 2009) and is then subjected to heat treatment to form rutile TiO_2 on its surface (Pattanayak et al., 2011a), the resultant product forms apatite on its surface in SBF (Pattanayak et al., 2011b) and exhibits a high level of osteoinduction (Fujibayashi et al., 2004; Takemoto et al., 2006; Fukuda et al., 2011b) as well as osteoconduction (Takemoto et al., 2005), as shown in Figure 11.24 (Tanaka et al., 2009).

The NaOH-, HCl-, and heat-treated porous Ti metal was applied to a spinal fusion device in a canine model (Takemoto et al., 2007). Based on this result, it was subjected to clinical trials as a spinal fusion device for five patients in the period of November 2008 to June 2009, as shown in Figure 11.25 (Fujibayashi et al., 2011). Conventional spinal fusion devices need an autograft for fixation to the surrounding bone, since they cannot bond to living bone, whereas the present device can be fixed without an autograft, as it does bond to living bone. All of the clinical cases to date have resulted in a successful outcome.

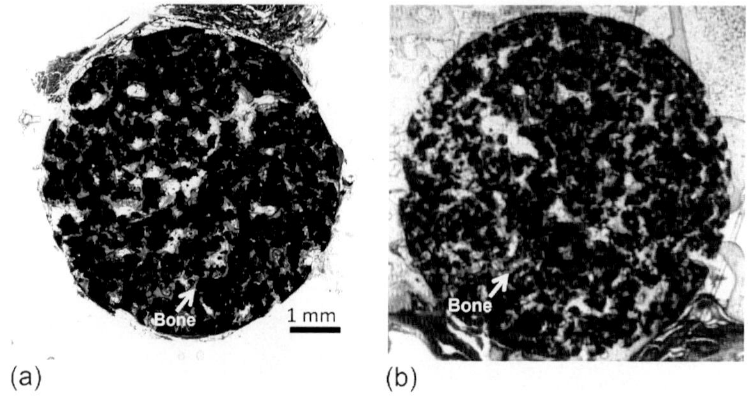

Figure 11.24 Optical micrographs of porous Ti metal subjected to NaOH, HCl, and heat treatments implanted into muscle of dog for 12 months (a) and into rabbit femur for 26 weeks (b). Reproduced from Takemoto et al. (2006), with permission Elsevier and after Tanaka et al. (2009).

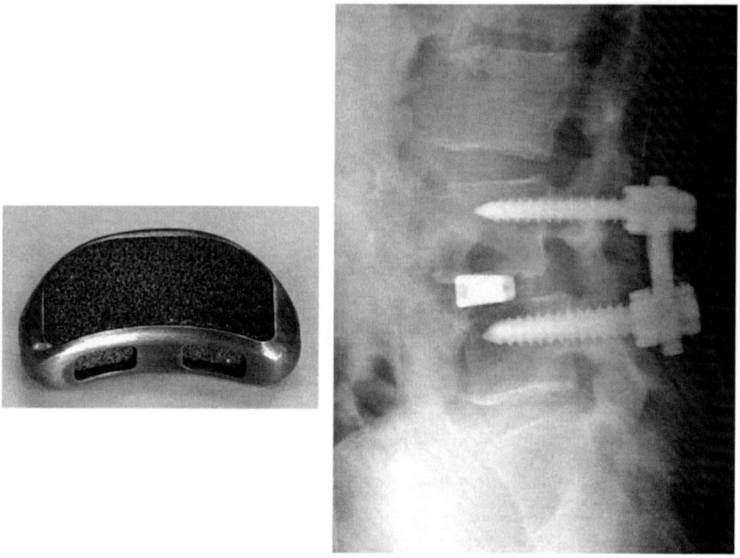

Figure 11.25 Spinal fusion device of porous Ti metal subjected to NaOH, HCl, and heat treatments (left-hand side) and its clinical application (right-hand side).
Reproduced from Fujibayashi et al. (2011), with permission Springer.

11.6 Future aspects of biomineralization on metals

The general principle for obtaining bone-bonding metals by using biomineralization has been demonstrated for Ti metal and its alloys.

The general principle for inducing bone-bonding ability via biomineralization in general is expected to be revealed even for metals other than just Ti metal and its alloys in the future.

Various functions besides bone-bonding ability are expected to be derived from metals by biomineralization.

A variety of bone-bonding ceramics have been developed and are being used clinically as important bone substitutes. However, they cannot be used under load-bearing conditions, since their fracture toughness is insufficient. The metals currently used to induce biomineralization exhibit not only bone-bonding properties but also high fracture toughness. Therefore, these metals are likely to find broad applications in orthopedics and dentistry.

References

Armitage, D.A., Mihoc, R., Tate, T.J., McPhail, D.S., Chater, R., Hobkirk, J.A., Shinawi, L., Jones, F.H., 2007. The oxidation of calcium implanted titanium in water: a depth profiling study. Appl. Surf. Sci. 253, 4085–4093.

Bjursten, L.M., Rasmusson, L., Oh, S., Smith, G.C., Brammer, K.S., Jin, S., 2010. Titanium dioxide nanotubes enhance bone bonding in vivo. J. Biomed. Mater. Res. 92A, 1218–1224.

Chen, X., Nouri, A., Li, Y., Lin, J., Hodgson, P.D., Wen, C., 2008. Effect of surface roughness of Ti, Zr, and TiZr on apatite precipitation from simulated body fluid. Biotechnol. Bioeng. 101, 378–387.

Chen, X.-B., Li, Y.C., Plessis, J.D., Hodgson, P.D., Wen, C., 2009. Influence of calcium ion deposition on apatite-inducing ability of porous titanium for biomedical applications. Acta Biomater. 5, 1808–1820.

Cooper, L.F., Zhou, Y., Takebe, J., Guo, J., Abron, A., Holmen, A., Ellingsen, J.E., 2006. Fluoride modification effects on osteoblast behavior and bone formation at TiO_2 grit-blasted c.p. titanium endosseous implants. Biomaterials 27, 926–936.

Diefenback, M., Muckley, T., Shrader, C., Schmidt, J., Zankovych, S., Bossert, J., Jandt, K.D., Faucon, M., Finger, U., 2011. The effect of plasma chemical oxidation of titanium alloy on bone-implant contact in rats. Biomaterials 32, 8041–8047.

Dong, W., Zhang, T., Epstein, J., Cooney, L., Wang, H., Li, Y., Jiang, Y.-B., Cogbill, A., Varadan, V., Tian, Z.R., 2007. Multifunctional nanowire bioscaffolds on titanium. Chem. Mater. 19, 4454–4459.

Ferraris, S., Spriano, S., Pan, G., Venturello, A., Bianchi, C.L., Chiesa, R., Faga, M.G., Maina, G., Verne, E., 2011. Surface modification of Ti-6Al-4V alloy for biomineralization and specific biological response: part 1, inorganic modification. J. Mater. Sci. Mater. Med. 22, 533–545.

Fujibayashi, S., Neo, M., Kim, H.-M., Kokubo, T., Nakamura, T., 2004. Osteoinduction of porous bioactive titanium metal. Biomaterials 25, 443–450.

Fujibayashi, S., Takemoto, M., Neo, M., Matsushita, T., Kokubo, T., Doi, K., Ito, T., Shimizu, A., Nakamura, T., 2011. A novel synthetic material for spinal fusion: a prospective clinical trial of porous bioactive titanium metal for lumbar interbody fusion. Eur. Spine J. 20, 1486–1495.

Fukuda, A., Takemoto, M., Saito, T., Fujibayashi, S., Neo, M., Yamaguchi, S., Kizuki, T., Matsushita, T., Niinomi, M., Kokubo, T., Nakamura, T., 2011a. Bone bonding bioactivity of Ti metal and Ti-Zr-Nb-Ta alloys with Ca ions incorporated on their surfaces by simple chemical and heat treatments. Acta Biomater. 7, 1379–1386.

Fukuda, A., Takemoto, M., Saito, T., Fujibayashi, S., Neo, M., Pattanayak, D.K., Matsushita, T., Sasaki, K., Nishida, N., Kokubo, T., Nakamura, T., 2011b. Osteoinduction of porous Ti implants with a channel structure fabricated by selective laser melting. Acta Biomater. 7, 2327–2336.

Gold, J.M., Schmidt, M., Steinmann, S.G., 1989. XPS study of amino acid adsorption to titanium surfaces. Helv. Phys. Acta 62, 246–249.

Karthega, M., Rajendran, N., 2010. Hydrogen peroxide treatment on Ti-6Al-4V alloy: a promising surface modification technique for orthopaedic application. Appl. Surf. Sci. 256, 2176–2183.

Kato, H., Nakamura, T., Nishiguchi, S., Matsusue, Y., Kobayashi, M., Miyazaki, T., Kim, H.-M., Kokubo, T., 2000. Bonding of alkali- and heat-treated tantalum implants to bone. J. Biomed. Mater. Res. B: Appl. Biomater. 53, 28–35.

Kawai, T., Kizuki, T., Takadama, H., Matsushita, T., Unuma, T., Nakamura, T., Kokubo, T., 2010. Apatite formation on surface titanate layer with different Na content on Ti metal. J. Ceram. Soc. Jpn. 118, 19–24.

Kawai, T., Takemoto, M., Fujibayashi, S., Neo, M., Akiyama, H., Yamaguchi, S., Pattanayak, D.K., Matsushita, T., Nakamura, T., Kokubo, T., 2012. Bone-bonding properties of Ti metal subjected to acid and heat treatments. J. Mater. Sci. Mater. Med. 23, 2981–2992.

Kawai, T., Takemoto, M., Fujibayashi, S., Akiyama, H., Yamaguchi, S., Pattanayak, D.K., Doi, K., Matsushita, T., Nakamura, T., Kokubo, T., Matsuda, S., 2013. Osteoconduction of porous Ti metal enhanced by acid and heat treatments. J. Mater. Sci. Mater. Med. 24, 1701–1715.

Kawai, T., Takemoto, M., Fujibayashi, S., Akiyama, H., Tanaka, S., Yamaguchi, S., Pattanayak, D.K., Doi, K., Matsushita, T., Nakamura, T., Kokubo, T., Matsuda, S., 2014. Osteoinduction on acid and heat treated porous Ti metal samples in canine muscle. PLoS One 9, e88366.

Kawanabe, K., Ise, K., Goto, K., Akiyama, H., Nakamura, T., Kaneuji, A., Sugimori, T., Matsumoto, T., 2009. A new cementless total hip arthroplasty with bioactive titanium porous-coating by alkaline and heat treatments: average 4.8-year results. J. Biomed. Mater. Res. B: Appl. Biomater. 90B, 476–481.

Kim, H.-M., Miyaji, F., Kokubo, T., Nakamura, T., 1996. Preparation of bioactive Ti and its alloys via simple chemical surface treatment. J. Biomed. Mater. Res. 32, 409–417.

Kim, H.-M., Miyaji, F., Kokubo, T., Nishiguchi, S., Nakamura, T., 1999. Graded surface structure of bioactive titanium prepared by chemical treatment. J. Biomed. Mater. Res. 45, 100–107.

Kim, H.-M., Takadama, H., Miyaji, F., Kokubo, T., Nishiguchi, S., Nakamura, T., 2000a. Formation of bioactive functionally graded structure on Ti-6Al-4V alloy by chemical surface treatment. J. Mater. Sci. Mater. Med. 11, 555–559.

Kim, H.-M., Takadama, H., Kokubo, T., Nishiguchi, S., Nakamura, T., 2000b. Formation of bioactive graded surface structure on Ti-15Mo-5Zr-3Al alloys by chemical treatment. Biomaterials 21, 353–358.

Kim, H.-M., Himeno, T., Kawashita, M., Lee, J.-H., Kokubo, T., Nakamura, T., 2003. Surface potential change in bioactive titanium metal during the process of apatite formation in simulated body fluid. J. Biomed. Mater. Res. A 67A, 1305–1309.

Kizuki, T., Takadama, H., Matsushita, T., Nakamura, T., Kokubo, T., 2010. Preparation of bioactive Ti metal surface enriched with calcium ions by chemical treatment. Acta Biomater. 6, 2836–2842.

Kizuki, T., Takadama, H., Matsushita, T., Nakamura, T., Kokubo, T., 2013. Effect of Ca contamination on apatite formation in a Ti metal subjected to NaOH and heat treatments. J. Mater. Sci. Mater. Med. 24, 635–644.

Kizuki, T., Matsushita, T., Kokubo, T., 2014. Antibacterial and bioactive calcium titanate layers formed on Ti metal and its alloys. J. Mater. Sci. Mater. Med. 25, 1737–1746.

Kokubo, T., Takadama, H., 2006. How useful is SBF in predicting in vivo bone bioactivity? Biomaterials 15, 2907–2915.

Kokubo, T., Pattanayak, D.K., Yamaguchi, S., Takadama, H., Matsushita, T., Kawai, T., Takemoto, M., Fujibayashi, S., Nakamura, T., 2010. Positively charged bioactive charged bioactive Ti metal prepared by simple chemical and heat treatments. J. R. Soc. Interface 7, S503–S513.

Lee, M.-H., Park, H.-S., Min, K.-S., Ahn, S.-G., Park, J.-M., Song, K.-Y., Park, C.-W., 2007. Evaluation of in vitro tests for surface-modified titanium by H_2SO_4 and H_2O_2 treatment. Met. Mater. Int. 13, 109–115.

Leeuwenburgh, S.C.G., Wolke, J.G.C., Jansen, J.A., de Groot, K., 2008. Calcium phosphate coatings. In: Kokubo, T. (Ed.), Bioceramics and Their Clinical Applications. Woodhead, Cambridge, pp. 464–484.

Li, P., Ohtsuki, C., Kokubo, T., Nakanishi, K., Soga, N., de Groot, K., 1994. The role of hydrated silica, titania, and alumina in inducing apatite on implants. J. Biomed. Mater. Res. 28, 7–15.

Li, Y., Zou, S., Wang, D., Feng, G., Bao, C., Hu, J., 2010. The effect of hydrofluoric acid treatment on titanium implant osseointegration in ovariectomized rats. Biomaterials 31, 3266–3273.

Liu, X., Zhao, Z., Leng, Y., 2007. Biomimetic calcium phosphate coatings on nitric-acid-treated titanium surfaces. Mater. Sci. Eng. C 27, 700–708.

Minagar, S., Wang, J., Barndt, C.C., Ivanova, E.P., Wen, C., 2013. Cell response of anodized nanotubes on titanium and titanium alloys. J. Biomed. Mater. Res. A 101A, 2726–2739.

Miyazaki, T., Kim, H.-M., Kokubo, T., Kato, H., Nakamura, T., 2000. Bioactive tantalum metal prepared by NaOH treatment. J. Biomed. Mater. Res. 50, 35–42.

Miyazaki, T., Kim, H.-M., Kokuo, T., Ohtsuki, C., Nakamura, T., 2001a. Apatite-forming ability of niobium oxide gels in a simulated body fluid. J. Ceram. Soc. Jpn. 109, 929–933.

Miyazaki, T., Kim, H.-M., Kokubo, T., Kato, H., Nakamura, T., 2001b. Induction and acceleration of bonelike apatite formation on tantalum oxide gel in simulated body fluid. J. Sol-Gel Sci. Technol. 21, 83–88.

Miyazaki, T., Kim, H.-M., Kokubo, T., Miyaji, F., Kato, H., Nakamura, T., 2001c. Effect of thermal treatment on apatite-forming ability on NaOH-treated tantalum metal. J. Mater. Sci. Mater. Med. 12, 683–687.

Miyazaki, T., Kim, H.-M., Kokubo, T., Ohtsuki, C., Kato, H., Nakamura, T., 2002. Enhancement of bonding strength by graded structure at interface between apatite layer and bioactive tantalum metal. J. Mater. Sci. Mater. Med. 13, 651–655.

Nayab, S.N., Jones, F.H., Olsen, I., 2007. Effects of calcium ion-implantation of titanium on bone cell function in vitro. J. Biomed. Mater. Res. 83A, 296–302.

Nishiguchi, S., Kato, H., Fujita, H., Kim, H.-M., Miyaji, F., Kokubo, T., Nakamura, T., 1999. Enhancement of bone-bonding strengths of titanium alloy implants by alkali and heat treatments. J. Biomed. Mater. Res. B: Appl. Biomater. 48, 689–696.

Nishiguchi, S., Fujibayashi, S., Kim, H.-M., Kokubo, T., Nakamura, T., 2003. Biology of alkali- and heat-treated titanium implants. J. Biomed. Mater. Res. 67A, 26–35.

Park, J.-W., Park, K.B., Suh, J.-Y., 2007. Effects of calcium ion incorporation on bone healing of Ti6Al4V alloy implants in rabbit tibia. Biomaterials 28, 3306–3313.

Park, J.-W., Kim, Y.-J., Jang, J.-H., Kwon, T.G., Bae, Y.-C., Suh, J.-Y., 2010. Effects of phosphoric acid treatment of titanium surface on surface properties, osteoblast response and removal of torque forces. Acta Biomater. 6, 1661–1670.

Pattanayak, D.K., Kawai, T., Matsushita, T., Takadama, H., Nakamura, T., Kokubo, T., 2009. Effect of HCl concentrations on apatite-forming ability of NaOH-HCl- and heat-treated titanium metal. J. Mater. Sci. Mater. Med. 20, 2401–2411.

Pattanayak, D.K., Yamaguchi, S., Matsushita, T., Kokubo, T., 2011a. Effect of heat treatments on apatite-forming ability of NaOH- and HCl-treated titanium metal. J. Mater. Sci. Mater. Med. 22, 273–278.

Pattanayak, D.K., Yamaguchi, S., Matsushita, T., Kokubo, T., 2011b. Nanostructured positively charged bioactive TiO_2 layer formed on Ti metal by NaOH, acid and heat treatments. J. Mater. Sci. Mater. Med. 22, 1803–1812.

Pattanayak, D.K., Yamaguchi, S., Matsushita, T., Nakamura, T., Kokubo, T., 2012. Apatite-forming ability of titanium in terms of pH of the exposed solution. J. R. Soc. Interface 9, 2145–2155.

Rautray, T.R., Narayanan, R., Kwon, T.-Y., Kim, K.-H., 2010. Surface modification of titanium and titanium alloys by ion implantation. J. Biomed. Mater. Res. B: Appl. Biomater. 93B, 581–591.

Shibata, Y., Suzuki, D., Omori, S., Tanaka, R., Murakami, A., Ktaoka, Y., Baba, K., Kamijo, R., Miyazaki, T., 2010. The characteristics of in vitro biological activity of titanium surfaces anodically oxidized in chloride solution. Biomaterials 31, 8546–8555.

So, K., Kaneuji, A., Matsumoto, T., Matsuda, S., Akiyama, H., 2013. Is the bone-bonding ability of a cementless total hip prosthesis enhanced by alkaline and heat treatments? Clin. Orthop. Relat. Res. 471, 3847–3855.

Sugino, A., Ohtsuki, C., Tsuru, K., Hayakawa, S., Nakano, T., Okazaki, Y., Osaka, A., 2009. Effect of spatial design and thermal oxidation on apatite formation on Ti-15Zr-4Ta-4Nb alloy. Acta Biomater. 5, 298–304.

Suh, Y.-T., 2003. The significance of the surface properties of oxidized titanium to the bone response: special emphasis on potential biochemical bonding of oxidized titanium implant. Biomaterials 24, 3893–3907.

Takadama, H., Kim, H.-M., Kokubo, T., Nakamura, T., 2001a. An X-ray photoelectron spectroscopy study of the process of apatite formation on bioactive titanium metal. J. Biomed. Mater. Res. 55, 185–193.

Takadama, H., Kim, H.-M., Kokubo, T., Nakamura, T., 2001b. TEM-EDX study of mechanism of bonelike apatite formation on bioactive titanium metal in simulated body fluid. J. Biomed. Mater. Res. 57, 441–448.

Takemoto, M., Fujibayashi, S., Neo, M., Suzuki, J., Kokubo, T., Nakamura, T., 2005. Mechanical properties and osteoconductivity of porous bioactive titanium. Biomaterials 26, 6014–6023.

Takemoto, M., Fujibayashi, S., Neo, M., Suzuki, J., Matsushita, T., Kokubo, T., Nakamura, T., 2006. Osteoinductive porous titanium implants: effect of sodium removal by dilute HCl treatment. Biomaterials 27, 2682–2691.

Takemoto, M., Fujibayashi, S., Neo, M., So, K., Akiyama, N., Matsushita, T., Kokubo, T., Nakamura, T., 2007. A porous bioactive titanium implant for spinal interbody fusion: an experimental study using a canine model. J. Neurosurg. Spine 7, 435–443.

Takeuchi, K., Saruwatari, L., Nakamura, H., Yang, J.-M., Ogawa, T., 2005. Enhanced intrinsic biomechanical properties of osteoblastic mineralized tissue on roughened titanium surface. J. Biomed. Mater. Res. A 72A, 296–305.

Tanaka, K., Takemoto, M., Fujibayashi, S., Kawanabe, K., Matsushita, T., Kokubo, T., Nakamura, T., 2009. Long term study of osteoconductivity of bioactive porous titanium metals: effects of sodium removal by dilute HCl treatment. Key Eng. Mater. 396–398, 353–356.

Tanaka, M., Takemoto, M., Fujibayashi, S., Kawai, T., Yamaguchi, S., Kizuki, T., Matsushita, T., Kokubo, T., Nakamura, T., Matsuda, S., 2014. Bone bonding ability of a chemically treated low elastic modulus Ti alloy: gum metal. J. Mater. Sci. Mater. Med. 25, 635–643.

Turkan, U., Guden, M., 2010. The effect of nitiric acid surface treatment on CaP deposition on Ti6Al4V open-cell foams in SBF solution. Surf. Coat. Tech. 205, 1904–1916.

Uchida, M., Kim, H.-M., Miyaji, F., Kokubo, T., Nakamura, T., 2002a. Apatite formation on zirconium metal treated with aqueous NaOH. Biomaterials 23, 313–317.

Uchida, M., Kim, H.-M., Kokubo, T., Tanaka, K., Nakamura, T., 2002b. Structural dependence of apatite formation on zirconia gels in a simulated body fluid. J. Ceram. Soc. Jpn. 110, 710–715.

Ueda, M., Ikeda, M., Ogawa, M., 2009. Chemical-hydrothermal combined surface modification of titanium for improvement of osteointegration. Mater. Sci. Eng. C 29, 994–1000.

Ueno, T., Tsukimura, N., Yamada, M., Ogawa, T., 2011. Enhanced bone-integration capability of alkali- and heat-treated nanopolymorphic titanium in micro-to-nanoscale hierarchy. Biomaterials 32, 7297–7308.

Wang, X.-X., Hayakawa, S., Tsuru, K., Osaka, A., 2002. Bioactive titania gel layers formed by chemical treatment of Ti substrate with H_2O_2/HCl solution. Biomaterials 23, 1353–1357.

Wang, X.J., Li, Y.C., Yamada, Y., Hodgson, P.D., Wen, C.E., 2008. In vitro bioactivity evaluation of titanium and niobium metals with different surface morphologies. Acta Biomater. 4, 1530–1535.

Whiteside, P., Matykina, E., Gough, J.E., Skeldon, P., Thompson, G.E., 2010. In vitro evaluation of cell proliferation and collagen synthesis on titanium following plasma electrolytic oxidation. J. Biomed. Mater. Res. A 94A, 38–46.

Wu, J.-M., Hayakawa, S., Tsuru, K., Osaka, A., 2004. Low-temperature preparation of anatase and rutile layers on titanium substrates and their ability to induce in vitro apatite deposition. J. Am. Ceram. Soc. 87, 1635–1642.

Xie, L., Liao, X., Yin, G., Huang, Z., Yan, D., Yao, D., Liu, W., Chen, X., Gu, J., 2011. Preparation, characterization, in vitro bioactivity, and osteoblast adhesion of multi-level porous titania layer on titanium by two-step anodization treatment. J. Biomed. Mater. Res. A 98A, 312–320.

Yamaguchi, S., Takadama, H., Matsushita, T., Nakamura, T., Kokubo, T., 2009. Cross-sectional analysis of the surface ceramic layer developed on Ti metal by NaOH-heat treatment and soaking in SBF. J. Ceram. Soc. Jpn. 117, 1126–1130.

Yamaguchi, S., Takadama, H., Matsushita, T., Nakamura, T., Kokubo, T., 2010. Apatite-forming ability of Ti-15Zr-4Nb-4Ta alloy induced by calcium solution treatment. J. Mater. Sci. Mater. Med. 21, 439–444.

Yamaguchi, S., Takadama, H., Matsushita, T., Nakamura, T., Kokubo, T., 2011. Preparation of bioactive Ti-15Zr-4Nb-4Ta alloy from HCl and heat treatments after an NaOH treatment. J. Biomed. Mater. Res. A 97A, 135–144.

Yamaguchi, S., Kizuki, T., Takadama, H., Matsushita, T., Nakamura, T., Kokubo, T., 2012. Formation of a bioactive calcium titanate layer on gum metal by chemical treatment. J. Mater. Sci. Mater. Med. 23, 873–883.

Yamaguchi, S., Matsushita, T., Kokubo, T., 2013a. A bioactive Ti metal with a Ca-enriched surface layer releases Mg ions. RSC Adv. 3, 11274–11282.

Yamaguchi, S., Matsushita, T., Nakamura, T., Kokubo, T., 2013b. Bioactive Ti metal with Ca-enriched surface layer able to release Zn ion. Key Eng. Mater. 529–530, 547–552.

Yamaguchi, S., Nath, S., Matsushita, T., Kokubo, T., 2014. Controlled release of strontium ions from a bioactive Ti metal with a Ca-enriched surface layer. Acta Biomater. 10, 2282–2289.

Yan, W.-Q., Nakamura, T., Kobayashi, M., Kim, H.-M., Miyaji, F., Kokubo, T., 1997. Bonding of chemically treated titanium implants to bone. J. Biomed. Mater. Res. 37, 267–275.

Zhang, L., Ayukawa, Y., LeGeros, R.Z., Matsuya, S., Koyano, K., Ishikawa, K., 2010. Tissue-response to calcium-bonded titanium surface. J. Biomed. Mater. Res. A 95A, 33–39.

Zhang, P., Zhang, Z., Li, W., Zhu, M., 2013. Effect of Ti-OH groups on microstructure and bioactivity of TiO_2 coating prepared by micro-ark oxidation. Appl. Surf. Sci. 268, 381–386.

Zhao, X., Liu, X., Ding, C., 2005. Acid-induced bioactive titania surface. J. Biomed. Mater. Res. A 75A, 888–894.

Zhao, X., Liu, X., You, J., Chen, Z., Ding, C., 2008. Bioactivity and cytocompatibility of plasma-sprayed titania coating treated by sulfuric acid treatment. Surf. Coat. Technol. 202, 3221–3226.

Zhao, L., Mei, S., Chu, P.K., Zhang, Y., Wu, Z., 2010. The influence of hierarchical hybrid micro/nano-textured titanium surface with titania nanotubes on osteoblast functions. Biomaterials 31, 5072–5082.

Zhou, W., Zhou, X., Wu, X., Yuan, L., Shu, Q., Xia, Y., Ostrikov, K., 2007. Plasma-controlled nanocrystallinity and phase composition of TiO_2: a smart way to enhance biomimetic response. J. Biomed. Mater. Res. A 81A, 453–464.

Zhou, R., Wei, D., Yang, H., Cheng, S., Feng, W., Li, B., Wang, Y., Jia, D., Zhou, Y., 2014. Osseointegration of bioactive microarc oxidized amorphous phase/TiO_2 nanocrystals composite coatings on titanium after implantation into rabbit tibia. Mater. Sci. Mater. Med. 25, 1307–1318.

Biomaterials for catalysed mineralization of dental hard tissues

12

R. Osorio, M. Toledano
University of Granada, Granada, Spain

12.1 Introduction

12.1.1 Necessity for enamel and dentine remineralization in restorative dentistry

Dental caries is a multifactorial disease caused by the interaction of dietary factors, dental biofilm and the host's dental tissue within the oral environment. Enamel and dentine are not static tissues, and constantly undergo dynamic transformation. Enamel, in the oral environment, is covered by microbial deposits with metabolic activity resulting in periods of pH decrease and demineralization; reactive pH increases are produced in the course of remineralization processes. Under normal physiological conditions (pH 7), saliva is supersaturated with calcium and phosphate ions, making the progress of demineralization slow and facilitating remineralization. Saliva also incorporates buffering capacity and defensive elements into its composition. However, frequently the physiological remineralization fails, demineralization proceeds, enamel breaks and cavitation take place. Dentine is then exposed to demineralization (Fejerskov and Clarkson, 1996). The reaction of dentine to caries is further complicated mainly due to the high organic content of this tissue, the presence of multiple enzymes and by non-collagenous proteins playing a crucial role in this infectious disease (Chaussain-Miller et al., 2006). At this stage, restorative materials are needed.

Aesthetic and plastic restorative materials in dentistry are now in great demand, and are used in dental clinics. Most of these materials are resin based, and require demineralization procedures to be performed on dentine and enamel prior to resin application. Infiltration of hydrophilic monomers into demineralized dental tissue will create a hybrid layer that will couple resinous restorative materials to the demineralized dentine. Resin infiltration is never complete, and monomers will not be effectively polymerized in the presence of water. Therefore, a defective interface susceptible to collagen and resin degradation will be formed, jeopardizing the longevity of bonded restorations (Spencer et al., 2010). Dentine matrix metalloproteinases (MMPs) play a crucial role in collagen degradation at the hybrid layer, with the subsequent presentation of secondary caries under restorations. If effective inhibitors of MMPs are included in resin–dentine bonding interfaces, they may protect the seed crystallite-sparse collagen fibrils of the scaffold from degradation, and they could be

Biomineralization and Biomaterials. http://dx.doi.org/10.1016/B978-1-78242-338-6.00013-2

remineralized. Reincorporation of mineral into the demineralized. dentine matrix is important since the mineral precipitated may become a constant site for further nucleation, and the remineralized subsurface of the tissue may be more resistant to subsequent acid attack (Liu et al., 2011b). Development of biomaterials able to catalyse remineralization of incompletely resin-infiltrated collagen matrices created by resin adhesives will represent a great advance in dental care.

12.1.2 Interaction of biomaterials with dental hard tissues for remineralization

Enamel remineralization is the process whereby calcium and phosphate ions are supplied to promote ion deposition into crystal voids in demineralized enamel, to produce net mineral gain. The ability of saliva to remineralize demineralized enamel crystals has been proved. It is made possible by its ability to supply calcium and phosphate ions. Ion stabilization will be produced by salivary phosphoproteins. Ion stabilization guarantees that precipitation will not occur and ions will remain available to diffuse into mineral, to allow for remineralization of demineralized enamel. However, the remineralization produced by saliva is small and is a slow process. An efficient remineralization system for enamel should supply stabilized bioavailable calcium, phosphate and fluoride ions that favour subsurface mineral gain rather than deposition only in the surface layer (Cochrane et al., 2010).

Demineralized dentine may be naturally remineralized. Two different models of *in vitro* mineralization can be found in the literature, classified as the top-down or classical and bottom-up or non-classical approaches. In the classical, dentine remineralization is obtained through the epitaxial growth of residual crystals, acting as nucleating sites. However, evidence exists indicating that it results in an extrafibrillar and non-functional remineralization. The bottom-up approach was suggested as alternative and does also require the use of synthetic substitutes for dentine matrix proteins. Two different analogues are needed: (1) a sequestering agent that facilitates amorphous calcium phosphate formation and (2) a dentine matrix phosphoprotein substitute that will facilitate positioning of hydroxyapatite at special sites of collagen, leading to intrafibrillar remineralization (Gu et al., 2011; Watson et al., 2014). Thus, a dual analogue biomimetic system functioning as sequestration analogue, and a phosphate-based templating analogue, may mimic the full role of non-collagenous proteins, resulting in a highly ordered intrafibrillar nanoapatite assembly (Li et al., 2013; Gu et al., 2011; Liu et al., 2011a).

Phosphoproteins are non-specific and may be substituted by other immobilized, naturally occurring phosphoproteins to induce apatite deposition within collagen (Gu et al., 2011). Many biominerals are formed by an amorphous precursor pathway mediated by non-collagenous proteins. The initial stage is the sequestration of the calcium and phosphate ions from solution into amorphous calcium phosphate nanoprecursors, the subsequent stage is templating the nucleation, and the growth of the nanoprecursors at the special sites of the collagen substrate by non-collagenous proteins (Li et al., 2013; Liu et al., 2011a).

12.2 Agents to facilitate remineralization through dental materials

Catalysed dentine bioremineralization will be attained through materials which ideally should be able to: (1) deliver mineral ions (calcium, phosphorus or silicon) in a sustained manner (bioavailability), (2) bind to collagen, acting both as sequestering/templating agent of calcium and phosphorus (bioactivity), and also perform as a nucleation promoter to aid in apatite crystallization, (3) protect collagen from degradation, and (4) provide an adequate pH to facilitate inhibition of collagenolityc enzymes and favour new mineral formation.

12.2.1 *Ionic dissolution products from bioactive dental materials*

Several inorganic materials have been shown to be bioactive, resorbable, and able to deliver remineralizing ions. However, the exact mechanisms of ion elution and interaction between the ionic dissolution products of such materials and dental tissues are not fully known. Due to the chemical similarity to the inorganic phase of dental tissue, different calcium phosphate-based compounds have been employed. The biodegradability of these materials depends on many parameters such as: crystallinity, porosity, chemical purity, surface roughness, pH in the media, other solubilized ions present in the biological fluid at the oral environment, etc. (Hoppe et al., 2011; Cochrane et al., 2010).

The main inorganic materials employed may be classified as: (1) Calcium phosphates or calcium sodium phosphosilicates, referred to as bioactive glasses – 45S5 Bioglass, Brushite, β-TCP or α-TCP. The main problem with applying these materials to the oral cavity to promote enamel or dentine remineralization is the poor solubility of the calcium phosphate phases. The dissolution of the calcium phosphate phase in saliva or at the dentine interface requires undersaturation of these ions with respect to that crystalline phase. In the normal pH range of biological oral fluid, these crystalline calcium phosphate phases would sparingly dissolve (Cochrane et al., 2010). (2) Unstabilized amorphous calcium phosphates and silicates where a calcium salt, phosphate salt, and silicates are delivered separately. Even when these amorphous calcium phosphates are unstable, they tend to be transformed into a stable crystalline phase. But, before phase transformation occurs, ions should be available to facilitate remineralization (Cochrane et al., 2010).

Ionic dissolution (mainly calcium, phosphorus and silicon) from these products is a key factor in understanding the behaviour of remineralizing materials. Since calcium and phosphorus are the main components of biological apatite, these ions play an essential role in remineralization. Other inorganic ions have been incorporated in new developed restorative materials due to their known effects in hard-tissue metabolism, as fluoride, zinc, strontium or magnesium (Sauro et al., 2013; Toledano et al., 2012).

The formula for stoichiometric hydroxyapatite is $Ca_{10}(PO_4)_6(OH)_2$. However, biological apatite is calcium deficient, and contains substantial amounts of carbonate. Carbonated apatite is a precursor of hydroxyapatite. In natural apatite, substitution by foreign ions occurs frequently. Apatite may be substituted by many different trace elements in small concentrations. These ions include carbon, fluoride or zinc, and formed minerals are carbohydroxyapatite, fluorapatite or scholzite (Fejerskov and Clarkson, 1996; Osorio et al., 2014b). These new minerals usually have some different properties to those of hydroxyapatite, being very resistant to dissolution in acid environments or having antibacterial properties. An ideal caries preventive material should release these ions in the oral environment (Longbottom et al., 2009).

The above-mentioned ions have been shown to exert other beneficial effects on remineralization. Silicon is able to stimulate collagen formation and induce hydroxyapatite precipitation. The silanol groups (Si–OH) can interact with calcium ions and induce the formation of apatite crystals. Magnesium, strontium, and zinc may act as crystal growth inhibitors facilitating amorphous calcium phosphate stabilization and intrafibrillar remineralization of collagen. Zinc stimulates remineralization as a direct activator of the enzyme alkaline phosphatase. Magnesium, strontium and zinc have also been shown to inhibit osteoclast activity, and induce new bone formation through regulation of osteoblast differentiation (Hoppe et al., 2011; Ma et al., 2013). The effects of these ions on odontoblast activity remain to be ascertained.

12.2.2 Phosphopeptides and other templating, stabilizing and nucleating agents

Solubility of calcium and phosphate ions in biological systems is tightly regulated by proteins. Biological fluids containing high concentrations of calcium and phosphate ions also contain proteins (mainly phosphopeptides) to avoid ion precipitation. These proteins have a distinct similarity of structure, all containing negatively charged residues. These residues interact with calcium and phosphate ion clusters to stabilize them in aqueous environments. It results in biological fluids with remineralization potential (Cochrane et al., 2010).

These phosphopeptides are characterized by organic phosphate groups covalently bound to structural macromolecules, and are found in mineralized tissues. In dentine, many proteins and ions exist at the location of apatite formation, and the nucleation and crystal growth are regulated by these templates and ions at the same time. Promotion and inhibition of crystal formation both coexist. Some proteins have nucleation/promotion and inhibition sites (i.e. in collagen the positive net charge close to the C-terminal end of collagen promotes the infiltration of amorphous calcium phosphate, while the charged amino acids form nucleation sites that control the conversion from amorphous phase into orientated apatite crystals). The carboxyl groups on the template molecules are also an important nucleation site. These acidic groups have a strong interaction with calcium ions, and then phosphate will be attracted by the adsorbed calcium ions. Other nucleation sites are carbonyl, hydroxyl, and amino groups (Gu et al., 2011; Ma et al., 2013).

Non-collagenous acidic proteins inhibit apatite nucleation for amorphous phase stabilization. This occurs with some well-identified analogues of these proteins, such as polyacrylic acid and polyaspartic acid. These molecules are adsorbed on the collagen fibres and retard the mineralization process. The mineralization of collagen is highly dependent upon the interaction of collagen with such acidic molecules. However, inorganic polyphosphates have to be used as a site-directing molecule in conjunction with a nucleation inhibitor for amorphous phase stabilization. The combination of polyphosphate salts and polyacrylic acid directs amorphous precursor phases into periodic apatite–collagen nanocomposites. The same mechanism of cooperative nucleation was proposed by studying the exposure procedure of collagen templates with polyaspartic acid, calcium, and phosphate ions (Zeiger et al., 2011). Polyaspartic acid is a biodegradable, water-soluble polyaminoacid with the potential to inhibit deposition of calcium phosphate salts and chelate metal ions. Polyaspartic acid and the polymer-induced liquid precursor (PILP) have been proposed as analogues for non-collagenous proteins and the biomimetic system for the natural process of mineralization of dentine and enamel (Burwell et al., 2012). The PILP process consists of adding liquid nanoprecursors encapsulated by polyanionic polymers. The anionic polymer sequesters calcium ions, which then build up a charge to sequester counter ions (phosphate or carbonate), which induces liquid–liquid phase separation in the crystallizing medium. The PILP process is also a pioneer work in bone mineralization (Olszta et al., 2003).

The presence of two phosphorylated non-collagenous proteins is important to attain dentine mineralization: (1) dentine phosphoprotein (DPP), in the presence of Ca^{2+}, tends to form a compact globular structure, with a specific affinity for collagen molecules. DPP localizes to the boundary between the gap, and overlap regions of collagen fibrils (acting as a crystal inhibitor); (2) dentine matrix protein 1 (DMP1) induces crystal nucleation, and regulates the phase transition from amorphous calcium phosphate to carbonated apatite. In conjunction, both lead to the formation of highly organized mineralized collagen fibrils. In contrast, in the presence of non-phosphorylated DPP and DMP1, no organized mineralization of collagen fibrils occurred. Phosphorylation is essential for proper collagen mineralization (Ma et al., 2013). Binding of biomimetic molecules containing multiple phosphate groups to collagen fibrils produces phosphorylated and negatively charged surfaces that attract calcium ions by electrostatic interaction. The phosphorylated collagen serves as a scaffold for homogeneous nucleation of apatite crystallites. Trimetaphosphate is one of the proposed analogues for phosphorylated non-collagenous proteins, as it may provide the interfacial linkage between mineral crystallites and collagen fibrils. Some of the phosphate groups immobilized on the collagen fibrils are also binding sites for calcium ions, and facilitate apatite nucleation onto collagen. Collagen and non-collagen proteins or their analogues, other ions and molecules have to work together in order to obtain formation of biological minerals (Gu et al., 2011).

Casein phosphopeptide (CPP), which is present in milk, is one of the most studied molecules. It has also been shown that its efficacy in remineralizing dental substrates is enhanced if combined with amorphous calcium phosphate and fluoride. CPP forms nanoclusters with amorphous calcium phosphate (CPP–ACP), thus providing a pool of calcium and phosphate which can maintain the supersaturation of saliva. The CPPs have a high binding affinity for apatite. The CPPs were shown to prefer binding to some faces of hydroxyapatite crystals, being able to change direction in crystal growth. CPP will release calcium, phosphate, and fluoride ions, and this process is promoted at low pH (Cochrane et al., 2010). Since CPP–ACP can stabilize calcium and phosphate in the solution, it can also help in the buffering of plaque pH, and so calcium and phosphate levels in plaque are increased. Therefore calcium and phosphate concentrations within the subsurface lesions are kept high, which results in dentine and enamel remineralization (Wilson, 2007). There is now a large body of scientific evidence demonstrating that CPP–ACP and CPP–ACP combined with fluoride can promote the remineralization of enamel subsurface lesions.

The structural domains of proteins in human saliva, as statherin, are partly responsible for the protection and recalcification of tooth enamel, and analogues of these proteins have also been synthesized. The abilities of these fragments to adsorb at hydroxyapatite surfaces and to inhibit its mineralization in supersaturated solutions are determining factors in facilitating enamel remineralization, as it occurs with proteins involved in the dentine remineralization process (Raj et al., 1992).

12.2.3 Collagen protectors: Metalloproteinase inhibitors, crosslinkers and antibacterial agents

The organic phase in dentine is a templating scaffold for mineral crystals to grow, and dentine may be naturally remineralized as long as seed crystallites are present and non-collagenous proteins can act as nucleation sites. Therefore, collagen protection from enzymatic degradation is crucial to obtain remineralization.

MMPs have been suggested to play an important role in the destruction of dentine organic matrix following demineralization by bacterial acids, and in the control or progression of carious decay. During the caries process, the mineral part of dentine is dissolved, exposing the organic matrix to breakdown by bacterially derived enzymes, as well as by host-derived enzymes such as the MMPs present within the dentine (Chaussain-Miller et al., 2006). Demineralized exposed collagen, which is unprotected by intrafibrillar minerals, due to caries or to therapeutic acid treatment before resin bonding, can undergo degradation by endogenous MMPs. Effective inhibitors of MMPs included in resin–dentine bonding interfaces may protect the seed crystallite-sparse collagen fibrils of the scaffold from degradation, before they can be remineralized (Liu et al., 2011b).

Most first-generation MMP inhibitors are based on zinc- or calcium-chelating groups. Doxycycline and chlorhexidine digluconate have been shown to be efficient zinc chelators. These and other compounds may have potent effects but lack selectivity because of the strong homology between catalytic sites of MMPs. Some of these MMPs should be active in promoting further remineralization. Several studies have shown inhibition of MMPs by metals, metal salts, and zinc present in dental products. Zinc concentration strongly reduced MMP-mediated collagen degradation in etched dentine. A collagen protector effect, exerted through binding at the collagen-sensitive cleavage sites of metalloproteinases, has been advocated (Osorio et al., 2011b), so MMPs will be able to act in the subsequent remineralization process. The presence of hydrophilic monomers at the bonded hybrid layer may also inhibit MMP-mediated collagen degradation, through adsorption of MMPs (Osorio et al., 2011a). The coordination of the hydroxyl group of HEMA with zinc that is present in the catalytic domain of MMPs has been suggested as a reversible inhibitory mechanism.

Dentine collagen cleavage by cysteine proteases (also present in dentine) has been questioned, as only Cathepsin K may act extracellularly and, at the moment, it has not been isolated in the extracellular dentine matrix. Cathepsin-mediated collagen hydrolysis is restricted to an acidic pH environment, and thus confined to the sub-osteoclastic and lysosomal compartments. It is not possible to reach this pH, extracellularly, at the hybrid layer due to the buffer capacity of dentine (Kunawarote et al., 2010).

The mechanical performance and stability of collagen can be improved by crosslinking, thereby facilitating collagen remineralization (Gu et al., 2011). It seems that crosslinking opens intramolecular spaces and minimizes the excluded volume of collagen fibrils (Bleek and Taubert, 2013). Attempts to introduce primers with collagen crosslinkers have been researched. Proven dentine collagen crosslinkers include glutaraldehyde, riboflavin and carbodiimide (Ma et al., 2013).

The elimination of bacteria from the dental tissue interface, and providing a bacteria-tight seal between the dental tissue and oral cavity, are requisites for long-term success in dental restorative treatments. Antibacterial agents have been incorporated into dental restorative materials to achieve this goal. Antibacterials have been added to resin materials mainly by two strategies: antibacterial filler incorporation or attaching some antibacterial agent to resin monomers. Some of the researched antibacterial nanoparticles have been silver, zinc oxide, calcium and phosphate ion-releasing fillers (amorphous calcium phosphate or hydroxyapatite), alone or combined with barium-, boro-, or fluor-aluminosilicate glass particles. The main advantage of these composite resins containing calcium phosphate fillers is that they can release calcium and phosphate when the pH is reduced from neutral to cariogenic pH. Furthermore, these materials can rapidly neutralize the acidic medium, increasing the pH. The pH increase may also help the antibacterial activity. Of the antibacterial monomers, methacryloyloxydodecylpyridinium bromide has been incorporated into a commercially available adhesive system. It is a quaternary ammonium compound that may be attached to a methacrylate group of the resin monomer, forming a non-agent-releasing antibacterial dental resin (Imazato et al., 2014).

12.2.4 pH modulation

Remineralization is a dynamic process in which amorphous phase formation, phase stabilization, and transition of calcium phosphate continuously occur. pH value has a significant effect on these processes. At low pH the equilibrium concentrations of calcium and phosphate are greater than at high pH. In the pH range between 5 and 6, the equilibrium concentrations approximate the values for physiological fluids. Above that value, precipitation of mineral may occur from a supersaturated solution; below that value, dissolution of mineral occurs in an undersaturated solution. Calcium phosphate may precipitate in various crystalline forms differing in solubility (Ten Cate and Featherstone, 1996).

The alkalinity potential of the remineralizing materials should also be considered. Some calcium phosphate compounds in solution are able to induce an essential increase in pH (pH ranging from 9 to 10.5). This alkalinity is due to the progression of calcium hydroxide formation in calcium phosphate cements or to a rapid release of

Na^+ or K^+, and the incorporation of H^+ or H_3O^+ into glass particles (Sauro et al., 2013). This alkalinizing potential will facilitate an antibacterial effect, and will inhibit dentine MMPs which are activated under acidic conditions but act at neutral pH (higher than 6, optimum pH 7.5) (Fields, 2013). This pH increase may be beneficial for remineralization whenever it is maintained below 9, as 8–9 is the optimum pH range for alkaline phosphatase to act (Ross et al., 1951).

12.3 Applied dental restorative materials for catalysed bioremineralization

Dentine remineralization through materials is not easy to achieve. Using the previously described matrix protein analogues clinically is not realistic, and it is challenging to apply. Incorporation of amorphous precursors able to bind collagen is, at the moment, a more applicable approach to clinical dentistry (Watson et al., 2014).

12.3.1 Calcium hydroxide cement

Since its introduction to dentistry (1928), calcium hydroxide has been widely used as a mineralizing agent as well as an effective antimicrobial medication. Some relevant characteristics are found in this material, which may facilitate dentine remineralization. Calcium hydroxide is able to liberate calcium, and has been shown to behave as a bioactive surface able to precipitate calcium carbonate and calcium phosphate. As carbonated apatite represents the biological apatite phase seen in bone, cementum, and dentine, these crystals may aid in initiation of dentinogenesis.

Calcium hydroxide induces matrix formation and mineralization through the release of bone morphogenetic proteins, cytokines, and some specific tissue growth factors. Calcium hydroxide may also stimulate alkaline phosphatase activity.

The mineralizing action of calcium hydroxide has been attributed in part to release of hydroxyl ions and the resultant alkaline pH that may also neutralize acids, produce bactericidal effects and cause denaturation of bacteria-derived products (Sangwan et al., 2013).

12.3.2 Glass ionomer cements

Glass ionomers were first introduced to dentistry in 1975. They are composed of fluoro-aluminosilicate powder, a polyalkenoic acid (polyacrilyc acid) and water to facilitate the setting of the cement (acid–base reaction).

When this cement is placed on dentine or enamel an ion exchange process is produced. Aluminium, silicon and fluoride leach out of the cement. Phosphorus and calcium may move out of the underlying dental tissue due to the created acidic environment. This process seems to aid in dentine and enamel remineralization (Watson et al., 2014). A ligand exchange mechanism within the formed gel at the dental interface endows glass fillers not only with the ability to release but also to recharge fluoride ions. The formation of calcium polycarboxylate has also been demonstrated in dentine and enamel, facilitating not only tissue remineralization but also chemical bonding, and it is a significant factor in the excellent long-term adhesion and mineralization ability of these materials (Falsafi et al., 2014).

The effect of polyacrylic acid as a stabilizing agent or mobilizing dentine biomolecule, facilitating remineralization, has not been fully explained. Some further glass ionomer modifications include incorporation of strontium, hydroxyapatite nanoparticles, liquid silica and other bioactive elements.

12.3.3 Calcium silicate cements

Portland cement was the first introduced calcium silicate cement. A similar material, mineral trioxide aggregate (MTA), was developed and recommended initially as a root-end filling material, and subsequently has been used for pulp capping, pulpotomy, apexogenesis, apical barrier formation in teeth with open apexes, repair of root perforations, and root canal-filling material. MTA powder contains fine hydrophilic particles that set in the presence of moisture.

This material is principally composed of di- and tricalcium silicate, tricalcium aluminate and tetracalcium aluminoferrite, and bismuth oxide. These cements set, after water mixing, through a hydration reaction. MTA is biocompatible, and a bioactive material with hard-tissue conductive and inductive properties. A high pH value (around 10) is obtained after mixing, and is maintained over time. The high pH value is due to the constant release of calcium forming calcium hydroxide. Both Portland cement and MTA contain tricalcium and dicalcium silicate, which on hydration produce calcium silicate hydrate gel, calcium hydroxide, and calcium carbonate, and thus are able to exert remineralizing effects (Parirokh and Torabinejad, 2010; Watson et al., 2014).

12.3.4 Advanced polymers for bioremineralization

New resin materials are being developed to facilitate dentin remineralization. Most of these incorporate inorganic fillers (bioglass, calcium phosphate and/or calcium silicate particles, hydroxyapatite and silicon nanoparticles) trying to produce calcium, silicon and phosphorus liberation in order to promote remineralization processes at the bonded interface (Sauro et al., 2013; Besinis et al., 2014).

Zinc (zinc oxide particles or zinc chloride) has also been successfully incorporated into dental resin adhesives, as an MMP inhibitor agent able to stabilize dentine collagen (Toledano et al., 2012; Osorio et al., 2014a) and promoting dentine remineralization at the adhesive interface (Toledano et al., 2013; Osorio et al., 2014b).

12.4 Future trends and research in bioactive dental restorative materials

An alternative strategy that is becoming the focus of research in the biomaterials field is the use of polymeric particles as calcium- and phosphate-sequestering materials (Osorio et al., 2014c). These polymers may also act as carriers of other biological factors for the management of tissue mineralization, permitting a controlled ion release rate (Leonor et al., 2009; Wu et al., 2011; Musyanovych and Landfester, 2014). Surfaces of these polymers exhibit a high concentration of functional groups (i.e. $COOH^-$) able to form complexes with metal cations, and exert a potent quelant effect (Li et al., 2013; Osorio et al., 2014c). These functional groups may also be conjugated with various targeting molecules including proteins, growth factors or any other specific molecules. Studies have also shown that self-assembling polymers and biomimetic peptides based on dentine phosphophorin, which contain multiple phosphoseryl residues, will nucleate hydroxyapatite (Hartgerink et al., 2001).

A polymeric sphere system able to combine bioactivity with a capacity for controlled protein/drug delivery is ideal for teeth and also for bone regeneration (Wu et al., 2011). Filler nanoparticles will not only induce mineralization but also may be used for loading/delivering some antibiotic drugs, osteoinductive agents, or other proteins useful in the dental environment.

Biocatalysed mineralization is desirable within the dentine–resin hybrid layer, potentially reinforcing or stabilizing this zone. Biocatalysts can remain active within a polymer; a dental resin can potentially carry the catalyst for subsequent formation of particles in order to create a mineral gradient. The principle is that particles need to be transported, or could be formed *in situ*, as is done in nature (Goldberg et al., 2009).

There is a need to delve much deeper into dentine. Dentine is a biologically active tissue which contains trapped bioactive molecules which might be liberated if the dentine matrix is demineralized or solubilized by external agents. These molecules are released into the immediate vicinity and affect any reparative processes. Precise knowledge of these active molecules, their function and regulation mechanisms is necessary, if we really want to aid nature in dentine remineralization with newly developed materials.

Acknowledgements

This work was supported by Grants MINECO/FEDER MAT2011#24551, MAT2014#52036-P and Grant CEIBiotic from the University of Granada.

References

Besinis, A., van Noort, R., Martin, N., 2014. Remineralization potential of fully demineralized dentin infiltrated with silica and hydroxyapatite nanoparticles. Dent. Mater. 30, 249–262.

Bleek, K., Taubert, A., 2013. New developments in polymer-controlled, bioinspired calcium phosphate mineralization from aqueous solution. Acta Biomater. 9, 6283–6321.

Burwell, A.K., Thula-Mata, T., Gower, L.B., Habelitz, S., Kurylo, M., Ho, S.P., Chien, Y.C., Cheng, J., Cheng, N.F., Gansky, S.A., Marshall, S.J., Marshall, G.W., 2012. Functional remineralization of dentin lesions using polymer-induced liquid-precursor process. PLoS One 7, e38852.

Chaussain-Miller, C., Fioretti, F., Goldberg, M., Menashi, S., 2006. The role of matrix metalloproteinases (MMPs) in human caries. J. Dent. Res. 85, 22–32.

Cochrane, N.J., Cai, F., Huq, N.L., Burrow, M.F., Reynolds, E.C., 2010. New approaches to enhanced remineralization of tooth enamel. J. Dent. Res. 89, 1187–1197.

Falsafi, A., Mitra, S.B., Oxman, J.D., Ton, T.T., Bui, H.T., 2014. Mechanisms of setting reactions and interfacial behavior of a nano-filled resin-modified glass ionomer. Dent. Mater. 30, 632–643.

Fejerskov, O., Clarkson, B.H., 1996. Dynamic of caries lesion formation. In: Fejerskov, O., Ekstrand, J., Burt, B.A. (Eds.), Fluoride in Dentistry. Munksgaard, Copenhagen, pp. 187–213.

Fields, G.B., 2013. Interstitial collagen catabolism. J. Biol. Chem. 288, 8785–8793.

Goldberg, A.J., Advincula, M.C., Komabayashi, T., Patel, P.A., Mather, P.T., Goberman, D.G., Kazemi, R.B., 2009. Polypeptide-catalyzed biosilification of dentin surfaces. J. Dent. Res. 88, 377–381.

Gu, L.S., Kim, Y.K., Liu, Y., Takahashi, K., Arun, S., Wimmer, C.E., Osorio, R., Ling, J.L., Looney, S.W., Pashley, D.H., Tay, F., 2011. Immobilization of a phosphonated analog of matrix phosphoproteins within cross-linked collagen as a templating mechanism for biomimetic mineralization. Acta Biomater. 7, 268–277.

Hartgerink, J.D., Beniash, E., Stupp, S.I., 2001. Self-assembly and mineralization of peptide-amphiphile nanofibers. Science 23, 1684–1688.

Hoppe, A., Güldal, N.S., Boccaccini, A.R., 2011. A review of the biological response to ionic dissolution products from bioactive glasses and glass-ceramics. Biomaterials 32, 2757–2774.

Imazato, S., Ma, S., Chen, J.H., Xu, H.H., 2014. Therapeutic polymers for dental adhesives: loading resins with bio-active components. Dent. Mater. 30, 97–104.

Kunawarote, S., Nakajima, M., Shida, K., Kitasako, Y., Foxton, R.M., Tagami, J., 2010. Effect of dentin pretreatment with mild acidic HOCl solution on microtensile bond strength and surface pH. J. Dent. 38, 261–268.

Leonor, I.B., Balas, F., Kawashita, M., Reis, R.L., Kokubo, T., Nakamura, T., 2009. Biomimetic apatite deposition on polymeric microspheres treated with a calcium silicate solution. J. Biomed. Mater. Res. B: Appl. Biomater. 91, 239–247.

Li, J., Yang, J., Li, J., Chen, L., Liang, K., Wu, W., Chen, X., Li, J., 2013. Bioinspired intrafibrillar mineralization of human dentine by PAMAM dendrimer. Biomaterials 34, 6738–6747.

Liu, L., Liu, J.Y., Kong, X.D., Cai, Y.R., Yao, J.M., 2011a. Porous composite scaffolds of hydroxyapatite/silk fibroin via two-step method. Polym. Adv. Technol. 22, 909–914.

Liu, Y., Tjäderhane, L., Breschi, L., Mazzoni, A., Li, N., Mao, J., Pashley, D.H., Tay, F.R., 2011b. Limitations in bonding to dentin and experimental strategies to prevent bond degradation. J. Dent. Res. 90, 953–968.

Longbottom, C., Ekstrandb, K., Zeroc, D., Kambarad, M., 2009. Novel preventive treatment options. Monogr. Oral Sci. 21, 156–163.

Ma, J., Wang, J., Ai, X., Zhang, S., 2013. Biomimetic self-assembly of apatite hybrid materials: from a single molecular template to bi-/multi-molecular templates. Biotechnol. Adv. 32 (4), 744–760. http://dx.doi.org/10.1016/j.biotechadv.2013.10.014.

Musyanovych, A., Landfester, K., 2014. Polymer micro- and nanocapsules as biological carriers with multifunctional properties. Macromol. Biosci. 14 (4), 458–477. http://dx.doi.org/10.1002/mabi.201300551.

Olszta, M.J., Odom, D.J., Douglas, E.P., Gower, L.B., 2003. A new paradigm for biomineral formation: mineralization via an amorphous liquid-phase precursor. Connect. Tissue Res. 44 (Suppl. 1), 326–334.

Osorio, R., Yamauti, M., Osorio, E., Ruiz-Requena, M.E., Pashley, D., Tay, F.R., Toledano, M., 2011a. Effect of dentin etching and chlorhexidine application on metalloproteinase-mediated collagen degradation. Eur. J. Oral Sci. 119, 79–85.

Osorio, R., Yamauti, M., Osorio, E., Ruiz-Requena, M.E., Pashley, D.H., Tay, F.R., Toledano, M., 2011b. Zinc reduces collagen degradation in demineralized human dentin explants. J. Dent. 39, 148–153.

Osorio, R., Cabello, I., Toledano, M., 2014a. Bioactivity of zinc-doped dental adhesives. J. Dent. 42, 403–412.

Osorio, R., Osorio, E., Cabello, I., Toledano, M., 2014b. Zinc induces apatite and scholzite formation during dentin remineralization. Caries Res. 48, 276–290.

Osorio, R., Osorio, E., Medina-Castillo, A.L., Toledano, M., 2014c. Polymer nanocarriers for dentin adhesion. J. Dent. Res. 93 (12), 1258–1263. http://dx.doi.org/10.1177/0022034514551608.

Parirokh, M., Torabinejad, M., 2010. Mineral trioxide aggregate: a comprehensive literature review – part I: chemical, physical, and antibacterial properties. J. Endod. 36, 16–27.

Raj, P.A., Johnsson, M., Levine, M.J., Nancollas, G.H., 1992. Salivary statherin. Dependence on sequence, charge, hydrogen bonding potency, and helical conformation for adsorption to hydroxyapatite and inhibition of mineralization. J. Biol. Chem. 267, 5968–5976.

Ross, M.H., Ely, J.O., Archer, J.G., 1951. Alkaline phosphatase activity and pH optima. J. Biol. Chem. 192, 561–568.

Sangwan, P., Sangwan, A., Duhan, J., Rohilla, A., 2013. Tertiary dentinogenesis with calcium hydroxide: a review of proposed mechanisms. Int. Endod. J. 46, 3–19.

Sauro, S., Osorio, R., Fulgencio, R., Watson, T.F., Cama, G., Thompson, I., Toledano, M., 2013. Remineralisation properties of innovative light-curable resin-based dental materials containing bioactive micro-fillers. J. Mater. Chem. B 1, 2624–2638.

Spencer, P., Ye, Q., Park, J., Topp, E.M., Misra, A., Marangos, O., Wang, Y., Bohaty, B.S., Singh, V., Sene, F., Eslick, J., Camarda, K., Katz, J.L., 2010. Adhesive/dentin interface: the weak link in the composite restoration. Ann. Biomed. Eng. 38, 1989–2003.

Ten Cate, J.M., Featherstone, J.D.B., 1996. Physicochemical aspects of fluoride-enamel interactions. In: Fejerskov, O., Ekstrand, J., Burt, B.A. (Eds.), Fluoride in Dentistry. Munksgaard Textbook, Copenhagen, pp. 187–213.

Toledano, M., Yamauti, M., Ruiz-Requena, M.E., Osorio, R., 2012. A ZnO-doped adhesive reduced collagen degradation favouring dentine remineralization. J. Dent. 40, 756–765.

Toledano, M., Sauro, S., Cabello, I., Watson, T., Osorio, R., 2013. A Zn-doped etch-and-rinse adhesive may improve the mechanical properties and the integrity at the bonded-dentin interface. Dent. Mater. 29, e142–e152.

Watson, T.F., Atmeh, A.R., Sajini, S., Cook, R.J., Festy, F., 2014. Present and future of glass-ionomers and calcium-silicate cements as bioactive materials in dentistry: biophotonics-based interfacial analyses in health and disease. Dent. Mater. 30, 50–61.

Wilson, N., 2007. Minimally Invasive Dentistry: The Management of Dental Caries. Quintessence, London, pp. 69–70.

Wu, C., Zhang, Y., Fan, W., Ke, X., Hu, X., Zhou, Y., Xiao, Y., 2011. $CaSiO_3$ microstructure modulating the in vitro and in vivo bioactivity of poly(lactide-co-glycolide) microspheres. J. Biomed. Mater. Res. A 98A, 122–131.

Zeiger, D.N., Miles, W.C., Eidelman, N., Lin-Gibson, S., 2011. Cooperative calcium phosphate nucleation within collagen fibrils. Langmuir 5, 8263–8268.

Designing biomaterials based on biomineralization for bone repair and regeneration

13

I.B. Leonor[1,2], *A.I. Rodrigues*[1,2], *R.L. Reis*[1,2]
[1]University of Minho, Guimarães, Portugal, [2]ICVS/3B's—PT Government Associate Laboratory, Braga/Guimarães, Portugal

When Nature finishes to produce its own species, man begins using natural things in harmony with this very nature to create an infinity of species.

Leonardo da Vinci (1452–1519)

13.1 Introduction

Nature is full of biological structures, from nanoscale to macroscopic level, with a variety of shapes and forms. These structures are crystals, clays, inorganic/organic composites, seashells, pearls, bones and teeth, wood, silk, horn and collagen, among others (Boskey, 1999; Cuisinier, 1996; Dove et al., 2003; Green et al., 2002; Hou et al., 2004; Lowenstam and Weiner, 1989; Mann, 2001; Rubner, 2003; Sommerfeldt and Rubin, 2001; Zhang, 2003).

For centuries, these structures have inspired scientists to learn more about them and find ways to create them synthetically. Studying these natural structures has generated a growing interest in materials science and in the adaptation of biological processes that may lead to significant advances in designing complex structures and controlling sophisticated processing routes that lead to the final shape. To date, it has not yet been possible to fully replicate these structures using nonbiological processing methods. Learning from biological concepts (bioinspiration) and mimicking biological surfaces (biomimetic) has led to important advances in the manufacture of synthetic materials in recent years.

For instance, the abalone shell nacre, one of the most beautiful and delicate structures in the natural world, had inspired several research groups in the development of biomimetic-advanced materials (Ball, 2001). These shells serve as protective barriers for invertebrates, are impenetrable to the crushing action of predators, are resistant to the pressures in the deep ocean around hydrothermal vents where mollusks proliferate, and have toughened to avoid shattering in intertidal areas due to wave action (Kaplan, 1998).

Biomineralization and Biomaterials. http://dx.doi.org/10.1016/B978-1-78242-338-6.00014-4

In the case of bone it is a complex living tissue in which a complex matrix is arranged in different hierarchical levels of organization going from macro- to nanoscales (Bonzani et al., 2006; Stevens and George, 2005). The nano arrangement refers to a nanocomposite structure (tough and flexible type I collagen fibers reinforced by hydroxyapatite crystals), which is responsible for bone's compressive strength and high fracture toughness (Bayliss et al., 2012; Bonzani et al., 2006; Stevens and George, 2005).

From these two examples, we can say that Nature continues to be an important source of inspiration for designing smart materials and solving technological problems associated with biomaterials.

So, faced with the challenge of designing new materials for bone regeneration/replacement, this chapter underlines the study of bone as a composite material, and sheds light on using this information to develop novel bioactive materials able to induce bone regeneration. We start by summarizing the most important aspects of bone biology in relation to the inorganic matrix, the apatite component, since the amount of information (Cohen, 2006; Dove et al., 2003; Lowenstam and Weiner, 1989; Simkiss and Wilbur, 1989; Weiner et al., 1999) on bone is so large that it warrants a separate chapter. This is followed by an explanation of the concept of a bioactive material, explaining the requirements for bone bonding as well as the mechanism of integration of bioactive materials with bone. Finally, we highlight the importance of understanding the surface chemistry of biomaterials, as well as the main mechanism for apatite formation and how we can design a three-dimensional (3D) structure with a properly functionalized surface to induce bone regeneration.

13.2 Clinical need for bone replacement/regeneration

Over the course of the twentieth century, the quality of human life has improved and also a rapid and unprecedented increase in human life expectancy has been observed. These two indicators are growing in opposite directions as people are living longer and their quality of life often starts to decline between the ages of 60 and 90 due to the gradual degradation of our connective tissues and bone density and strength (Hench et al., 2000). The need to replace or repair deteriorating bones and joints has pushed scientists to look for new designs and processing methods to develop innovative biomaterials for replacement/regeneration of tissues. Every year millions of people suffer from severe bone fractures that constitute a major worldwide health problem responsible for the expenditure of several billions of euros by national health care systems (O'Keefe and Mao, 2011).

Currently, the therapeutic strategies used for bone replacement/regeneration are based on the bone grafting (autografts, allografts, demineralized bone matrix, and bone substitutes) with all the associated complications and drawbacks (Drosse et al., 2008; Giannoudis et al., 2005; Schroeder and Mosheiff, 2011; Willie et al., 2010). Autografts and allografts are the most common treatments in the field; however, they present several weaknesses. In the case of autografts there is a potential risk of donor-site morbidity, mismatch in size between the donor and the recipient sites due to the limited amount of biological materials available for grafting, and the surgical procedure can lead to infection and chronic pain (Drosse et al., 2008; Giannoudis et al., 2005). Allografts are especially attractive when a larger segment of bone needs to be replaced, although it also has limitations, namely the possibility of immune rejection and pathogen transfer from donor to host (Giannoudis et al., 2005).

Currently, a large number of bone graft substitutes are commercially available for orthopedic use; however, so far, they are not in favor compared to autografts and allografts (Giannoudis et al., 2005; Willie et al., 2010).

The search for new bone regeneration strategies is thus a priority driven by the debilitating pain associated with bone damage and increasing human life expectancy (Drosse et al., 2008; Stevens, 2008).

Tissue engineering (TE) of autologous bone is a promising alternative for surgically derived autologous bone grafting. However, to date, available bone grafts are very limited with respect to size and appropriate 3D architecture, so healthy bone has to be removed and used as an autograft in bone reconstruction, with all the associated disadvantages. All the properties of an autogenous bone graft should be part of the ideal bone implant and most of the available biomaterials tend to be either predominantly osteogenic or osteoinductive, or simply osteoconductive (Keating and McQueen, 2001).

In the past years, considerable efforts have been made to develop synthetic bone grafts, such as calcium phosphate (CaP) ceramics, as an alternative to patients' own bone and bone from donors. These ceramics, due to their chemical similarity to the inorganic phase of bone, are one of the main sources of bone graft substitutes (Hoppe et al., 2011; Samavedi et al., 2013). Also, several research groups have recently moved from the development of inert polymers, which mimic the biomechanical properties of native tissue, to bioactive materials able to promote tissue self-healing (He et al., 2010; Rios et al., 2009; Xia et al., 2014; Wang et al., 2011; Lee et al., 2012; Yeo and Kim, 2012). One of the challenges of bone TE consists of designing a bioactive matrix capable of mimicking the natural properties of bone while providing a temporary replacement for the injured tissue with a suitable biodegradation rate, and thus sustaining tissue repair and regeneration.

In the following sections, the importance and use of simulated body fluid (SBF) as an *in vitro* model to study apatite formation or precipitation on the surfaces of different materials will be explained, and then some of the research work related to surface functionalization to induce the formation of an apatite layer on their surfaces will be described.

13.3 Bone structure and function

Biology has an enormous influence on the demanding quest to develop biomedical technologies, as the most modern strategies for designing advanced materials are inspired by Nature. It is fascinating to observe bone structure, one of the most remarkable materials in Nature, a hard composite material with a complex structure that, in association with cells, has different functions in a living organism such as movement, protection of organs, as well as red and white blood cell production and mineral storage (Cohen, 2006; Bayliss et al., 2012; Stevens and George, 2005; Reznikov et al., 2014).

Bone is a dynamic and highly vascularized tissue that undergoes a continuous remodeling process throughout the lifetime of an individual (Bonzani et al., 2006; Stevens, 2008). It is a highly complex hierarchical structure with several levels of embedded structures that extend from the molecular (micro scale) to the macroscopic scale (Rho et al., 1998; Weiner et al., 1999; Reznikov et al., 2014). At the microscale, bone apatite-reinforced collagen forms individual lamellae that range in size from the nanometer to the micrometer scale, while at the macroscale, interstitial bone is composed of osteons ranging in size from micrometers to millimeters (Rho et al., 1998; Wang, 2003). This hierarchically organized structure has an irregular, yet optimized, arrangement and orientation of the components, making the bone material heterogeneous and anisotropic (Bonfield et al., 1998).

With regard to bone structure (Figure 13.1), an organic framework (mostly collagen) can be observed in which mineral (hydroxyapatite or apatite) plate-shaped crystals are precipitated along the collagen fibrils (Isaac and Green, 1991; Marks and Hermey, 1996; Posner, 1985; Wenk and Heidelbach, 1999; Dunlop and Fratzl, 2010). Herewith it can be said that the bone tissue can be seen as a two-component composite material composed primarily of a collagen, organic matrix, and a mineral, hydroxyapatite, inorganic matrix.

With respect to the organic phase, type I collagen constitutes approximately 95% of the organic matrix; the remaining 5% is composed of proteoglycans and numerous

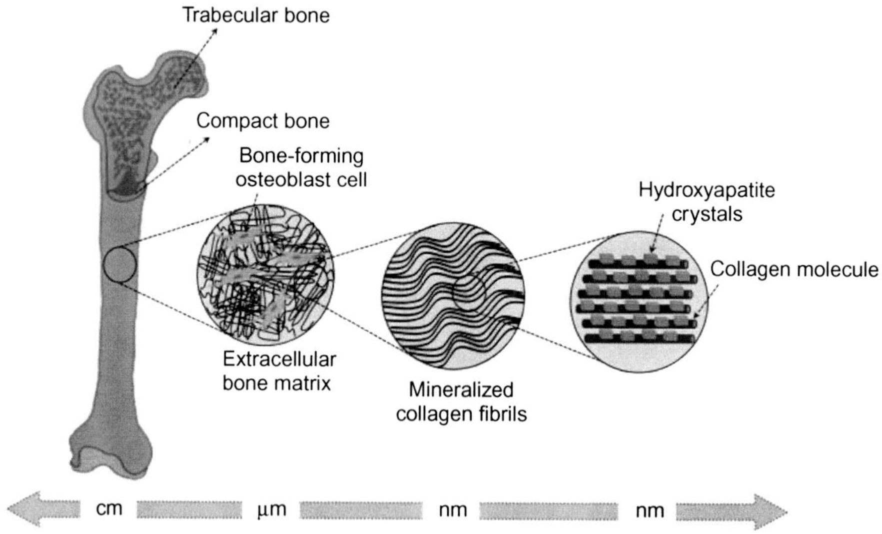

Figure 13.1 Schematic illustration showing the different levels of the hierarchical structure of bone.
Source: Data adapted from Stevens (2008) and Taton (2001).

noncollagenous proteins (Bayliss et al., 2012; Marks and Hermey, 1996). The collagen constitutes the main component of a 3D matrix into which, and in some cases onto which, the mineral forms (Weiner, 1986; Reznikov et al., 2014). Also, it confers to bone high fracture toughness (Bayliss et al., 2012; Bonzani et al., 2006; Stevens, 2008; Dunlop and Fratzl, 2010).

The mineral phase constitutes two-thirds of the total bone matrix and is nanocrystalline apatite, structurally disordered, and compositionally nonstoichiometric with variable amounts of carbonate (CO_3^{2-}), fluoride (F^-), citrate ($C_6H_5O_7^{4-}$), and hydroxyl (OH^-) substitutions (Bayliss et al., 2012; Bonfield et al., 1998; Cho et al., 2003). This mineral matrix is responsible for bone's compressive strength (Bayliss et al., 2012; Bonzani et al., 2006; Stevens, 2008; Dunlop and Fratzl, 2010).

The third major component is water, located within the collagen fibrils, in the gaps between triple-helical molecules (Weiner, 1986; Reznikov et al., 2014). The major components are intimately associated into an ordered structure, mineralized collagen fibrils (Weiner, 1986; Reznikov et al., 2014). Their proportions, however, can vary considerably between family members of bone.

Weiner (1986) and Weiner and Wagner (1998) described them, three decades ago, as highly complex structures in terms of up to seven hierarchical levels of organization for lamellar bone. Nevertheless, Reznikov et al. (2014) have recently reported that lamellar bone could be arbitrarily divided into nine hierarchical levels; however, in the case of woven bone the number of hierarchical levels is fewer. The mechanical properties of bone are derived from the organized mineralization of hydroxyapatite within a matrix of collagen fibrils, glycoproteins, and many other types of protein (Boskey, 1999; Weiner and Addadi, 1997). Such combination of inorganic and organic components provides an increased toughness compared with hydroxyapatite alone (Boskey, 1999; Weiner and Addadi, 1997). These properties, together with their capacity to rapidly mobilize mineral stores on metabolic demand, make it the ultimate smart material, which is, in fact, a very important goal in designing and engineering composite materials (Weiner and Addadi, 1997).

Engineering materials for bone replacement/regeneration are a complex process and require a multidisciplinary approach to generate a hierarchically organized implant that is able to replace the traditional bone grafts.

The following sections were written with this goal in mind and are by no means exhaustive, but intended to motivate the community to think more broadly about the source materials for the development of bioactive materials, and also work as a guide to some examples of the work developed in this field.

13.4 Bioactive ceramics for bone replacement/regeneration

The real challenge of the bone replacement/regeneration research field lies in the development of a material with a natural structure, in which features of different length scales can be hierarchically organized to obtain fast tissue formation and integration within the body. In this sense bone implants should cover three essential aspects: osteoconduction (support of bone growth and promotion of the ingrowth of surrounding bone), osteoinduction (capable of promoting the differentiation of progenitor cells into an osteoblastic lineage), and a potential for osseointegration (integration into surrounding bone) (Matsuno et al., 2010; Giannoudis et al., 2005; Schroeder and Mosheiff, 2011).

In the last three decades, a new class of biomaterials designated as surface-active biomaterials have been widely used in clinical applications, particularly musculoskeletal and dental applications (periodontal diseases, bone fillers, and ossicular reconstruction) (Pashuck and Stevens, 2012; Hench and Wilson, 1984).

These surface-active biomaterials allow the formation of direct physicochemical bonds between the bone tissue and the implant, without formation of a fibrous capsule (Hench, 1988). The interfacial bond with bone is generally stronger than either the bone or the implant (Hench, 1991; Hench and Wilson, 1984). Also, it has often been assumed that the interfacial stability achieved using bioactive fixation would ensure an improvement in implant survivability (Hench, 1998). These so-called bioactive materials or bone-bonding materials have the capacity to form an interfacial bond with adjacent tissue through a layer. The chemical composition of this layer usually includes phosphorus and calcium, and is thus similar to the mineral composition of bone tissue (Hench, 1991; Hench and Wilson, 1984). These bioactive materials include dense hydroxyapatite, bioactive glasses, bioactive glass-ceramics, and other CaPs (Hench and Wilson, 1984; Salinas et al., 2013; Vallet-Regi, 2001; Jones, 2013). Hydroxyapatite, a native component of bone, is a popular bone substitute as a coating or filler due to its documented ability to promote mineralization. In spite of that, another material that is referenced in the field of bioactive ceramics and will continue to be an inspiration to develop bioactive materials for repair and regenerate bone tissues is Bioglass® 45S5. This original bioactive glass, 45S5, with a single phase and only four components (Na_2O, CaO, P_2O_5, and SiO_2), has been used as a model for most surface studies for 40 years, largely because no other bioactive glass composition has been found with better biological properties (Jones, 2013; Hench, 1991). In the next sections there will be a strong emphasis on 45S5 Bioglass®, since most of the work developed in our research group for bone applications have been inspired by this bioactive glass.

Larry Hench invented this bioactive glass in 1969 by demonstrating an artificial material forming a chemical bond with bone, as mentioned above (Jones, 2013; Hench and Paschall, 1973). This important bone-grafting material provides an ideal environment for colonization, proliferation, and differentiation of osteoblasts to form new bone (Xynos et al., 2000a; Hench, 2009). With the discovery of this glass the concept of bioactivity was introduced, defined as a condition that elicits a specific biological response at the interface of the material, which results in the formation of a bond between the tissues and the materials. The most interesting properties in this bioactive glass were the fact that it could bond to both hard and soft tissues (Hench and Wilson, 1984). Wilson et al. (1981) had proved that collagen fibers, which are a constituent of soft tissues, became embedded and bonded within the growing silica-rich and hydroxycarbonate apatite (HCA) layer on the 45S5. This discovery is especially important, due to the fact that many implants require a stable interface with both soft and hard connective tissues (Hench, 1988; Hench et al., 1971). Furthermore, an important advantage of bioactive fixation is the fact that a bioactive bond to bone has strength equal to or greater than bone after 3–6 months (Hench, 1998). The high strength of bonding between bioactive implants and both hard and soft tissues is due to the growth of mineralized collagen fibrils, where nanometer-scale HCA crystals are embedded into the collagen fibrils (Hench, 1998).

These bioactive ceramics have the capacity to form a mechanically strong bond to bone through a biologically active HCA layer formed on their surface that is chemically and structurally similar to the mineral phase in bone when they are implanted (Hench, 1991; Legeros, 2002; Neo et al., 1992b; Ducheyne and Qiu, 1999). Furthermore, the analysis of the bone–implant interface revealed that the presence of hydroxyapatite is one of the key features in the bonding zone (Ducheyne and Qiu, 1999). So, in other words, it can be said that one of the essential requirements for the artificial material to bond to living bone is the formation of the biologically active apatite layer on their surfaces (Hench et al., 1971; Neo et al., 1992a,b, 1993; Vallet-Regi, 2001).

All the bioactive glasses have the capacity to bond with living tissue through an intervening apatite layer (Vallet-Regi, 2001; Jones, 2013; Hench, 1991; Neo et al., 1992b, 1993). However, each glass has its own mechanisms of bone bonding, which is dependent on their chemical composition, relative crystallization and solubility, e.g. the bioactivity of the materials (Neo et al., 1992b). Hench (1988, 1991) proposed five stages for HCA formation on the surface of most bioactive glasses in body fluid *in vivo* or in SBF *in vitro* (see Table 13.1). After all these years, different characterization techniques had been used to demonstrate that these stages already occur (Jones, 2013).

These five reaction stages (Table 13.1) that occur on the glass interface do not depend on the presence of tissues. In fact, they can occur in acellular SBFs with ion concentrations nearly equal to those of human blood plasma (ISO, 2014; Kokubo et al., 1990; Kokubo and Takadama, 2006). The surface reactions lead to a biochemical adsorption of growth factors, and then the synchronized sequence of cellular events led to a rapid formation of new bone within 7–10 days (Hench, 1995; Hench and Wilson, 1984). These stages are required for the bond between implant and tissues to occur. The large difference in the rate of bone bonding and proliferation at the interface of bioactive implants indicates that different biochemical factors are occurring at the implant–tissue interface, which depends on the composition of the material and the rate of surface reactions (Hench, 1988, 1995).

Each surface-active ceramic has its own characteristics in regard to the formation of the apatite layer, and its process and kinetics can be affected by bulk factors such as density and surface area, as well as by surface factors such as composition and structure. For example, the formation and growth of the CaP on the crystallized Bioglass® 45S5 differ from those on hydroxyapatite. In order to clarify this process, the *in vitro* bioactivity of partially crystallized 45S5 as a function of immersion time in SBF was investigated by *in situ* observation using atomic force microscopy (Leonor et al., 2002). 45S5 was previously sintered at 800 °C to obtain a material that is less resorbable, but still bioactive and stiffer than standard 45S5. Subsequently, the results obtained for the crystallized 45S5 were compared to those of hydroxyapatite *c*- and *a*-faces. In Figure 13.2, it can be seen that a CaP layer grows on the crystallized 45S5 via the following three stages:

Table 13.1 Reaction stages of a bioactive implant (Hench, 1988, 1991)

Stage	Reaction
1	Fast ion exchange of alkali ions with hydrogen ions from body fluids, where it is usually controlled by diffusion Rapid exchange of Na^+ or K^+ with H^+ or H_3O^+ from solution: $\mathrm{Si-O-Na^+ + H^+ + OH^- \rightarrow Si-OH^+ + Na^+(solution) + OH^-}$
2	Network dissolution, where it is usually controlled by interfacial reaction Loss of soluble SiO_2 in the form of $Si(OH)_4$ to the solution, resulting from breaking of Si–O–Si bonds and formation of Si–OH (silanols) at the glass solution interface: $\mathrm{Si-O-Si + H_2O \rightarrow Si-OH + OH-Si}$
3	Condensation and repolymerization of a silica (SiO_2)-rich layer on the surface depleted in alkalis and alkaline-earth cations: 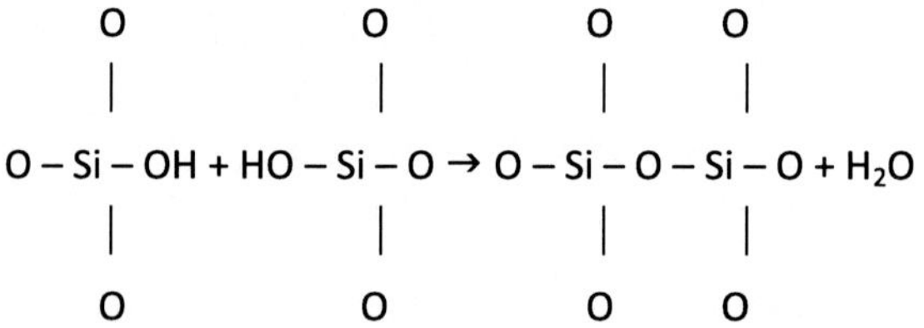
4	Migration of Ca^{2+} and PO_4^{3-} groups to the surface through the SiO_2-rich layer, forming a $CaO–P_2O_5$-rich film on top of the SiO_2-rich layer, followed by growth of the amorphous $CaO–P_2O_5$-rich film by incorporation of soluble calcium and phosphates from solution
5	Chemisorption and crystallization of the HCA layer Crystallization of the amorphous $CaO–P_2O_5$ film by incorporation of OH^-, CO_3^{2-}, or F^- anions from solution to form a mixed hydroxyl, carbonate, apatite (HCA) layer

Stage 1: Rapid formation of island structures on the surface ($0 < T < 15$ min).
Stage 2: Increase in the number of these islands (15 min $< T <$ 4 h).
Stage 3: Fusion of the islands (4 h $< T$).

The surface of the partially crystallized 45S5 was almost fully covered with a CaP layer after 24 h. The CaP layer grows on the partially crystallized 45S5 by multiple 2D nucleation and fusion of these islands after an incubation period. CaP formation on the partially crystallized 45S5 arises from a low surface energy of the surface layer and/or an effect of the layer to lower the resistance when the growth unit of CaP is incorporated into the growing island.

In this sense, a detailed insight into the mechanism of apatite formation on bioactive ceramics is considered crucial for developing bioactive materials with enhanced functions, namely physical, chemical, and biological.

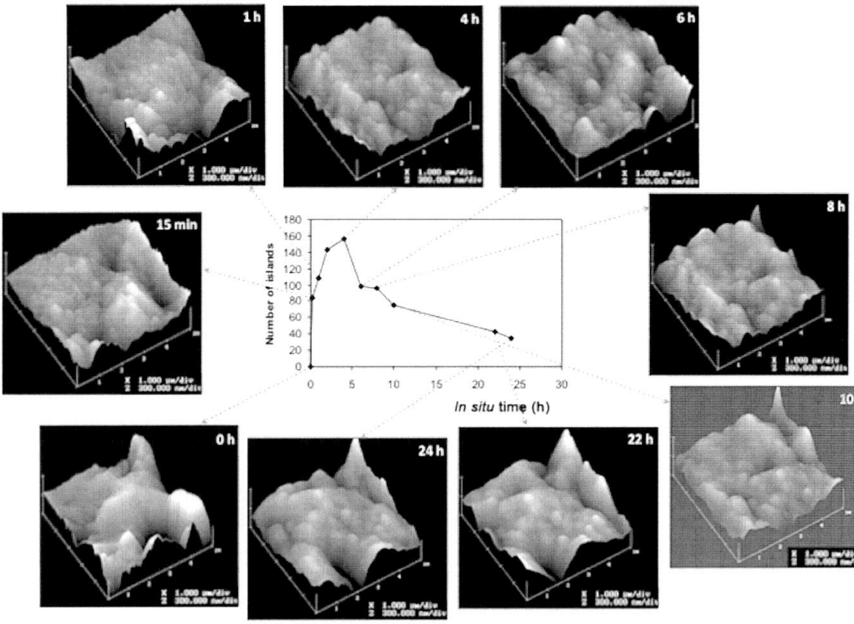

Figure 13.2 The relations between the changes in the number of islands and the formation of CaP layer on the surface of partially crystallized 45S5 Bioglass® as a function of immersion time in SBF by *in situ* observation using atomic force microscopy. After 24 h in SBF the surface of the 45S5 was almost fully covered with a CaP layer. This CaP layer grows by multiple two-dimensional nucleation and fusion of these islands after an incubation period. Data adapted from the results of Leonor et al. (2002).

13.5 Assessment of *in vitro* bioactivity

When bioactive ceramics interact with body fluid the formation of an apatite layer on their surfaces is expected, which is considered as one of the first events. However, each surface-active ceramic has its own characteristics in regard to the formation of the apatite layer.

The concept of bioactivity is defined as a surface property of an implant material, which allows it to form a chemical bond with living tissues (Kim, 2001, 2003; Kokubo et al., 2003, 2004; Kokubo and Takadama, 2006). As mentioned before, a bioactive material can be defined as the one that elicits a specific biological response at the interface of the material, which results in the formation of a bond between the tissues and the materials (Kim, 2001, 2003; Kokubo et al., 2003, 2004; Kokubo and Takadama, 2006; Hench and Paschall, 1973). There are few materials that directly bond to living bone without the formation of detectable apatite on their surfaces.

The body fluid is supersaturated with proteins, cells, and inorganic ions that may be involved in the biomineralization and thus is too complex to mimic *ex vivo* for designing bioactive materials (Kim, 2003). In this sense, Kokubo et al. (ISO, 2014; Kokubo et al., 1990; Kokubo and Takadama, 2006) proposed a protein-free and acellular SBF with pH 7.40 and ionic composition nearly equal to that of human blood plasma (see Table 13.2). Thus, the bioactivity of a material can be evaluated by observing the formation *per se* of apatite on its surface when immersed in an SBF solution. In 2003, conventional SBF (1.0 SBF) with the refined recipe was proposed to the Technical Committee ISO/TC150 of the International Organization for Standardization as a solution for *in vitro* measurement of the apatite-forming ability of implant materials (Kokubo and Takadama, 2006; Oyane et al., 2003b). In 2012, the conventional SBF solution was approved by the ISO-TC as a solution for testing the apatite-forming ability of materials and in 2014 it was revised (ISO, 2014). Furthermore, this solution can be used for screening bioactive materials before animal testing and therefore to reduce the number of animals used and the duration of animal experiments, which can assist in the efficient development of new types of bioactive materials (Kokubo and Takadama, 2006).

In recent years, SBF has been used for the *in vitro* assessment of bioactivity of artificial materials by examining their apatite-forming abilities in this solution. On the other hand, the SBF provided significant information on the compositional and structural dependence of apatite formation on ceramic surfaces (Kim, 2001). Moreover, the SBF is a powerful source to prepare a highly bioactive apatite layer on different materials, through which bioactive materials with different functions may be fabricated. It can be said from these tests that a material able to induce the formation of an apatite layer on its surface in SBF can bond to living bone through the apatite formed on its surface in the living body, as long as the material does not contain any substance that induces toxic or antibody reactions (Kokubo and Takadama, 2006). For more information about the history of SBF and the correlation of the ability of apatite to form on various materials in SBF with *in vivo* bone bioactivities, we suggest the manuscript of Kokubo and Takadama (2006).

Table 13.2 Nominal ion concentrations of SBF in comparison with those in human blood plasma (ISO, 2014; Kokubo et al., 1990; Kokubo and Takadama, 2006)

	Ion concentrations (mM)	
Ion	**Blood plasma (pH 7.2–7.4)**	**Simulated fluid (pH 7.40)**
Na^+	142.0	142.0
K^+	5.0	5.0
Mg^{2+}	1.5	1.5
Ca^{2+}	2.5	2.5
Cl^-	103.0	147.8
HCO_3^-	27.0	4.2
HPO_4^{3-}	1.0	1.0
SO_4^{2-}	0.5	0.5

Nevertheless, there has been criticism of the correlation of bioactivity tests with apatite-forming ability (Bohner and Lemaitre, 2009; Pan et al., 2010). The main concern is that bioactivity tests should not be used to predict the formation of an apatite layer on a substrate by only soaking in SBF. These are based on several aspects, namely the inability of SBF to mimic physiological conditions, the fact that the content of carbonate is not controlled, the procedure of preparing the solution is long and difficult, among others (Bohner and Lemaitre, 2009; Pan et al., 2010). Bohner and Lemaitre (2009) have compared different SBF compositions from the thermodynamic viewpoint, paying special attention to maintaining the equilibrium with the physiological temperature and partial CO_2 pressure. Also, they proposed a simplified alternative to prepare the SBF solution (Bohner and Lemaitre, 2009).

Despite these criticisms, it is important to bear in mind that SBF is a simple and rapid test to investigate the mineralization potential of a material *in vitro*; however, there is no guarantee that this material, able to induce the formation of an apatite layer *in vitro*, will bond to bone in an *in vivo* situation. This effect is related to the number of factors that are involved when a material is implanted and therefore it is a complex process to precisely reproduce.

In the case of Bioglass®, A–W glass–ceramic, and alumina we can say that there is a correlation between bioactivity tests and *in vivo* tests, where the first two materials bond directly to bone through an intervening apatite layer and in the last material there was no direct contact between bone and alumina (Neo et al., 1992a,b; Oonishi et al., 1999, 2000). In these cases the bioactive materials elicited a specific biological response at the interface of the material, resulting in the formation of a bond between the tissues and the materials. As pointed out by Bohner and Lemaitre (2009), it is important that when the bioactivity of a material is tested there should be a correlation of its *in vitro* behavior with the *in vivo* behavior.

Summarizing, the principle of this bioactivity test is to form an apatite layer on several substrates, which are not bioactive by themselves, from the viewpoint of stability as well as similarity to blood plasma in ion concentrations. In this regard, controlling the composition and structure of apatite produced in SBF in an approach analogous to bone mineral is of great interest in the development of new types of bioactive materials.

13.6 The influence of material properties in rendering self-mineralized polymers

Several aspects in terms of surface properties, namely composition, topography, and surface energy, have been explored for developing bioactive materials, such as tailoring the chemical bioactivity of ceramics by incorporating ions with a specific function. This section will present and describe some research work related to the development of bioactive polymers by *in situ* functionalization. The idea is to demonstrate the importance of functional groups for apatite nucleation and explain their mechanism on the polymers that are not intrinsically bioactive.

Our current knowledge is that functional groups that are able to induce apatite nucleation can develop a negative charge when immersed in water and can therefore lead to nucleation, meaning the acidity of a functional group perhaps works as a nucleating agent for apatite formation.

Kim et al. (2003, 2005) have used electrophoresis to investigate changes in surface composition and structure, and rationalize them in terms of change in surface potential, the process and kinetics of apatite formation on hydroxyapatite. This work demonstrated that the initialization of apatite nucleation involves an electrostatic interaction between the functional groups that are initially negative and calcium ions. Thus, functional groups that are able to become negatively charged at the blood plasma pH ($\approx$7.4) are assumed to be potentially effective for apatite nucleation in an *in vivo* environment.

Sulfonic ($-SO_3H$) groups, in terms of acidity, are worthy of research as nucleating agents for apatite formation. To demonstrate this, four types of biocompatible polymeric surfaces, namely polyethylene terephthalate (PET), polycaprolactam (Nylon 6), high-molecular-weight polyethylene (HMWPE), and ethylene-vinyl alcohol copolymer (EVOH), were subjected to sulfonation by soaking in sulfuric acid (H_2SO_4) or chlorosulfonic acid ($ClSO_3H$) at different concentrations (Leonor et al., 2007a). The selection of these polymers was based on their high chemical stability, good mechanical properties and biocompatibility, and also for their being well known in the medical field. This work clearly showed the successful incorporation of $-SO_3H$ groups onto the surface of HMWPE and EVOH through sulfonation and stabilization by soaking in a calcium hydroxide ($Ca(OH)_2$) saturated solution. The surface-modified samples are able to form a dense and uniform apatite layer after soaking for up to 7 days in SBF, as can be seen in Figure 13.3. When the sulfonated samples are soaked in $Ca(OH)_2$ saturated solution, it is expected that the $-SO_3H$ groups become negatively charged and interact with the positively charged calcium (Ca^{2+}) ion from the solution, as shown in Figure 13.3.

Nevertheless, using PET and Nylon 6 it was not possible to incorporate $-SO_3H$ groups. For PET, the amount of hydrophilic functionalities in the polymer surface had to be increased prior to sulfonation. This modification of PET involves chain cleavage, leading to new chain ends in the polymer surface and causing degradation of polymer specimens. Reaction conditions had to be tuned in order to optimize the concentration of functional groups and minimize degradation of the polymer. In the present study, all tested conditions lead to the incorporation of $-SO_3H$ groups in a very low concentration and their degradation, which makes these specimens unsuitable for the desired applications. Similar features were found for Nylon 6; the sulfonation process allows the incorporation of $-SO_3H$ groups in very low concentrations together with the wearing down of the surface layer on the specimens.

The efficacy for apatite nucleation of the $-SO_3H$ groups attached to different polymeric surfaces, HMWPE and EVOH, and combined with Ca^{2+} ions, was clearly seen in this work. This report confirms that the proposed sulfonation reactions only transform the chemical composition and structure of the outer surface layer, while the substrate bulk properties remain unaltered.

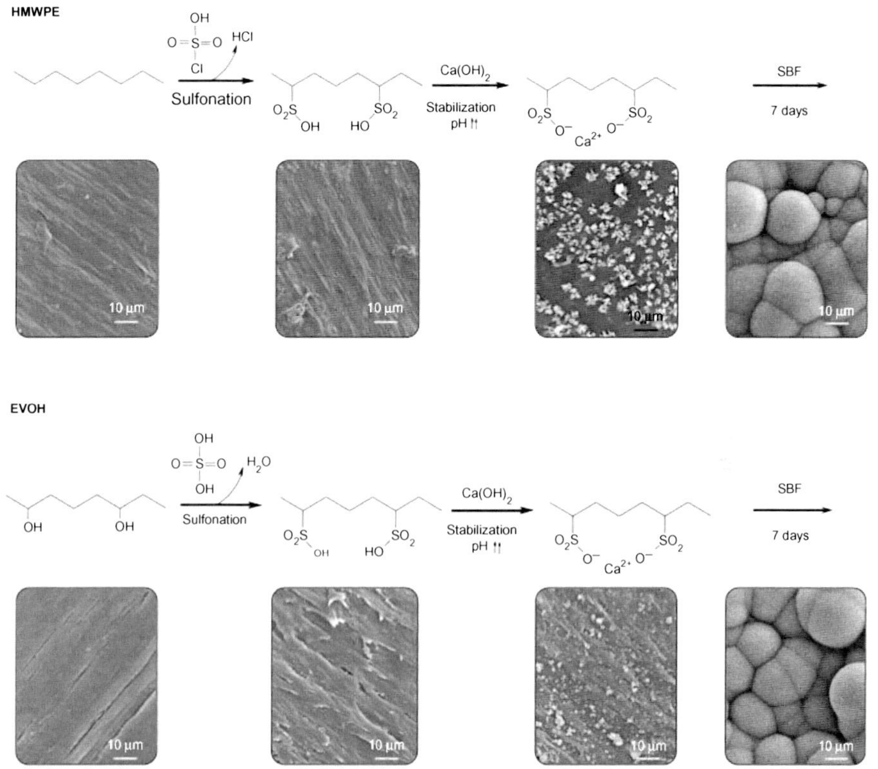

Figure 13.3 Development of polymer-based bioactive materials by the incorporation of sulfonic ($-SO_3H$) groups onto the surface of HMWPE and EVOH through sulfonation and stabilization by soaking in a calcium hydroxide ($Ca(OH)_2$) saturated solution. Scanning electron microscopy (SEM) micrograph images of the surfaces of these polymers, which were subjected to sulfonation treatment, to subsequent $Ca(OH)_2$ treatment, and then immersed in simulated body fluid (SBF) for 7 days, are shown. The proposed method allows the formation of an apatite layer onto the surface of polymeric materials that are not intrinsically bioactive. Data adapted from the results of Leonor et al. (2007a).

Tanahashi and Matsuda (1997) have demonstrated that the complex process of apatite formation is due to the electrostatic interaction between the substrate surface and specific ions in the fluid. Biomineralization, such as apatite formation, mainly occurs by calcium ion adsorption and complexation with a negatively charged group of the artificial material and its subsequent complexation with phosphate ion (Tanahashi and Matsuda, 1997). As a result, an amorphous CaP with low Ca/P ratio will form, which is phase metastable, and then will eventually be transformed into a stable apatite layer (Kokubo et al., 2003, 2004).

So, to clarify the mechanism of apatite nucleation of the $-SO_3H$ groups and their dependence on $Ca(OH)_2$ treatment, we decided to investigate the potential surface change of the bioactive polyethylene using laser electrophoresis as a function of immersion time in SBF (Leonor et al., 2007c). This work proved that formation of the

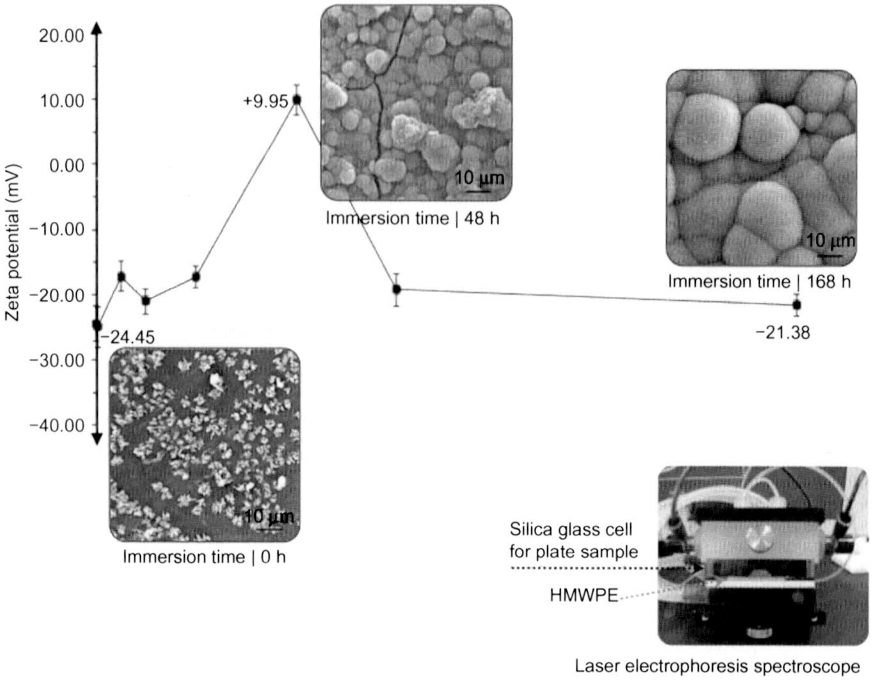

Figure 13.4 Zeta potential measurement of HMWPE surface by laser electrophoresis spectroscopy. The surface of HMWPE was subjected to sulfonation treatment and subsequent $Ca(OH)_2$ treatment. Afterwards, the variation of the surface potential of HMWPE was evaluated as a function of soaking time in SBF. These results demonstrated that the formation of apatite on bioactive polyethylene in SBF is due to electrostatic interaction of the polymer surface and ions in the fluid.
Data adapted from the results of Leonor et al. (2007c).

apatite layer on bioactive HMWPE in SBF is due to electrostatic interaction of the polymer surface and ions in the fluid. This process is illustrated in Figure 13.4 and can be described as follows: (1) formation of $-SO_3H$ groups with a negative charge by the release of Ca^{2+} ions from the HMWPE; (2) formation of an amorphous calcium sulfate with a positive charge by combining the negatively charged $-SO_3H$ groups with the positively charged Ca^{2+} ions in the SBF; (3) formation of an amorphous CaP by combining the positively charged calcium sulfate with the negatively charged phosphate ions in the SBF; and (4) formation of an apatite film with negative charge by transforming the CaP into crystalline apatite.

Finally, it is worth noting that laser electrophoresis spectroscopy can be a very useful tool to understand the mechanism of apatite formation by measuring the zeta potential of the substrate surface and therefore improving the design of high-performance bioactive polymers.

In these studies, we demonstrated the incorporation of $-SO_3H$ groups by covalent attachment on different polymeric surfaces, allowing us to confirm the assumption of the role of this functional group as well as to develop a bioactive surface *in situ* by functionalization. This combination should allow the development of bioactive apatite–polymer composites.

Reis and Cunha (1995) proposed starch-based polymers as alternative biomaterials for temporary biomedical applications. In several published works (Gomes et al., 2001, 2003, 2006; Marques et al., 2002, 2004, 2005; Salgado et al., 2007; Santos et al., 2007), it has been demonstrated that these polymers have good potential to be used as alternative biomaterials for temporary biomedical applications, as their surface properties can be controlled to achieve a favorable interaction with the biological system. In fact, one of the advantages of these polymers is the possibility to chemically tune their properties, such as producing surfaces that are more hydrophilic and capable of carrying functional groups. The structure of corn starch with ethylene-vinyl alcohol biodegradable blend (SEVA-C) contains hydroxyl (–OH) groups, a reactive functional group (Leonor et al., 2007b). Nonetheless, the abundance of –OH groups in its structure does not confer bioactive behavior, i.e. it cannot induce the formation of an apatite layer by itself. On the other hand, the presence of these groups allow the incorporation of other polar groups such as carboxyl (–COOH), a negatively charged group effective in apatite nucleation (Tanahashi and Matsuda, 1997). Thus, to allow starch-based biodegradable polymers to mineralize by themselves, we used wet chemistry modification to functionalize their surfaces (Leonor et al., 2007b). For this purpose, we used two different types of activation: (i) $Ca(OH)_2$ and (ii) sodium hydroxide (NaOH) solution (Figure 13.5).

These treatments led to the formation of carboxylic acid-rich SEVA-C surfaces. This simple functionalization via etching and hydrolysis provides a polymeric surface with precursors of apatite, prior to the soaking in SBF, and therefore decreases the period required for apatite deposition from SBF. The two alkaline treatments have some significant differences. $Ca(OH)_2$ treatment was more efficient for the induction of an apatite layer than NaOH treatment. These results are attributed to the formation of –COOH groups, and also the water uptake capability of this polymer allowed the material to absorb higher quantities of Ca^{2+} ions from the calcium solution. As result, these groups, together with Ca^{2+} ions, lead to an increase of the ionic activity product of the apatite and accelerates apatite nucleation. The NaOH treatment allowed SEVA-C to absorb large quantities of sodium (Na^{+}) ion, and when soaked in SBF, these ions are released into the solution, leading to the formation of sodium carbonate (Na_2CO_3). This is due to the fact that sodium ions have great capacity to bind with carbonate (CO^{3-}) ions from the SBF solution. Consequently, the amount of CaP deposited on the NaOH-treated SEVA-C is less, i.e. the formation of the apatite layer was delayed due to the –COOH groups being busy with Na^{+} ions and not being available for bonding with Ca^{2+} ions to induce apatite nucleation.

These results suggest that these simple treatments are effective for surface functionalization and subsequent mineral nucleation and growth on biodegradable polymers to be used for bone-related applications. Moreover, due to its simplicity, the present surface modification has advantages in terms of cost and processing time and it can be applied to biodegradable scaffolds.

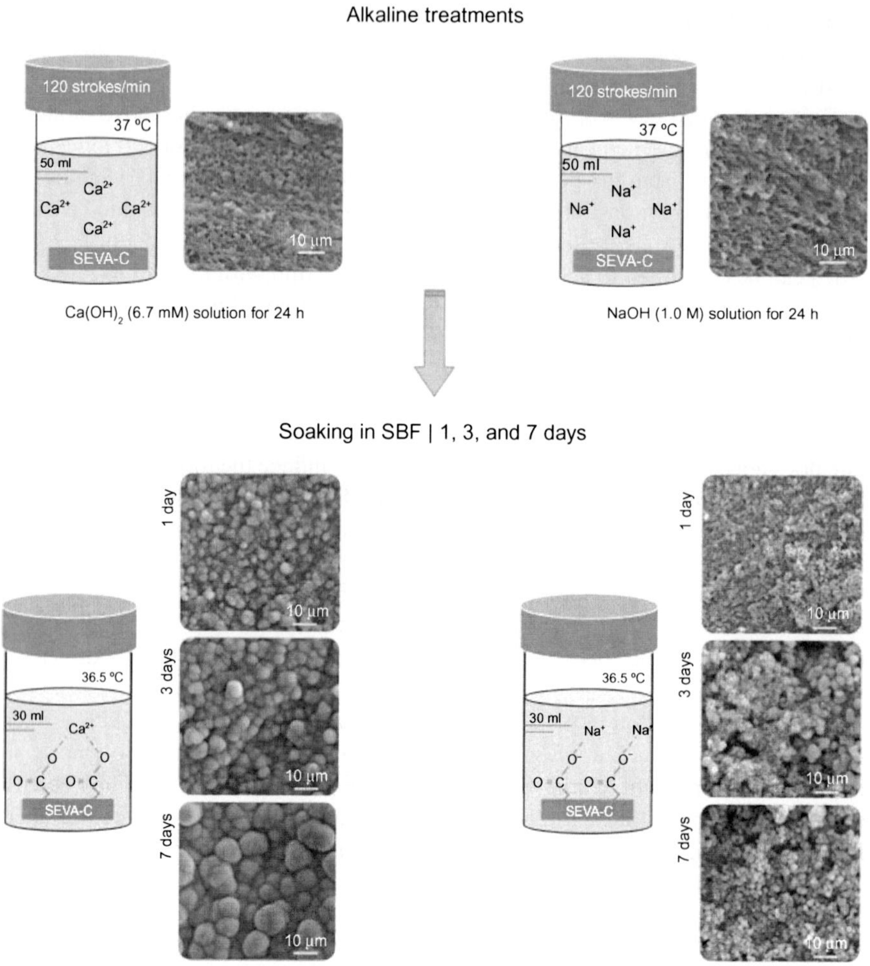

Figure 13.5 Straightforward alkaline treatments to render starch-based degradable polymers self-mineralizable. In this work a blend of corn starch with ethylene-vinyl alcohol, 50/50 wt% (SEVA-C), was submitted to two different treatments, $Ca(OH)_2$ and NaOH, for the incorporation of carboxyl (–COOH) functional groups. To evaluate the formation of an apatite layer the treated samples were soaked in simulated body fluid (SBF) for 1, 3, and 7 days. Data adapted from the results of Leonor et al. (2007b).

All studies described in this section share a common finding that a surface with an organized arrangement of functional groups can act as a template for apatite nucleation. Such materials can be developed by combining an inorganic and an organic phase, and in this way mimicking the bone structure that has been developed over millions of years of evolution (Chai et al., 2012). On the basis of these fundamental findings, several kinds of bone-bonding materials with different mechanical properties can be developed, and it is expected that a properly functionalized surface will lead to a desirable cellular response.

13.7 Engineering a bioactive 3D structure for bone TE

As mentioned earlier, three essential elements of bone regeneration, along with a stable and strong interface between the implant and bone tissue, are osteoconduction, osteoinduction, and osteogenesis. In this way, we are able to overcome the numerous complications associated with bone-grafting strategies. Bone TE success strongly depends on our ability to develop new materials by combining these three crucial elements.

TE of artificial bone grafts is a recent promising approach, which combines the state of the art of materials science with the concepts of biology/life sciences (Langer and Vacanti, 1993). Generally, the published works on TE consist of seeding cells onto a synthetic or natural scaffold (Gurtner et al., 2007) and bioactive scaffolds that contain or recruit growth factors (Kempen et al., 2009; Kruyt et al., 2009). For example, polymers are frequently used due to their versatility in terms of chemical and physical properties; however, their limited bioactivity provides only partial cues to guide tissue regeneration (Laurencin et al., 1999). So far, polymeric scaffolds (naturally derived or synthetic) have shown promising potential to be used as artificial bone grafts due to their favorable mechanical properties, namely elasticity (Kanczler and Oreffo, 2008; Salerno et al., 2010). However, bone TE is currently compromised by its inability to produce load-bearing scaffolds, with osteoconductivity, and osteoinductivity and biological cues to ensure the creation of a microenvironment to guide tissue regeneration. Although CaP mineral-based scaffolds dominate the commercial bone-substitute materials market, they lack the fibrillar protein component and are brittle, typically failing at compressive loads of less than 5 MPa (Collins et al., 2009).

Knowing the advantages and weaknesses of ceramics and natural polymers, it is possible to synergize the best properties of both materials for developing bioactive 3D structures for use in bone regeneration applications. To accomplish that, we proposed to engineer a material with inorganic and organic phases, mimicking natural bone matrix, by means of combining wet-spinning technologies with a calcium silicate solution as a coagulation bath for different applications in the bone TE field (Rodrigues et al., 2012, 2014; Leonor et al., 2011; Requicha et al., 2014a,b).

The wet-spinning process and calcium silicate solution allow us to develop an *in situ* functionalization methodology to obtain 3D wet-spun fiber meshes with silanol (Si–OH) groups through a simple, economic, and reliable process, as described in Figure 13.6 (Leonor et al., 2011; Requicha et al., 2014a,b; Rodrigues et al., 2012, 2014). This process has the advantage that no further CaP coating or chemical modification of the fiber mesh is required to render bioactive behavior as in the classic ceramic materials described in Section 13.3, such as an organized arrangement of Si–OH groups.

Li et al. (1992, 1993a,b) have confirmed the catalytic effect of Si–OH groups for apatite nucleation on pure silica gel, prepared by a sol–gel method, to form an apatite layer on its surface in SBF. Previous studies also showed that the presence of Si–OH groups in polymeric materials is crucial to induce apatite formation in SBF solution (Leonor et al., 2008; Oyane et al., 2003a).

Some studies (Carlisle, 1981; Hench, 2009; Hing et al., 2006; Pietak et al., 2007) suggest that silicon ions are an essential mineral and play a significant role in bone

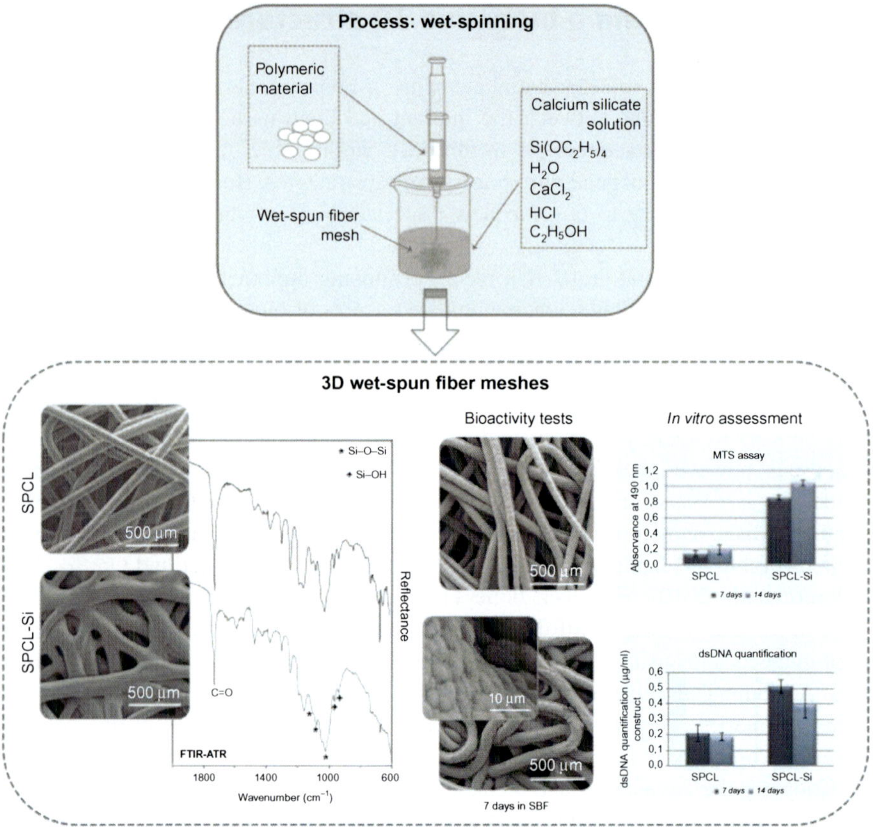

Figure 13.6 Generation of bioactive materials from a calcium silicate solution: 3D wet-spun fiber meshes were fabricated via a one-step synthesis, in which the wet-spun polymer acts as a template mesh for incorporation of silanol (Si–OH) groups while it is precipitated in the solution. Scanning electron microscopy (SEM) micrograph images of fiber mesh scaffolds produced by wet-spinning in methanol solution (SPCL—control) and calcium silicate solution (SPCL-Si), and the respective Fourier transform infrared spectroscopy with attenuated total reflection (FTIR–ATR) spectra of both fiber meshes, are shown. Evaluation of *in vitro* bioactivity of wet-spun fiber mesh scaffolds: after 7 days of immersion in SBF a dense and uniform apatite layer was formed on SPCL-Si scaffolds. At higher magnifications a finer structure is seen where needle-like crystals are agglomerated. For the control, SPCL scaffolds, no formation of an apatite layer was observed. Biological assessment of wet-spun fiber mesh scaffolds: cell viability (MTS assay) and cell proliferation (DNA quantification) of GBMSCs seeded onto SPCL and SPCL-Si scaffolds after 7 and 14 days in osteogenic medium.
Data adapted from the results of Leonor et al. (2011) and Rodrigues et al. (2012).

and cartilage systems, acting on the physiological system most prominently during the growth and development of the skeleton. Moreover, silicon and calcium ions act on the skeletal system during its growth and development by regulating and activating genes in osteoprogenitor cells (Hench, 2009). Thus, the main aim of these works was to control the architecture and surface chemistry of the scaffold, through the incorporation of functional groups, with the expectation of regulating cell behavior and, simultaneously, guiding new tissue formation. In this case, a scaffold is considered not as an inert 3D substrate, but as a conduit for dynamic extracellular signaling, chemical, biological, and mechanical cues, thus facilitating progenitor cell recruitment, growth, and differentiation (Dawson and Oreffo, 2008). Therefore, the development of these scaffolds could play a direct role in tissue regeneration as they better simulate bone, and the presence of cells is likely to have an important role in interaction and communication within the native bone environment.

With wet-spinning technologies it is possible to create hybrid materials with high levels of organization, with a particular array of chemical and physical properties, and also avoiding the thermal degradation of natural origin polymers compared to other techniques, such as the melt-spinning technique (Tuzlakoglu and Reis, 2009).

For this work a blend of corn starch with polycaprolactone was used (30/70 wt%, starch with polycaprolactone, SPCL) to produce 3D fiber mesh scaffolds by the wet-spinning technique.

In Figure 13.6, we can observe SPCL fiber mesh scaffolds with Si–OH groups (SPCL-Si) with a large surface area and highly interconnected pores. The porosity was around 56.84% (the pores are completely interconnected) and the fiber diameter was around 100 μm. As for SPCL, without Si–OH groups, the micrographs revealed a typical SPCL fiber mesh structure with a random distribution of the fibers within the mesh and highly interconnected pores. The porosity was about 85% and the fiber diameter was around 100 μm, as determined by microcomputed tomography analysis (Leonor et al., 2011).

As mentioned earlier, the bioactivity of artificial materials is commonly evaluated by examining the formation of an apatite layer on the surface of the scaffold after immersion in an SBF solution. After soaking in SBF for 7 days, an apatite layer was formed in SPCL-Si scaffolds, while SPCL scaffolds could not induce the formation of an apatite layer (Figure 13.6). Moreover, at higher magnifications a finer structure is seen, where needle-like crystals agglomerate.

Fourier transform infrared spectroscopy with attenuated total reflection (FTIR–ATR) and X-ray photoelectron spectroscopy (XPS) analysis clearly demonstrated that by precipitating a polymeric solution, in this case SPCL, into the calcium silicate coagulation bath, we were able to incorporate Si–OH groups into the surface of 3D wet-spun fiber meshes.

By FTIR–ATR analysis, the components typical of PCL at 1725 cm^{-1} (C=O) and 1170 cm^{-1} (C–O) decreased in intensity, while a new shoulder band appearing at 1210 cm^{-1} was observed (Elzein et al., 2004; Pashkuleva et al., 2008). The decrease in stretching vibrations of carbonyl and ether groups can be related to the incorporation of silicon from the precipitation of SPCL polymeric solution into the calcium silicate solution. Also, several characteristic bands of sol–gel-derived silica were observed. The presence of Si–OH groups comes from the hydrolyzed and incomplete polycondensation reaction of TEOS, which can lead to the formation of hydrogen-bonding interactions with the carbonyls of PCL. Also, the presence of a new shoulder can be attributed to the stretching vibration of carbonyl groups of PCL that are bonded to the silanol groups from the calcium silicate solution. Several research works have demonstrated these hydrogen-bonding interactions between the carbonyls of PCL and Si–OH of the silica network in hybrid materials (Eglin et al., 2003; Nie et al., 2005; Taddei et al., 2005).

The results obtained from XPS analysis are in good agreement with the FTIR–ATR data (Leonor et al., 2011; Rodrigues et al., 2012), confirming the incorporation of Si ions in the SPCL fiber mesh, using the calcium silicate solution as a coagulation bath, since these ions were not detected in the control samples, SPCL. In the C_{1s} core-level XPS spectra for SPCL fiber mesh scaffolds three mains peaks were observed, one at 285 eV corresponding to the main backbone carbon peak (polymeric C–C or CH_2–CH_2) and another which is at about 286.4 eV for –C–O–. The third peak at 288 eV corresponds to the carbonyl/carboxyl bonded carbons (–C(=O)O–) (Pashkuleva et al., 2008). In the case of SPCL-Si fiber mesh scaffolds, a decrease of the intensity of the carbon signal was clearly observed, especially for the bonds (–C–O–) and (–C(=O)O–). This decrease could be related to the incorporation of silicon and oxygen, which was confirmed by FTIR–ATR measurements. This hypothesis was also confirmed by the presence of a new peak around 103.35–103.5 eV in the Si_{2p} core-level XPS spectra that is typically attributed to the Si–O bond present in SiO_2. As for the O_{1s} region in SPCL-Si, two peaks were detected with binding energies of 532.26 and 531.68 eV, respectively.

Concerning the mechanical properties, dynamic mechanical analysis shows that incorporation of silicon from the precipitation of SPCL polymeric solution in the calcium silicate solution leads to a higher storage modulus, i.e. SPCL-Si possess a higher stiffness than the SPCL control samples. The incorporation of inorganic phase into the organic phase leads to a restriction of the SPCL chain mobility and therefore increases the stiffness of SPCL-Si structures. It would thus appear that when a polymeric solution, SPCL solution, is precipitated into the coagulation bath containing calcium silicate solution, it incorporates Si–OH groups that impose restrictions on the flexible SPCL chains. Some researchers have reported that the mobility of PCL chains was restricted when this polymer was part of the ceramer (Tian et al., 1998, 1999).

An *in vitro* biological assessment of wet-spun fiber mesh scaffolds of SPCL with or without Si–OH groups was performed to evaluate their osteoconductivity and osteogenic properties using goat bone marrow stromal cells (GBMSCs). From this study, it was observed that the cell viability and proliferation were higher in cells seeded onto scaffolds with silanol groups (Figure 13.6). The higher alkaline phosphatase (ALP) activity and matrix production indicated that Si–OH groups improve cellular functionality toward the osteoblastic phenotype.

Moreover, culturing cells under dynamic conditions have been shown to provide an important stimulus for the proliferation and differentiation of cells and, furthermore, to mimic *in vivo* pressure gradients. Thus, the effect of the incorporation of Si–OH groups into SCPL scaffolds on the osteogenic differentiation of human adipose stem cells (hASCs) cultured under dynamic and static conditions was also assessed (Rodrigues et al., 2012). Once more, the functionalized materials exhibit the capacity to sustain cell proliferation and induce their differentiation into the osteogenic lineage. The formation of mineralization nodules was observed in cells cultured onto the functionalized materials. Culture under dynamic conditions using a flow perfusion bioreactor was shown to enhance hASC proliferation and differentiation and a better distribution of cells within the material.

Overall, the results obtained from previous works (Leonor et al., 2011; Requicha et al., 2014a,b; Rodrigues et al., 2012, 2014) demonstrated that polymeric wet-spun fiber meshes incorporating bioactive functional Si–OH groups has the potential to work as a supporting matrix for osteogenic differentiation of mesenchymal stem cells. The positive effect of these groups may be related to the biological relevance of silicon in natural metabolic processes of bone due to its influence in the gene upregulation of osteoblastic cells (Hench, 2009; Xynos et al., 2000a,b), underlying the potential of SPCL-Si scaffolds and their intrinsic properties to sustain *in vitro* osteogenic features.

This innovative procedure developed by our group (Leonor et al., 2011; Rodrigues et al., 2012) allows us to design bioactive scaffolds by surface functionalization using functional groups instead of a CaP coating as is often described in scientific publications (Yeung et al., 2013). The fact that this *in situ* functionalization is performed under mild conditions allows the incorporation of biologically active molecules. This methodology allowed the development of 3D wide-net meshes by assembling these thin wet-spun fiber meshes, and allowed their structure to be adjusted according to the location of implantation and the size of the bone defect.

13.8 Concluding remarks and future perspectives

Structures of various shapes and functions found in Nature are not only compelling for their beauty and diversity but also clearly inspire the design of new architectures (Hecht, 2005). Nevertheless, it is necessary to take into account the fact that biomineralization has evolved over millions of years, while materials science has progressed only over the past century (Hou et al., 2004). It is certain that there is still much to be learned from the way that Nature assembles its many biologically important structures. Research efforts have recently moved from the development of inert materials that mimic the biomechanical properties of native tissue to bioactive materials, which promote tissue self-healing.

The research work presented in this chapter highlights the importance of understanding the surface chemistry of biomaterials, as well as the main mechanism for apatite formation, which provides the fundamental principles and tools for designing new bioactive materials. These fundamental mechanisms of bioactive ceramics are believed to open a new avenue for future breakthroughs for clinical biointeractive materials with novel physical, chemical, and biological functions. Here, it was demonstrated that any organic polymer can also exhibit bioactivity if its surface is modified with a functional group effective in apatite nucleation. These new types of bioactive materials are expected to play an important role in bone replacement and regeneration. Therefore, it can be concluded that the key point lies in designing an organized functionalized surface to control the mechanisms of apatite nucleation. Additionally, we highlight that controlling the material properties by functionalization can work as an approach to control cell proliferation and differentiation.

A thorough understanding of the ways in which these bioactive materials cooperate in regenerating bone provide us with novel combinatorial therapies that can be used to engineer an implant with tailored functional properties able to create an appropriate environment for guiding cell recruitment and driving bone tissue formation or to enhance bone regeneration of aging people by delivering a stimulus to reactivate the osteoprogenitor cells or reverse the deterioration of bone. However, the mechanism of how inorganic ions, released from those bioactive ceramics or polymeric structures, activate and regulate the osteogenic process is currently unknown (Samavedi et al., 2013; Chai et al., 2012; Hoppe et al., 2011). To overcome this, there is a need for a comprehensive understanding of cell signaling pathways and molecular mechanisms that underlie the process of bioactive material-mediated osteogenesis and angiogenesis for effective bone defect repair. Also, several pertinent questions continue to be unanswered, namely how can cells sense the chemical and physical changes in materials, such as surface topography? What chemical and mechanical cues provided by a scaffold are needed to drive cells, in particular the differentiation of stem cells toward osteoblastic lineages? What are the best properties of materials to promote vascularization?

The answers to these questions require collaborative work of scientists with different academic backgrounds ranging from bioengineering, cell and molecular biology, chemistry, material sciences, medicine, and physics. Recent advances in the development of nanotechnology, a window of new technologies and materials, have shed new light on the importance of deepening the interface between biology and the synthetic world, as cells have a unique ability to sense and respond to signals. Nanotechnology can play a significant role in the bone TE field by helping to recreate the cellular microenvironmental niche in the development of nanostructured biomaterials, incorporating bioactive cues and topographical features to direct stem cell differentiation (Saiz et al., 2013; Zhang and Webster, 2009). However, it is important to keep in mind that it is necessary to monitor the cost of these solutions since accessible, simple, and reasonably priced approaches are crucial (Schroeder and Mosheiff, 2011).

In this way, major breakthroughs in the rational design of future smart bioactive materials are expected and will be beneficial for patients needing treatment for bone defects after fracture, osteoporosis, or after operation on malignant bone tumors, and more importantly that are able to last for the lifespan of the patient.

Acknowledgments

The authors gratefully acknowledge the support provided by the Portuguese Foundation for Science and Technology (FCT): OSTEOSYNTHESIS Project (PTDC/CTM-BIO/0814/2012) and SFRH/BD/69962/2010 (A.I. Rodrigues doctoral grant).

References

Ball, P., 2001. Life's lessons in design. Nature 409, 413–416.

Bayliss, L., Mahoney, D.J., Monk, P., 2012. Normal bone physiology, remodelling and its hormonal regulation. Surgery (Oxford) 30, 47–53.

Bohner, M., Lemaitre, J., 2009. Can bioactivity be tested in vitro with SBF solution? Biomaterials 30, 2175–2179.

Bonfield, W., Wang, M., Tanner, K.E., 1998. Interfaces in analogue biomaterials. Acta Mater. 46, 2509.

Bonzani, I.C., George, J.H., Stevens, M.M., 2006. Novel materials for bone and cartilage regeneration. Curr. Opin. Chem. Biol. 10, 568–575.

Boskey, A., 1999. Mineralization, structure and function of bone. In: Seibel, M., Robins, S., Bilezikian, J. (Eds.), Dynamics of Bone and Cartilage Metabolism. Academic Press, San Diego.

Carlisle, E.M., 1981. Silicon—a requirement in bone-formation independent of vitamin-D1. Calcif. Tissue Int. 33, 27–34.

Chai, Y.C., Carlier, A., Bolander, J., Roberts, S.J., Geris, L., Schrooten, J., van Oosterwyck, H., Luyten, F.P., 2012. Current views on calcium phosphate osteogenicity and the translation into effective bone regeneration strategies. Acta Biomater. 8, 3876–3887.

Cho, G., Wu, Y., Ackerman, J.L., 2003. Detection of hydroxyl ions in bone mineral by solid-state NMR spectroscopy. Science 300, 1123–1127.

Cohen, M.M., 2006. The new bone biology: pathologic, molecular, and clinical correlates. Am. J. Med. Genet. A 140A, 2646–2706.

Collins, A.M., Skaer, N.J.V., Cheysens, T., Knight, D., Bertram, C., Roach, H.I., Oreffo, R.O.C., Von-Aulock, S., Baris, T., Skinner, J., Mann, S., 2009. Bone-like resorbable silk-based scaffolds for load-bearing osteoregenerative applications. Adv. Mater. 21, 75–78.

Cuisinier, F.J.G., 1996. Bone mineralization. Curr. Opin. Solid State Mater. Sci. 1, 436.

Dawson, J.I., Oreffo, R.O.C., 2008. Bridging the regeneration gap: stem cells, biomaterials and clinical translation in bone tissue engineering. Arch. Biochem. Biophys. 473, 124–131.

Dove, P.M., Deyoreo, J.J., Weiner, S., 2003. Biomineralization. Reviews in Mineralogy and Geochemistry. Mineralogical Society of America, Washington, DC.

Drosse, I., Volkmer, E., Capanna, R., De Biase, P., Mutschler, W., Schieker, M., 2008. Tissue engineering for bone defect heating: an update on a multi-component approach. Injury—Int. J. Care Inj. 39, S9–S20.

Ducheyne, P., Qiu, Q., 1999. Bioactive ceramics: the effect of surface reactivity on bone formation and bone cell function. Biomaterials 20, 2287–2303.

Dunlop, J.W.C., Fratzl, P., 2010. Biological composites. Annu. Rev. Mater. Res. 40 (40), 1–24.

Eglin, D., Ali, S.A.M., Perry, C.C., 2003. A statistical study of poly(epsilon-caprolactone) crystallinity in poly(epsilon-caprolactone)-silica sol–gel materials and their in vitro calcium phosphate-forming ability. Polym. Int. 52, 1807–1819.

Elzein, T., Nasser-Eddine, M., Delaite, C., Bistac, S., Dumas, P., 2004. FTIR study of polycaprolactone chain organization at interfaces. J. Colloid Interface Sci. 273, 381–387.

Giannoudis, P.V., Dinopoulos, H., Tsiridis, E., 2005. Bone substitutes: an update. Injury—Int. J. Care Inj. 36, 20–27.

Gomes, M.E., Reis, R.L., Cunha, A.M., Blitterswijk, C.A., De Bruijn, J.D., 2001. Cytocompatibility and response of osteoblastic-like cells to starch-based polymers: effect of several additives and processing conditions. Biomaterials 22, 1911–1917.

Gomes, M.E., Sikavitsas, V.I., Behravesh, E., Reis, R.L., Mikos, A.G., 2003. Effect of flow perfusion on the osteogenic differentiation of bone marrow stromal cells cultured on starch-based three-dimensional scaffolds. J. Biomed. Mater. Res. 67A, 87–95.

Gomes, M.E., Holtorf, H.L., Reis, R.L., Mikos, A.G., 2006. Influence of the porosity of starch-based fiber mesh scaffolds on the proliferation and osteogenic differentiation of bone marrow stromal cells cultured in a flow perfusion bioreactor. Tissue Eng. 12, 801–809.

Green, D., Walsh, D., Mann, S., Oreffo, R.O., 2002. The potential of biomimesis in bone tissue engineering: lessons from the design and synthesis of invertebrate skeletons. Bone 30, 810–815.

Gurtner, G.C., Callaghan, M.J., Longaker, M.T., 2007. Progress and potential for regenerative medicine. Annu. Rev. Med. 58, 299–312.

He, J.W., Genetos, D.C., Leach, J.K., 2010. Osteogenesis and trophic factor secretion are influenced by the composition of hydroxyapatite/poly(lactide-co-glycolide) composite scaffolds. Tissue Eng. Part A 16, 127–137.

Hecht, S., 2005. Construction with macromolecules. Mater. Today 8, 48.

Hench, L.L., 1988. Bioactive ceramics. In: Ducheyne, P., Lemons, J.E. (Eds.), Bioceramics: Material Characteristics Versus In Vivo Behaviour. Academy of Sciences, New York.

Hench, L.L., 1991. Bioceramics: from concept to clinic. J. Am. Ceram. Soc. 74, 1487–1510.

Hench, L.L., 1995. Bioactive implants. Chem. Ind. 3, 547–550.

Hench, L.L., 1998. Biomaterials: a forecast for the future. Biomaterials 19, 1419–1423.

Hench, L.L., 2009. Genetic design of bioactive glass. J. Eur. Ceram. Soc. 29, 1257–1265.

Hench, L.L., Paschall, H.A., 1973. Direct chemical bond of bioactive glass-ceramic materials to bone and muscle. J. Biomed. Mater. Res. 7, 25–42.

Hench, L.L., Wilson, J., 1984. Surface-active biomaterials. Science 226, 630–636.

Hench, L.L., Splinter, R.J., Allen, W.C., Greenlee, T.K., 1971. Bonding mechanisms at the interface of ceramic prosthetic materials. J. Biomed. Mater. Res. 5, 117–141.

Hench, L.L., Polak, J.M., Xynos, I.D., Buttery, L.D.K., 2000. Bioactive materials to control cell cycle. Mater. Res. Innov. 3, 313–323.

Hing, K.A., Revell, P.A., Smith, N., Buckland, T., 2006. Effect of silicon level on rate, quality and progression of bone healing within silicate-substituted porous hydroxyapatite scaffolds. Biomaterials 27, 5014–5026.

Hoppe, A., Guldal, N.S., Boccaccini, A.R., 2011. A review of the biological response to ionic dissolution products from bioactive glasses and glass-ceramics. Biomaterials 32, 2757–2774.

Hou, D.F., Zhou, G.S., Zheng, M., 2004. Conch shell structure and its effect on mechanical behaviors. Biomaterials 25, 751–756.

Isaac, D.H., Green, M., 1991. Preferred orientation in bone. Interf. Med. Mech. 2, 76.

ISO, 2014. Implants for Surgery—In Vitro Evaluation for Apatite-Forming Ability of Implant Materials. ISO—International Organization for Standardization.

Jones, J.R., 2013. Review of bioactive glass: from Hench to hybrids. Acta Biomater. 9, 4457–4486.

Kanczler, J.M., Oreffo, R.O., 2008. Osteogenesis and angiogenesis: the potential for engineering bone. Eur. Cell. Mater. 15, 100–114.

Kaplan, D.L., 1998. Mollusc shell structures: novel design strategies for synthetic materials. Curr. Opin. Solid State Mater. Sci. 3, 232.

Keating, J.F., McQueen, M.M., 2001. Substitutes for autologous bone graft in orthopaedic trauma. J. Bone Joint Surg. Br. Vol. 83B, 3–8.

Kempen, D.H., Kruyt, M.C., Lu, L., Wilson, C.E., Florschutz, A.V., Creemers, L.B., Yaszemski, M.J., Dhert, W.J., 2009. Effect of autologous bone marrow stromal cell seeding and bone morphogenetic protein-2 delivery on ectopic bone formation in a microsphere/poly(propylene fumarate) composite. Tissue Eng. Part A 15, 587–594.

Kim, H.M., 2001. Bioactive ceramics: challenges and perspectives. J. Ceram. Soc. Jpn. 109, S49–S57.

Kim, H.M., 2003. Ceramic bioactivity and related biomimetic strategy. Curr. Opin. Solid State Mater. Sci. 7, 289.

Kim, H.M., Himeno, T., Kawashita, M., Lee, J.H., Kokubo, T., Nakamura, T., 2003. Surface potential change in bioactive titanium metal during the process of apatite formation in simulated body fluid. J. Biomed. Mater. Res. A 67, 1305–1309.

Kim, H.M., Himeno, T., Kokubo, T., Nakamura, T., 2005. Process and kinetics of bonelike apatite formation on sintered hydroxyapatite in a simulated body fluid. Biomaterials 26, 4366–4373.

Kokubo, T., Takadama, H., 2006. How useful is SBF in predicting in vivo bone bioactivity? Biomaterials 27, 2907–2915.

Kokubo, T., Kushitani, H., Sakka, S., Kitsugi, T., Yamamuro, T., 1990. Solutions able to reproduce in vivo surface-structure changes in bioactive glass-ceramic A-W. J. Biomed. Mater. Res. 24, 721–734.
Kokubo, T., Kim, H.M., Kawashita, M., 2003. Novel bioactive materials with different mechanical properties. Biomaterials 24, 2161–2175.
Kokubo, T., Kim, H.M., Kawashita, M., Nakamura, T., 2004. Bioactive metals: preparation and properties. J. Mater. Sci. Mater. Med. 15, 99–107.
Kruyt, M.C., Delawi, D., Habibovic, P., Oner, F.C., Van Blitterswijk, C.A., Dhert, W.J., 2009. Relevance of bone graft viability in a goat transverse process model. J. Orthop. Res. 27, 1055–1059.
Langer, R., Vacanti, J.P., 1993. Tissue engineering. Science 260, 920–926.
Laurencin, C.T., Ambrosio, A.M., Borden, M.D., Cooper Jr., J.A., 1999. Tissue engineering: orthopedic applications. Annu. Rev. Biomed. Eng. 1, 19–46.
Lee, H.J., Ahn, S.H., Kim, G.H., 2012. Three-dimensional collagen/alginate hybrid scaffolds functionalized with a drug delivery system (DDS) for bone tissue regeneration. Chem. Mater. 24, 881–891.
Legeros, R.Z., 2002. Properties of osteoconductive biomaterials: calcium phosphates. Clin. Orthop. Relat. Res. (February), 81–98.
Leonor, I.B., Ito, A., Onuma, K., Kanzaki, N., Zhong, Z.P., Greenspan, D., Reis, R.L., 2002. In situ study of partially crystallized Bioglass and hydroxylapatite in vitro bioactivity using atomic force microscopy. J. Biomed. Mater. Res. 62, 82–88.
Leonor, I.B., Kim, H.-M., Balas, F., Kawashita, M., Kokubo, T., Reis, R.L., Nakamura, T., 2007a. Functionalization of different polymers with sulfonic groups as a way to coat them with a biomimetic apatite layer. J. Mater. Sci. Mater. Med. 18, 1923–1930.
Leonor, I.B., Kim, H.M., Balas, F., Kawashita, M., Reis, R.L., Kokubo, T., Nakamura, T., 2007b. Alkaline treatments to render starch-based biodegradable polymers self-mineralizable. J. Tissue Eng. Regen. Med. 1, 425–435.
Leonor, I.B., Kim, H.-M., Balas, F., Kawashita, M., Reis, R.L., Kokubo, T., Nakamura, T., 2007c. Surface potential change of polyethylene modified with sulfonic functional groups in simulated body fluid. J. Mater. Chem. 17, 4057–4063.
Leonor, I.B., Baran, E.T., Kawashita, M., Reis, R.L., Kokubo, T., Nakamura, T., 2008. Growth of a bonelike apatite on chitosan microparticles after a calcium silicate treatment. Acta Biomater. 4, 1349–1359.
Leonor, I.B., Rodrigues, G.T., Gomes, M.E., Reis, R.L., 2011. In situ functionalization of wet-spun fibre meshes for bone tissue engineering. J. Tissue Eng. Regen. Med. 5, 104–111.
Li, P.J., Ohtsuki, C., Kokubo, T., Nakanishi, K., Soga, N., Nakamura, T., Yamamuro, T., 1992. Apatite formation induced by silica-gel in a simulated body-fluid. J. Am. Ceram. Soc. 75, 2094–2097.
Li, P., Ohtsuki, C., Kokubo, T., Nakanishi, K., Soga, N., Nakamura, T., Yamamuro, T., 1993a. Effects of ions in aqueous media on hydroxyapatite induction by silica gel and its relevance to bioactivity of bioactive glasses and glass-ceramics. J. Appl. Biomater. 4, 221–229.
Li, P., Ohtsuki, C., Kokubo, T., Nakanishi, K., Soga, N., Nakamura, T., Yamamuro, T., 1993b. Process of formation of bone-like apatite layer on silica-gel. J. Mater. Sci. Mater. Med. 4, 127–131.
Lowenstam, H., Weiner, S., 1989. On Biomineralization. Oxford University Press, New York.
Mann, S., 2001. Biomineralization: Principles and Concepts in Bioinorganic Materials Chemistry. Oxford University Press, Oxford.
Marks, S.C., Hermey, D.C., 1996. The structure and development of bone. In: Bilezikian, J.P., Raisz, L.G., Rodan, G.A. (Eds.), Principles of Bone Biology. Academic Press, New York.

Marques, A.P., Reis, R.L., Hunt, J.A., 2002. The biocompatibility of novel starch-based polymers and composites: in vitro studies. Biomaterials 23, 1471–1478.

Marques, A.P., Reis, R.L., Hunt, J.A., 2004. Cytokine secretion from mononuclear cells cultured in vitro with starch-based polymers and poly-L-lactide. J. Biomed. Mater. Res. A 71, 419–429.

Marques, A.P., Reis, R.L., Hunt, J.A., 2005. An in vivo study of the host response to starch-based polymers and composites subcutaneously implanted in rats. Macromol. Biosci. 5, 775–785.

Matsuno, T., Omata, K., Hashimoto, Y., Tabata, Y., Satoh, T., 2010. Alveolar bone tissue engineering using composite scaffolds for drug delivery. Jap. Dent. Sci. Rev. 46, 188–192.

Neo, M., Kotani, S., Fujita, Y., Nakamura, T., Yamamuro, T., Bando, Y., Ohtsuki, C., Kokubo, T., 1992a. Differences in ceramic-bone interface between surface-active ceramics and resorbable ceramics: a study by scanning and transmission electron microscopy. J. Biomed. Mater. Res. 26, 255–267.

Neo, M., Kotani, S., Nakamura, T., Yamamuro, T., Ohtsuki, C., Kokubo, T., Bando, Y., 1992b. A comparative study of ultrastructures of the interfaces between four kinds of surface-active ceramic and bone. J. Biomed. Mater. Res. 26, 1419–1432.

Neo, M., Nakamura, T., Ohtsuki, C., Kokubo, T., Yamamuro, T., 1993. Apatite formation on three kinds of bioactive material at an early stage in vivo: a comparative study by transmission electron microscopy. J. Biomed. Mater. Res. 27, 999–1006.

Nie, K.M., Zheng, S.X., Lu, F., Zhu, Q.R., 2005. Inorganic–organic hybrids involving poly(epsilon-caprolactone) and silica network: hydrogen-bonding interactions and isothermal crystallization kinetics. J. Polym. Sci. Part B Polym. Phys. 43, 2594–2603.

O'Keefe, R.J., Mao, J., 2011. Bone tissue engineering and regeneration: from discovery to the clinic—an overview introduction. Tissue Eng. B Rev. 17, 389–392.

Oonishi, H., Hench, L.L., Wilson, J., Sugihara, F., Tsuji, E., Kushitani, S., Iwaki, H., 1999. Comparative bone growth behavior in granules of bioceramic materials of various sizes. J. Biomed. Mater. Res. 44, 31–43.

Oonishi, H., Hench, L.L., Wilson, J., Sugihara, F., Tsuji, E., Matsuura, M., Kin, S., Yamamoto, T., Mizokawa, S., 2000. Quantitative comparison of bone growth behavior in granules of bioglass, A-W glass-ceramic, and hydroxyapatite. J. Biomed. Mater. Res. 51, 37–46.

Oyane, A., Kawashita, M., Nakanishi, K., Kokubo, T., Minoda, M., Miyamoto, T., Nakamura, T., 2003a. Bonelike apatite formation on ethylene-vinyl alcohol copolymer modified with silane coupling agent and calcium silicate solutions. Biomaterials 24, 1729–1735.

Oyane, A., Kim, H.M., Furuya, T., Kokubo, T., Miyazaki, T., Nakamura, T., 2003b. Preparation and assessment of revised simulated body fluids. J. Biomed. Mater. Res. A 65A, 188–195.

Pan, H., Zhao, X., Darvell, B.W., Lu, W.W., 2010. Apatite-formation ability—predictor of "bioactivity"? Acta Biomater. 6, 4181–4188.

Pashkuleva, I., Azevedo, H.S., Reis, R.L., 2008. Surface structural investigation of starch-based biomaterials. Macromol. Biosci. 8, 210–219.

Pashuck, E.T., Stevens, M.M., 2012. Designing regenerative biomaterial therapies for the clinic. Sci. Transl. Med. 4, 160sr4.

Pietak, A.M., Reid, J.W., Stott, M.J., Sayer, M., 2007. Silicon substitution in the calcium phosphate bioceramics. Biomaterials 28, 4023–4032.

Posner, A.S., 1985. The mineral of bone. Clin. Orthop. Relat. Res. 200 (November), 87–99.

Reis, R.L., Cunha, A.M., 1995. Characterization of two biodegradable polymers of potential application within the biomaterials field. J. Mater. Sci. Mater. Med. 6, 786–792.

Requicha, J.F., Moura, T., Leonor, I.B., Martins, T., Munoz, F., Reis, R.L., Gomes, M.E., Viegas, C.A., 2014a. Evaluation of a starch-based double layer scaffold for bone regeneration in a rat model. J. Orthop. Res. 32, 904–909.

Requicha, J.F., Viegas, C.A., Munoz, F., Azevedo, J.M., Leonor, I.B., Reis, R.L., Gomes, M.E., 2014b. A tissue engineering approach for periodontal regeneration based on a biodegradable double-layer scaffold and adipose-derived stem cells. Tissue Eng. Part A 20 (17–18), 2483–2492.

Reznikov, N., Shahar, R., Weiner, S., 2014. Bone hierarchical structure in three dimensions. Acta Biomater. 10, 3815–3826.

Rho, J.Y., Kuhn-Spearing, L., Zioupos, P., 1998. Mechanical properties and the hierarchical structure of bone. Med. Eng. Phys. 20, 92–102.

Rios, C.N., Skoracki, R.J., Miller, M.J., Satterfield, W.C., Mathur, A.B., 2009. In vivo bone formation in silk fibroin and chitosan blend scaffolds via ectopically grafted periosteum as a cell source: a pilot study. Tissue Eng. Part A 15, 2717–2725.

Rodrigues, A.I., Gomes, M.E., Leonor, I.B., Reis, R.L., 2012. Bioactive starch-based scaffolds and human adipose stem cells are a good combination for bone tissue engineering. Acta Biomater. 8, 3765–3776.

Rodrigues, M.T., Leonor, I.B., Groen, N., Viegas, C.A., Dias, I.R., Caridade, S.G., Mano, J.F., Gomes, M.E., Reis, R.L., 2014. Bone marrow stromal cells on a 3D bioactive fiber mesh undergo osteogenic differentiation in the absence of osteogenic media supplements: the effect of silanol groups. Acta Biomater. 10 (10), 4175–4185.

Rubner, M., 2003. Materials science: synthetic sea shell. Nature 423, 925–926.

Saiz, E., Zimmermann, E.A., Lee, J.S., Wegst, U.G.K., Tomsia, A.P., 2013. Perspectives on the role of nanotechnology in bone tissue engineering. Dent. Mater. 29, 103–115.

Salerno, A., Guarnieri, D., Iannone, M., Zeppetelli, S., Netti, P.A., 2010. Effect of micro- and macroporosity of bone tissue three-dimensional-poly(epsilon-caprolactone) scaffold on human mesenchymal stem cells invasion, proliferation, and differentiation in vitro. Tissue Eng. Part A 16, 2661–2673.

Salgado, A.J., Coutinho, O.P., Reis, R.L., Davies, J.E., 2007. In vivo response to starch-based scaffolds designed for bone tissue engineering applications. J. Biomed. Mater. Res. A 80A, 983–989.

Salinas, A.J., Esbrit, P., Vallet-Regi, M., 2013. A tissue engineering approach based on the use of bioceramics for bone repair. Biomater. Sci. 1, 40–51.

Samavedi, S., Whittington, A.R., Goldstein, A.S., 2013. Calcium phosphate ceramics in bone tissue engineering: a review of properties and their influence on cell behavior. Acta Biomater. 9, 8037–8045.

Santos, M.I., Fuchs, S., Gomes, M.E., Unger, R.E., Reis, R.L., Kirkpatrick, C.J., 2007. Response of micro- and macrovascular endothelial cells to starch-based fiber meshes for bone tissue engineering. Biomaterials 28, 240–248.

Schroeder, J.E., Mosheiff, R., 2011. Tissue engineering approaches for bone repair: concepts and evidence. Injury—Int. J. Care Inj. 42, 609–613.

Simkiss, K., Wilbur, K.M., 1989. Biomineralization. Academic Press, San Diego.

Sommerfeldt, D.W., Rubin, C.T., 2001. Biology of bone and how it orchestrates the form and function of the skeleton. Eur. Spine J. 10 (Suppl. 2), S86–S95.

Stevens, M.M., 2008. Biomaterials for bone tissue engineering. Mater. Today 11, 18–25.

Stevens, M.M., George, J.H., 2005. Exploring and engineering the cell surface interface. Science 310, 1135–1138.

Taddei, P., Tinti, A., Reggiani, M., Fagnano, C., 2005. In vitro mineralization of bioresorbable poly(epsilon-caprolactone)/apatite composites for bone tissue engineering: a vibrational and thermal investigation. J. Mol. Struct. 744, 135–143.

Tanahashi, M., Matsuda, T., 1997. Surface functional group dependence on apatite formation on self-assembled monolayers in a simulated body fluid. J. Biomed. Mater. Res. 34, 305–315.

Taton, T.A., 2001. Nanotechnology—boning up on biology. Nature 412, 491–492.
Tian, D., Blacher, S., Dubois, P., Jerome, R., 1998. Biodegradable and biocompatible inorganic-organic hybrid materials—2. Dynamic mechanical properties, structure and morphology. Polymer 39, 855–864.
Tian, D., Blacher, S., Jerome, R., 1999. Biodegradable and biocompatible inorganic-organic hybrid materials: 4. Effect of acid content and water content on the incorporation of aliphatic polyesters into silica by the sol–gel process. Polymer 40, 951–957.
Tuzlakoglu, K., Reis, R.L., 2009. Biodegradable polymeric fiber structures in tissue engineering. Tissue Eng. Part B Rev. 15, 17–27.
Vallet-Regi, M., 2001. Ceramics for medical applications. J. Chem. Soc. Dalton Trans. 2, 97–108.
Wang, M., 2003. Developing bioactive composite materials for tissue replacement. Biomaterials 24, 2133–2151.
Wang, G.C., Zheng, L., Zhao, H.S., Miao, J.Y., Sun, C.H., Ren, N., Wang, J.Y., Liu, H., Tao, X.T., 2011. In vitro assessment of the differentiation potential of bone marrow-derived mesenchymal stem cells on genipin-chitosan conjugation scaffold with surface hydroxyapatite nanostructure for bone tissue engineering. Tissue Eng. Part A 17, 1341–1349.
Weiner, S., 1986. Organization of extracellularly mineralized tissues: a comparative study of biological crystal growth. CRC Crit. Rev. Biochem. 20, 365–408.
Weiner, S., Addadi, L., 1997. Design strategies in mineralized biological materials. J. Mater. Chem. 7, 689–702.
Weiner, S., Wagner, H.D., 1998. The material bone: structure mechanical function relations. Annu. Rev. Mater. Sci. 28, 271–298.
Weiner, S., Traub, W., Wagner, H.D., 1999. Lamellar bone: structure-function relations. J. Struct. Biol. 126, 241–255.
Wenk, H.R., Heidelbach, F., 1999. Crystal alignment of carbonated apatite in bone and calcified tendon: results from quantitative texture analysis. Bone 24, 361–369.
Willie, B.M., Petersen, A., Schmidt-Bleek, K., Cipitria, A., Mehta, M., Strube, P., Lienau, J., Wildemann, B., Fratzl, P., Duda, G., 2010. Designing biomimetic scaffolds for bone regeneration: why aim for a copy of mature tissue properties if nature uses a different approach? Soft Matter 6, 4976–4987.
Wilson, J., Pigott, G.H., Schoen, F.J., Hench, L.L., 1981. Toxicology and biocompatibility of bioglasses. J. Biomed. Mater. Res. 15, 805–817.
Xia, Z., Villa, M.M., Wei, M., 2014. A biomimetic collagen-apatite scaffold with a multi-level lamellar structure for bone tissue engineering. J. Mater. Chem. B 2, 1998–2007.
Xynos, I.D., Edgar, A.J., Buttery, L.D.K., Hench, L.L., Polak, J.M., 2000a. Ionic products of bioactive glass dissolution increase proliferation of human osteoblasts and induce insulin-like growth factor II mRNA expression and protein synthesis. Biochem. Biophys. Res. Commun. 276, 461–465.
Xynos, I.D., Hukkanen, M.V.J., Batten, J.J., Buttery, L.D., Hench, L.L., Polak, J.M., 2000b. Bioglass (R) 45S5 stimulates osteoblast turnover and enhances bone formation in vitro: implications and applications for bone tissue engineering. Calcif. Tissue Int. 67, 321–329.
Yeo, M.G., Kim, G.H., 2012. Preparation and characterization of 3D composite scaffolds based on rapid-prototyped PCL/beta-TCP struts and electrospun PCL coated with collagen and HA for bone regeneration. Chem. Mater. 24, 903–913.
Yeung, W.K., Reilly, G.C., Matthews, A., Yerokhin, A., 2013. In vitro biological response of plasma electrolytically oxidized and plasma-sprayed hydroxyapatite coatings on Ti-6Al-4V alloy. J. Biomed. Mater. Res. B Appl. Biomater. 101B, 939–949.
Zhang, S., 2003. Building from the bottom up. Mater. Today 6, 20.
Zhang, L.J., Webster, T.J., 2009. Nanotechnology and nanomaterials: promises for improved tissue regeneration. Nano Today 4, 66–80.

Bio-inspired calcium phosphate materials for hard-tissue repair

14

E. Cunningham[1], *G. Walker*[2], *F. Buchanan*[1], *N. Dunne*[1]
[1]Queen's University, Belfast, UK, [2]University of Limerick, Limerick, Ireland

14.1 Introduction

Usually observed in their dried state, bones are often thought of as non-living immutable structures, when in fact they are flexible living tissue, capable of remodelling to meet the mechanical requirements demanded of them. Made up of an organic matrix of collagen fibres and calcium and phosphate minerals, in the form of hydroxyapatite (HA), they form a relatively hard and lightweight composite (Toole, 1995). Despite their reputation of being an inert and static material, they have the ability to adapt mass and morphology to a variety of functional demands. With the facility for self-repair and a capacity to rapidly mobilize mineral stores on metabolic demand, they are in fact the ultimate "smart" material (Sabolinski et al., 1996).

Bones' unique capacity to adapt geometry and microstructure to specific local mechanical environments and to detect and eliminate damages and micro-crack formation, inevitable in the physiological environment, depends heavily on maintaining a delicate balance between bone resorption by osteoclasts and bone formation by osteoblasts.

This "remodelling" and maintenance of the skeletal system can be disrupted in a variety of ways that can lead to a number of bone diseases and disorders. These include genetic abnormalities, such as osteogenesis imperfecta and osteopetrosis, developmental defects that occur at or before birth, along with diseases such as osteoporosis and Paget's disease. While these directly affect the health of bone there are many disorders that have an indirect influence by interfering with mineral metabolism. Problems in the endocrine system's release of oestrogen and androgen, hormones which play a pivotal role in male and female bone growth, can disrupt the balance of osteoclast and osteoblast action. Other causes include the use of certain medication, lifestyle behaviours and skeletal complications of malignancies.

As we age, the balance is inexorably tipped in favour of osteoclasts so that resorption exceeds formation, rendering bones brittle and prone to fracture (Rodan and Martin, 2000). It is now estimated that only 31–36% of people over 70 years of age have a normal bone mass (Srouji and Livne, 2005).

Osteoporosis, by far the most prevalent skeletal disease with an estimated 200 million sufferers worldwide, is escalating, expecting to double by 2050 (Huang and Kung, 2006). Defined as "low bone mineral density (BMD) with micro-architectural alterations of bone, increased bone fragility and a greater risk of fractures" (Bianchi, 2007), osteoporosis is said to represent a major public health

Biomineralization and Biomaterials. http://dx.doi.org/10.1016/B978-1-78242-338-6.00015-6

problem (Hamerman, 2005). While the reduction of BMD and the corresponding fragility, it brings depends heavily upon many factors such as ethnicity, geography, and overall health, the impact of resulting osteoporotic fractures to the individual, their family and society is significant, inducing adverse changes in quality of life, deficits in self-care, mobility and ambulation.

In 2000, there were an estimated 9 million new osteoporotic fractures, of which 1.6 million were at the hip, 1.7 million were at the forearm and 1.4 million were clinical vertebral fractures (Johnell and Kanis, 2006). Considering gains in life expectancy and UN projections for the over-60 population, estimated to reach 1.9 billion by 2050, the economic impact and burden on the healthcare system is expected to rise significantly (Rousculp et al., 2007; Becker et al., 2010).

Non-union of fractures incurred, especially in instances of extensive bone loss due to trauma, may exceed the body's regenerative capabilities and require surgical intervention to bridge the skeletal defect. Management of a non-union is an often prolonged, challenging and costly exercise, emphasizing the importance for further bone tissue engineering (TE) strategies.

At present bone (or attempted facsimiles) is second only to blood as the material most often implanted in the body. The demand is highlighted by the fact that there are between 2 and 4 million operations involving bone grafting or bone substitutes performed annually around the world (Brydone et al., 2010; Giannoudis et al., 2005; Vaccaro, 2002). A variety of materials are used, ranging from the traditional autografts, allografts and xenografts to the alternative demineralized bone matrix, collagen, bioceramics (CaP, HA and TCP), coralline HA, bone marrow aspirates, bone morphogenetic proteins, bioactive glasses and biodegradable polymers (Giannoudis et al., 2005). All these materials are intended to provide support, fill voids and enhance biologic repair of skeletal defects. They are used by orthopaedic surgeons, neurosurgeons, craniofacial surgeons and periodontists to repair skeletal deficiencies which arise from bone diseases, tumours, trauma or abnormal development.

To date, autografts, defined as "tissue grafted into a new position in or on the body of the same individual", remain the gold standard for these procedures providing all three interdependent elements necessary to maximize bone-forming ability (Giannoudis et al., 2005; Vaccaro, 2002):

1. Scaffolding which will support osteoconduction (new bone growth).
2. Growth factors which promote the stimulation and growth of the surrounding tissue providing osteoinductive potential.
3. Progenitor cells within the graft which have the potential to develop into bone-forming osteoblasts, providing osteogenic potential.

Associated drawbacks include quantity limitations, morbidity and donor-site complications which increase recovery time, disability and chronic pain (Trentz, 2003; Vaccaro, 2002).

Regarded as a surgeon's second choice, allografts are transplanted tissue from a genetically non-identical member of the same species. Their use has increased 15-fold in the past decade and accounts for around one-third of bone grafts performed (Vaccaro, 2002). Their lower success rate, when compared with autografting, is attributed to the fact they are acellular. The lack of cellular material triggers an immune response, giving the graft a characteristic known as immunogenicity. The increased risks of immunogenicity, coupled with the risks of bacterial contamination and viral transmission, yield variable clinical results (Vaccaro, 2002).

Having been investigated since the 1960s, bone xenografts, transplanted from a donor species, can also be described as a "traditional" grafting technique (Blom, 2007). Bovine bone, in particular, has been shown to have a high level of biocompatibility with human osteoblasts; however, risks of disease transmission and reported intense inflammatory response have limited their application (Blom, 2007).

Hard-tissue reconstruction procedures stemming from trauma, tumour, deformity, degeneration and an ageing population have dramatically increased, triggering a high demand on the advancement of bone implant technology (Wu et al., 2014). Coupled with the ongoing realization of the limitations of traditional bone repair, tissue engineers have a responsibility to develop viable alternatives that will enhance the functional capabilities of bone graft substitutes and eliminate the need for auto-, allo- or xenografts (Parikh, 2002).

At present, the use of alternative bone grafts accounts for ~10% of all graft procedures performed (Vaccaro, 2002). Their limited use is due to a perceived inferiority to autografts and allografts for reasons including poor resorbability, the inclusion of processed animal components, inferior handling characteristics and cost (Lewandrowski et al., 2000; Vaccaro, 2002). It has also been stated that bone regeneration, in the presence of synthetic grafts, can take a considerable length of time to achieve complete anatomical and functional recovery, if at all (Mastrogiacomo et al., 2006).

Bone regeneration using three-dimensional (3D) "scaffold" constructs has been described as one of the most promising TE developments, with the potential to contribute a genuine alternative to the gold standard autogenic bone transplants (Hutmacher and Garcia, 2005).

Providing essential support for the proliferation and colonization of cellular material, depending on their design, composition and mechanical stability, bone scaffolds can be utilized for various skeletal functions around the body.

14.2 Biomimetics

Nature has a vast pool of inventions that have passed the harsh test of practicality and durability in a constantly changing environment (Bar-Cohen, 2006). Recognizing that its capabilities are significantly ahead of modern-day technologies, engineers continue to mimic, with varying degrees of success, the highly effective and efficient biological mechanisms found in organisms around the world (Vaccaro, 2002). Solutions arrived at by millennia of evolution are a source of inspiration and an excellent starting point in the search for answers to modern-day problems.

Often biological systems have extraordinary mechanical properties exceeding those attainable using the same synthetic materials. For example, the nacre of the abalone shell has been reported to have a work of fracture over 3000 times that of its mineral constituent (Jackson et al., 1990). Spiders web is lightweight, has a tensile strength similar to high-grade steel, is insoluble and resistant to wind and sunlight (Bar-Cohen, 2006; Jackson et al., 1990). The compressive stress of a shark tooth bite, measured at 600 MPa, exceeds that of the individual mineralized portions of the tooth (Jackson et al., 1990). The gecko, achieving dry adhesion via van der Waals forces, can run on polished glass and attach its body to the wall using a single toe to support its body weight (Bar-Cohen, 2006).

The field known as biomimetics, bionics or biogenesis applies methods and systems found in nature to the study, design and construction of new engineering systems, materials, chemical compounds and modern technology (Dorozhkin, 2012). In terms of recent scientific applications it has reached astonishing levels simulating synaptic processes in the brain (Matsumoto et al., 2014), shark skin on the surface of ship propellers (Zhao et al., 2012), the hearing of insects (Reeve et al., 2007), flight patterns of eagles, silk production of spiders and even the traffic control of ants (Kammoun et al., 2014).

The study of biological systems as structures dates back to the early parts of the twentieth century, and D'Arcy W. Thompson can be considered to be one of the progenitors of biomimetics (Meyers et al., 2008). In the development of three-dimensionally porous architectures nature provides abundant inspiration (Figure 14.1). Rigid biological materials, shells, bone and sponge spicules, have proved attractive as models for synthetic structural composites due to their efficient combinations of mechanical properties, such as strength, stiffness and toughness (Mayer, 2005). In these instances, as is often the case with biomimetic strategies, the aim is not to directly copy the structures of biological materials but to abstract key concepts from the biological systems that can be adapted within a synthetic context (Green et al., 2002).

In the context of TE, a more appropriate definition for a biomimetic material would be *any scaffolding material that mimics one or more characteristic of the natural extracellular matrix* (Ma, 2008).

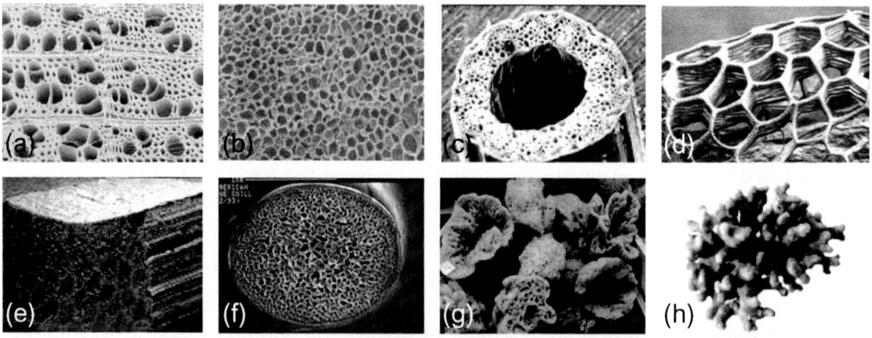

Figure 14.1 Detailed structures of: (a) yellow poplar wood (Kultikova, 1999), (b) cork (from Brazinha et al., 2013, with permission), (c) plant stem (from Dawson and Gibson, 2007, with permission), (d) *Castilleja coccinea* flower seed (Robertson, 2005; courtesy of Robertson), (e) bamboo culm (Gibson et al., 2010), (f) porcupine quill (from Gibson, 2005, with permission), (g) marine sponges and (h) calcified alga.

The prevailing theory embedded in the majority of scaffold production techniques is that the probability of successful integration is strengthened if the structure of the material used mimics that of natural bone (Schliephake, 1991).

14.3 Balancing the requirements for successful *in vivo* integration

TE has evolved from the use of biomaterials to repair or replace diseased or damaged tissue to using controlled 3D scaffolds to guide the proliferation and spread of cells *in vitro* and *in vivo* (Vats et al., 2003).

In a typical TE application, cells harvested from the patient are seeded onto a scaffold, creating a living biocomposite (Jones and Hench, 2003). This biocomposite should subsequently function in tandem with the rest of the body without risk of rejection or complications (Leong et al., 2003). Its successful integration into the body depends heavily on balancing interlinked and often opposing structural and biological requirements (Byrne et al., 2007).

14.3.1 Host response

Today, biocompatibility is defined as *the ability to perform with an appropriate host response in a specific application* and over the past 30–40 years many attempts have been made to identify the key material and surface characteristics that control it (Ellinsen and Lyngstadaas, 2003; Best et al., 2008). In the context of implanted biomaterials, this is not easily accomplished due to the complexity of the physiological environment.

With regard to the replacement of bone tissue, the earliest attempts (in the late 1970s to early 1980s) aimed only to restore functionality, eliciting minimal biological response. These materials were classed as *bioinert* and a perceived successful outcome was one without a toxic response (Best et al., 2008). By the mid-1980s, the focus switched to the development of *bioactive* materials that would induce a controlled reaction at the implant site (Hench and Polak, 2002). To date, materials which have been successfully investigated as TE scaffolds have fallen under the umbrellas of (1) natural polymers, such as collagens, glycosaminoglycan, starch, chitin and chitosan; (2) synthetic polymers, based on polylactic acid, polyglycolic acid and their co-polymers (polylactic-co-glycolic acid, PLGA) and (3) ceramics such as calcium phosphates (compounds containing calcium ions (Ca^{2+}) together with orthophosphates (PO_4^{3-}), metaphosphates (PO_3^{-})), or pyrophosphates ($P_2O_7^{4-}$) and occasionally hydrogen (H^+) or hydroxide ions (OH^-), bioactive glasses and glass ceramics (Buckley and O'Kelly, 2004).

Of these purported alternatives, calcium phosphates have gained clinical acceptance as a valuable bone substitute over the last 20 years, in large part due to compositional and crystallographic similarities between the calcium phosphate (CaP) compounds and the mineral component of natural human bone (Bohner and Baroud, 2005; Jarcho, 1979). Several authors have documented that these similarities yield a high degree of biocompatibility, providing them with an inherent ability to bond directly with bone minerals and promote new bone formation by osteoconduction (Kasten et al., 2003; Jun et al., 2003). It must be noted, however, that the degree of biocompatibility and bioactivity can vary considerably. Subtle differences in the chemical composition, density, porosity and crystallinity may have a profound effect on their behaviour *in vivo* (Yamamuro, 2005). Despite this, numerous *in vivo* and *in vitro* assessments have demonstrated that calcium phosphates, in bulk, coated, powder or porous forms, in both crystalline and amorphous phases, have always supported the attachment, differentiation and proliferation of osteoblasts and mesenchymal cells (Rezwan et al., 2006).

Despite the grafting potential of calcium phosphate products being studied for over 80 years, β-tricalcium phosphate (β-TCP), ($Ca_3(PO_4)_2$), and HA, ($Ca_{10}(PO_4)_6(OH)_2$), or a biphasic combination of the two, have dominated recent medical studies (Jun et al., 2003). Both are excellent carriers of osteoinductive growth factors and osteogenic cell populations, which promote the growth of new bone (von Doernberg et al., 2006). They comprise a large proportion of commercially available TE scaffolds (Table 14.1).

As the material most similar to bone, HA is said to have much greater osteoconductive properties than other calcium phosphates. Classed as a surface-active material, it binds to bone via the formation of an apatite layer consisting of carbonate-apatite crystals, similar to those found in bone (Fujita et al., 2003; Handschel et al., 2002). This apatite layer subsequently takes on its own active role, adsorbing acellular proteins which serve as growth factors, making new bone formation easier (Hench and Wilson, 1984). The bonding mechanisms are said to include direct ionic bonds, electrostatic ionic bonds, hydrogen bonds and Van der Waals bonds (Chen et al., 2004).

Within the scientific and orthopaedic community, the consensus is that the ideal material for bone TE will provide initial mechanical strength and support, with gradual resorption and replacement by newly synthesized host tissue (Clarke et al., 2011). However, it is advisable to match the rate of resorption with the expected rate of bone

Table 14.1 Commercially available ceramic scaffolds

Material	Trade name	Manufacturer
HA	ApaPore	ApaTech, UK
	Ostim	Ostim Heraeus, Germany
	Bonefil	Pentax, Japan
	BoneSource	Stryker Orthopaedics, USA
	Endobon	Biomet, USA
	Pro Osteon 500	Biomet, USA
β-TCP	Bioresorb	Sybron Implant Solutions, Germany
	Biosorb	SBM S.A., France
	Cerasorb	Curasan, Germany
	Conduit	DePuy Spine, USA
	Vitoss	Orthovita, USA
HA + β-TCP	Biosel	Bioland, France
	Bonesave	Stryker Orthopaedics, USA
	Bonesource	Stryker Orthopaedics, USA
	Tribone	Stryker, Europe

ingrowth (Best et al., 2008). Due to the highly dense and crystalline structure of HA, it is essentially non-degradable with resorption rates of 5–15% per year (Giannoudis et al., 2005). As a result, implants of porous HA may prevent newly formed bone within its pores from experiencing the mechanical stresses required for bone remodelling (Giannoudis et al., 2005).

TCP, on the other hand, specifically the α and β polymorphs, have been shown to be resorbable in the physiological environment *in vivo* with new bone replacing the implanted material (Lin et al., 2007). The apatite formation that occurs on the surface of implanted HA does not occur on implanted TCP, leaving it open for resorption and replacement (Fujita et al., 2003). This is said to impart a significant advantage over other biomedical materials, as the presence of Ca and PO_4 ions released through dissolution may enhance bone growth (Miranda et al., 2006). It has been reported, however, that resorption only occurs in a mechanically loaded environment (Handschel et al., 2002; Neo et al., 1998; Schliephake et al., 1997). Without mechanical stimulus, osteocytes do not produce the signalling molecules required to modulate the activities of osteoblasts and osteoclasts (Klein-Nulend et al., 2005). As TCP is known to have low mechanical stability, this has severely hindered its use in clinical applications (Lin et al., 2007).

It must be noted that when the solubility of a calcium phosphate is higher than the rate of tissue regeneration, it is logical that any potential orthopaedic use will be limited to bone cavity and defect filling (Best et al., 2008).

14.3.2 Porosity

Defined as *the percentage of void space within a solid* (Karageorgiou and Kaplan, 2005), porosity is an essential requirement for the integration of any bone substitute scaffold, but the optimum level cited in the literature varies considerably. While Best et al. recommend relatively high porosities of 70–80%, scaffolds with porosity between 45% and 90% have been reported to have successfully induced bone growth (Best et al., 2008). This can be attributed to the fact that the porosity within cortical and cancellous bone (Figure 14.2) itself varies between 3–12% and 50–90%,

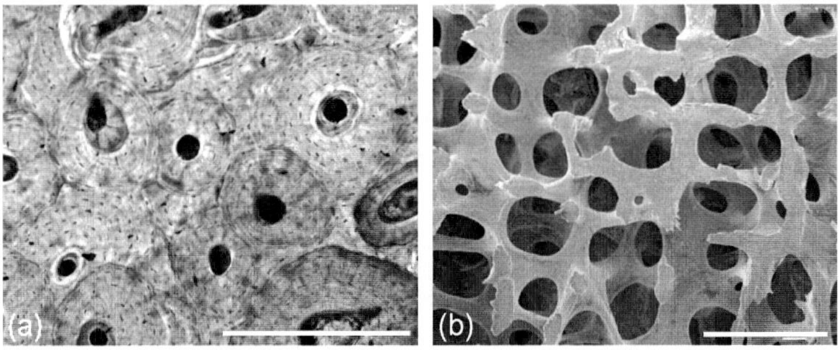

Figure 14.2 Porosity in: (a) cortical bone and (b) cancellous bone. Note the scale bar on both images represents 1 mm.
Courtesy of Tim Arnett, University College London.

respectively (Karageorgiou and Kaplan, 2005). With regard to cancellous bone, there are well-established advantages to having porosities at both ends of this spectrum. High levels of porosity result in greater bone ingrowth while low levels increase mechanical stiffness and strength (Byrne et al., 2007). Porosity is a prime example of a scaffold characteristic, for which a healthy balance must be found (Byrne et al., 2007).

14.3.3 Pore size distribution

As with porosity there is a plethora of opinion as to the required pore size distribution for complete integration of scaffolds into the body. Karageorgiou and Kaplan (2005) cite the development of several HA scaffolds, with pore sizes ranging from 150 to 800 μm, all of which have heralded some level of bone ingrowth. While von Doernberg et al. (2006) concur with this range, a number of authors including Sopyan et al. (2007), Giannoudis et al. (2005) and Engin and Tas (1999) cite narrower ranges of 100–400, 150–500 and 250–300 μm, respectively.

As far as pinpointing an optimum pore size within the ranges cited, the literature provides no clear conclusion. Von Doernberg et al. (2006) discuss several contradictory investigations where increasing the pore size (within the 150–800 μm range) has seen: (i) increased bone ingrowth, (ii) decreased bone ingrowth and (iii) no effect. The fact that these investigations do not yield an optimum elevates the importance of the other requirements discussed.

There is, however, a general consensus that the minimum pore size required to facilitate cell and ion transport, initiating osteogenesis, is 50–100 μm (Woodard et al., 2007; Chang et al., 2000; Gauthier et al., 1998; Le Huec et al., 1995; von Doernberg et al., 2006).

The presence of microporosity (pores $\leq$20 μm) is also necessary according to Hing et al., providing increased surface area for protein adsorption, increasing ionic solubility and providing attachment points for osteoblasts (Woodard et al., 2007). This is reaffirmed by Woesz et al. (2005), who state that the nutrition of the cells within any structure must be ensured for viability, and that this is only possible via the flow of serum through both micro- and macropores (pores $\geq$50 μm). Yang (2001) provided a complete breakdown of pore size requirements for complete bone regeneration (Table 14.2).

Table 14.2 Breakdown of scaffold pore size requirements for bone regeneration (Yang, 2001)

Scaffold pore size (μm)	Observations
5	Neovascularization
15–40	Fibroblast ingrowth
40–100	Osteoid ingrowth
200–350	Significant bone ingrowth
>500	Rapid vascularization

14.3.4 Interconnectivity

Numerous studies of cancellous bone have shown that both the bone and marrow constituents are fully interconnected, and that neither exist in a fully isolated manner (Odgaard and Gundersen, 1993). Logically then, the successful integration of any bone substitute scaffold depends heavily on achieving a high level of interconnectivity between the pores. This will allow for the provision of sufficient blood and cellular material throughout the scaffold, enhancing bone deposition, nourishment of new bone and removal of waste products (Woodard et al., 2007; Bignon et al., 2003; Cyster et al., 2005).

It also plays an important role in ensuring the original surrounding bone remains healthy. The vascularization of the implanted scaffold allows the migration of osteoblasts and osteoclasts to continue as normal, meaning the remodelling process will be unaffected (Mastrogiacomo et al., 2006).

Bohner et al. go as far as to downplay the relevance of pore size distribution, stating that bone ingrowth should not be affected by pore size as long as they are fully interconnected through openings larger than 50 μm (von Doernberg et al., 2006). This may account for the large variability in the optimum pore size distribution cited in the reviewed literature.

For such an important characteristic of 3D scaffolds, interconnectivity is often not quantified (Bignon et al., 2003). Instead, average pore size and level of porosity are often cited and interconnection between the pores is inferred (Bignon et al., 2003).

14.3.5 Surface porosity

If bone ingrowth and vascularization are to be achieved, having appropriate pore size distribution and interconnectivity means nothing without adequate surface porosity. Not only will it provide access to the core of a scaffold, it is also said to improve mechanical interlocking between the implanted scaffold and the surrounding healthy bone, enhancing mechanical stability at this vital interface (Karageorgiou and Kaplan, 2005).

14.3.6 Mechanical properties

The primary function of any bone substitute is to provide support and stimulate integration into the site of trauma. Obtaining the optimum porosity, pore size distribution and interconnectivity to facilitate integration, however, should not be achieved at the expense of structural integrity. At the very least, a scaffold should have sufficient stability to maintain its architecture during *in vitro* culturing and handling throughout implantation. After implantation, the scaffold must possess similar mechanical properties to the once healthy bone it has been designed to replace (Table 14.3) (Babis and Soucacos, 2005). Once integration has occurred the implanted scaffold should function in tandem with neighbouring healthy bone and assume a shared load, reducing the risk of rejection and failure. If successful, the stability of the implant should increase, as reported by Sopyan et al. (2007), who found the compressive strength of porous HA implants rose from 2 to 20 MPa 3 months after implantation.

Table 14.3 **Mechanical properties of cancellous bone**

Anatomical location	Young's modulus (GPa)		Compressive strength (MPa)		References
	Range	Average	Range	Average	
Femur	0.044–1.531	0.389	0.56–22.9	7.36	Keaveny (1998)
Lumbar spine	0.001–0.110	0.023	0.05–8	1.55	Keaveny (1998)
Proximal tibia	0.061–1.174	0.445	0.68–14.1	5.33	Keaveny (1998) and Karageorgiou and Kaplan (2005)
Proximal femoral		0.441		6.8	Karageorgiou and Kaplan (2005)
General	0.01–0.5	0.255	4–12	7	Yang (2001) and Rezwan et al. (2006)

When engineering substitute bone, unlike the majority of engineering endeavours, the pursuit of mechanical strength in excess of what is required should be avoided. Wolff's law suggests that if a substitute is introduced with the capability of exceeding these values, stress shielding may be induced, resulting in atrophy or tissue loss (Black, 2005).

14.4 Natural solutions

While the focus of this section is on the development of bio-inspired calcium phosphate TE scaffolds, it is important to provide a brief overview of the production techniques currently employed. In the last few decades a multitude of bone scaffold manufacturing techniques have been developed in an attempt to address the biological, architectural and mechanical requirements described. It will become clear that their success varies considerably and in the majority of techniques only a select number of requirements are addressed. The vast majority of these techniques employ ceramics, polymers and combinations of the two, including solvent casting and particulate leaching, gas foaming, fibre meshes and fibre bonding, phase separation, melt moulding, emulsion freeze drying, solution casting, freeze drying, fused deposition modelling, rapid prototyping, membrane lamination, microsphere sintering and supercritical fluid technology. Table 14.4 provides a glimpse of the plethora of scaffold manufacturing techniques currently available and the pros and cons, average pore size distribution, average porosity encountered, along with an example micrograph.

It is evident that, without the aid of computer control, conventional scaffold fabrication techniques fail to allow precision control over pore size distribution, pore geometry, distribution of pores and construction of interconnected channels (Sachlos and Czernuszka, 2003). In particular, the inability to attain complete interconnectivity will restrict the migration of cells beyond the first 500 μm of the scaffold

Table 14.4 **Advantages, disadvantages, pore size distribution and porosity of current scaffold production techniques (Buckley and O'Kelly, 2004; Yang, 2001; Hutmacher, 2000; Rezwan et al., 2006; Karageorgiou and Kaplan, 2005)**

Fabrication process	Achievable pore distribution (μm)	Porosity (%)	Advantages	Disadvantages	Example SEM	SEM references
Solvent casting/ particle leaching	30–300	20–50	Controlled porosity	Limited thickness (3 mm), use of organic solvents, poor control of internal architecture	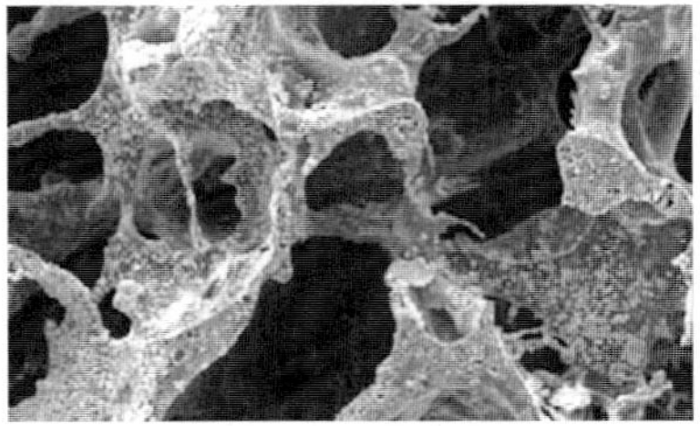	Rezwan et al. (2006)
Fibre bonding	20–100	Max 95	High porosity	Residual solvents, lack of mechanical strength	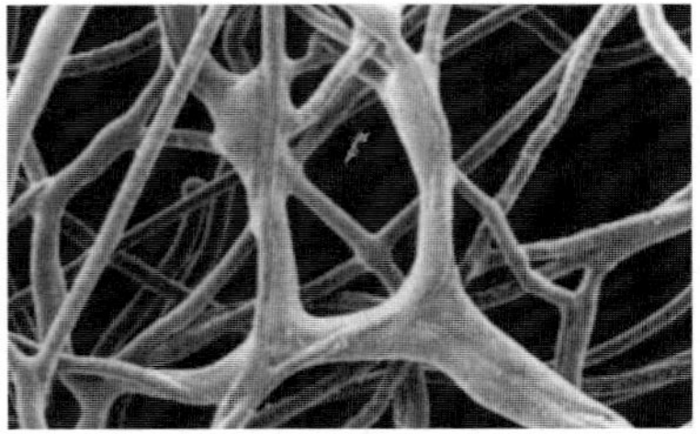	Gibson (2005)
Phase separation	Max 200	Max 97	High porosity	Poor control of internal architecture, limited pore range	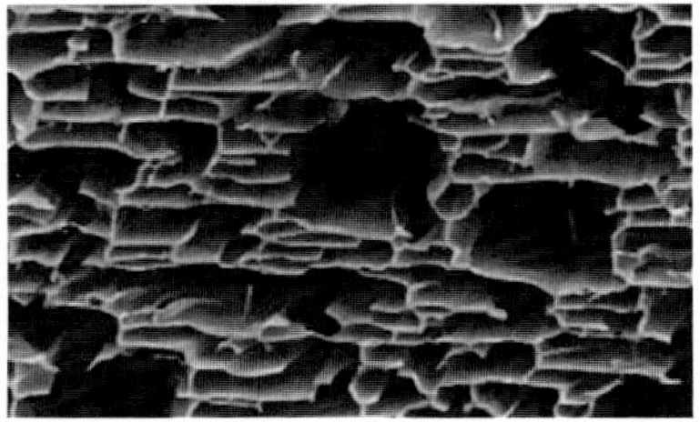	Rezwan et al. (2006)

Continued

Table 14.4 **Continued**

Fabrication process	Achievable pore distribution (μm)	Porosity (%)	Advantages	Disadvantages	Example SEM	SEM references
Solid freeform fabrication (SFF)	Min 150	Max 80	100% Inter-connected	Poor resolution at the micro scale	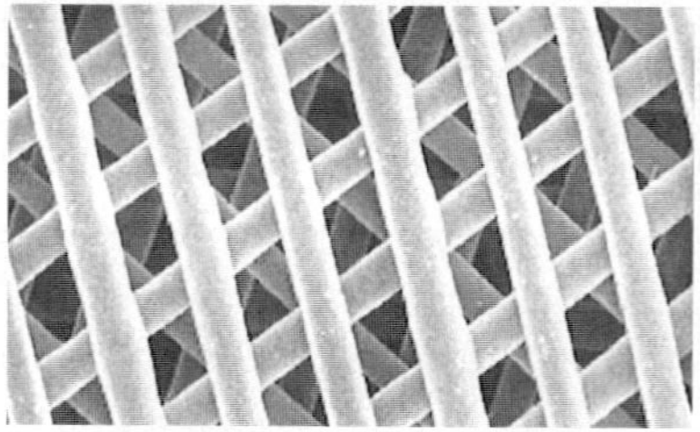	Hutmacher (2000)
Melt moulding	50–500	Max 80	Good control of pore size and porosity	Residual solvents	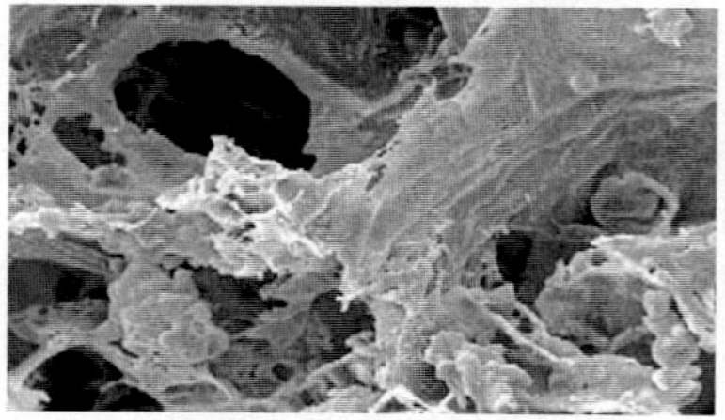	Oh et al. (2003)
Emulsion freeze drying	Max 200	Max 97	Highly inter-connected	Limited to small pores	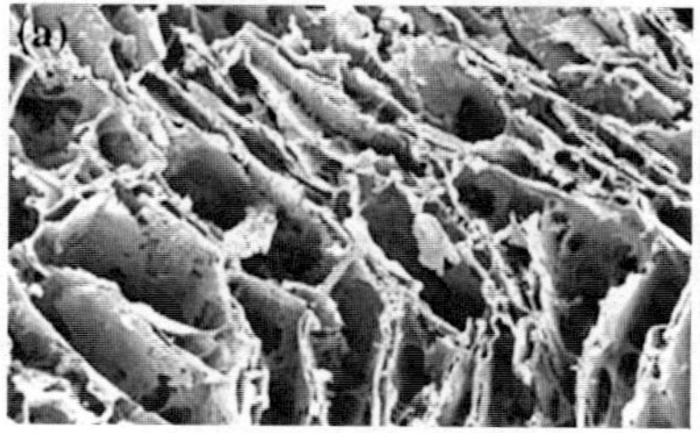	Sultana and Wang (2008)

Freeze drying						
Collagen/HA	300–500		Highly inter-connected	Limited to smallpores	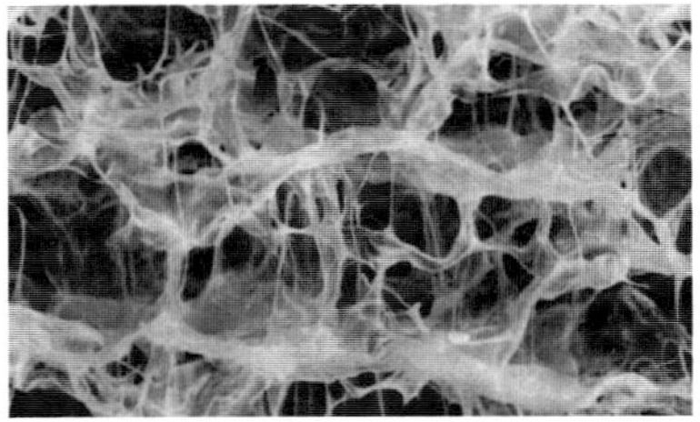	Gibson (2005)
Chitosan/HA	30–300	49–85				
Membrane lamination	30–300	Max 85	Control of porosity and pore size	Limited inter-connectivity, lack of mechanical strength	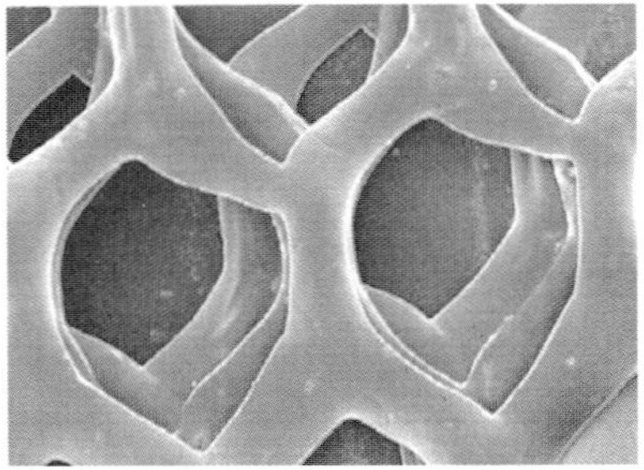	Vozzi et al. (2003)
Microsphere sintering	90	40	Control of porosity, can be fabricated into complex shapes	Limited inter-connectivity, use of organic solvents	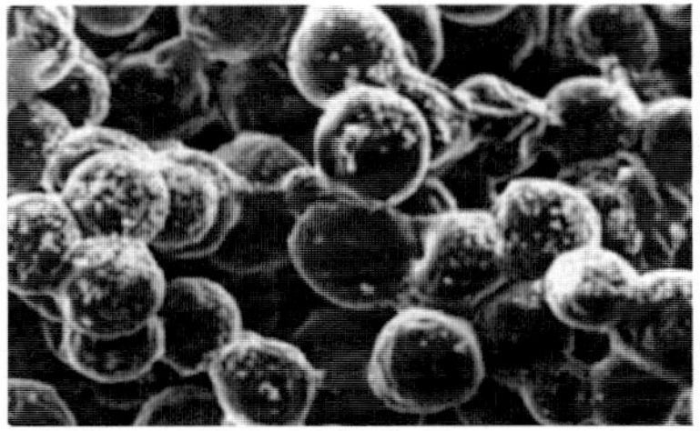	Rezwan et al. (2006)

Continued

Table 14.4 **Continued**

Fabrication process	Achievable pore distribution (μm)	Porosity (%)	Advantages	Disadvantages	Example SEM	SEM references
Supercritical fluid technology	Micropores <40 macropores <400	Max 97	Inter-connected macropore structure	High volume of non-interconnected micropore structure	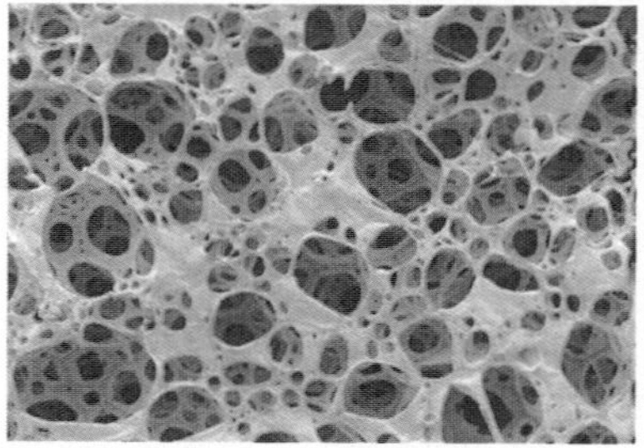	Kim et al. (2006)

(Sachlos and Czernuszka, 2003). Mineralization at the outer surface can limit the transfer of oxygen, nutrients and waste products, diminishing the chance for complete integration into the body (Sachlos and Czernuszka, 2003).

While techniques that do incorporate computer control, under the umbrella term of "solid freeform fabrication", reportedly offer advantages over conventional techniques, there are issues which should be addressed. Despite a greater control over the macro-architecture, poor micro-resolution, difficulty in support powder removal, high-temperature processing preventing the use of biological molecules and inflexible processing parameters are all issues that remain today (Rezwan et al., 2006; Sachlos and Czernuszka, 2003).

There is also a concern within a number of techniques arising from the use of organic solvents such as chloroform and methylene chloride. Their toxic and carcinogenic properties can have a devastating effect on cellular material (Sachlos and Czernuszka, 2003).

These shortcomings continue to push tissue engineers to develop new techniques for the production of scaffolds which meet all the requirements for integration into the body.

14.4.1 Replication of cancellous bone

In pursuit of a production method that will yield architectural characteristics similar to cancellous bone, a logical starting point would be to study techniques utilizing the material itself.

Tancred et al. (1998) describe a multi-stage process for the macrostructural replication of bovine cancellous bone.

The first step in the process is to remove the organic content from within specimens of around $35 \times 12 \times 8\,mm^3$. This is achieved via thermal ashing (burning off the organic constituents) or chemical extraction. As thermal ashing can result in some unwanted structural breakdown, chemical extraction of organic matter using ethylenediamine (1,2-diaminoethane, ED) and a Wheaton soxhlet extraction kit is recommended (Tancred et al., 1998).

The use of ethylenediamine as an extraction solvent is well established. As far back as 1954, Williams and Irvine (1954) describe how ethylenediamine meets the requirements for any organic solvent: (1) vigorously breaking down the organic constituent of bone at as low a temperature as possible; (2) easily volatized, allowing it to be readily distilled from the construct and (3) achieving extraction with minimum erosion of inorganic particulates.

Once deproteinization is achieved, a negative mould is created by submerging the specimens in molten wax and removing the bone through immersion in a 10% hydrochloric acid solution for 48 h. To aid this decalcification process the specimens are pretreated in boiling sodium hypochlorite, resulting in the deposition of a crystalline coating sealing the microporosity. While the replication of the microporosity would be desirable, its presence makes complete decalcification difficult (Tancred et al., 1998).

The negative mould is then infiltrated with a ceramic slurry made up of combinations of HA and β-TCP and fired in a furnace, leaving what is claimed to be a replica of the original macrostructure (Tancred et al., 1998).

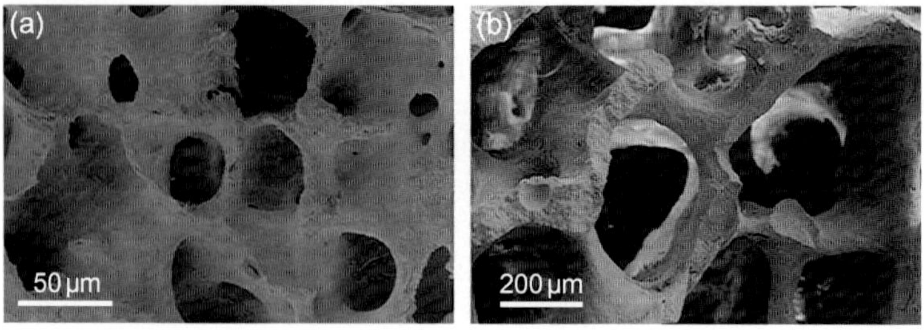

Figure 14.3 Micrographs of: (a) cancellous bone and (b) HA replicated bone (from Tancred et al., 1998, with permission).

There is, however, a lack of quantitative evidence on pore size distribution and interconnection between pores is merely inferred from micrographs and a high porosity (Tancred et al., 1998). The success of replication cannot be verified as the supplied micrographs (Figure 14.3) are of different samples. In any case, the average pore sizes within the supplied micrographs are considerably different.

While organic free bovine bone is commercially available (Endobon®-Biomet, UK Ltd), the use of processed animal components can elicit fears of disease transmission and immunorejection (Woesz et al., 2005). Despite maceration and decalcification, scaffolds produced using the method described by Tancred et al. may raise similar concerns.

The obvious solution is to replicate human cancellous bone, but this too has restrictions. Aside from the risk of bacterial contamination, viral transmission and immunogenicity, there are logistical issues (Vaccaro, 2002; Hing et al., 2007). According to Blom (2007), the supply of human bone simply cannot meet the demands of hip revision surgery, let alone for all other operations requiring bone.

While there are unanswered questions within the investigation of Tancred et al. aspects of the procedure including the chemical extraction and ceramic slurry production may be adapted and used for a variety of scaffold production techniques.

14.4.2 Polymer foam replication

With concerns over contamination and disease transmission associated with processes using natural bone, researchers have looked at other porous structures to mimic in their pursuit of developing the perfect bone substitute.

In 1963, Brenzy and Green patented a polymeric foam replication technique (Muhamad Nor et al., 2008). The process, adopted by numerous authors (Chen et al., 2006; Sopyan et al., 2007; Muhamad Nor et al., 2008; Saiz et al., 2007; Ramay and Zhang, 2003; Kim et al., 2004, 2005; Callcut and Knowles, 2002), involves the coating of open-celled polymer foams with ceramic slurry followed by removal of the polymer through sintering. Despite subtle variations in processing techniques, the scaffolds produced are highly porous ($\geq$70%) with characteristically round pores. The flexibility that comes from varying the characteristics of the polymer sponge allows for the production of scaffolds with controllable pore size (Ramay and Zhang, 2003). Further manipulation of the structural and mechanical properties can be achieved by varying

Table 14.5 Overview of structural characteristics and mechanical properties of scaffolds produced using the polymer foam replication technique

Material	Porosity (%)	Pore size distribution (μm)	Compressive strength (MPa)	References
45S5 Bioglass®	89–92	510–720	0.27–0.42	Chen et al. (2006)
Glass reinforced HA	85–97.5	420–560	0.01–0.175	Callcut and Knowles (2002)
HA	86	420–560	0.21	Chen et al. (2006)
HA	69	490	0.03	Chen et al. (2006)
HA	86	1130	0.29	Chen et al. (2006)
HA	–	100–200	–	Saiz et al. (2007)
HA	70–77	200–400	0.55–5	Ramay and Zhang (2003)
HA	82–86	500–700	0.21–0.41	Kim et al. (2005)
HA	87	150–200	0.16	Kim et al. (2004)
HA	–	400–600	1.3–10.5	Sopyan et al. (2007)

the solid content of the slurry and the sintering temperature (Muhamad Nor et al., 2008; Sopyan et al., 2007).

Due to the isotropic (uniform in all directions) nature of the preform polymer sponges, the pore size distribution in ceramic replicas is usually within tight bands of 200 μm at most (Table 14.5). According to the literature cited these distributions will be insufficient to allow the necessary processes, from neovascularization to rapid vascularization, required for complete integration into the body (Yang, 2001).

In the techniques cited, interconnectivity between pores is inferred from high levels of porosity, fracture surface micrographs or pore window sizes (Saiz et al., 2007; Chen et al., 2006; Sopyan et al., 2007; Callcut and Knowles, 2002; Kim et al., 2004, 2005; Muhamad Nor et al., 2008; Ramay and Zhang, 2003). While claims of complete interconnectivity can be understood in some cases (Figure 14.4a–c), in others it would appear to be a massive assumption (Figure 14.4d–f).

With the exception of Chen et al. (2006), who state that pore windows of 30–120 μm are *considerably smaller* than those required for osteoblast penetration, the authors cited fail to predict the effects of both pore size distribution and interconnectivity on the potential vascularization of their scaffolds (Sopyan et al., 2007; Callcut and Knowles, 2002; Chen et al., 2006; Kim et al., 2004, 2005; Muhamad Nor et al., 2008; Ramay and Zhang, 2003; Saiz et al., 2007). There is, in fact, little evidence to suggest that scaffolds produced using this method have been successfully integrated into skeletal defects.

One explanation for this may be their poor mechanical properties. On average, the compressive strength of scaffolds manufactured via polymer sponge replication is between 0.2 and 0.4 MPa (Table 14.5). This suggests their application would be severely limited.

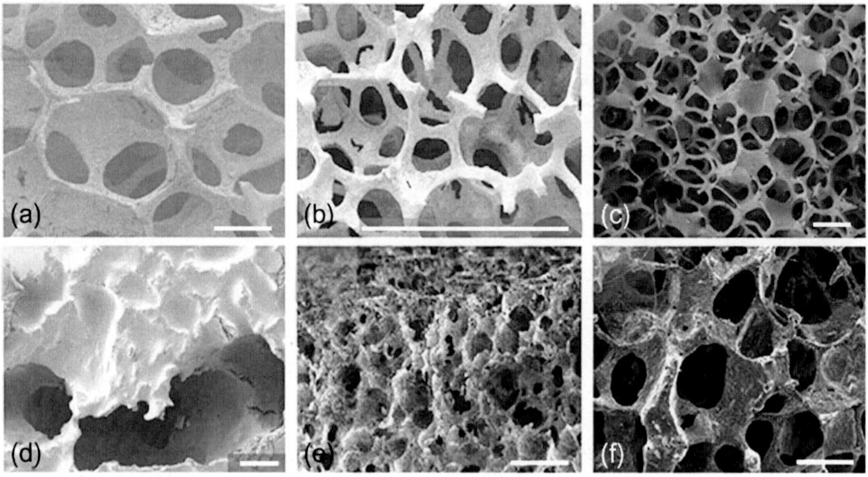

Figure 14.4 Fracture surfaces of HA scaffolds produced via polymer sponge replication provided as evidence of pore interconnectivity by: (a) Kim et al. (2005); (b) Kim et al. (2004); (c) Callcut and Knowles (2002); (d) Sopyan et al. (2007); (e) Saiz et al. (2007) and (f) Chen et al. (2006). White scale bars are all 500 µm.

In a study in 2007, Muhammad Nor et al. attempted to improve the mechanical stability of ceramic foams (for non-biomedical applications) by varying the density of the ceramic slurry prior to infiltration. Despite producing scaffolds with a compressive strength of ca. 23 MPa, the increase in mechanical strength caused a corresponding reduction in porosity to 20% (Muhamad Nor et al., 2008). While this extreme case would of course be unsuitable for use as a bone substitute, Muhamad Nor et al. (2008) do show the importance of achieving an optimum ceramic slurry density.

Some authors, such as Ramay and Zhang (2003) and Kim et al. (2005), have incorporated sol–gel and poly(ε-caprolactone) coating techniques in an attempt to increase mechanical stability, with mixed success. Ramay and Zhang produced scaffolds with a compressive strength of ~5 MPa while Kim et al. saw no significant increase in compressive strength. Kim et al. (2005) did, however, report an increased capacity for energy absorption, defined as *the energy necessary to deform a specimen to a strain* (ε).

As with the replication of bone itself, there are elements of the polymer sponge replication technique which limit its potential as a scaffold manufacturing technique. The failure to yield a pore size distribution that lends itself to integration into the body and poor mechanical strength are major hurdles to be overcome.

If these problems can be solved, however, the process itself has an inherent flexibility which would be extremely useful in manufacturing bone substitute scaffolds for specific locations within the body.

14.4.3 Replication of commercial cellulose sponge

In what would appear to be a logical progression from replicating polymer foam, a number of researchers have reported the use of cellulose sponge as a viable preform in the production of porous ceramics with pore size distributions mimicking that of cancellous bone (Guicciardi et al., 2001; Landi et al., 2003; Mastrogiacomo et al., 2006). Due to the nature of their production they exhibit a wider and non-isotropic pore size distribution when compared to PU foam, evident in Figure 14.5.

Originally patented by DuPont in the early twentieth century, production involves moulding and heating a suspension of cellulose, sodium crystals, cut hemp fibres and dye. As the mixture heats the cellulose hardens while the sodium crystals melt, leaving pores in the finished sponge. It is the size and grade (fine to coarse) of these crystals which ultimately allows control of pore characteristics.

In a technique similar to polymer foam replication, Guicciardi et al. (2001), Landi et al. (2003) and Tampieri et al. (2001) coat cellulose sponges with HA slurry followed by removal via sintering. The results from these investigations are summarized in Table 14.6. Note that (L), (M) and (H) denote the use of low (20%), medium (45%) and high (80%) crystalline HA.

While a greater pore size distribution has been achieved in the majority of scaffolds produced, only those created by Guicciardi et al., using 20% and 45% crystalline HA, have both a porosity and compressive strength close to the required standard. It must be noted, however, that the pore size distribution exceeds the 50–800 μm range cited as optimum for bone ingrowth (Yang, 2001) and a scaffold with a compressive strength of 0.8 MPa would have limited applications.

Interconnectivity between pores is, once again, characterized via SEM analysis of fracture surfaces. In one of the more thorough evaluations, Mastrogiacomo et al. use micrographs to describe two types of interconnection between pores, which they term *tunnels*, and *contact points*, represented in Figure 14.6a by *T* and *C*, respectively. On average the tunnels *T* and contact points *C* are ~200 and 100 μm, respectively.

As validation, Mastrogiacomo et al. (2006) confirmed their SEM findings by analysing histological sections after 8 weeks *in vivo* (Figure 14.6b). Despite minor variations, the results from both methods showed good correlation. This also provided an

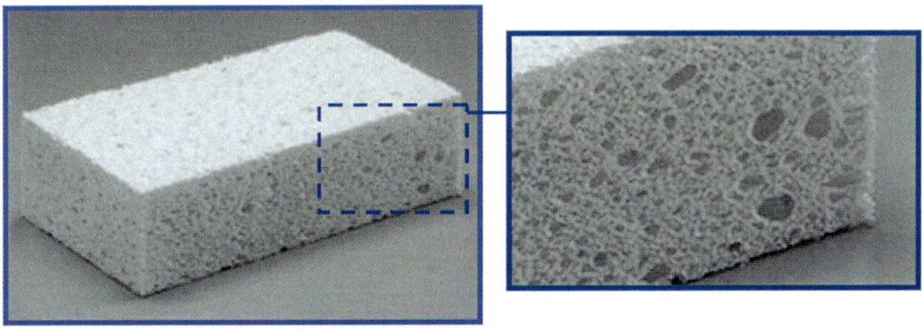

Figure 14.5 Typical commercial cellulose sponge.

Table 14.6 Overview of structural characteristics and mechanical properties of scaffolds produced using the cellulose sponge replication technique

Macro porosity (>50 μm)	% Micro porosity (<10 μm)	Total porosity (%)	Compressive strength (MPa)	Author and references
50 μm to 2 mm	19.9	56	16.1	(L) Guicciardi et al. (2001)
50 μm to 2 mm	23.8	67	0.8	(M) Guicciardi et al. (2001)
50 μm to 2 mm	8.8	30	32.6	(H) Guicciardi et al. (2001)
100–200 μm	9	45	6 ± 0.5	Landi et al. (2003)
500–1000 μm	"Present" (not quantified)	65 ± 5	N/A	Mastrogiacomo et al. (2006)

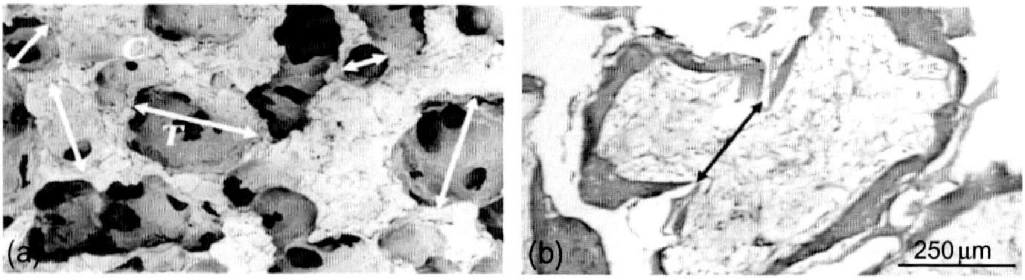

Figure 14.6 (a) Distribution of interconnection pore size analysed using SEM before implantation. (b) Histological evaluation after 8 weeks *in vivo* (from Mastrogiacomo et al., 2006, with permission).

opportunity to observe bone matrix deposition (stained in pink) and the presence of blood vessels within the pores (Mastrogiacomo et al., 2006).

According to the literature cited the wider pore size distribution attainable via the replication of cellulose sponges (Table 14.6) should yield greater bone ingrowth than scaffolds produced via polymer foam replication. The range of mechanical properties attainable is also superior with a range of 0.8–16.1 MPa, compared to the 0.03–10.5 MPa cited for polymer foam replication in Section 14.4.2. Despite this improvement, however, further progress may still be made.

14.4.4 Replication of natural sponges

Rambo et al. (2004) and Sieber (2005) have both commented on the aptitude for biological materials such as wood, bamboo and grass to facilitate the transportation of fluid, and have attempted various replication techniques. While the focus of their work was the creation of biomorphous oxide ceramics for a number of applications such as sensors, filters and high-strength lightweight structural components, they posited that the resulting open porous structures would be of particular interest for applications as bioceramic scaffolds in bone TE (Rambo et al., 2004; Sieber, 2005). In a related investigation, Rambo et al. (2006) used *Hippospongia lachne* in the development of calcium phosphate-coated scaffolds. Again the porous structures were not specifically developed to be used as bone substitute scaffolds (Rambo et al., 2006). While the constructs produced by Rambo et al. do not meet the pore distribution criterion required, and definitive comment cannot be made on interconnectivity or mechanical stability, the proposal to utilize the porous architecture of marine sponges as preforms in the production of ceramic scaffolds is one that deserves further inspection. Their functional design, which has evolved over millennia, mirrors the goal of a bone scaffold substitute. Specifically geared to utilize water flow, sponges efficiently filter required organic material through a highly developed system of pores (Müller et al., 2004). With over 5000 species around the world, each with various porous architectures, sponges represent one of the highest biodiversities in the world (Müller et al., 2004). Recognizing this potential it is important to fully understand their functional architecture.

Lacking the distinct tissues and organs that make up more complex animals, sponges of the phylum *Porifera* (from the Latin *porus* "pore" and *ferre* "to bear") are said to possess the most primitive morphology of all animals (Sethmann and Worheide, 2008). Predominantly marine (ocean dwelling), these sessile (without a stalk) organisms, known as filter feeders, syphon and consume suspended organic particles and micro-organisms found in sea water (Sethmann and Worheide, 2008). While they have existed and evolved for over 800 millennia their architecture can be categorized into three types, Asconoid (Figure 14.7a), Syconoid (Figure 14.7b) and Leuconoid (Figure 14.7c), increasing in complexity from Figure 14.7a–c.

A simple perforated bottle-shaped structure, the Asconoid is characterized by numerous pores, known as porocytes. These pores form the openings of unbranched canals which lead to a central body known as the spongocoel. The inner lining of the spongocoel is lined with collar cells, choanocytes and protruding flagella, which

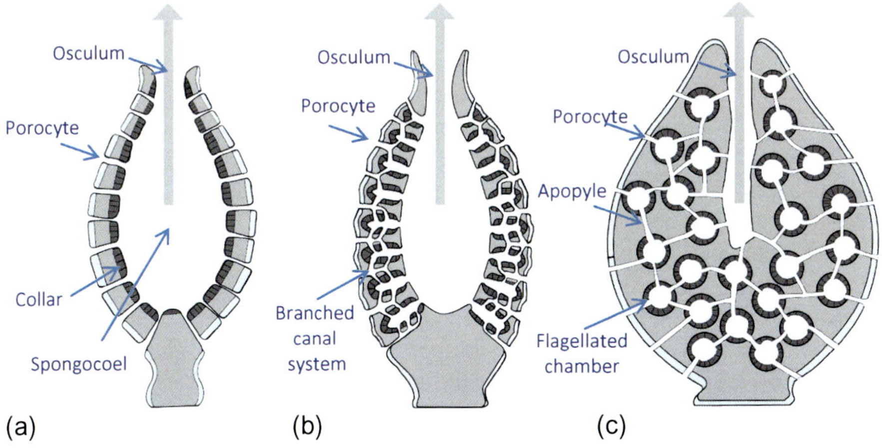

Figure 14.7 Schematic of: (a) Asconoid structure; (b) Syconoid and (c) Leuconoid structure. Image adapted from WikiMedia.org.

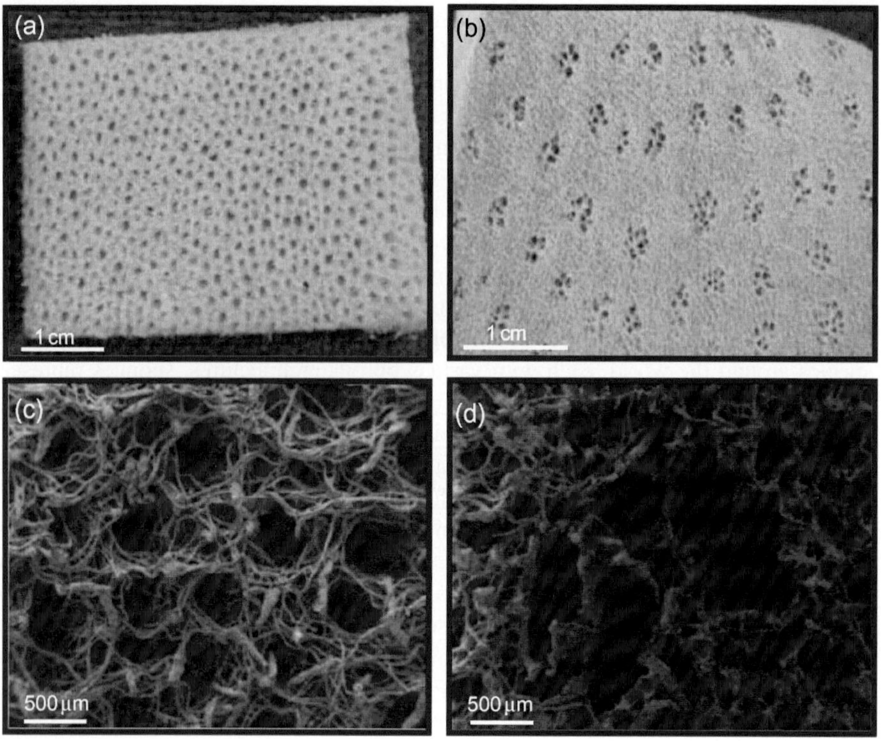

Figure 14.8 (a) Inhalant surface of *Spongia agaricina*. (b) Exhalent surface of *Spongia agaricina*. (c) Low-resolution micrograph of inhalant surface. (d) Typical pore cluster on exhalent surface of *Spongia agaricina*.
All images from Cunningham (2008).

continuously flick to produce water current which flows from the exterior through the ostia and canals exiting the sponge at the Osculum.

The Syconoid sponges, often larger than the Asconoid sponges, are slightly more complex, with the body wall (Mesohyl) continuously folding in on itself (Figure 14.8). The wall is thicker with longer pore channels branching off into a system of canals, each of which is lined with choanocytes, collar cells and flagellum. This increases the overall surface area/volume ratio, improving the sponge's ability to filter the necessary organic material from sea water.

The final and predominant morphological group are the Leuconoid sponges (Figure 14.7c). These show the greatest complexity, with longer canals and further branching. Unlike the Asconoid or Syconoid structures the canals lead to chambers lined with choanocytes. There is also no spongocoel present. Instead, water is filtered through to a larger exit canal known as the apopyle (Figure 14.7c).

Cunningham et al. (2009, 2011) have demonstrated the ability to replicate a number of species exhibiting a Leuconoid structure. After developing an 80 wt% (55.9 vol%) solid-loaded HA slip with a viscosity of 126 MPa using predominantly spherically shaped particles three species of natural sponge, Dalmata Fina (*Spongia officinalis*

Linnaeus), Fina Silk (*Spongia zimocca*) and Elephant Ear (*Spongia agaricina*), were vacuum infiltrated, squeezed to remove excess HA and sintered at 1400 °C. Of the marine sponge replica scaffolds produced, *Spongia agaricina* (Figure 14.8) was said to have the greatest potential for success as a TE scaffold. Characterized by intertwined fibrous struts (Figure 14.8c), which are on average $16 \pm 8\,\mu m$ thick and 20–30 μm apart, they are composed of a fibrin mesh with regular macro pores of ca. $470 \pm 120\,\mu m$ on the inhalant surface (Figure 14.8a) and regular clusters of 5–7 pores of $510 \pm 140\,\mu m$ on the exhalent surface (Figure 14.8d).

The sponge replica (Figure 14.9) scaffolds demonstrated a normal pore distribution, with over >90% within the 1–500 μm range. Moreover, >99% of the pores were classified as interconnected, and both permeability ($k = 16.8 \times 10^{-10}\,m^2$) and compressive properties measured were in agreement with values reported for healthy cancellous bone. In early cytotoxicity studies, they exhibited the ability to support adherence and proliferation of both primary endothelial cell types (human umbilical vein endothelial cells, HUVEC; and human dermal microvasular endothelial cells, HDMEC) and the osteoblast-like cell line MG63 (Cunningham et al., 2011).

Further development of this work, carried out by Newe et al. (2012), have attempted to determine the impact of HA:β-TCP ratio, structural properties and pH of the physiological environment on the dissolution rate in an attempt to create predictive *in vivo* models.

14.4.5 Coral and coralline algae

Discounting sponges briefly, the marine environment remains a rich source of inspiration for 3D porous structures with architectural and compositional similarities to those of bone. Indeed, a plethora of research has stemmed from similarities encountered in species of coral and coralline (coral-like) algae.

In the early 1970s a number of researchers, including Chiroff et al., Roy et al. and White et al., recognized that certain species of coral, in particular *Goniopora* and *Porites*, have an interconnected porous architecture similar to that of bone (Chiroff et al., 1975; Roy and Linnehan, 1974).

With a *trabecular pattern* similar to cancellous bone, *Goniopora* is characterized by 500–600 μm pores with 220–260 μm interconnections. *Porites*, having a greater density, have an architecture similar to cortical bone, and are characterized by 200–250 μm pores with interconnections of 190 μm (Truumees and Herkowitz, 1999).

These architectural features coupled with a skeletal composition of calcium carbonate have generated great interest in orthopaedics.

From the work of Chiroff et al., Roy et al. and White et al., a number of techniques have been developed to exploit these architectures in the production of bone substitute scaffolds.

Chiroff et al. (1975) utilize a technique first developed by White et al. in 1972. Known as *replamineform* (Chiroff et al., 1975), the process involves machining coral into a desired shape and firing it in a furnace to remove carbon dioxide (CO_2) from the calcium carbonate ($CaCO_3$), leaving a calcite body with the original microstructure (Gyger et al., 2007; Hench, 1991). Using a process that would later influence Tancred et al., the calcite body is impregnated with one of a number of materials, including

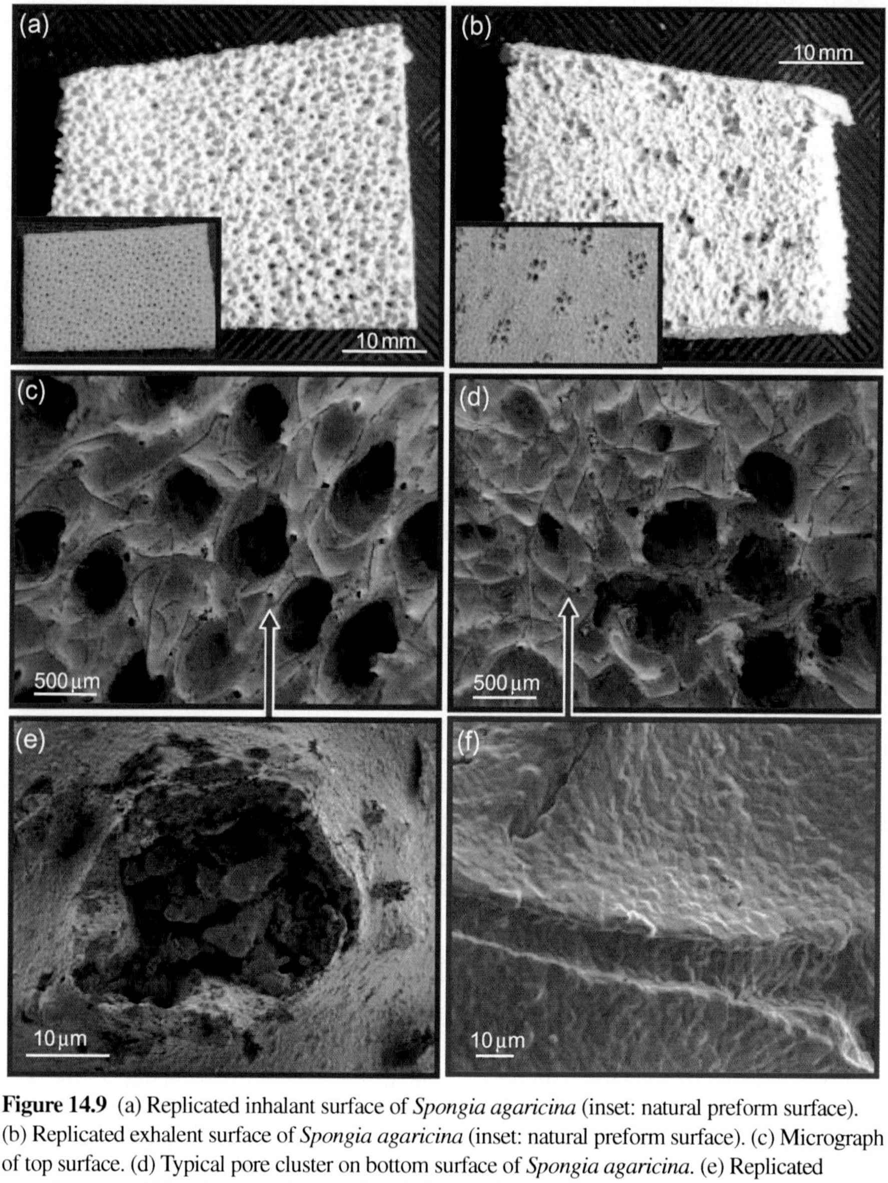

Figure 14.9 (a) Replicated inhalant surface of *Spongia agaricina* (inset: natural preform surface). (b) Replicated exhalent surface of *Spongia agaricina* (inset: natural preform surface). (c) Micrograph of top surface. (d) Typical pore cluster on bottom surface of *Spongia agaricina*. (e) Replicated spongin struts within micro openings on the inhalant surface. (f) High-magnification grain structure of replica elephant ear showing typical track mark found across the surface of the construct. All images from Cunningham et al. (2011).

alumina, titania, polyurethane, silicone rubber or calcium phosphate. Finally the calcite body is removed using a dilute hydrochloric solution, leaving a negative mould of the original coral with interconnected porosity (Hench, 1991). Using replicas composed of alumina, Cr–Co–Mo alloy and HA, Chiroff et al. (1975) reported almost complete bone ingrowth after 8 weeks *in vivo* in the proximal tibiae and distal femora of a dog.

According to Hench et al. and Chiroff et al., the advantage of the replamineform technique is the ability to create constructs with uniform pore size and complete interconnectivity between pores (Hench, 1991). It must be noted that the authors are assuming complete interconnectivity as a consequence of removing the coral using the dilute hydrochloric solution. No actual quantification of interconnectivity was carried out.

An approach developed by Roy and Linnehan (1974) employs a hydrothermal process to convert $CaCO_3$ within the coral to HA. The conversion is achieved by placing the $CaCO_3$ in a phosphate solution at a temperature and pressure of 260 °C and 103 MPa, respectively. As with the replamineform techniques the microstructure of the coral is not affected. Ultimately this research has led to the commercial development of coral-based orthopaedic products such as Pro-Osteon™ 200R and Pro-Osteon™ 500R.

While Roy and Linnehan reported complete conversion to HA, current commercially available materials manufactured via the hydrothermal process, such as Interpore 200™ and Interpore 500™, intentionally achieve only partial conversion (Ben-Nissan, 2003). This, however, creates a mismatch of degradation rates which compromises the scaffold's durability, its ability for tissue integration and, as a result, longevity (Ben-Nissan, 2003). Despite its fragility, however, coralline HA has been successfully used in clinical situations, although only as a space-filling material in non-load-bearing applications such as maxillofacial and dental surgery (Blom, 2007).

Alternatively, commercially available Biocoral™ uses the original $CaCO_3$ skeleton directly. Even in its natural state, coral mineral (aragonite or calcite forms of $CaCO_3$) has the ability to form chemical bonds with bone and soft tissue (Ben-Nissan, 2003). However, due to its high dissolution rate and poor mechanical stability, it is unsuitable for the majority of implant purposes (Ben-Nissan, 2003).

In summary, the ability to choose the scaffold material gives the replamineform technique an inherent advantage over both the hydrothermal technique and using coral in its natural state. Ultimately, this versatility allows greater control over a scaffold's mechanical properties, a characteristic which is lacking when using the other two techniques (Ben-Nissan, 2003; Blom, 2007).

Unlike coral, the architecture of coralline algae has not been utilized as a scaffold or bone substitute. Instead, their calcium carbonate (in the form of calcite) has been converted to calcium phosphate using similar hydrothermal conversion methods as those used by Roy et al. In 2008, Walsh et al. developed a production technique for the conversion of *Corallina officinalis* alga to HA using a two-stage process involving pyrolysis (to remove any organic constituents) and chemical synthesis. The calcite–HA material produced successfully maintained the interconnecting porous morphology of the precursor material (Kusmanto et al., 2008; Walsh et al., 2008).

14.4.6 Harvesting and sustainability

With regard to the utilization of human cancellous bone as a precursor for replication, while it may appear to be the logical answer to attain the perfect scaffold architecture, there are inherent drawbacks. Aside from the risk of bacterial contamination, viral transmission and immunogenicity, there are logistical issues (Vaccaro, 2002; Hing et al., 2007). According to Blom (2007), the supply of human bone simply cannot meet the demands of hip revision surgery, let alone for all other operations requiring bone.

For the biomimetic replication of naturally occurring species, sponges, coral, algae or otherwise, prior to harvesting and exploitation material scientists and engineers have a responsibility to determine the effect on their long-term sustainability. Protocols must be in place to limit the ecological impact either through lab-based cultivation or in situ replenishment.

For example, in order to meet the market demand, Biomet Interpore (manufacturers of Pro-Osteon™) harvest 2–4 tonnes of coral per year from the Pacific and Indian Oceans. This level of removal coupled with the slow growth rate of coral, reported by Smith et al. to be on average 0.8–26 mm/year (Roth, 1979), has created serious environmental concerns. In response to these concerns, all species of stony coral have been listed under the Convention on International Trade in Endangered Species treaty (Harriott, 2003).

Similarly, current demand for select species of marine sponges has resulted in their inclusion on endangered and protected species lists. Greeks and Tunisians have been harvesting marine sponges as far back as 4000–5000 years ago. Resistant to acids, and tougher and easier to clean than synthetic sponges, they are still used today in pottery, painting, polishing, filtering and even in surgery (Josupeit, 1990). While their commercial exploitation peeked in the 1930s, with ~70% being used for industrial applications, the development of synthetic alternatives and an increase in price has led to their predominant use as luxury bath sponges (Josupeit, 1990).

Researchers such as Gleason (1996), and more recently Louden et al. (2007), have attempted to address these issues and demonstrated the potential of aquaculture to provide a sustainable supply of bath and industrial sponges and alleviate the environmental effects of over-harvesting. Indeed, sponges are said to be well suited to cultivation; even small cut pieces of sponge will regenerate under the appropriate conditions (Castro and Huber, 2005). The use of laboratory or natural settled larvae together with asexual recruits, such as coral branches, colony fragments and entire small colonies, provides a feasible approach for the rehabilitation of damaged reefs (Gleason, 1996).

14.5 Ceramic processing

The shaping and forming of bioceramics into increasingly complex 3D shapes are pushing engineers to reinvestigate old manufacturing techniques and develop new ones. Ultimately, the intended application, whether complete hard-tissue replacement or integration with surrounding healthy tissue, will heavily influence any production technique employed. In general, two types of processing technologies are applicable where calcium phosphates are the predominant constituent (i.e. discounting polymer composites, which may use CaPs as a functional filler/additive): (1) employment of a high-solid content, low-viscosity slip/slurry of CaP powders for shaping and subsequent firing and (2) application of self-setting and self-hardening water-wet moulded powders (calcium phosphate cements).

Key to the scaffold production techniques discussed here is the development of a free-flowing HA slurry with optimum rheological properties. The aim is to achieve a colloidal suspension (a dispersion of solid particles in a liquid) with as high a solid content as possible while maintaining a low viscosity which will easily penetrate the pores of sacrificial constructs. According to Studart et al., a viscosity of between 1 and 6 Pa s is suitable to penetrate pores of between 1 and 2 mm (Brown and Green, 1994). As pore sizes in the sacrificial preforms discussed are ~30–1000 μm, and with little information available as to the maximum viscosity that will penetrate these pores, one can assume that it will be <1 Pa s. Achieving this requires an understanding of the methods of dispersion and the effects of particle characteristics (specific surface area and size distribution), solid content and mixing methods (Sadeghian et al., 2006).

Assuming that, at room temperature, the thermal energy (repulsive energy) of the particles is less than the attractive potential energy created via the van der Waals bonds, if particles in a suspension collide, they will stick together (Rahaman, 2007). As particles in a colloidal suspension are said to exhibit Brownian motion (the random movement of particles suspended in a liquid or gas) the assumption, therefore, is that they will collide, stick together and agglomerate. According to Zhang et al. (2006), this is the limiting factor in the production of a stable castable slurry. These agglomerations can be broken down and dispersed by inducing particle repulsion using electrosteric stabilization. This has been reported extensively to be the most effective method for achieving dispersion of HA powder (Sadeghian et al., 2006; Yasuda et al., 2000; Zhang et al., 2006, 2007; Lelievre et al., 1996; Tsetsekou et al., 2001). Most often this involves using a polymer with an ionizable group which dissociates (splits up into ions), giving rise to charged polymer chains. Known as polyelectrolytes, these chains are adsorbed onto the surface of HA particles, providing steric stabilization while the dissociated ions yield electrostatic repulsion (Figure 14.10) (Rahaman, 2007).

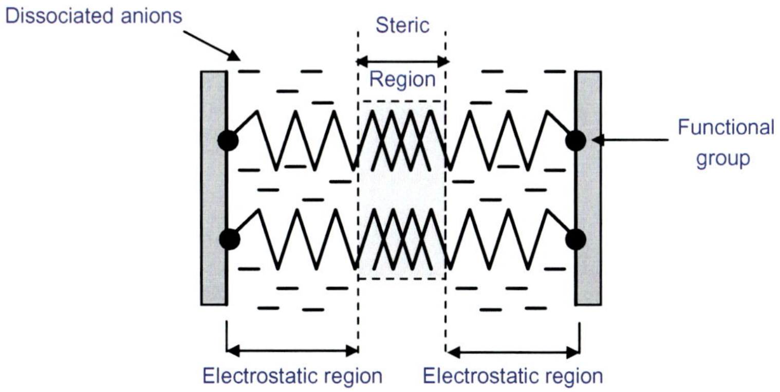

Figure 14.10 Electrosteric stabilization using an anionic polyelectrolyte via interaction of dissociated anions anchored to uncharged polymer chains (Cunningham, 2008).

An extensive study carried out by Lelievre et al. (1996) determined that the most efficient dispersing agent for HA was an anionic polyelectrolyte. In this case, the polymer's ionizable group splits up into anions (negative ions) which readily attach to the positively charged surface of HA (Sadeghian et al., 2006). The choice of an anionic polyelectrolyte has been reported by numerous authors, adding further weight to its selection as dispersing agent (Zhang et al., 2007; Yasuda et al., 2000). It should be noted that there is an optimum concentration of dispersing agent. Too little will result in insufficient electrostatic repulsion while too much will cause compression of the electrical double layer, nullifying its effects (Sadeghian et al., 2006).

Along with dispersion, particle size distribution will affect the ability of the HA particles to efficiently pack together. Enhancing this *packing ability* will increase the density of the pre-sintered construct, reducing the shrinkage incurred during sintering and improving the mechanical stability of the scaffold (Hsu et al., 2007; Tsetsekou et al., 2001; Cyster et al., 2005). Theoretically, if a slip is composed of spherical particles of equal diameter in a face-centred cubic arrangement the porosity will be 26% (Hampton et al., 1988). Consequently, it has been reported that either a bimodal particle size distribution with a mixture of *coarse* ($\approx$6 μm) and *fine* ($\approx$0.7 μm) particles (Hampton et al., 1988) or a broad continuous distribution of sizes (Rahaman, 2007) will enhance packing ability and aid the development of a high-solid-content/low-viscosity slip.

Particle morphology will also affect the packing ability of a colloidal suspension (Rahaman, 2007). While acicular (elongated or rod-like) particles have been reported to allow the production of high-density sintered constructs if adequate particle alignment is achieved (Rahaman, 2007), some authors have documented difficulty in achieving this (Lelievre et al., 1996). One explanation for this is the increased surface tension corresponding to increased contact area between rod-like particles. Logically, spherical particles will have the least surface tension as contact between particles is restricted to a single point (Janney, 1999).

Along with an appropriate particle size distribution and optimum choice and quantity of dispersant, the solid content can be further increased by adding the HA in ever decreasing amounts over a prolonged period (Woesz et al., 2005; Cyster et al., 2005). To aid homogenization of the mix and enhance particle packing, some form of mechanical agitation should be applied continually until use. This can be in the form of stirring, milling (without milling media, e.g. agate balls) or ultrasonic agitation (Rodríguez-Lorenzo et al., 2001).

14.5.1 Sintering

Predominantly, the final stage in the scaffold production techniques described is sintering. Defined as *the bonding of a powdered mass of adjacent particles by heating*, sintering involves three stages: initial, intermediate and final (Rahaman, 2007). Within these stages matter is rearranged via five processes, shown schematically in Figure 14.11a, including: (1) surface diffusion, (2) lattice diffusion from the particle

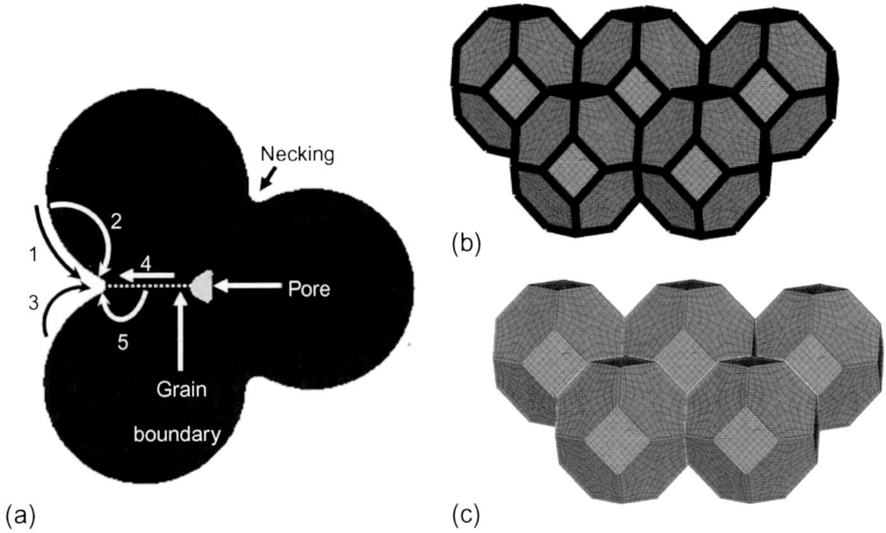

Figure 14.11 (a) Five mechanisms contributing to sintering of crystalline powder particles. (b) Intermediate stage of sintering. (c) Final stage of sintering.
Adapted from Rahaman (2007).

surface, (3) vapour transport, (4) grain boundary diffusion and (5) lattice diffusion from the grain boundary (Rahaman, 2007).

The initial stage, comprising of processes (1)–(3), begins at the onset of atomic mobility and is characterized by the formation of concave necks between individual particles. Driven by a reduction of surface free energy, these merging particles continue to unite throughout the intermediate stage. The curvatures formed in the initial stage disappear and a 3D microstructure of *tetrakaidecahedron* grains begins to take shape (Figure 14.11b). At the edges of these tetrakaidecahedron grains there are continuous *channel-like* pores with circular cross-sections represented by the black lines in Figure 14.11b (Rahaman, 2007). Densification via grain boundary diffusion and lattice diffusion from the grain boundary results in grain growth and a corresponding reduction in pore size. In the final stage, there is a further increase in density as these pores continue to shrink (Figure 14.11c) (Rahaman, 2007).

The relevant literature, which points to the use of a broad continuous particle size distribution with predominantly spherical particles ranging between ~2 and 13 μm as effective in the production of a high-solid content homogeneously packed slip (Rahaman, 2007; Lelievre et al., 1996), shows that the same characteristics yield high-density sintered constructs. The increased particle contact that arises in an efficiently packed slip reduces the shrinkage incurring during densification. Ultimately, the consequence is increased mechanical stability in the sintered construct (Hsu et al., 2007; Tsetsekou et al., 2001; Cyster et al., 2005; Rahaman, 2007).

At a certain temperature during sintering the density of a ceramic construct will begin to rise. After an initial phase of gradual densification, a second accelerated phase sees the density rise to ~90–95% of the theoretical value, before inevitably slowing and stopping. The composition, particle size, particle size distribution and therefore particle packing ability of the HA, coupled with the rate of heating and sintering atmosphere, will ultimately determine the maximum density attainable (Rahaman, 2007). Achieving a high density may, however, be hindered by the domination of *coarsening* (steps (1)–(3) described previously) over densification. If this does occur, grain growth is accompanied by increased porosity around the grain boundaries, compromising mechanical stability (Rahaman, 2007). Described as a key requirement in the development of a high-density construct, the suppression of coarsening mechanisms is achievable by determining and using an optimum sintering temperature (Benaqqa et al., 2005; He et al., 2008; Saiz et al., 2007; Rahaman, 2007).

While it is difficult to ascertain a definitive HA sintering regime from the techniques discussed in Section 14.4, cited in Table 14.7, there is certainly a pattern. Common to all is the application of multistage sintering leading to a peak temperature of between 1000 and 1350 °C. As the sintering temperature for most ceramics is typically between 0.5 and 0.8 of the melting temperature (Rahaman, 2007), this is a logical range for HA, the melting point of which is between 1570 and 1650 °C (depending on crystallinity) (Cruz et al., 2006).

The peak temperature is reached, in all cases, via a controlled heating rate and maintained for a period of between 1 and 3 h. During this plateau the majority of densification and microstructural development occurs (Rahaman, 2007). Controlling the rate of temperature increase (ramp rate) is essential to ensure uniform densification occurs without cracking (Rahaman, 2007).

While Ruys et al. (1995) and Zhang et al. (2007) have reported that sintering temperatures of between 1200 and 1450 °C can induce the compositional breakdown of HA to anhydrous calcium phosphates such as TCP, Jokanovic et al. (2008) have suggested that the process can begin as early as 900 °C. This decomposition, which arises due to irreversible dehydroxylation (removal of hydroxyl (OH) groups), is said to trigger higher dissolution rates (Zhang et al., 2007). It is also said to suppress densification and is accompanied by a decrease in mechanical stability (Muralithran and Ramesh, 2000). Consequently, the chemical composition of constructs produced via the sintering of HA at these temperatures should be analysed to ensure compositional breakdown has not occurred.

Also common in the regimes cited is the introduction of a temperature plateau of between 1 and 3 h, prior to reaching the sintering temperature. This is used in order to fully pyrolyse any polymer or organic phases present. Ramay and Zhang (2003) report the use of thermogravimetric analysis to study the pyrolysis of PU foam and use it to determine the optimum temperature to carry out this pyrolysis. The same could also be applied for any potential organic removal techniques.

Each regime cited allows for the production of constructs with >90% density. While the final mechanical properties of a scaffold will of course be dependent on its architecture, the aim of any sintering regime developed should be to, at the very least, match this value.

Table 14.7 **Sintering regimes utilized in various scaffold production techniques**

Technique	Material	Sintering regime	References
Replication of bovine bone	HA/β-TCP	1 °C/min to 350 °C → 4 °C/min to 1200 °C → 3 h @ 1200 °C → 4 °C/min cooling	Tancred et al. (1998)
PU foam replication	HA/monomers	1 °C/min to 600 °C → 1 h @ 600 °C → 3 °C/min to 1250 °C → 2 h @ 1350 °C	Ramay and Zhang (2003)
PU foam replication	HA PCL	1 °C/min to 600 °C → 3 h @ 600 °C → 3 h @ 1200 °C	Kim et al. (2005)
PU foam replication	HA	Between 1000 and 1200 °C	Saiz et al. (2007)
PU foam replication	Glass reinforced HA	1 °C/min to 600 °C → 5 °C/min 1250 °C → 8 h @ 1250 °C	Callcut and Knowles (2002)
Cellulose sponge replication	(L)(M)(H) HA	1 h @500 °C → 50 °C/h → 1 h @ 1200 °C	Guicciardi et al. (2001)
Cellulose sponge replication	HA	1 h @560 °C → 1 h @ 900 °C	Landi et al. (2003)
Cellulose sponge replication	HA	2 h @ 600 °C → 1 h @ 1280 °C	Mastrogiacomo et al. (2006)

Note: (L), (M) and (H) denote the use of low (20%), medium (45%) and high (80%) crystalline HA.

14.6 The future of biomimetic-inspired calcium phosphate materials

The array of interconnected porous preforms that nature currently possesses continues to inspire tissue engineers to utilize these intricate morphologies for hard-tissue replacement. These inorganic skeletons, whether in a raw state, hydrothermally converted to HA or replicated in a biphasic ceramic have been shown to support bone tissue and maintain their function, providing structural environments analogous to bone, as well as the high macroporosity required for vascularization, bone cell invasion and angiogenesis (Green et al., 2002). There is, however, potential for major advancement through continued discovery and utilization of new and existing naturally occurring structures coupled with the advancement of bioceramics. The challenge will be to combine apparently disparate chemical, biological and clinical disciplines to provide structures and materials that can positively impact the health of so many (Jun et al., 2003). As collaboration continues, individual strategies must merge into a single, focused approach. In the future, the composition, microstructure and molecular surface chemistry of various types of calcium orthophosphates should be tailored to match the specific biological and metabolic requirements of tissues (Dorozhkin, 2010). Moreover, as nanotechnology advances, the ability to control fabrication becomes routine, these structures should allow for the encapsulation of biologically active proteins, molecules and programmed autologous cells (Green et al., 2002). As pressure rises on an increasingly ageing society the next generation of calcium orthophosphate bioceramics based on biomimetic architectures should enhance the quality of life of millions of people.

References

Babis, G.C., Soucacos, P.N., 2005. Bone scaffolds: the role of mechanical stability and instrumentation. Injury 36, S38–S44.

Bar-Cohen, Y., 2006. Biomimetics – using nature to inspire human innovation. Bioinspir. Biomim. 1, 1.

Becker, D.J., Kilgore, M.L., Morrisey, M.A., 2010. The societal burden of osteoporosis. Curr. Rheumatol. Rep. 12, 186–191.

Benaqqa, C., Chevalier, J., Saädaoui, M., Fantozzi, G., 2005. Slow crack growth behaviour of hydroxyapatite ceramics. Biomaterials 26, 6106–6112.

Ben-Nissan, B., 2003. Natural bioceramics: from coral to bone and beyond. Curr. Opin. Solid State Mater. Sci. 7, 283–288.

Best, S.M., Porter, A.E., Thian, E.S., Huang, J., 2008. Bioceramics: past, present and for the future. J. Eur. Ceram. Soc. 28, 1319–1327.

Bianchi, M.L., 2007. Osteoporosis in children and adolescents. Bone 41, 486–495.

Bignon, A., Chouteau, J., Chevalier, J., Fantozzi, G., Carret, J.P., Chavassieux, P., Boivin, G., Melin, M., Hartmann, D., 2003. Effect of micro- and macroporosity of bone substitutes on their mechanical properties and cellular response. J. Mater. Sci. Mater. Med. 14, 1089–1097.

Black, J., 2005. Stress Shielding. Biological Performance of Materials: Fundamentals of Biocompatibility, fourth ed. CRC Press, Boca Raton, FL.

Blom, A., 2007. (V) Which scaffold for which application? Curr. Orthop. 21, 280–287.

Bohner, M., Baroud, G., 2005. Injectability of calcium phosphate pastes. Biomaterials 26, 1553–1563.

Brazinha, C., Fonseca, A.P., Pereira, H., Teodoro, O.M.N.D., Crespo, J.G., 2013. Gas transport through cork: modelling gas permeation based on the morphology of a natural polymer material. J. Membr. Sci. 428, 52–62.

Brown, D.D., Green, D.J., 1994. Investigation of strut crack formation in open cell alumina ceramics. J. Am. Ceram. Soc. 77 (6), 1467–1472.

Brydone, A.S., Meek, D., Maclaine, S., 2010. Bone grafting, orthopaedic biomaterials, and the clinical need for bone engineering. Proc. Inst. Mech. Eng. H: J. Eng. Med. 224, 1329–1343.

Buckley, C.T., O'Kelly, K.U., 2004. Regular scaffold fabrication techniques for investigations in tissue engineering. In: Prendergast, P.J., McHugh, P.E. (Eds.), Topics in Bio-Mechanical Engineering. Trinity Centre for Bio-Engineering and National Centre for Biomedical Engineering Science, Dublin.

Byrne, D.P., Lacroix, D., Planell, J.A., Kelly, D.J., Prendergast, P.J., 2007. Simulation of tissue differentiation in a scaffold as a function of porosity, Young's modulus and dissolution rate: application of mechanobiological models in tissue engineering. Biomaterials 28, 5544–5554.

Callcut, S., Knowles, J.C., 2002. Correlation between structure and compressive strength in a reticulated glass-reinforced hydroxyapatite foam. J. Mater. Sci. Mater. Med. 13, 485–489.

Castro, P., Huber, M., 2005. Marine animals without a backbone. In: Schreiber, F. (Ed.), Marine Biology. fifth ed. McGraw-Hill, New York.

Chang, B.-S., Lee, C.-K., Hong, K.-S., Youn, H.-J., Ryu, H.-S., Chung, S.-S., Park, K.-W., 2000. Osteoconduction at porous hydroxyapatite with various pore configurations. Biomaterials 21, 1291–1298.

Chen, Q.Z., Wong, C.T., Lu, W.W., Cheung, K.M.C., Leong, J.C.Y., Luk, K.D.K., 2004. Strengthening mechanisms of bone bonding to crystalline hydroxyapatite in vivo. Biomaterials 25, 4243–4254.

Chen, Q.Z., Thompson, I.D., Boccaccini, A.R., 2006. 45S5 Bioglass derived glass ceramic scaffolds for bone tissue engineering. Biomaterials 27, 2414–2425.

Chiroff, R.T., White, E.W., Weber, K.N., Roy, D.M., 1975. Tissue in growth of replamineform implants. J. Biomed. Mater. Res. 9, 29–45.

Clarke, S.A., Walsh, P., Maggs, C.A., Buchanan, F., 2011. Designs from the deep: marine organisms for bone tissue engineering. Biotechnol. Adv. 29, 610–617.

Cruz, F.J.A.L., Lopes, J.N.C., Calado, J.C.G., 2006. Molecular dynamics simulations of molten calcium hydroxyapatite. Fluid Phase Equilibr. 241, 51–58.

Cunningham, E., 2008. Development of Bone Substitutes Based on Biomimetic Structures. Queen's University, Belfast.

Cunningham, E., Dunne, N., Walker, G., Maggs, C., Wilcox, R., Buchanan, F., 2009. Hydroxyapatite bone substitutes developed via replication of natural marine sponges. J. Mater. Sci. Mater. Med. 21 (8), 2255–2261. http://dx.doi.org/10.1007/s10856-009-3961-4.

Cunningham, E., Dunne, N., Clarke, S., Choi, S.Y., Buchanan, F., Walker, G., Wilcox, R., Unger, R., Kirkpatrick, C.J., 2011. Comparative characterisation of 3-D hydroxyapatite scaffolds developed via replication of synthetic polymer foams and natural marine sponges. J. Tissue Sci. Eng. S1.

Cyster, L.A., Grant, D.M., Howdle, S.M., Rose, F.R.A.J., Irvine, D.J., Freeman, D., Scotchford, C.A., Shakesheff, K.M., 2005. The influence of dispersant concentration on the pore morphology of hydroxyapatite ceramics for bone tissue engineering. Biomaterials 26, 697–702.

Dawson, M.A., Gibson, L.J., 2007. Optimization of cylindrical shells with compliant cores. Int. J. Solids Struct. 44, 1145–1160.

Dorozhkin, S.V., 2010. Calcium orthophosphates as bioceramics: state of the art. J. Funct. Biomater. 1, 22–107.

Dorozhkin, S., 2012. Calcium Orthophosphates Applications in Nature, Biology and Medicine. Pan Stanford, Singapore.

Ellinsen, J.E., Lyngstadaas, S.P., 2003. Bio-Implant Interface. Improving Biomaterials and Tissue Reactions. CRC Press, Boca Raton, FL.

Engin, N.O., Tas, A.C., 1999. Manufacture of macroporous calcium hydroxyapatite bioceramics. J. Eur. Ceram. Soc. 19, 2569–2572.

Fujita, R., Yokoyama, A., Nodasaka, Y., Kohgo, T., Kawasaki, T., 2003. Ultrastructure of ceramic-bone interface using hydroxyapatite and β-tricalcium phosphate ceramics and replacement mechanism of β-tricalcium phosphate in bone. Tissue Cell 35, 427–440.

Gauthier, O., Bouler, J.-M., Aguado, E., Pilet, P., Daculsi, G., 1998. Macroporous biphasic calcium phosphate ceramics: influence of macropore diameter and macroporosity percentage on bone ingrowth. Biomaterials 19, 133–139.

Giannoudis, P.V., Dinopoulos, H., Tsiridis, E., 2005. Bone substitutes: an update. Injury 36, S20–S27.

Gibson, L.J., 2005. Biomechanics of cellular solids. J. Biomech. 38, 377–399.

Gibson, L.J., Ashby, M.F., Harley, B.A., 2010. Cellular Materials in Nature and Medicine. Cambridge University Press, Cambridge.

Gleason, M.G., 1996. Coral recruitment in Moorea, French Polynesia: the importance of patch type and temporal variation. J. Exp. Mar. Biol. Ecol. 207, 79–101.

Green, D., Walsh, D., Mann, S., Oreffo, R.O.C., 2002. The potential of biomimesis in bone tissue engineering: lessons from the design and synthesis of invertebrate skeletons. Bone 30, 810–815.

Guicciardi, S., Galassi, C., Landi, E., Tampieri, A., 2001. Rheological characteristics of slurry controlling the microstructure and the compressive strength behavior of biomimetic hydroxyapatite. J. Mater. Res. 16, 163–170.

Gyger, J.L.S., Kulkarni, P., Bruck, H.A., Gupta, S.K., Wilson, J.O.C., 2007. Replamineform inspired bone structures (RIBS) using multi-piece molds and advanced ceramic gelcasting technology. Mater. Sci. Eng. C 27, 646–653.

Hamerman, D., 2005. Bone health across the generations: a primer for health providers concerned with osteoporosis prevention. Maturitas 50, 1–7.

Hampton, J.H.D., Savage, S.B., Drew, R.A.L., 1988. Experimental analysis and modeling of slip casting. J. Am. Ceram. Soc. 71 (12), 1040–1045.

Handschel, J., Wiesmann, H.P., Stratmann, U., Kleinheinz, J., Meyer, U., Joos, U., 2002. TCP is hardly resorbed and not osteoconductive in a non-loading calvarial model. Biomaterials 23, 1689–1695.

Harriott, V.J., 2003. Can corals be harvested sustainably? Ambio 32, 130–133.

He, L.-H., Standard, O.C., Huang, T.T.Y., Latella, B.A., Swain, M.V., 2008. Mechanical behaviour of porous hydroxyapatite. Acta Biomater. 4, 577–586.

Hench, L.L., 1991. Bioceramics: from concept to clinic. J. Am. Ceram. Soc. 74, 1487–1510.

Hench, L., Polak, J.M., 2002. Third generation biomedical materials. Science 295, 1014–1017.

Hench, L., Wilson, J., 1984. Surface active biomaterials. Science 226, 630–636.

Hing, K.A., Wilson, L.F., Buckland, T., 2007. Comparative performance of three ceramic bone graft substitutes. Spine J. 7, 475–490.

Hsu, Y., Turner, I., Miles, A., 2007. Mechanical characterization of dense calcium phosphate bioceramics with interconnected porosity. J. Mater. Sci. Mater. Med. 18, 2319–2329.

Huang, Q.-Y., Kung, A.W.C., 2006. Genetics of osteoporosis. Mol. Genet. Metab. 88, 295–306.

Hutmacher, D.W., 2000. Scaffolds in tissue engineering bone and cartilage. Biomaterials 21, 2529–2543.

Hutmacher, D.W., Garcia, A.J., 2005. Scaffold-based bone engineering by using genetically modified cells. Gene 347, 1–10.

Jackson, A.P., Vincent, J.F.V., Turner, R.M., 1990. Comparison of nacre with other ceramic composites. J. Mater. Sci. 25, 3173–3178.
Janney, M.A., 1999. Attaining high solids in ceramic slurries. In: Jillavenkatesa, A., Onoda, G.Y. (Eds.), Advances in Process Measurements for the Ceramics Industry. Wiley, New York.
Jarcho, M., et al., 1979. Synthesis and fabrication of beta-TCP (whitlockite) ceramics for potential prosthetic applications. J. Mater. Sci. 14, 142–150.
Johnell, O., Kanis, J.A., 2006. An estimate of the worldwide prevalence and disability associated with osteoporotic fractures. Osteoporos. Int. 17, 1726–1733.
Jokanovic, V., Jokanovic, B., Markovic, D., Zivojinovic, V., Pasalic, S., Izvonar, D., Plavsic, M., 2008. Kinetics and sintering mechanisms of hydro-thermally obtained hydroxyapatite. Mater. Chem. Phys. 111, 180–185.
Jones, J.R., Hench, L.L., 2003. Regeneration of trabecular bone using porous ceramics. Curr. Opin. Solid State Mater. Sci. 7, 301–307.
Josupeit, H., 1990. Sponges: World Production and Markets. South Pacific Aquacultural Development Project.
Jun, Y.-K., Kim, W.H., Kweon, O.-K., Hong, S.-H., 2003. The fabrication and biochemical evaluation of alumina reinforced calcium phosphate porous implants. Biomaterials 24, 3731–3739.
Kammoun, H.M., Kallel, I., Casillas, J., Abraham, A., Alimi, A.M., 2014. Adapt-Traf: an adaptive multiagent road traffic management system based on hybrid ant-hierarchical fuzzy model. Transport. Res. C-Emer. 42, 147–167.
Karageorgiou, V., Kaplan, D., 2005. Porosity of 3D biomaterial scaffolds and osteogenesis. Biomaterials 26, 5474–5491.
Kasten, P., Luginbühl, R., Van Griensven, M., Barkhausen, T., Krettek, C., Bohner, M., Bosch, U., 2003. Comparison of human bone marrow stromal cells seeded on calcium-deficient hydroxyapatite, β-tricalcium phosphate and demineralized bone matrix. Biomaterials 24, 2593–2603.
Keaveny, T.M., 1998. Cancellous bone. In: Black, J., Hastings, G. (Eds.), Handbook of Biomaterial Properties, first ed. Chapman & Hall, London.
Kim, H.-W., Knowles, J.C., Kim, H.-E., 2004. Hydroxyapatite/poly(ε-caprolactone) composite coatings on hydroxyapatite porous bone scaffold for drug delivery. Biomaterials 25, 1279–1287.
Kim, H.-W., Knowles, J.C., Kim, H.-E., 2005. Hydroxyapatite porous scaffold engineered with biological polymer hybrid coating for antibiotic Vancomycin release. J. Mater. Sci. Mater. Med. 16, 189–195.
Kim, T.K., Yoon, J.J., Lee, D.S., Park, T.G., 2006. Gas foamed open porous biodegradable polymeric microspheres. Biomaterials 27, 152–159.
Klein-Nulend, J., Bacabac, R.G., Mullender, M.G., 2005. Mechanobiology of bone tissue. Pathol. Biol. 53, 576–580.
Kultikova, E.V., 1999. Structure and properties relationships of densified qood. M.S. Thesis, Virginia Polytechnic Institute and State University.
Kusmanto, F., Walker, G., Gan, Q., Walsh, P., Buchanan, F., Dickson, G., McCaigue, M., Maggs, C., Dring, M., 2008. Development of composite tissue scaffolds containing naturally sourced mircoporous hydroxyapatite. Chem. Eng. J. 139, 398–407.
Landi, E., Celotti, G., Logroscino, G., Tampieri, A., 2003. Carbonated hydroxyapatite as bone substitute. J. Eur. Ceram. Soc. 23, 2931–2937.
Le Huec, J.C., Schaeverbeke, T., Clement, D., Faber, J., Le Rebeller, A., 1995. Influence of porosity on the mechanical resistance of hydroxyapatite ceramics under compressive stress. Biomaterials 16, 113–118.

Lelievre, F., Bernache-Assollant, D., Chartier, T., 1996. Influence of powder characteristics on the rheological behaviour of hydroxyapatite. J. Mater. Sci. Mater. Med. 7, 489–494.

Leong, K.F., Cheah, C.M., Chua, C.K., 2003. Solid freeform fabrication of three-dimensional scaffolds for engineering replacement tissues and organs. Biomaterials 24, 2363–2378.

Lewandrowski, K.-U., Gresser, D., Wise, D.L., Trantolo, D.J., 2000. Bioresorbable bone graft substitutes of different osteoconductivities: a histologic evaluation of osteointegration of poly(propylene glycol-co-fumaric acid)-based cement implants in rats. Biomaterials 21, 757–764.

Lin, K., Chang, J., Lu, J., Wu, W., Zeng, Y., 2007. Properties of β-$Ca_3(PO_4)_2$ bioceramics prepared using nano-size powders. Ceram. Int 33, 979–985.

Louden, D., Whalan, S., Evans-Illidge, E., Wolff, C., De Nys, R., 2007. An assessment of the aquaculture potential of the tropical sponges *Rhopaloeides odorabile* and *Coscinoderma* sp. Aquaculture 270, 57–67.

Ma, P.X., 2008. Biomimetic materials for tissue engineering. Adv. Drug Deliv. Rev. 60, 184–198.

Mastrogiacomo, M., Scaglione, S., Martinetti, R., Dolcini, L., Beltrame, F., Cancedda, R., Quarto, R., 2006. Role of scaffold internal structure on in vivo bone formation in macroporous calcium phosphate bioceramics. Biomaterials 27, 3230–3237.

Matsumoto, Y., Ii, S., Shiozaki, S., Sugiyama, K., Takagi, S., 2014. Toward the multi-scale simulation for a human body using the next-generation supercomputer. Procedia IUTAM 10, 193–200.

Mayer, G., 2005. Rigid biological systems as models for synthetic composites. Science 310, 1144–1147.

Meyers, M.A., Chen, P.-Y., Lin, A.Y.-M., Seki, Y., 2008. Biological materials: structure and mechanical properties. Prog. Mater. Sci. 53, 1–206.

Miranda, P., Saiz, E., Gryn, K., Tomsia, A.P., 2006. Sintering and robocasting of β-tricalcium phosphate scaffolds for orthopaedic applications. Acta Biomater. 2, 457–466.

Muhamad Nor, M.A.A., Hong, L.C., Arifin Ahmad, Z., Md Akil, H., 2008. Preparation and characterization of ceramic foam produced via polymeric foam replication method. J. Mater. Process. Technol. 207, 235–239.

Müller, W.E.G., Batel, R., Schröder, H.C., Müller, I.M., 2004. Traditional and modern biomedical prospecting: part I—the history sustainable exploitation of biodiversity (sponges and invertebrates) in the Adriatic Sea in Rovinj (Croatia). Evid. Based Complement. Alternat. Med. 1, 71–82.

Muralithran, G., Ramesh, S., 2000. The effects of sintering temperature on the properties of hydroxyapatite. Ceram. Int. 26, 221–230.

Neo, M., Herbst, H., Voigt, C.F., Gross, U.M., 1998. Temporal and spatial patterns of osteoblast activation following implantation of β-TCP particles into bone. J. Biomed. Mater. Res. 39, 71–75.

Newe, C., Cunningham, E., Buchanan, F., Walker, G., Prendergast, P., Lennon, A., Dunne, N., 2012. Static and dynamic degradation of sintered calcium phosphate ceramics. Key Eng. Mater. 493–494, 861–865.

Odgaard, A., Gundersen, H.J.G., 1993. Quantification of connectivity in cancellous bone, with special emphasis on 3-D reconstructions. Bone 14, 173–182.

Oh, S.H., Kang, S.G., Kim, E.S., Cho, S.H., Lee, J.H., 2003. Fabrication and characterization of hydrophilic poly(lactic-co-co-glycolic acid)/poly(vinyl alcohol) blend cell scaffolds by melt-molding particulate-leaching method. Biomaterials 24, 4011–4021.

Parikh, S., 2002. Bone graft substitutes: past, present, future. Postgrad. Med. 48, 142–148.

Rahaman, M.N., 2007. Ceramic Processing. CRC Taylor & Francis, Boca Raton, FL.

Ramay, H.R., Zhang, M., 2003. Preparation of porous hydroxyapatite scaffolds by combination of the gel-casting and polymer sponge methods. Biomaterials 24, 3293–3302.

Rambo, C.R., Cao, J., Seiber, H., 2004. Preparation and properties of highly porous, biomorphous YSZ ceramics. Mater. Chem. Phys. 87, 345–352.

Rambo, C.R., Müller, F.A., Müller, L., Sieber, H., Hofmann, I., Greil, P., 2006. Biomimetic apatite coating on biomorphous alumina scaffolds. Mater. Sci. Eng. C 26, 92–99.

Reeve, R., Van Schaik, A., Jin, C., Hamilton, T., Torben-Nielsen, B., Webb, B., 2007. Directional hearing in a silicon cricket. Biosystems 87, 307–313.

Rezwan, K., Chen, Q.Z., Blaker, J.J., Boccaccini, A.R., 2006. Biodegradable and bioactive porous polymer/inorganic composite scaffolds for bone tissue engineering. Biomaterials 27, 3413–3431.

Robertson, K., 2005. Honeycomb-Like Seed Coat [Online]. Available at: http://www.inhs.uiuc.edu/~kenr/prairiephotos/castcocc.seeds.SEM2.jpg (accessed 05.05.08).

Rodan, G.A., Martin, T.J., 2000. Therapeutic approaches to bone diseases. Science 289, 1508–1514.

Rodríguez-Lorenzo, L.M., Vallet-Regí, M., Ferreira, J.M.F., 2001. Colloidal processing of hydroxyapatite. Biomaterials 22, 1847–1852.

Roth, A.A., 1979. Coral reef growth. Origins 6 (2), 88–95.

Rousculp, M.D., Long, S.R., Wang, S., Schoenfeld, M.J., Meadows, E.S., 2007. Economic burden of osteoporosis-related fractures in Medicaid. Value Health 10, 144–152.

Roy, D.M., Linnehan, S.K., 1974. Hydroxyapatite formed from coral skeletal carbonate by hydrothermal exchange. Nature 247, 220–222.

Ruys, A.J., Wei, M., Sorrell, C.C., Dickson, M.R., Brandwood, A., Milthome, B.K., 1995. Sintering effects on the strength of hydroxyapatite. Biomaterials 16, 409–415.

Sabolinski, M.L., Alvarez, O., Auletta, M., Mulder, G., Parenteau, N.L., 1996. Cultured skin as a "smart material" for healing wounds: experience in venous ulcers. Biomaterials 17, 311–320.

Sachlos, E., Czernuszka, J.T., 2003. Making tissue engineering scaffolds work. Review on the application of solid freeform fabrication technology to the production of tissue engineering scaffolds. Eur. Cell. Mater. 5, 29–40.

Sadeghian, Z., Heinrich, J.G., Moztarzadeh, F., 2006. Influence of powder pre-treatments and milling on dispersion ability of aqueous hydroxyapatite-based suspensions. Ceram. Int. 32, 331–337.

Saiz, E., Gremillard, L., Menendez, G., Miranda, P., Gryn, K., Tomsia, A.P., 2007. Preparation of porous hydroxyapatite scaffolds. Mater. Sci. Eng. C 27, 546–550.

Schliephake, H., 1991. Influence of pore dimensions on bone ingrowth into porous hydroxylapatite blocks used as bone graft substitutes. A histometric study. Int. J. Oral Maxillofac. Surg. 20, 53–58.

Schliephake, H., Redecker, K., Kage, T., 1997. Reconstruction of calvarial defects by bioresorbable ceramics: an experimental study in rats. Br. J. Oral Maxillofac. Surg. 35, 219.

Sethmann, I., Worheide, G., 2008. Structure and composition of calcareous sponge spicules: a review and comparison to structurally related biominerals. Micron 39, 209–228.

Sieber, H., 2005. Biomimetic synthesis of ceramics and ceramic composites. Mater. Sci. Eng. A 412, 43–47.

Sopyan, I., Mel, M., Ramesh, S., Khalid, K.A., 2007. Porous hydroxyapatite for artificial bone applications. Sci. Technol. Adv. Mater. 8, 116–123.

Srouji, S., Livne, E., 2005. Bone marrow stem cells and biological scaffold for bone repair in aging and disease. Mech. Ageing Dev. 126, 281–287.

Sultana, N., Wang, M., 2008. Fabrication of HA/PHBV composite scaffolds through the emulsion freezing/freeze-drying process and characterisation of the scaffolds. J. Mater. Sci. Mater. Med. 19, 2555–2561.

Tampieri, A., Celotti, G., Sprio, S., Delcogliano, A., Franzese, S., 2001. Porosity-graded hydroxyapatite ceramics to replace natural bone. Biomaterials 22, 1365–1370.

Tancred, D.C., McCormack, B.A.O., Carr, A.J., 1998. A synthetic bone implant macroscopically identical to cancellous bone. Biomaterials 19, 2303–2311.

Toole, G., 1995. Structure of bone. In: Toole, S. (Ed.), Understanding Biology, third ed. Stanley Thorne, Ellenborough House.

Trentz, O., 2003. Osteoblast response to bone substitutes in vitro. Eur. Cell. Mater. 5, 55.

Truumees, E., Herkowitz, H., 1999. Alternatives to autologous bone harvest in spine surgery. Univ. Penn. Orthop. J. 12, 77–88.

Tsetsekou, A., Agrafiotis, C., Milias, A., 2001. Optimization of the rheological properties of alumina slurries for ceramic processing applications. Part I: slip-casting. J. Eur. Ceram. Soc. 21, 363–373.

Vaccaro, A., 2002. The role of the osteoconductive scaffold in synthetic bone graft. Orthopaedics 25, s571–s578.

Vats, A., Tolley, N.S., Polak, J.M., Gough, J.E., 2003. Scaffolds and biomaterials for tissue engineering: a review of clinical applications. Clin. Otolaryngol. 28, 165–172.

Von Doernberg, M.-C., Von Rechenberg, B., Bohner, M., Grünenfelder, S., Van Lenthe, G.H., Müller, R., Gasser, B., Mathys, R., Baroud, G., Auer, J., 2006. In vivo behavior of calcium phosphate scaffolds with four different pore sizes. Biomaterials 27, 5186–5198.

Vozzi, G., Flaim, C., Ahluwalia, A., Bhatia, S., 2003. Fabrication of PLGA scaffolds using soft lithography and microsyringe deposition. Biomaterials 24, 2533–2540.

Walsh, P.J., Buchanan, F.J., Dring, M., Maggs, C., Bell, S., Walker, G.M., 2008. Low-pressure synthesis and characterisation of hydroxyapatite derived from mineralised red algae. Chem. Eng. J. 137, 173–179.

Wikimedia.org. *Porifera Types* [Online]. Available at: http://upload.wikimedia.org/wikipedia/commons/thumb/b/b6/Porifera_Types-fr.svg/800px-Porifera_Types-fr.svg.png (accessed 10.06.08).

Williams, J.B., Irvine Jr., J.W., 1954. Preparation of the inorganic matrix of bone. Science 119, 771–772.

Woesz, A., Rumpler, M., Stampfl, J., Varga, F., Fratzl-Zelman, N., Roschger, P., Klaushofer, K., Fratzl, P., 2005. Towards bone replacement materials from calcium phosphates via rapid prototyping and ceramic gelcasting. Mater. Sci. Eng. C 25, 181–186.

Woodard, J.R., Hilldore, A.J., Lan, S.K., Park, C.J., Morgan, A.W., Eurell, J.A.C., Clark, S.G., Wheeler, M.B., Jamison, R.D., Wagoner Johnson, A.J., 2007. The mechanical properties and osteoconductivity of hydroxyapatite bone scaffolds with multi-scale porosity. Biomaterials 28, 45–54.

Wu, S., Liu, X., Yeung, K.W.K., Liu, C., Yang, X., 2014. Biomimetic porous scaffolds for bone tissue engineering. Mater. Sci. Eng. R. Rep. 80, 1–36.

Yamamuro, T., 2005, September. Bioactive ceramics and their clinical use. In: Paper presented at the Professor Larry L Hench Retirement Symposium: Bioactive Ceramics and their Clinical Use. Imperial College, London.

Yang, S., Leong, K.F., Du, Z., Chua, C.K., 2001. The design of scaffolds for use in tissue engineering. Part I. Traditional factors. Tissue Eng. 7 (6), 679–689.

Yasuda, H.Y., Mahara, S., Umakoshi, Y., Imazato, S., Ebisu, S., 2000. Microstructure and mechanical property of synthesized hydroxyapatite prepared by colloidal process. Biomaterials 21, 2045–2049.

Zhang, J., Maeda, M., Kotobuki, N., Hirose, M., Ohgushi, H., Jiang, D., Iwasa, M., 2006. Aqueous processing of hydroxyapatite. Mater. Chem. Phys. 99, 398–404.

Zhang, J., Tanaka, H., Ye, F., Jiang, D., Iwasa, M., 2007. Colloidal processing and sintering of hydroxyapatite. Mater. Chem. Phys. 101, 69–76.

Zhao, D.-Y., Huang, Z.-P., Wang, M.-J., Wang, T., Jin, Y., 2012. Vacuum casting replication of micro-riblets on shark skin for drag-reducing applications. J. Mater. Process. Technol. 212, 198–202.

Mineralization of fibers for bone regeneration

15

A. El-Fiqi, S.-J. Seo, H.-W. Kim
Institute of Tissue Regeneration Engineering (ITREN), Dankook University, Cheonan, South Korea

15.1 Introduction

The structural and morphological similarities between fibrous scaffolds and native extracellular matrix (ECM) have attracted significant attention in tissue engineering. The ECM is a nanofibrous network of primarily proteins and polysaccharides, having both regulatory and structural functions. Interactions between cells and ECM, via cell-surface receptors and cell-adhesive ligands, regulate many cellular behaviors, such as adhesion, proliferation, migration, and differentiation (Martins et al., 2007; Han and Gouma, 2006; Ma et al., 2005a). Several studies have been devoted to exploring the *in vitro* and *in vivo* biological responses of fibrous scaffolds for soft- and hard-tissue regeneration (Chong et al., 2007; Choi et al., 2008a; Chen et al., 2008; Liu et al., 2010; Kumbar et al., 2008a; He et al., 2005a,b; Aghdam et al., 2013; Sreerekha et al., 2013; Shih et al., 2006; Mo et al., 2004; Yoshimoto et al., 2003; Bolgen et al., 2005, 2007; Shin et al., 2004; Fu et al., 2012; Ji et al., 2012). Fibrous scaffolds that mimic the topography and scale of native ECM have shown better cellular responses than nonfibrous equivalents (Xu et al., 2004; Kim and Park, 2006; Sell et al., 2007; Woo et al., 2003, 2007).

As the bone ECM is a type of composite constituted mainly of collageneous fibers embedded with hydroxyapatite (HA) nanocrystallites, the development of fibrous nanocomposites mimicking bone composition and structure is desirable. The biological characteristics of HA such as its specific affinity to many adhesive proteins and its direct involvement in the bone cell differentiation and mineralization processes make it a potential candidate for use in bone regeneration (Zhou and Lee, 2011; Rajzer et al., 2014; Thomas et al., 2006; Wang et al., 2007; Phipps et al., 2011). A rapidly growing strategy combining an electrospinning technique and biomimetic deposition of HA mineral on the fiber surface to closely mimic the structure and composition of bone ECM is currently considered as a promising approach to improve the biological performance of fibrous scaffolds. The surface modification of scaffolds with HA mineral has a significant effect in controlling the vital cellular events at early stages, e.g. protein adsorption, cell attachment, proliferation, and tissue differentiation. Moreover, the presence of HA mineral on the fiber surface facilitates surface tailoring with functional proteins, e.g. fibronectin or growth factors (Suarez-Gonzalez et al., 2012; Lee et al., 2014; Kim et al., 2014a). The effects of HA within the synthetic polymers on the biological function of bone-associated cells, such as adhesive protein adsorption, cell anchorage, signal transduction, osteoblastic differentiation, and bone matrix synthesis,

Biomineralization and Biomaterials. http://dx.doi.org/10.1016/B978-1-78242-338-6.00016-8

have been investigated in many *in vitro* studies (Peng et al., 2011; Zhao et al., 2011; Lee et al., 2010a; Lu et al., 2013).

A variety of bone-like scaffolds have been developed either by direct introduction of HA nanoparticles into polymeric matrices or by the mineralization of HA on the surface of polymeric substrates (Kim et al., 2006a, 2007; Yang et al., 2008a, 2009; Lao et al., 2011; Liao and Cui, 2004; Zhang et al., 2009a; Liu et al., 2009a; Kretlow and Mikos, 2007; Song et al., 2005; Mavis et al., 2009; Takeuchi et al., 2003; Oyane et al., 2005a). Several methods for fiber surface modification have also been proposed, e.g. hydrophilic coating with natural polymers (Qu et al., 2007; He et al., 2005b; Duan et al., 2007), tethering the surface with cell-adhesive proteins (Ma et al., 2005b; Horne et al., 2010; Chen and Su, 2011), mussel-inspired surface modification by polydopamine (Xie et al., 2012a; Rim et al., 2012; Jiang et al., 2014; Cho et al., 2014; Tsai et al., 2014), and biomimetic mineralization (Xie et al., 2013; Rodriguez et al., 2011; Cai et al., 2011; Ma et al., 2011; Li et al., 2008). These mineralization strategies of fibrous scaffolds and the effects on *in vitro* and *in vivo* responses are the main focus of this chapter.

15.2 Fibers as scaffolds for bone regeneration

Fibrous scaffolds which closely mimic the architecture of native tissue have been identified as more favorable for cell adhesion, proliferation, and differentiation (Xu et al., 2004; Kim and Park, 2006; Sell et al., 2007; Woo et al., 2003, 2007), and this is correlated with fibrous scaffold topography and their extremely high surface area-to-volume ratio. Moreover, the typical features of fibrous scaffolds such as high porosity, thoroughly interconnected micro/nanoscale pore-networks, and formability into a wide variety of sizes and shapes have made them potential scaffolds for tissue engineering. Engineered scaffolds that can mimic the nanofibrous morphology, structure, composition, and biochemical cues of native bone ECM are currently needed for effective bone regeneration.

15.2.1 Processing techniques of fibers

Various techniques have been used to prepare polymeric micro/nanofibers; in the following text we focus mainly on two techniques, namely microfluidics and electrospinning.

15.2.1.1 Microfluidics

Microfluidics is emerging as an effective tool for tissue engineering and biology. The rapid development of microfluidic devices has resulted in a variety of microfluidic applications, i.e. microfluidics has recently begun to be realized as a promising tool for the preparation of highly uniform polymer microfibers and monodisperse microparticles (Nunes et al., 2014; Cheng et al., 2014; Jun et al., 2013; Lee et al., 2010b; Mazzitelli et al., 2011; Shin et al., 2007; Matthews et al., 2002; Yang et al., 2005;

Li et al., 2003; Min et al., 2004; Jegal et al., 2011; Pham et al., 2006; Xu et al., 2005, 2009; Duncanson et al., 2012; Breslauer et al., 2010; He et al., 2011; Choi et al., 2009). Advances in microfluidics technology have revolutionized biomedical fields due to significant advantages such as miniaturization, very small volume consumption, and multiplexing. Microfluidic fiber fabrication is processed via a very promising route at the micro- and nanoscale, providing precise control over fiber shape, size, chemical anisotropy, and biological activity (Nunes et al., 2014; Cheng et al., 2014; Jun et al., 2013; Lee et al., 2010b; Mazzitelli et al., 2011; Shin et al., 2007).

Hwang et al. (2008) developed a method to produce poly(lactic-*co*-glycolic acid) (PLGA) microfibers within a microfluidic chip for the generation of 3D tissue-engineered scaffolds. The synthesis of PLGA fibers was achieved by using a polydimethylsiloxane-based microfluidic spinning device in which linear streams of PLGA dissolved in dimethyl sulfoxide (DMSO) were precipitated in a glycerol-containing water solution. PLGA fibers comprised of a dense outer surface and a highly porous interior were formed with diameters ranging from 20 to 230 μm on changing the flow rate of 1–50 μl min^{-1} with a sheath flow rate of 250 or 1000 μl min^{-1}. A modified setup comprised of a microfluidic device and a fiber winding system is depicted in Figure 15.1.

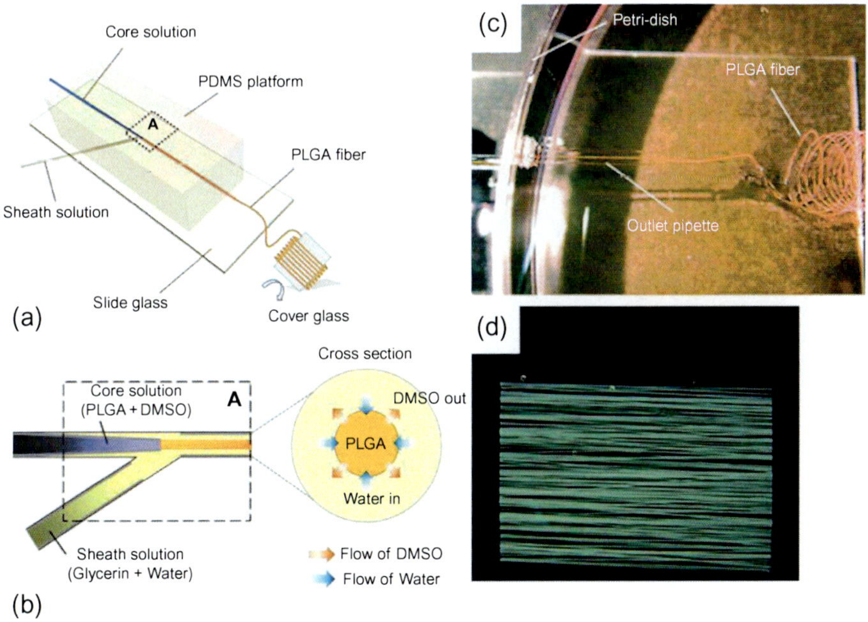

Figure 15.1 Exemplary process of microfibers (microfluidics). (a) Schematic showing microfluidic chip and cover glass winding device for the aligned PLGA fibers. (b) Principle of the phase inversion process during polymer precipitation, PLGA spinning with a microfluidic chip. (c) Fiber generation with the chip and (d) PLGA fiber scaffold for cell culture wound around a coverslip (18 × 18 mm^2) during spinning (at 60 rpm).

Adapted with permission from American Chemical Society, Copyright (2008).

The end of a core pipet (30 μm in diameter) was cut and ground smoothly using a microforge. The core solution (10% PLGA in DMSO) and sheath solution (mixture of glycerin and distilled water, 50% (v/v)) were introduced into each respective inlet, and the two fluids merged (around the dotted rectangle of Figure 15.1). The sheath fluid surrounding the tip of the core glass is merged while the core fluid is extruded through the core glass to form a stable coaxial flow because of microfluidic phenomena. At the interface between the core PLGA solution and sheath fluid, the exchange of DMSO with water occurs and the polymer in the liquid phase is solidified.

15.2.1.2 Electrospinning

Electrospinning is widely used for the fabrication of fibers with diameters ranging from micrometers down to a few nanometers in a straightforward and cost-effective manner (Matthews et al., 2002; Yang et al., 2005; Li et al., 2003; Min et al., 2004; Jegal et al., 2011; Pham et al., 2006). As depicted in Figure 15.2, a polymer solution in a typical electrospinning process is extruded through a syringe to a metal nozzle, and a pendent drop is formed at the tip of the syringe nozzle. Then, a high voltage is applied between a tip and a grounded collection target. When the electric field strength overcomes the surface tension of the droplet, a polymer solution jet is initiated and accelerated toward the collection target. As the jet travels through the air, the solvent evaporates and a nonwoven polymeric fabric is formed on the target. Figure 15.2 shows an example of a fibrous scaffold made of poly(ε-caprolactone) (PCL) and gelatin (GEL) using electrospinning (El-Fiqi and Kim, 2014).

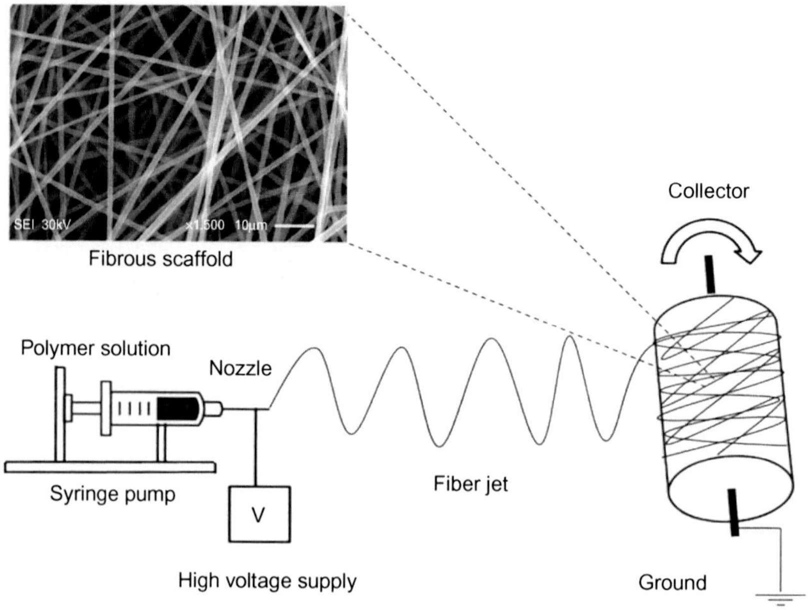

Figure 15.2 Exemplary process of nanofiber fabrication (electrospinning).

In this process, a PCL–GEL polymeric solution was prepared by mixing PCL (10%, w/w) with GEL (10%, w/w) solutions in 2,2,2-trifluroethanol (TFE) in a 1:1 ratio. The PCL–GEL solution was introduced into a plastic syringe with a 23-gage needle tip and then the syringe was connected to an injection pump adjusted at a flow rate of 1 ml h^{-1}. The electrospun fibers were fabricated at 10.5 kV using a high-voltage power supply. The electrospun nanofibers were collected on an aluminum-foil sheet attached to a rotating cylinder located a distance of 14.5 cm from the tip of the syringe needle. The SEM image in Figure 15.2 shows a nonwoven fibrous mat made of PCL–GEL with a smooth fiber surface.

15.2.2 Types of fibers

Based on the chemical composition, fibers can be classified into organic, inorganic, or organic–inorganic composite/hybrid fibers. Organic fibers are mainly based on the natural and synthetic polymers, while inorganic-based fibers contain metallic oxide ceramics, carbon, or glasses. The combinations of organic–inorganic phases yield a broad category of fibers with varying properties.

15.2.2.1 Polymer fibers

A variety of synthetic and natural polymers have been investigated in terms of their electrospinnability (fiber production ability). A wide range of biodegradable and biocompatible synthetic polymers and their copolymers, such as poly(lactic acid) (PLA), poly(glycolic acid) (PGA), PCL, and PLGA, have been successfully electrospun into fibrous scaffolds (Park and Lee, 2011; Boland et al., 2001; Reneker et al., 2002; Kumbar et al., 2008b). Barnes et al. (2007) electrospun PLA in chloroform, methylene chloride, and 1,1,1,3,3,3-hexafluoro-2-propanol (HFP) at concentrations ranging from 1:7 to 1:10 (w/v ratios), resulting in nanofiber creation with diameters ranging from 100 nm to about 10 μm. Uniaxial tensile tests of these PLA electrospun mats revealed moduli of the order of 180 MPa. PLA was also electrospun by Yang et al. (2005) to produce aligned fibers for the study of neurite outgrowth and differentiation of neural stem cells. Boland et al. electrospun PGA in HFP to produce fibers with diameters ranging from ~1.5 μm to 200 nm (Boland et al., 2004). PLGA nanofibers were also electrospun with diameters ranging from 500 to 800 nm for tissue engineering applications (Li et al., 2002). The electrospinning of pure PCL was reported by Reneker et al. (2002) using a PCL solution in acetone at 14–18 wt%. Kumbar et al. (2008b) reported the fabrication of PLGA mats with different fiber diameters in the range of 150–6000 nm via electrospinning for skin tissue engineering. The fabricated mats were found to have a tensile modulus from 39.23 ± 8.15 to 79.21 ± 13.71 MPa which falls in the range for normal human skin.

Natural polymers such as collagen (COL), GEL, and chitosan (CS) have also been electrospun, and generally showed good biocompatibility (Rho et al., 2006; Huang et al., 2004; Homayoni et al., 2009). Rho et al. (2006) reported electrospinning of type I COL in HFP to fabricate a biomimetic nanofibrous ECM for tissue engineering. The average diameter of COL nanofibers electrospun from 8% COL solution in HFP was 460 nm (range 100–1200 nm). Zhang et al. (2005a) produced electrospun webs of

GEL in TFE at various ratios of weights to volume to determine the optimal concentration of GEL for electrospinning. The GEL solution prepared at 10% (w/v) produced uniform GEL fibers with a tensile strength of 2.5 MPa. Huang et al. (2004) electrospun GEL into nanofibers with diameters ranging from 100 to 340 nm using TFE as a solvent and a mass concentration in the range of 5–12.5%. The GEL nanofiber mat that had no beads on the surface obtained from 7.5% mass concentration exhibited the largest tensile modulus and strength, compared to those produced from mass concentrations of 5%, 10%, and 12.5% solutions.

15.2.2.2 Inorganic fibers

Fabrication of micro/nanofibers from inorganic materials by electrospinning was not easy due to the difficulty in preparing inorganic-based solutions with viscosity suitable for electrospinning. Electrospinning of inorganic-based solutions is possible by using sol–gel solutions of the inorganic materials mixed with a polymeric binder, e.g. poly(vinyl pyrrolidone) (PVP), poly(vinyl butyral) (PVB), or poly(vinyl alcohol) in order to adjust the viscosity to a level suitable for electrospinning (Kim et al., 2006b; Xie et al., 2012b; Lu et al., 2009; Xia et al., 2007). The combination of electrospinning and a sol–gel process enabled direct production of inorganic fibers such as silica (SiO_2) (Qiu et al., 2013; Ravichandran et al., 2013), titania (TiO_2) (Ding et al., 2008), HA (Kim and Kim, 2006), and bioactive glass (BG) nanofibers (Kim et al., 2006b, 2011; Asgharnia and Alizadeh, 2013; Gao et al., 2011; Huang et al., 2012; Hong et al., 2010a,b).

For example, BGs were mainly produced in the form of micro-sized powders, granules, or spheres, but not in nanofiber form. The advantages of the nanofiber form, such as ECM-mimicking architecture, high surface area-to-volume ratio, and high porosity, encourage researchers to produce nanofibrous BG. Kim et al. (2006b) first reported fabrication of electrospun BG nanofibers from a silica glass-based solution mixed with PVB. The glass nanofibers produced showed rapid formation of a bone mineral-like apatite phase in contact with simulated body fluid (SBF), which was attributed to the extremely large surface area of the glass nanofibers. The nanofibrous form of BGs supported a population of rat bone marrow-derived stem cells and osteogenic differentiation to a level significantly higher than that on dense sintered BGs or PCL polymer nanofibers, indicating the potential of BG nanofibers in terms of both morphological and compositional benefits.

In spite of their attractive bone bioactivity, electrospun nanofibers of bioactive inorganic materials, including calcium phosphates and BGs, may have limitations in their use as tissue regeneration matrices on account of their brittleness. Also, post heat treatment limits their potential as drug carriers. However, BG nanofibers can be applied as nanofillers for the production of nanocomposite scaffolds with degradable polymers (Kim et al., 2006c, 2008a; Jo et al., 2009; Lee et al., 2008; Noh et al., 2010), e.g. BG nanofibers were homogenized with COL or a poly-L-lactic acid (PLLA) solution to produce uniform scaffolds and membranes, ultimately improving the bone bioactivity of the organic phase and osteogenic differentiation and cellular mineralization. This approach, which aimed to combine the bone bioactivity of the inorganic component with the shape formability of the organic phase, highlights the useful application of electrospun inorganic nanofibers as a bone-bioactive nanocomponent.

15.2.2.3 Composite/hybrid fibers

Combining organic polymers with inorganic materials in the course of electrospinning is considered a fascinating and reasonable way of generating nanofibers with suitable properties targeted for bone regeneration. The inorganic phase may act to improve the biological properties of polymeric nanofibers, such as cell compatibility and bone-forming processes, involving the osteogenic differentiation and calcification of bone matrix (Kim et al., 2006c, 2008a; Jo et al., 2009; Lee et al., 2008; Noh et al., 2010). The introduction of a polymeric phase should provide mechanical flexibility and formability.

Bone ECM is a type of composite constituted mainly of collagenous fibers embedded with HA nanocrystallites, which necessitates the development of fibrous nanocomposites mimicking bone structure (Jegal et al., 2011; Venugopal et al., 1917; Jang et al., 2009; Asran et al., 2010; Frohbergh et al., 2012). Therefore, electrospun scaffolds of biodegradable polymers with nanoparticles of HA have received much interest in bone tissue engineering (Jegal et al., 2011; Venugopal et al., 1917; Jang et al., 2009; Asran et al., 2010; Frohbergh et al., 2012; Zhang et al., 2008; Nilza et al., 2014; Zheng et al., 2013; Thomas et al., 2007). The electrospun COL/HA (Nilza et al., 2014; Thomas et al., 2007; Teng et al., 2008; Zhang et al., 2010) and GEL/HA (Jegal et al., 2011; Kim et al., 2005; Li et al., 2014; Venugopal et al., 2008; Zheng et al., 2009) composite fibers have been explored to repair bone defects with a similar composition and structure to natural bone tissue. A biomimetic nanocomposite nanofibrous scaffold made of HA/CS was prepared by Zhang et al. (2008). HA/CS nanocomposite with an HA mass ratio of 30 wt% was synthesized through the co-precipitation method to attain a homogeneous dispersion of the spindle-shaped HA nanoparticles (~100–30 nm) within the CS matrix. By using a small amount (10 wt%) of ultrahigh-molecular-weight poly(ethylene oxide), continuous HA/CS nanofibers with a diameters of 214 ± 25 nm were electrospun successfully with HA nanoparticles incorporated within. *In vitro* culture with human fetal osteoblast (hFOB) cells for up to 15 days demonstrated that the incorporation of HA nanoparticles into CS nanofibrous scaffolds led to significant bone formation compared to pure electrospun CS scaffolds. Therefore, composite scaffolds containing bioactive inorganic particles are expected to improve interactions with bone cells. Both *in vitro* and *in vivo* studies have demonstrated that organic–inorganic composite scaffolds support attachment, differentiation, and proliferation of osteoblasts or mesenchymal stem cells (MSCs) and facilitate bone healing (Peng et al., 2011; Lu et al., 2013; Liao and Cui, 2004; Jo et al., 2009; Ko et al., 2008).

Organic–inorganic hybrid materials organized on a nanoscale mimic the bone ECM and potentially provide attractive matrices for bone regeneration. Electrospun hybrid nanofibers of polymers and inorganic materials, e.g. silica and BGs, were reported by several groups (Song et al., 2008; Castano et al., 2014; Hong et al., 2010c; Toskas et al., 2011, 2013; Gao et al., 2013; Allo et al., 2013; Chen et al., 2009;

Wakita et al., 2009, 2011; Obata et al., 2010; Mahony et al., 2010; Poologasundarampillai et al., 2010). Silica hybrid materials favored cell attachment as evidenced by culture of MSCs on a silica/GEL hybrid (Mahony et al., 2010). Toskas et al. (2011) fabricated CS-containing poly(ethylene oxide) (PEO)/silica electrospun hybrid nanofibers prepared from a modified silica solution containing tetraethoxysilane and 3-glycidyloxypropyltriethoxysilane (GPTES). The hybrid nanofibers promoted the attachment and proliferation of murine osteoblast-like 7F2 cells. The combination of these hybrid nanofibers made of biocompatible polymers CTS/PEO and silicate facilitated the advantages of both materials and offered tunable properties for creating bioactive scaffolds for bone repair. Gao et al. (2013) prepared hybrid scaffolds of GEL and BG with a fibrous microstructure by a combined sol–gel and electrospinning technique. The fibers consisting of GEL and BG were covalently linked by GPTES to form a homogeneous phase. The GEL/BG hybrid scaffold showed excellent bioactivity through formation of an HA-like layer on the surface of the fibers within 12 h of immersion in SBF. The GEL/BG hybrid scaffold showed improved mechanical properties where a tensile strength of 4.3 ± 1.2 MPa and an elongation to failure of $168 \pm 14\%$ were obtained for the GEL/BG hybrid scaffold compared to values of 0.5 ± 0.2 MPa and $63 \pm 2\%$ for GEL scaffolds. Furthermore, the GEL/BG hybrid scaffolds supported the proliferation of osteoblastic MC3T3-E1 cells, alkaline phosphatase (ALP) activity, and mineralization during *in vitro* culture, showing good biocompatibility. Gupta et al. (2009) combined the electrospinning and electrospraying techniques and found that HA nanoparticles sprayed on the surface of the electrospun scaffold could not only enhance the attachment and proliferation of hFOBs, but also improved the mechanical properties compared to the electrospun scaffold blended with HA nanoparticles.

15.3 Biomimetic mineralization of fibers

Mineralization of biomaterials through biomimetic processes has been widely explored for bone regeneration (Xie et al., 2013; Rodriguez et al., 2011; Cai et al., 2011; Ma et al., 2011; Li et al., 2008; Kim et al., 2006d; Azami et al., 2012; Chen et al., 2005, 2006a, 2010a, 2012; Liu et al., 2009b; Strange and Oyen, 2011; Arafat et al., 2011; Oyane et al., 2003, 2005a,b; Qu et al., 2007; Yokoyama et al., 2007; Ning et al., 2008; Lee et al., 2013). Biomineralization is a process by which mineral crystals are deposited in an organized fashion in the matrix (either cellular or extracellular) of living organisms (Boskey, 1998). Bone biomineralization involves a well-orchestrated process in which crystals of calcium phosphate are deposited on the matrix produced by bone-forming cells (Weiner et al., 1999). The biomimetic synthesis of biomaterials provides control over the morphology, crystal orientation, and distribution of the inorganic phase within the organic matrix during the mineralization process. Recently, there is a rapidly growing trend to combine the bulk properties of polymers with the surface properties of biomimetic mineralized HA in the design of functional scaffolds for bone tissue engineering (Xie et al., 2013; Rodriguez et al., 2011; Cai et al., 2011; Ma et al., 2011; Li et al., 2008; Kim et al., 2006d; Azami et al., 2012;

Chen et al., 2005, 2006a, 2010a, 2012; Liu et al., 2009b; Strange and Oyen, 2011; Arafat et al., 2011; Oyane et al., 2003, 2005a,b; Qu et al., 2007; Yokoyama et al., 2007; Ning et al., 2008; Lee et al., 2013).

Indeed, there are several methods for coating polymeric surfaces with HA such as electrodeposition, electrospraying, and laser ablation; however, biomimetic mineralization, which allows easy deposition of HA under physiological conditions, is more controllable with respect to microstructural features, such as crystal size and shape, and the stoichiometry of bone-like HA. Apatite deposition on a substrate using a biomimetic process was first developed by Kokubo and his group (Kokubo et al., 1990; Abe et al., 1990; Kokubo, 1995; Kim et al., 1999). Kokubo et al. proposed a biomimetic process involving a reaction between a bioactive CaO–SiO_2-based glass and SBF with ion concentrations nearly equal to those of human blood plasma. In this process, an organic substrate is placed in the vicinity of BG immersed in SBF. Apatite deposition is initiated by the heterogeneous nucleation of apatite induced by dissolved silicate and calcium ions released from BG. The dissolved silicate ions provide a heterogeneous nucleation site for apatite on the substrate, and dissolved calcium ions increase the degree of supersaturation of the fluid with respect to the apatite. Once the apatite nuclei are formed on the substrate, they can grow spontaneously by consuming calcium and phosphate/carbonate ions from the surrounding fluid. This process has attracted much attention because a bone-like apatite layer can be coated onto organic substrates. Apatite crystals were grown to form a layer on the surface of a polymer substrate by immersion in 1.5× SBF, which has an ion concentration 1.5 times higher than that of normal SBF (Abe et al., 1990; Kokubo, 1995; Kim et al., 1999).

This finding provides a fundamental idea that apatite deposition on a polymeric substrate can be triggered both by the existence of specifically effective functional groups for heterogeneous nucleation of apatite, and by an increased concentration of calcium ions in the surrounding fluid. Thereby, the bone-like apatite layer can be formed on the surface of substrates in a solution mimicking body fluid when some functional groups are introduced in the substrates. This process is called a biomimetic process. The formed apatite, so-called bone-like apatite, is more similar to natural bone with a low crystallinity, composition, and nanocrystal size, which is also an important issue in its degradation. Moreover, the conditions used in this process are very mild and allow application of a variety of polymeric surfaces. Forming apatite on the surface of polymeric substrates through a biomimetic process was studied by many investigators (Xie et al., 2013; Rodriguez et al., 2011; Cai et al., 2011; Ma et al., 2011; Li et al., 2008; Kim et al., 2006d; Azami et al., 2012; Chen et al., 2005, 2006a, 2010a, 2012; Liu et al., 2009b; Strange and Oyen, 2011; Arafat et al., 2011; Oyane et al., 2003, 2005a,b; Qu et al., 2007; Yokoyama et al., 2007; Ning et al., 2008; Lee et al., 2013). The classical biomimetic process developed by Kokubo usually involves immersion of substrates in SBF or 1.5× SBF for about 1–4 weeks (Abe et al., 1990; Kokubo, 1995; Kim et al., 1999). However, this is a time-consuming process, taking several weeks to form ideal mineralized layers. Taguchi et al. (1998, 1999a,b) developed an alternate soaking method to rapidly form HA on a substrate under near physiological conditions. In one cycle of alternate soaking, a substrate was soaked for a short time in a calcium chloride solution, rinsed with water, soaked in a sodium phosphate dibasic solution, and then rinsed again. The pH of both solutions was maintained at pH 7.4 and the process was typically performed at 37 °C. An alternate soaking method developed by Taguchi et al. offered several advantages such as being simple, efficient, and faster than other methods for HA precipitation.

15.3.1 Mineralization of polymer fibers

Natural bone is a type of composite in which inorganic apatite nanocrystals are deposited on organic COL fibers woven into a three-dimensional structure. The development of scaffolds capable of mimicking the nanofibrous morphology, structure, function, and the biochemical complexities of the natural bone ECM is desirable for effective bone regeneration. HA-mineralized nanofibrous scaffolds have been proposed as promising scaffolds for bone regeneration due to their ability to mimic both nanoscale features and chemical composition of natural bone ECM. Many bone-like electrospun fibrous scaffolds have been developed either by the direct introduction of HA nanoparticles into the polymeric matrices (Peng et al., 2011; Zhao et al., 2011; Lee et al., 2010a; Lu et al., 2013; Frohbergh et al., 2012; Zhang et al., 2008, 2010; Nilza et al., 2014; Zheng et al., 2013; Thomas et al., 2007; Teng et al., 2008; Kim et al., 2005, 2006e; Li et al., 2014; Madurantakam et al., 2009; Ma et al., 2011; Teo et al., 2011; Hongju et al., 2012; Sui et al., 2007; Sonseca et al., 2012; Choi et al., 2008b; Heydarkhan-Hagvall et al., 2008; Schneider et al., 2008; Nirmala et al., 2010) or by the biomimetic mineralization of bone-like HA on the surface of polymeric fibers (Chen et al., 2006b, 2010b; Yang et al., 2008b, 2011; Fu et al., 2013; Chang et al., 2013; Cui et al., 2010a; Buczynska et al., 2011; Zhang et al., 2009b; Patel et al., 2009; Liao et al., 2008; Takeuchi et al., 2003; He et al., 2010, 2014; Abdal-hay et al., 2014; Kolluru et al., 2013; Kim et al., 2014b; Karaman et al., 2013; Xie et al., 2010; Prosecka et al., 2012; Choi and Kim, 2012). The biological characteristics of HA, such as its specific affinity to many adhesive proteins and its direct involvement in the bone cell differentiation and mineralization processes, make HA a potential biomaterial for utilization in bone regeneration. The presence of HA on the surface of polymeric scaffolds offers several benefits: it renders synthetic polymer scaffolds bioactive and osteoconductive; it changes the chemistry as well as topography at the surface and it provides osteophilic surface to implants for bonding with natural bone after implantation.

Biomimetic HA deposition on fibrous scaffolds can be achieved using several different methods. Examples are as follows: (i) direct incorporation of HA particles into polymeric solutions (Peng et al., 2011; Zhao et al., 2011; Lee et al., 2010a; Lu et al., 2013; Frohbergh et al., 2012; Zhang et al., 2008, 2009b, 2010; Nilza et al., 2014; Zheng et al., 2013; Thomas et al., 2007; Teng et al., 2008; Kim et al., 2005, 2006e, 2014b; Li et al., 2014; Madurantakam et al., 2009; Ma et al., 2011; Teo et al., 2011; Hongju et al., 2012; Sui et al., 2007; Sonseca et al., 2012; Choi et al., 2008b; Heydarkhan-Hagvall et al., 2008; Schneider et al., 2008; Nirmala et al., 2010; Chen et al., 2006b, 2010b; Yang et al., 2008b, 2011; Fu et al., 2013; Chang et al., 2013; Cui et al., 2010a; Buczynska et al., 2011; Patel et al., 2009; Liao et al., 2008; Takeuchi et al., 2003; He et al., 2010, 2014; Abdal-hay et al., 2014; Kolluru et al., 2013; Karaman et al., 2013; Xie et al., 2010; Prosecka et al., 2012) in which poor dispersion and agglomeration of HA particles may occur; (ii) the SBF soaking method developed by Kokubo et al. (Abe et al., 1990; Kokubo, 1995; Kim et al., 1999), which uniformly precipitates a bone-like HA layer over the polymer fiber surface but requires several days for HA deposition; and (iii) alternate soaking in calcium and phosphate solutions developed by Taguchi et al. (1998, 1999a,b), which offered several advantages such as being simple, efficient, and faster. All these cases should involve either the active surface chemistry to capture calcium or phosphate ions, and/or the high supersaturation capacity of the medium to induce mineral deposition. Therefore, approaches to enhance mineralization of fiber have mostly focused on either of these: (i) activating the fiber surface either by changing composition or surface modification/functionalization and (ii) increasing supersaturation capacity by increasing ionic concentrations, mainly calcium and/or phosphate, or alternative/combinatory coating methods to improve nucleation.

One exemplar study activated the surface of synthetic polymer fibers followed by a combinatory mineralization method. Yu et al. (2009) produced surface mineralized nanofibers of PCL. The PCL was first electrospun into a nanofibrous mat, and the surface was further mineralized with HA as summarized in Table 15.1 using a series of surface-modification steps involving the activation of nanofibers in an alkaline solution (2 N NaOH) to generate carboxylic groups, followed by alternate dipping in Ca- and P-rich solutions (150 mM of Ca^{2+} and HPO_4^{2-}) to allow mineral nucleation, and then further soaking in a Ca–P pseudo-saturated solution (SBF).

The surface of the mineralized PCL nanofiber was observed to be almost fully covered with nanocrystalline apatite (Figure 15.3). Through mineralization, the wettability of the nanofiber matrix was greatly improved. Murine-derived osteoblastic cells were shown to attach and grow actively on the apatite-mineralized nanofibrous substrate. The mineralized substrate significantly stimulated the expression of bone-associated

Table 15.1 Example of mineralization process of synthetic polymer fibers

Fiber mineralization process		
Surface activation	**Mineral induction**	**SBF soaking**
Dipping into alkaline solution: 1. Fiber dipping into 2 M NaOH solution with shaking at 80 rpm for 6 h 2. Fiber removal and washing with distilled water	Alternate dipping into Ca- and P-solutions: 1. Dipping into 150 mM $CaCl_2$ solution for 30 s 2. Washing with distilled water for 5 s 3. Dipping into 150 mM Na_2HPO_4 solution for 30 s 4. Washing with distilled water for 5 s 5. Repeating the alternate dipping cycle six times and finally drying	Incubation of the treated fiber for 7 days into SBF solution containing: 142 mM NaCl 4.2 mM $NaHCO_3$ 5.0 mM KCl 1.0 mM $K_2HPO_4{\cdot}3H_2O$ 1.5 mM $MgCl_2{\cdot}6H_2O$ 2.5 mM $CaCl_2$ 0.5 mM Na_2SO_4

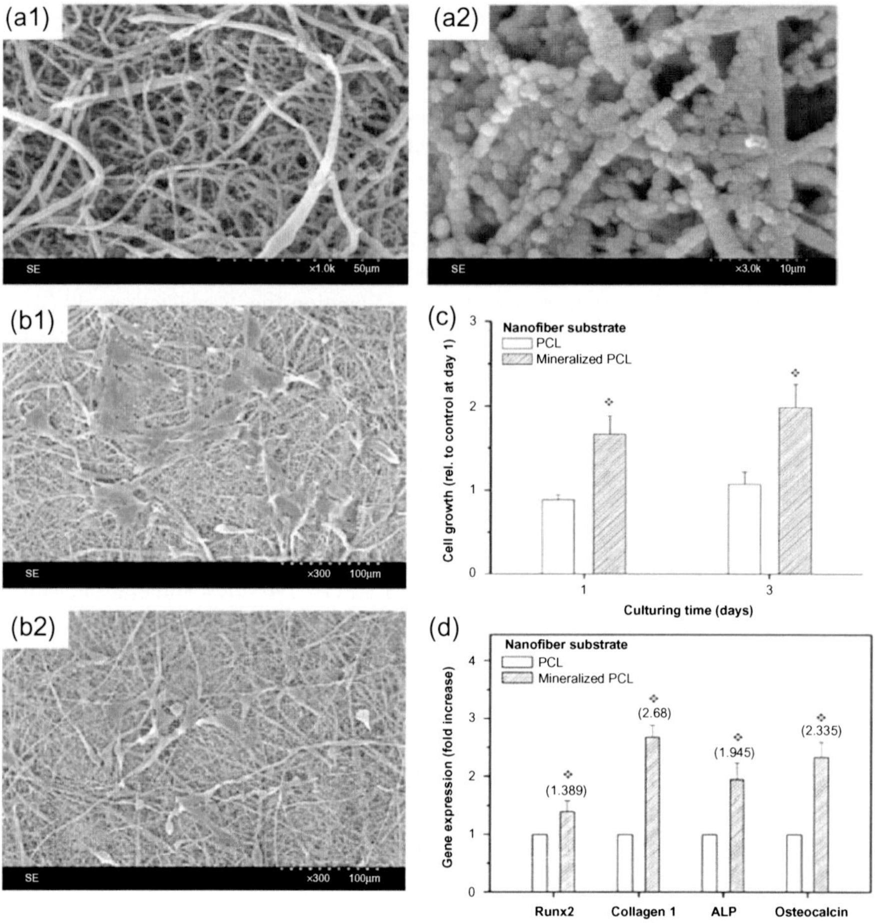

Figure 15.3 Exemplary study of polymer fiber mineralization. (a1 and a2) Morphology of the surface-mineralized PCL nanofibrous matrix obtained by a series of solution-mediated mineralization steps involving activation in an alkaline solution, alternate soaking in Ca- and P-solutions, and immersion in SBF. (b1 and b2) Initial osteoblastic cell response showing better spreading behavior on the mineralized PCL nanofiber (b1) than on pure PCL nanofiber (b2) at 8 h of culturing. (c) Cell viability and (d) expression of bone-associated genes was significantly enhanced by surface mineralization. Adapted with permission from Yu et al. (2009), Wiley Interscience Co.

genes, including Runx2, COL type I, ALP, and osteocalcin (OCN), when compared with the substrate without mineralization.

For the case of PCL, in order to make the surface active, a plasma surface treatment was also used instead of alkaline solution. Yang et al. (2008a) reported mineralization of PCL fibrous scaffolds after plasma treatment. After activation through plasma treatment, the PCL fibers were immersed in highly concentrated SBF (10× SBF) for varying time periods. After 2 h mineralization started, and after 6 h minerals covered the surface completely. However, the coating phase was a mixture of apatite and dicalcium phosphate dihydrate (DCPD). A prolonged soaking changed the phase carbonate apatite, which was similar to biological apatite. For PLLA, which is more hydrophilic and degradable than PCL, water hydrolysis is enough to activate the surface. Yuan et al. (2001) reported mineralization of PLLA fibers in 1.5× SBF after hydrolysis in water. After 15 days in 1.5× SBF, a thick apatite layer was formed on the surface of the fibers. This apatite had a Ca/P ratio similar to that of natural bone.

Cui et al. (2010b) examined more systematically the effects of surface functional groups pretreated on the fiber surface on the mineralization behaviors. They functionalized PDLA fibers with carboxyl, hydroxyl and amino groups and their combinations. Electrospun fibers with higher densities of carboxyl groups, combination of hydroxyl and carboxyl groups in a ratio of 3:7, and combination of amino, hydroxyl, and carboxyl groups in a ratio of 2:3:5 were favorable for HA nucleation and growth in SBF, resulting in higher content and lower crystal size of formed HA mineral.

A polydopamine has also recently been used to activate fiber surfaces for mineralization. Xie et al. (2013) examined mineralization on the surface of polydopamine-coated PCL fibers. The mineralization was controlled by adjusting the concentration of HCO_3 and dopamine in the mineralized solution, the reaction time, and the surface composition of the fibers. The deposited HA coatings through this mineralization gave PCL electrospun fibers much higher stiffness, ultimate tensile strength and toughness, which could be closer to the mechanical properties of natural bone. Liu et al. (2011) used CS and heparin prior to mineralization. The PLGA fibrous mats were immobilized with CS and heparin, and then immersed in 10× SBF. The mineral coating had a significant impact on the mechanical properties of the fibers, such as elastic modulus.

A combination of the above approaches was also used. Li et al. (2009) used plasma surface treatment and/or GEL coating on electrospun PLGA and PCL nanofibers coupled with mineralization in 10× SBF. A linear gradient in calcium phosphate was achieved across the surface of nanofiber mats. The gradient had functional consequences with regard to stiffness and biological activity. Specifically, the gradient of mineral content resulted in a gradient in the stiffness of the scaffold and further influenced the activity of preosteoblastic cells. These nanofiber-based scaffolds were proposed for repairing the tendon-to-bone insertion site via a tissue engineering approach.

The use of natural polymers mixed with synthetic polymers to improve the mineralization ability has also been reported (Meng et al., 2013; Ngiam et al., 2009; Lin et al., 2014). One example of mineralized PLGA and PLGA/GEL electrospun nanofibers was reported by Meng et al. (2013) using a concentrated SBF method, supersaturated calcification solution method, and alternate soaking method. A large amount of CaP composed of DCPD, HA, and OCP was deposited on the surface of nanofibers in short times. The mineralized PLGA/GEL nanofibers had higher CaP deposition than PLGA nanofibers, where GEL acted as nucleation site for the formation of CaP. Ngiam et al. (2009) also prepared PLGA and PLGA/COL (50% PLGA +50% COL) nanofibers using electrospinning. A bone-like HA was deposited better on the PLGA/COL nanofibers than on the PLGA nanofibers. The PLA/GEL composite fibers were also investigated by Cai et al. (2011) in 5× SBF under continuously bubbled CO_2 to keep the pH value stable at 6.4. Similarly, a mineralization-accelerating effect of GEL was noticed.

Lin et al. (2014) coated CS at different concentrations on the PLA nanofiber, and then mineralized in 10× SBF. The CS, as a cationic polysaccharide, is used to promote more nucleation sites and growth of CaP. All CS-coated PLA fiber mats displayed better mineralization behaviors than pure PLA. In particular, the incorporation of 4% CS resulted in the most significant improvement in CaP crystal formation. The composition of the deposited minerals was a mixture of dicalcium phosphate dehydrate and apatite. Li et al. (2008) coated GEL onto PCL fibers via a layer-by-layer approach, and then mineralized in 10× SBF. After 2 h, the deposited mineral phase was found to be a mixture of dicalcium phosphate dehydrate (a precursor to apatite) and apatite.

15.3.2 Mineralization of inorganic fibers

Bioactive materials, such as CaP and BG/BG-ceramics, are a potential class of materials for use in the regeneration of hard tissues. The processing of such materials into nanoscale fibers with high surface area greatly influences their physicochemical properties and especially their biomineralization behavior. The mineralization of BG nanofibers in SBF is simple, straightforward, and easier than the mineralization of polymeric fibers due to their intrinsic bioactive compositions and their capability of releasing ions, e.g. Si, Ca, and P ions. Mineralization of polymeric fibers requires initial activation of the fiber surface prior to mineralization in SBF solution. The activation of the fiber surface, e.g. by alkali hydrolysis, introduces functional groups, e.g. COOH, which can facilitate the deposition and nucleation of HA when fibers are immersed in SBF solution.

Kim et al. (2006b) first fabricated BG nanofibers based on SiO_2–CaO–P_2O_5 three-component composition using a sol–gel method coupled with an electrospinning process. The group examined the *in vitro* biomineralization of the prepared BG nanofibers in SBF and the results showed that BG nanofibers possessed excellent bioactivity compared to the traditional BG. The prepared nanofibers showed excellent *in vitro* bone-like HA deposition when immersed in SBF. Figure 15.4a and b shows the degree of HA mineral formed on BG nanofibers' surface when immersed in SBF for 1 and 3 days as observed by transmission electron microscopy (TEM).

Mesoporous BG (MBG) fibers using the combination of a sol–gel process and a high-velocity spray technique were then prepared by Yi et al. (2008). The *in vitro* biomineralization of MBG fibers showed the growth of nanoparticles on the fiber surface after soaking in SBF for 8 h; however, when the soaking time was extended to 24 h, nanoparticles with rod-like morphology similar to that of HA were observed covering the entire surface of MBG fibers. After immersing in SBF for 72 h, the HA layer deposited on the surface of fibers was thickened while the fibrous 3D macrostructure of the scaffold was still maintained. The MBG fibers induced much faster precipitation of HA than conventional BG fibers, indicating that the MBG fibers had excellent *in vitro* biomineralization ability.

Xia et al. (2007) used PVP and Pluronic P123 (EO20–PO70–EO20) as templates to produce BG nanofibers. The *in vitro* biomineralization of such BG nanofibers showed formation of CaP nanoparticles on the surface of the nanofibers after immersion in SBF for 6 h. Then, nanoparticles gradually grew into apatite flowers after 12 h in SBF.

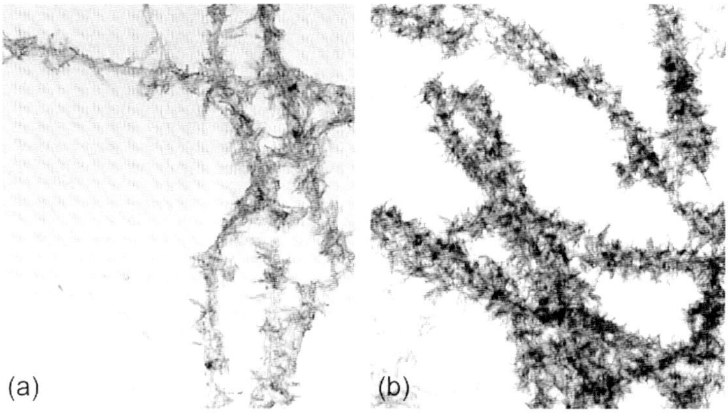

Figure 15.4 Mineralization of the glass nanofiber (d_{ave} = 84 nm) after incubation in a simulated body fluid: TEM images for: (a) 1 and (b) 3 days.
Reprinted from authors' own publication.

The nanofibers were entirely wrapped by thick apatite layers after 24 h, indicating higher *in vitro* biomineralization ability than conventional BG fibers. Similarly, Hong et al. (2010a) produced ultrathin MBG hollow fibers fabricated by combining an electrospinning technique with a sol–gel process using a phase-separation-inducing agent, PEO. Immersion of the prepared fibers in SBF rapidly resulted in the development of a layer of enamel-like apatite mesocrystals at the fiber surfaces and apatite nanocrystals inside the hollow cores. The effects of different Ca/Si molar ratios were investigated by Seol et al. (2009). The biomineralization ability was found to increase with increasing Ca to Si molar ratio and low-crystalline apatite crystals were formed on the surface after immersion in SBF for 7 days.

The coating with biopolymers of the BG nanofibers was also effective in mineralization. Kim et al. (2011) coated BG nanofibrous membrane with degradable biopolymers, such as PCL and PLA. A thin layer of the biopolymers was covered on the glass nanofiber through a capillary-driven infiltration of the polymer solution into the small-sized nanofibrous mesh. The biopolymer-coated glass nanofibers showed significantly enhanced mechanical properties and rapid induction of apatite mineralization in SBF.

An interesting approach to mineralized nanofibers was found in biomolecular conjugation, as performed by Kim et al. (2008b). This group used *in situ* HA-mineralized BG nanofibers to introduce fibronectin biomolecules on to the nanofiber surface. Cell-adhesive fibronectin was effectively coupled with apatite mineral on the surface of BG nanofibers, which demonstrated significant enhancement in initial osteoblast adhesion and spreading.

15.3.3 Mineralization of composite/hybrid fibers

Using the principles of biomimetic mineralization, composite/hybrid nanofibrous scaffolds can also be mineralized by the growth of CaP crystals and formation of bone-like apatite coating on nanofiber surfaces to mimic structure, physical architecture, and chemical composition of natural bone ECM. Surface biomimetic mineralization of composite/hybrid scaffolds is currently one of the growing strategies in research on bone tissue engineering.

Inorganic nanoparticles of HA have been incorporated within a variety of polymeric matrices to improve polymer mineralization ability (Jaiswal et al., 2012, 2013a,b). Jaiswal et al. (2012) prepared PLLA/HA and PLLA/GEL/HA electrospun nanofibrous scaffolds by electrospinning, followed by mineralization via alternate soaking in CaP solutions. Each alternate soaking cycle leads to a continuous growth of HA on the fibers. Leaching study showed that HA in PLLA/GEL/HA acted as a binder of GEL, preventing the GEL leaching from scaffolds and thus eliminating the use of toxic crosslinking agents. The same group (Jaiswal et al., 2013a) compared the effects of surface- or internal-existing HA in the PCL–GEL fiber system. HA nanopowder was either incorporated within the fiber or HA mineral phase was induced on the surface by an alternate soaking method. The alternate soaking proved to be rapid and efficient for surface precipitation of HA over the scaffold while blending of HA powders with PCL–GEL solution failed to display HA particles on the surface of the scaffold. The microstructure of fibers showed HA particles on the surface of the PCL–GEL scaffold and embedded HA particles in the interior of the PCL–GE fibers. Alternately soaked nanofibers showed better mechanical strength than PCL–GEL, but the blended fibers exhibited higher Young's moduli and tensile strength as compared to mineralized fibers. In the alternate soaking process, the addition of GEL was shown to improve mineralized composition similar to native bone matrix (Jaiswal et al., 2013b); 66.8% of HA was achieved, which is similar to that of native human bone (60%).

As well as mixing inorganic nanoparticles with polymeric compositions, solution-based organic–inorganic hybrid compositions were also investigated. GEL/siloxane hybridized nanofibers with various compositions (GEL/siloxane 0.5, 1, and 2 by weight ratio) were also prepared by Song et al. (2008). Hybrids were synthesized by grafting GEL to GPTMS via hydrolysis and polycondensation reactions. This can yield a siloxane network where GPTMS molecules bridge the GEL chains. An aqueous solution of GEL was mixed with polysilane GPTMS at various ratios (siloxane/GEL = 0.5, 1, and 2) containing a small concentration (2.5 wt%) of $CaCl_2$, which was hydrolyzed, condensed, and then electrospun into nanofibers. The involvement of siloxane groups within the GEL significantly improved the chemical stability of GEL by forming linkages with amide groups of GEL to produce a hybridized network. The hybridized nanofibers significantly enhanced osteoblastic cell differentiation, as shown in Figure 15.5.

On the hybridized nanofiber matrices, osteoblastic cells were observed to attach, spread, and populate actively. Incorporation of Ca within the hybridized nanofibers improved the bone cell activity, since Ca ions were shown to enhance the *in vitro* apatite-forming ability within SBF when added to the GEL–siloxane gel matrices. The viability of the cells assessed at day 3 was better on the hybridized nanofibers than on the pure GEL equivalent and was even better on the Ca-incorporated one. The ALP activity at day 10 was significantly higher on the Ca-incorporated nanofibers than on the pure GEL nanofibers.

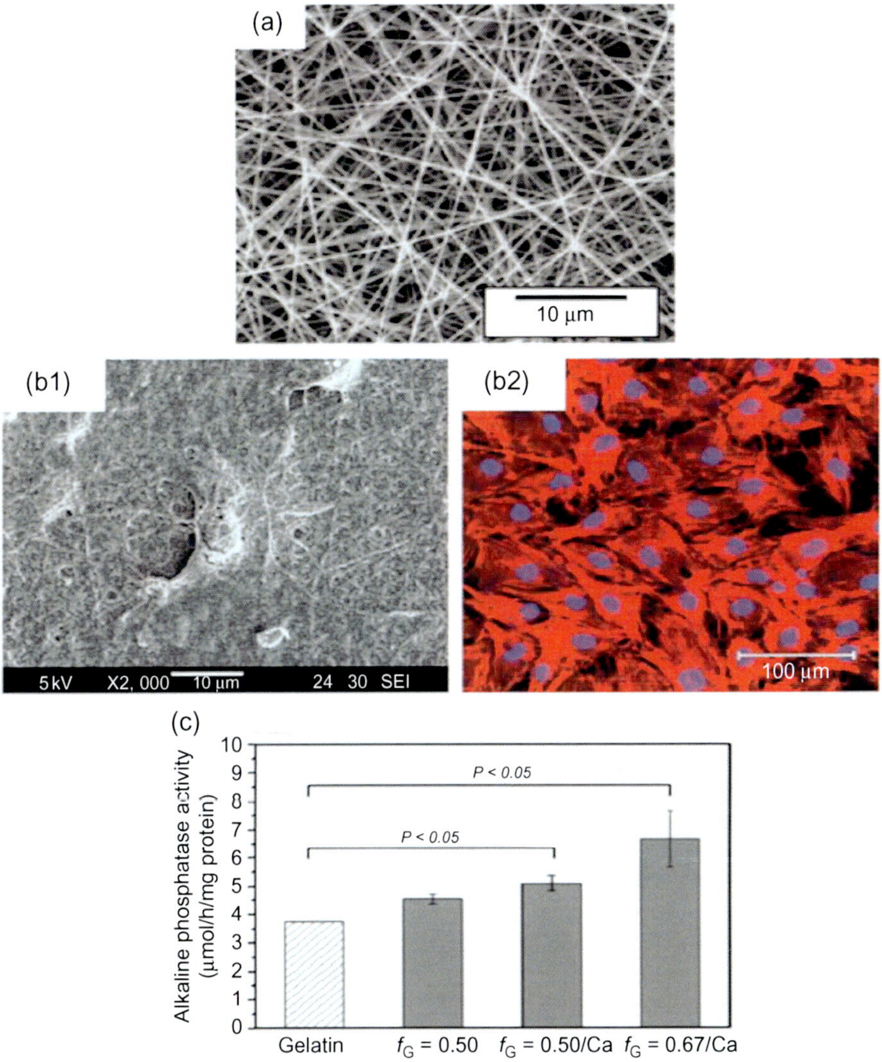

Figure 15.5 Mineralization of hybrid nanofibers. (a) Morphology of gelatin–siloxane hybridized nanofibers (gelatin:polysilane = 1:1, $CaCl_2$ addition = 2.5 wt%). (b1 and b2) MC3T3-E1 growth on the hybridized nanofibers after 3 days (scanning electron microscopy (b1) and confocal laser scanning microscopy (b2)). (c) Alkaline phosphatase activity of osteoblastic cells on the nanofibers showing a significantly higher level on the hybridized ones (ANOVA). Reprinted with permission from Song et al. (2008), Wiley Interscience Co.

15.4 Biological effects of mineralized fibers

Surface modification of scaffolds is a crucial factor in regulating cell functions for the repair of hard tissues, such as bones and teeth. Bone-like apatite coating of polymeric substrates by means of biomimetic mineralization is a possible way to provide favorable environments for osteoblasts/osteoprogenitor cells and to enhance the bone-bonding ability. Surface-mineralized fibrous scaffolds have largely shown better biological responses (Li et al., 2008, 2009; Ma et al., 2011; Chang et al., 2013; Cui et al., 2010a; Buczynska et al., 2011; Yang et al., 2011; He et al., 2014; Yu et al., 2009; Luickx et al., 2014; Ngiam et al., 2009; Lin et al., 2014; Jaiswal et al., 2012, 2013a,b; Araujo et al., 2008; Gandhimathi et al., 2013; Ravichandran et al., 2012), as detailed in the following.

15.4.1 In vitro *cell responses*

The initial biological responses to the surface of synthetic polymers, such as protein adsorption and cell anchorage, are significantly limited because they are inherently hydrophobic and have no cell-adhesive domains. Considerable efforts have been made to improve the poor surface properties and weak cell-adhesive potential of synthetic polymers. For example, hydrophilic polymers such as COL and GEL were introduced within the synthetic polymers to produce polymer composite nanofibers with improved biological responses (Chong et al., 2007; He et al., 2005a,b; Kim and Park, 2006; Duan et al., 2007; Ma et al., 2005b; Zhang et al., 2005a,b; Choi et al., 2008b; Hong and Kim, 2013; Fu et al., 2014). Moreover, the incorporation of bioactive inorganic phases such as HA into the polymeric nanofibers has also enhanced osteoblastic activity (Rajzer et al., 2014; Thomas et al., 2006; Wang et al., 2007; Peng et al., 2011; Hongju et al., 2012; Kim et al., 2006e; Sui et al., 2007; Sonseca et al., 2012).

Because anchorage-dependent cells initially recognize the outermost surface of nanofiber, it is believed that the bioactive inorganic composition, particularly HA mineral phase, can be properly tailored upon the surface of the polymer nanofibers (Chen et al., 2006b, 2010b; Yang et al., 2008b, 2011; Fu et al., 2013; Chang et al., 2013; Cui et al., 2010a; Buczynska et al., 2011; Zhang et al., 2009b; Patel et al., 2009; Liao et al., 2008; Takeuchi et al., 2003; He et al., 2010, 2014; Abdal-hay et al., 2014; Kolluru et al., 2013; Kim et al., 2014b; Karaman et al., 2013; Xie et al., 2010; Prosecka et al., 2012). Surface-mineralized polymer fibers are considered to be an appropriate matrix for cultivating stem cells and stimulating the biological function of bone-associated cells, such as adhesive protein adsorption, cell anchorage, osteoblastic differentiation, and bone matrix synthesis (Li et al., 2008, 2009; Ma et al., 2011; Chang et al., 2013; Cui et al., 2010a; Buczynska et al., 2011; Yang et al., 2011; He et al., 2014; Yu et al., 2009; Luickx et al., 2014; Ngiam et al., 2009; Lin et al., 2014; Jaiswal et al., 2012, 2013a,b). Mineralized scaffolds promote osteogenic cellular activities and mineral deposition (Li et al., 2008, 2009; Ma et al., 2011; Chang et al., 2013; Cui et al., 2010a; Buczynska et al., 2011; Yang et al., 2011; He et al., 2014; Yu et al., 2009; Luickx et al., 2014; Ngiam et al., 2009; Lin et al., 2014; Jaiswal et al., 2012, 2013a,b). Mineralized scaffolds have shown better attachment and proliferation of osteoblasts, which facilitates differentiation of MSCs towards osteoblastic lineage (Li et al., 2008, 2009;

Ma et al., 2011; Chang et al., 2013; Cui et al., 2010a; Buczynska et al., 2011; Yang et al., 2011; He et al., 2014; Yu et al., 2009; Luickx et al., 2014; Ngiam et al., 2009; Lin et al., 2014; Jaiswal et al., 2012, 2013a,b). The proliferation rate of the cells on mineralized scaffolds has been shown to be significantly higher (by 1.9-fold) as compared to that on pristine fibrous scaffolds (Li et al., 2008). Improved attachment, proliferation, and differentiation of osteoblasts have been observed on HA-coated carboxymethyl CS scaffold as compared to noncoated scaffold (Budiraharjo et al., 2012). Also, HA mineralization on electrospun poly(DL-lactide) fibers acted as an excellent cell support to maintain desirable cell–substrate interactions, to provide favorable conditions for cell proliferation and to stimulate osteogenic differentiation (Cui et al., 2010b).

As to the responses of osteoblasts/osteoblastic cell lines, several studies have reported improvements in cell adhesion, phenotype expressions, matrix syntheses, and mineralization. Pre-osteoblastic cell-line MC3T3-E1 viability, adhesion, and proliferation on mineralized electrospun PCL fibrous scaffolds confirmed improved cell affinity when compared to the unmineralized PCL samples (Luickx et al., 2014). Furthermore, the ALP activity of human osteosarcoma cells (MG-63) was found to be the highest on PLLA/GEL/HA compared to PLLA and PLLA/HA scaffolds. Apatite-coated PCL nanofibers produced favorable physical and biological environments, by increasing ECM deposition of osteoblastic cells such as high levels of OCN and ALP expressions (Mavis et al., 2009). A synergistic effect of GEL and biomimetic deposited HA layer on PLLA/GEL fibrous scaffolds has also been reported (Jaiswal et al., 2013b). When the PCL–GEL fibrous scaffolds were mineralized, MG-63 cell adhesion, proliferation, and mineralization were significantly improved (Jaiswal et al., 2013a). Similar behaviors on MC3T3-E1 cells were also noticed on the mineralized electrospun PCL fibers (Li et al., 2008).

In the case of stem cells, similar behaviors to osteoblastic cells have also been documented. Yu et al. (2009) examined the effects of apatite mineralization of PCL nanofiber surfaces on the growth behavior and osteogenic differentiation of rat bone marrow stromal cells. BMSCs isolated from adult rats were seeded on mineralized PCL nanofibers and on nonmineralized fibers as a control, and then cultured for up to 28 days. The cells showed better adherence to the mineralized PCL nanofibers than to the control, as shown in Figure 15.6a–d, indicating favorable cell affinity to the mineralized nanofibrous surface. The secretion of COL by the cells was shown to increase with culturing time on both types of nanofiber. The cell viability was similar at day 7 but was higher on the control at a prolonged period of day 14. However, ALP activity was noticed to a significantly higher level on the mineralized PCL nanofibers than on the control at days 14 and 28, as shown in Figure 15.6e. Overall, the surface-mineralized PCL nanofibrous substrate was shown to support the adhesion and growth of BMSCs and to stimulate the differentiation into an osteogenic lineage.

15.4.2 *In* vivo *tissue reactions*

While most studies on the effects of surface mineralization of fibrous scaffolds have been conducted *in vitro*, very few works have explored the *in vivo* performance. *In vivo* bone formation of nonwoven electrospun fibers of PLLA coated and/or blended with

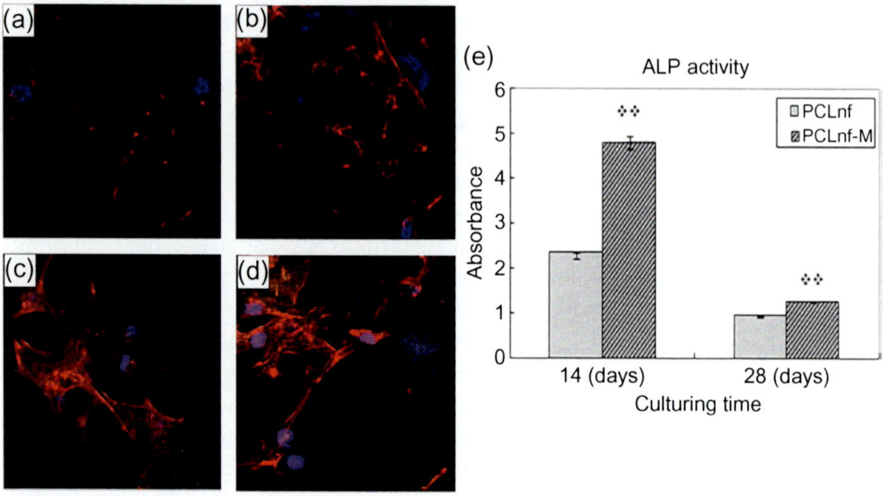

Figure 15.6 Cell responses stimulated by nanofiber mineralization. Confocal laser scanning microscopy (CLSM) images of the stained cells on (a and b) PCLnf and (c and d) PCLnf-M substrates after culturing for (a and c) 3 and (b and d) 7 days. The nuclei and cytoskeletal extensions are stained blue and red, respectively. ALP activity of the cells on the nanofibers without (PCLnf) and with the mineralization treatment (PCLnf-M) are shown. A significant difference between the samples was observed at 14 and 28 days ($^{**}p<0.01$). ALP activity of the cells on the nanofibers without (PCLnf) and with the mineralization treatment (PCLnf-M) is shown in (e).

HA nanocrystals was evaluated by Yanagida et al. (2011). The *in vivo* study was performed by covering rat calvarium defects with HA-coated or HA-mixed PLLA fabrics. Typical histological sections after 1-month implantation of samples in rat calvaria defects showed that the nontreated PLLA sample surface was covered with thin connective tissue, but there was no new bone observed inside the calvarium defect. Proliferation of connective tissues was clearly observed inside the defects covered with HA-coated or HA-mixed PLLA. The proliferated connective tissues contained bone tissues, including osteocytes. The bone tissue in the connective tissue had a lamellar structure resembling the existing bone tissue, but newly formed small bone trabeculae were also observed in the proliferated connective tissue adjacent to the edge of the defect covered with HA-coated or HA-mixed PLLA. In this way, the HA coating or HA mixing promoted the formation of new bone inside the defects covered with PLLA fabrics; however, the HA nanocrystal surface coating was more effective in improving *in vivo* bone compatibility as compared to simple HA mixing, and this was attributed to the uniform and effective covering of the PLLA surface with HA. Therefore, the presence of the HA nanocrystal coating on the bone defect material accelerated its remodeling at the subacute stage.

In vivo subcutaneous implantation of mineralized PLLA nanofibrous scaffolds in mice (Seyedjafari et al., 2010) showed that the HA coating on the surface of PLLA nanofibers induced bone formation under the skin (ectopic bone formation) of mice in the absence of exogenous inductive agents or cells. Marked ossification, calcification, and formation of trabeculi were observed in the HA/PLLA scaffold when compared to

PLLA (Seyedjafari et al., 2010). Microscopic examinations of implanted PLLA scaffolds showed no evidence of calcium deposition while HA/PLLA implants showed deposition of calcium and bone formation. Moreover, trabeculi and bone marrow elements were identified in ossification areas.

A recent elegant study by Lee et al. (2014) proposed a strategy in which biomimetic mineralization was used to engineer a biointerface with cell-attractive and osteogenic potential. They modified the surface of a PCL–GEL electrospun fibrous scaffold with HA mineral through biomimetic mineralization. The mineralization of the biopolymer surface facilitated tethering of a fusion protein of fibronectin 9-10 domain (FNIII9-10) and OCN (FNIII9-10–OCN) through a molecular recognition of OCN to the HA crystal lattice. The FNIII9-10–OCN anchorage to the HA-mineralized fiber was observed to be highly specific and tightly bound to preserve stability over a long period. *In vitro* investigations showed that the initial cell adhesion levels as well as the spreading shape and process of MSCs within 24 h were strikingly different between the fibers linked with and without fusion protein. *In vivo* study in a rat calvarium showed new bone formation on HA-coated PCL–GEL fibrous scaffold. A higher quantity of new bone formation in the fiber linked with fusion protein was noticed, and furthermore a significant increase was noticed when the MSCs were tissue engineered with the fusion protein-linked fiber.

The *in vivo* osteogenic effects of a biomimetic CaP-coated tubular electrospun nanofiber mesh loaded with platelet-rich plasma (PRP)/bone morphogenetic protein-7 (BMP-7) was reported by Berner et al. (2012) in critical sized segmental bone defects. Bilateral full-thickness diaphyseal segmental defects were created in rats and nanofiber mesh tubes were placed around the defect. Defects received either treatment with a CaP-coated nanofiber mesh tube, an uncoated nanofiber mesh tube, a CaP-coated nanofiber mesh tube with PRP, or a CaP-coated nanofiber mesh tube in combination with BMP-7 and PRP. Significantly higher bone volume, rate of bone regeneration, and biomechanical properties were obtained for the BMP group compared to the control groups.

An *in vivo* implantation of HA-, BG-, and tricalcium phosphate (TCP)-coated electrospun PLLA nanofibrous mats was also performed in rat critical-size defects (Dinarvand et al., 2011). A higher level of bone formation was observed in the animals receiving HA-, BG-, and TCP-coated scaffolds compared to an untreated control group. Furthermore, HA–BG-coated nanofibers induced the highest bone formation among all groups with sufficient osseointegration, suggesting the *in vivo* role of surface HA mineral phase in bone formation.

15.5 Conclusions

Fibrous scaffolds have been an attractive 3D matrix for cell culture and tissue engineering. A range of compositions have been generated into micro/nanofibers through the methods of electrospinning or microfluidics. For the potential use of fibrous scaffolds, surface mineralization is a rational approach to provide at the biointerface environments that are favorable for cellular anchorage, migration, and subsequent differentiation into hard-tissue lineage. Many *in vitro* studies have demonstrated the effects of the surface-mineralized fiber scaffolds on cell adhesion and osteogenic stimulation, and some *in vivo* findings have also supported the role of surface mineral phase in new bone induction. More recently, some biomolecules have also been combined with the surface mineral phase to improve the therapeutic functions of the mineralized fiber scaffolds.

References

Abdal-hay, A., Oh, Y.S., Yousef, A., Pant, H.R., Vanegas, P., Lim, J.K., 2014. In vitro deposition of CaP nanoparticles on air jet spinning Nylon 6 nanofibers scaffold for bone tissue engineering. Appl. Surf. Sci. 307, 69–76.

Abe, Y., Kokubo, T., Yamamuro, T., 1990. Apatite coating on ceramics, metals and polymers utilizing a biological process. J. Mater. Sci. Mater. Med. 1, 233–238.

Aghdam, R.M., Shakhesi, S., Najarian, S., Mohammadi, M.M., Ahmadi Tafti, S.H., Mirzadeh, H., 2013. Fabrication of a nanofibrous scaffold for the in vitro culture of cardiac progenitor cells for myocardial regeneration. Int. J. Polym. Mater. Polym. Biomater. 63, 229–239.

Allo, B.A., Lin, S., Mequanint, K., Rizkalla, A.S., 2013. Role of bioactive 3D hybrid fibrous scaffolds on mechanical behavior and spatiotemporal osteoblast gene expression. ACS Appl. Mater. Interfaces 5, 7574–7583.

Arafat, M.T., Lam, C.X.F., Ekaputra, A.K., Wong, S.Y., Li, X., Gibson, I., 2011. Biomimetic composite coating on rapid prototyped scaffolds for bone tissue engineering. Acta Biomater. 7, 809–820.

Araujo, J.V., Martins, A., Leonor, I.B., Pinho, E.D., Reis, R.L., Neves, N.M., 2008. Surface controlled biomimetic coating of polycaprolactone nanofiber meshes to be used as bone extracellular matrix analogues. J. Biomater. Sci. Polym. Ed. 19, 1261–1278.

Asgharnia, S., Alizadeh, P., 2013. Synthesis and characterization of SiO_2–CaO–P_2O_5–MgO based bioactive glass and glass-ceramic nanofibres by electrospinning. Mater. Lett. 101, 107–110.

Asran, A.S., Henning, S., Michler, G.H., 2010. Polyvinyl alcohol–collagen–hydroxyapatite biocomposite nanofibrous scaffold: mimicking the key features of natural bone at the nanoscale level. Polymer 51, 868–876.

Azami, M., Moosavifar, M.J., Baheiraei, N., Moztarzadeh, F., Ai, J., 2012. Preparation of a biomimetic nanocomposite scaffold for bone tissue engineering via mineralization of gelatin hydrogel and study of mineral transformation in simulated body fluid. J. Biomed. Mater. Res. A 100, 1347–1355.

Barnes, C.P., Sell, S.A., Boland, E.D., Simpson, D.G., Bowlin, G.L., 2007. Nanofiber technology: designing the next generation of tissue engineering scaffolds. Adv. Drug Deliv. Rev. 59, 1413–1433.

Berner, A., Boerckel, J.D., Saifzadeh, S., Steck, R., Ren, J., Vaquette, C., Zhang, J.Q., Nerlich, M., Guldberg, R.E., Hutmacher, D.W., Woodruff, M.A., 2012. Biomimetic tubular nanofiber mesh and platelet rich plasma-mediated delivery of BMP-7 for large bone defect regeneration. Cell Tissue Res. 347, 603–612.

Boland, E.D., Wnek, G.E., Simpson, D.G., Pawlowski, K.J., Bowlin, G.L., 2001. Tailoring tissue engineering scaffolds using electrostatic processing techniques: a study of poly (glycolic acid) electrospinning. J. Macromol. Sci. A 38, 1231–1243.

Boland, E.D., Telemeco, T.A., Simpson, D.G., Wnek, G.E., Bowlin, G.L., 2004. Utilizing acid pretreatment and electrospinning to improve biocompatibility of poly(glycolic acid) for tissue engineering. J. Biomed. Mater. Res. B Appl. Biomater. 71, 144–152.

Bolgen, N., Menceloglu, Y.Z., Acatay, K., Vargel, I., Piskin, E., 2005. In vitro and in vivo degradation of non-woven materials made of poly(ε-caprolactone) nanofibers prepared by electrospinning under different conditions. J. Biomater. Sci. Polym. Ed. 16, 1537–1555.

Bolgen, N., Vargel, I., Korkusuz, P., Menceloglu, Y.Z., Piskin, E., 2007. In vivo performance of antibiotic embedded electrospun PCL membranes for prevention of abdominal adhesions. J. Biomed. Mater. Res. B Appl. Biomater. 81, 530–543.

Boskey, A.L., 1998. Biomineralization: conflicts, challenges, and opportunities. J. Cell. Biochem. Suppl. 30–31, 83–91.

Breslauer, D.N., Muller, S.J., Lee, L.P., 2010. Generation of monodisperse silk microspheres prepared with microfluidics." Biomacromolecules. Biomacromolecules 11, 643–647.

Buczynska, J., Pamula, E., Blazewicz, S., 2011. Mechanical properties of (poly(l-lactide-co-glycolide))-based fibers coated with hydroxyapatite layer. J. Appl. Polym. Sci. 121, 3702–3709.

Budiraharjo, R., Neoh, K.G., Kang, E.T., 2012. Hydroxyapatite-coated carboxymethyl chitosan scaffolds for promoting osteoblast and stem cell differentiation. J. Colloid Interface Sci. 366, 224–232.

Cai, Q., Xu, Q., Feng, Q., Cao, X., Yang, X., Deng, X., 2011. Biomineralization of electrospun poly(l-lactic acid)/gelatin composite fibrous scaffold by using a supersaturated simulated body fluid with continuous CO_2 bubbling. Appl. Surf. Sci. 257, 10109–10118.

Castano, O., Sachot, N., Xuriguera, E., Engel, E., Planell, J.A., Park, J.H., Jin, G.Z., Kim, T.H., Kim, J.H., Kim, H.W., 2014. Angiogenesis in bone regeneration: tailored calcium release in hybrid fibrous scaffolds. ACS Appl. Mater. Interfaces 6, 7512–7522.

Chang, W., Mu, X., Zhu, X., Ma, G., Li, C., Xu, F., Nie, J., 2013. Biomimetic composite scaffolds based mineralization of hydroxyapatite on electrospun calcium-containing poly(vinyl alcohol) nanofibers. Mater. Sci. Eng. C Mater. Biol. Appl. 33, 4369–4376.

Chen, J.P., Su, C.H., 2011. Surface modification of electrospun PLLA nanofibers by plasma treatment and cationized gelatin immobilization for cartilage tissue engineering. Acta Biomater. 7, 234–243.

Chen, Y., Mak, A.F., Li, J., Wang, M., Shum, A.W., 2005. Formation of apatite on poly(α-hydroxy acid) in an accelerated biomimetic process. J. Biomed. Mater. Res. B Appl. Biomater. 73, 68–76.

Chen, Y., Mak, A.F., Wang, M., Li, J., 2006a. Composite coating of bonelike apatite particles and collagen fibers on poly L-lactic acid formed through an accelerated biomimetic coprecipitation process. J. Biomed. Mater. Res. B Appl. Biomater. 77, 315–322.

Chen, J., Chu, B., Hsiao, B.S., 2006b. Mineralization of hydroxyapatite in electrospun nanofibrous poly(l-lactic acid) scaffolds. J. Biomed. Mater. Res. A 79, 307–317.

Chen, J.-P., Chang, G.-Y., Chen, J.-K., 2008. Electrospun collagen/chitosan nanofibrous membrane as wound dressing. Colloid. Surf. A Physicochem. Eng. Asp. 313–314, 183–188.

Chen, L.-J., Liao, J.-D., Lin, S.-J., Chuang, Y.-J., Fu, Y.-S., 2009. Synthesis and characterization of PVB/silica nanofibers by electrospinning process. Polymer 50, 3516–3521.

Chen, J., Nan, K., Yin, S., Wang, Y., Wu, T., Zhang, Q., 2010a. Characterization and biocompatibility of nanohybrid scaffold prepared via in situ crystallization of hydroxyapatite in chitosan matrix. Colloids Surf. B: Biointerfaces 81, 640–647.

Chen, J., Li, X., Cui, W., Xie, C., Zou, J., Zou, B., 2010b. In situ grown fibrous composites of poly(dl-lactide) and hydroxyapatite as potential tissue engineering scaffolds. Polymer 51, 6268–6277.

Chen, J., Yu, Q., Zhang, G., Yang, S., Wu, J., Zhang, Q., 2012. Preparation and biocompatibility of nanohybrid scaffolds by in situ homogeneous formation of nano hydroxyapatite from biopolymer polyelectrolyte complex for bone repair applications. Colloids Surf. B: Biointerfaces 93, 100–107.

Cheng, Y., Zheng, F., Lu, J., Shang, L., Xie, Z., Zhao, Y., Chen, Y., Gu, Z., 2014. Bioinspired multicompartmental microfibers from microfluidics. Adv. Mater. 26, 5184–5190.

Cho, H.J., Perikamana, S.K., Lee, J.H., Lee, J., Lee, K.M., Shin, C.S., Shin, H., 2014. Effective immobilization of BMP-2 mediated by polydopamine coating on biodegradable nanofibers for enhanced in vivo bone formation. ACS Appl. Mater. Interfaces 6, 11225–11235.

Choi, M.O., Kim, Y.J., 2012. Fabrication of gelatin/calcium phosphate composite nanofibrous membranes by biomimetic mineralization. Int. J. Biol. Macromol. 50, 1188–1194.

Choi, J.S., Leong, K.W., Yoo, H.S., 2008a. In vivo wound healing of diabetic ulcers using electrospun nanofibers immobilized with human epidermal growth factor (EGF). Biomaterials 29, 587–596.

Choi, J.S., Lee, S.J., Christ, G.J., Atala, A., Yoo, J.J., 2008b. The influence of electrospun aligned poly(ε-caprolactone)/collagen nanofiber meshes on the formation of self-aligned skeletal muscle myotubes. Biomaterials 29, 2899–2906.

Choi, C.-H., Jung, J.-H., Hwang, T.-S., Lee, C.-S., 2009. In situ microfluidic synthesis of monodisperse PEG microspheres. Macromol. Res. 17, 163–167.

Chong, E.J., Phan, T.T., Lim, I.J., Zhang, Y.Z., Bay, B.H., Ramakrishna, S., Lim, C.T., 2007. Evaluation of electrospun PCL/gelatin nanofibrous scaffold for wound healing and layered dermal reconstitution. Acta Biomater. 3, 321–330.

Cui, W., Li, X., Xie, C., Chen, J., Zou, J., Zhou, S., Weng, J., 2010a. Controllable growth of hydroxyapatite on electrospun poly(DL-lactide) fibers grafted with chitosan as potential tissue engineering scaffolds. Polymer 51, 2320–2328.

Cui, W., Li, X., Xie, C., Zhuang, H., Zhou, S., Weng, J., 2010b. Hydroxyapatite nucleation and growth mechanism on electrospun fibers functionalized with different chemical groups and their combinations. Biomaterials 31, 4620–4629.

Dinarvand, P., Seyedjafari, E., Shafiee, A., Jandaghi, A.B., Doostmohammadi, A., Fathi, M.H., Farhadian, S., Soleimani, M., 2011. New approach to bone tissue engineering: simultaneous application of hydroxyapatite and bioactive glass coated on a poly(l-lactic acid) scaffold. ACS Appl. Mater. Interfaces 3, 4518–4524.

Ding, Y., Zhang, P., Long, Z., Jiang, Y., Xu, F., Lei, J., 2008. Fabrication and photocatalytic property of TiO_2 nanofibers. J. Sol-Gel Sci. Technol. 46, 176–179.

Duan, Y., Wang, Z., Yan, W., Wang, S., Zhang, S., Jia, J., 2007. Preparation of collagen-coated electrospun nanofibers by remote plasma treatment and their biological properties. J. Biomater. Sci. Polym. Ed. 18, 1153–1164.

Duncanson, W.J., Zieringer, M., Wagner, O., Wilking, J.N., Abbaspourrad, A., Haag, R., Weitz, D.A., 2012. Microfluidic synthesis of monodisperse porous microspheres with size-tunable pores. Soft Matter 8, 10636–10640.

El-Fiqi, A., Kim, H.-W., 2014. Mesoporous bioactive nanocarriers in electrospun biopolymer fibrous scaffolds designed for sequential drug delivery. RSC Adv. 4, 4444–4452.

Frohbergh, M.E., Katsman, A., Botta, G.P., Lazarovici, P., Schauer, C.L., Wegst, U.G.K., Lelkes, P.I., 2012. Electrospun hydroxyapatite-containing chitosan nanofibers crosslinked with genipin for bone tissue engineering. Biomaterials 33, 9167–9178.

Fu, S., Ni, P., Wang, B., Chu, B., Peng, J., Zheng, L., Zhao, X., Luo, F., Wei, Y., Qian, Z., 2012. In vivo biocompatibility and osteogenesis of electrospun poly(ε-caprolactone)–poly(ethylene glycol)–poly(ε-caprolactone)/nano-hydroxyapatite composite scaffold. Biomaterials 33, 8363–8371.

Fu, S., Yang, L., Fan, J., Wen, Q., Lin, S., Wang, B., Chen, L., Meng, X., Chen, Y., Wu, J., 2013. In vitro mineralization of hydroxyapatite on electrospun poly(ε-caprolactone)-poly(ethylene glycol)-poly(ε-caprolactone) fibrous scaffolds for tissue engineering application. Colloids Surf. B: Biointerfaces 107, 167–173.

Fu, W., Liu, Z., Feng, B., Hu, R., He, X., Wang, H., Yin, M., Huang, H., Zhang, H., Wang, W., 2014. Electrospun gelatin/PCL and collagen/PLCL scaffolds for vascular tissue engineering. Int. J. Nanomedicine 9, 2335–2344.

Gandhimathi, C., Venugopal, J., Ravichandran, R., Sundarrajan, S., Suganya, S., Ramakrishna, S., 2013. Mimicking nanofibrous hybrid bone substitute for mesenchymal stem cells differentiation into osteogenesis. Macromol. Biosci. 13, 696–706.

Gao, C., Gao, Q., Bao, X., Li, Y., Teramoto, A., Abe, K., 2011. Preparation and in vitro bioactivity of novel mesoporous borosilicate bioactive glass nanofibers. J. Am. Ceram. Soc. 94, 2841–2845.

Gao, C., Gao, Q., Li, Y., Rahaman, M.N., Teramoto, A., Abe, K., 2013. In vitro evaluation of electrospun gelatin-bioactive glass hybrid scaffolds for bone regeneration. J. Appl. Polym. Sci. 127, 2588–2599.

Gupta, D., Venugopal, J., Mitra, S., Giri Dev, V.R., Ramakrishna, S., 2009. Nanostructured biocomposite substrates by electrospinning and electrospraying for the mineralization of osteoblasts. Biomaterials 30, 2085–2094.

Han, D., Gouma, P.I., 2006. Electrospun bioscaffolds that mimic the topology of extracellular matrix. Nanomedicine 2, 37–41.

He, W., Yong, T., Teo, W.E., Ma, Z., Ramakrishna, S., 2005a. Fabrication and endothelialization of collagen-blended biodegradable polymer nanofibers: potential vascular graft for blood vessel tissue engineering. Tissue Eng. 11, 1574–1588.

He, W., Ma, Z., Yong, T., Teo, W.E., Ramakrishna, S., 2005b. Fabrication of collagen-coated biodegradable polymer nanofiber mesh and its potential for endothelial cells growth. Biomaterials 26, 7606–7615.

He, C., Xiao, G., Jin, X., Sun, C., Ma, P.X., 2010. Electrodeposition on nanofibrous polymer scaffolds: rapid mineralization, tunable calcium phosphate composition and topography. Adv. Funct. Mater. 20, 3568–3576.

He, T., Liang, Q., Zhang, K., Mu, X., Luo, T., Wang, Y., Luo, G., 2011. A modified microfluidic chip for fabrication of paclitaxel-loaded poly(l-lactic acid) microspheres. Microfluid. Nanofluid. 10, 1289–1298.

He, C., Jin, X., Ma, P.X., 2014. Calcium phosphate deposition rate, structure and osteoconductivity on electrospun poly(l-lactic acid) matrix using electrodeposition or simulated body fluid incubation. Acta Biomater. 10, 419–427.

Heydarkhan-Hagvall, S., Schenke-Layland, K., Dhanasopon, A.P., Rofail, F., Smith, H., Wu, B.M., Shemin, R., Beygui, R.E., MacLellan, W.R., 2008. Three-dimensional electrospun ECM-based hybrid scaffolds for cardiovascular tissue engineering. Biomaterials 29, 2907–2914.

Homayoni, H., Ravandi, S.A.H., Valizadeh, M., 2009. Electrospinning of chitosan nanofibers: processing optimization. Carbohydr. Polym. 77, 656–661.

Hong, S.G., Kim, G.H., 2013. Mechanically improved electrospun PCL biocomposites reinforced with a collagen coating process: preparation, physical properties, and cellular activity. Bioprocess. Biosyst. Eng. 36, 205–214.

Hong, Y., Chen, X., Jing, X., Fan, H., Gu, Z., Zhang, X., 2010a. Fabrication and drug delivery of ultrathin mesoporous bioactive glass hollow fibers. Adv. Funct. Mater. 20, 1503–1510.

Hong, Y., Chen, X., Jing, X., Fan, H., Guo, B., Gu, Z., Zhang, X., 2010b. Preparation, bioactivity, and drug release of hierarchical nanoporous bioactive glass ultrathin fibers. Adv. Mater. 22, 754–758.

Hong, K.S., Kim, E.C., Bang, S.H., Chung, C.H., Lee, Y.I., Hyun, J.K., Lee, H.H., Jang, J.H., Kim, T.I., Kim, H.W., 2010c. Bone regeneration by bioactive hybrid membrane containing FGF2 within rat calvarium. J. Biomed. Mater. Res. A 94, 1187–1194.

Hongju, P., Zi, Y., Huanhuan, L., Xiao, C., Bei, F., Huihua, Y., Bo, S., Hongwei, O., Yanzhong, Z., 2012. Electrospun biomimetic scaffold of hydroxyapatite/chitosan supports enhanced osteogenic differentiation of mMSCs. Nanotechnology 23, 485102.

Horne, M.K., Nisbet, D.R., Forsythe, J.S., Parish, C.L., 2010. Three-dimensional nanofibrous scaffolds incorporating immobilized BDNF promote proliferation and differentiation of cortical neural stem cells. Stem Cells Dev. 19, 843–852.

Huang, Z.-M., Zhang, Y.Z., Ramakrishna, S., Lim, C.T., 2004. Electrospinning and mechanical characterization of gelatin nanofibers. Polymer 45, 5361–5368.

Huang, S., Kang, X., Cheng, Z., Ma, P., Jia, Y., Lin, J., 2012. Electrospinning preparation and drug delivery properties of Eu3+/Tb3+ doped mesoporous bioactive glass nanofibers. J. Colloid Interface Sci. 387, 285–291.

Hwang, C.M., Khademhosseini, A., Park, Y., Sun, K., Lee, S.H., 2008. Microfluidic chip-based fabrication of PLGA microfiber scaffolds for tissue engineering. Langmuir 24, 6845–6851.

Jaiswal, A.K., Chandra, V., Bhonde, R.R., Soni, V.P., Bellare, J.R., 2012. Mineralization of nanohydroxyapatite on electrospun poly(l-lactic acid)/gelatin by an alternate soaking process: a biomimetic scaffold for bone regeneration. J. Bioact. Compat. Polym. 27, 356–374.

Jaiswal, A.K., Chhabra, H., Soni, V.P., Bellare, J.R., 2013a. Enhanced mechanical strength and biocompatibility of electrospun polycaprolactone-gelatin scaffold with surface deposited nano-hydroxyapatite. Mater. Sci. Eng. C Mater. Biol. Appl. 33, 2376–2385.

Jaiswal, A.K., Kadam, S.S., Soni, V.P., Bellare, J.R., 2013b. Improved functionalization of electrospun PLLA/gelatin scaffold by alternate soaking method for bone tissue engineering. Appl. Surf. Sci. 268, 477–488.

Jang, J.-H., Castano, O., Kim, H.-W., 2009. Electrospun materials as potential platforms for bone tissue engineering. Adv. Drug Deliv. Rev. 61, 1065–1083.

Jegal, S.H., Park, J.H., Kim, J.H., Kim, T.H., Shin, U.S., Kim, T.I., Kim, H.W., 2011. Functional composite nanofibers of poly(lactide-co-caprolactone) containing gelatin-apatite bone mimetic precipitate for bone regeneration. Acta Biomater. 7, 1609–1617.

Ji, W., Yang, F., Seyednejad, H., Chen, Z., Hennink, W.E., Anderson, J.M., van den Beucken, J.J., Jansen, J.A., 2012. Biocompatibility and degradation characteristics of PLGA-based electrospun nanofibrous scaffolds with nanoapatite incorporation. Biomaterials 33, 6604–6614.

Jiang, J., Xie, J., Ma, B., Bartlett, D.E., Xu, A., Wang, C.H., 2014. Mussel-inspired protein-mediated surface functionalization of electrospun nanofibers for pH-responsive drug delivery. Acta Biomater. 10, 1324–1332.

Jo, J.H., Lee, E.J., Shin, D.S., Kim, H.E., Kim, H.W., Koh, Y.H., Jang, J.H., 2009. In vitro/in vivo biocompatibility and mechanical properties of bioactive glass nanofiber and poly(ε-caprolactone) composite materials. J. Biomed. Mater. Res. B Appl. Biomater. 91, 213–220.

Jun, Y., Kim, M.J., Hwang, Y.H., Jeon, E.A., Kang, A.R., Lee, S.H., Lee, D.Y., 2013. Microfluidics-generated pancreatic islet microfibers for enhanced immunoprotection. Biomaterials 34, 8122–8130.

Karaman, O., Kumar, A., Moeinzadeh, S., He, X., Cui, T., Jabbari, E., 2013. Effect of surface modification of nanofibres with glutamic acid peptide on calcium phosphate nucleation and osteogenic differentiation of marrow stromal cells. J. Tissue Eng. Regen. Med. http://dx.doi.org/10.1002/term.1775.

Kim, H.W., Kim, H.E., 2006. Nanofiber generation of hydroxyapatite and fluor-hydroxyapatite bioceramics. J. Biomed. Mater. Res. B Appl. Biomater. 77, 323–328.

Kim, T.G., Park, T.G., 2006. Biomimicking extracellular matrix: cell adhesive RGD peptide modified electrospun poly(D,L-lactic-co-glycolic acid) nanofiber mesh. Tissue Eng. 12, 221–233.

Kim, H.M., Kishimoto, K., Miyaji, F., Kokubo, T., Yao, T., Suetsugu, Y., Tanaka, J., Nakamura, T., 1999. Composition and structure of the apatite formed on pet substrates in SBF modified with various ionic activity products. J. Biomed. Mater. Res. 46, 228–235.

Kim, H.W., Song, J.H., Kim, H.E., 2005. Nanofiber generation of gelatin–hydroxyapatite biomimetics for guided tissue regeneration. Adv. Funct. Mater. 15, 1988–1994.

Kim, S.S., Sun Park, M., Jeon, O., Yong Choi, C., Kim, B.S., 2006a. Poly(lactide-co-glycolide)/hydroxyapatite composite scaffolds for bone tissue engineering. Biomaterials 27, 1399–1409.

Kim, H.W., Kim, H.E., Knowles, J.C., 2006b. Production and potential of bioactive glass nanofibers as a next-generation biomaterial. Adv. Funct. Mater. 16, 1529–1535.

Kim, H.-W., Song, J.-H., Kim, H.-E., 2006c. Bioactive glass nanofiber–collagen nanocomposite as a novel bone regeneration matrix. J. Biomed. Mater. Res. A 79A, 698–705.

Kim, S.S., Park, M.S., Gwak, S.J., Choi, C.Y., Kim, B.S., 2006d. Accelerated bonelike apatite growth on porous polymer/ceramic composite scaffolds in vitro. Tissue Eng. 12, 2997–3006.

Kim, H.W., Lee, H.H., Knowles, J.C., 2006e. Electrospinning biomedical nanocomposite fibers of hydroxyapatite/poly(lactic acid) for bone regeneration. J. Biomed. Mater. Res. A 79, 643–649.

Kim, S.S., Ahn, K.M., Park, M.S., Lee, J.H., Choi, C.Y., Kim, B.S., 2007. A poly(lactide-co glycolide)/hydroxyapatite composite scaffold with enhanced osteoconductivity. J. Biomed. Mater. Res. A 80, 206–215.

Kim, H.W., Lee, H.H., Chun, G.-S., 2008a. Bioactivity and osteoblast responses of novel biomedical nanocomposites of bioactive glass nanofiber filled poly(lactic acid). J. Biomed. Mater. Res. A 85A, 651–663.

Kim, H.-W., Lee, H.-H., Knowles, J.C., 2008b. Nanofibrous glass tailored with apatite-fibronectin interface for bone cell stimulation. J. Nanosci. Nanotechnol. 8, 3013–3019.

Kim, J.-J., Won, J.-E., Shin, U.S., Kim, H.-W., 2011. Improvement of bioactive glass nanofiber by a capillary-driven infiltration coating with degradable polymers. J. Am. Ceram. Soc. 94, 2812–2815.

Kim, T.H., Kim, J.J., Kim, H.W., 2014a. Basic fibroblast growth factor-loaded, mineralized biopolymer-nanofiber scaffold improves adhesion and proliferation of rat mesenchymal stem cells. Biotechnol. Lett. 36, 383–390.

Kim, J.J., Bae, W.J., Kim, J.M., Kim, J.J., Lee, E.J., Kim, H.W., Kim, E.C., 2014b. Mineralized polycaprolactone nanofibrous matrix for odontogenesis of human dental pulp cells. J. Biomater. Appl. 28, 1069–1078.

Ko, E.K., Jeong, S.I., Rim, N.G., Lee, Y.M., Shin, H., Lee, B.K., 2008. In vitro osteogenic differentiation of human mesenchymal stem cells and in vivo bone formation in composite nanofiber meshes. Tissue Eng. A 14, 2105–2119.

Kokubo, T., 1995. Apatite formation on organic polymers by a biomimetic process. Eur. J. Solid State Inorg. Chem. 32, 819–827.

Kokubo, T., Kushitani, H., Sakka, S., Kitsugi, T., Yamamuro, T., 1990. Solutions able to reproduce in vivo surface structure changes in bioactive glass-ceramic A-W. J. Biomed. Mater. Res. 24, 721–734.

Kolluru, P.V., Lipner, J., Liu, W., Xia, Y., Thomopoulos, S., Genin, G.M., Chasiotis, I., 2013. Strong and tough mineralized PLGA nanofibers for tendon-to-bone scaffolds. Acta Biomater. 9, 9442–9450.

Kretlow, J.D., Mikos, A.G., 2007. Mineralization of synthetic polymer scaffolds for bone tissue engineering. Tissue Eng. 13, 927–938.

Kumbar, S.G., James, R., Nukavarapu, S.P., Laurencin, C.T., 2008a. Electrospun nanofiber scaffolds: engineering soft tissues. Biomed. Mater. 3, 034002.

Kumbar, S.G., Nukavarapu, S.P., James, R., Nair, L.S., Laurencin, C.T., 2008b. Electrospun poly(lactic acid-co-glycolic acid) scaffolds for skin tissue engineering. Biomaterials 29, 4100–4107.

Lao, L., Wang, Y., Zhu, Y., Zhang, Y., Gao, C., 2011. Poly(lactide-co-glycolide)/hydroxyapatite nanofibrous scaffolds fabricated by electrospinning for bone tissue engineering. J. Mater. Sci. Mater. Med. 22, 1873–1884.

Lee, H.H., Yu, H.S., Jang, J.H., Kim, H.W., 2008. Bioactivity improvement of poly(ε-caprolactone) membrane with the addition of nanofibrous bioactive glass. Acta Biomater. 4, 622–629.

Lee, J.H., Rim, N.G., Jung, H.S., Shin, H., 2010a. Control of osteogenic differentiation and mineralization of human mesenchymal stem cells on composite nanofibers containing poly[lactic-co-(glycolic acid)] and hydroxyapatite. Macromol. Biosci. 10, 173–182.

Lee, K.H., Shin, S.J., Kim, C.B., Kim, J.K., Cho, Y.W., Chung, B.G., Lee, S.H., et al., 2010b. Microfluidic synthesis of pure chitosan microfibers for bio-artificial liver chip. Lab Chip 10, 1328–1334.

Lee, J.H., Park, J.-H., Yun, Y.-R., Jang, J.-H., Lee, E.-J., Chrzanowski, W., Wall, I.B., Kim, H.-W., 2013. Tethering bi-functional protein onto mineralized polymer scaffolds to regulate mesenchymal stem cell behaviors for bone regeneration. J. Mater. Chem. B 1, 2731–2741.

Lee, J.H., Park, J.H., El-Fiqi, A., Kim, J.H., Yun, Y.R., Jang, J.H., Han, C.M., Lee, E.J., Kim, H.W., 2014. Biointerface control of electrospun fiber scaffolds for bone regeneration: engineered protein link to mineralized surface. Acta Biomater. 10, 2750–2761.

Li, W.J., Laurencin, C.T., Caterson, E.J., Tuan, R.S., Ko, F.K., 2002. Electrospun nanofibrous structure: a novel scaffold for tissue engineering. J. Biomed. Mater. Res. 60, 613–621.

Li, D., Wang, Y., Xia, Y., 2003. Electrospinning of polymeric and ceramic nanofibers as uniaxially aligned arrays. Nano Lett. 3, 1167–1171.

Li, X., Xie, J., Yuan, X., Xia, Y., 2008. Coating electrospun poly(ε-caprolactone) fibers with gelatin and calcium phosphate and their use as biomimetic scaffolds for bone tissue engineering. Langmuir 24, 14145–14150.

Li, X., Xie, J., Lipner, J., Yuan, X., Thomopoulos, S., Xia, Y., 2009. Nanofiber scaffolds with gradations in mineral content for mimicking the tendon-to-bone insertion site. Nano Lett. 9, 2763–2768.

Li, D., Sun, H., Jiang, L., Zhang, K., Liu, W., Zhu, Y., Fangteng, J., Shi, C., Zhao, L., Sun, H., Yang, B., 2014. Enhanced biocompatibility of PLGA nanofibers with gelatin/nano-hydroxyapatite bone biomimetics incorporation. ACS Appl. Mater. Interfaces 6, 9402–9410.

Liao, S.S., Cui, F.Z., 2004. In vitro and in vivo degradation of mineralized collagen-based composite scaffold: nanohydroxyapatite/collagen/poly(l-lactide). Tissue Eng. 10, 73–80.

Liao, S., Murugan, R., Chan, C.K., Ramakrishna, S., 2008. Processing nanoengineered scaffolds through electrospinning and mineralization suitable for biomimetic bone tissue engineering. J. Mech. Behav. Biomed. Mater. 1, 252–260.

Lin, C.C., Fu, S.J., Lin, Y.C., Yang, I.K., Gu, Y., 2014. Chitosan-coated electrospun PLA fibers for rapid mineralization of calcium phosphate. Int. J. Biol. Macromol. 68, 39–47.

Liu, C., Han, Z., Czernuszka, J.T., 2009a. Gradient collagen/nanohydroxyapatite composite scaffold: Development and characterization. Acta Biomater. 5, 661–669.

Liu, X., Smith, L.A., Hu, J., Ma, P.X., 2009b. Biomimetic nanofibrous gelatin/apatite composite scaffolds for bone tissue engineering. Biomaterials 30, 2252–2258.

Liu, X., Lin, T., Fang, J., Yao, G., Zhao, H., Dodson, M., Wang, X., 2010. In vivo wound healing and antibacterial performances of electrospun nanofibre membranes J. Biomed. Mater. Res. A 94, 499–508.

Liu, W., Yeh, Y.-C., Lipner, J., Xie, J., Sung, H.-W., Thomopoulos, S., Xia, Y., 2011. Enhancing the stiffness of electrospun nanofiber scaffolds with a controlled surface coating and mineralization. Langmuir 27, 9088–9093.

Lu, H., Zhang, T., Wang, X.P., Fang, Q.F., 2009. Electrospun submicron bioactive glass fibers for bone tissue scaffold. J. Mater. Sci. Mater. Med. 20, 793–798.

Lu, L.X., Zhang, X.F., Wang, Y.Y., Ortiz, L., Mao, X., Jiang, Z.L., Xiao, Z.D., Huang, N.P., 2013. Effects of hydroxyapatite-containing composite nanofibers on osteogenesis of mesenchymal stem cells in vitro and bone regeneration in vivo. ACS Appl. Mater. Interfaces 5, 319–330.

Luickx, N., Van den Vreken, N., D'Oosterlinck, W., Van der Schueren, L., Declercq, H., De Clerck, K., Cornelissen, M., Verbeeck, R., 2014. Optimization of the activation and nucleation steps in the precipitation of a calcium phosphate primer layer on electrospun poly(ε-caprolactone). J. Biomed. Mater. Res. A. http://dx.doi.org/10.1002/jbm.a.35191.

Ma, Z., Kotaki, M., Inai, R., Ramakrishna, S., 2005a. Potential of nanofiber matrix as tissue-engineering scaffolds. Tissue Eng. 11, 101–109.

Ma, Z., He, W., Yong, T., Ramakrishna, S., 2005b. Grafting of gelatin on electrospun poly (caprolactone) nanofibers to improve endothelial cell spreading and proliferation and to control cell Orientation. Tissue Eng. 11, 1149–1158.

Ma, Z., Chen, F., Zhu, Y.J., Cui, T., Liu, X.Y., 2011. Amorphous calcium phosphate/poly (D,L-lactic acid) composite nanofibers: Electrospinning preparation and biomineralization. J. Colloid Interface Sci. 359, 371–379.

Madurantakam, P.A., Rodriguez, I.A., Cost, C.P., Viswanathan, R., Simpson, D.G., Beckman, M.J., Moon, P.C., Bowlin, G.L., 2009. Multiple factor interactions in biomimetic mineralization of electrospun scaffolds. Biomaterials 30, 5456–5464.

Mahony, O., Tsigkou, O., Ionescu, C., Minelli, C., Ling, L., Hanly, R., Smith, M.E., Stevens, M.M., Jones, J.R., 2010. Silica–gelatin hybrids with tailorable degradation and mechanical properties for tissue regeneration. Adv. Funct. Mater. 20, 3835–3845.

Martins, A., Araujo, J.V., Reis, R.L., Neves, N.M., 2007. Electrospun nanostructured scaffolds for tissue engineering applications. Nanomedicine 2, 929–942.

Matthews, J.A., Wnek, G.E., Simpson, D.G., Bowlin, G.L., 2002. Electrospinning of collagen nanofibers. Biomacromolecules 3, 232–238.

Mavis, B., Demirtaş, T.T., Gümüşderelioğlu, M., Gündüz, G., Çolak, Ü., 2009. Synthesis, characterization and osteoblastic activity of polycaprolactone nanofibers coated with biomimetic calcium phosphate. Acta Biomater. 5, 3098–3111.

Mazzitelli, S., Capretto, L., Carugo, D., Zhang, X., Piva, R., Nastruzzi, C., 2011. Optimised production of multifunctional microfibres by microfluidic chip technology for tissue engineering applications. Lab Chip 11, 1776–1785.

Meng, Z.X., Li, H.F., Sun, Z.Z., Zheng, W., Zheng, Y.F., 2013. Fabrication of mineralized electrospun PLGA and PLGA/gelatin nanofibers and their potential in bone tissue engineering. Mater. Sci. Eng. C 33, 699–706.

Min, B.-M., Lee, G., Kim, S.H., Nam, Y.S., Lee, T.S., Park, W.H., 2004. Electrospinning of silk fibroin nanofibers and its effect on the adhesion and spreading of normal human keratinocytes and fibroblasts in vitro. Biomaterials 25, 1289–1297.

Mo, X.M., Xu, C.Y., Kotaki, M., Ramakrishna, S., 2004. Electrospun P(LLA-CL) nanofiber: a biomimetic extracellular matrix for smooth muscle cell and endothelial cell proliferation. Biomaterials 25, 1883–1890.

Ngiam, M., Liao, S., Patil, A.J., Cheng, Z., Yang, F., Gubler, M.J., Ramakrishna, S., Chan, C.K., 2009. Fabrication of mineralized polymeric nanofibrous composites for bone graft materials. Tissue Eng. A 15, 535–546.

Nilza, R., Susana, R.S., Clemens, A.V.B., Lorenzo, M., Fernando, J.M., 2014. A biocomposite of collagen nanofibers and nanohydroxyapatite for bone regeneration. Biofabrication 6, 035015.

Ning, C., Cheng, H., Zhu, W., Yin, Z., Chen, H., Zheng, H., Lei, S., Yin, S., Tan, G., 2008. In vitro mineralization of surface-modified porous polycaprolactone scaffolds in simulated body fluid. Appl. Surf. Sci. 255, 429–431.

Nirmala, R., Nam, K., Navamathavan, R., Park, S.-J., Kim, H., 2010. Hydroxyapatite mineralization on the calcium chloride blended polyurethane nanofiber via biomimetic method. Nanoscale Res. Lett. 6, 1–8.

Noh, K.-T., Lee, H.-Y., Shin, U.-S., Kim, H.-W., 2010. Composite nanofiber of bioactive glass nanofiller incorporated poly(lactic acid) for bone regeneration. Mater. Lett. 64, 802–805.

Nunes, J.K., Wu, C.-Y., Amini, H., Owsley, K., Di Carlo, D., Stone, H.A., 2014. Fabricating shaped microfibers with inertial microfluidics. Adv. Mater. 26, 3712–3717.

Obata, A., Hotta, T., Wakita, T., Ota, Y., Kasuga, T., 2010. Electrospun microfiber meshes of silicon-doped vaterite/poly(lactic acid) hybrid for guided bone regeneration. Acta Biomater. 6, 1248–1257.

Oyane, A., Kawashita, M., Nakanishi, K., Kokubo, T., Minoda, M., Miyamoto, T., Nakamura, T., 2003. Bonelike apatite formation on ethylene-vinyl alcohol copolymer modified with silane coupling agent and calcium silicate solutions. Biomaterials 24, 1729–1735.

Oyane, A., Uchida, M., Choong, C., Triffitt, J., Jones, J., Ito, A., 2005a. Simple surface modification of poly(ε-caprolactone) for apatite deposition from simulated body fluid. Biomaterials 26, 2407–2413.

Oyane, A., Uchida, M., Yokoyama, Y., Choong, C., Triffitt, J., Ito, A., 2005b. Simple surface modification of poly(ε-caprolactone) to induce its apatite-forming ability. J. Biomed. Mater. Res. A 75, 138–145.

Park, J.-Y., Lee, I.-H., 2011. Controlled release of ketoprofen from electrospun porous polylactic acid (PLA) nanofibers. J. Polym. Res. 18, 1287–1291.

Patel, P.A., Eckart, J., Advincula, M.C., Goldberg, A.J., Mather, P.T., 2009. Rapid synthesis of polymer-silica hybrid nanofibers by biomimetic mineralization. Polymer 50, 1214–1222.

Peng, F., Yu, X., Wei, M., 2011. In vitro cell performance on hydroxyapatite particles/poly (l-lactic acid) nanofibrous scaffolds with an excellent particle along nanofiber orientation. Acta Biomater. 7, 2585–2592.

Pham, Q.P., Sharma, U., Mikos, A.G., 2006. Electrospun poly(ε-caprolactone) microfiber and multilayer nanofiber/microfiber scaffolds: characterization of scaffolds and measurement of cellular infiltration. Biomacromolecules 7, 2796–2805.

Phipps, M.C., Clem, W.C., Catledge, S.A., Xu, Y., Hennessy, K.M., Thomas, V., Jablonsky, M.J., Chowdhury, S., Stanishevsky, A.V., Vohra, Y.K., Bellis, S.L., 2011. Mesenchymal stem cell responses to bone-mimetic electrospun matrices composed of polycaprolactone, collagen I and nanoparticulate hydroxyapatite. PLoS One 6, e16813.

Poologasundarampillai, G., Ionescu, C., Tsigkou, O., Murugesan, M., Hill, R.G., Stevens, M.M., Hanna, J.V., Smith, M.E., Jones, J.R., 2010. Synthesis of bioactive class II poly(g-glutamic acid)/silica hybrids for bone regeneration. J. Mater. Chem. 20, 8952–8961.

Prosecka, E., Buzgo, M., Rampichova, M., Kocourek, T., Kochova, P., Vyslouzilova, L., Tvrdik, D., Jelinek, M., Lukas, D., Amler, E., 2012. Thin-layer hydroxyapatite deposition on a nanofiber surface stimulates mesenchymal stem cell proliferation and their differentiation into osteoblasts. J. Biomed. Biotechnol., 428503. http://dx.doi.org/10.1155/2012/428503.

Qiu, P., Qu, X., Brackett, D.J., Lerner, M.R., Li, D., Mao, C., 2013. Silica-based branched hollow microfibers as a biomimetic extracellular matrix for promoting tumor cell growth in vitro and in vivo. Adv. Mater. 25, 2492–2496.

Qu, X., Cui, W., Yang, F., Min, C., Shen, H., Bei, J., Wang, S., 2007. The effect of oxygen plasma pretreatment and incubation in modified simulated body fluids on the formation of bone-like apatite on poly(lactide-co-glycolide) (70/30). Biomaterials 28, 9–18.

Rajzer, I., Menaszek, E., Kwiatkowski, R., Chrzanowski, W., 2014. Bioactive nanocomposite PLDL/nano-hydroxyapatite electrospun membranes for bone tissue engineering. J. Mater. Sci. Mater. Med. 25, 1239–1247.

Ravichandran, R., Venugopal, J.R., Sundarrajan, S., Mukherjee, S., Ramakrishna, S., 2012. Precipitation of nanohydroxyapatite on PLLA/PBLG/collagen nanofibrous structures for the differentiation of adipose derived stem cells to osteogenic lineage. Biomaterials 33, 846–855.

Ravichandran, R., Gandhi, S., Sundaramurthi, D., Sethuraman, S., Krishnan, U.M., 2013. Hierarchical mesoporous silica nanofibers as multifunctional scaffolds for bone tissue regeneration. J. Biomater. Sci. Polym. Ed. 24, 1988–2005.

Reneker, D.H., Kataphinan, W., Theron, A., Zussman, E., Yarin, A.L., 2002. Nanofiber garlands of polycaprolactone by electrospinning. Polymer 43, 6785–6794.

Rho, K.S., Jeong, L., Lee, G., Seo, B.M., Park, Y.J., Hong, S.D., Roh, S., Cho, J.J., Park, W.H., Min, B.M., 2006. Electrospinning of collagen nanofibers: Effects on the behavior of normal human keratinocytes and early-stage wound healing. Biomaterials 27, 1452–1461.

Rim, N.G., Kim, S.J., Shin, Y.M., Jun, I., Lim, D.W., Park, J.H., Shin, H., 2012. Mussel-inspired surface modification of poly(l-lactide) electrospun fibers for modulation of osteogenic differentiation of human mesenchymal stem cells. Colloids Surf. B: Biointerfaces 91, 189–197.

Rodriguez, K., Renneckar, S., Gatenholm, P., 2011. Biomimetic calcium phosphate crystal mineralization on electrospun cellulose-based scaffolds. ACS Appl. Mater. Interfaces 3, 681–689.

Schneider, O.D., Loher, S., Brunner, T.J., Uebersax, L., Simonet, M., Grass, R.N., Merkle, H.P., Stark, W.J., 2008. Cotton wool-like nanocomposite biomaterials prepared by electrospinning: in vitro bioactivity and osteogenic differentiation of human mesenchymal stem cells. J. Biomed. Mater. Res. B Appl. Biomater. 84, 350–362.

Sell, S., Barnes, C., Smith, M., McClure, M., Madurantakam, P., Grant, J., McManus, M., Bowlin, G., 2007. Extracellular matrix regenerated: tissue engineering via electrospun biomimetic nanofibers. Polym. Int. 56, 1349–1360.

Seol, Y.J., Kim, K.H., Kang, Y.M., Kim, I.A., Rhee, S.H., 2009. Bioactivity, pre-osteoblastic cell responses, and osteoconductivity evaluations of the electrospun non-woven SiO_2-CaO gel fabrics. J. Biomed. Mater. Res. B Appl. Biomater. 90, 679–687.

Seyedjafari, E., Soleimani, M., Ghaemi, N., Shabani, I., 2010. Nanohydroxyapatite-coated electrospun poly(l-lactide) nanofibers enhance osteogenic differentiation of stem cells and induce ectopic bone formation. Biomacromolecules 11, 3118–3125.

Shih, Y.R., Chen, C.N., Tsai, S.W., Wang, Y.J., Lee, O.K., 2006. Growth of mesenchymal stem cells on electrospun type I collagen nanofibers. Stem Cells 24, 2391–2397.

Shin, M., Yoshimoto, H., Vacanti, J.P., 2004. In vivo bone tissue engineering using mesenchymal stem cells on a novel electrospun nanofibrous scaffold. Tissue Eng. 10, 33–41.

Shin, S.-J., Park, J.-Y., Lee, J.-Y., Park, H., Park, Y.-D., Lee, K.-B., Whang, C.-M., Lee, S.-H., 2007. "On the fly" continuous generation of alginate fibers using a microfluidic device. Langmuir 23, 9104–9108.

Song, J., Malathong, V., Bertozzi, C.R., 2005. Mineralization of synthetic polymer scaffolds: a bottom-up approach for the development of artificial bone. J. Am. Chem. Soc. 127, 3366–3372.

Song, J.H., Yoon, B.H., Kim, H.E., Kim, H.W., 2008. Bioactive and degradable hybridized nanofibers of gelatin-siloxane for bone regeneration. J. Biomed. Mater. Res. A 84, 875–884.
Sonseca, A., Peponi, L., Sahuquillo, O., Kenny, J.M., Giménez, E., 2012. Electrospinning of biodegradable polylactide/hydroxyapatite nanofibers: study on the morphology, crystallinity structure and thermal stability. Polym. Degrad. Stabil. 97, 2052–2059.
Sreerekha, P.R., Menon, D., Nair, S.V., Chennazhi, K.P., 2013. Fabrication of electrospun poly (lactide-co-glycolide)–fibrin multiscale scaffold for myocardial regeneration in vitro. Tissue Eng. A 19, 849–859.
Strange, D.G., Oyen, M.L., 2011. Biomimetic bone-like composites fabricated through an automated alternate soaking process. Acta Biomater. 7, 3586–3594.
Suarez-Gonzalez, D., Barnhart, K., Migneco, F., Flanagan, C., Hollister, S.J., Murphy, W.L., 2012. Controllable mineral coatings on PCL scaffolds as carriers for growth factor release. Biomaterials 33, 713–721.
Sui, G., Yang, X., Mei, F., Hu, X., Chen, G., Deng, X., Ryu, S., 2007. Poly-l-lactic acid/hydroxyapatite hybrid membrane for bone tissue regeneration. J. Biomed. Mater. Res. A 82, 445–454.
Taguchi, T., Kishida, A., Akashi, M., 1998. Hydroxyapatite formation on/in poly(vinyl alcohol) hydrogel matrices using a novel alternate soaking process. Chem. Lett. 27, 711–712.
Taguchi, T., Kishida, A., Akashi, M., 1999a. Apatite formation on/in hydrogel matrices using an alternate soaking process: II. Effect of swelling ratios of poly (vinyl alcohol) hydrogel matrices on apatite formation. J. Biomater. Sci. Polym. Ed. 10, 331–339.
Taguchi, T., Kishida, A., Akashi, M., 1999b. Apatite formation on/in hydrogel matrices using an alternate soaking process (III): effect of physico-chemical factors on apatite formation on/in poly(vinyl alcohol) hydrogel matrices. J. Biomater. Sci. Polym. Ed. 10, 795–804.
Takeuchi, A., Ohtsuki, C., Miyazaki, T., Tanaka, H., Yamazaki, M., Tanihara, M., 2003. Deposition of bone-like apatite on silk fiber in a solution that mimics extracellular fluid. J. Biomed. Mater. Res. A 65, 283–289.
Teng, S.-H., Lee, E.-J., Wang, P., Kim, H.-E., 2008. Collagen/hydroxyapatite composite nanofibers by electrospinning. Mater. Lett. 62, 3055–3058.
Teo, W.E., Liao, S., Chan, C., Ramakrishna, S., 2011. Fabrication and characterization of hierarchically organized nanoparticle-reinforced nanofibrous composite scaffolds. Acta Biomater. 7, 193–202.
Thomas, V., Jagani, S., Johnson, K., Jose, M.V., Dean, D.R., Vohra, Y.K., Nyairo, E., 2006. Electrospun bioactive nanocomposite scaffolds of polycaprolactone and nanohydroxyapatite for bone tissue engineering. J. Nanosci. Nanotechnol. 6, 487–493.
Thomas, V., Dean, D.R., Jose, M.V., Mathew, B., Chowdhury, S., Vohra, Y.K., 2007. Nanostructured biocomposite scaffolds based on collagen coelectrospun with nanohydroxyapatite. Biomacromolecules 8, 631–637.
Toskas, G., Cherif, C., Hund, R.D., Laourine, E., Fahmi, A., Mahltig, B., 2011. Inorganic/organic (SiO_2)/PEO hybrid electrospun nanofibers produced from a modified sol and their surface modification possibilities. ACS Appl. Mater. Interfaces 3, 3673–3681.
Toskas, G., Cherif, C., Hund, R.D., Laourine, E., Mahltig, B., Fahmi, A., Heinemann, C., Hanke, T., 2013. Chitosan(PEO)/silica hybrid nanofibers as a potential biomaterial for bone regeneration. Carbohydr. Polym. 94, 713–722.
Tsai, W.B., Chen, W.T., Chien, H.W., Kuo, W.H., Wang, M.J., 2014. Poly(dopamine) coating to biodegradable polymers for bone tissue engineering. J. Biomater. Appl. 28, 837–848.
Venugopal, J., Prabhakaran, M.P., Zhang, Y., Low, S., Choon, A.T., Ramakrishna, S., 1917. Biomimetic hydroxyapatite-containing composite nanofibrous substrates for bone tissue engineering. Philos. Trans. A Math. Phys. Eng. Sci. 368, 2065–2081.

Venugopal, J.R., Low, S., Choon, A.T., Kumar, A.B., Ramakrishna, S., 2008. Nanobioengineered electrospun composite nanofibers and osteoblasts for bone regeneration. Artif. Organs 32, 388–397.

Wakita, T., Obata, A., Kasuga, T., 2009. New fabrication process of layered membranes based on poly(lactic acid) fibers for guided bone regeneration. Mater. Trans. 50, 1737–1741.

Wakita, T., Nakamura, J., Ota, Y., Obata, A., Kasuga, T., Ban, S., 2011. Effect of preparation route on the degradation behavior and ion releasability of siloxane-poly(lactic acid)-vaterite hybrid nonwoven fabrics for guided bone regeneration. Dent. Mater. J. 30, 232–238.

Wang, H., Li, Y., Zuo, Y., Li, J., Ma, S., Cheng, L., 2007. Biocompatibility and osteogenesis of biomimetic nano-hydroxyapatite/polyamide composite scaffolds for bone tissue engineering. Biomaterials 28, 3338–3348.

Weiner, S., Traub, W., Wagner, H.D., 1999. Lamellar bone: structure–function relations. J. Struct. Biol. 126, 241–255.

Woo, K.M., Chen, V.J., Ma, P.X., 2003. Nano-fibrous scaffolding architecture selectively enhances protein adsorption contributing to cell attachment. J. Biomed. Mater. Res. A 67, 531–537.

Woo, K.M., Jun, J.H., Chen, V.J., Seo, J., Baek, J.H., Ryoo, H.M., Kim, G.S., Somerman, M.J., Ma, P.X., 2007. Nanofibrous scaffolding promotes osteoblast differentiation and biomineralization. Biomaterials 28, 335–343.

Xia, W., Zhang, D., Chang, J., 2007. Fabrication and in vitro biomineralization of bioactive glass (BG) nanofibres. Nanotechnology 18, 135601.

Xie, J., Li, X., Lipner, J., Manning, C.N., Schwartz, A.G., Thomopoulos, S., Xia, Y., 2010. "Aligned-to-random" nanofiber scaffolds for mimicking the structure of the tendon-to-bone insertion site. Nanoscale 2, 923–926.

Xie, J., Michael, P.L., Zhong, S., Ma, B., MacEwan, M.R., Lim, C.T., 2012a. Mussel inspired protein-mediated surface modification to electrospun fibers and their potential biomedical applications. J. Biomed. Mater. Res. A 100, 929–938.

Xie, J., Blough, E.R., Wang, C.H., 2012b. Submicron bioactive glass tubes for bone tissue engineering. Acta Biomater. 8, 811–819.

Xie, J., Zhong, S., Ma, B., Shuler, F.D., Lim, C.T., 2013. Controlled biomineralization of electrospun poly(ε-caprolactone) fibers to enhance their mechanical properties. Acta Biomater. 9, 5698–5707.

Xu, C., Inai, R., Kotaki, M., Ramakrishna, S., 2004. Electrospun nanofiber fabrication as synthetic extracellular matrix and its potential for vascular tissue engineering. Tissue Eng. 10, 1160–1168.

Xu, S., Nie, Z., Seo, M., Lewis, P., Kumacheva, E., Stone, H.A., Garstecki, P., Weibel, D.B., Gitlin, I., Whitesides, G.M., 2005. Generation of monodisperse particles by using microfluidics: control over size, shape, and composition. Angew. Chem. Int. Ed. Engl. 44, 724–728.

Xu, J.H., Li, S.W., Tostado, C., Lan, W.J., Luo, G.S., 2009. Preparation of monodispersed chitosan microspheres and in situ encapsulation of BSA in a co-axial microfluidic device. Biomed. Microdevices 11, 243–249.

Yanagida, H., Okada, M., Masuda, M., Narama, I., Nakano, S., Kitao, S., Takakuda, K., Furuzono, T., 2011. Preparation and in vitro/in vivo evaluations of dimpled poly(l-lactic acid) fibers mixed/coated with hydroxyapatite nanocrystals. J. Artif. Organs 14, 331–341.

Yang, F., Murugan, R., Wang, S., Ramakrishna, S., 2005. Electrospinning of nano/micro scale poly(l-lactic acid) aligned fibers and their potential in neural tissue engineering. Biomaterials 26, 2603–2610.

Yang, F., Wolke, J.G.C., Jansen, J.A., 2008a. Biomimetic calcium phosphate coating on electrospun poly(ε-caprolactone) scaffolds for bone tissue engineering. Chem. Eng. J. 137, 154–161.

Yang, D., Jin, Y., Zhou, Y., Ma, G., Chen, X., Lu, F., Nie, J., 2008b. In situ mineralization of hydroxyapatite on electrospun chitosan-based nanofibrous scaffolds. Macromol. Biosci. 8, 239–246.

Yang, F., Both, S.K., Yang, X., Walboomers, X.F., Jansen, J.A., 2009. Development of an electrospun nano-apatite/PCL composite membrane for GTR/GBR application. Acta Biomater. 5, 3295–3304.

Yang, D., Yu, K., Ai, Y., Zhen, H., Nie, J., Kennedy, J.F., 2011. The mineralization of electrospun chitosan/poly(vinyl alcohol) nanofibrous membranes. Carbohydr. Polym. 84, 990–996.

Yi, J., Wei, G., Huang, X., Zhao, L., Zhang, Q., Yu, C., 2008. Sol–gel derived mesoporous bioactive glass fibers as tissue-engineering scaffolds. J. Sol-Gel Sci. Technol. 45, 115–119.

Yokoyama, Y., Oyane, A., Ito, A., 2007. Biomimetic coating of an apatite layer on poly(l-lactic acid); improvement of adhesive strength of the coating. J. Mater. Sci. Mater. Med. 18, 1727–1734.

Yoshimoto, H., Shin, Y.M., Terai, H., Vacanti, J.P., 2003. A biodegradable nanofiber scaffold by electrospinning and its potential for bone tissue engineering. Biomaterials 24, 2077–2082.

Yu, H.S., Jang, J.H., Kim, T.I., Lee, H.H., Kim, H.W., 2009. Apatite-mineralized polycaprolactone nanofibrous web as a bone tissue regeneration substrate. J. Biomed. Mater. Res. A 88, 747–754.

Yuan, X., Mak, A.F., Li, J., 2001. Formation of bone-like apatite on poly(l-lactic acid) fibers by a biomimetic process. J. Biomed. Mater. Res. 57, 140–150.

Zhang, Y., Ouyang, H., Lim, C.T., Ramakrishna, S., Huang, Z.-M., 2005a. Electrospinning of gelatin fibers and gelatin/PCL composite fibrous scaffolds. J. Biomed. Mater. Res. B Appl. Biomater. 72B, 156–165.

Zhang, Y.Z., Venugopal, J., Huang, Z.M., Lim, C.T., Ramakrishna, S., 2005b. Characterization of the surface biocompatibility of the electrospun PCL-collagen nanofibers using fibroblasts. Biomacromolecules 6, 2583–2589.

Zhang, Y., Venugopal, J.R., El-Turki, A., Ramakrishna, S., Su, B., Lim, C.T., 2008. Electrospun biomimetic nanocomposite nanofibers of hydroxyapatite/chitosan for bone tissue engineering. Biomaterials 29, 4314–4322.

Zhang, P., Hong, Z., Yu, T., Chen, X., Jing, X., 2009a. In vivo mineralization and osteogenesis of nanocomposite scaffold of poly(lactide-co-glycolide) and hydroxyapatite surface-grafted with poly(l-lactide). Biomaterials 30, 58–70.

Zhang, H., Liu, J., Yao, Z., Yang, J., Pan, L., Chen, Z., 2009b. Biomimetic mineralization of electrospun poly(lactic-co-glycolic acid)/multi-walled carbon nanotubes composite scaffolds in vitro. Mater. Lett. 63, 2313–2316.

Zhang, Y., Reddy, V.J., Wong, S.Y., Li, X., Su, B., Ramakrishna, S., Lim, C.T., 2010. Enhanced biomineralization in osteoblasts on a novel electrospun biocomposite nanofibrous substrate of hydroxyapatite/collagen/chitosan. Tissue Eng. A 16, 1949–1960.

Zhao, H., Wang, G., Hu, S., Cui, J., Ren, N., Liu, D., Liu, H., Cao, C., Wang, J., Wang, Z., 2011. In vitro biomimetic construction of hydroxyapatite-porcine acellular dermal matrix composite scaffold for MC3T3-E1 preosteoblast culture. Tissue Eng. A 17, 765–776.

Zheng, X., Zhou, S., Xiao, Y., Yu, X., Feng, B., 2009. In situ preparation and characterization of a novel gelatin/poly(d,l-lactide)/hydroxyapatite nanocomposite. J. Biomed. Mater. Res. B Appl. Biomater. 91, 181–190.

Zheng, F., Wang, S., Wen, S., Shen, M., Zhu, M., Shi, X., 2013. Characterization and antibacterial activity of amoxicillin-loaded electrospun nano-hydroxyapatite/poly(lactic-co-glycolic acid) composite nanofibers. Biomaterials 34, 1402–1412.

Zhou, H., Lee, J., 2011. Nanoscale hydroxyapatite particles for bone tissue engineering. Acta Biomater. 7, 2769–2781.

Index

Note: Page numbers followed by *f* indicate figures and *t* indicate tables.

P

R

S

CPI Antony Rowe
Eastbourne, UK
October 23, 2015